AF526763

Neem and Pest Management

Ram Prakash Srivastava
Principal Scientist
Central Institute For Subtropical Horticulture,
Rehmankhera, P.O.- Kakori,
Lucknow-227 017 (U.P.)

International Book Distributing Co.
(Publishing Division)
Chaman Studio Building, 2nd Floor,
Charbagh, LUCKNOW-226 004, U.P. (INDIA)
Tel. : Off. : (0522) 450004, 450007, 459058; Res. : (0522) 343274, 342829.
Fax : 0522-458629

Published by
International Book Distributing Co.
(Publishing Division)
Chaman Studio Building, 2nd Floor,
Charbagh, LUCKNOW-226 004, U.P. (INDIA)
Tel. : Off. : (0522) 450004, 450007, 459058; Res. : (0522) 343274, 342829.
Fax : 0522-458629. E-mail : sushil@lw1.vsnl.net.in

First Edition : 2001

ISBN 81-85860-58-0

Composed and designed at :
Panacea Computers
33, Nehru Road, Sadar Cantt.
Lucknow - 226 002 (U.P.)
Tel. : (0522) 481164, 480546
E-Mail : prasgupt@hotmail.com

Printed at :
Army Printing Press,
33, Nehru Road, Sadar Cantt.
Lucknow - 226 002 (U.P.)
Tel. : (0522) 481164, 480546

A small step towards
conserving India's rich
bio-heritage in the
New Millennium

Dedicated to

My beloved wife

(Late) Smt. Anandeshwari Srivastava

Contents

DR. R. S. PARODA
SECRETARY
&
DIRECTOR-GENERAL

GOVERNMENT OF INDIA
DEPARTMENT OF AGRICULTURAL RESEARCH & EDUCATION
AND
INDIAN COUNCIL OF AGRICULTURAL RESEARCH
MINISTRY OF AGRICULTURE, KRISHI BHAVAN,
NEW DELHI-110 001
Tel. 3382629; Fax 91-11-3387293;
E-mail : @icar.delhi.nic.in

FOREWORD

Use of natural products in agriculture has assumed greater practical importance with the current thrust on environmental sustainability and organic farming. *Neem* (*Azadirachta inidca* A. Juss) tree, which is native to India, has been identified as one of the most potent source of products for environmentally safe production and agricultural development. On account of its medicinal and insecticidal properties, it has been intricately interwoven with culture, life and civilization of India.

In last few years *neem* has attracted global attention due to its strong and safe insecticidal properties. Intensive search during the past decade has led to identification of *neem* as one of the most potent alternatives to toxic chemical pesticides and fertilizers. The neem products are effective biodegradable, relatively less toxic and easily available.

Although, the *neem* tree has been used for centuries by Indian farmers to control the crop pests, both in the field and in storage, yet its chemical properties have been patented in India very recently on a limited scale. Considering the socio-economic importance of *neem* and its use in integrated pest management system, it is imperative that effective formulations of *neem* are developed for controlling insect, disease and nematode pests of agricultural crops.

I congratulate Dr. R. P. Srivastava, Principal Scientist, CISH, Lucknow for bringing out this valuable book on "**Neem and Pest Management**" that addresses the important issues like antifeedant, repellent, deterrent effects, and toxic effects of neem products etc. on the pests of agricultural crops. I am sure that the book will be useful to all those research and extension workers who are engaged in R & D activities on *neem* and its products.

(R. S. PARODA)
Secretary, Department of Agricultural Research & Education
and Director General, Indian Council of Agricultural Research,
Ministry of Agriculture, Krishi Bhavan, New Delhi-110 001

JULY, 24, 2000

DR. R. S. PARODA
SECRETARY
&
DIRECTOR GENERAL

GOVERNMENT OF INDIA
DEPARTMENT OF AGRICULTURAL RESEARCH & EDUCATION
AND
INDIAN COUNCIL OF AGRICULTURAL RESEARCH
MINISTRY OF AGRICULTURE, KRISHI BHAVAN
NEW DELHI 110 001
Tel.: [illegible], Fax: [illegible]
E-mail: [illegible]

F O R E W O R D

Use of natural products in agriculture has assumed greater significance with the current thrust on environmental sustainability and organic farming. Neem (*Azadirachta indica* A. Juss) tree, which is native to India, has been identified as one of the most potent source of products for environmentally safe production and agricultural development on account of its medicinal and insecticidal properties. It has been intricately interwoven with the culture, life and civilization of India.

In last few years neem has attracted global attention due to its strong and safe insecticidal properties. Intensive research during the past decade has led to identification of neem as one of the most potent alternatives to toxic chemical pesticides and fertilizers. The neem products are effective, biodegradable, relatively less toxic and easily available.

Although, the neem tree has been used for centuries by Indian farmers to control the crop pests, both in the field and in storage, yet its chemical properties have been patented in India very recently on a limited scale. Considering the socio-economic importance of neem and its use in integrated pest management system, it is imperative that effective formulations of neem are developed for controlling insect, disease and nematode pests of agricultural crops.

I congratulate Dr. R. P. Srivastava, Principal Scientist, CISH, Lucknow for bringing out this valuable book on "Neem and Pest Management" that addresses the important issues like antifeedant, repellent, deterrent, efficacy and toxic effects of neem products etc. on the pests of agricultural crops. I am sure that the book will be useful to all those research and extension workers who are engaged in R & D activities on neem and its products.

(R. S. PARODA)
Secretary, Department of Agricultural Research & Education,
and Director General, Indian Council of Agricultural Research,
Ministry of Agriculture, Krishi Bhavan, New Delhi 110 001

JULY 24, 2000

CENTRAL FOR RESEARCH AND DEVELOPMENT OF WASTE AND MARGINAL LAND
B-1, J Road, Mahanagar Extension, Lucknow-226 006 (U.P.), INDIA

Dr. M. D. Pathak
Chairman & Ex. D.G.,
UPCAR

FOREWORD

Role of neem in pest management has emerged as a practical, safe, and indigenous method in recent years.

Neem is an indigenous tree of India, and is well recognized for its ability to grow even under sub-optimum environmental and edaphic conditions. Neem tree's luxuriant, evergreen growth with medicinal and pesticidal attributes have made it a very popular plant with considerable international interest. There are about 15 million neem trees growing in India with annual seed yield of nearly half a million tonnes. Recognizing its potential, several countries have started bringing large areas under neem plantations. The information on isolation and identification of biological ingredients from various parts of neem and their properties have also become available now.

Dr. R. P. Srivastava, a well known Entomologist with many years of experience in pest management on several agricultural and horticultural crops, has done outstanding job in bringing out the book, "**Neem and Pest Management**". It is a comprehensive treatise on various aspects of neem including its origin, geographical distribution, pests and diseases, biologically active ingredient, biological effects and chemistry of neem compound and various uses and properties, especially the pest control efficiency of neem-based pesticides using innovative bioassays and application methods. The author also incorporates much needed information on the effect of neem on bio-control agents, and other non-targeted organisms. He also evaluates the efficacy of neem-based pesticides *vis-à-vis* chemicals, other plant derived pesticides and bio-control agents.

In view of these, neem and neem-based pesticides are poised to play a major role in integrated pest management system. Dr. R. P. Srivastava's book on neem is very timely and of great significance and would be very useful to teachers, researchers and students, etc.

I congratulate Dr. Srivastava for writing such a comprehensive book on this important multi-utility plant.

M. D. Pathak

(M. D. PATHAK)

CENTRAL FOR RESEARCH AND DEVELOPMENT OF WASTE AND MARGINAL LAND
B-4, J Road, Mahanagar Extension, Lucknow-226 006, U.P. INDIA

Dr. M. D. Pathak
Chairman & Ex.D.G.
UPCAR

FOREWORD

Role of neem in pest management has emerged as a practical, safe and indigenous method in recent years.

Neem is an indigenous tree of India, and is well recognised for its ability to grow even under sub-optimum environmental and edaphic conditions. Neem tree's toxicant, evergreen growth with medicinal and pesticidal attributes have made it a very popular plant with considerable international interest. There are about 18 million neem trees growing in India with annual seed yield of nearly half a million tonne. Recognizing its potential, several countries have started bringing large areas under neem plantations. The information on isolation and identification of biological ingredients from various parts of neem and their properties have also become available now.

Dr. R. P. Srivastava, a well known Entomologist with many years of experience in pest management on several agricultural and horticultural crops, has done outstanding job in bringing out the book "Neem and Pest Management". It is a comprehensive treatise on various aspects of neem including its origin, geographical distribution, pests and diseases, biologically active ingredient, biological effects and chemistry of neem compound and various uses and properties, especially the pesticidal efficiency of neem-based pesticides using innovative bioassays and application methods. The author also incorporates much needed information on the effect of neem on bio-control agents and other non-targeted organisms. He also evaluates the efficacy of neem based pesticides vis-a-vis chemicals, other plant derived pesticides and bio-control agents.

In view of these, neem and neem-based pesticides are poised to play a major role in integrated pest management system. Dr. R.P. Srivastava's book on neem is very timely, and of great significance and would be very useful to teachers, researchers and students etc.

I congratulate Dr. Srivastava for writing such a comprehensive book on this important multi-utility plant.

(M. D. PATHAK)

PREFACE

Neem (*Azadirachta indica* A. Juss) is an attractive evergreen tree which is native to the Indian sub-continent but cultivated throughout Pakistan, Sri Lanka, South East Asia, Africa (Somalia, Nigeria, Mauritania, Togo etc.), East and Sub Sahelion Africa, Fizi, Mauritius, many countries in Central and South America, the Caribbean, Puerto Rico, the Virgin Islands and Australia. There are host of products obtained from every part of the neem tree, many of which have potential use in pest management programme. Therefore, the tree has gained worldwide attention and recognition. Neem tree is indispensable to the Indian Sub-continent having cultural, medicinal and pesticidal values.

Historically, commercial uses of neem tree are shrouded in mystique and lore of Vedic period of India, which began about 4000 B.C. In Ayurved, a system of medicine, neem bark and leaves are being used in dermal problems, flower as tonic and stomachic and fruits as purgative and emollient.

No systemic work has been undertaken on neem tree which has a numerous qualities and usage is truly called as "wonder tree". It is, therefore, felt that the workers engaged in research on neem may consider the following thrust areas for high quality neem production :

1. It is now important to establish Field Gene Bank of neem cultivars found within the Country and abroad at a suitable location. There is a need for genetic improvement of neem of local species and for generating more information on various uses, its production and suitability systems.

2. An integrated process of cleaning, decortication and crushing of neem seed for extracting neem oil needs to be developed.

3. It is imperative to have international neem improvement programme network. The long term objectives of the network comprise improvement of genetic quality and adaptability of neem and its utilization world wide as pesticides, medicine, etc.

4. Increasing awareness towards eco-friendly technologies and efforts to enhance economic value of neem and its products, are likely to provide much desired impetus and sustainability to their utilization. Therefore, it is imperative that effective formulations of high biological activity of neem are developed in order to minimize the use of chemical formulations.

5. Standardization and development of neem coated urea based fertilizers, for their slow nutrient release properties is required to be expedited.

6. Sustainable IPM programmes, having neem as one of the important component, needs to be standardized and implemented at a large scale.

With the continued robust growth of the global biopesticide market, azadirachtin is uniquely positioned to become a key insecticide. The broad-spectrum activities of azadirachtin at low use rates coupled with the insect growth regulator activity and unique mode of action (ecdysone disruptor) make it an ideal product with additional advantage of prevention of insecticide resistance development in case of pests. Therefore, it could become integral component of integrated pest control and organic pest control programmes. Azadirachin has been exempted from residue tolerance requirements by the US Environmental Protection Agency for food crop applications. It has been found to exhibit good efficacy against key pests such as white flies, leafminers, gnats, thrips, aphids and many leaf-eating caterpillars, fungi, bacteria and viruses. Azadirachtin has minimal impact on non-target organisms and is compatible with other bio-control agents. The world's largest azadirachtin extraction facility has been commissioned in India to process over 1000 tonnes neem seeds per annum. This will ensure adequate availability of technical grade azadirachtin material in the future. At present, it can be justifiably asserted that full potential of neem has still not been exploited.

The current manuscript has been organized systematically in 11 chapters. The chapters deal with the detailed information on neem and control of insect pests, diseases and nematodes of agricultural crops. Each chapter has upto date information on the subject and contains exhaustive bibliography.

I am expressing my deep gratitude to Dr. R. S. Paroda, Secretary, DARE, Govt. of India and Director General, ICAR, New Delhi and Dr. M. D. Pathak, Ex. Director General, U.P. Council of Agricultural Research, Lucknow for writing foreword to this book and extending their blessings. I am extremely thankful to Dr. R.K. Pathak, Technical Co-ordinator, Diversified Agricultural Support Project, Lucknow (Now Director, Central Institute of Subtropical Horticulture, Lucknow) for giving very useful suggestions. I thank to Director, CISH, Lucknow for giving encouragement and support to undertake this task. I also extend my sincere thanks to Dr. Anand Akhila, Sr. Scientist, Central Institute of Medicinal and Aromatic Plants, Lucknow for critically going through the chapter on Biological active ingredient, biological effects and chemistry of neem compounds.

My sincere thanks are due to Dr. R. K. Ojha and Dr. Ajay Varma, Senior Scientists, CISH, Lucknow for their valuable help in the preparation of this book, "**Neem and Pest Management**" which is before you. I very sincerely thank Shri R. P. Shankhwar, Technical Officer, CISH, Lucknow, who has helped me a lot in preparing this book. The guidance and help rendered by Shri Prem Kumar, Technical Officer, and Shri Habib-Ur Rehman, Artist are highly appreciated and I thank them very much.

The moral and physical support extended by Sarvesh Mathur and Alok Mathur, my sons-in-laws; daughters Mrs. Rashmi, Mrs. Rachna and son Rajeev Srivastava, daughter-in-law Mrs. Veethi Srivastava are duly and affectionately acknowledged. I extend my thanks to publisher of this book for his keen interest and sincere devotion in bringing out this book in shortest possible time.

I hope that this book would be of immense value to graduate, post graduate students and research scholars working on neem.

I shall feel highly obliged to all the readers who spare some time in providing critical suggestions, if any, for the improvement of this book in any of its prospective editions.

RPSrivastava

(Ram Prakash Srivastava)

Chapter 1

Neem *Azadirachta indica* A. Juss

Neem, the versatile tree having many good and useful qualities belong to the order Rutales and the family Meliaceae (Fig. 1.1). This genus *Azadirachta indica* was described by A. Juss in 1830. Later it was described synonyms with *Melia azedarach* L, *Melia indica* Brandis and *Antelaea azadirachta* (L) Adelb.

Neem tree is indegenous to India from where it has spread to many Asian and African countries. viz. Pakistan, Bangladesh, Myanmar (Burma), Srilanka, Thailand, Indonesia, Malaysia, Singapore, Iran, Yemen, Australia, Papua New Guinea, Fizi, Nigeria, Tanzania, Cameroon, Madagascar, U.S.A., Latin America, Germany, France, Portugal, Spain, and United Kingdom. The earliest authentic record of its medicinal use is available in Arthashastra of Kautilya in 4th century B.C. Various parts of the tree have been used in Ayurveda of the famous Hindi Physician Susruta (1000 AD). For centuries, the tree has been held in esteem by Indian folk because of its medicinal and insecticidal values. The Sanskrit name Nimba and Arishta indicates that this tree was in use since time immemorial. There is a story behind this tree, it is said that when Amrita (heavenly Nector) was being taken to heaven, a few drops fell on neem tree. It is because of this fact the tree is being planted in front of every house, worshipped and used in various rituals.

Geographical Distribution :

The origin of *A. indica* is not very clear. Some says that it is from Myanmar (Burma) and others that it is from South India. Many consider its origin from south eastern and southern Asia and from Indonesia to Iran.

In Indonesia, neem exists mainly in the low lying northern and eastern parts of Java. In Philippines it was introduced from India, Africa and Indonesia. In China *A. Indica* was planted from the seeds obtained from Togo in Africa. The Indian neem tree is scattered in Thailand. In Myanmar neem is very common in the central part of the country. This is very common in the central part of Bangladesh and Pakistan. Ketkar (1967) reported more than 14 million trees in India. It is also said that there are more than 20 million trees available in entire India. In Nepal, it is found in southern, low lying areas, while in Srilanka it spreads in northern parts.

In Africa, neem was introduced from India. Now a days neem exists along the east African coast in Somalia, Kenya and Tanzania. In Egypt around 40 thousand trees are planted in recent years. In Ethiopia it grows well in northern parts. This tree is very common in other parts of Nigeria especially in Benin, Togo, Burkina, Faso, Ghana and Ivory coast. Neem also exists in Mali. In northern America, neem trees occurs in Miami and other parts of Florida. In Brazil, neem tree was planted in this tropical parts of the country. Neem is also reported from Surinam and Guyana. In Medagascar and Mauritius it is common in the dry regions of the south west. In Egypt, approximately 4000 trees were planted mainly on Nile delta. In Sudan it is grown on large scale especially in irrigated region of the central Sudan. In Ethiopia it grows well in low lying north western parts. In central Africa neem is grown in lake Chad basin. The famous neem trees belt in Niamey, the capital of Niger is well known. Neem also exists in Mali especially along river Niger is well known. Neem is also grown in Southern parts of Mauretania, in Gambia, in the Guineas and in Sierra Leone. There exists some old trees in the island of Gran Canaria in Las Palmas (Schmutterer, 1995).

Ecology :

Neem is fast growing sclerophyllous tree and grows well in climates from semi and to semi-humid. It thrives well in places with less than 500 mm. rains in a year. Normally, it is a drought resistance. In case, where severe drought is there, the leaves are shed partially or some times completely. Neem also grows in the area where rainfall is upto 2,500 mm. in a year but with poor fruit yeild. The soil requirements are modest and it grow best on well drained deep, sandy soils. It also grows equally well on poor, shallow, sandy or stony ground. In India and Sudan it thrive well on black cotton soils. In Nigeria high clay contents, in Africa on lateritic soils. In Caribbean on calcarious soils, in Madagascar stony shallow soil. In Somalia it grows very well in saline ground water.The pH value between 6.2 and 7.0 seems to be best for this tree but pH 5-9 and 10 may also be tolerated. In Australia, neem trees on soils with pH value from slightly alkaline to slightly acidic (Rice, 1993), water logged soils are unsuitable for neem as they may lead to rot of the taperoot.

Neem thrive well between 21 to 32°C temperature but it tolerate upto 50°C during summer. When temperature falls below 4°C shedding of leaves do occur. Neem also thrive well at the altitudes upto 700-800 m. The young plants are normally raised under shades during unfavourable conditions while mature trees needs lot of light. According to Gruber (1991), the net rate of photosynthesis, measured in Australia was 10-17 umol CO_2 m-2s-1 which is intermediate to high compared with tropical fruit trees.

Neem Tree :

In India neem is called as limba, limo, nim, nimb, nimba, vembu, vepa, veppam etc. In Pakistan , Nimmi, in Myanmar, tamarkha, in Australia, neem, in Yemen, megaimarah, in Fizi, neem, in Nigeria, Baboyaro, Dogon yaro, in Medagascar and Latin America Nim, in U.S.A. neem, in Germany niem, nim, niembaum, in France Azadirac, Margousier, in Spain and Portugal Margosa, in United Kingdom Indian Lilac, neem.

Young plants have to be protected against competetion from weeds, especially during first year. The famous neem tree has the reputation of possessing a large number of biological activities such as insecticidal, nematicidal, bactericidal, spermicidal, antiviral, diuretic and antipyretic. However, the centuries old knowledge has not been explored in India on large scale. The pioneering efforts of the Khadi and Village Industries Commission in the past two decades on various aspects of processing the fruit and seeds and utilization of neem cake neem oil etc are only instrumental in developing this resource.

For healthy growth of the neem tree, pH, water table, rainfall, irrigation, nutrients, inter and intraspecific competetion, mycorrhiza temperature and genetic factors play an important role. It is observed that in the region of dry areas, High altitude, very cool temperature and the area of high rainfall, the general health of the plant is very poor and growth is retarded. At the altitude of less than 1000 m the growth is poor and which ultimately effect poor yield. The areas where rainfall is very poor, the growth of the tree is also hampered. Frost may also hamper the growth of young neem plants but this was observed in northern Pakistan near Himalayan mountains and in Florida (U.S.A.)

Radwanski (1977) observed that in Nigeria in first three years nearly 66 percent growth takes place and the plant reaches to the height of 4-7m. The observations are that in Africa, the trees reach a height of 1.5m. in first year, 2 m. in second year and 4.5 m. in fourth year. In Nicaragua, neem grown in plantations at the age of 8 years at five locations had an average height of 7.7-9.6 m (Gruber 1991).

As regards yield of fruit is concerned, no reliable information is available. Normally the first fruits appears at the age of 3-6 years after planting. In norther hemisphere of Indian sub continent, the flowering takes place in January and february and in south of the equator in November and December. In west Africa two flowering season are present. In Azua Valley in the

Dominican Republic, three flowering seasons are found.

Neem tree has attracted world wide attention due to its wide ranging capacity as a biocide. There is a need to create enough encouragement for more cultivation of neem sò that low cost plant protection prcducts may be prepared. Neem plantations may be planted on both side of the roads as has been done in many states of our country (Fig. 1.2). There is a need for proper planning to grow neem plantations in larger areas on important germplasms. Neem is one tree which can be grown in any type of land. To promote mass cultivation of neem, the growers may be given some incentives and proper education for neem cultivation and allied problems. There is a need to have Cooperative Societies to develop this trade. To expand the neem plantations, it is also desirable to have better export potential. This neem farming is called Margoculture. Neem can grow very well between 0°C to 47.5° C but at low temperature the small plants should be protected from frost. The rainfall from 450-1200 mm is good for its growth but as little as 130 mm rainfall, the neem can tolerate and grow. It can grow on all kind of soils, clay, saline and alkaline soil but it grows very well on black cotton soil.

Dlwaulle (1979) recommended general spacing of 5x5 m in rainfed zones. However, Hoamuangkaew *et al.* (1990) reported at Thailand that 2x4 m spacing yielded a great volume than that of the spacing at 4x4 m.

Kazmi (1980) observed that *A. indica* is also known as 'merimeri' and this plant can tolerate 250 mm annual rain fall and dry saline soils. Ahmad *et al.* (1985) observed that *A. indica* and *Melia azedarach* were grown on coastal sand using 10-30 percent dilution of saline water for irrigation. Plants irrigated with saline water showed an increase in moisture contents and decrease in chlorophyll, sugar and protein contents. Under saline water irrigation, *M. azedarach* showed more rapid growth than *A. indica*.

Vijayan and Rehill (1987) reported the occurrence of twin seedlings in *A. indica* for the first time. Seedlings with equal and unequal growth were observed and frequency of occurrence was 11-27 percent.

Tampubalon and Alrasyid (1989) reported the neem tree and its developmental prospects in rainfed zones in Indonesia. They suggested the use of the neem tree in Taungya systems, greening schemes, shelteobelt and road side plantings. Ahmad (1989) reported large area planting of neem in Saudi Arabia. The plantation was done in 10 km^2 area using 50,000 trees. However, Gruber (1996) reported that from 1986 to 1995 about 600,000 neem trees were planted in Nicarangua. The seed size (1.7 cm length) has been the only selection criterion for nurseries. Best average from yield 30 kg/tree was obtained from trees grwon at a distance of 8x8 m on sandy to loamy soils in

Fig. 1.1

Fig. 1.3

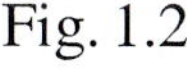

Fig. 1.2

Fig. 1.1 Neem tree

Fig. 1.2 Photograph showing neem tree grown on both sides of the road as shadow tree (Courtesy : Shri Alok Mathur and Smt. Rachna Mathur, Madhya Pradesh)

Fig. 1.3 Neem at flowering stage

Fig. 1.4

Fig. 1.5

Fig. 1.6

Fig. 1.4 Leaves
Fig. 1.5 Fruiting neem twig
Fig. 1.6 Fruits bunch with heavy fruiting

warm dry regions but with at least 1200 mm annual rainfall.

Burman *et al*. (1991) observed that average application of 46 litres of water per plant for 2 weeks intervals led to maximum growth and biomass production with no mortality. According to Brenner *et al*. (1991), windbreaks are being recommended for many areas of dryland Africa. The study shows that a linear relationship exists between the leaf area and stem basal area of young neem trees. Betanus and Morrobel (1994) observed that the best plant growth was obtained with the mixture of rice husks and peat, followed by 30 percent coconut fibre + 70 percent peat.

Jhorar *et al*. (1996) transplanted 9-12 month old saplings of *A. indica* in desert ecosystem in 0.6x0.6 m pits filled with field soil. FYM and pond silt/ heavy soils in the ratio of 1:0.5:0.5 in 6x6 m spacing. In each plant, 10 ltres water was irrigated. The percentage survival was 95 percent.

According to Gogati and Gujar (1996), the shape of the crown may be mushroom, conical, or inverted conical and the occurrence of these shapes in the population is 39, 38 and 23 percent, respectively. Crown volume varies from 471.08 to 2095.64 m^3 and there is correlation between crown size and bole diameter. Flowering behaviour is not uniform; out of 331 trees under study, a total of 292 trees flowered and fruited in the 1992 season. Fruiting starts in late April and fruit fall is from mid-May to mid-July. Fruit production varies from 7.5 to 17.325 kg (dry weight) per tree, the average being 12.795 kg. Fruit production is also consonance with crown size and bole diameter. Selections for specified traits and concerted efforts at international and national levels are reccommended for genetic improvement.

Kumar (1998) reported that the barley (Jagrati) and oat (OPO-250) showed between the rows of neem plantation gave yield margin of 2.76 and 1.95 q/ ha, respectively and fully precluded from insects, pests and soil born diseases during rabi season in both the year. It is worthwhile to mention here that the green manuring dhaincha given in kharif season of both the years under agroforestry system during its decomposition period, no incidece of insect pests was observed probably due to standing of neem plants in the fields as well as natural fall of their leaves.

Propagation :

In general, neem plants are being propogated by using seeds as it can be easily collected and sown, but seed propagated plants have lot of variability and therefore may be discouraged. To obtain early maturity crop, true to type stock and better uniformity of the growth characteristics, vegetative propagated plants may be used for mass plantation. Since vegetative propagation is not very common among the farmers, it is desirable to select good quality seeds from the good mother plants having good vigour and age, better crown shape, well distributed branching, having big size seed with distributed branching, and with heavy fruiting and disease free plants. It is also observed that the seeds collected at the begining and at the end are not useful but greenish yellow skin fruit is good for plantation. Seeds may be sown in drills 15 cm apart at a depth of 2.5 cm. The seeds normally germinates within 1 to 2 weeks and promissing seedlings are picked out when they are 5 cm tall.

Neem is proved to be very good when propogated vegetatively through root suckers as it is possible to propogate through root cuttings. Stem cuttings of neem are successfully propogated. Hard wood cuttings are treated with 1000 ppm IAR and IBA. Softwood cuttings could be used by treating with naphthol or salicylic acid. Air layering is a better way of vegetative propogation as they carry roots and appears to be sure method of establishment. To develop rooting it is necessary to treat the air layers with 0.1 percent IBA or NAA (Shanmugavelu, 1967). However, air layering is best method.

Pong-Anat *et al*. (1989) observed 85-90% success by whip (tongue) method of grafting, 75% splice grafting and 65% Cleft grafting. They observed that optimal temperature and high humidity yeild better results. Cannell (1983) gave details of management practice in the cultivation of neem plantations for obtaining high quality seeds. The parameters are growth patterns, floral initiation, pollination, fruit shedding and irregular bearing. The juvenile vegetative period can be enhanced by increasing fertility, mulching and irrigation and removing competition. The grafting may be done on flower bearing roots stocks. He suggested that for better floral initiation, intensive prunning all over the crown for enough light availability for good floral initiation may be done. Since neem is a cross pollinated crop, good pollination is ensured by controlling the air current by wind breaks. To take care of fruit shedding, proper irrigation, application of fertilizer and growth regulators, removal of shades and removal of shoot tips may be regulated. To minimize irregular bearing, prune half of the tree or a portion and keep it as off year and the left over portion as on year crop. It is also essential to develop hybrids having regular bearing tendencies.

Solanki and Gupta (1999) observed that very high variability has been reported for azadirachtin contents in seed kernels from neem tree from differennt agroclimatic regions which ranges from 0.14 to 1.66 percent. It is still not clear that this variability in the germplasm is due to genetical factors or environmental factors. It is, therefore, essential to plant the germplasm in a single location for real assessment and evaluation. A minimum of 6 years are required to plant the germplasm at one place and the tree starts fruiting at 6-8 years of age.

The National Research Centre for agroforestry, Jhansi (U.P.), India, has standardized an air layering technique for neem seedling propagation. Such plants are genetically identical to their mother plants. It was also observed that flowering and fruiting phenomena in air layered seedlings occur every year during May-June which help to evaluate the germplasm even in second year. In this process of air layering, rooting in the treated branches and survival in polybags are cent percent.

Protocols for vegetative propagation of neem using traditional methods viz. grafting annd budding, air layering, rooting branch cuttings, procurement of juvenile cuttings, rooting hard wood cuttings, rooting semihard wood cuttings, rooting softwood cuttings and the newly developed techniques of tissue culture are now available. In clonal seed orchards meant for quality seed production and controlled breeding, the use of mature clonal propagation is beneficial as it results in early flowering and seed set but rapid multiplication of superior clonal material for raising fast growing clonal plantations of neem necessitates use of juvenile cuttings.

The wide distribution of neem and its growth under a variety of climates is indicative of the presence of large genetic diversity the selective use of which can be made for yeild improvement of this species. One of the main variability in neem is seed weight, size, leaf length, height and growth of the seedlings. Jacobson (1981) reported that neem seed of Indian origin contains 1-3 percent aza (azadirachtin) where as seeds from African origin has 5-6 percent aza compound and even upto 9 percent aza was recovered. The cultivar of neem (var *siamensis*) from Thailand have their leaf margins larger and smoother than Indian neem.

Selection procedure for the utilization of genetic diversity through clonal approaches are an important aspect. To obtain good clonal collection, selection of superior seed sources are important. Once from these seeds, superior clones are identified, mass multiplication of tested clones must be done. It is also important to develop hybridization of superior clones using the hybridization methods and also by developing genetic transformation technique. The strategies for application of clonal approaches on an operational scale will be on selection of clones which includes height and diameter growth, stem and crown form, foliage, bark, seed, oil content, aza contents of seed and oil, salanin content, tanin contents and quality of bark. It is also necessary to establish clone culture techniques. The need to establish the clonal seed orchard for the production of superior quality seed is important aspect. Besides clone testing and raising superior clonal plantations is urgent need to obtain better quality seeds.

Almost every part of the tree is bitter and has found application in indigenous medicine. The seed oil (Margosa oil) is acrid, yellow, bitter in taste and has a disagreeable garlic like odour. It is used in skin diseases such as Scrofula, indolent ulcers and sores and ringworm. It is also applied in cases of rheumatism as a liniment. It is also reported to possess anthelmentic, antiinflammatory, antipyorrhoeic ,insecticidal properties, antifeedant, deterrent, growth inhibition, anti-viral, antiscabic, cardiac, juvenile hormones, affect moulting, ovipositional deterrent, repellent sterilizing effects, spermicidal, fungicidal larvicidal, piscicidal and nematicidal properties.

Micropropagation

It has been observed that propagation of neem plants true to type is very difficult by conventional technique. Since two decades, many research workers are busy in the application of plant protoplast, cell and tissue culture for the mass propagation of neem. To establish successfull tissue culture plants, it is best to start with the explants from indeterminate growth zone. The work done on micropropagation is given below.

Sanyal (1981) observed *in vitro* hormones induced chemical and histological differentiation in stem callus of neem, *Azadirachta indica*. Neem bark contains 0.04 percent bitter triterpenes, including the medicinal compound nimbin Callus tissue derived from young stem bark and devoid of nimbin was grown in modified Murashige and Skoog (MS) media containing various concentrations (0.1-0.5 mg/lt) and combinations of IAA and BA, in darkness. IAA at 0.5 mg/lt induced rooting in 40 percent of tissue cultures. The presence of BA stimulated production of tracheary elements, the maximum number being produced with 0.5 mg IAA + 0.25 mg BA. Cultures without growth substance made very poor growth. Nimbin (0.25% of dry weight) was detected only in cultures differentiating roots.

Narayan and Jaiswal (1985) collected leaflets from a 15-yr old tree of neem from Uttar Pradesh, India. Five mm diameter explants were cut from with a cork borer and cultured on Murashige and Skoog medium supplemented with various concentrations of 2,4-D, BAP [benzyladenine] and NAA at 25°C with 12 h illumination daily at 4000 lx. Explants produced light green friable callus after 6 weeks culture in medium supplemented with 0.5 mg/lt 2,4-D and 0.05 mg/lt benzyladenine. The calli were subcultured on media containing 0.05-1 mg/lt NAA or benzyladenine separately or combined. Root formation from calli occurred at all concentrations of NAA tested, in the absencce of benzyladenine; shoot formation followed the reverse patter. Addition of a low concentration of NAA (0.05 mg/ml) to medium containing benzyladenine increased the frequency of shoot formation. Callus derived shoots produced roots and developed into plantlets when transferred to liquid medium supplemented with 1.0 mg/ml IAA.

Sanyal and Datta (1986) collected the callus raised from young stem bark of *A. indica* and mainntained for 1.5 years was completely devoid of nimbin, an important secondary metabolite (triterpene). Redifferentiation of compounds and organs by adjusting the nutritional balance was attempted. After several trials with Murashige and Skoog inorganic salts plus various auxins, cytokinins, vitamins and natural supplements for organogenesis and plantlet formation, high BA and low NAA was shown to produce good shoot buds while the reverse induced roots. IBA (0.5 mg/lt) when added to the best shoot-producing combination of substances, produced shoots and roots together. Nimbin synthesis increased markedly only with organogenesis. Root formation was accompanied by the least quantity of nimbin, shoot formation by more nimbin synthesis, and differentiation of root and shoot together by the highest nimbin synthesis, but this was still less than in the cotyledons.

Rao *et al*. (1988) studied induction and growth of callus in *Azadirachta indica*. Cotyledons, leaves and stem segments were cultured on MS medium supplemented with different concentrations of various growth hormones. Data indicated that fresh cotyledons were the best source followed by leaves and that the combination of 2,4-D and IAA was the most effective in producing calli.

Ramesh and Pankya (1990) reported *in vitro* propagation of neem from leaf discs. Excised leaf discs of *A. indica* cultured on Wood and Braun's medium supplemented with kinetin (Kn) and benzyladenine (BAP), produced 10-12 adventitious shoot buds within four weeks. Addition of adenine sulfate to the culture medium increased the number of buds induced to 18-20. Histological studies revealed that each shoot bud originated from a group of epidermal cells. Divisions of these cell initials produced a meristemoid which developed into a shoot bud. Each isolated shoot bud grown on medium containing gibberellic acid (GA3) developed into a healthy shoot. When shoots grown on medium containing IAA (6 μM) for two weeks were transferred to basal medium they produced tap roots. This procedure permitted 12-16 plantlets to be produced within six months from a single leaf disc of the mature neem tree. After acclimatization to natural environmental conditions, plantlets were transferred to pots.

Gautam *et al*. (1993) Studied conducted on the development of shoots and roots in another derived callus of *A. indica*. Callus originated in microsporangial wall layer, and connective tissues of anthers containing uninucleate microspores on Nitsch or MS medium supplemented with growth regulators. A higer percentage of cultures produced callus on Nitsch medium containing 10 μM IAA + 1 μM 6-benzyladenine (BA). After 13-15 weeks, green nodular structures and prominent roots developed in 25 percent of the cultures on MS medium supplemented with 10 μM NAA + 1 μM kinetin. Multiple shoots were induced in this anther-derived callus when subcultured on MS medium augumented with 4.44 μM BA + 0.53 μM NAA along with 18.75 μM polyvinylpyrrolidone. Excised shoots formed roots after subculturing on MS medium containing 4.9 μM IBA + 18.75 μM polyvinylopyrrolidone, thus developing complete plantlets. Examination of anthers producing callus also revealed 2- to multi-celled pollen masses with intact exine.

Allan *et al.* (1994) observed that callus produced from the leaves of a Ghanaian strain of *Azadirachta indica* produced azadirachtin when grown in a defined medium. The azadirachtin was isolated by standard procedure of solvent partition and column chromatography monitored by supercritical fluid chromatography, and identified by SFC, TLC and HPLC. In antifeedant tests using *Schistocerca gregaria*, it was 100 percent effective at <0.4 mg/lt. The yield of azadirachtin was 0.0007 percent based on the dry weight of callus.

According to Joshi and Thengane (1996), clonal propagation of neem from non-woody branch explants (from 2-5 yr old trees) was achieved on MS medium supplemented with 6-benzylaminopurine [benzyladenine] at 0.5 mg/lt and kinetin at 1.0 mg/lt. Two to three shoots per explant was observed after 20 days. An increase in multiplication frequency of shoots per explant and average shoot length (4.7 cm) was observed after 14-16 weeks on medium containing 3/4 dilution of MS macronutrients, full strength MS micronutrients, kinetin (0.05 mg/lt), benzyladenine (0.1 mg/lt), calcium pantothenate and biotin (both 0.05 mg/lt). Rooting of these shoots was achieved on half strength MS liquid medium supplemented with IAA (2 mg/lt). The soil regenerated plantlets were successfully establilshed in soil.

According to Gosal *et al.* (1996), Nodal segments excised from shoots of mature trees *A. indica* were cultured on modified MS medium. In *A. indica*, the nodal segments taken from basal portion of coppice shoots gave maximum 80.1 percent shoot bud regeneration. The shoots were rooted on MS medium and transferred to pots with 90 percent survival. In addition, propagation was also achieved through somatic embryogenesis in *A. indica* by using leaf squares on Linsmaier and Skoog medium supplemented with BAP, proline and maltose. The regenerated embryos developed roots on 0.5 MS medium. The developed technology involves efficient propagation through tissue culture in mature trees resulting in higher sustained production of forests by increasing production/unit area/unit time.

Akula and Drew (1996) studied somatic embryogenesis of *A. indica*. Efficient induction of somatic embryogenesis was achieved in four selected genotypes of neem, *A. indica*, either directly from root and nodal cuttings or indirectly from leaf explants after a callus phase. Induction of somatic embryogensis occurred directly both from nodal and root cuttings within 4 weeks of culture on MS basal medium supplemented with Ig/L casein hydrolysate, 100 mg/L adenine sulphate and 100 mg/L glutamine. The induction of embryos from leaf explants was quite different. Callus initially formed along the mid rib of the leaf explants, 10days after incubating on the medium plus 0.5 mg/L thidiazuron (TDZ) and 0.1 mg/L 2,4-D. Callus was maintained on the same medium without subculturing and produced clusters of globular somatic embryoids after 60 days incubation. The origin of embryoids from all 3 explants was from peripheral cell/cells without any vascular connections to the original explant, and they eventually separated from the callus. These embryoids showed a high rate of multiplication 3-4 weeks after subculturing onto fresh medium devoid of growth hormones. Embryoids left on the same medium passed through a series of developmental stages and grew to complete plantlets with a well developed taproot system.

Weweizer (1996) studied the callus cultures (CC) from *A. indica* and the productin of azadirachtin. Callus initiation from bark tissue was much more rapid which takes 3 weeks than from leaf tissue which takes 5 weeks. Callus of *A. indica* are capable of producing azadirachtin depending on their age and the source of the explant. Callus subcultured thrice produced 2-3 times more azadirachtin than CC subcultured only twice. Highest yields were obtained by CC derived from Nicaraguan leaves.

Morgan *et al*. (1996) studied biotechnological processes for the production of active insecticidal compounds from the neem and related tree species are being explored as a possible alternative to harvesting the seeds and extracting an active fraction. Callus cultures of *A. indica* and *A. excelsa* (Jack) from different plant explants have been created. *A. indica* cultures have been shown to express azadirachtin and other related compounds. Towards the optimisation of suspension cultures for the production of azadirachtin, different culture parameters have been tested on the growth kinetics and desired target product expression.

Ramesh and Padhya (1996) observed that flavonoids are the natural protective compounds which form an array of characterstic secondary metabolites. This study describes the biosynthesis of these compounds in neem cell and tissue cultures which can be enhanced by altering the culture conditions. The product accumulation/synthesis was correlated with the biomass yield where the synthesis occurred when cells attain the stationary phase.

Flavonoids, in particular, kaempfero, and quercetin were found maximum in flowers (1.93 and 1.43%) and their cell cultures (0.84 and 0.68%). Sucrose level at 3 percent enhanced their quantities *in vitro*. Low levels of inorganic nitrogen (30 µM/lt) in WB medium favoured the biomass yield and flavonoid synthesis, while MS medium rich in nitrogen source (60 µM/lt) was found inhibitory.

Islam *et al*. (1996) described methodss for regeneration of plants *in vitro* from cotyledon and nucellus of *A. indica*. High frequency somatic embryogenesis was achieved from cotyledon explants in MS medium containing different concentrations of NAA. The embryoids developed into shoots when the embryogenic calli were transferred to cytokinin - supplemented media. However, the freqency of embryo conversion was very low. A high concentration of BA (0.5-2.5 mg/lt) along with a comparatively low one (0.1-0.2 mg/lt) of NAA induced cotyledon and nucellus explants to produced adventitious shoots. Shoot organogenesis best occurred with 2 mg/lt BA and 0.2 mg/lt NAA.

Joarder *et al*. (1996) studied micropropagation of neem through axillary bud culture. A micropropagation method is described using nodal explants from field-grown adult *A. indica* tree. The explants were established by culturing two weeks in hormone-free MS followed by four weeks in cytokinin supplemented-medium. Established explants upon transfer to medium containing low concentration (0.05-0.3 mg/lt) of cytonin prduced shoots. Multiplication of shoots was achieved by culturing nodal segments from *in vitro* proliferated shoots. Roooting was induced at a high frequency by subculturing individual shoots on medium with 0.5 mg/lt IBA. Plantlets were acclimated and successfully established in soil.

Nirmala Kumari *et al*. (1996) did experiments to study the *in vitro* response of various explants viz., leaf, stem, bark and seed cotyledon for callus induction and shoot-tip and cotyledons for micropropagation on MS medium with different levels of auxins and cytokinins. The leaf explants produced nodular and friable calli on MS + 2,4-D (2.0 mg/lt) + Ki (0.5 mg/lt) _ Sucrose (3%). Shoot tip and cotyledons were inoculated on MS + BAP (1.1 mg/lt) + CH (200 mg/lt) + Sucrose (2%) + Sorbitol (1%). Cotyledons produced small shoot buds after 30 days of inoculation, which gave 15-20 leafy shoots per cotyledon. Shoot tips directly produced 6-7 shoots. One month old callli of leaf and stem were subcultured on MS + BAP (2.0 mg/lt) + IAA (0.5 mg/lt) + Sucrose (3%). They differentiated for embryoid formation annd produced leafy shoots. The authors indicated the possibility of micropropagation of neem trees for large scale cultivation.

Thiagarajan and Murali (1996) studied high frequency regeneration of neem *in vitro* from immature embryo explants. A study was earlier undertaken to optimize conditions to establish plantlets of neem culturing seed embryos. In this study, they reported high frequency regenration of neem plantlets from excised embryos, with intact cotyledon, grouped according to their size from immature fruits collected from trees 30 days after flowering. The embryos were cultured in basal MS medium supplemented with various combinations of sugars, nitrates, cytokinins, and auxins. The embryos elicited interesting responses with increasing concentraions of indole acetic acid (IAA) during the regenration of plantlets. The optimum condition for high frequency regeneration of neem plantlets from intact embryos was in a medium containing 3 percent sucrose supplemented with 0.1 mg/lt naphthalene acetic acid and 0.3 mg/lt 6-benzyl amino purine. The regenerated plantlets have been successfully established in soil.

Joshi and Thengane (1996) initiated studied for the propagation of neem by organogenesis and somatic embryogenesis. The induction of somatic embryos was achieved directly on explant tissue or from callus. MS basal medium supplemented with auxins, indole 3-acetic acid (IAA), or 2,4-D, along with cytokinins, 6-benzyl amino purine, and kinetin, were used. Adventitious multiple shoots (10-15) were induced from excised axillary and terminal buds of neem on medium containing BAP, kinetin, and sucrose. The individual shoots were transferred to sand:soil (1:1) mixture for direct rooting and grown in polyhouse, where 90 percent survival was observed. Somatic embryos were induced via callus, as well as directly on the cotyledonary explants. However, type of the auxin was found to be a deciding factor for embryo induction pathway. Direct embryo induction was observed on medium supplemented with 2,4-D, BAP, and casein hydrolysate, while indirect embryogenesis was observed when IAA was used. Seventy percent embryos showed germination. Survival of the plantlets was 80-90 percent.

Drew (1996) developed a technique for production of micro-cuttings *in vitro* for clonal propagation of neem lines with high yields of azadirachtin. Neem shoots were established in cluture from nodal bud explants of galsshouse-grown plants. Apically dominant shoots were grown on hormone-free MS medium and multiplied by dissection into nodal sections. The use of cytokinins in this medium caused excess callus production on cut sections of the explants. Good root systems were initiated whenactively grwoing axillary shoots were removed from nodal sections and cultured on MS medium plus 10 μM (IBA) for four days, with subsequent root development on hormone-free medium. Longer exposure to IBA reduced rooting percentages, increased time to initiate roots, and produced excess callus at the base of the explants.

Rajagopal and Ramaswamy (1996) undertook investigations to isolate compounds having therapeutic value in neem through the production of secondary metabolites in callus and suspension cultures. Two explants (bark and root) were used for callus induction, cell suspension, regeneration and isolation of secondary metabolites. Among the two geotypes studied, one genotype (G1) responed well to callus induction in both the explants and a callus induction to 97.5 percent and 92.5 percent in bark and root, respectively. The best medium combiation for callus induction in both the explants was MS + 2,4-D (2 mg/lt) + Kin (0.5 mg/lt) + Sucrose (3%).

Weiven *et al*. (1997) studied induction of somatic embryogenesis in *A. indica*. Embryogenic calli were initiated from cotyledons or hypocotyles using an MS agar medium supplemented with 0.5 mg NAA, 1 mg 6-benzylaminopurine [benzyladenine] (BA), 1 g casein hydrolysate, and 50 g sucrose per litre. When transferred to a liquid medium similar to the agar medium but with NAA replaced by 0/5 mg IAA per litre, calli formed globular structures which developed rudimentary roots after 4-5 weeks incubation. When transferred into a hormone-free MS medium containing 1 g casein hydrolysate and 50 g sucrose per litre, active embryo masses started to appear after 1-2 weeks. Embryo production was found to improve more than 2-fold by adding 0.2 mg zeatin per litre to the medium.

Tawfik (1997) developed a micropropagation system to overcome the difficulty in conventional propagation techniques used for the neem tree. Four types of explants (shoot-tips, stem nodes, petiole and leaf segements) were used. Shoot tips and stem explants were cultured on woody plant medium (WPM) supplemented with kinetin and benzyladenine (BA) each at 0, 0.5 and 1 mg/lt. Four to five shoots per shoot tip explants were observed after 4-5 weeks in culture. Shoots excised from stem node explants produced 3-6 more shoots within 4 weeks when cultured on the same fresh medium. Petiole and leaf explants were cultured on WPM supplemented with BA, kinetin and thidiazuron (TDZ) alone or in a combination with NAA. Petiole segments produced callus when cultured on a medium containing BA, kinetin or TDZ each at 0/5 mg/lt. Regenerated shoots formed roots in a medium containing 0/5 or 1 mg/lt IBA. The plantlets were transplanted into soil potting medium, and were acclimatized gradually over 60 days.

Wewetzer *et al*. (1997) subcultured leaf and bark segments from 2-3 year-old *A. indica* plants from Nicaraguan and Togo twice and analysed for azadirachtin content. Nicaraguan cell lines had a significantly higher azadirachtin content than those from Togo. Calli from Togo only yielded azadirachtin when derived from leaf explants and cultured on White's medium, though no differentiation was observed on this medium. Azadirachtin content was 3 times higher on media supplemented with 15 g sucrose per litre compared to those with 30 g sucrose per litre.

Eeswara *et al*. (1997) reported a standard procedure for the micropropagation of the neem tree. Micropropagated shoots were initiated from leaf explants of neem, taken from potted plants (6 or 18-months old) grown from seed. Regardless of their seed origin (5 sources from Niger, 2 from Ghana, and 1 from Sri Lanka), shoots were successfully produced by culturing leaf explants on MS medium containing benzylaminopurine [benzyladenenine] (1 mg/lt), kinetin (0.8 mg/lt) and adenine sulfate (6 mg/lt) in complete darkness. These shoots were further multiplied on MS medium containing benzyladenenine (0.1 mg/lt), kinetin (0.08 g/lt) and adenine sulfate (0.6 mg/lt). Within 32 weeks, 80 shoots could be produced from a single leaf explant of size 10x10 mm. Some 55 percent of these shoots rooted on MS medium containing IBA (1 mg/lt) and all of these grew on transfer to soil.

Zypman *et al*. (1997) initiated light yellow friable callus and shoots from hypocotyl explants of *A. indica* on MS medium supplemented with 0.5 mg IAA and 1 mg benzylaminopurine/lt. Shoots were regenerated from nodal buds and calli. Approximately 60 percent of shoots developed roots when transferred to MS medium without BA. The best medium for enhanced root development was MS medium with 5 μM IAA. After 3 months of incubation in the same medium, somatic embryogenesis of callus culture was observed.

George and Kulkarni (1997) studied explants of leaflets, rachis, inflorescence axis and cotyledons from locally growing trees in Maharashtra callused with high frequency in an *in vitro* culture medium of MS + 5 percent coconut water + 2 μM BAP in 2-3.5 weeks, and differentiated into shoots in 2.5-5.5 months. The frequency of shoot initiation and their number per callus was higher in calluses derived from reproductive than vegetative parts of the trees. The shoots initiated roots were transferred to medium containing IAA and NAA, but the frequency was higher with IAA. The resulting plantlets could be successfully acclimatized to pot growth with 85 percent survival value. Plants in pots could be obtained within a period of 195-288 days from the date of initiation of the callus, depending on the explants used. The inflorescence axis and cotyledons yielded plantlets earlier than the rachis and leaflet lamina.

Lei-GuangFu *et al*. (1998) studied callus formation in explants of *A. indica* and the azadirachtin contents. All calli induced from root, leaf, stem and bark tissue were capable of synthesizing azadirachtin. Callus growth rate and azadirachtin content were highest in leaf explants. Azadirachtin content was 0.40, 0.50, 0.57 and 0.63 mg/g dry weight for bark, stem, root and leaf explants, respectively. Low ammonium salt medium had the highest callus growth rate. High ammonium salt MS medium had the greatest callus azadirachtin content. White medium, without ammonium salt, resulted in a slow rate of growth and low azaidirachtin content. The best callus growth and azdirachtin formation was found at the second or third subculture.

Venkateswarlu *et al.* (1998) standardized a micropropagation protocol for a mature tree of neem. Nodal explants were collected from actively growing young shoots of a 22-yr-old tree in Andhra Pradesh (India) with a straight stem, 50 kg seed yield/year, 25 percent oil in seeds and 0.6 percent azadirachtin in the kernels. Successful shoot bud induction and multiplication could be achieved on MS medium with suitable concentrations of cytokinin. Elongated shoots could be rooted *in-vitro* using 1/2 strength MS medium with 2.5 mg/lt IAA. Rooted plantlets were transferred to a soil.

Murthy and Saxena (1998) studied somatic embryogenesis and plant regeneration of neem. Somatic embryos were initiated from mature seeds of neem when cultured on MS medium supplemented with thidiazuron (TDZ). Regeneration occurred via somatic embryogenesis, direct embryo formation and through an intermediary callus phase. TDZ was very effective and induced somatic embryogenesis across a wide range of concentraions (1-50 μM). However, somatic embryogenesis was accompanied by callus formation at concentrations of 20 μM and above. Cell suspension cultures were established with the TDZ-induced callus and groups of large cell clumps were formed within 2-3 weeks. Plants were regenerated from both directly formed somatic embryos and somatic embryos derived from cell suspensions plated on semisolid medium devoid of growth regulators. Regenerated plantlets continued to grow in soil and were similar phenotypically to zygotic seedlings. This simple regeneration system may be beneficial for mass propagation of selected elite clones of neem.

Srividya *et al.* (1998) developed callus with 100 percent response from immature embryos of *A. indica* on MS medium supplemented with 2.0 mg NAA + 0.5 mg BA, 2 mg BA and 1.0 mg NAA + 0.5 mg BA/lt. For optimum multiple shoot formation, medium containing 2.0 mg BA or 0.5 mg BA + 0.1 mg NAA was used. These shoots rooted in half-strength MS medium supplemented with 0.5 mg IBA/lt. Roots contained 4 ppm azadirachtin and shoots contained 8 ppm azadirachtin and 3 ppm nimbin. Callus had negligible amounts of azadirachtin.

Veeresham *et al.* (1998) established callus cultures from leaves and flowers of *A. inidca*. Twelve week old callus produced 2.46 percent azadirachtin. Wewetzer (1999) established callus from leaf and bark. The azadirachtin contents were also obtained by TLC and HPLC methods.

Neem Leaf :

Leaf is one of the main non wood product of neem. An average tree of 7.5 to 8 m. tall gives about 350 kg of leaves (Fig. 1.4). Neem trees are good fodder for livestock when mixed with other fodder. In Andhra Pradesh farmers feed neem leaves to cattle and goats immediately after parturition for increasing milk secretion.

Neem is an attractive evergreen plant and reaches to the height between 15-40 m. and diameter between 15-20 m. It is observed that under normal circumstances, leaves remain green but sometime due to extended dry period, the leave are shed. The leaves are 20-40 cm long, pinnate, unpaired and the terminal leaflet is often missing. The petiole are short. The young leaves are reddish to purplish in color. The leaves possess insecticidal properties and thus used in medicine formulations. The leaves contains12.40-18.27 % crude Protien, 11.40-23.08 % crude fibre, 43.32-66.60 % N. free extract, 2.27-6.24 % ether extract. 7.75-18.37 % total ash, 0.89-3.96 % calcium and 0.10-0.30 % phosphorus. The leaves are used as antiseptic antihelmintic and also in oriticaria, earache, glandular tumer, postural eruption, jaundice, prurigo, eruption of skin causing itching, boils ulcers, liver ailments, psora and other postural eruption, leucoderma, diabetes etc. Sauce prepared from leaves is eaten in South India and is considered good for health. The dried leaves are used. to protect our woolen cloths and against stored grain pests. A rare kind of neem tree with leaves of having sweet taste has been reported from temple campus of Maganahalli in Karnataka. Surprisingly, however, all other trees in the vicinity are bitter.

Sarma *et al*. (1992) observed that major venation pattern was pinnate eucamptodromous in *A. indica* while pinnate craspedodromous in *Melia azedarach*. Patel and Singh (1996) reported that the number of leaves retained per plant was highest in *Pithecellobium dulce* followed by *M. azedarach*. The total leaf area per plant was highest in *M. azedarach* followed by *A. indica*.

Bark :

The bark is regarded as a bitter tonic, astringent, antipyretic, antiperiodic and febrifuge. It is also used in thirst, nausea, vomiting, ulcer as well as in skin diseases. It is said to contain a resinous bitter principle and is usually prescribed in the form of a tincture or an infusion. It is also very effective against malarial fever. The bark is moderately thick with longitudinal or oblique furrows on the outer surface. It is dark grey outside and reddish inside.The trees grows wild in dry forests of the Decan and in the open scrub forests of the dry one of Burma. It thrives best in the drier climate of the north Western parts, where normal rainfall varies from 18-45 inches and maximum shade temperature may be as high as 120°F. It grows on all kinds of soil viz. soline, alkaline and lavacidic soil but does well on black cotton soil. It is well suited for afforestation in dry situations and is a common avenue tree.

Root :

Studies were conducted on the root development and its relation with the grwoth of *A. indica,* in 2 age groups of trees (1 and 9 yr old). Each tree was measured for various shoot and root characteristics (roots to a depth of 1 m.). Soil samples (0-20, 20-40, 40-60, 60-100 cm depth) were collected for physico-chemical analysis. For all tree species, survival, growth and root weight were significantly higher on the sites where performance of trees was generally good than where it was poor. Lateral root twisting was the most common root deformation observed; a one sided root system was observed less frequently. Kinked tap roots were only observed on 1-yr old trees and L-shaped tap roots on 9-yr old trees. For both ages most of the roots were observed in the surface soil layers (0-20 cm). Also, there was a higher percentage of roots in the surface soil layers for trees on the poor performance sites than on good performance sites. The amount of roots (0-40 cm soil layer) had a positive and significant reltion with tree growth while root deformation was independent of tree growth. Over all sites, there were significant differences in some soil properties between those with poor and good tree performance.

Pal *et al.* (1992) investigaged on the effect of indole butyric acid (IBA), 1-naphthol, 2-naphthol carbolic acid and salicylic acid on the rooting of leafy softwood branch cuttings, and of IBA and 1-naphthyl acetic acid (NAA) on rooting of semi-hardwood branch cuttings of neem. Cuttings were collected in July from the lower part of the crown of 6-yr old trees. Auxin treated and untreated semi-hardwood cuttings failed to root. About 30 percent of the leafy softwood cuttings rooted even without chemical treatment. IBA, as well as the phenols, when applied alone, stimulated adventitious root regeneration on the cuttings. A combination of IBA with the phenols inhibited rooting.

Toky and Bisht (1992) observed the rooting patterns of some agroforestry trees in an arid region of North-Western India and found that in *M. azedarach*, the roots were confined to the top 80 cm of the soil profile. Reddy *et al.* (1993) conducted experiments to find out the rooting responses of shoot cuttings of *Inga dulcis* and *A. indica* to 10 rooting media showed that ordinary sand was the best medium, followed by vermiculite. The least rooting percentage was obtained in quartz sand for both the species. Other rooting media tested were peat moss, compost, soilrite, perlite, green moss, sawdust and soil.

Pal *et al*. (1994) conducted rooting trials using terminal and sub-terminal leafy softwood binodal segments (shoot cuttings) taken from actively growing branches of neem in the middle portion of the canopy of 6-yr old trees. The terminal segments rooted easily even without auxin (1000 or 5000 ppm IBA) treatment. The segments taken from the sub-terminal greenwood zone were only moderately easy to root but their rooting response improved on treatment with potassium permanganate (1000 ppm) and ascorbic acid (5000 ppm), but not with cobalt cholride (1000 ppm).

Verma *et al*. (1996) studied the effect of auxins on rooting of neem cuttings. The effects of IBA and NAA were studied on the rooting of neem (*A. indica*) cuttings. The cuttings were treated with 100, 500 and 1000 ppm of the auxins in February-March. The treatments were given by dipping the lower 1/3 protion of cuttings into auxin solution for 24 hr with a water treatment acting as control. All IBA and NAA treatments stimulated rooting and sprouting of neem cuttings, while control cuttings did not root or sprout. The 100 ppm dosage was best for both NAA and IBA, but IBA was more effective than NAA at all dosages.

Kamaluddin and Ali (1996) studied the effects of leaf area and auxin on rooting and growth of rooted stem cuttings of neem. Green stem cuttings, obtained from 2-yr old managed stock plants of neem with 30, 50 or 100 percent leaf area were rooted in a non-mist propagator following treatment with 0, 0.2, 0.4 or 0.8 percent IBA solution. The rooted cuttings were transferred to polyethylene pots and grown in a nursery for 12 weeks. Rooting percentages of IBA treated cuttings did not significantly differ from those of control cuttings. Likewise leaf area did not significantly influence the percentage of cuttings that rooted. The effects of leaf area or IBA treatment in the propagation unit were reflected on the growth of potted, rooted cuttings. The number of roots increased with the increase in IBA concentration and this response was enhanced by increasing leaf area. Root biomass increased with the increase in leaf area, even in control cuttings. Root development was highly influenced by leaf area and IBA, and hence the growth of rooted cutting. Cuttings treated with 0.2 or 0.4 percent IBA at 100 percent leaf area yielded the best performing rooted cuttings.

Shanthi *et al*. (1996) studied rooting of cotyledonary scions in *Azadirachta indica* seedlings. Seedlings 7 and 20 days old were decapitated in the hypocotyl region just below the point of cotyledon attachment, and the resulting scions planted in sand medium for rooting. After 15 days the acion from the 7day old seedlings had developed a good root system, and the stock had also produced healthy new shoots. The scion from the 20 day old seedlings developed no roots and died after 4 days, although the stock continued to produce new shoots. The stock and surviving scions were transplanted to polybags after 8 weeks, and planted out after 3 months. They grew well, reaching a height of 2 m and collar diameter of 24 cm after 2 years.

Stephen *et al*. (1996) determined the growth pattern of pot-grown neem seedlings in light and dark and the response of these seedlings to spray application of brassinolide (BR), gibberellic acid A3 (GA), benzyl aminopurine (BAP), edylene as chloroethylphosphonic acid (ET), and cycocel (CCC) either alone or in combination. One centimetre hypocotyl and epicotyl segments were excised and treated with the above plant growth regulators to observe elongation and developmental changes. Segments were also treated with indole butyric acid (IBA) to promote rooting in vermiculite. Both in dark and in lilght seedling growth was not continuous but sigmoidal with a plateau phase setting in after 10 days.

Gera *et al*. (1997) did rooting trial of *A. indica* through shoot cuttings. Branch cuttings 22 cm long and of 1.5 cm diameter were collected from young trees (5-10 yr old) in the first week of March 1996. The cuttings were treated with 0.1 percent Bavistin (Carbendazim) for 4-5 min and then washed thoroughly. The basal ends were then treated by the quick dip method with 1 of 6 treatments : water (control), $KMnO_4$ (1000 ppm) + IBA (1000 ppm), ascorbic acid (5000 ppm) + IBA (1000 ppm), or IBA (1000 ppm). The top ends of the cuttings were sealed with wax, and they were then planted in coarse sand at 15x15 cm spacing and grown under intermittent mist at 30°C and RH 70-80 percent. Rooting was recorded after 70 days. All treatments enhanced rooting over that recorded in the control (23%) but the best treatment was IBA alone, which gave 70 percent rooting, and also the largest number, length and dry weight of roots.

Palanisamy and Kumar (1997) studied the effect of position size of cuttings and environmental factors on adventitious rooting in neem. Variation in rooting was studied of branch cuttings of neem collected from 10 suprior 25 yr old trees at Jabalpur, Madhya Pradesh, India. Cuttings were from proximal, middle and distal positions of the branch and of different lengths (5, 12 and 25 cm) and diameters (distal cuttings 0.5 cm, middle 1 cm and proximal 2 cm); all were treated with 1000 ppm IBA before planting in sand beds in a mist chamber. Cuttings from the distal end of the branch rooted better than those from the proximal end, probably due to higher endogenous auxin level. Cuttings 25 cm long and 0.5 cm diameter from the distal end gave 100 percent rooting. Percentage rooting decreased significantly when the length of the cutting was reduced, and no rooting was observed in 5 cm long cuttings irrespective of the position, suggesting a sub-optimal level of endogenous auxin and other root-inducing factors in small cuttings.

Palanisamy *et al*. (1998) studied adventitious rooting in shoot cuttings of neem (*Azadirachta indica*) and karanj (*Pongamia pinnata*). Adventitious rooting was investigated in shoot cuttings of neem and karanj, collected from 10 yr old trees near Jabalpur, Madhya Pradesh, India. The influence of auxins (IAA, IBA, NAA) and B-vitamins on neem only was evaluated. Maximum rhizogenesis coincided with the emergence of new sprouts i.e. in Febrauary on neem and in March on karanj. Considering both rooting percentage and root biomass, IBA was the most effective auxin tested. Compared with auxin free controls, IBA (1000 ppm in neem and 800 ppm in karanj) significantly increased adventitious rooting percentage to 80 and 100 percent, respectively, numbers of roots and root biomass for both species. In neem, B-vitamin treatment gave 40-50 percent rooting of cuttings. In neem, some of the cuttings treated with 2000-3000 ppm IBA had well developed root systems without any sprouts.

Flowers :

The flowers are white and arranged in axillary normally more or less drooping panicles which are up to 25 cm long. The inflorescence (Fig. 1.3) bear 150-250 flowers. Protandric bisexual flowers and male flowers exists on the same individual. The ovary is trilocular. There are ten glabrous anthers. The number of chromosomes is 2n=28,30 n=14. The flower appears from March-May and the fruits ripen during June-August. They are greenish yellow when ripe and usually contain only one seed. The flower yeild a glucoside, nimbosterin, nimbosterol, nimbecetin and fatty acids. An infusion of the flower is the stimulant tonic and stomachic. It controls nervous headaches and skin diseases.

Karmaker and Chatterjee (1994) isolated two allergenically active components present in *A. indica* from whole pollen extract by sequential ammonium sulfate precipitation (0-90%), and DEAE-Sephadex A-50 ion-exchange chromatography followed by gel filtration through Sephadex G-200. The allergenicity of fractionated materials was tested by the skin prick test and ELISA inhibition, and showed that AlaI and AlaIVb are the major allergens.

Gupta *et al*. (1996) studied the reproductive biology of neem and suggested that neem (*A. indica*) may be a self pollinated tree species. They studied pollen dehiscence time, flower anthesis, pollen germination, fruit setting in selfed and open pollinated flowers and progeny performance of selfed and open pollinated seeds.

Datta and Chaturvedi (1997) observed that the concentration of airborn pollen of *A. indica* was lower than that of *Holoptelea integrifolia*. However, the distribution of *A. indica* was wider than *H. integrifolia* and extended in to southern limits of the country.

The floral biology, breeding behaviour and breeding strategy in neem was studied by Singh, *et al*. (1996). New vegetative growth appeared in late March, immediately followed by flower initiation. There was a considerable variation in the level of flowering ranging from its absence to profuse flowering. Flowering continued till the end of April. Flowers open in the morning from 7 to 11 am. Usually 3 to 5 flowers open daily in an inflorescence which is a cymose panicle consisting of hermaphrodite and male flowers. Thus inflorescence is andromonoceous in nature. This is the main reason for predominantly cross pollination in this species. Hermaphrodite flowers varied from 20 to 24 percent depending upon season. Fruit set was 3.0 (1995) to 9.0 percent (1994) of total flowers. Period of peak flowering was about 22 days. Allogamous nature of neem was durther confirmed through selfing where fruit set was from 0.0 (1994) to 0.15 percent (1995) of total flowers compared to more than 3 percent under open pollination.

Neem Oil :

Neem oil is a major non wood product of neem. The oil is greenish yellow to brown in color. It is a fixed acrid bitter oil having disagreeable odour. It is extracted from the seed by boiling on pressure. The oil is generally extracted by crushing of kernels or seeds or dried fruits. It is observed that 2-5 months old seed gives maximum oil. The oil has high copherol contents (1.17 mg/g) made up almost equally at gamma and delta form with just a trace of beta form. This oil has a high specific gravity 0.9235 at 15.5°C. The oil was easily soluble in ether, chloroform Carbon bisulphide, benzole etc. The oil contains 0.427 percent Sulphur. The quantity of sulphur is reduced to 0.109 percent if the oil is agitated with alcohol. Recently a process has been devoloped at Hyderabad to produce colourless and debittered neem oil that can be used for edible purpose.This oil is most important for commercial point of view.The oil yield is sometimes as high as 50 percent of the weight of Kernel. It has been estimated that India produces 3.5 million tons of kernals each year which produces 700,000 of 150,000 tons. It has been observed that neem seeds must be harvested during wet season and in the absence of drying facilities get the contaminations of *Aspergillus flavus* which produces the toxin aflatoxin. The khadi and village Industries commission has devised and popularised simple methods of developing, drying and decorticating neem products.The neem oil contains the following acids namely, 61.9 percent oleic acid, 14.9 percent palmitic acid 14.4 percent stearic acid, 7.5 percent linolic acid and 1.3 percent arachidic acid. The following glycerides are avilable in neem oil viz 0.6 percent fully saturated glycerides, 22.0 percent tri-unsaturated glycerides, 34 percent stearodiolein, 26.0 percent palmito-diolein,12% oleopalmitostearin, 5.0 percent oleodipalmitin. The oil also contains 1.2 -

1.6 percent nimbidin, 0.1 percent Nimbin and 0.01 percent Nimbinin and 0.2 - 1.0 percent Azadirachtin. The neem oil contents and the quantity of azadirachtin presence in the neem seed of different regions of the world were also observed by Ermel (1986). The details are given in the following table-1.

Table-1 : Average azadirachtin and oil contents of neem seed kernel samples from different countries :

Country	No. of Samples	Oil (% ± S.E.)	Azadirachtin (mg/g ± S.E.)
Dominican Rep.	44	46.6±6.4	3.43±0.74
Haiti	16	46.5±4.6	3.05±0.59
Honduras	1	45.3	4.20 -
Equador	26	46.6±3.1	3.99±1.19
Guinea Bissa	1	48.2	-2.40 -
Mali	2	54.1	-2.05 -
Senegal	22	44.3±4.0	3.30±0.63
Gambia	6	45.5±1.5	2.98±0.88
Niger	15	48.1±2.6	3.40±0.67
Togo	3	46.8	-5.40 -
Benin	57	47.4±4.4	3.75±0.94
Sudan	9	44.8±4.0	2.53±0.60
Somalia	19	46.4±2.4	2.90±0.88
Zanzibar	1	44.5	-4.80 -
Madagascar	1	38.2	-2.20 -
Iran	4	45.4±1.9	2.75±1.65
Yemen	7	49.7±2.5	4.44±0.9
India	9	47.6±5.5	5.14±1.80
Sri Lanka	3	50.1±5.0	3.40±0.34
Myanmar(Burma)	3	48.4	-6.10±0.70
Thailand	6	45.0±5.0	5.20±1.10
Australia	1	48.1	-4.90 -

Neem Cake :

The neem cake is the main by product of neem seeds. It contains 1.07-1.30 percent sulphur. This is one of the very useful fertilizer for our crop. This is both organic nitrogen manure and nitrification inhibitor. The average nutrient contents of neem cake are 5.2 percent nitrogen, 1.0 percent P_2O_5 and

1.4 percent K_2O. It has been experimented that neem cake when mixed with urea retards the nitrification rate for about two weeks. Thus more nitrogen is available as ammonium for a longer period.

Ketkar (1967) studied the effect and economics of using neem cake with one or more than one fertilizer like urea, superphosphate etc. He found 25-50 percent saving of urea nitrogen coupled with improve yeilds. The neem cake are very efficacious in the management of insects and other pests. The cake contains salanin, nimbin, azadirachtin and azadiradione as the major components. The azadirachtin repels the insects, inhibit feeding and affect their harmonal balance. It also protect the crop from the attack of termite. The latest research indicates that more than 200 insects species could be controlled effeciently. The neem cake contains 13- percent 5 percent crude protein, 26-50 percent Carbohydrate, 8-26 percent Crude fiber, 5-18 percent ash and 1-17 percent Acid insoluble ash.

Gum and Resins :

The gum which exudes from the bark occurs in the form of clear, bright amber coloured tears or fragments. It contains water 13-15 ash 3 ; galactans 12 ; pentosans 26 percent and some albumins and oxidase. A sweet liquid, neem toddy is occasionally obtained as an exudation from the upper parts of some trees. It has the foul smell of neem and contains sugars 6.5 and albuminous and gummy matter about 6.5 percent.

Arumugasamy *et al*. (1996) observed that resin canals and ducts in *A. indica* are embryonic in origin. Their initials are mostly confined to the peripheral regions of the embryo and endosperm. They are distributed naturally in the fruit walls, shoot apices, cortex of young stem, and midrib regions. These resin canals are unbranched, elongated, tube-like structures without an accompanying well-organized epithelium of secretory cells. At a few places, the canals exactly resemble resin ducts due to their discontinuity. The resin canals contain many dense, globular bodies inside the lumen and their resinous nature is revealed by fluorescence and electromicroscopic studies. However, these resin canals differ from the resin glands and nectaries of Meliaceae both in structure and in nature of the secretory products.

Selvi and mohan (1999) reported for the first time, the presence of a new flavonoid viz., 8-prenyl-5,7-dihydroxy-3-(3-hydroxy-3, 3-dimethylbutyl) - 4-methoxy-flavanone, in the resinous exudate of the glands of the tender leaves and shoot apices of *A. indica*. Besides, two maxine sterols viz., fucosterol and (24E)-isopropenyl cholesterol were isolated from the same parts of the plants.

Wood :

The wood is moderately heavy, medium textured and narrowly interlocked-grained. The sap wood is greyish-white. The heart wood is red when first exposed and later turns reddish brown. It is not very lustrous and is slightly aromatic. Anatomically it is featured by ditinct growth rings, medium sized vessels occluded by reddish brown gum, parenchyma cells with deposits of gum, and fine rays. The timber is used in house building and for making boards, panels toys, ploughs and in furniture.

Parasitic Plants :

Browne (1968) reported *Dendrophtoe falcata, Tapinanthus* sp. and *Loranthus* sp. as parasites of neem. Vyas (1989) reported parasite *Ficus religiosa* Linn for the first time as epiphytic on the main branch of a roadside tree of *A. indica* in Gujrat, India. Kalkulu (1993) reported epiphytic moss *Palytrichum juniperinum* on neem tree.

Karunaichamy *et al*. (1993) reported *D. falcata* on neem. They observed that the host and the parasite had a similar diurnal pattern of photosynthesis although net photosynthesis was greater in *D. falcata*. Stomata of both plants opened quickly in the morning, then gradually closed. Level of K and Mg were greater in leaves of *D. falcata* than that of *A. indica* whereas N, P and Ca were similar in both.

Schmutterer (1998) reported the semi-parasite *Cassytha filiformis* on neem and found that the plant is able to kill mature trees growing under unfavourable conditions.

Nutritional Aspects :

In the waste land areas, fertilizer application is a must to grow neem plantation. Jordan (1985) suggested that neem can grow in poor fertility soils. However, it is a established fact that to aim in getting better yeild of fruits, nutrient supply is a must. Kandaswamy et al (1988) observed that application of bio-fertilizer like VAM (consists of *Glomus fasciculatum* - a fungi) is an important step to grow good seedlings. This VAM enhance nutrient supply and water up take and protection against diseases. Bala Subramanian (1992) emphasized the need to treat the seedlings with Frankia and *Glomus*.

Fruits and Seeds :

Neem at the age of 5-6 years reaches to reproductive maturity but good yeild is obtained when it is 10-12 years old. A medium size tree produces 37-55 kg of fruits while at the initial stage it yielded 25 kg fruits per tree. Yellow green fruits are the best for good germination but mature yellow fruits are equally good. The seeds must be dried under shade soon after harvest. One should take precautions not to allow *Aspergillus flavus* to grow other wise the seed quality will be affected. Maximum oil yield is recovered after 3 months of storage. It would be better to put tarpauline / polythene sheets / gunny bag sheets below the tree to collect dust free good quality seeds (Fig. 1.7).

The oil contents of the seed is 35-45%. The effective ingredients are present in all parts of the tree but most highly concentrated in the seeds. The effective substances of neem loose their potency in sunny conditions being broken down by ultra violet radiation. It is therefore best to apply neem preprations in the evening. So far neem preparations seem to have had no, or almost none, negative effects on Hymenoptera family whose members are useful pollinators and parasites of insect pest and also on bees. (Schmutterer 1995). The fruits are glabrous olive like which vary in shape from elongate oval to nearly roundish and when ripe are 1.4-2.8 x 1.0-1.5 cm. They are green when young and yellowish green to yellow when mature (Fig. 1.5, 1.6). The white hard shell enclosed one or sometimes two seed kernals. The seed is very rich in oil. Azadirachtin an active ingredient of neem has been isolated from the seeds. From neem seeds neem oil and its by product neem cake is produced. During 1991-92 about 3220-20 tons of neem fruits costing Rs. 2648/ton was made available. Again in 1992-93 about 2471 tons of neem fruit costing Rs 2725/ton was made available. In other parts of India the kernal is costing 3000-4000 rupees per ton while neem fruit costs Rs 1200-2000/ton.

Tropical recalcitrant seeds can be stored for longer duration by maintaing high seed moisture contents and a certain amount of gas exchange. These seeds are sensitive to low temperature. Chaisurisri (1986) successfully stored the seeds at 15°C and moisture contents 46.18 percent. The seed under these conditions retains germination upto 62%. Seeds with relatively high moisture content requires high O_2 concentration for higher retention of viability, perforated containers are therefore better for maintaining low CO_2/O_2 ratio and this in fact prolongs the viability.

Ketkar (1976) reported that 8m tall tree produces about 37-55 kg fruits per year. Radwansky (1977) reported average production of 20.5 kg fruits per tree in Nigeria. In Nicaragua, the average fruit yield was 19.2 kg from 6-9 years old trees. Hedge (1993) recorded 25 kg yield per mature tree per year. In Kenya even 100 kg seeds have been obtained from a single neem tree.

Ezumah (1986) stored neem seeds with 16-18 percent moisture at 26°C for 56 days having 35 percent viability of seeds. Roederer and Bellefontaine (1989) reported that seed with endocarp stored at the centre technique Forestier Tropical shared 42 percent germination capacity even after more than 5 years of storage.

It is, therefore, important for doing large scale research on (i) testing of the different combinations of temperatures, containers and seed moisture content for long term storage of neem seed (ii) Development of techniques for cryogenic storage of excised embryos for neem seeds for long term germ plasm conservation.

Maithani *et al.* (1989) observed that physiologically mature seeds with maximum germination capacity and longevity at 10-12 week after flowering (late July-early Aug.) when some of the fruits turned yellow and started to fall. This peirod coincided with peak green weight, size, and a fully grown embryo in a hard fibrous endocarp. Seeds were collected in early Aug. 1986 from a further 5 trees and seeds extracted and dried for 4 days as before. They were stored at room temperatre and at 15 or 5°C in sealed and perforated polythene bags, perforated cardboard boxes (seeds wrapped in blotters) and over silica gel in desiccators. Both sealed containers and low temperature (5°C) caused rapid deterioration of seeds gave complete loss of germination capacity in 1-4 months, while the use of aerated containers at room temperature or 15°C inhibited deterioration so that about 15 percent seed still germinated after 6 months storage, compared with a initial germination of about 60 percent.

Roederer and Bellefontaine (1989) reported that neem seed after 2-6 months storage had a germination percentage at or near zero. However, seeds with endocarp stored at CTFT for several years showed 42 percent germination after 75 years storage.

Venkatesh *et al.* (1990) studied the viability of neem seeds. Mature fruits of neem were collected from Karnataka, India. The fruits were peeled, washed in running water, sun dried and stored in cloth bags at room temperature. Sampling for germiation tests was done at 3-day intervals for 15 days, and thereafter at 15-day intervals up to 4 months after collection. Germination was carried out in sand. Germination of fresh seeds was 91 percent, but this value decreased to 50 percent after 15 days storage and 10 percent after 4 months storage.

Ponnuswamy *et al.* (1991) studied on seed longevity and the effect of fruit grades in neem. Depulped neem fruits (Fig. 1.8) (collected from a 20-yr old tree in Tamil Nadu) were graded by the flotation method, which gave 16 percent floaters. These weighed comparatively less than the sinkers, possessed a higher endocarp content and recorded low viability in terms of maximum germination and vigour. In contrast the sinkers recorded an increase of 35 percent in viability and 92 percent in vigour index. In a separate study, seeds stored at an ambient temperature of 33.8°C recorded a germination of 8 percent at the end of 3 months compared with 90 percent when freshly collected. This is attributed to desiccation injury as the moisture content dropped from an initial 30.8 to a low of 15.5 percent at the end of 3 months. In contrast, seeds stored in earthen pots burried in a moist sand bed manifested little loss in seed moisture content and recorded a germination of 62 percent at the end of 3 months. Density grading for maximizing viability and vigour, and moist storage in earthen pots for prolonging longevity, are recommended for the species.

Kumaran *et al.* (1994) studied the effect of growth regulators on seed germination and seedling growth of neem. The effects were determined with GA (gibberellic acid), indole acetic acid (IAA), kinetin (KN) and chlorocholine-chloride (CCC) at 200 and 400 ppm on seed germination and seedling growth of *A. indica*. Seeds were soaked in the solutions for 24 and 48 hrs; water soaked seeds served as control. Results indicated that germination was enhanced by KN, with the highest germination at 200 ppm KN. Seedling growth in terms of root and shoot length and number of leaves was greatly ennhanced by CCC, the effect being higher at 400 ppm concentration with a 48 hrs soaking period.

Kumaran *et al.* (1996) collected seeds from 28 one-parent families of neem (*A. indica*) in seven agroclimatic zones of Tamil Nadu, India and studied for the variation and heritability of seed length, seed breadth, seed length-breadth ratio, on 100 seed weight, and oil content. Results showed significant differences among the one-parent families for all the seed parameters studied. Hundred seed weight recorded moderate genotypic (12.9%) and phenotypic (15.8%) coefficients of variance. Heritability was the highest for oil content (99.8%). Seed length showed high gennetic gain. From these results, it is concluded that hundred seed weight may serve as a reliable selection index by virtue of its high genotypic coefficient of variance, high genetic advance, and moderate heritability.

Karivaratharaju *et al.* (1996) studied the effect of seed moisture and drying method on neem seed quality. The study revealed that drying methods do not significantly influence the germination percent, root length, epicotyl length, hypocotyl length, dry matter production and vigour index of the neem seeds. Study also revealed that the neem seeds can be dried to 10 percent moisture content by different drying methods namely shade, sun and oven drying methods without much damage to the seed viability.

Kumar *et al*. (1996) observed that average germination percentage of seeds plucked from the trees was significantly higher than that of seeds collected from the ground under the trees. They also observed that the average germination of depulped seeds was more than 7 times higher than that of seeds with pulp.

Rai (1996) collected fruits at the yellowish tinge stage i.e. they were neither over- not under-ripe, depulped immediately after collection, dried overnight, treated with charcoal ash and stored in gunny bags. These seeds were brought to the nursery and sown in plastic trays and polybags. Seeds raised in plastic trays in a glasshouse gave significantly higer germintation (70.06%) than seeds raised in polybags (35.27%), and there were signifcant differences between provenances.

Viswanathan *et al*. (1996) observed the physical properties of neem nut. The length of neem nuts, measured at moisture contents (MC) from 7.6 to 21 percent ranged from 12.87 to 16.20 mm and stem-end diameter ranged from 6.86 to 8.52 mm. Crushing strength along the longitudinal and diametral axes decreased as MC increased, while 1000-nut weight, kernel mass as a percentage of total mass and angle of repose increased as MC increased. Porosity, bulb density and particle density decreased linearly as MC increased. Coefficient of static friction on various surfaces also increased as MC increased and was highest on plywood, followed by mild steel, galvanized iron and glass.

Bharathi *et al*. (1996) studied the effect of drupe maturity on seed germination and beedling vigour in neem. A study was carried out with seeds obtained from a 20-yr old neem tree at Coimbatore, India at different drupe maturity stages - green (G), greenish yellow (GY) and yellow (Y). Germination percentage, root and shoot lengths, dry matter production (DMP), vigour index (VI) and germination value were assessed. The results showed that among the maturity stages, GY drupes registered superiority over Y and G drupes. Hence, for maximum vigour and viability of neem seeds, the drupes should be harvested when they form GY in colour.

Gamene *et al*. (1996) studied storage behaviour of neem seeds from Barkina Faso. The initial germination capacity, expressed as percentage normal seedlings, was 87 percent. To determine the effect of moisture content, seeds with intact endocarp were stored in drums with air of a constant temperature (20°C) and different relative humidities. Durig 9 weeks of storage, the germination capacity remained highest (70% or more) at relative humidities of 55 percent (moisture content about 9% on a fresh weight basis) and 75 percent (moisture content about 13%) and was markedly lower at either lower (32 and 20-25%) or higher (95%) relative humidities. The effect of storage temperature was examined in 2 experiments. In the first, seeds with 8.9 percent moisture were stored at 3, 20, 30 or 50°C for 2 weeks. The germination capacity of these seeds was lowered at the highest temperature only. In the second, seeds with 4.5, 9.2 or 12.9 percent moisture were stored at both 3 and 30°C for up to 5 weeks. Irrespective of seed moisture content, germination capacity was lost most rapidly at the lowest termperature.

Kumaran *et al.* (1996) studied the effect of presowing chemical treatment on germination and seedling growth in neem. Seeds of neem collected from mature trees in Tamil Nadu, India were soaked in potassium chloride (KCl), potassium dihydrogen phosphate (KH_2PO_4), potassium nitrate (KNO_3) and diammonium phosphate (DAP) solutions, each at two concentrations (1.0 and 2.0%) for 24 hrs. Seeds soaked in water served as controls. The treated seeds and controls were sown in nursery beds, and germination and seedling growth at 1 month was recorded. Soaking in 2 percent KH_2PO_4 gave maximum germination, shoot length and number of leaves, and 1 percent KH_2PO_4 gave maximum root length. The KCl treatments performed similarly to the controls with the other treatments promoting intermediate performances.

Shanthi *et al.* (1996) studied the genetic significance of late flowering in neem and its effect on the germination of seeds. *A. indica* normally flowers in April (in Tamil Nadu, India) and the fruits are ready for harvesting by July, although the fruiting season sometimes extends to August. The seeds have a short period of viability i.e. only a few months after harvest so late flowering would be useful in solving seed procurement problems for later planting. Such a genotype was identified in Tirupur which flowered in September and produced fruits in December. Seed germination and seedling characteristics were compared from the 2 periods; geminnation was 83 percent in the early flowering type and 67 percent in the later flowering type. Seedling growth was similar, but there were more abnormal seedlings in the later flowering type, probably due to selfing. Late flowering types would be useful for breeding studies.

Dumet *et al.* (1997) studied desiccation tolerance and cryopreservation of embryonic axes of recalcitrant seeds. Because recalcitrant seeds are generally large and show very curtailed longevity even if stored hydrated cryopreservation of their embryonic axes seems to be the most obvious way to create a gene bank. The aim of this investigation was to determine the minimal water content tolerated by embryos of 5 recalcitrant species and their subsequent survival after cryopreservation. As sucrose is known to improve desiccation- and cryopreservation-tolerance of many plant tissues, experiments were performed with both non-pre-treated and sucrose pre-treated embryonic axes. Embryonic axes of *Camellia sinensis* and *Azadirachta indica* tolerated dehydration to very low water contents, 0.14 and 0.23 g water per g of dry weight, respectively. In such dry states most of the axes survived cryopreservation.

Dod *et al*. (1997) studied the effect of storage container on the viability of neem seeds. Seeds of *A. indica* (9% moisture content) were stored in cloth, polylined jute, paper or polyethylene bags for 2 months under ambient conditions of Vidarbha region, India. Seeds lost viability with storage period. Seeds stored in cloth bags maintained significantly higher viability, 37.9 percent germination after 60 days of storage compared with the other treatments giving 19.03-30.37 percent germination after 60 days storage.

Sridharan *et al*. (1997) collected seed samples from 12 locations in Tamil Nadu, India and determined the oil contents and azadirachtin. The authors found variation in the seeds collected from different locations. Climatic factors at the different locations were correlated with azadirachtin and oil contents of seeds, with a significant positive correlation of sunshine hours during September-March with seed oil content, and a negative correlation between the total number of rainy days during the fruiting season (April-August) with seed azadirachtin content.

Singh *et al*. (1997) tested two methods of drying neem seeds (after depulping the fruits) sun drying and shade drying. Moisture loss was more rapid during sun drying, dropping from an initial 26.6 to 6.0 percent in 6 days compared with 11 percent in shade dried seeds. Drying did not reduce germination much : and germination was as high as 60 percent at 6 percent moisture content. Storing seeds with a high moisture contennt (>25%) under deep freeze conditions had a deleterious effect on viability. Storage of such seeds under ambient or refrigerated conditions gave better results, but viability loss was less if seeds were stored under ambient or refrigerated conditions with a moisture content <10 percent.

Sacande *et al*. (1997) tested the germinability of developing seeds and cell cycle events and found that germination could occur after 8 weeks of seed development but normal seedlings resulted only after 10 weeks or more. The change in colour from green to yellow takes place after 12 weeks of seed development. A strong beta-tubulin signal was observed in tips of young embryonic radicles and cotyledons, but non-detectable signal was found in 9 week old ones and in those from green-mature and yellow fruits. Because DNA replication and beta-tubulin level were almost negligible at seed maturity, it is suggested that these factors are not involved in the intermediate storage behaviour of neem seeds.

According to Sacande *et al*. (1998), *A. indica* seeds of 1 Sri Lankan and 4 Burkina Faso provenances were dried to equilibrium moisture content (MC) at different RH at 20°C, then placed in open storage at 20°C or hermetically sealed in packets and stored at temperatures ranging from -20 to +20°C for 2 years. There was hardly any difference in storage behaviour between seed lots, regardless of provenance. Seeds originating from mature yellow fruits had greater longevity than seeds from green or brown fruits. In all storage experiments with seeds having MC 10 percent, viability was preserved best at 10-15°C, indicating that neem seed is chilling and freezing sensitive. There was no survival longer than 2 years under these conditions. At MC of 4-8 percent, seeds were considerably more tolerant of low temperature storage and had 40-60 percent viability after 2 years at all temperatures tested (-20 to + 20°C). However, the seeds were sensitive to imbibitinal stress, which could be alleviated by imbibition at temperatures of 25-30°C or above.

Eeswara *et al*. (1998) stored neem seed extracted from fruit harvested at two stages of maturity i.e. ripe and fully mature were stored at three seed moisture contents (MC) (undried, <15 and <10% MC) and in three storage conditons; an air-conditioned room (22°C, 76% RH), refrigerator (3°C), and freezer (-20°C) in sealed thick polythene bags. The germination of the undried ripe seeds was high (86%) and remained above 80 percent when seeds were dried to 13.2 and 7.1 percent MC whereas the low germination (46%) of the fully mature seed fell to around 30 percent after drying to 14.2 and 8.1 percent MC. At all seed MCs the loss of germination during storage occurred most slowly in the refrigerator followed by the air-conditioned room. With the exception of seed having <10 percent MC, storage in the freezer was highly damaging. Thus after 5 weeks storage at 13.2 percent MC, the ripe seeds had germinations of 65, 52 and 0 percent when stored in the refrigerator, air-conditioned room and freezer, respectively. Reducing the seed MC prolonged the storage life of both ripe and fully mature seeds in all three storage conditions.

However, Karivaradaraju *et al*. (1999) observed that the drying methods did not significantly influence the germination percentage root length, epicotyl length, hypocotyl length, dry matter production and vigour index of the seeds. It is also observed that at 10 percent moisture, neem seed can be stored without much damage to seed viability. Singh *et al*. (1999) recommended the collection practices, picking of neem fruit, depulping, cleaning and drying, decortication and storage methods.

Advantages and Disadvantages of neem

Advantages :

Neem is one tree where every part of the tree is used for many useful purposes. This is used for the treatment of human diseases since times immemorial. The uses are :

(i) Neem an eco-friendly pesticide : A number of insects,plant pathogens and nematodes are being controlled by neem products. About 50 neem products are being manufactured by various private agencies and government agencies for the control of pests and nematodes. They are called eco-friendly pesticides as they do not normally harms mammals, birds and useful insect, bioagents and pollinators.

(ii) Neem useful in Ayurveda, Unani and Homeopathy : In Ayurveda medicines, green leaves are used against epistaxis, eye trouble, small pox, leprosy, lack of appetite and to control intestinal worms. The old leaves are used against ulcers and boils. The flowers helps in bile control and in the control of intestinal worms. The young twigs relieve coughs, asthama, piles and control intestinal worms . The unripe fruit is effective against piles, intestinal worms and urinary troubles. The ripe neem seeds control epistaxis and eye trouble. The seed kernel helps in controlling leprosy and intestinal worms. The bark controls skin diseases when paste is prepared but the liquid extract controls fever.

Some times in old trees cracks develops in the bark and from the stem liquid oozes out which we call as Neem Ka Neera or Neem Ki Toddy. The liquid once started oozing, continues for months together. The place from which this liquid comes out, produce a sound which in later period becomes light. The toddy is a useful medicine. It is colorless, bitter in smell, unacceptable with light sweet in taste. The pH is 3.5-5.00, density 0.997-1.0525, water availability 89.7-92% and boiling point 97'C. The liquid has high amino acid and sugar contents. This also contains gum, phosphorus and protein. This is used against a number of diseases viz. skin disease like sun burn, leprosy, weakness, blood pressure, wounds, ringworms and ulcers.

The Unani medicines are preapred against leprosy , itches, pimples and blood purifying medicines. It also serves as excellent digestive. In Homeopathy, as many as 67 drugs have been prepared from neem. These medicines were introduced in Germany, Europe and America.

Gum of neem is generally excluded out from the bark during spring and summer season. It is yellowish in color and contains lot of protien. The Indian neem gum is also called as 'East India Gum'. This gum is used to color silken items and also to color thread. When eaten, it gives mental and physical strength.

The leaves of neem are very useful for a number of diseases. These contains Moisture 59.4%, Protein 7.1% Fat 1.0%, Carbohydrate 22.9% Mineral elements 3.4% and threads 6.2%. The nutrient material consists of calcium 510 mg/100 gm, Phosphorus 80510 mg/100 gms, Iron 17 mg/100gm, Thiamin 0.04/100 g, Niasin 1.4 mg /100 g, Vitamin C 218 mg/100 g, and Karotin 2.00 mg/100 g. Considering these nutrient minerals, it is observed that neem leaves are nutritionally better than other leafy vegetables. Neem leaves and tumeric paste prepared in the ratio of 4:1 cures all the skin diseases.

The neem brush is very useful in cleaning our teeth. The use of neem brush helps in killing bacteria, give nutrition to teeth and tooth gum, and in removing bad smell of the mouth. Neem leaves are used to remove plaque diseases of the teeth which is like Pyrohaea. This disease increases with the increase of the bacteria in the teeth. The liquid and enzymes oozes out of this bacteria spoils our gum and we call it gingvitis disease, later the bones of the gum are destroyed and then it is called as Pyorohea or Pariodontis disease. A number of toothpaste, antibiotic, antibacterias, enzymes, urea and floride compounds are in use but juice of the neem (1%) is the best in controlling the above disease. The flower of neem are used for eye ailments, for helminth parasites problems and for gall bladder problems. The inflorescence of neem when soaked in water over night and then taken as drink cures skin diseases and weakness.

(iii) Use of neem in agricultural production : The neem seed and the kernel is one of the very important part of the neem plant. The azadirachtin oil and neem cake is produced from this. All these products are commercially very important. If the temperature is above 70°C, the kernel looses oil production, properties, medicinal properties and azadirachtin quality. It is therefore important that neem seeds should not be dried on floor, or polythene sheet or by hot air blowers. The kernel contains insecticidal properties, antifeedant properties etc. The neem cake is used in agriculture as fertilizer, to reduce the nematode infestation in soil and also for nitrogen conservation. Lately the cake-water solution is also used to control insects.

The kernel is also used in extracting oil which is used for soap making. About 80 percent of the oil in India is used for soap making. In an estimate in 1975, about 18000 metric ton neem oil was used for soap making. Soap prepared from neem oil is one which has medium solubility, medium hardness, more fog and more cleaness of cloth.

(iv) Neem in birth control : The National Institute of Imunology has prepared a cream called Praneem from neem seeds which is very efficacious in controlling sex diseases and also in aids control. The neem seed has antifertility effects. Women in south-west Medagascar chew neem leaves everyday to prevent conception. Womens in Ghana and Gambia eat neem leaves for abortion purposes for 2-3 months continously. The neem oil kills the sperms of the human being in 30 seconds at the lethal concentration of 1000 mg percent. It is found that Nim-76 is the active ingredient which kills the sperms. Now excelsior company of Kanpur (U.P.) India has prepared a neem product called sensal which has antifertility effect. This is also used as antiseptic and vaginal toner. This National Institute of Immunology has developed a method for termination of unwanted pregnancy by oral intake of neem extract and published first report on antifertility affects by neem components in human beings and rats.

Disadvantages :

(i) When sunlight is exposed to neem products it loose its effectiveness. It is therefore necessary to have sun screens to avoid spoilage. These material when sprayed on plants remains effective for 2-3 weeks and can be stored at low temperatures. Simmilarly neem seed may be stored at low temperature in hygenic condition in the area where low humidity is available.

(ii) The neem seed has a very short viability and hence it is difficult to introduce the seed in any new location. Many times the seed germinates well but sooner it starts dieing. These are some of the major limitations in neem tree planting in our country and abroad.

(iii) The neem tree is very sensitive to frost and it can be grown in tropical and subtropical regions. It is observed that under dry condition the growth and yeild of seeds can be erratic. If the place is very dry, there are chances for high pest and disease infestation.

(iv) The neem tree planting has still not been done in methodical way for its commercial planting. We do not see any big area where neem is planted. However, neem trees are grown in the courtyard of the compound in village houses, road side trees and in two sides of the canals. During planting one do not foresee any problem from the livestock while human beings do create some problem by plucking the twigs for tooth brush etc. Therefore, for at least 3-4 years it is very important to protect the trees.

(v) We normally use to discuss a lot that the neem products are safe and very easy to apply but one should also consider that we do not have a fool proof method for mechanizing the process of collecting, storing and handling of seeds and the products. Neem is one crop in India which give fruiting only once and that too the fruits mature and drops from the trees in mansoon season when there are chances of more spoilage of the seeds and thus the supply is effected. Gathering the neem seed and then processing to prepare the product takes lot of time and efforts.

(vi) Although we say that there is no hazards in the use of neem products but no systematic studies have so far been made to know the residues and its side effect on human being and other mammals. There are reports of advrse effects of neem products on mammals and hence in some countries, the use of these products are limited. It is said that if the seeds are stored well after drying under shades and the moisture contents are below 14 percent, the infestation of *Aspergillus flavus* which produce a toxin, aflatoxins gets minimized, and as such the products prepared do not have side effects on human being and mammals. The presence of the fungus is therefore causes degradation of the various constituent of the neem seed kernal which ultimately looses its pesticidal effect and in turn the toxin aflatoxin causes hazards. It is therefore necessary to take care of the harvested seeds properly and avoid planting the neem trees in the humid areas.

(vii) Another set back which has came to the notice that by using neem products, we do get damage to same type of crops. It is another point to note that potato plants do not take up azadirachtin where as bean do accept. Hence this uncertainty is to be cheked in each plant species. In certain green house or field trials, it has been found that these neem products damage the plants like cabbages, onions and tomatoes. It is also considered that crude preparations and the extracts prepared scientifically project different actions. However pure extracts gave a better performance in the field.

(viii) Time of application of neem is very important factor but in general the extracts are very effective if spraying is done in morning and evening. It is always better to avoid spraying when the Sun is very hot. Another point which is important for us is to spray after knowing fully the life cycle, habits, and other details. A slight mistake in the operation at wrong timings leads to failure of crop. It is therefore of paramount importance to trained the farmers for its use, otherwise the effect of the neem will be minimized.

(ix) Another strange point in neem to be considered by us is that although neem protects a large number of insect pests in various crops, but this do have a high infestation of various pests damaging various parts of the neem tree. There are reports that the neem tree is being killed by these pests in northern Cameroon, Chad, north eastern Nigeria and eastern Niger. The scale insect *Aonidiella orientalis* is a very serious pest in Central and south India. These scales suck the sap of the plants and cause damage. Simillarly the neem leaves although have the fungicidal and antibacterial qualities, but still different pathogens do attack different parts of the plant and cause damage. The leaves of the neem are susceptible to leaf spot *Cercospora subsessilis* and bacterial blight *Pseudomonas azadirachtae.*

(x) The neem products have a little slow action, as compared with the chemical pesticides. In general the insect die when treated with neem a little late and therefore do not possess the required knock down effect. Although some tall claims are avialable when it is said that some products kills the insect within 24 hours. This slow action of the neem is another setback for promoting neem products in pest control. It is therefore essential to promote and do further research to develop neem extracts which could give quick effects. It is also to be noticed that neem products do have adverse effect on *Gambusia* sp. fish when neem products were applied in water. It is therefore very important that sincere efforts may be made to do toxicological research with neem products. Although neem products are safer to beneficial insects but a few cases are available when neem gave adverse effect. e.g. larvae of hover flies. When bees and butterflies drink nectar from the flowers of the neem sprayed plant, these may carry the traces of neem and cause poor reproducing character.

(xi) It is also observed that registration of the neem products is very difficult as the authorities are psychologically not sound to accept the product as an alternative to pesticides. Not only this there are other problems like poor standardization of products and synergestic mixture of other ingredients are another important points which are disadvantages to accept neem registration.

(xii) The Pharmacological effect of the neem are being discussed in many old literature but its overall clear effect on the diseases are not yet been critically analyzed. There are examples in west Africa where neem leaves causes kideney damage when taken over for a long period. Neem extracts were found to be toxic to guinea pigs and rabbits where as leaves when fed to goats and guinea pigs reduce their growth rate. The use of neem oil as an antivaginal spermicide or male contraceptive is still unclear.

References

Ahmad, R.; Khan, D. Ismall, S. (1985). Growth of *Azadirachta indica* and *Melia azedirach* at coastal sand using highly saline water for irrigation. *Pakistan Journal of Botany*, **17**(2) : 229-233.

Ahmed, S.; Salem Bamofleh and Ma'toug Munshi (1989). Cultivation of neem (*Azadirachta indica*, Meliaceae) in Saudi Arabia. *Economic Botany*, **43**(1) : 35-38.

Allan, E.J.; Eeswara, J.P.; Johnson, S.; Mordue, A.J.; Morgan, E.D. and Stuchbury, T. (1994). The production of azadirachtin by *in-vitro* tissues of neem, *Azadirachta indica*. *Pesticide Science*, **42**(3) : 147-152.

Arumugasamy, K.; Udaiyan, K.; Manian, S.; Balasubramanian, C. and Mohan, P.S. (1996). Studies on the resin canals and ducts of *Azadirachta indica* A. Juss. (Meliaceae). *Neem and Environment*, **1** : 151-158.

Balasubramanian, A. (1992). Role of microorganisms in agro-forestry extension and training. Directorate of extension and education TNAU Coimbatore : 43.

Bharathi, A.; Umarani, R.; Karivaratharaju, T.V.; Vanangamudi, K. and Manonmani, V. (1996). Effect of drupe maturity on seed germination and seedling vigour ih neem. *Journal of Tropical Forest Science,* **9**(2) : 147-150.

Brenner, A.J.; Jarvis, P.G. and VanDenBeldt, R.J. (1991). Transpiration from a neem windbreak in the Sahel. *Proceedings International Workshop*, Niamey, February, 1991.

Burman, U.; Kathju, S.; Garg, B.K. and Lahiri, A.N. (1991). Water management of transplanted seedlilngs of *Azadirachta indica* in arid areas. *Forest Ecology and Management*, **40**(1-2) : 51-63.

Cannell, M.G.R. (1983). Plant Management in Agroforestry Manipulation of trees, population densities and mixtures of trees and herbaceous crops. Plant Research and Agroforestry (ed. P.A. Huxley) CCRAF Nairobi Kenya.

Chaisurisri, K.B. Ponay and Wasuwanich. P. (1986). Storage of *Azadirachta indica* A. Juss seeds. *Embryon* **2** : 19-27.

Chander, Harish; Singh, R.R.; Mandal, B.S. and Chander, H. (1996). Effect of auxin on number of leaves per cutting and length of main branch in neem (*Azadirachta indica*). *Annals of Biology*, Ludhiana, **12**(1) : 57-61.

Datta, K. and Chaturvedi, M. (1997). Distribution of airborne pollen of *Azadirachta* and *Holoptelea* in India. *Journal of Palynology*, **33**(1-2) : 83-94.

Delwaulle, J.C. (1979). Forest plantation in dry tropical Africa. Techniques and species to use [IV]. *Boise of Forets des Tropiques*, No. **184** : 45-59.

Dhar, A. and Drew, R. (1996). Somatic embryogenesis of *Azadirachta indica* (A. Juss) and HPLC analysis for *in vitro* azadirachtin production. *Neem Newsletter*, **13**(2) : 26-27.

Dod, V.N.; Gabhane, V.V.; Patil, B.N. and Pagar, P.C. (1997). Storability of neem (*Azadirachta indica* A. Juss) seeds as influenced by storage containers. *Crop Research*, Hisar, **14**(2) : 289-292.

Drew, R.A. (1996). Clonal propagation of neem by tissue culture. *Neem and Environment*, **2** : 999-1006.

Dumet, D.; Berjak, P.; Ellis, R.H. (ed.); Black, M. (ed.); Murdoch, A.J. (ed.) and Hong, T.D. (1997). Desiccation tolerance and cryopreservation of embryonic axes of recalcitrant species. *Proceedings Fifth International Workshop on Seeds*, held at Reading, U.K. on 10-15 September, 1995 : 771-776.

Eeswara, J.P.; Allan, E.J. and Powell, A.A. (1998). The influence of stage of seed maturity, moisture content and storage temperature on the survival of neem (*Azadirachta indica*) seed in storage. *Seed Science and Technology*, **26**(2) : 299-308.

Eeswara, J.P.; Stuchbury, T.; Allan, E.J. and Mordue (Luntz), A.J. (1997). A standard procedure for the micropropagation of the neem tree (*Azadirachta indica* A. Juss). *Plant Cell Reports*, **17**(3) : 215-219.

Ermel, K; Pahlich, E; Sehmutterer, H. (1986). Azadirachtin content of neem Kernels from different geographical locations and its dependence on temperature, relative humidity and light. Proceeding third international Neem Conference, Nairobi, Kenya 10-15 July : 171-184.

Ezumah, B.S. (1986). Germination and storage of neem (*Azadirachta indica*) seeds. *Seed Science, Technology* **14** : 593-600.

Gamene, C.S.; Kraak, H.L.; Pijlen, J.G. van; Vos, C.H.R. de; Van Pijlen, J.G. and De Vos, C.H.R. (1996). Storage bahaviour of neem (*Azdirachta indica*) seeds from Burkina Faso. *Seed Science and Technology*, **24**(3) : 441-448.

Gautam, V.K.; Nanda, K. and Gupta, S.C. (1993). Development of shoots and roots in anther-derived callus of *Azadirachta indica* A. Juss - a medicinal tree. *Plant Cell Tissue and Organ culture*, **34**(1) : 13-18.

George, I.A. and Kuulkarni, A.R. (1997). Micropropagation of neem - *Azadirachta indica* A. Juss. *Annals of Forestry*, **5**(1) : 50-54.

Gera, Mohit; Gera, N. and Meena, S.L. (1997). Rooting trial of *Azadirachta indica* A. Juss through shoot cuttings. *Indian Forester*, **123**(9) : 860-862.

Gill, R.I.S.; Gosal, S.S.; Dieters, M.I. (ed.); Matheson, A.C. (ed.); Nikles, D.G. (ed.); Harwood, C.E. (ed.) and Walker, S.M. (1996). Micropropagation of economically important tropical forest trees. *Tree improvement for sustainable tropical forestry, QFRI IUFRO Conference,* Caloundra, Queensland, Australia, 27 October- 1 November, 1996 Vol. 1 : 230-233.

Gogate, M.G. and Gujar, D.R. (1996). An assay of genetic variability through phenological studies on neem (*Azadirachta indica* A. Juss.) plantation. *Neem and Environment*, **1** : 123-128.

Gruber, A.K. (1991). Wachstum Fruchtertrag und Azadirachtingehalt der Samen von *Azadirachta indica* A. Juss auf Verschiedenen - Standorten in Nicaragua. *Doctor Thesis*, Techn Univ. of Berlin, Germany.

Gruber, A.K. (1996). Management of neem plantations and development of neem-based insecticides in Nicaragua. *Neem Newsletter*, **13**(3) : 30-31.

Gupta, V.K.; Solanki, K.R.; Gupta, R.; Kumar, R.V. and Datta, A. (1996). Reproductive biology of neem (*Azadirachta indica* A. Juss). *Range Management and Agroforestry*, **17**(2) : 187-192.

Hedge, N.G. (1993). Neem production and development constraints at grass root level. *World Neem Conference* (Bangalore, India). *Souvernir*, pp. 1-5.

Hoamuangkaew, W.; Medema, L. and Chuntanaparb, L. (1990). Yield and financial analysis of *Azadirachta indica* A. Juss. *Journal of Tropical Science*, **3**(1) : 72-79.

Inamdar, J.A.; Subramanian, R.B. and Mohan, J.S.S. (1986). Studies on the resin glands of *Azadirachta indica* A. Juss. (Meliaceae). *Annals of Botany*, **58**(3) : 425-429.

Islam, R.; Joarder, N.; Joarder, O.I.; Hossain, M.; Rahman, S.M. and Naderuzzaman, A.T.M. (1996). Somatic embryogenesis and organogenesis in neem (*Azadirachta indica* A. Juss). *Neem & Environment*, **2** : 955-960.

Jacobson, M. (1981). Neem research in the U. S. Department of Agriculture, Chemical, BIological and Cultural Aspects. *Proceedings First International Neem Conference*, FRG, Rottach, Egern.

Jha, Mohan; Choudhary, L.D. and Jha, M. (1990). Trial on stump planting of some tree species. *Indian Forester*, **116**(4) : 283-285.

Jhorar, B.S.; Dhilon, R.S. and Singh, V.P. (1996). Comparative evaluation of neem (*Azadirachta indica* A. Juss.) and Khejri (*Prosopis cineraria*) in desert ecosystem. *Neem Newsletter*, **13**(4) : 59.

Joarder, O.I.; Naderuzzaman, A.T.M.; Islam, R.; Hossain, M.; Joarder, N.

and Biswas, B.K. (1996). Micropropagation of neem through axillary bud culture. *Neem & Environment*, **2** : 961-966.

Jordan C.F. (1985) Nutrient cycling in Forest Ecosystems John Willey and Sons. London.

Joshi, M.S. annd Thengane, S.R. (1996). In vitro propagation of *Azadirachta indica* A. Juss (neem) by shoot proliferation. *Indian Journal of experimental biology*, **34**(5) : 480-482.

Joshi, M.S. annd Thengane, S.R. (1996). Potential application of *in vitro* methods for propagation of neem (*Azadirachta indica* A. Juss). *Neem and Environment*, **2** : 967-974.

Jussieu, A.De (1830). Memoire Sur le groupe des Meliacees. *Mem. Mus. Hist., Nat.*, Paris, 19 : 220.

Lei GuangFu; Zhu XiRu; Zhang YunKai; Zahng HaiBao; Liu Wei; Lei, G.F.; Zhu, X.R.; Zhang, Y.K.; Zhang, H.B. and Liu, W. (1998). Callus formation in explants of *Azadirachta indica* and the azadirachtin content. *Journal of Tropical and Subtropical Botany*, **6**(3) : 267-270.

Kakulu, S.E. (1993). Biological monitoring of atmospheric trace metal deposition in North-Eastern Nigeria. *Environmental Monitoring and Assessment*, **28**(2) : 137-143.

Kamaluddin, M. and Ali, M. (1996). Effects of leaf area and auxin on rooting and growth of rooted stem cuttings of neem. *New Forests*, **12**(1) : 11-18.

Karmakar, P.R. and Chatterjee, B.P. (1994). Isolation and characterization of two IgE-reactive proteins from *Azadirachta indica* pollen. *Molecular and Cellular Biochemistry*, **131**(1) : 87-96.

Kandaswamy D.M.; Rangarajan, G.; Obliswamy, R.; Narayanan and C. Ponnaiah (1988). Mycorrhizal occurrence and their effect on certain forest plant species. *National Workshop on Mycorrhizae*, J.N.U., New Delhi, 13-15 March, 1987 : 111-118.

Karivaratharaju, T.V.; Bharathi, A. and Umarani, R. (1996). Effect of seed moisture and drying methods on neem seed quality. *Neem Newsletter*, **13**(2) : 25.

Karivaradaraju, T.V.; Bharathi, A. and Umarani, R. (1999). Effect of seed moisture and drying methods on seed quality of neem (*Azadirachta indica*, A. Juss). *Azadirachta indica A. Juss*, ed. Singh and Saxena : 277-284.

Karunaichamy, K.S.T.K.; Paliwal, K. and Natarajan, K. (1993). Diurnal course of leaf gas exchange of mistletoe (*Dendrophthoe falcata*) and its host (*Azadirachta indica*) in a semi-arid region of southern India. *Proceed-*

ings Indian National Science Academy, Part B, Biological Sciences, **59**(5) : 505-510.

Kazmi, S.M.A. (1980). 'Geed-hindi' - *Melia azadirachta* L. - a most common cultivated tree in Somalia. *Somali Range Bulletin*, No. **9** : 20-23.

Ketkar, C.M. (1976). Utilization of neem (*Azadirachta indica*) and its by-procucts. *Khadi and Village Industries Commission*, Bombay, India.

Khedkar, M.H. and Subramanian, K. (1996). Introduction of root trainer nursery technology in forestry sector-Maharashtra. *Indian Forester*, **122**(3) : 199-211.

Kijoti, W.E. and Chamshama, S.A.O. (1990). Root development and plantation performance of some tree species at Dodoma, Tanzania. *Record Faculty of Forestry*, Sokoine University of Agriculture, No. **47**.

Kumaran, K.; Palani, M.; Jerlin, R. and Surendran, C. (1994). Effects of growth regulators on seed germination and seedling growth of neem (*Azadirachta indica*). *Journal of Tropical Forest Science*, **6**(4) : 529-532.

Kumaran, K.; Surendran, C. and Palani, M. (1996). Effect of presowing chemical treatment on germination and seedling growth in neem (*Azadirachta indica* A. Juss.). *Indian Journal of Forestry*, **19**(1) : 87-88.

Kumaran, K.; Surendran, C. and Rai, R.S.V. (1996). Variation studies and heritable components of seed parameters in neem (*Azadirachta indica* A. Juss.). *Neem and Environment*, **1** : 167-172.

Maithani, G.P.; Bahuguna, V.K.; Rawat, M.M.S. and Sood, O.P. (1989). Fruit maturity and interrelated effects of temperature and container on longevity of neem (*Azadirachta indica*) seeds. *Indian Forester*, **115**(2) : 89-97.

Morgan, E.D.; Van der Esch, S.A.; Jarvis, A.P.; Maccioni, O.; Giagnacovo, G. and Vitale, F. (1996). Production of natural insecticides from *Azadirachta* species by tissue culture. *Neem Newsletter*, **13**(3) : 31-32.

Munshi Kumar; Singh, R.A. and Warsi, A.S. (1998). Neem : A potential tree for agroforestry and its eco-friendly contribution in pest management. Abs. *ICPPMSA*, 11-13 December, 1998, CSAU A&T, Kanpur (India) : 160-161.

Murthy, B.N.S. and Saxena, P.K. (1998). Somatic embryogenesis and plant regeneration of neem (*Azadirachta indica* A. Juss). *Plant Cell Reports*, **17**(6-7) : 469-475.

Narayan, P. and Jaiswal, V.S. (1985). Plantlets regeneration from leaflet callus of *Azadirachta indica* Juss. *Journal of Tree Science*, **4**(2) : 65-68.

Nirmala Kumari, A.; Ramaswamy, N.M. and Sree Rangaswamy, S.R. (1996). Tissue culture studies in neem (*Azadirachta indica* A. Juss). *Neem and Environment*, **2** : 981-992.

Palanisamy, K.; Ansari, S.A.; Pramod Kumar and Gupta, B.N. (1998). Adventitious rooting in shoot cuttings of *Azadirachta indica* and *Pongamia pinnata*. *New Forests*, **16**(1) : 81-88.

Pal, Mohinder; Adarsh Kumar; Meena Bakshi; Bhandari, H.C.S. and Pankaj Khullar (1994). Cheap non-auxinic chemicals for rooting nodal segments of neem (*Azadirachta indica*). Special issue : Focus on Vegetative Propagation. *Indian Forester*, **120**(2) : 138-141.

Pal, Mohinder; Badola, K.C. and Bhandari, H.C.S. (1992). Stimulation of adventitious root regeneration on leafy shoot cuttings of neem (*Azadirachta indica*) by auxin and phenols. *Indian Journal of Forcstry*, **15**(1) : 68-70.

Palanisamy, K. and Pramod Kumar (1997). Effect of position, size of cuttings and environmental factors on adventitiouus rooting in neem (*Azadirachta indica* A. Juss). *Forest Ecology and Management*, **98**(3) : 277-280.

Patel, N.L. and Singh, S.P. (1996). Dynamics of growth in some agroforestry tree species under South Saurashtra region of Gujrat. Special issue : *Agroforestry, Indian Forester*, **122**(7) : 570-576.

Pong-Anant, K.C.; Wongamance, P.; Wasuwanich and S. Kirkar (1989). Grafting of *Azadirichta indica* A. Juss, Muak-lek, Saraburi, Thailand Asean-Canada Forest tree seed centre (Thailand).

Ponnuswamy, A.S.; Rai, R.S.V.; Surendran, C. and Karivaratharaju, T.V. (1991). Studies on maintaining seed longevity and the effect of fruit grades in neem (*Azadirachta indica*). *Journal of Tropical Forest Science*, **3**(3) : 285-290.

Radwanski, S.A. (1977). Neem Tree-2 : Use and Potential Use. *World Crops Livestock*, **29** : 111-113.

Rajagopal, B. and Ramaswamy, N.M. (1996). Studies on *in vitro* culture and production of secondary metabolites in neem (*Azadirachta indica* A. Juss). *Neem Newsletter*, **13**(2) : 28.

Rajiv Rai ((1996). To study the efficacy of different nursery raising techniques in neem (*Azadirachta indica*). *Journal of Tropical Forestry*, **12**(4) : 183-188.

Rakesh Kumar; Bangarwa, K.S. and Kumar, R. (1996). Seed collection and pre-sowing handling methods in *Azadirachta indica* A. Juss. *Annals of Biology*, Ludhiana, **12**(1) : 67-70.

Ramesh, K. and Padhya, M.A. (1990). *In vitro* propagation of neem, *Azadirachta indica* (A. Juss), from leaf discs. *Indian Journal of experi-*

mental Biology, **28**(10) : 932-935.

Ramesh, K. and Padhya, M.A. (1996). De novo synthesis of secondary metabolites in cultured cells of neem. *Neem & Environment*, **1** : 173-186.

Rao, M.V.S.; Rao, Y.V.; Rao, Y.S. and Manga, V. (1988). Induction and growth of callus in A. Juss. *Crop Improvement*, **15**(2) : 203-205.

Reddy, K.K.; Venkaiah, K. and Praveena, P.L.R.J. (1993). Rooting media *vis-a-vis* rooting percentage. *Indian Journal of Forestry*, **16**(2) : 103-106.

Rice, M. (1993). Development of neem research and industry in Australia. *World Neem Conference* (Bangalore, India), *Souvenir* : 8-24.

Roederer, Y and Bellefontaine (1989). Can neem seeds be expected to keep their germinative capacity for several years after collection ? *Forest Genetics Res*ources Information, No. **17** : 30-33.

Sacande, M.; Groot, S.P.C.; Hoekstra, F.A.; Castro, R.D. de and Bino, R.J. (1997). Cell cycle events in developing neem (*Azadirachta indica*) seeds : are they related to intermediate storae behaviour? *Seed Science Research*, **7**(2) : 161-168.

Sacande, M.; Hoekstra, F.A.; Pijlen, J.G.van; Groot, S.P.C. and van Pijlen, J.G. (1998). A multifactorial study of conditions influecnce longevity of neem (*Azadirachta indica*) seeds. *Seed Science Research*, **8**(4) : 473-482.

Sanyal, M.; Das, A.; Banerjee, M. and Datta, P.C. (1981). *In vitro* hormone induced chemical and histological differentiation in stem callus of neem *Azadirachta indica* A. Juss. *Indian Journal of Experimental Biology*, **19**(11) : 1067-1068.

Sanyal, M. and Datta, P.C. (1986). Nutritional environment and redifferentiation from cultured tissues of *Azadirachta*. *Acta Horticulture*, **188** : 99-103.

Sarma, V.; Rao, S.R.S. and Beena, C.H. (1992). Leaf architecture in relation to taxonomy : Meliaceae. *Feddes Repertorium*, **103**(7-8) : 535-542.

Schmutterer, H. (1995). **The Neem Tree**. Pub. VCH Verlagsgesellschaft weinheim (Federal Republic of Germany) and VCH Publishers Inc., New York NY (USA), pp. 696.

Schmutterer, H. (1998). Some arthropod pests and a semi-parasitic plant attacking neem (*Azadirachta indica*) in Kenya. *Anzeiger fur Schadlingskunde, Pflazenschutz, Umweltschutz*, **71**(2) : 36-38.

Shanmugarelu, K.G. (1967). A note on air-layering of *Eugenia* (*Eugenia gambalana* Lam) and neem (*Azadirachta india* A. Juss). *South indian*

Horticulture **15** : 70-71.

Shanthi, K.; Mahadevan, N.P.; Geetha, S. and Singh, B.G. (1996). Rooting of cotyledonary scions in *Azadirachta indica* A. Juss. seedlings. *Indian Forester*, **122**(4) : 346-347.

Shanthi, K.; Manimuthu, L. and Singh, B.G. (1996). Genetic significacne of late flowering forms in neem (*Azadirachta indica* A. Juss) as reflected by germination studies. *Indian Forester*, **122**(3) : 263-264.

Singh, B.G.; Mahadevan, N.P.; Shanthi, K.; Manimuthu, L. and Geetha, S. (1997). Effect of moisture content on the viability and storability of *Azadirachta indica* A. Juss seeds. *Indian Forester*, **123**(7) : 631-636.

Singh, R.P.; Lehari, A. and Singh, D.P. (1999). Standardization of methodology of collection practices for neem seed and their storage. Abs. National Workshop on Neem Based Industry, Techno. Commercial Aspects and Exposition, 26-27 October, 1999 organized by Fragrance and Flavour Development Centre, Kannauj at Kanpur.

Singh, V.P.; Dhillon, R.S. and Jhorar, B.S. (1996). Floral biology, breeding behaviour and breeding strategy in neem (*Azadirachta indica* A. Juss). *Neem Newsletter*, **13**(4) : 56-57.

Solanki, K.R. and Gupta, V.K. (1999). Breeding neem germplasm faster through air-layering for evaluation of their azadirachtin content. *ICAR Newsletter*, **5**(4) : 6-7.

Sridharan, S.; Venugopal, M.S.; Dhaliwal, G.S. (ed.); Arora, R. (ed.); Randhawa, N.S. (ed.) and Dhawan, A.K. (1998). Effect of environmental conditions on the yield of azadirachtin and oil in neem. *Proceedings International conferennce on Ecological Agriculture : Towards Sustainable Development*, Chandigarh, India, 15-17 Novermber, 1997 : 510-518.

Srividya, N.; Devi, B.P. and Satyanarayana, P. (1998). Azadirachtin and nimbin content in *in vitro* cultured shoots and roots of *Azadirachta indica* A. Juss. *Indian Journal of Plant Physiology*, **3**(2) : 129-134.

Stephen, A.; Muruganandam, B.; Ravichandran, P. and Dayanandan, P. (1996). Neem seedlings - Growth potential of different organs. *Neem and Environment*, **1** : 159-166.

Su WelWen; Hwang Wening; Kim SeYoung; Sagawa, Y.; Su, W.W.; Hwang, W.I. and Kim, S.Y. (1997). Induction of somatic embryogenesis in *Azadirachta indica. Plant Cell Tissue and Organ Culture*, **50**(2) : 91-95.

Tampubolon, A.P. and Alrasyid, H. (1989). The neem tree and its developmental prospects in rainfed zones in Indonesia. *Duta Rimba*, **15**(109-

110) : 3-12.

Tawfik, A.A. (1997). Micropropagation and plant regeneration of neem tree (*Azadirachta indica* Juss). *Assiut Journal of Agricultural Sciences*, **28**(2) : 1-10.

Thiagarajan, M. and Murali, P.M. (1996). High frequency regeneration of neem *in vitro* (*Azadirachta indica* A. Juss.) from immature embryo explants. *Neem & Environment*, **2** : 975-980.

Toky, O.P. and Bisht, R.P. (1992). Observations on the rooting patterns of some agroforestry trees in an arid region of north-western India. *Agroforestry Systems*, **18**(3) : 245-263.

Veeresham, C.; Kumar, M.R.; Sowjanya, D.; Kokate, C.K.; Apte, S.S. and Raj Kumar, M. (1998). Production of azadirachtin form callus cultures of *Azadirachta indica*. *Fitoterapia*, **69**(5) : 423-424.

Venkatesh, D.A.; Dundaiah, M. and Mahadevamurthy, S. (1990). Viability of neem seeds. *Myforest*, **26**(4) : 369-370.

Venkateswarlu, B.; Katyal, J.C.; Jayjayanti Choudhuri, Mukhopadhyay, K. and Choudhuri, J. (1998). Micro-propagation of plus neem (*Azadirachta indica* A. Juss) and evaluation of field transferred plants. *Indian Forester*, **124**(7) : 537-543.

Verma, R.C.; Dhillon, R.S. and Singh, V.P. (1996). Effect of auxins on rooting of neem cuttings in spring seasons. *Annals of Bilology*, Ludhiana, **12**(1) : 52-56.

Vijayan, A.K. and Rehill, P.S. (1987). Occurrence of twin seedlings in *Azadirachta indica* A. Juss. *Indian Journal of Forestry*, **10**(3) : 214-216.

Viswanathan, R.; Palanisamy, P.T.; Gothandapani, L. and Sreenarayanan, W. (1996). Physical properties of neem nut. *Journal of Agricultural Engineering Research*, **63**(1) : 19-25.

Vyas, K.J. (1989). *Azadirachta indica* A. Juss., the host of *Ficus religiosa* Linn. *Indian Forester*, **115**(4) : 276.

Wewetzer, A. (1996). Production of azadirachtin by callus cultures from *Azadirachta indica* A. Juss. *Neem Newsletter*, **13**(2) : 27-28.

Wewetzer, A. (1999). Production of azadirachtin by callus cultures from *Azadirachta indica* A. Juss. **Azadirachta indica A. Juss**, Ed. Singh and Saxena : 247-258.

Wewtzer, A.; Kleeberg, H. (ed.) and Zebitz, C.P.W. (ed.) (1997). Tissue culture of *Azadirachta indica* (A. Juss) and its potential for the production of azadirachtin. *Proceedings 5th Workshop Wetzlar*, Germany, 22-25 Jan. 1996 : 243-249.

Fig. 1.7

Fig. 1.8

Fig. 1.9

Fig. 1.7 Ripe fruits of neem (*A. indica*)
Fig. 1.8 Depulped seed of neem (*A. indica*)
Fig. 1.9 Seed kernel (*A. indica*)

Fig. 3.7

Fig. 3.8

Fig. 3.7 Attack of fungal pathogen on neem seed (*Aspergillus* sp.)
Fig. 3.8 Black leaf spot on neem leaves caused by *Colletotrichum* sp.
Fig. 3.9 Brown leaf spot on neem leaves caused by *Cercospora* sp.

Wongmanee, C.; Onnom, W. and Piewluang, C. (1990). Spacing for stump preparatin and stump sizes for planting *Azadirachta indica* A. Juss. var. *siamensis* Valeton. *Embryon*, **3**(1) : 11-15.

Zypman, S.; Ziv, M.; Applebaum, S.; Altman, A. (ed.) and Ziv, M. (1997). Tissue culture methods and cloning of the neem tree (*Azadirachta indica*) for bioinsecticide production. *Acta Horticulturae*, **447** : 235-236.

Chapter 2

Biological Active Ingredient, Biological Effects and Chemistry of Neem Compounds

The biological effects of neem extracts were determined for the last two decades and attracted the attention of many chemist, biologists all over the world. In the beginning the crude extracts of different parts of the neem plants were prepared and tested for its efficacy against a number of insects. These extracts posses antifeedent, insecticidal, nematicidal, fungicidal and insect growth retarder qualities.

Neem research on chemical constituents originated from National Chemical Laboratory Pune, India. Later Narayanan *et al*. (1967) concentrated mostly on the biological active ingredient and on the structure of bitter principle neem oil nimbin was established. The bitter principle of neem oil have been investigated by several workers who have obtained the bitter principle the nimbidin by extraction with alcohol. This compound is highly bitter and contain sulphur. On hydrolysis it gives nimbidinic acid which is generally bitter and retains sulpur. Nimbidin preparations are reported to be free from the unpleasant smell of the oil and to be efficacious in a variety of skin diseases septic sores and ulcers. This is also useful in bleeding gums and phyrrhoea.

More than 100 triterpenoid compounds have been isolated and identified. The study of the biological activity of neem was restricted till 1975 to neem oil and the extracts of different parts of the plants. The development of neem research in India was initiated by Pradhan *et al*. (1962) when they demonstrated that water suspensions of neem kernal powder against locusts have good antifeedent activity.

At least nine compounds possess insect growth regulatory activity. Besides, azadirachtin, Meliantrial (Lavie *et al*., 1967) and Salannin (Henderson *et al*., 1968) also deter insect feeding. A new compound deacetylazadirachtinol isolated from fresh fruits is an potent feeding deterrent as azadirachtin Siddiqui *et al*. (1986) isolated two new tetranortriterpenoid gamma-hydroxybutenolides desacetylnimbinolide and desacetylisonimbinolide along with desacetylnimbin from fresh green twings. Structures were determined by chemical and spectral means.

Siddiqui *et al.* (1986) isolated new triterpenoid (nimbocinone) and two sterols (sitosteril and stigmasterol) were isolated from fresh undried winter leaves collected in Pakistan. Vepaol with antifeedent and growth inhibitory effects and two new compounds (isove paol and nimibidin) have been isolated from neem seed (Sankaran et al (1987).

Rembold *et al.* (1987) studied structure and biological activity of azadirachtin A and B. The activity of these compounds were tested in laboratory assays using the chrysomelid *Epilachna* sp. Chemical structures were compared with biological activity and it is postulated that epoxide ring at C-13/C-14 is the essential structural element interfering with hormonal control of *Epilachna* sp. growth and development. A new growth inhibitor 7-deacetyl 17 β-hydroxy -azadiradione has also ben isolated from neem. (Lee *et al.*, 1988). Four new tricyclic diterpenoids-nimosome, nimbosme, methyl nimbiol and methyl nimbione, with many Phenolic deterpane sugiol have been obtained from neem bark.

Siddiqui *et al.* (1988) isolated two new diterpenoids, nimbionone and nimbionol from the acidic fraction of the ethanolic extract of the bark and their structures as 12-hydroxy-13-methoxypodocarpa-8, 11, 13-trien-3, 7-dione and 3, 12-dihydroxy-13-methoxypodocarpa-8, 11, 13-trien-7-one, respectively, through chemical and spectral studies. The acid fraction, which ultimately yielded nimbionone and nimbionol, showed significant antibacterial activity against the Gram positive bacteria, *Bacillus subtilis, Staphylococcus epidermis* and *S. aureaus*, and the Gram negative bacterium *Klebsiella ozaenae*.

Ley *et al.* (1988) studied the chemistry of insect antifeedents from *A. indica.* Azadirachtin (a tetranortriterpenoid isolated from the neem trees, *A. indica*) can be converted to the natural product 22, 23-dihydro-23β-methoxyazdirachtin via selective bromomethoxylation of the C-22,23 enol ether double bond and tri-n-butyltin hydride reduction; the corresponding acetic acid adduct on pyrolysis affords azadirachtin in high yield. These compounds possess hypocholesterolemic antitamor atilenkemic and antibiotic activity and also cause plant cell expansion, inhibit cell division and are insecticidal. Ara *et al.* (1988a,b) isolated two compounds namely isomeric diterphenoids nimbonane and nimbonone from the stem bark of neem along with methyl.grevillate.

TABLE 1 : LC 50% value of neem seed oil and its constituents and their yield/kg oil

	Approximate yield (g)/kg oil	L C50%
Hexane extract (oil)*	-	0.674
Ethanol-soluble part	160	0.328
Ethanol-insoluble part	-	Non toxic
Nonterpenoid	1	0.104
Epinimbin	5	Non toxicity at 0.3%
Nimbin	5	Non toxicity at 0.3%
Salannin	16	0.055
Desacetylnimbin	0.25	Not tested
Desacetylsalannin	0.25	Not tested
Salannin derivative	0.5	0.096
Azadirachtin	Traces	Not tested

* 1kg seed kernel yielded 450g of oil.

Rojatkar *et al*. (1989) isolated two new tetranortriterpenoids, an aldehyde named nimbanal and the 3-acetyl derivative of salannol, from a seed extract of *A. indica*. The structure of nimbanal was established by spectral data while that of salannol-3- acetate was proved by its correlation with salannin.

Ara *et al*. (1989a,b,c,d,e) studied the neutral fraction of stem bark of *A. indica* which yielded the new diterpenoids margolone, margolonone and isomargolonone. The structures of these compounds were established as 12-methyl-7-oxopodocarpa-8, 11-13-triene-13-carboxylic acid, 12-methyl-3, 7-dioxopodocarpa-8, 11, 13-triene-13-carboxylic acid, and 13-methyl-3, 7-dioxopodocarapa-8, 11,13-triene-12-carboxylic acid, respectively, through chemical transformations and spectral studies. These diterpenoids showed antibacterial activity against various gram-positive and gram-negative organisms and represent the first isolation of podocarpane derivatives with carbon substiuents at both C-12 and C-13.

Kraus *et al*. (1989) isolated 1,3-Diacety-11, 19-deoxa11-oxa-meliacarpin, a possible intermediate in the biosynthesis of the insect antifeedant azadirachtin, from methanolic extracts of A. *indica* seeds. A structure is proposed on the basis of 1H and 13C NMR data. The neem seed oil was found to have appreciable contact toxicity [(Singh *et al.*, 1988). They even identified compounds responsible for toxicity of oil. The details are given in table.1].

Siddiqui *et al*. (1992) isolated triterpenoids from the fresh fruit coats of *A. indica*. Three novel degraded triterpenoide: deefurano-azadiradione (7 α-acetoxy-4,4, 8-trimethyl-6 α-(13αlMe)- androsta-1, 14-dien-3, 16-dione); 7α-acetoxy-4,4, 8-trimethyl-5α (13 αlMe)-17-oxa-androsta-1, 14-dien-3, 16-dione and 7α-acetoxy-4,4, 8-trimethyl-5α-17-oxa-androsta-1, 14-dien-3, 16-dione wer isolated from the neutral fraction of the ethanolic extract of the fresh, undried, uncrushed, ripe fruit coats of neem. Their structures were elucidated from spectroscopies analysis. This is the first report of the isolation of such degraded triterpenoids of this skeleton from any plant source.

Rani and Akhila (1994) observed biosynthetic relationship between Nimocinol and Nimocinolide in *Azadirachta Indica*. Fresh undried winter leaves of *Azadirachta Indica* were fed with [2-^{14}C, 4R-$^{3}H_{1}$] MVA: The cut shoots of A. *Indica* were immersed in a sterile, bacteria free solution containing 10 μci of ^{14}C and Ca5O μci of ^{3}H as a mixturte of [2-^{14}C] and [4R-$^{3}H_{1}$] MVA and left under natural conditions to allow the uptake of the substrate. After 36 hrs, the leaves were seperated from the twig and washed free of substrate. The isolation of limonoids were carried out. Atomic and isotopic ratios in nimocinol and nimocinolide isolated showed the loss of one ^{14}C from side chain and four tritium atoms — one from side chain and three from apotetracyclic triterpene skeleton. The biosynthetic pathway suggesting the involvement of euphol (Δ8,9 compounds) which is isomerised to butyrospermol (Δ7,8 compounds) followed by an apo-rearrangment during which one methyl migrates from C-14 to C-8 and double bond from C-7 to C-14.

Senniappand *et al*. (1996) studied chemical examination of the resinous exudate from A. *indica* some of the volatile compounds of trichome secretions serve as repllents. Recently, it has been reported, for the first time, the presence of a new flavonoid, 8-prenyl-5, 7-dihydro 3' (3-hydroxy-3, 3-dimethyl-butyl)-4-1-methoxy-flavanon in the tender leaves and shoot apices of A. *indica*.

Kumar *et al*. (1996) isolated four new compounds, 11-hydroxyazadirachtin-B, 1-tigloyl-3-acetylazadiractinin, 1,2-diacetyl-7-tigloyl-12-hydroxyvilasinin and 23-desmethyllimocin-B were isolated from seeds of A. *indica*. There structures were characterized from spectral data.

Govinderchari *et al*. (1996) isolated and quantified the azadirachtin and other major triterpenoids in the seeds oil of A. *indica* and major azadirachtins

A, B, D, H have been isolated, identified and estimated. Estimation of the different components in neem oil that we analysed shows that it contains salannin, nimbin, azadiradione, 6-deacetylnimbin and epoxyazadiradione in 1.4%, 0.5%, 0.2%, 0.5% and 0.15% respectively while the misture of azadirachtins constitute about 0.3 percent.

Rojatkar *et al.* (1996) reported 1-α-Destigloyl-1-α-benzoylazadirachtin, 11-β-Hydroxyazadirachtinin, and 4-α-benzoyl nimbandiol from the methanol extract of seeds of *A. indica*. These structures were proposed on the basic of H- NMR spectral data.

Akhila and Kumar (1996) observed that Azadirachtin, salannin, meliantriol, nimbin and nimbidin are the best known limonoids. The biogenesis of limonoids apparently starts from a tetracyclic triterpenoid precursor of the tirucallol-type. Little evidence is available regarding the biogenetic pathway to these limonoids. The involvement of a terpenoid intermediate, possibly euphol with a double bond at the C-8 position in biosynthesis of nimbolide has been established in the leaves of *A. indica* by feeding [2-14-2. 4R-3H1], mevalonic acid lactone. Seeing the structure of meliantriol and nimbocinone, it appears that no intermediate with a double-bond at C-9 would be involved during the biosynthesis of meliantriol and nimbocinone. However, an atomic ratio of 6:5 (^{14}C:^{3}H) in meliantriol obtained from fresh fruits which were fed with [$2^{14}C^{4}R^{3}H1$] mevalonic acid clearly indicated that a double bond at C-9 is formed at some initial stage of biosynthesis. Doubly labelled (^{14}C:^{3}H 6:5) meliantriol when fed to fresh winter leaves produced nimbocinone ($^{14}C^{3}H2$:1), indicating the loss of two more tritium atoms obviously from C-17 and C-3, thus concluding that meliantriol is the precursor of nimbocinone. This confirms, that euphol is isomerised to butyrospermol which in turn produces meliantriol and nimbocinone sequentially.

Sundaram and Curry (1996) studied photostabilization of neem-based azadirachtin (aza) applied on to glass surfaces in the presence of 3 UV absorbers, 2,4-diydroxy benzophenone (Uvinul M-400, UM) 4-aminobenzoic acid (PBA) and Fluorescent brightener-28 (FB-28), a stilbene disulfonic acid derivative. The UV absorber, UM, provided excellent protection, increasing the dissipation half-life (DT50) of pure AZ-A from 3.87 to 22.54 days. Photostabilization due to PBA was marginal. The UV absorber, FB-28, acted as an effective photosensitizer, reducing the DT50 of AZ-A from 3.87 to 0.31 days. Similar trends in DT50 were observed when the 3 UV absorbers were added to 4 formulations containing AZ-A. It was found that for effective photostabilizatoin, both AZ-A and the UV abosorber must have matching UV spectra with similar lambdmax. The mechanism of photostabilization was probably due to either an energy transfer from AZ-A to the YV absorber and/

or competitive absorption of UV photons by the absorber. Photoinstability of AZ-A in the presence of FC-28 was due to energy transfer from the activated FB-28 to AZ-A molecules. Based on the UV spectral data, UV protectants can be selected and matched to stabilize UV-labile pesticides.

Akhila (1996) reported biogenesis of azadirachtin in *A. indica*. The azadirachtin molecule seems to be formed with the two fragments, a decalin fragment and another hydroxyfuran fragment. Scientists have synthesised these fragments separately but in the plant it apperas to be a long biosynthetic pathway strating from euphol -> butyrospermol->Apo-rearangement-cleavage of C-12-C-13 bond -> oxidation of C-29, C-1, C-19, C-12 -> epoxidation at ^{13}O and formation of hydroxyfuran ring. This chain-of-reactoin can be proved by feeding ^{14}C and ^{3}H bearing precursors at various positions in sodium acetate, MVA and squalene. In one of the preliminary experiments he successfully incorporated ^{14}C into azadiractin by feeding aqueous solutions of ^{14}C-acetates and mevalonates to the twigs of neem tree bearing green/ripe fruits. 3-Deacetyl-3-cinnamyl-azadirachtin, 3-Deacetyl-3-cinnamyl-azadirachtin, 3-acetyl-11-methoxyazadirachtin, 22-23-dihydro-23β methoxyazadirachtin, 3-tigloylazadirachtin, closely related to azadiracthin, seem to be either intermediate precursors or the metabolic products of azadirachtin.

Niranjan *et al.* (1996) reported a new tetranortriterpenoid, 11-epi-azadirachtin H, was isolated from the extract of seeds of *A. indica* (collected from Tamil Nadu, India). Its structure was proposed on the basis of various spectral analyses.

Cohen *et al.* (1996) observed that neem seed preparations contain not only azadirachtin as the active insect antifeedant or growth regulator but also a variety of other limonoids, some of which are cytotoxic to N1E-115 neuroblastoma (mouse), 143B.TK-osteosarcoma (human) and *Spodoptera frugiperda* S19 cultured cell lines. The most potent of these limonoids is nimbolide with an IC50 ranging from 4 to 10 μM and averaging 6 uM for the 3 cell lines. Other limonoids of decreasing potency and their average IC50 values (uM) are epoxyazadiradione 27 μM, salannin 112 μM, and nimbin, deacetylnimbin and azadiracthin each >200 μM.

Dong and Zhao (1996) reported toxic effects and mode of action of azadirachtin. Azadirachtin has antifeedant, stomach poison, contact poison, and growth inhibition effects on many insects. The main mode is disruption of endocrinic activities in insect. Azadirachtin attacks, neurosecretory cells and *Corpus cardiacum* prothoracic glands, causing interruption of synthesis and release of prothoracicotropic hormone (PPTH) in the brain, and redcution of receptivity of prothoracic glands to PTTH which accounts for the shortage of 20-hydroxyecdysone [ecdysteron] in synthesis and release. For these reasons, growth and metamorphosis of insects are disturbed.

According to Sidhu and Behl (1996), *A. indica* produces seeds during the rainy season in India. Phenotypes that produced seeds out-of-season in November-December were identified. Qualitative and quantitative variations in azadirachtin content were determined in seeds produced in the rainy season (July) and in winter. Seed produced during July contained higher concentrations of crude azadirachtins (1.53%) than those produced in the winter (1.26%). Concentrations of azadirachtin A, B and F were determined by HPLC using an improved solvent-eluting system for better resolution. Azadirachtin A was the major metabolite in July seeds. Azadirachtins A and B were in nearly equal proportions in winter seeds. The concentration of azadirachtin F was>2-fold higher in winter seeds than in July seeds. The sum of the peak area of the azadirachtin S, B and F was higher in seeds produced in winter. Winter stress appeared to favour synthesis of azadirachtins B and F. Such phenotypes could be a material of choice for clonal multiplication of *A. indica*.

Chirathamjaree *et al.* (1997) collected the azadirachtin (aza A) contents of 95 neem seed samples, from 17 provinces in Thailand in 1995, were determined from methanol extracts by using high performance liquid chromatograhpy (HPLC). Samples of Indian neem (*Azadirachta indica*) collected from Rayong province contain the highest amount of aza. The contents of the a.i. range from 4.7 to 7.8 mg aza/g kernel. In contrast of aza A in Thai neem seeds is 2.4 kg/g kernel. The figure of aza A is low when compared with other surveys. Thai neem seeds also contain less oil. the average oil yields of *A. indica* samples is 40.6% (w/w), whereas Thai neem seeds contain only 34.0% (w/w) oil. The composition of substances in the neem seed samples was determined by using gradient HPLC. Thai neem samples show fewer compounds, 1-tigloyl-3-acetylazadirachtol are the dominating tetranortreiterpenoids in Thai neem samples. The latter one is considered as a marker peak for the Thai neem seeds because this compound could not be detected in *A. indica* neem samples. On the other hand, *A. indica* samples contain salannin and nimbin as major tetranotriterpenoids, and these less polar compounds are nearly missing in the Thai neem. No correlation is found between aza A content and size of tree. The result clearly shows that the Indian neem is superior to the Thai neem regarding active ingredinents.

Jarvis *et al.* (1997) observed photoxidation of nimbin and salannin, tetranortriterpenoids from neem tree. Nimbin and salannin, major triterpenoids accompanying azadirachtin in extracts of *A. indica* seeds, were photooxidized by ultraviolet light in the presence of oxygen to more polar, unstable intermediates that rearranged on silica gel to 2 final products in which the furan ring had been oxidized to isomeric hydroxybutenolides. The isomeric hydroxybutenolides were also readily formed when a crude extract of triterpenoids from neem seeds was irradiated, and both isomers of salannin were isolated from seeds. Photooxidation of nimbin and salannin proceeded much faster than that of azadirachtin.. All photoproducts showed some biological activity against *Spodoptera littoralis, Locusta migratoria* and *Schistocerca gregaria*. Isonimbinolide was as potent as azadirachtin at inhibiting feeding in all 3 species, and it also inhibited the growth of *S. littoralis*. Isolsalanninolide showed potent antifeedant and growth inhibitory activity against *S. littoralis*.

Suresh *et al*. (1997). Nimonol was obtained from the fresh green whole leaves of *Azadiractha indica* and its structure has been reassigned on the basis of COSY, NOESY and 1H NMR spectral data. 6-Acetylnimonol was identifed as 6-acetoxyazadirone and dihydronimonol was shown to be indentical to isomeldemnin. The melting point and 1H NMR spectra of 6-acetylnimocionol were markedly different from those of 6-acetoxyazadirone showing that nimocinol and nimonol have different structures.

Nair *et al*. (1997) obtained α-Linolenic acid form fresh *A. indica* leaves. An optimized procedure for the isolation of the limonoids nimbolide and 28-deoxonimbolide is also described. Leaves dried in the shade for 48 h were extracted inhexane. Low polarity compounds were removed. Separatoin of the limonids was carried out by converting them to their methyl esters followed by chroatography on $AgNO_3$-impregnated silica gel.

Stark (1997) stated that the principal ingredient in neem is azadirachtin, a highly oxidized limonoid. Although other limonoids present in neem are toxic or repellent, or antifeedant to insest. Azadirachtin acts on insect chemoreceptors to deter feeding and oviposition and directly affects overall fitness through subtle yet deleterious effects on mitosis, muscle and gut physiology and biological rhythms. Its main mode of action is inhibition of release of morphogenetic peptide hormones, resulting in disruption of ecdysteroid and juvenile hormone concentration in the haemolymph which in turns effects moulting, metamorphosis and reproduction. Bidmon *et al*. (1987) povided enough evidence for the growth regulating effects of azadirachtin. It was observed that infection of azadirachtin into blowfly larveal caused distinct alteration larval and pupal peaks of ecdysteroids, which in control larvae show sharp, brief surges peaks in the ecdysteroid content in treated larvae were delayed and did not decline rapidly.

Although neem pesticides exert broad-spectrum action, they are reported to be less toxic to natural enemics of insect pests. The insecticidal toxicity action of neem is still disputed inspite of the fact that large number of references are available which clearly give evidences that azadirachtine does exhibit contract toxicity to test organisms. The azadirechtins do show symptoms of increased development time, distended elytra, misshappen abdomens, non-functional wings and other morphological abnormalities which prevented flight and mating. It must be realized that all the neem insecticides are not the same and it is the formulation which matters.

According to Klaus *et al.* (1997), Thailand has 3 species of the neem trees, *Azadirachta siamensis* (Velton) (Thai neem tree or sadan), *A. excelsa* (Jack) (locally known as Tiam tree) and *A. indica* (A. Juss) (the Indian neem tree). The *A. indica* do not produce enough seeds in Thailand. The climate condition is such that rainy period is extended in Thailand and probably do not favour the fruit production. The Thai neem is indiginous to Thailand and grows in most part of the country. The Thai neem seeds contain less azadirachtin A (0.5 to 4.6 mg/g kernel) in comparison to Indian neem seeds (4.7 to 7.8 mg/g kernel). According to Sombatsiri *et al.* (1995), the presence of chlorophyll and water in the outer layer of the seed kernels may have an impact on the less contents of azadirachtin A.

The Tiam Neem, *A. excelsa* has big seed which is also characterized by the presence of chlorophyll in the other layer. They contain azadirachtin and marrangin, it is also found that both the species *A. excelsa* and *A. siamensis* contain another bioactive compound which is not occuring in Indian neem.

Harvest and Post harvest measures are essential for the quality neem seed and the kernel. The good quality of kernel is of green or greenish yellow color. The seed will then turn brown and finally block, showing increasing oxidation and finally the complete decomposition of aza A. It is also known that combination of high temperature and high humidity cause the rapid degradation. Therefore the following points may be kept in mind to obtain with contents of aza A from the seeds.

1. Harvest the neem fruits as soon as possible
2. Depulp the fruit to obtain the seeds
3. Dry the seeds rapidly to a moisture content of about 10% or less.
4. Pervent the seeds from taking up moisture after drying.
5. Process the seeds as soon as possible after harvest. However there is not much degradatioin during one year period because the seeds were properly dried before storage.

Kalinowski *et al.* (unpublished results) obtained from Thai neem, another compound punified for the first time as 1-tigloyl-3-acetyl-azadirachtol a derivative of aza B. However, this derivatives was not detected in Indian neem.

To obtain good quality azadirachtin from seeds the following quality parameters may be considered.

(i) A suitable harvest and post-harvest technology should be developed.

(ii) Equipment to pick up the fruits from the tree or to collect them from the ground.

(iii) A depulper for neem frutis.

(iv) A seed (kernel) dryer.

(v) Appropriate storage facilities.

(vi) Implementation of extraction technology to enrich the active ingredient.

(vii) Production of a storage stable formulation.

(viii) selection and Micropropagaton of effective ecotype.

(ix) Cell and tissue culture for the production of active ingredients.

(x) Genetic engineering to create high productive plants.

Govindachari *et al.* (1997) isolated a new tetranotriterpenoid, 13, 14-desepoxyazadirachtin-A, from neem kernel extract by HPLC and its structure established by spectroscopic methods.

Selvi and Mohan (1997) isolated, a marine sterol namely (24xi)-Isopropenyl, cholesterol and its biosynthetic precursor Fucosterol were isolated for the first time from the species *Azadirachta indica* and chemically characterized by spectroscopic methods.

Tonthu bthimthong and Tasakorn (1997) reported that azadirachtin is a natural insecticide, non-toxic to mammals. It is extracted from defatted ground neem seed kernel using methanol. Subsequentluy, 100 ml of methanol extract containing 2mg azadirachtin was found when agitated with 2.33 g of activated carbon powder with avergae size of 516 um for 1 h. The activated carbon can remove 97% of azadirachtin from the solution. The adsorbed azadirachtin can be stored for 1 month with negligible degradation. However, after 5 months, it declines to only 65 percent.

Jarvis *et al.* (1998) observed that the natural insecticide azadirachtin is most stable in mildy acidic aqueous solutions between pH 4 and 6 at room temperature. It is unstable in mildly alkaline and strongly acidic solutions. It is stable when stored in neutral organic solvents at room temperature for months. While it is relatively stable to heating in the seeds or as a pure solid, it is rapidly destroyed or altered by heating in aqueous solution and methanol. In methanol at 90°C, it is quantitatively converted to 3-acetyl-1-tigloylazadirachtinin.

Cohen *et al.* (1998) observed that neem seeds preparation from *Azadirachta indica*) contains not only azadirachtin as the active insect antifeedant or growth regulator but also a variety of other limonoids, some of which are cytotoxic to N1E-115 neuroblastoma (mouse), 143B. TK-osteosarcoma (human and *Spodoptera frugiperdea*) Sf9 cultured cell lines. The most potent of these limonoids is nimbolide with an IC50 ranging from 4 to 10 fM and averaging 6 fM for the 3 cell lines. Other limoinods of decreasing potency and their average IC50 values (fM) are epoxyazadiradione 27 fM, salannin 113 fM, and nimbin, deacetylnimbin and azadirachtin each >200 fM.

Niranjan *et al*. (1998) studied a new tetranortriterpene, which could have potential as an insect antifeedant, 11-epi-azadirachtin D (L-tigloyl-3-acetyl-11 α-hydroxy-4β-methyl-meliacarpin) was isolated from the methanolic extracts of *Azadirachta indica* seeds. Its structures have been proposed on the basis of various spectral analyses.

Johnson and Morgan (1998) investigated the selective extraction of nimbin, salannin, azadirachtin and oil from *Azadirachta indica* seeds, using supercritical carbon dioxide and methanol, was investigated under different conditions. Extraction for 30 min with 100% carbon dioxide using a capillary restrictor to maintain the critical pressure removed only small amounts of oil and triterpenoids. Slightly higher levels of oil and triterpenoids were removed when a back-pressure regulator was used. Over a 150 min period, in 30 min intervals, 100% carbon dioxide extracted all the nimbin and salannin from the seeds, whilst leaving some azadirachtin behind.

Akhila *et al* (1998) reported the production of Radioactive azadirachtin nimbin and salanin in the seeds kernels of *A.Indica* by using ^{14}C mevalonate and acetate as precursors under *in vivo / in vitro* conditions. The funding of radio-lablled precursors was done on fresh green yellow and ripe fruits. The incubation ranged from 24 hrs. to 3 weeks. It was observed that maximum incorporation was obtained in yellow fruits at room temperature in the dark after two weeks.

Akhila and Rani (1999) published the latest review on the chemistry of neem tree (*A.Indica* A Juss.) In their review they discussed 17 major groups of limonoids present in neem with their molecular formula, molecular weight, the plant part, yield. Specific rotation UV, IR, ^{1}H and ^{13}C NMR, NOESY and mass spectroscopy, derivatives and their spectra biological activity and synthesis.

According ot Jarvis *et al*. (1999), a dichloromethane extract of the triterpenoids from seed kernels of the neem tree (*Azadirachta indica* from Bangalore in India) was fractionated rapidly and with economic use of solvents using the Biotage TM flash chromatography system. The fractions were analysed for 11 tetranortiterpenoids, including the natural insecticide azadirachtin, by means of thin layer chromatrography (TLC), or quantitatively by supercritical fluid chromatography or HPLC. In the Biotage flash column, pure compounds (including salannin and azadirachtin) were obtained directly although these compounds comprised <0.3% of the total mass of the seeds extract and were present in a complex mixture of similar substances.

Azadirachtin and its analogues

Due to continuous awareness, Butterworth and Morgan (1968) were the first to isolate azadirachtine from neem seeds. They in 1971 developed a simplified method ot isolate azadirachtin through solvent partition followed by Column and preparative thin layer chromatography. By taking 2 kg of neem kernal 1.5 g azadirachtin (more than 90 per cent phase azadirachtin) was isolated by this method. Later this method was further modified through repeated reversed phase high performance chromatography and this yielded 8.7 g aza from 48.2 kg seeds. Schroeder and Nakanishi (1987) got 5 g of aza from 2 kg of seeds by vacuum liquid chromatorgrahpy followed by crystallization from carbon tetrachloride and flash chromatography. Lee and Klocke (1987) isolated 364.8 mg of aza of 7.99 percent purity from 1.5 kg of seeds by florisil column displacement chromatography and droplet counter current chromatography.

Efforts to chemical synthesis for commercial exploitation of azadirachtin failed. It is after this, work on the constituents and biological activity of azadirachtin gained special significance. Azadirachtin is most important neem compound for commercial point of view. Aza is found in neem kernels (0.2-1.0%). In India 1 kg of neem kernel yield about 5 grms of aza. This compound is so efficacious that 50 gms of this is sufficient to control insect pest in one hectare area. The quality and efficacy of the commercial products of neem depends on the amount of aza available in them. In two formulation of American company namely Margosan 'O' and Alin contains aza 0.3 percent and respectively. In India Achook, Nimbicidin, Bioneem Margocide 'O' k, Neemark I.D.9-Repelin contains 0.03 percent aza while Rakshak Margocide ok and Shkrina had aza 0.15 percent.

Aza has the following biological characters

1. It demonstrate the feeding deterrent properties and the crops treated with this were protected from the attack of insects. It is presumed that the neem extract stimulate the neurone which is responsible for antifeedent activities. Such compounds may be termed as "deterrent neurone". Schoonhoven (1981) reported that different neurone are stimulated by aza in case of lepidopterous insects viz., *Pieris brassicae, Lymantria dispar Hypsipyla grandella*. These deterrent neurons are being stimu-

lated by aza in case of *Spodoptera exempta, S. littoralis, S. frugiperda, Heliothis virescens, Helicoverpa armigera* and *Mamestra. brassicae.*

2. The aza, when goes in side the host causes growth disturbances and act as insect growth regulator. The larval adult development is disrupted due to which no moulting takes place and insect dies in between. Experiments with *E. varivestis* indicates that Aza interferes at a dose of 1.25 ppm exclusively with larval-pupal and 0.25ppm with pupal- adult moultings. Many times the survived pupae, adults were observed malformed with shrivelled wings. Aza acts on the hormonal regulation of insect growth, development and reproduction. The most important hormones are ecdysone and 20-hydroxy ecdysome as moulting hormones and the juvenile hormones 0-III is envolved in many metabolic pathway. Beside about 100 biologically active peptides have been identified which are acting as hormones. It has been observed that in aza treated insects, oviposition period is delayed by 5-7 days and only a few eggs are layed.
3. Aza does not have bad effects on mammals, birds and useful insects.
4. Aza has insecticidal, nematicidal, bactericidal and fungicidal effects.

The azadirachtin is one of the important biological active ingredient isolated from neem. It appeans to cause some 90 percent effect on most of the pests. In fact, it is so potent that a mere trace of its presence prevents some insects from even touching the plants. Azadirachtin is structurally similar to insect harmones called ecdysones which control the process of metamorphosis. It also affects the corpus cardicaum, an organ similar to the human pituitory which controls the secretion of hormones. Azadirachtin seems to be ecdysone blocker, this breaks the life cycle and biological activity of azadirachtin gained special significance. In U.S.A. and west Germany, research work on the antifeedant ovicidal and aphidicidal activities of the active ingredient was taken up in large scale. As the active principles were present in small quantities, two active rich fraction namely Neemric I and Neemrich II were obtained. The process is given below.

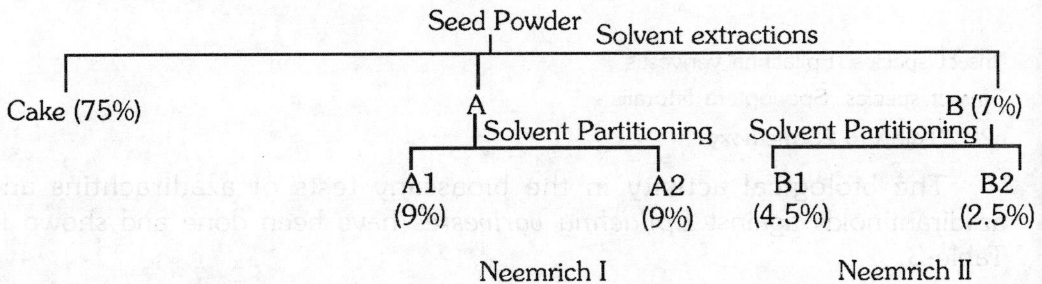

The fraction Neemrich I consists of Lipids and exhibit insecticidal activity against sucking insects and soft bodied insects e.g. aphids, jassids white flies etc. The compound Neemrich II contains azadirechtine and related tetranortriterperiods which exhibit antifeedent insect growth retarder activity against tobacco caterpillar, gram pod berer, ctlten ball worm etc. the compound isolated is demonstrated in Table-2.

Table 2 : Bio-efficacy of Azadirachtin and its naturally occurring derivatives

Compound	R1	R2	R3	R4	R5	G1*	AF	PPM
					LC_{50} (ppm)		Index	
Azadirachtin A	Tig	Ac	COOMe	OH	COOMe	1.66	99	1
Azadirachtin B	Ac	Tig	COOMe	H	COOMe	1.30		
Aradirachtin C	Tig	Ac	?	?	?	1.26		
Azadirachtin D	Tig	Ac	Me	OH	COOMe	1.57		
Azadirachtin E	H	Ac	COOMe	OH	COOMe	2.80		
Azadirachtin F	H	Tig	COOMe	OH	COOMe	1.15		
3-Tigloyl Azadirachtol	H	Tig	COOMe	H	COOMe	7.69	97	1
Compound 1	H	Tig	COOMe	H	COOMe			
Compound 2	H	Ac	Me	H	COOMe			
Compound 3	Tig	Fer	COOMe	OH	COOMe			
Compound 4	H	Ac	COOMe	OH	H			
Compound 5	Tig	Ac	COOMe	OH	H			
Compound 6	Tig	H	COOMe	OH	COOMe			
Compound 7	Tig	Ac	COOMe	OH	COOMe		81	1

*Insect species: Epilachna verivestis

**Insect species: Spodoptera littoralis

a22-23-dihydro 23 methoxy

The biological activity in the bioassany tests of azadirachtins and azadirantinoids against *Epilachna varivestes* have been done and shown in Table-3.

Table-3 : Biological activity in the Epilachna varivestis bio-assay of a selected number of azadirachtins and azadirachtinoids.

Compound	LC_{50} ppm	Reference (s)
3-De-(2-methyl-2-cis-butenoyloxy)-3-hydroxy-azadirachtin B	0.1	1,2
22,23-Dihydro-azadirachtin B	0.3	1,2
Azadirachtin I	0.3	3
Azadirachtin L	0.3	7
3-Deacetoxy-3-hydroxy-azadirachtin A	0.4	1,2
2',3', 22,23-Tetrachydro-azadirachtin B	0.5	1,2
AzadirachtinH	0.5	3
Azadirachtin F	1.1	1,2
Azadirachtin K	1.1	4
Azadirachtin A	1.3	1,2
22,23-Dihydro-azadirachtin A	1.3	1,2
Azadirachtin B	1.3	1,2
Azadirachtin D	1.6	1,2
Azadirachtin E	2.8	1,2
20-O-(Methylthiomethyl)-22,3-dilhydro-azadirachtin A	20	6
11,20-Di-O-methyl-azadirachtin A	30	5
3-Deacetoxy-1-de-(2-methyl-2-cis-butenoyloxy)-1,3 -dihydroxy-22,23-dihydro-11,20-di-O-methyl-azadirachtin A	50	6
22,23-Dihydro-7, 11-di-O-methyl-azadirachtin A	50	6
11-O-Acetyl-azadirachtin A	60	5
Salannin	>100	1,2
R,S-11-(22,23-dihydro)-azadirachtinin	>100	5
22,23-Dhydro-11,20-di-O-methyl-azadirachtinin	>100	6
7,11,20-Tri-O-methyl-azadiracthin A	>100	5
3-Deacetoxy-l-de-(2-methyl-2-cis-butenoyloxy)1,3-dihydroxy-22,23-dihydro-7,11,20-tri-O-methyl-azadiracthin A	>100	6

1=FORSTER (1987), 2=REMBOLD (1989), 3=GOVINDACHARI et al. (1991), 4=GOVINDACHARI *et al.* (1992) 5=REMBOLD and PUHLMANN (1993), 6=Unpubl. results, 7=ERMEL et al . (1991), KALINOWSKI *et al.* (1993).

[Source Rembold, H. and Puhlmann. I (1995)]

Rangasamy and Parmar (1997) investigated azadirachtin -A contents of the seed kernels from 11 of the 21 agroecological regions of India in relation to climate, growth period and soil. They observed that Azadirachtin-A contents ranged from 0.14-2.02 percent w/w with an average of 0.68 percent Further, is was observed that in coastal-arid and semi-arid ecosystem showed <0.72 percent and in subhumid ecosystem it is as low as 0.27 percent. The ecotypes growing in red soil had higher azadirachtin-A as compared to other soils.

The efficacy of these two products were tested against cotton and sorghum crop along with insecticides (Table-4) and it was observed that Neemric I has given good productivity of the crop against sucking insects etc. Therefore, this product is one which could be considered as an alternative fraction for pest management in compensation to chemical insecticides.

Table 4 : Some bioefficacy data of field trails with Neemrich-I and II

Treatment	Cotton	Sorghum	
	Yield/ha	Grain q/ha	Fodder q/ha
Neemrich-II	18.90	16.18	8.72
Dimethoate	20.27	16.58	8.81
Endosulfan	20.28	16.46	8.62
Neemrich-II & Dimethoate	21.65	17.20	9.15
Neemrich-II & Endosulfan	21.82	17.42	9.38
Neemrich-I	16.31	12.63	6.81
Inert material	13.68	11.66	6.30
Untreated control	11.78	8.64	5.42

The azadirachtin compound extracted from neem has developed special status due to its antifeedent and insect growth inhibition properties. A number of aza derivatives with modified ester groups attached to C-1 and C-3 have been obtained. The 3-deacetyl azadirachtin obtained from aza by methanolic neem extracts 1-detigloylazadirachitin (azadirachtin) is a hydrolysis product of aza. 3-deacetyl Cinnamoyl azadirachtin obtained from leaf extracts. Besides 1-dihydrotigloyl, 1-detigloyl-1-isobutyroyl, 1-detigloyl-1 isovaleroyl azadirachtin and 1-detigloyel-1-isocaproyl-3-deacetyl-3-epoxymethacroylazadiracthin were obtained from seed extracts. Two dihydromethoxy derivatives of aza, vepaol, and isovepaol had been Known biologically active compounds. A biodegradation product of aza 11-demethoxy Carbonylaza (azadirachtin H) is also obtained from seed extracts. Marrangin (azadirachtin L) was also isolated (Fig. 2.1).

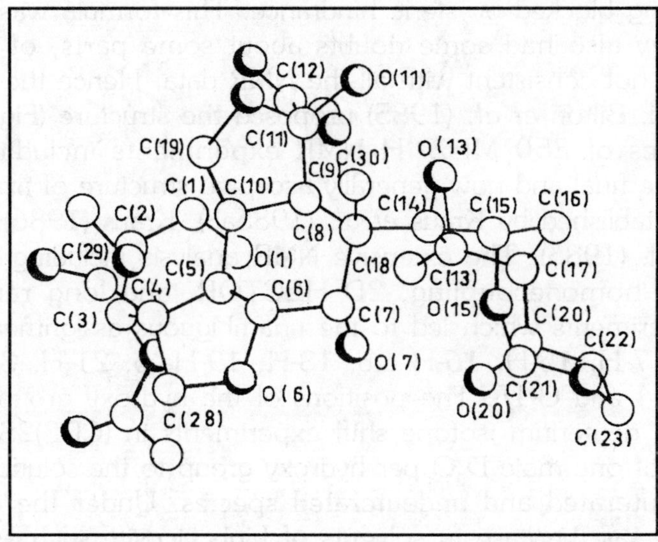

Fig.2.1 Molecular structure of azardirachtin (Broughton, et al. (1986)

There was much controversy about the chemical structure of azadirachtin but Broughton *et al*. (1986) through X-ray Crystallography of a detigloylated dihydro derivatives of azadirachtin have proposed the following molecular structure of azadirachtin (C35 H44 016).

Although purified aza is commerically available but cent percent purity is still not proposed. A number of derivatives including deacetylazadirachtin dihdroazadirachtin, tetrahydroazadirachtin, detigloyl-dihydro azadirachtin, detigoylpyruvoylazadirachtin and some acetylated compounds obtained by chemical modification.

Nakanishi (1975) and Zanno *et al*. (1975) derived structural formula (Fig. 2.2) from NMR investigation of the 14-acetate, obtained from aza by acetylation presumably via the 7-acetate followed by acyl migration, the further C-7

Fig. 2.2

Fig. 2.3

acetylation being blocked by steric hindrance. This formula was accepted for 10 years. They also had some doubts about some parts, of the structure because it was not consistent with all the NMR data. Hence the investigations were continued. Bilton *et al*. (1985) proposed the structure (Fig. 2.3) for aza based on series of 250 MHZ 'H NMR experiments including difference decoupling. The final and now generally accepted structure of aza as shown in (Fig. 4) was established by Kraus *et al*. (1985ab), Kraus (1986), Kraus *et al*. (1987a), Pohnl, (1985). The extensive NMR analysis including FT difference NOE 13C-1H-homodecoupling, 2D HECTOR, and long range coupling (COLOC) experiments which led to the unambiguous assignment of the crucial signals 0f 7-H, 15-H, 16-H a,b, 18-H, 19-Ha,b, 21-H, 30-H, C-7, C-11, C-13, C-14 and C-15. The positions of the hydroxy groups were determined by 13C deuterium isotope shift experiments in (CD3)2SO as the solvent: addition of one mole D_2O per hydroxy group to the solution gives a 1:1 mixture of deuterated and undeuterated species. Under the conditions of slow exchange, as the case in solvents of high vicosity such as DMSO, the signals of both isotopomers ROD and ROH can be observed in the 13C NMR spectrum. β-isotope shifts were found for C-7 (δ-73.82/73.71), C-20 (δ-81.78/81.69) and C-11 (δ-104.07/103.97), but not for C-13 C-14 indicating that the earlier assignments of 7-OH and 20-OH were correct, whereas the third off group is not attached to C-13 or C-14 but to C-11 which is now hemiacetol carbon.

Wehrli and Nishida (1979) confirmeced the hemiacetal structure (Fig. 2.4) by long rang coupling C-11/11-OH and by NOE experiments 11-OH 19-H and 11-OH/30-H. The remaining oxygen atom was assigned to an epoxide oxygen connected to C-13-C-14 on the basic of the chemical shifts of C-13 and C-14 which are typical for quarternary oxirane carbon atoms.

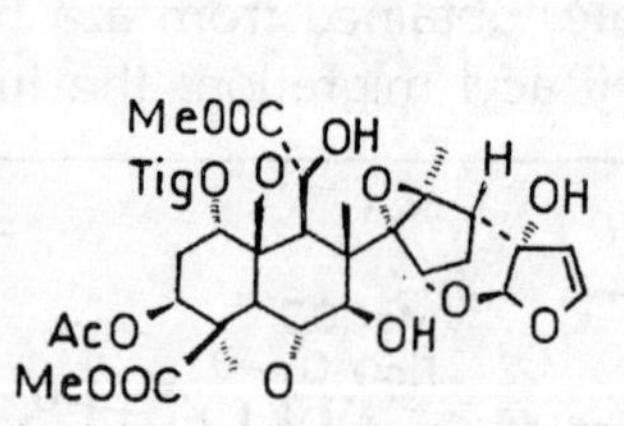

Fig. 2.4 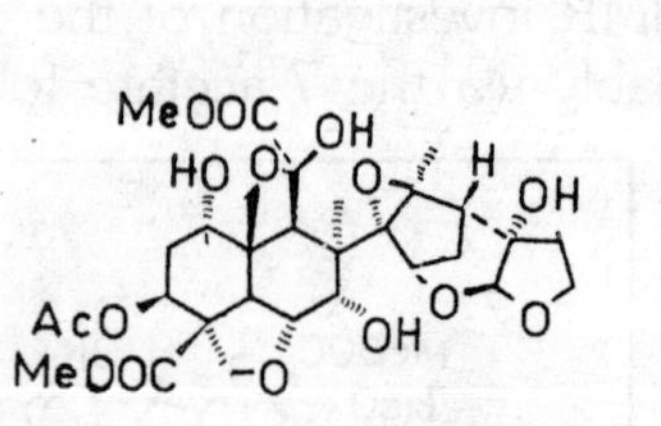 Fig. 2.5

The NMR derieved structure (Fig. 2.4) for aza was latter confirmed by X-ray diffraction analysis carried out on 1-detigloyl, 22-23, dihydroazadirachtin (Fig. 5) (Broughton *et al*., 1986 and Bilton *et al*., 1987). Many constituents related to azadirachtin have been isolated which can be kept in three major groups viz., azadirachtols, azadirachtins and meliacarpins. This nomenclature is better than the one given by Rembold *et al.* (1987 a,b) who suggested aza derivatives as aza-A, aza-B, aza-C etc. but overall this nomenclature is not accepted.

Azadirachtol Group

Kubo *et al.* (1984) reported a new neem derivatives deacetyl azadirachtinol and suggested the following structure (Fig. 2.6). In this an additional hydroxyl group was attached to C-13. Later studies made with NMR spectra and the structure 3-tigloylazadirachtol was named. A few azadirachtol derivatives were also indicated as 2-3 dihydrotigloylazadirachtol. The Azadirachtol (3-detiglyoylazadirachtin B) (Fig.2.7), dihydro azadirachtols and tetrahydroazadirachtols and besides this, a number of other derivatives were biologically active against many insects (table I) Another structure 3-tigloyl-13, 14-desepoxy-17 hydroxy-azadirachtol (azadirachtin G) were also obtained from seeds. One more product 1-tigloyl-3-acetylazadirachtol was also obtained from *A. excelsa* seed oil.

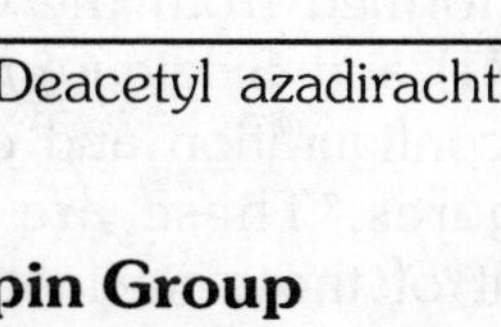

Fig. 2.6 Deacetyl azadirachtinol

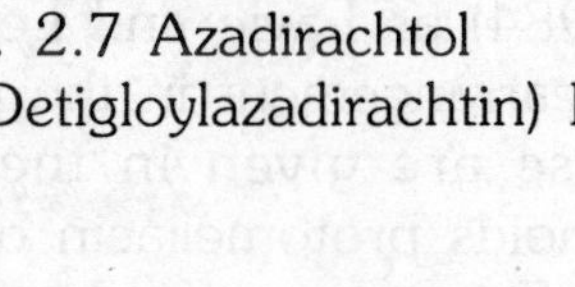

Fig. 2.7 Azadirachtol (3-Detigloylazadirachtin) B)

Meliacarpin Group

These compounds were obtained both from *M. azedarach* and *A. indica*. The carbon skeleton of the base structures of the meliacarpins is same as that of aza. 1-Tigloyl-3-acetyle-11 hydroxymeliacarpin (azadirachtin D) was reported to be biologically active ingredient of neem. (Kraus *et al.*, 1991 and Zhou Ha -Wart, 1993). Besides the bio-degradable products such as 1-tigloyl-3 acetyl-11-hydroxy-11 demthoxycarbonyl meliacarpin (azadirachtin I) and 11-demethoxy-carbonyl-11-oxomeliacarpin were also found in Meliacarpin group.

To obtain good quality of triterpenoids, fresh material is always taken. If the material is wet, it may be dried at ambient temperature under shade.

The solvents generally used for extraction of triterpenoids are ethanol, methanol, acetone, diethyl ether, petroleum ether, hexane etc. Most of the extraction process were aimed to isolating azadirachtin (Aza) and ethanol was normally used. (Butterworth and Morgan 1968, 1971). Warthen *et al*. (1984) used methanol for extraction method. According to Uebel *et al*. (1979), acetone is not very suitable for such extraction. Ethanol was found as good solvent for the extraction of terpenoidal and nonterpenoidal compounds. (Ara *et al*., 1990a,b) However, Yamasaki *et al*. (1988) obtained good results with hexane while from petroleum ether was found as a good extraction material (Ara *et al*., 1989b).

Two groups of triterpenoids are available in neem. (1) Euphal (Euphal 20 + Hβ) (Fig. 2.8), (2) Tirucallol (Tirucallol 20 + Hα) (Fig. 2.9). The other derivatives Δ7- Euphal (20 +Hβ) and Δ7- Tirucallol (20+Hα) (Fig. 2.10) are

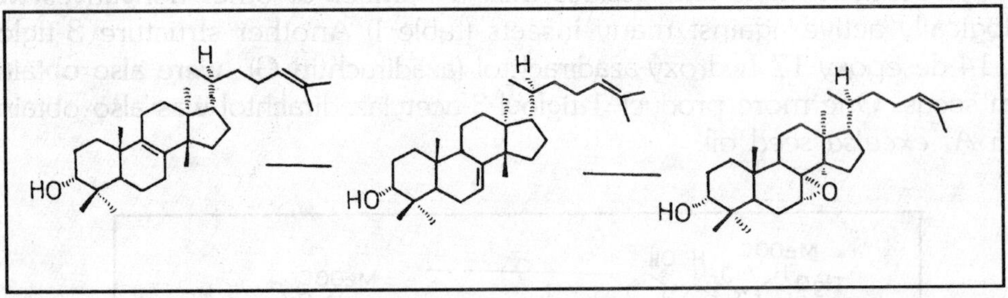

Fig. 2.8
(i) Euphol (20-H_β)
(ii) Tirucallol (20-H_α)

Fig. 2.9
(i) Δ^7-Euphol (20-H_β)
(ii) Δ^7-Tirucallol (20-H_α)

Fig. 2.10
(i) (3a) 20-H_β
(ii) (3b) 20-H_α

formed from the Euphal and Tirucallol by double-bond and shift. Similarly apo-euphal and apo-tirucallol derivatives are formed from the earlier one. Dreyer (1984) as Lavie and Levy (1968, 1971) and Siddiqui *et al*. (1988) obtained rearrangement in the presence of β-configuration and α configuration. These are given in the following figures. These are termed as protolimonoids protomeliacin or melianes, out of these meliantrial was reported as efficacious antifeedent. Another compound isolated are azadirachtal, azadirol Limocin. These tetranortriterpenoids are formed from the Δ7- euphal/tirucallol and apo-euphal/tirucallol by the oxidative degradation of the C-17 side chain resulting in loss of 4 carbon atomas. The degredation occurs by direct formation of the C-17 furanring from apo-euphal/apo-trirucallol or by the formation of hemiacetals of type (22) which turned to hydroxy butenolides 23+24 which finally become furane derivatives 21. These Limonoids occurring in neem are related to 9 different basic structures azadirone (Fig. 2.11), amoorastatin (Fig. 2.12), Vepinin (Fig. 2.13) and Vilasinin (Fig. 2.14). The others are Gedunin (D-seco) (Fig. 2.15), nimbin (Fig. 2.16), nimbolinin (Fig. 2.17) and salannin (C-seco) (Fig. 2.18).

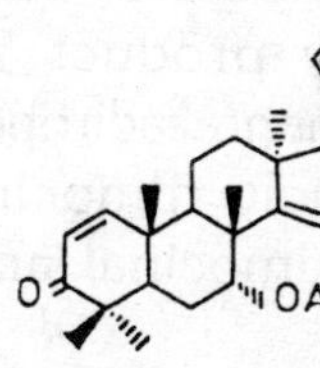
Fig. 2.11

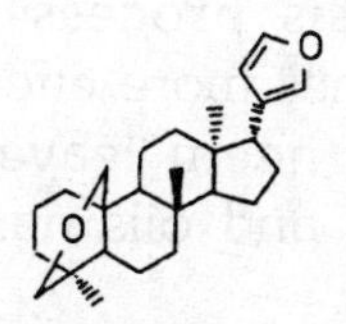
Fig. 2.12

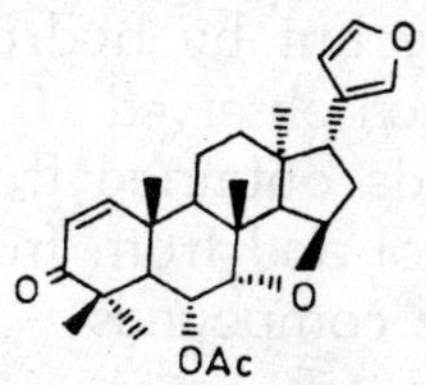
Fig. 2.13

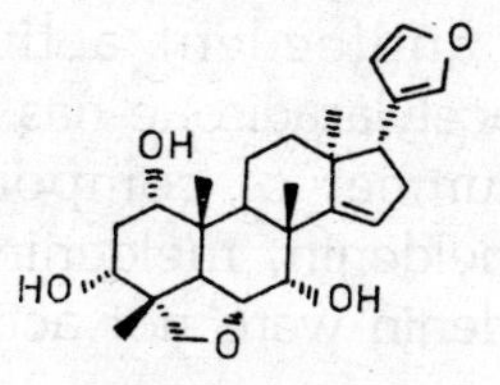
Fig. 2.14

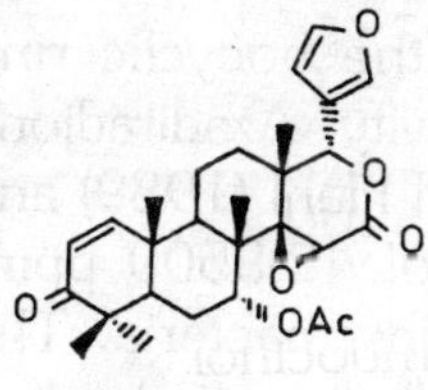
Fig. 2.15

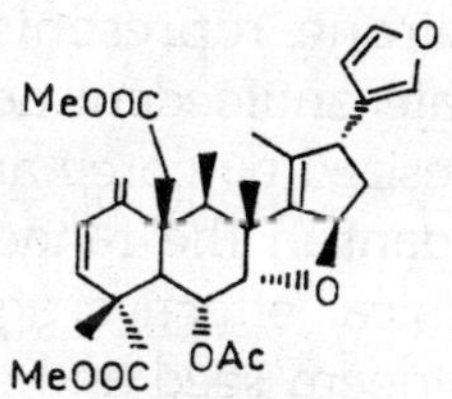
Fig. 2.16

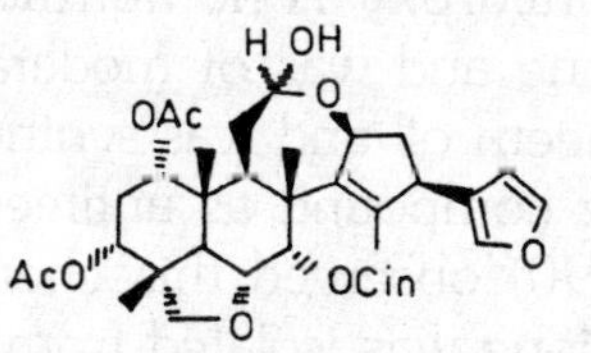
Fig. 2.17

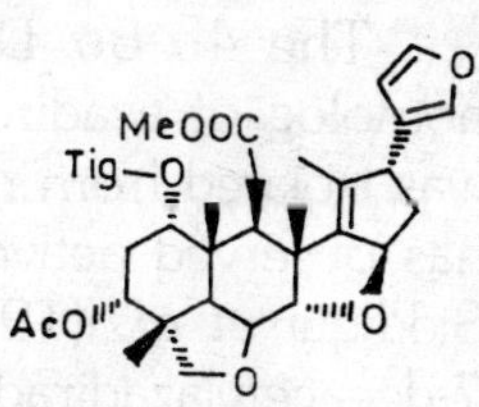
Fig. 2.18

Fig. 2.11 Azadiron, Fig. 2.12 Amoorastatin, Fig. 2.13 Vipinin, Fig 2.14 Vilasinin, Fig. 2.15 Gedunin, Fig. 2.16 Nimbin, Fig. 2.17 Nimbolinin, Fig. 2.18 Salannin

Azadirone Group

This can be grouped in to three Azadirone, Azadiradione (Fig. 2.19) and expoxyazadiradione (Fig. 2.20) and these are biogenetically related to each

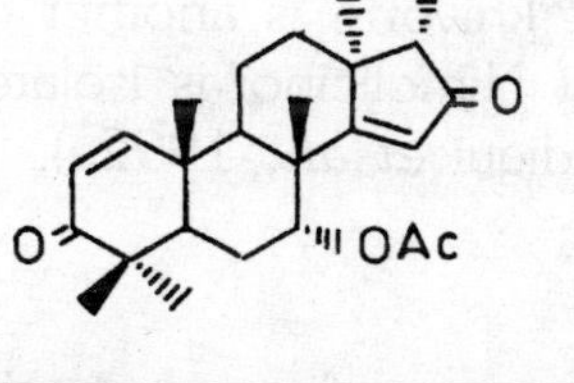
Fig. 2.19 Azadiradione

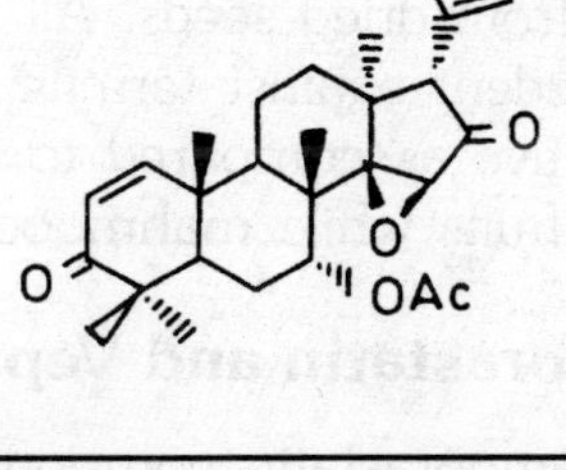
Fig. 2.20 Epoxyazadiradione or Nimbinin

other by oxidation process. Azadirone obtained from neem oil possess very low antifeedent activity but by hydrolysis process the by product 7-decacetylazadirone has been observed 10 times more effective than azadirone. A number of compounds obtained from neem leaves namely nimocin, isomeldenin, meldenindiol and from fruits and oils namely nimocinal and meldenin were not active compounds.

Devakumar and Sukhdev (1993) observed that from dried neem seed azadirone at C-14/15 leads to the formation of 7- acetyl neutrichilenone which possess antitumor activity.

The 4α-6α Dihydroxy A-homoazadirone represents the bocyclic ring homolog of azadirone and was of moderate antifeedent activity. Azadiradione was isolated from neem oil and was synthesized by corey and Hahl (1989) and has observed active compound as antifeedant in the range of 42-5500 ppm. Siddiqui *et al*. (1990) observed this chemical effective against bacteria. The 7-decacetylazadiradione was isolated from neem seed and called as nimbocinol and is much less active than azadiradione.

Two more campounds reacted to Azadiradione was extracted from neem seed. These are 17-B hydroxyazadiradione and 17-B hydroxynimbocinol. Both these campounds are moderate antifeedant but very efficacious against termite species (Ishida *et al*., 1992).

The compound Epoxyazadiradione was isolated from neem oil (Siddiqui, 1992b) and it was called nimbinin. This product is a weak antifeedent but showed high bactericidal activity. Nimocinolide and isonimocinolide isolated from green leaves possess larvicidal properties. The derivative isonimocinolide, and isonimbolide showed insect growth disrupting proprieties.

Gedunim Group

This is the group in which the compounds are formed from azadirone and its derivatives. In this group Gedunin and 7-deacetyldedunin were isolated from neem seed kernal and bark where as benzoyl derivative was isolated from dried seeds. All these compounds are biologically active and act as antifeedent against termite (Ishida *et al*., 1992). khivorin is another effective derivative as compared to gedunin. The product Nimolicinol is isolated from fresh fruits while mahmoodin from seed oil (Siddiqui *et al*., 1992a).

Amoorastatin and Vepinin groups

Amoorastatin compounds are common in some meliaceae species. The compoud, Azadirachtanin has been isolated from fresh neem leaves. The vepinin was isolated from neem oil (Narayanan *et al*., 1969).

Vilasinin group

Vilasinin was isolated from green neem leaves the vilasinin derivatives, nimbolin (1,3-diacetly 7- cinnamoyl vilasinin) was isolated from heart wood. 1-Tigloyl-3-acetylvilosin and its isomer 1-acetyl-tigloylvilaxinin were extracted from seeds. Nimbidinin was found in seed oil, where as 1-tigloyl-3-acetyl-12-α-acetoxy vilasinin, 1-3-diacetyl-12-α-acetoxy vilasinin, 7-tigloyl-12-α, acetoxy vilasinin and 1-acetyl-7-tigloynimbidinin from dried seeds (Kraus *et al*., 1987a, 1993a; Thilele, 1991).

Nimbin group

Siddiqui (1942) for the first time isolated nimbin compounds. These nimbins are formed from 12-hydroxynimocinol or sendanol type precursors.

The Nimbin has been isolated from the seeds and also from leaves (Harris *et al*., 1968; Bokel *et al*., 1990). The 6-Deacetylnimbin was obtained from seeds, leaves and bark (Bokel *et al*., 1990; Narayanan *et al*., 1967a) and according to Bokel *et al*. (1990), Kigodi *et al*. (1989) and Kraus *et al*. (1987a), this contains 2-3 dihydronimbie acid, nimbolide, 28-deoxonimbolide and 6-deacetylnimbinal. Besides nimbanal and nimbinol have been isolated from seeds. Hydroxybutenolides are also present in the nimbin group. Nimbinolide was recently isolated from the seed extracts and named as isophotonimbin. Insect antifeedant activity, bactericidal activity, cytotoxic activity were observed with these compounds. Nimbolide is very low toxic against rodents but very active against *Staphyloccus* strains of bacterium and protozoan parasite. They effect potato virus X, Vaccinia virus and fowl pox virus. These compounds open up a new vista to control these and other viral diseases of crops and livestock.

Salannin group

This is one of the important triterpenoid isolated from neem. This compounds is very powerful antifeedent. Yamasaki *et al*. (1988) developed analytic isolation methods for isolating salanin. Besides 2-3-Dehydrosalannol and 3-deacetyl salanin was observed in the leaf extracts. It is also seem that gama-hydroxybutenolids are also known in salanin. Garg and Bhakuni (1985) obtained from seeds salannolide which is identical to photosalannin. Siddiqui *et al*. (1986a) obtained isoazadirolide from fresh seem leaves. Margosinolide and isomargosinolide are also important structures. Salannol actame-21 and Salannolactame-23 are the other limonoids isolated from neem. (Kraus *et al*., 1987d).

References

Akhila, A. (1996). Biogenesis of azadirachtin in *Azadirachta indica* A. Juss. *Neem Newsletter*, **13**(2) : 24.

Akhila, A.; and Kumar, S. (1996). Biosynthesis of limonoids in *Azadirachta indica*, meliantriol and nimbocinone. *Neem Newsletter*, **13**(2) : 19-20.

Akhila, A and Rani K (1999). Chemistry of the Neem tree (*Azadirachta Indica* A.juss) *Fortschritte der chemie organischer Naturstoffe,* Progress in the Chemistry of organic Natural products-78 Edited by W.Herz, Hfalk, G.W.Kirby, R.E. Moore, Ch. Tamm. Springerwien New York pp 47-149.

Akhila, Anand, Srivastava Meenu and Rani Kusum (1998). Production of radioactive azadirachtine in the seed kernels of *Azadirachta indica* (The Indian Neem Tree). *Natural product letters*, **11** : 105-110.

Ara, I.; Siddiqui, B.S.; Faizi, S.; Siddiqui, S. (1988a). Tricyclic diterpenoids from the stem bark of *Azadirachta indica. J. Nat. Prod.*, **51** : 1054-1061.

Ara, I.; Siddiqui, B.S.; Faizi, S.; Siddiqui, S. (1988b). Tetrapenoids from the stem bark of *Azadirachta indica. Phytochemistry*, **27** : 1801-1804.

Ara, I.; Siddiqui, B.S.; Faizi, S.; Siddiqui, S. (1989a). Diterpenoids from the stem bark of *Azadirachta indica. Phytochemistry*, **28** : 1177-1180.

Ara, I.; Siddique, B.S.; Faizi, S.; Siddiqui, S. (1989b). Diterpenoids from the root bark of *Azadirachta indica. Z. Naturforsch*, **44b** : 1279-1282.

Ara, I.; Siddique, B.S.; Faizi, S.; Siddiqui, S. (1989c). Two new terpenoids from root bark of *Azadirachta indica. J. Nat. Prod.*, **52** : 1209-1213.

Ara, I.; Siddique, B.S.; Faizi, S.; Siddiqui, S. (1989d). Structurally novel diterpenoid constituents from the stem bark of *Azadirachta indica* (Meliaceae). *Journal of the Chemical Society-Perkin Transactions,* **I**(2) : 343-345.

Ara, I.; Siddique, B.S.; Faizi, S.; Siddiqui, S. (1989e). Margosinone and margosinolone, two new polyactetate derivatives from *Azadirachta indica. Fitoterapia*, **60** : 519-523.

Ara, I.; Siddique, B.S.; Faizi, S.; Siddiqui, S. (1990a). Three new deterpenoids from the stem bark of *Azadirachta indica. J. Nat. Prod.*, **53** : 816-820.

Ara, I.; Siddique, B.S.; Faizi, S.; Siddiqui, S. (1990b). Tricyclic diterpenes from the stem bark of *Azadirachta indica. Planta Med.*, **56** : 84-86.

Balasubramanian, C.; Mohan, P.S.; Arumugasamy, K. and Udaiyan, K. (1994). Flavonoid from resin glands of *Azadirachta indica. Phytochemistry*, **34**(4) : 1194-1195.

Bidmon, J.H.J.; Kauser. A.; Mobus, P. and Koolman, J. (1987). Effect of azadirachtin on blowfly larvae and pupae. In Natural Pesticides From the Neem Tree (Azadirachta indica A. Juss.) and Other Tropical Plants, H. Schmutterer and K.R.S. Ascher, eds. *Proc. Third 3rd Intternational Neem conference*, Nairobi, Eschborn, GTZ : 141-150.

Bilton, J.N.; Broughton, H.B.; Ley, S.V.; Lidert, Z.; Morgan, E.D.; Rzepa, H.S. and Sheppard, R.N. (1985). Structural reappraisal of the limonoid insect antifeedant azadirachtin. *J. Chem. Soc., Chem. Commum.* : 968-971.

Bilton, J.N.; Jones, P.S.; Ley, S.V.; Robinson, N.G. and Sheppard, R.N. (1988). Chemistry of insect antifeedant from *Azadirachta indica* (part I) : Conversion from the azadirachtin to the azadirachtin skeletons. *Tetrahedron Letters*, **29** : 1849-1852.

Bokel, M.; Cramer, R.; Gutzeit, H.; Reeb, S. and Kraus, W. (1990). Tetranortriterpenoids related to nimbin and nimbolide from *Azadirachta indica* A. Juss (Meliaceae). *Tetrahedron*, **46** : 775-782.

Broughton, H.B.; Ley, S.V.; Slawin, A.M.Z.; Williams, D.J. and Morgan, E.D. (1986). X-ray crystallographic structure determination of detigloyldihydroazadirachtine and reasignment of the structure of the limonoid insect antifeedant azadirachtin. *J. Chem. Soc., Chem. Commun.* : 46-47.

Butterworth, J.H. and Morgan, E.D. (1968). Isolation of a substance that suppresses feeding in locusts. *J. Chem. Soc., Chem. Commun.* : 23-24.

Butterworth, J.H. and Morgan, E.D. (1971). Investigation of the locust feeding inhibition of the seeds of the neem tree, *Azadirachta indica. J. Insect Physiol.*, **17** : 969-977.

Chirathamjaree, C.; Ermel, K. and Sangwanich, A. (1997). Azadirachtin content of neem seed kernels from selected locations in Thailand. *Proceedings first International Symposium on Biopesticide*, held at Naresuan University, Phitsanulok, Thailand, Oct 27-31, 1996 : 192.

Cohen, E.; Quistad, G.B. and Casida, J.E. (1996). Cytotoxicity of nimbolide, epoxyazadiradione and other limonoids from neem insecticide. *Life Science* , **58**(13) : 1075-1081.

Cohen, E.; Quistad, G.B. and Casida, J.E. (1998). Cytotoxicity of nimbolide, epoxyazadiradione and other limonoids from neem insecticide. *Neem Newsletter*, **15**(2) : 55-56.

Corey, E.J. and Hahl, R.W. (1989). Synthesis of a limonoid, azadiradione. *Tetrahedron Letters*, **30** : 3023-3026.

Devakumar, C. and Sukh Dev (1993). Chemistry. In Randhawa, N.S.; Parmar, B.S. (eds.), *Neem Research and Development*. Society of Pesticide Science, New Delhi, India : 63-96.

Dong Li Xiao and Zhao ShanHuan (1996). The toxic effects and mode of azadirachtin on insects. *Journal of South China Agricultural University*, **17**(1) : 118-122.

Dreyer, D.L. (1984). Biogenetic relationships of degraded triterpenes in the Rutales. In Nes, W.D.; Fuller, G. and Tsai, L.-S. (eds.). *Isopentenoids in Plants, Biochemistry and Function*. Marcel Dekker, New York, NY : 247-265.

Ermel, Klaus; Chirathamjaree, C. and Sangwanich, A. (1997). *Processing of Thai Neem (Azadirachta siamensis Valeton) and the Quality Problems. Proceedings First International Symposium on Biopesticides*, held at Naresuan University, Phitsanulok, Thailand, Oct. 27-31, 1996 : 171-184.

Garg, H.S. and Bhakuni, D.S. (1985). 2',3'-Dehydrosalannol, a tetranortriterpenoid from *Azadirachta indica* leaves. *Phytochemistry*, **24** : 866-867.

Govindachari, T.R. and Geetha Gopalakrishana (1997). 13,14-Desepoxyazadirachtin-A, a tetranortriterpenoid from *Azadirachta indica*. *Phytochemistry*, **45**(2) : 397-399.

Govindachari, T.R.; Gopalakrishnan, G.; Raghunathan, R. and Rajan, S.S. (1994). Crystallisation of azadirachtin-A. *Current Scince*, **66**(4) : 295-297.

Govindachari, T.R.; Gopalakrishnan, G. and Suresh, G. (1996). Isolation and quantification of azadirachtins and other major triterpenoids in the seed oil of *Azadirachta indica* A. Juss. *Neem Newsletter*, **13**(2) : 21-22.

Harris, M.; Henderson, R.; McCrindle, R.; Overton, K.H. and Turner, D.W. (1968). Tetranortriterpenoids -VIII. The constituents and stereochemistry of nimbin. *Tetrahedron*, **24** : 1517-1523.

Henderson, R.; McCrindle, R.; Melera, A. and Overton, K.H. (1968). Tetranortriterpenoids -IX. The constituents and stereochemistry of salannin. *Tetrahedron*, **24** : 1525-1528.

Ishida, M.; Serit, M.; Nakata, K.; Juneja, L.R.; Kim, M. and Takahashi, S. (1992). Several antifeedant from neem oil, *Azadirachta indica* A. Juss., against *Reticulitermes speratus* Kolbe (Isoptera: Rhinotermitidae). *Biosci. Biotech. Biochem.*, **56** : 1835-1838.

Jacobson, M. (1986). Pharmacological annd toxicological effects of neem and chinaberry on warm-blooded animals. *Neem Newsletter*, **3**(4) : 39-43.

Jarvis, A.P.; Johnson, S. and Morgan, E.D. (1998). Stability of the natural insecticide azadirachtin in aqueous and organic solvents. *Pesticide Science*, **53**(3) : 217-222.

Jarvis, A.P.; Johnson, S.; Morgan, E.D.; Simmonds, M.S.J. and Blaney, W.M. (1997). Photoxidation of nimbin and salannin, tetranortriterpenoids from the neem tree (*Azadirachta indica*). *Journal of Chemical Ecology*, **23**(12) : 2841-2860.

Jarvis, A.P.; Morgan, E.D. and Edwards, C. (1999). Rapid separation of triterpenoids from neem seed extracts. *Phytochemical Analysis*, **10**(1) : 39-43.

Johnson, S. and Morgan, E.D. (1998). Supercritical fluid extraction of oil and triterpenoids from neem seeds. *Neem Newsletter*, **15**(4) : 48.

Kigodi, P.G.; Blasko, G.; Thebtaranonth, Y.; Pezzuto, J.M. and Cordell, G.A. (1989). Spectroscopic and biological investigation of nimbolide and 28-deoxonimbolide from *Azadirachta indica. J. Nat. Prod.*, **52** : 1246-1251.

Kraus, W. (1986). Constituents of neem and related species. A revised structure of azadirachtin. *Stud. Org. Chem.*, **26** : 237-256.

Kraus, W.; Baumann, S.; Bokel, M.; Bruhn, A.; Cramer, R.; Ehhammer, B.; Gutzeit, H.; Herr, B.; Kaufmann-Horlacher, I.; Keller, U.; Klingele, M.; Schwinger, M.; Thiele, S.; Vogler, B.; Zhou, Y.; Soellner, R.; Wendisch, D.; Steffens, R. and Wachendorff, U. (1991). Constituents of neem and other Meliaceae species in pest control. In Ahmed, S. (ed.) Neem (*Azadirachta indica*) for pest control and rural development in Asia and the Pacific. *17th Pacific Science Congress* (Honolulu, HI, 1991), in press.

Kraus, W.; Baumann, S.; Bokel, M.; Cramer, R.; Grimminger, W.; Hendlmeier, R.; Keil, E.; Keller, U.; Klenk, A.; Klingele, M.; Pohnl, H. and Schwinger, M. (1987a). Insect feeding and development controlling constituents of Meliaceae. *Proc. 1st Princess Chulabhorn Sci. Congr. Int. Congre. on Nat. Products* (Bangkok, Thailand), Vol. **II** : 554-563.

Kraus, W.; Baumann, S.; Bokel, M.; Keller, U.; Klenk, A.; Klingele, M.; Pohnl, H. and Schwinger, M. (1987b). Control of insect feeding and development by constituents of *Melia azedarach* and *Azadirachta indica*. In Schmutterer, H.; Ascher, K.R.S. (eds.) Natural Pesticides from the Neem Tree and Other Tropcial Plants. *Proc. 3rd Int. Neem Conf.* (Nairobi, Kenya, 1986) : 149-160.

Kraus, W.; Bokel, M.; Cramer, R.; Klenk, A. and Pohnl, H. (1985a). Constituents of neem and related species. A revised structure of azadirachtin.

Third Int. Conf. on Chemistry and Biotechnology of Biologically Active Natural Prducts (Sofia, Bulgaria, 1985), Abstr. : 446-450.

Kraus, W.; Bokel, M.; Klenk, A. and Pohnl, H. (1985b). The structure of azadirachtin and 22,23-dihydro-23 β-methoxyazadirahctin. *Tetrohedron Letters*, **26** : 6435-6438.

Kraus, W.; Bokel, M.; Schwinger, M.; Vogler, B.; Soellner, R.; Wendisch, D.; Steffens, R. and Wachendorff, U. (1993a). The chemistry of azadirachtin and other insecticidal constituents of Meliaceae. In Van Beek, T. and Breteler, H. (eds). *Phytochemistry and Agriculture*, Oxford University Press : 18-39.

Kraus, W.; Gutzeit, H. and Bokel, M. (1989). 1,3-Diacetyl-11,19-deoxa-11-oxo-meliacarpin, a possible precursor of azadirachtin, from *Azadirachta indica* A. Juss (Meliaceae). *Tetrahedron Letters*, **30**(14) : 1797-1798.

Kraus, W.; Klenk, A.; Bokel, M. and Vogler, B. (1987d). Tetranortriterpenoid-lactame mit insektenfrafBhemmender Wirkung aus *Azadirachta indica* A. Juss (Meliaceae). *Liebigs Am. Chem.* : 337-340.

Kubo, I.; Matsumoto, T.; Matsumoto, A. and Shoolery, J.N. (1984). Structure of deacetylazadirachtinol, application of 2D^{1}H-^{1H} and ^{1}H-^{13}C shift correlation spectroscopy. *Tetraheron Letters*, **25** : 4729-4732.

Kumar, C.S.S.R.; Malladi Srinivasa and Shirish Yakkundi (1996). Limonoids from the seeds of *Azadirachta indica. Phytochemistry*, **43**(2) : 451-455.

Lavie, D.; Jain, M.K. and Kirson, I. (1967). Terpenoids. Part VI. The complete structure of melianone. *J. Chem. Soc. (C)* : 1347-1351.

Lavie, D. and Levy, E.C. (1968). Studies on epoxides IV. Rearrangements in triterpenoids. *Tetrahedron Lett.* : 2097-2100.

Lavie, D. and Levy, E.C. (1971). Meliane-meliacin relationship. *Tetrahedron*, **27** : 3941-3947.

Lee, S.M. and Klocke, J.A. (1987). Combined florisil, droplet counter current, and high performance liquid chromatographies for preparative isolation and purification of azadirachtin from neem (*Azadirachta indica*) seeds. *J. Liq. Chromatogr.*, **10** : 1151-1163.

Lee, S.M.; Olsen, J.I.; Schwelzer, M.P. and Klocke, J.A. (1988). 7-Deacetyl-17-β-hydroxy azadiradione, a new limonoid insect growth inhibitor from *Azadirachta indica. Phytochemistry*, **27** : 2773-2775.

Ley, S.V.; Anderson, J.C.; Blancy, W.M.; Lidert, Z.; Morgan, E.D.; Robinson, N.G. and Simmonds, M.S.J. (1988). Chemistry of insect antifeedants from *Azadirachta indica* (part 3): reactions on the C-22,23 enol ether double bond of azadirachtin and conversion to 22,23-dihydro-23βmethoxyazadirachtin. *Tetrahedron Letters*, **29**(42) : 5433-5436.

Nair, M.S.; Sindhu Gopal; Deepa Issac and Gopal, S. (1997). Optimised isolation procedure for biologically active compounds nimbolide and 28-deoxonimbolide from *Azadirachta indica* leaves. *Phytochemistry*, **46**(7) : 1177-1178.

Nakanishi, K. (1975). Structure of the insect antifeedant azadirachtin. *Rec. Adv. Phytochem.*, **9** : 283-298.

Narayanan, C.R.; Pachapurkar, R.V. and Sawant, B.M. (1967). Nimbinin: a new tetranortriterpenoid. *Tetrahedron Lett.* : 3563-3565.

Narayanan, C.R.; Pachapurkar, R.V.; Sawant, B.M. and Wadia, M.S. (1969). Vepinin, a new constituent of neem oil. *Indian J. Chem.*, **7** : 187.

Narayanan, C.R. and Iyer, K.N. (1967a). Isolation and characterisation of deacetylnimbin. *Indian J. Chem.*, **5** : 460.

Niranjan, R.; Venkatakrishnan, K. and Madyastha, K.M. (1996). 11-Epi-azadirachtin D : an epimeric azadirachtin analogue from *Azadirachta indica*. *Phytochemistry*, **49**(1) : 265-267.

Niranjan, R.; Venkatakrishnan, K.; Madyastha, K.M. and Ramji, N. (1998). 11-Epi-Azadirachtin H from *Azadirachta indica*. *Phytochemistry*, **42**(2) : 561-562.

Pohnl, H. (1985). Neue Verbindungen der Azadirachtin-Reihe und andere bilogisch aktive Tetranortriterpenoids aus *Melia azedarach* L. - Isolierung und Strukturaufklarung. *Doctor Thesis*, Univ. of Hohenheim, Germany.

Pradhan, S.; Jotwani, M.G. and Rai, B.K. (1962). Neem seed deterrent to locusts. *Indian Farming*, **12**(8) : 7-11.

Rangasamy, S. and Parmar, B.S. (1997). Azadirachtin-A contents of seeds of neem ecotypes in relation to the agro-ecological region of India. *Neem Newsletter*, **14**(1) : 19.

Rani, Kusum and Anand, Akhila (1994). Biosynthetic relationship between nimocinol and nimocinolide in *Azadirachta indica*. *Natural product letter*, **4** : 179-182.

Rembold, H.; Forster, H.; Czoppelt, C.; Schmutterer, H. (ed.) and Ascher, K.R.S. (1987). Structure and biological activity of azadirachtins A and B. In Schmutterer, H. and Ascher, K.R.S. (eds.) *National Pesticides from the Neem Tree and Other Tropical Plants. Proceedings of the 3rd International Neem conference*, Nairobi, Kenya, 10-15 July, 1986 : 149-160.

Rembold, H.; Forester, H. and Sonnenbichler (1987b). Structure of azadirachtin B. *Z. Naturforsch*, **42c** : 4-6.

Rojatkar, S.R.; Bhat, V.S.; Kulkarni, M.M.; Joshi, V.S. and Nagasampagi,

B.A. (1989). Tetranortriterpenoids from *Azadirachta indica. Phytochemistry*, **28**(1) : 203-205.

Rojatkar, S.R.; Sawaikar, D.D. and Nagasampagi, B.A. (1996). New tetra and pentanortriterpenoids from *Azadirachta indica* A. Juss. *Neem and Environment*, **1** : 203-206.

Sankaram, A.V.B.; Murthy, M.M.; Bhaskaraiah, K.; Subramanyam, M.; Sultana, N.; Sharma, H.C.; Leuschner, K.; Ramaprasad, G.; Sitaramaiah, S.; Rukmini, C. and Rao, P.U. (1987). Chemistry, biological activity and utilization aspects of some promising neem extractives. In Schmutterer, H. and Ascher, K.R.S. (eds.). Natural Pesticides from the Neem Tree and Other Tropical Plants. *Proc. 3rd Int. Neem Conf*. (Nairobi, Kenya, 1986) : 127-148.

Schoonhoven, L.M. (1981). Perception of azadirachtin by some lepidopterous larvae. *Proceedings First International Neem Conference*, Rottach, Egern, 1980, GTZ Eschborn : 105-108.

Schroeder, D.R. and Nakanishi, K. (1987). A simplified isolation procedure for azadirachtin. *J. Nat. Prod*., **50** : 241-244.

Selvi, S.T. and Mohan, P.S. (1997). Chemical examination of the resinous exudate from *Azadirachta indica* A. Juss. *Indian Journal of Chemistry Section B, Organic including Medicinal*, **36** (11) : 1082-1083.

Senniappan Thamarai, Selvi;Chinnusamy Balasubramanian and Palathurai Subramaniam Mohan (1996). Chemical examination of the resinous exudate from *Azadirachta indica* A. Juss. *Neem Newsletter*, **13** (2) : 21.

Siddiqui, B.S.; Ghiasuddin; Faizi, S. and Siddiqui, S. (1992a). Triterpenoids from the fresh fruit coats of *Azadirachta indica. Phytochemistry*, **31**(12) : 4275-4278.

Siddiqui, S. (1942). A note on the isolation of three new bitter principles from the nim oil. *Curr. Sci*., **11** : 278-279.

Siddiqui, S.; Faizi, S.; Siddiqui, B.S. and Ghiasuddin (1992b). Consituents of *Azadirachta indica*: Isolation and structure elucidation of a new antibaceterial tetranortriterpenoid, mahmoodin, and a new protolimonoid, naheedin. *J. Nat. Prod*., **55** : 303-310.

Siddiqui, S.; Iffat Ara; Shaheen Faizi; Tariq Mahmood and Siddiqui, B.S. (1988). Phenolic tricyclic diterpenoids from the bark of *Azadirachta indica. Phytochemistry*, **27**(12) : 3903-3907.

Siddiqui, S.; Mahmood, T.; Siddiqui, B.S. and Faizi, S. (1986). Two new tetranortriterpenoids from *Azadirachta indica. Journal of Natural Products*, **49**(6) : 1068-1073.

Siddiqui, S.; Siddiqui, B.S.; Faizi, S. and Mahmood, T. (1986a). Isoazadirolide, a new tetranortriterpenoid from *Azadirachta indica* A. Juss (Meliaceae). *Heterocycles*, **24** : 3163-3167.

Siddiqui, S.; Siddiqui, B.S.; Faizi, S. and Mahmood, T. (1988). Tetracyclic triterpenoids and their derivatives from *Azadirachta indica. J. Nat. Prod.*, **51** : 30-43.

Siddiqui, S.; Siddiqui, B.S.; Ghiasuddin and Faizi, S. (1990). Triterpenoids from kernel of *Azadirachta indica. Proc. Pak. Acad. Sci.*, **27** : 333-348.

Siddiqui, S.; Tariq Mahmood; Siddiqui, B.S. and Shaheen Faizi (1986). Isolation of a triterpenoid from *Azadirachta indica*. *Phytochemistry*, **25**(9) : 2183-2185.

Sidhu, O.P. and Behl, H.M. (1996). Seasonal variation in azadirachtins in seeds of *Azadirachta indica*. *Current Science*, **70**(12) : 1084-1086.

Singh, R.P.; Dev Kumar, C. and Dhingra, S. (1988). Activity of neem seed kernel extract against the mustard aphid, *Lipaphis erysimi*. *Phytoparasitica*, **16**(3) : 225-229.

Sombatsiri; Ermel, K. and Schmutterer, H. (1995). The Thai neem tree, *Azadirachta siamensis* (Val.). In H. Schmutterer (ed.), The Neem Tree *Azadirachta indica* A. Juss. and Other Meliaceous Plants, VCH Verlogsgesellschaft mbH, Weinheim Germany : 585-597.

Stark, J.D. (1997). Risk assessment of neem insecticides : Persistence in the environment and potential impact on non-target organisms. *Proceedings First International Symposium on Biopesticide* held at Naresuan University, Phitsanulok, Thailand, Oct. 27-31, 1996 : 69-74.

Sundaram, K.M.S. and Curry, J. (1996). Effect of some UV light absorbers on the photostabilization of azadirachtin, a neem-based biopesticide. *Chemosphere*, **32**(4) : 649-658.

Suresh, G.; Narasimhann, N.S. and Palani, N. (1997). Structure of nimonol from fresh whole green leaves of *Azadirachta indica*. *Phytochemistry*, **45**(4) : 807-810.

Thiele, S. (1991). Isolierung und Strukturaufklarung neuer Tetranortriterpenoide aus *Azadirachta indica* A. Juss (neem tree, Meliaceae) und Untersuchung der nematiziden Wirkung von Neem-Extrakten. *Doctor Thesis*, Univ. of Hohenheim, Germany.

Tonthubthimthong, P. and Tasakorn, P. (1997). Concentration of crude azadirachtin extract by adsorption on active carbon. *Proceedings First International Symposium on Biopesticide* held at Naresuan University, Phitsanulok, Thailand, Oct. 27-31, 1996 : 169.

Uebel, E.C.; Warthen, J.D.Jr. and Jacobson, M. (1979). Preparative reversed-phase liquid chromatographic isolation of azadirachtin from neem kernels. *J. Liq. Chromatogr.*, **2** : 875-882.

Warthen, J.D.Jr.; Stokes, J.B.; Jacobson, M. and Kozempel, M.F. (1984). Estimation of azadirachtin content in neem extract and formulations. *J. Liq. Chromatogr.*, **7** : 591-598.

Wehrli, F.W. and Nishida, R. (1979). The use of carbon-13 nuclear magnetic resonance spectroscopy in natural products chemistry. In Zechmeister, L.; Herz, W.; Griesebach, H. and Kirby, G.W. (eds.). *Fortschr. Chem. Org. Naturst*, **36**, Springer Verlag, New York, NY : 1-229.

Yamasaki, R.B.; Ritland, T.G.; Barnby, M.A. ad Klocke, J.A. (1988). Isolation and purification of salannin form neem seeds and its quantification in neem and chinaberry seeds and leaves. *J. Chromatogr.*, **447** : 277-283.

Zanno, P.R.; Miura, I.; Nakanishi, K. and Elder, D.L. (1975). Structure of the phagorepellent azadirachtin. Application of PRFT/CWD carbon-13 nuclear magnetic resonance. *J. Am. Chem. Soc.*, **97** : 1975-1977.

Zhou-Halwart, Y. (1993). Isolierung und Strukturaufklarung neuer Tetranortriterpenoide aus *Azadirachta indica* A. Juss (Meliaceae, Neem-Baum) mit insektenfraBhemmender Wirkung. *Doctor Thesis*, Univ. of Hohenheim, VErlag Koster, Berlin, Germany.

Chapter 3

Pests and Diseases

Neem is one tree in which every part of the tree has one or the other properties such as antifeedant, deterent, growth inhibition, oviposi tional deterrent, repellent, insecticidal properties, fungicidal properties and nematicidal properties.

In the presence of these properties, very few pests and diseases thrive on this tree. However, few of them are known to cause considerable damage to seedlings and young plants. The pests and diseases reported on neem tree are given below.

Pests

Coleopterous pests

The coleopterous beetles reported on neem are *Siderodactylus sagitarius* Oliver, *Sternocera interrupta* Oliver, *Epitragus sallei, Doliopygus unispinosus* Schedl, *Bostrychopsis jesvita* Fab., *Anoplophora chinensis* Forster, *Phryneta spinator* Fabr. *Apate monachus* Fabr, *Apate terebrans* Pallos, *Holotrichia consanguinea, H. serrata, H. insularis* (Browne, 1968; Gupta, 1973). The infestation of beetle *Cryptocephalus ovulum* is common in Indian region. Both the larvae and adults feed on neem leaves but rearely cause any significant injuries in Northern India and Pakistan.

Other beetles feed on neem leaves are *Apirocalus mus, Hypomeces squamosus, Myllocerus dorsatus,* (Fig. 3.5, Fig. 3.6) *Myllocerus* spp. and the

Fig. 3.6 : Adult of *Myllocerus dorsatus*

borer, *Araecerus fasciculatus* (Uthamasamy *et al*., 1973). Rohilla *et al*. (1981) reported *Holotrichia serrata* (F) as predominant species feeding on neem. The grub is large brown and yellowish white. The beetle emerges during pre-monsoon rains in June and oviposit in soil at the onset of monsoon in July. The larvae after emergence start feeding on roots and cause heavy mortality in young seedlings. The adults feed on foliage at night, hiding in the soil during day time. Brar and Sandhu (1982) reported that *Holotrichia consanguinea* Blanchard were attracted to neem but they did not feed on the leaves while Vora and Ramakrishnan (1991) observed neem as most attractive plants. Khan and Ghai (1982) discribed *Holotrichia akolana* sp. n from adult males collected on *Azadirachta indica*. Ahmed and Seema Kumar (1990) reported *M. dorsatus* and *M. lactivirens* on neem leaves. Tewari (1992) reported *Araecerus sutularis* Boheman on neem tree. The beetles are the regular borer of neem berries on the tree. The eggs are generally laid on the berries and the larvae on hatching bore into the seed and hollow it out. Pupation takes place inside the seed. The emergence of beetle continues from March to December. Besides two more species namely, *Crioceris hamsoni* Jacoby, *Cryptocephalus aegualis* Clav have been recorded feeding as minor pests of neem. The adults feed on the foliage of neem causing very little damage.

Seema Kumar *et al*. (1996) and Seema Kumar *et al*. (1997) reported *Myllocerus lactivirens* Marshall as new pest record on neem in arid zone of India. The adults are small, 3-4 mm long, green with yellow tinge, scales on the elytra are on intervals of two types, the predominant circular without pedicel and the less predominant circulars, with narrow small pedicel, both having metallic green colour with yellowish tinge, scales on the spiral margins are cylindrical, green with yellow tinge. The maximum defoliating activity was recorded during July to October. The adults are voracious feeders and were found feeding on young tender as well as old leaves of neem seedling stage. The adults adhered on the ventral surface of the leaves. The adult weevil made holes but gradually started feeding the leaf laminae except midrib. Egg, grubs and pupal stages passed in the soil.

Fried (1999) reported wood beetle *Tristaria grouvellei*, stem weevil *Alcides bubo* Fab., seed weevil *Araecerus fasciculatus*, chrysomelid *Monolepta* sp. on neem tree.

Pests on seed

The stored neem seeds and kernels were found to be infested by coleopterous beetles, *Oryzaephilus acuminatus*, *Oryazaephilus surinamensis* (Fig. 3.1) (Linn.) and *Rhizopertha dominica* (Fabr.) (Fig. 3.2) (Sarup and

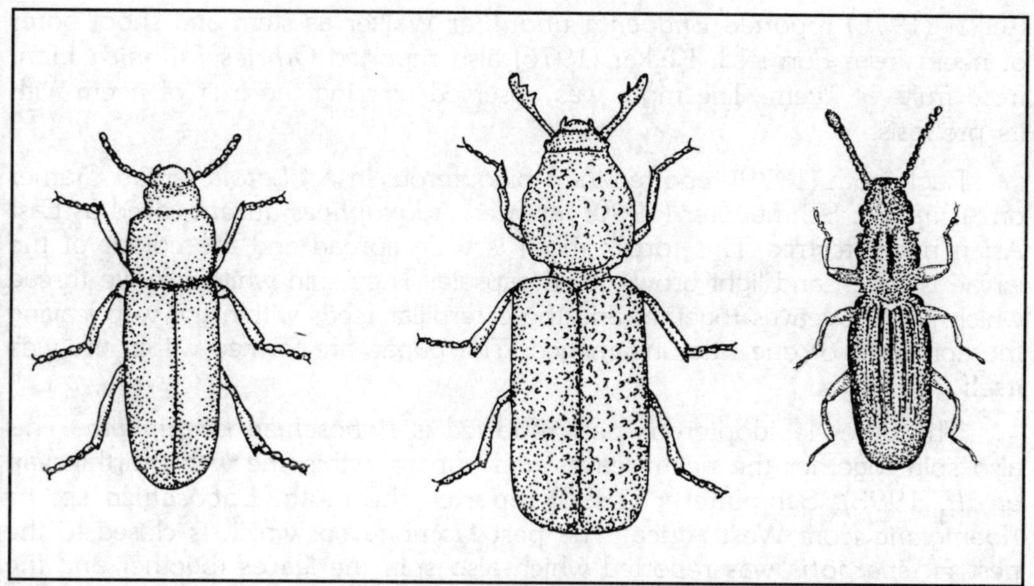

Fig. 3.1 Adult of *Tribolium Castaneum* Fig. 3.2 Adult of *Rhyzopertha dominica* Fig. 3.3 Adult of *Orzaephilus surinamensis*

Srivastava, 1971; Srivastava and Bhanotar, 1980; Thomas and Woodruff, 1983). O.O.L.T. (1987) reported *Rhyzopertha dominica* and *Stegobium paniceum* on stored neem seeds. Zongo (1990) found *Oryzaephilus surinamensis* infesting seeds of neem. Besides, the larvae of *Carpophilus dimidiatus* was reported feeding on neem seed kernels. The attack of *Tribolium castaneum* (Fig. 3.3) and *T. confusum* was reported on old neem seeds (Saxena *et al*., 1989).

Lepidopterous pests

Fletcher (1933) reported *Lasperyesia koenigiana* Fabr. from neem. The lepidopterous pests *Odites atmopha* Meyrick and *Leptoceneria reducta* Walker was also reported on neem.

Bessan (1941) reported stem and shoot borer, *Endoclita undulifer* Walker on neem. Besides, he reported *Ascotis selenaria* Hubner, *Boarmia variegata* Moore and *Cleora cornaria* Guenee. However, Nayar *et al*. (1989) reported *Eurema* sp. Uthamasamy *et al*. (1973) reported *Lasperyesia aurantiana* Meyr.

Browne (1968) reported *Leptocenaria reducita* Walker, an Australilan white ceder moth feeding on leaves of neem. The larvae feed in the night and cluster on stem durig the day. Ahmed and Murugesan (1975) reported semilooper, *Achaea janata* Linnaeus a new pests record on neem (*A. indica*).

Ketkar (1976) reported *Endoclita undulifer* Walker as stem and shoot borer of neem from Formosa. Ketkar (1976) also reported *Othries fullonica* Linn. from fruits of neem. The moth was observed piercing the fruit of neem with its probosis.

Raghunath (1979) reported the lepidopterous insect *Latoia lepida* Cramer on *A. indica*. Schmutterer (1990) reported *Adoxophyes aurata* called as East Asian neem tortrix. This tortrix moth is wide spread and the colour of the larvae is green and light brown head capsule. They spin white silk like thread which join the leaves together and the caterpillar feeds within the web leaving the epidermis of one side undamaged. The pupae are formed within the web itself.

The other lepidopterous pest reported is *Loboschiza koenigiana*. This also spin together the neem leaves and pupate within the web (Karthikeyan *et al.*, 1993). Schmutterer (1990) reported the moth, *Loboschiza* sp. *nr Koenigiana* from West Africa. The pest *Homona* sp. which is closed to the pest *H. spargotis* was reported which also spin the leaves together and the caterpillar feeds within the web.

Occasional lepidopterous pests of neem leaves reported form India are *Eurema* sp., *Ascotis selenaria* Hubner, *Boarmia variegata* Moore, *Cleora cornaria* Guenee, *Lasperyesia aurantiana* Meyr., *L. koenigiana* Fab. and *Odites atmopha* Meyrick. Friend (1993) reported adults of fruit piercing moth, *Othreis fullonica* which suck the sweet juice from the ripe fruits of neem. Besides, the white ceder moth, *Leptocneria reducta* sometimes found as a serious pest of *Melia azedrach*. The larvae of this dingy moth have been recorded feeding on leaves of neem *A. indica, Melia azedarach* in Australia. The fully grown larva is rusty brown with dorsal reddishh tubercles bearing grey setae. The larvae feed in the night and cluster in large number on the stem of the host during day (Browne, 1968).

Rice (1993) reported that young shoots of neem are attacked by the larvae of *Hypsipyla robusta* in South Australia but Chaudhary (1997) reported a new record of shoot borer, *Hypsipyla* sp. damaging all plants and infesting 97.75 percent of total shoots in India.

Friend (1996) reported white ceder moth *Lymantria reducta* (Walk.) feeding on the leaves of neem. Sundararaj and Murugesan (1997) noticed a severe infestation of *Thosea bipartita* on neem. This is a new record of the pest on neem in India. This pest was earlier reported from Thailand feeding on young plants of neem.

Heteropteran bug pests

The tea mosquito bug, *Helopeltis antonii* Signoret is wide spread in

Asia and is considered as serious pest of neem in India (Ayyar, 1983). These are very active narrow bodied insects with long limbs. This bug has a mixed red, black and white colour having peculiar erect knobbed process on its back. The damage is done to tender shoots and foliage. The eggs are burried singly in the tissues of young shoots. The reddish brown coloured first instar nymphs hatch after a week.

Pillai and Gopi (1990) observed drying up of the distal parts of the shoots and foliage of neem during December to April in South India. The insect pest, *Helopeltis antonii* has been observed to cause the drying symptoms. The other casual feeder pest reported is *Laspeyresia aurantiana*. Later, Annamalai *et al.* (1996) observed this bug as the most serious causing more than 90 percent loss in Periyar district during October to February. The biology of *H. antonii* was studied by Onkarappa and Kumar (1997) in Karnataka, India.

Eggs were laid within the tender shoots and the incubation period and total nymphal developmenet varied from 9 to 12 days and 13 to 20 days, respectively. The duration of the 1st, 2nd, 3rd, 4th and 5th instars was 3.5, 3.2, 2.5, 2.3 and 5.0 days, respectively. Hatching varied from 58.3 to 94.0 percent. The duration of the life cycle from egg to adult emergence varied from 23 to 30 days. Mean fecundity was 176.3 eggs/female. Mean longevity of males and females was 37.1 and 44.6 days, respectively. The pre-oviposition, oviposition and post-oviposition periods varied from 3 to 8, 17 to 61 and 0 to 3 days, respectively. The female to male ratio was 1 : 11.4. The insect completed 12 generations a year on neem shoots under laboratory conditions. Sundraraju and Babu (1998) studied the relative multiplication of *Helopeltis antonii* on its three main host plants (neem, guava and cashew). The net reproductive rate indicating the total female births (RO) was 65.63, 40.21 and 22.17; the population increased with an infinitesimal rate (RM) of 0.1255, 0.0857 and 0.1215; and finite rate (lambda) of 1.134, 1.09 and 1.129 female per day and a generation was completed in 33.35, 43.12 and 25.51 days on neem, guava and cashew, respectively. The stable age distribution (%) of eggs, nymphs and adults on these three hosts ranged from 65.883-68.499, 27.675-29.233 and 4.146-4.888, respectively.

Nezara viridula, green stink bug, the polyphagous insect (pentatomidae) is reported on neem by Valdix in 1994 in Nicaragua. These bugs feed on youg shoots and leaves.

Homopterous pests

Aphis spiraecola, the citrus aphids (Aphididae) are reported on neem in nurseries in Philippines and Dominicans Republic. They cause slight damage to young trees.

Browne (1968) reported grape mealy bug, *Pseudococcus maritimus* Ehrhorn on *Azadirachta indica* and *Melia azedarach* in Egypt. The nymphs feed principally on the sap of fruit or at the base of the shoot. There are generally two generations in a year.

Other soft scales reported on neem trees are *Macropulvinaria jacksoni* (from Africa and Asia), *Megapulvinaria maxima* (from India and Myanmar), *Vinsonia stellifera* (from Caribbean), *Parasaiseetia nigra* (from Malaysia), *Milviscutulus mangiferae* (from Papua and New Guinea), *Ceroplastes floridensis* (from Haiti and Madagascar), *C. ceriferus* (from India and Myanmar), *Coccus hesperidium* (from Medagascar), *Chloropulvinaria psidii* (from Philippines), *Pulvinaria azadirachtae* Green and *Aspidiotus hederae* Vallot. *Kerria lacca* (Lacciferidae) was also reported on neem in Southern Asia. The other scale pests reported are *Lecaniodiapsis azadirachta, Lecanium* sp. *Lepidosaphes meliae* (Tewari, 1992). Among mealy bugs, *Pseudococcus gilbertensis* was reported as an important neem pest in South India (Karthikeyan *et al*., 1993). Recently, mango mealy bug, *Drosicha mangiferae* has been reported on neem tree (Fig. 3.4) (Srivastava, 2000, unpublished). *Aonidiella orientalis*, oriental scale (Diaspididae) is one of the dangerous species of scale on neem (Balachowsky, 1956). Lale (1998) reported extent of the threat posed to neem plantation by *Aonidiella orientalis*, the scale insect. Another sap sucker *Megacoelum esmedorae* Baliard was observed feeding on green leaves and shoots of neem tree.

Meenakshi (1998) reported mealy bug, *Pulvinaria maxima* Green infesting on neem in Delhi region. There is a threat that this pest may effect the growth and development of neem tree. *Chrysomphalus aonidum*, Egyptian black scale (Diaspididae) is recorded on neem from Nicaragua, Papua, New Guinea and Madagascar (Schmutterer, 1990) and cause yellowing and dropping of leaves. *Parlatoria crypta*, mango white scale (Diaspididae) is a main pest of mango but recorded on neem. *Pinnapsis strachani*, lesser snow scale (Diaspididae) was reported as pest of neem. The other scales reported on neem are *Chrysomphalus dictyospermi* from Papua, New Guinea, *Selenaspidus articulatus* from Africa, Papua, New Guinea, Fizi; *Hemiberlesia lataniae* from Fizi; *Parlatoria proteus* from Madagascar; *P. orientalis* from India; *Borchseniaspis palmae* from Papua, New Guinea and *Lepidosaphes tapleyi* from Senegal (Schmutterer, 1990). Friend (1999) reported *Aonidiella aurantii* on neem.

Thysanopterous pests

Bhasin *et al*. (1958) reported the thrip *Taemothrips longistylus* Karny feeding on the sap of the leaves of neem. Ananthakrishnan (1973) has reported the thrip *Erythrothrips asiaticus* (Ramk and Margab) feedings on flowers of neem. Tewari (1992) reported the thrips, *Scirtothrips dorsalis* Hood and *Heliothrips haemorrhoidalis* and *Elaphrothrips* sp. as minor pests of neem in India.

White fly :

Beesan (1941) reported citrus whitefly *Dialeurodes citri* Ashmead infesting the foliage of *Melia azedarach*. This pest has also been reported on neem in USA. The other species reported on neem is *Dialuerodes armatus*.

Short horned and long horned gasshoppers

Uthamasamy *et al.* (1973) reported short horned hopper, *Orthacris simulans* feeding on neem leaves in India whereas Gruber (1992) reported short horned hopper, *Melanoplus* sp. damaging young leaves of neem in Nicaragua. Menendez and Perez (1994) reported long horned hopper, *Anurogryllus obortivus* feeding on leaves in Cuba. Chockalingam and Noorjahan (1984) observed the feeding behaviour of adults of *Microcentrum retinerve* on leaves of neem (*Azadirachta indica*) in the laboratory. The assimilation efficiency was 63.7 percent in males feeding on young leaves and 67.2 percent in females; the conversion efficiencies were very low relative to those of other Orthoptera. The tettigoniid longhorned grasshopper had higher feeding, assimilation annd conversion rates when reared on young leaves than on mature leaves. Young neem leaves contained higher concentrations of protein and total nitrogen, and lower amounts of phenol and lipid, than mature leaves. It is possible that the tettigoniid is able to detoxify the phagodeterrents which prevent other insects from feeding on neem.

Hymenopterous pests

Raghunath *et al.* (1982) reported the thousands of fire ants, *Solenopsis* sp. damaging neem in Andhra Pradesh, India and observed that plants of 3 to 6 years of age were completely defoliated. The attack was confined to neem only although there were many wild shrubs available in the vicinity. They reported it as the first record on neem. The ants feed voraciously on the leaves, thus leaving only the petiole and midrib of the leaves. Young rachis was also susceptible to the infestation.

The hemipteran ants *Acromyrmex* spp. and *Atta* spp. defoliate neem trees were reported by Schmutterer (1990). These ants are major pests of tree crop, ornamentals and other plants in Latin America. These ants defoliate in the night leaving untouched only the midribs and other strong veins.

Besides, the harvester ant, *Messar* spp. was found to collect neem seeds to their nests. The other hymenopterous pest *Xylocopa* sp. was also recorded on neem.

Isopterous pests

The termite, *Microtermes thoracalis* has been reported in Sudan, whereas in Nigeria, the termite, *Macrotermes bellicosus* (Rambur) (Isoptera : Termitidae) has been reported feeding on neem trunk. In India, *Odontotermes obesus* (Rambur) has been reported on young seedlings and sometimes on the trunk and branches of older trees, but no serious damage have been reported due to these termites (Radwanski & Wickens, 1981; Ahmad and Ahmed, 1992). Varma (1989) observed the infestation of the termite, *Odontotermes redemanni* on *Azadirachta inidca*. Delate and Grace (1995) reported the termite, *Coptotermes formosanus* Shir on neem.

Bioagents of neem pests

Khan and Shafee (1979) reported *Neochrysocharis pubipennis* sp. n. reared from *Pulvinaria maxima* Green on *A. indica*. Again Khan and Shafee (1981) reported the parasite *Tetrastichus ajmerensis* sp. n. from females and a male taken from *Pulvinaria* sp. on neem in Uttar Pradesh, India.

Karthikeyan *et al*. (1998) observed ten natural enemies attacking insect pests of *Azadirachta indica* in Tamil Nadu, India. Among them, the coccinellids, *Jauravia pallidula, Chilocorus nigrita* and spiders, *Platythomisus* sp. and *Cheiracanthium* sp. were found feeding on *Parlatoria orientalis* and *Pseudococcus gilbertensis*. *Podothrips* sp. was found to be predacious on *Parlatoria orientalis*. *Olios* sp. was found to attack *Myllocerus viridanus*. The braconids, *Apanteles maochaeralis* and *Fornicia* sp. were found to parasitize *Loboschiza koenigiana* and *Aphendala cana*. The egg parasitoid, *Anastatus* sp. was observed parasitizing eggs of *Hylas dentatus*. All the above mentioned natural enemies were recorded for the first time in the neem ecosystem.

Murugesan *et al*. (1998) collected four species of *Trichogramma, T. achaea, T. chilonis, T. roi* [*T. raoi*] and *T. japonicum* in Rajasthan, India, from eggs of *Corcyra cephalonica, Achaea janata, Laspeyresia koenigana* [?*Cydia koenigana*], *Boarmia variegata* and *Eurema* sp. All these parasitoids were recorded for the first time from *Azadirachta indica, Acacia milotica* and *Dalbergia sissoo* plantations.

However, Seema Kumar (1998) observed that *Fusarium* spp. (which causes root rot in new seedling) infecting land slug, a pest of neem and controls it.

Non-insect pest of neem :

(I) Mite pests : Among mites, *Schizotetranychus hindustanus* (Tetranychidae), *Tetranychus fici* Hirst and *T. urticae*, the spider mite attacks the lower part of the leaves and casues damage. Uthamasamy *et al.* (1973) reported an eriophyid mite, *Calepitrimerus azadirachtae* Channabasavanna (Acarina : Eriophyidae) attacking on young leaves. This eriophid mite prefer only the tender leaves for their feedig and cause very minute scars on them which in older leaves, become brown or white dots. The mite brings about deformity and the twigs and leaves dry out causing severe damage. Ahmed and Seema Kumar (1998) reported two species of phytophagous mites namely *Tetranychus* sp. and *Eutetranychus orientalis* (Acarina : Tetranychidae). Yousuf and Chouhan (1998) reported *Eutetranychus phaseoli* on neem plants (*A. indica*) while *E. maximae* attacks *M. azedarach*.

(ii) Other non-insect pests : Among non-insect pests, Indian koel *Eudynamys cyanocephala*, apostle bird *Struthidea cinerea*, greater bowerbird *Chlamydera nuchalis*. Village weaver *Textor cucullatus*, the white vented bulbul *Pycnonotus barbatus*, the long tailed starling *Lamprotornis caudatus*, the mouse bird *Urocolinus macrourus* and *Turdus pelios* have been recorded feeding on neem fruits.

Eidolon helvum, strain coloured flying fox feeds at dawn and during night on the pulp of ripe neem fruits. *Epomorphus gambianus,* another fruit bat species feeds on neem fruit in tropical Africa. *Camelus dromedarius*, camel feeds on neem leaves in Africa and India. *Capra hircus*, the goat feeds on leaves when no other fodder is available. These goats do not bite neem seeds into pieces but swallow them whole. Friend (1999) reported pale headed rosella, *Platycercus adscitus*, snail *Mollusca domestica* Linnaeus, slug *Stylommotophora* spp., toad *Bufo marinus*, B-tailed possum *Trichosurus vulpecula*, kangaroo *Macropus robustus*, cows *Bas taurus* L. and fruit bat sp. ex-Megachiroptera as pest of neem.

Diseases

Fungal disease

Narayanasamy and Ramakrishnan (1971) recorded a new pathogenic fungus, *Oidium azadirachtae* sp. nov. inefecting leaves of *A. indica*.

Sinniah *et al.* (1983) reported various species of the genus, *Aspergillus* from stored neem seeds (Fig. 3.7) or kernels in Asia, Africa, South and Central America. The *Aspergillus flavus* is the most dangerous fungus on neem kernels as it produces carcinogenic mycotoxin aflatoxin. Other species reported are *A. niger, A. tamarii, A. versicolor* and *A. quercinus* [=*A. mellus*).

The other fungi reported from seeds and kernels are *Alternaria palandui, A. tenuis, Cladosporium herbarum, Cephalosporium* sp., *Culvinaria lunata, Drechslera tetramera, Fusarium moniliforme, Monodyctis natans, Penecillium citrinum, Rhizopus* sp. and *Phoma* sp.

Iqbal and Chohan (1984) reported *Phoma jolyana* a new fungal pathogen on *Azadirachta indica* near Ludhiana, Punjab (India). They observed severe twig cankers and shot holes in the leaves. The pathogenecity was proved by inoculation method. *Hendersonula* sp attacks old neem trees in Sudan.

Castellani and Mohamed (1984) reported a new leaf spot disease caused by *Pseudocercospora subsessilis*. The incubation and reproduction periods were 25 and 50 days, respectively. Later Jamaluddin *et al*. (1988) described the symptoms of a blight disease caused by *Pseodocercospora subsessilis*, occurring on *Melia azadirachta* [*Azadirachta indica*] (bakayan neem) in Madhya Pradesh. The infection was observed both in plantations and in the nursery, and was most severe in Dec.-March. It started on the leaves, and spread to youg shoots and fruits, but was most severe on leaves and fruits, leading to premature shedding of both. Seeds of infected fruits were not viable.

Sankaran *et al*. (1986) reported two leaf spot diseases caused by *Colletotrichum capsici* (Fig. 3.8) and *Cercospora subsessilis*, (Fig. 3.9) a leaf blight and stem rot (*Sclerotium rolfsii*), a web blight (*Rhizoctonia solani*) and a wilt (*Fusarium solani*) for the first time in July-Dec. in 1983-84 on 2 to 3 month-old *A. indica* seedlings in Kerala. Although widespread of the leaf spots did not cause much damage, whereas the other diseases resulted in 20-30 percent mortality.

Later on Sankaran *et al*. (1988) reported a new fungal pathogen, *Colletotrichum gloeosporioides*, state of *Glomerella cingulata* on *Azadirachta indica*. The fungus, *Hendersonula* sp. was observed attacking old neem trees in Sudan.

Mehrotra (1989) gave an account of leaf web blight of *Melia azedirach*, (*Azadirachta indica*) *Cassia nodosa, Ceiba pentandra* and *Derris robusta*, caused by the *Rhizoctonia solani* an amorph of *Thanatephorus cucumeris*, and of leaf blight of *Engelhardtia spicata* caused by *Sclerotinia sclerotiorum*. The web blight was highly destructive in the nursery and killed heavily infected young seedlings. *M. azedarach* and *A. indica* were the most susceptible species, with 80-90 percent foliage damage, and mortality rates of 10-30 percent. In *A. indica*, mortality reached 60 percent where infection by *R. solani* was associated with infection by *Fusarium* sp. Growth of infected seedlings of these two species was very poor. The characteristics of isolates of *R. solani* from the different hosts indicated the existence of more than one strain at Burnihat. Recommended control measures for the web blight are : avoidance of dense sowing, and immediate removal of infected seedlings. In the case of *M. azedarach* and *A. indica*, raising in polypots instead of beds prevents the spread of infection (via the soil) in the heavy rains which occur in Assam, polypotted seedlings of *A. indica* at Burnihat were found to be almost free of the disease.

Mehrotra (1990) reported foliar disease caused by *Rhizoctonia solani* in forest nurseries. The pathogen has a wide host range causing top flagging leading to mortality of *Melia azadarach* and *A. indica*. Three biotypes of the fungus were distinguished by Sclerotial characteristics. Hiremath *et al*. (1991) observed *Melia azadirachta* [*A. indica*] was infected by *Cercospora leucosticta* and *Colletotrichum gloeosporioides* [*Glomerella cingulata*] both in forest nurseries in Karnataka.

Mehrotra and Pandey (1991) described leaf spotting and blight of neem caused by *Colletotrichum gloeosporioides* and *Alternaria alternata*. The infected leaves in case of *Colletotrichum* present a blighted appearance are eventually shed. The disease incidence (100%) and premature defoliation to the extent (32%) have been recorded. The disease is of short duration of about 3 to 4 weeks. In case of *Alternaria* leaf spotting and blight, the disease appears late in the last week of October or early November. The damage caused to about 80-100 percent leaf area. The fungus is known to attack the leaves when they become old and contain less soluble sugars.

Tewari *et al*. (1992) reported *Ganoderma* root rot disease in neem caused by *Ganoderma lucidum*. The paling of leaves and drying of branches appear after the maximum damage to the root system is done by the pathogen. The fungus attacks the sap wood and causes white spongy rot. The other disease reported from Bikaner, Rajasthan, India was pink disease caused by *Corticium salmonicolor*. The incidence was about 100 percent. The disease is characterized by the phenomenon of dieback and formation of pink incrustation and cankers on the infected stem and branches which are eventually killed.

Shukla (1992) observed seedlings of neem raised from seeds of 10 provenance at Dehra Dun, Uttar Pradesh, India were found to be infected by *Fusarium solani*. The pathogen caused blackening of the hypocotyl, followed by root rot, wilting and mortality of the seedlings. Pathogenicity of the isolated fungus was confirmed in tests with 2-month-old seedlings. Chemical control tests were undertaken in vitro with 6 fungicides. Ziram, Shield [Mancozeb], Karathene [Dinocap] and Bavistin [Carbendazim] were the most effective. Khan (1992) reported a disease caused by *Phomopsis* sp. at Dehradun. The fungus attacks the twigs which dry up showing dieback symptoms.

The other fungi reported on neem are *Corticum salmonicolor* stem and twig blight. The fungus, *Colletotrichum* state of *Glomerella cingulata* is reported on twig (Kaushik *et al*., 1993). Alam (1993) reported other leaf spot fungi namely, *Curvularia* sp. and *Exserohillem* sp. The other fungi are *Sclerotium rolfsii, Rhizoctonia solani,* leaf spot disease *Colletotrichum capsici,* wilt disease *Fusarium solani, Gonoderma lucidum*.

Kaushik *et al*. (1996) recorded *Colletotrichum* state of *Glomerella cingulata* (Stomen) spaned and Sct. causing blight of twigs and young branches of neem seedlings. Pre and post inoculation moist period contributed for high infection under screen house conditions. A temperature range of 24°C to 32°C combined with high relative humidity (>70%) increased disease intensity. Infected seedlings displayed minute circular to oblong, brown to black lesions on the succulent portion of the twigs or main shoot of 2-3 months old seedlings. The infection advanced and finally caused girdling and constriction of the entire region which extended towards the base, resulting in death of seedlings. Acervuli were observed submerged in the host tissue, they are scattered and black in colour. The first symptoms of disease appeared in the third week of July when the temperature and relative humidity ranged from 24.5 to 33.2°C and 60.9 to 86.0 percent, respectively. The disease progressed moderately up to first week of August. The disease becomes in severe form from August to first week of September when the temperature starts declining and the relative humidity starts increasing.

Kahn (1994) reported *Sarcinella fumosa* infecting *A. indica*. Sharma *et al*. (1996) reported Pseudocercospora meliacearum on *A. indica*, collected from forests of Madhya Pradesh, India. Da-Silva (1996) described fungal pathogen, *Corticium rolfsii* for the first time on neem (*A. indica*) in the state of Maranhao, Brazil. For confirmation, pathogenecity tests were carreied out on 30-day-old seedlings.Uniyal and Uniyal (1996) reported the mycoflora associated with the seeds of neem, *Azadirachta indica* from different locations in India. Species representing 6 fungal genera (*Fusarium, Aspergillus, Penicillium, Cephalosporium, Alternaria* and *Pythium*). *Mycelia sterilia* and bacteria were isolated. Among these, the dominant species were *Aspergillus niger, A. fumigatus* and *Fusarium solani*.

Sateesh *et al.* (1997) and Bhat *et al.* (1998) reported a new fungal pathogen, *Phomopsis azadirachtae* infecting *A. indica* in Karnataka, India. Friend (1999) reported invading ground fungus, *Phellinus noxius* on neem.

Stored neem seed dieseases

The disease of stored neem seed and seed kernel is causing lot of problem in storing the seeds/kernels in a better way as they destroy the active ingredients such as azadirachtin. *Aspergillus flavus* is the most dangerous fungus on neem kernels. The other species found are *A. niger, A. tamarii, A. versicolor and A. quercinus* (Sinniah *et al.*, 1983). Besides, the other fungi attack on seed and seed kernels are *Alternaria palandui, A. tenuis, Cladosporium herbarum, Cephalosporium* sp., *Curvularia lunata, Drechslera tetramera, Fusarium moniliforme, Monodyctis natans, Penicillium citrinum, Phoma* sp. and *Rhizopus* sp. (Srivastava, 1978).

Bacterial disease

The bacterial disease, *Pseudomonas azadirachtae* was reported from India by Desai *et al.* (1966). This causes angular leaf spots. The affected leaves become yellowish, dry up and fall off. The petiols are also susceptible (Srivastava and Patel, 1969; Chakravarti and Gupta, 1975). The other bacteria, *Xanthomonas azadirachtii* was also reported from India on neem by Chakravarti and Gupta (1975). Due to infection, the leaves become chlorotic, dry up and fall off prematurely.

Naina *et al.* (1989) studied genetic transformation and regeneration of transgenic neem (*A. Indica*) plants using *Agrobacterium tumefaciens*. Tumours were induced on *in vitro* grown seedlings of *A. indica* using *A. tumefaciens*, strains K12 X 562E and K12 X 167, which contain a recombinant derivative of the plasmid pTiA6. Transformed plantlets were regenerated from infected regions of the seedlings and also from the tumour callus. The transformed callus, shoots and plantlets showed autotrophic growth in culture and synthesized octopine; they were grown on a medium containing kanamycin.

Diatloff (1993) reported bacterial wilt of neem caused by *Pseudomonas solanocearum*. The *P. solanocearum* biovar 3 was isolated from neem trees (*A. indica*) in Queensland, Australia, affected by slow wilt following a prolonged dry period. Black to brown discoloration occurred in the roots and stems with a collar rot developing in the advanced stages. The isolates were pathogenic to neem tree and tomato but not to native white cedar (*Melia azedarach*). It is concluded that white cedar could possibly be used as a wilt-resistant root stock. This is the first record of bacterial wilt in neem tree.

References

Ahmed, S.I. (1992). Neem : Insect pest problems and their management in Arid Zone Report I. *AZFR Jodhpur* : 1-3.

Ahmed, S.I. and Murugesan, S. (1995). *Achaea janata* Linn. (Noctuidae : Lepidoptera), a new pest record on neem (*Azadirachta indica*) - its management through nuclear polyhedrosis virus. *Indian Forester*, **121**(1) : 63.

Ahmed, S.I. and Seema Kumar (1998). Prominent biological threats to neem. Abstract in *National Seminar on Entomology in 21st Centruy* held at Udaipur, April 30 to May 2, 1998 : pp. 133.

Alam, M.S. (1993). Some fungal diseases of leaf of *Azadirachta indica*. Abst. in *World Neem conference* held at Bangalore, India : pp. 25.

Anantha Krishnan, T.N. (1973). Thrips : Biology and control. *Macmillan India*, Delhi : 120.

Annamalai, R.; Surendran, C.; Thirumurthi, S.; Gunasekharan, V.; Nair, K.S.S.; Sharma, J.K. (ed.) and Varma, R.V. (1996). Insect damage to neem in Tamil Nadu, India. *Proceedings of the IUFRO Symposium*, Peechi, India, 23-26 November, 1993 : 349-355.

Ayyar, T.V.R. (1963). Hand Book of Economic Entomology for South India. Govt. Press, Madras, India.

Balachowsky, A.S. (1956). [Les cochenilles ducontinent Africain noir, Vol. I, Aspidiotini] (*Lere partie*) Ann. Mus. Roy Congo, Blege, Nouv. Ser. 4, Science Zoology, Vol. 3.

Beesan, C.F. (1941). The ecology and control of the forest insects of India and the neighbouring countries. *Forest Research Institute and College*, Dehradun : 1007.

Beesan, C.F. (1941). The ecology and control of the forest insects of India and the neighbouring countries. Vasant Press, Dehradun.

Bhasin, G.D.; Roonwal, M.L. and Singh, B. (1958). A list of insect pests of forest plants in India and adjacent countries. *Part III Indian Forest Bull. n.s.* NO. **171**(2) (Ent.)

Bhat, S.S.; Sateesh, M.K. and Devaki, N.S. (1998). A new destructive disease of neem (*Azadirachta indica*) incited by *Phomopsis azadirachtae*. *Current Science*, **74**(1) : 17-19.

Brar, K.S. and Sandhu, G.S. (1982). Field biology of the white grub, *Holotrichia consanguinea* Blanchard (Scarabaeidae : Coleoptera) in Punjab. *Journal of Soil Biology and Ecology*, **2**(1) : 32-35.

Browne, F.G. (1968). Pests and diseases of plantation trees. Claredon Press, Oxford, U.K.

Castellani, E. and Mohamed Ali, M.I. (1984). Diseases of *Azadirachta indica* in Somalia (1) Cercospora disease. *Rivista-di-Agricoltura Subtropicale-e-Tropicale*, **78**(3-4) : 733-739.

Castellani, E. and Mohamed, M.I. (1984). Malattie dell *Azadirachta indica* riscontrate in Somalia. *Cercosperiosi Riv. Agric. Subtrop. Trop.* **78** : 733-739.

Chakravarti, B.P. and Gupta, D.K. (1975). A nonpigmented strain of *Xanthomonas azadirachtii* causing leaf spot of neem (*Azadirachta indica* A. Juss.). *Current Science*, **44** : 240-241.

Chaudhary, O.P. (1997). New record of a shoot borer on neem. *Insect Environment*, **3**(1) : 8

Chockalingam, S. and Noorjahan, A. (1986). Consumption and utilization of neem leaves by *Microcentrum retincrvc* (Orthoptera . Tettigoniidae). *Annals of Entomology*, **2**(1) : 73-77.

Da-Silva, G.S. (1996). Occurrence of Sclerotium folfsii in neem (*Azadirachta indica*). *Fitopatologia Brasileira*, **21**(4) : 522.

Desai, S.G.; Gandhi, A.B.; Patel, M.K. and Kotasthanae, W.V. (1966). A new bacterial leaf spot and blight of *Azadirachta indica* A. Juss. *Indian Phytopathology*, **19** : 322-323.

Delate, K.M. and Grace, J.K. (1995). Susceptibility of neem to attack by the Formosan subterranean termite, *Coptotermes formosanus* Shir. (Isoptera : Rhinotermitidae). *Journal of Applied Entomolgoy*, **119**(2) : 93-95.

Diatloff, A.; Wood, B.A. and Wright, D.G. ((1993). Bacterial wilt of neem tree caused by *Pseudomonas solanacearum*. *Australasian Plant Pathology*, **22**(1) : 1.

Fletcher, T.B. (1933). Life histories of Indian Microlepidoptera (second series) Alucitidae (Pterophoridae) Tortricinae and Gelechidae). *Scientific Monograph*, Council of Agricultural Research India, No.2 : 58.

Friend, J.A. (1999). Pest and deseases of neem (*Azadirachta indica* A. Juss.) in Australia and associated biota. *Azadirachta indica* A. Juss. Ed. Singh and Saxena : 21-34.

Friend, J.A. (1996). Pests and deseases of neem in Australia. *Neem Newsletter*, **13**(3) : 47-48.

Gruber, A.K. (1991). Wachstum Fruchtertragund Azadirachtingehalt der Samen Von Azadirachta indica A Juss au Verschiedenn Standorten in Nicaragua. Doctor, *Thesis*, Techn. Univ. of Berlin, Germany.

Gruber, A.K. (1992). [Biologia y ecologia del arbal, de nim (*Azadirachta indica* A. Juss) extruccion medicion, toxicidad y potential decrear resistencia]. *Ceiba*, 33 : 249-256.

Gupta, K.M. (1973). Neem leaves attract white grub beetles. *Indian Journal of Entomology*, **35** : 276.

Harsh, N.S.K.; Tiwari, C.K. and Nath, V. (1989). Foliage diseases in forest nurseries and their control. *Journal of Tropical Forestry*, **5**(1) : 66-69.

Hegde, N.G. (1993). Neem production and development constraints at grass root level. *World Neem Conference* held at Bangalore, India. *Souvenir* : pp. 5.

Hiremath, S.V.; Srikant Kulkarni; Hegde, R.K.; Pall, B.S.; Bhat, M.N.; Naik, K.S.; Raghuchandar, T.; Hiremath, P.C. and Kulkarni, S. (1991). New host records of fungi. *Current Research*, University of Agricultural Sciences, Bangalore, **20**(5) : 93-94.

Iqbal Singh; Chohan, J.S. (1984). *Phoma jolyana*, a new pathogen on neem (*Azadirachta indica*). *Indian Forester*, **110**(10) :1058-1060.

Jamaluddin; Dadwal, V.S. and Sima Sahare (1988). *Cercospora* blight disease of *Melia azadirachta. Indian Forester*, **114**(12) : 894-895.

Jussieu, A.DE (1830). [Memoire Sur Le groupe des Meliacees.]. *Mem. Mus. Hest. Nat. Paries*, **19** : 220.

Karthikeyan, K; Rangarajan, A.V. and Muthuswami, M. (1998). New records of natural enemies on insect pests of neem tree, *Azadirachta indica* A. Juss. *Journal of Biological Control*, **11**(1-2) : 81-83.

Karthikeyan, K; Rangarajan, A.V. and Velusamy, R. (1993). Major pest of neem and their management in southern Tamil Nadu. World Neem Conference held at Bangalore, India. Abst. pp. 25.

Kaushik, J.C.; Bisla, S.S. and Singh, D.P. (1993). Twig blight - A new disease of neem seedlings. World Neem Conference held at Bangalore, India. Abst. pp. 42.

Kaushik, J.C.; Bisla, S.S. and Singh, D.P. (1996). Twig blight - A new disease of neem seedlings. *Neem and Environment*, **2** : 1007-1012.

Ketkar, C.M. (1976). Final Technical Report. Utilisation of neem (*Azadirachta indica*) and its by-products. Nana Dengle, Sadhna Press, India **III** : 47-56.

Ketkar, C.M. (1976). Utilisation of neem (*Azadirachta indica*) and its by-products, Khadi and Village Industries Commission, Bombay, India.

Khan, K.M. and Ghai, S. (1982). Taxonoic status of the genus *Holotrichia* Hope (Melolonthini : Melolonthinae : Scarabaeidae) with descriptions of five new species from India along with redescriptions of two poorly described species and a key to species. *Bulletin of Entomology*, **23**(1-2) : 28-45.

Khan, M.K. (1994). New host for some folilicolous hypomecetes from India.

Indian Phytopathology, **47**(4) : 441-442.

Khan, M.Y. and Shafee, S.A. (1979). Three new species of the genus *Neochrysocharis Kurdyumov* (Eulophidae : Entedontinae)) from India. *Journal of the Bombay Natural History Society*, **76**(3) : 480-490.

Khan, M.Y. and Shafee, S.A. (1981). On some new species of *Tetrastichus ajmerensis* (Hymenoptera : Eulophidae). *Journal of the Bombay Natural History Society*, **78**(2) : 337-343.

Lale, N.E.S. (1998). Neem in the Conventional Lake Chad Basin area and the threat of Oriental yellow scale insect (*Aonidiella orientalis* Newstead) (Homoptera : Diaspididae). *Journal of Arid Environments*, **40**(2) : 191-197.

Meenakshi (1998). Wonder tree neem in the grip of mealy bug. Abst. in *International Conference on Pest and Pesticidal Management for Sustainable Agriculture* held at CSAUAT, Kanpur, 11-13, Dece., 1998 : pp. 190.

Mehrotra, M.D. (1989). Leaf blight of some hardwood species in Assam and Meghalaya and its control in the nursery. *Indian Forester*, **115**(6) :378-384.

Mehrotra, M.D. (1990). *Rhizoctonia solani*, a potentially dangerous pathogen of khasi pine and hardwoods in forest nurseries in India. *European Journal of Forest Pathology*, **20**(6-7) : 329-338.

Mehrotra, M.D. and Pandey, P.C. (1991). Some important nursery diseases of *Azadirachta indica* and their control (Communicated).

Menendez, J.M.T. and Perez, R.P. (1994). [Comparazion preliminar del comportamiento de tres]. Procedencias de (*Azadirachta indica* A. Juss) en Cuba. Ler. Congo. Latinarmercano y del Caribe Sobre Nim y otros Insecticidas Vegetables. Sto. Domingo Republ. Dominicanna (Abstr.)

Murugesan, S.; Sundararaj, R.; Shivesh Kumar and Kumar, S. (1998). Survey, evaluation and attraction of *Trichogramma*, an egg parasitoid of Lepidoptera of forestry tree spcies. *Indian Journal of Environment and Toxicology*, **8**(1) : 47-48.

Naina, N.S.; gupta, P.K. and Mascarenhas, A.F. (1989). Genetic transformation and regeneration of transgenic neem (*Azadirachta indica*) plants using *Agrobacterium tumefaciens*. *Current Science*, **58**(4) : 184-187.

Narayanaswamy, P. and Ramakrishnan, K. (1971). Powdery Mildew of Coimbatore. *Journal of Madras University*, **38** : 84-99.

Nayar, K.K.; Ananthakrishnan, T.N. and David, B.V. (1989). **General and Applied Entomology**, Green Offset Press, Shahadra, Delhi.

Onkarappa, S. and Kumar, C.T.A. (1997). Biology of the tea mosquito bug,

Helopeltis antonii Sign. (Miridae : Hemiptera) on neem. *Mysore Journal of Agricultural Sciences*, **31**(1) : 36-40.

Oo, L.T. (1987). Neem Tree Research (Terminal Report) Burmese-German Plant Protection and Rodent Control Project, Rangoon, Tana Press, Bankok, Thailand, 105 pp.

Pillai, S.R.M. and Gopi, K.C. (1990). Seasonal drying up of the distal shoots of neem (*Azadirachta indica* A. Juss.) and important insect pests associated with it. *Myforest*, **26**(1) : 33-50.

Radwanski, S.A. (1977). Neem Tree, World Crops. **29** : 22-66, 111-113, 167-168, 222-224.

Radwanski, S.A. and Wickens, G.E. (1981). Vegetative follows and potential value of the neem tree (*Azadirachta indica*) in the tropics. *Economic Botany*, **35** : 398-414.

Raghunath, T.A.V.S. (1979). Additinal records of new insect pests in Andhra Pradesh. *Current Research*, **8**(6) : 103-104.

Raghunath, T.A.V.S.; Allam, M.A. and Venkaiah, K. (1982). Fire ant (*Solenopsis* sp) damaging neem (*Azadirachta indica* Juss.). *Indian Forester*, **108**(5) : 375.

Rice, M. (1993). Development of neem research and industry in Australia. Abstract in *World Neem Conference* held at Bangalore, India. *Souvenir* : 8-24.

Rohilla, H.R.; Singh, H.V. and Yadava, T.P. (1981). White grubs - increasing menace in Haryana. *Haryana Agricultural University Journal of Research*, **11**(2) : 256-258.

Roxburgh, W. (1874). Description of Indian plants. Today and Tomorrow, New Delhi (reprinted from Carey's Edition, 1832) : pp.763.

Sankaran, K.V.; Balasundaran, M. and Sharma, J.K. (1986). Seedling diseases of *Azadirachta indica* in Kerala, India. *European Journal of Forest Pathology*, **16**(5-6) : 324-328.

Sankaran, K.V.; Florence, E.J.M. and Sharma, J.K. (1988). Foliar diseases of some forest trees in Kerala - new records. *Indian Journal of Forestry*, **11**(2) : 104-107.

Sarup, P. and Srivastava, V.S. (1971). Observation on the damage of neem (*Azadirachta indica* A. Juss) seed kernel in storage of various pests and efficacy of the damaged kernel as an antifeedent against the desert locust, *Schistocera gregaria* Forsk. *Indian Journal of Entomology*, **33** : 228-230.

Sateesh, M.K.; Bhat, S.S. and Devaki, N.S. (1997). *Phomopsis azadirachtae* sp. nov. from India. *Mycotaxon*, **65** : 517-520.

Saxena, R.C.; Jilani, G. and Kareem, A.A. (1989). Effects of neem on stored grain insects. In Jacobson M. (Ed.) 1988. Focus on Phytochemical Pesticides, Vol. I. The Neem Tree, CRC Press, Boca Raton, E.L. : 97-111.

Schmutterer, H. (1990). [Beobachtungen an schadlingen von *Azadirachta indica* (Niembaum) und von verschiedenen, *Melia Arten*.] *Journal of Applied Entomology*, **109** : 390-400.

Schmutterer, H. (1995). The Neem Tree, *Azadirachta indica* A. Juss and other Meliaceous plants, sources of unique natural products for integrated pest management medicine, industry and other purposes. *V.C.H. Weinheim* : pp. 696.

Seema Kumar; Ahmed, S.I.; Chouhan, Sahadev; Kumar, S. and Chouhan, S. (1996). *Myllocerus laetivirens* Marshall (Coleoptera : Curculionidae), a new pest record on neem in arid zone of India. *Indian Journal of Entomology*, **58**(2) : 195-197.

Seema Kumar; Ahmed, S.I.; Chouhan, Sahadev and Kumar, S. (1997). *Myllocerus laetivirens* Marshall (Coleoptera : Curculionidae), a new pest record on neem in arid zone of India. *Annals of Agricultural Research*, **18**(1) : 102-103.

Seema Kumar; Srivastava, K.K. and Ammen Ullah (1998). New record of fungal infection - *Fuasarium* sp. in the culture of land slug *Laevicaulis alte* (Ferussac)- a severe pest of *Azadirachta indica* (neem) seedlings in Rajasthan and Gujarat. *Indian Forester*, **124**(5) ; 351-353.

Seema, C.D.; Gadpandey, K.K.; Rai, A.N. and Vyas, K.M. (1996). Some new hyphomycetes associated with leaf spots of forest plants in Madhya Pradesh, India. *Mycological Research* **100**(11) :1388-1392.

Sharma, C.D.; Gadpandey, K.K.; Rai, A.N. and Vyas, K.M. (1996). Some new hyphomycetes associated with leaf spots of forest plants in Madhya Pradesh, India. *Mycological Research*, **100**(11) : 1388-1392.

Shukla, A.N. (1992). Seedling blight and root rot in neem (*Azadirachta indica* A. Juss.). *Indian Journal of Forestry*, **15**(3) : 266-268.

Singh, I. and Chohan, J.S. (1984). *Phoma jolyana*, a new pathogen on neem (*Azadirachta indica*). *Indian Forestry* : 58-60.

Sinniah, D.; Varghese, G.; Baskaran, G. and Koo, S.H. (1983). Fungal flora of neem (*Azadirachta indica*) seeds and neem oil toxicity. *Malaysian Applied Biology*, **12**(1) : 1-4.

Srivastava, R.C. (1978). Mycoflora associated with *Azadirachta* seeds. *Fert. Technol*. **15** : 56.

Srivastava, S.K. and Patel, P.N. (1969). Epidemiology of bacterial leaf spot, blight and shot-hole disease of neem in Rajasthan. *Indian Phytopathology*, **22** : 238-244.

Srivastava, Y.N. and Bhanotar, R.K. (1980). Efficacy of neem powder infested with *Oryzaephilus surinamensis. Indian Journal of Forestry*, **3** : 353-356.

Sundararaj, R. and Murugesan, S. (1997). Record of *Thosea bipartita* (Limacodidae : Lepidoptera) on *Azadirachta indica* (neem) from Jodhpur, India. *Indian Forester*, **123**(3) : 261.

Sundararaju, D. and Babu, P.C.S. (1998). Life table studies of *Helopeltis antonii* Sign. (Heteroptera : Miridae) on neem, guava and cashew. *Journal of Entomological Research*, **22**(3) : 241-244.

Tewari, D.N. (1992). Monograph on neem. International Book Distributors, Dehradun, India : pp.278.

Thakur, M.S.; Pathak, N. and Saksena, S.B. (1978). Seasonal variation in leaf surface mycoflora of *Azadirachta indica. Indian Phytopathogloy*, **31**(2) : 255-256.

Thomas, M.C. and Woodruff, R.E. (1983). First records on a stored product pest, *Oryzaephilus acuminatus* Holstead from the western hemisphere. *Fla. Dep. Agric. Consum. Serv*. Div. Plant Ind. Entomolol. Circ. No. 257.

Uniyal, K. and Uniyal, D.P. (1996). Mycoflora of neem seeds. *Indian Forester*, **122**(5) : 423-427.

Uthamasamy, S.; Rao, P.V.S.; Mohanasundaram, M. and Subramaniam, T.R. (1973). A note on some new pests attacking the neem tree (*Azadirachta indica* A. Juss) in Tamil Nadu. *Science Culture*, **39** : 399-400.

Valdix, J. (1994). [EL'arbol Nim (Azadirachta indica) Un Cultivo nuevo en Nicaragua-Su susceptibilidad a plagas y Su function como hospedero de insectors Ler. Congr. Latinamericano y del Caribae.] *Sobre Nem y Otros Insecticidos Vegetale* Sto. Domingo, Republ. Dominicana (Abstr.).

Verma, S.C. (1989). Termite pests of forestry and agriculture in Haryana, India. *Indian Journal of Forestry*, **12**(1) : 1-6.

Vora, V.J. and Ramakrishnan, N. (1991). Attraction of the beetle, *Holotrichia consanguinea* Blanchard (Scarabaeidae : Coleoptera) to light and host plant. *Indian Journal of Entomology*, **53**(1) : 45-49.

Yousuf, M. and Chouhan, S. (1998). Infestation of four species of *Eutetranychus* (Acarina : Tetranychidae) on forest tree species in Rajasthan. *Annals of Forestry*, **6**(1) : 79-83.

Zongo, J.O. (1990). Insect pests of nuts and leaves of neem. *Sahel PV Info*. **27** : 19-20.

Fig. 3.4

Fig. 3.5

Fig. 3.4 Adult of *Drosicha mangiferae*
Fig. 3.5 Damage by *Myllocerus dorsatus*

Fig. 4.10

Fig. 4.11

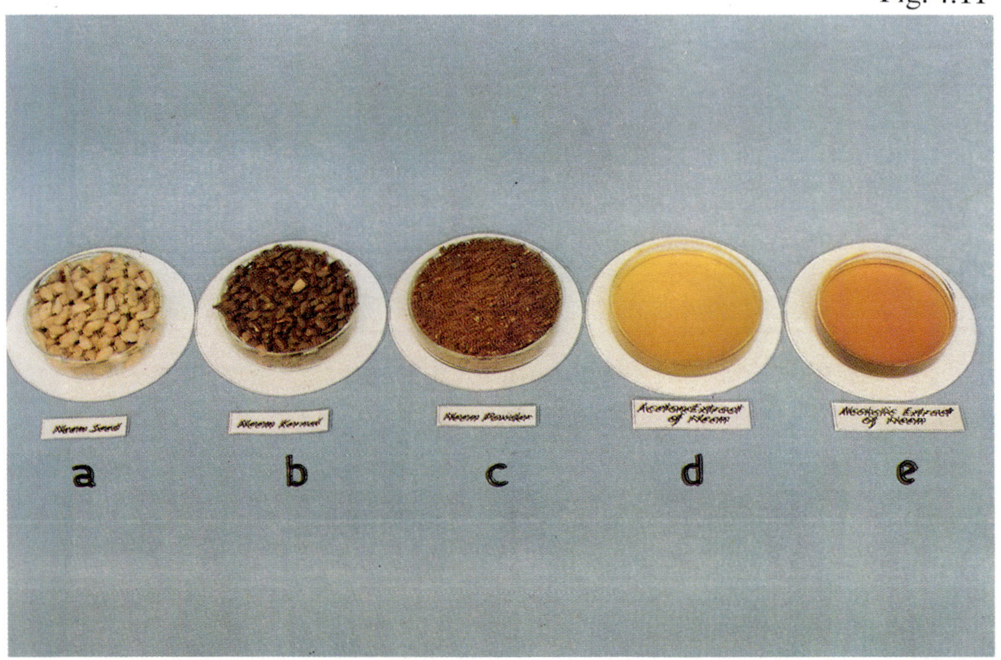

Fig. 4.10 Kernel powder (*A. indica*)

Fig. 4.11 a) Seed, b) Neem kernel, c) Neem powder, d) Acetone extract of neem, e) Alcoholic extract of neem

Chapter 4

Toxic effect of Neem products on Insects

The modern agriculture is based on judicious use of chemicals particu larly complex synthetic pesticides. Although these chemical pesticides are highly effective at the inital stage, but later create some major disadvantages such as high toxicity to non-targeted organisms, environmental pollution, and residual hazards which pose a serious threat to human livelihood. These chemicals also kill our eco-friendly natural enemies, cause resistance problem among insects to insecticides and also upset the natural balance of pest in nature. The high cost of these pesticides, labour and maintenance of equipments and resurgence of the treated population are some other defects in the use of chemical pesticides.

In recent years, there has been a growing interest in exploring plant-based chemicals for use in agriculture as an alternative to chemcial pesticides. The plant chemicals have received interest primarily because these are easily biodegradable, easily available and very cheap source of pest management.

Neem products have shown diverse biological effects agaisnt insects, pathogens, plant nematodes and mites. A number of neem products are ovicidal. The neem products also act as larvicide. Somė neem products when topically applied on adults, cause death, while other cause morphogenetic abnormalities on adults. Recently it has been established that neem is a good source of several alkaloids which have pesticidal properties. Some developed and developing countries have started taking interest in the plant chemicals to use them as pesticides. During the last two decades, neem has became a centre of interest for the scientific community all over the world.

Inspite of the wide recognition that many plant possess insecticidal properties, only a handful of pest controling products are in use in developed countries. The demonstrated effeicacy of the neem products and their recent approval for use in USA, Germany etc. has stimulated research and development programme on the insecticidal effect of neem.

Over 350 limonoids are known to date and about 150 are known to occur in neem, many of which cause insecticidal effects. The insecticidal efficacy of neem against about 350 species of arthropods are reported but still a very few products have been formulated. Many of these commercial formulations give high toxicity while other do not give the toxicity to the desired level. Thus failure is due to inconsistent field results. It is true that the principal barriers to commercialization of new botanicals are (i) scarcity of the natural resources (ii) standardization and quality control and (iii) problem of registration.

It is also observed that neem products when mixed at sublethal doses of chemical insecticides give synergistic effect or additive effects.

The toxic effects in various groups of insects and being discussed below.

1. Coleoptera

Qadri and Rao (1977) studied the insecticidal action of custard apple (*Annona squamosa*) in combination with neem (*Azadirachta indica*) against *Collosobruchus chinensis, Rhyzopertha dominica* and *Musca domestica nebulo* F. in the laboratory. The insects of each species were exposed for 24 hrs to extracts of the seeds of neem and custard apple and to DDT as a standard of comparision, in the form of dry deposits on glass. Statistical analysis of the results showed that neem extract alone was less toxic than either custard-apple extract or DDT, but that it had a synergistic action when combined with custard-apple extract; the combined seed extract, containing 4-5 parts of neem to 1 of custard apple, was as toxic as DDT to *C. chinensis* and about one-fifth as toxic as DDT to *R. dominica*.

The toxicity LC50 values of the custard apple alone to *C. chinensis*, *R. dominica* and *M. domestica nebulo* were 141.30, 524.80 and 120.20, respectively, while with neem, it was 891.30, 3311.00 and 169.80 but it was highest with DDT viz. 8.32, 16.60 and 7.08 micron g/dish, respectively. The relative toxicity with custard apple was 0.058, 0.032 and 0.041, with neem it was 0.009, 0.005 and 0.058. However, with DDT it was 1.00, 1.00 and 1.00, respectively, for the above three insects. In case of mixing the two botanicals, the synergistic action observed in terms of custard apple extract / neem seed extract ratio was 1:2, 1:3, 1:4, 1:5, 1:6. The synergistic values become twice, thrice, fourtimes, fivetimes and sixtimes, respectively in case of *C. chinensis* while in case of *R. dominica*, the toxicity was twice, fourtimes, fivetimes, sixtimes more against the ratio 1:2, 1:4, 1:5 and 1:6, respectively. In case of *M. domestica nebulo*, toxicity was fourtimes, fivetimes and sixtimes more toxic against the ratio 1:4, 1:5 and 1:6, respectively.

Fagoone and Lauge (1981) observed the toxic effect of neem extracts against *Crocidolomia binotalis* Zell. The potential of extracts of neem (*A.*

indica) to control *C. binotalis*, a pest of cabbage, was investigated in the laboratory. It was shown that in addition to being an antifeedant at concentrations as low as 0.001 percent of the methanol extract, neem was also very toxic, larval mortality was high and adult emergence was poor. Other effects of the extract included disruption of normal development, significant delays in larval moulting, the appearance of permanent larvae in the 3rd and 6th moults and abnormal pupation.

Borah and Soharia (1982) observed toxicity of neem and other insecticides against *Henosepilachna* [*Epilachna*]*vigintioctopunctata* F. A field-plot trial was carried out in Assam, India, during the kharif season of 1977 to determine the effectiveness of some insecticides against larvae of *H. vigintioctopunctata* on ridge gourd (*Luffa acutanglua*). All the compounds tested (malathion, carbaryl and fenitrothion at 1.0 kg a.i./ha, endosulfan at 0.035 kg a.i./ha, diazinon at 0.03 a.i./ha and dichlorvos at 0.05 a.i./ha) greatly reduced the larval population 2 days after spraying and showed similar reductions 7 days after treatment. The most effective compounds were malathion, reducing the larval population by 98.6 and 92.8 percent after 2 and 7 days, respectively, and carbaryl, reducing the population by 97.3 and 96.9 percent. The effectiveness of all the insecticides had fallen considerably by 15 days after treatment. Neem expeller cake powder (prepared from *Azadirachta indica*) was less effective than the other insecticides tested, with a population reduction of 57.2 and 59.5 percent 2 and 7 days after treatment, respectively.

Radhakrishnan *et al*. (1983) tested 11 compounds for the control of *C. chinensis* on stored seeds of pigeon pea (*Cajanus cajan*) in 1975-78 in Tindivanam, India. Over 92 percent bruchid mortality was obtained with dust formulations of 5 percent malathion, DDT or HCH (BHC), emulsions containing phoxim or pyrethrins and dusts of the fungicides fentin chloride (triphenyltin chloride) and fentin hydroxide (triphenyltin hydroxide), while 68.9 percent mortality was given by neem kernel powder; all these treatments permitted over 70.3 percent germination. Seed weight loss increased rapidly with storage duration.

Ali *et al*. (1983) studied the effectiveness of oils from seeds of various plants at various dosages in laboratory in India against eggs, larvae and adults of *C. chinensis* attacking seeds of gram (*Vigna radiata*) and their effects on oviposition, adult survival and subsequent population build-up. All the treatments (oils of plants including neem [*A. indica* (*Melia azadirachta*)], rape, coconut, mustard, mahua (*Bassia latifolia*) and oil palm were less effective at 0.05 ml/100 g seed than at 1.0 ml/100 g seed against all stages of the pest. At the higher dosage, the oils of neem, coconut, mustard, mahua and sesame caused 100 percent egg mortality, while the neem and coconut oils caused 100 percent larval mortality and 100 percent adult mortality was observed 3 days after treatmenet with neem, coconut, mahua, sesame and palm oils. Oviposition was prevented, and subsequent population increase inhibited, in seeds treated with oils of neem, coconut, mahua, sesame and palm oil.

Reed and Reed (1984) conducted experiments to assess the efficacy of crude formulations of neem seed against economically important insect pests of muskmelon in the greenhouse, and sweet maize, egg-plant (aubergine), potatoes and cabbage in the field. In tests with the chrysomelid *Acalymma vittatum* on muskmelon, a 20 percent dust of ground neem seed in kaolin gave protection for 6 days after application and was not phytotoxic. Against the noctuid *Heliothis zea* on sweet maize, neem in 2 and 4 percent spray formulations gave significantly better control than no treatment. Against the chrysomelid *Epitrix fuscula* and larvae of *Leptinotarsa decemlineata* on aubergine, neem as a 2 percent spray gave significantly better control than no treatment, but as a 20 percent dust it was not effective. On potatoes, larvae of *L. decemlineata* were controlled by neem spray, but adults were not controlled. On cabbage, neem gave significantly greater control of the noctuid *Trichoplusia ni* than no treatment, but as in the other field tests it was not as effective as Ammo (cypermethrin) at 0.06 lb. a.i./acre.

Jaipal *et al.* (1984) report the results of laboratory studies in India on the residual effect of various leaf extracts of neem (*A. indica*) on *R. dominica*, a serious pest of stored wheat. Alcohol extracts proved more toxic than acetone benzene, chloroform, ethyl acetate, hexane or petroleum ether extracts. Extracts equivalent to 1 or 2 g of leaf were allowed to dry in beakers at room temperature. Ten pairs of adults of *R. dominica* were released in 25 g wheat grains transferred to each beaker. Mortality was recorded daily for 10 days. In the beakers treated with the alcohol extract, all the adults died within 4 days of release, and a further 10 pairs released on the 5th day also died within 4 days of release.

Rao *et al.* (1984) studied the effect of neem and other insecticides against white grub. At Tirupati in Andhra Pradesh, India, soil treatment with granular formulations of 4 insecticides, seed treatment with 3 compounds and soil treatment with incorporated neem cake before sowing were applied to rain-fed groundnut and evaluated both for the control of *Holotrichia consanguinea* and for their effect on the growth and yield of the plant. Granules of phorate, carbofuran or quinalphos applied at 1.5 kg toxicant/ha and seed treatment with isofenphos at 25 g toxicant/kg seed all increased the pod yield as compared with no treatment, and also resulted in significantly more rapid leaf growth rate and duration. None of the treatments interfered with root nodule production. Phorate and carbofuran gave particularly good economic results. Neem, however, has no effect on the *H. consanguinea*.

Das (1986) tested pesticidal efficacy of some indeginous oils against *C. chinensis* L. The effect of different plants oils on the mortality and fecundity of adults of *C. chinensis* was investigated in the laboratory. Adults were introduced to containers of chick pea (*Cicer arietinum*) seeds mixed with either neem (*A. indica*), coconut, sesame, mustard or soyabean oil at 10 ml/kg seed. All adults died within 4 days of release, while none died in untreated seeds. Oviposition was completerly inhibted when stored seed was treated with the neem, sesame and coconut oils, and very few eggs were found on seeds treated with soyabean and mustard oils. After 1 and 3 months of storage, there was no infestaion in oil-treated seeds, while 25.9 and 98.5 percent, respectively, of control seeds were infested. The viability of the oil-treated seeds remained unaffected.

Singh *et al*. (1987) studied insecticidal properties of some oils against *R. dominica*. Six insecticides of plant origin were tested in films against adults of the stored products pest *R. dominica* in the laboratory. On the basis of LC50s, extracts of neem (*A. indica*), *Bassia* [*Madhuka*] *longifollia* and *Pongamia glabra* were highly toxic, being 1.7660, 1.2705 and 1.679 times as toxic, respectively, as the standard, castor oil.

Parmar and Dutta (1987) observed synergistic effects when oils including neem are mixed with malathion. Several inedible oils were tested as synergists for malathion against 15-day-old individuals of *Tribolium castaneum*. Oils of neem (*A. indica*), *Bassia* [*Madhuka*] *latifolia, Pongamia pinnata, Calophyllum inophyllum, Garcinia indica, G. morella, Vateria indica* and rubber (*Hevea brasiliensis*) were synergistic when applied with malathion in ratios of malathion to oil of 1:1 or 1:5.

Das (1987) examined the toxicity of neem (*A. indica*) kernel oil to eggs and larvae of the stored product pests *Callosobruchus chinensis* in the laboratory on chickpeas (*Cicer arietinum*) at 31-35.5°C and 81-86 percent RH. Treatment of seeds with 4, 6, 8 or 10 ml neem oil/kg increased egg mortality significantly, mortality being 87.2 percent at the highest concentration. Larval mortality was also greater following treatment with neem oil.

Varma and Joshi (1988) treated the collar zone of apple trees in Uttar Pradesh with neem cake at the rate of 250 g/tree, neem cake at 500 g/tree, charcoal cake at 250 g/tree, grease bandage with 100 g/aldrin dust per tree. Aldrin dust 250 g/tree, waste mobile oil 100 g/tree, ground plant material of *Acorus calamus* 50 g/tree, drenching with carbaryl 0.2 percent, pine wood saw dust 500 g + heptachlor dust 100 g/tree and lime 250 g/tree against coleopteran pest, *Dyscerus fletcharie* Marshall. The result indicates that all the treatments were proved significantly superior to untreated control. The aldrin dust 250 g/tree proved best follwed by neem cake 250 g and 500 g/tree.

Gujar and Mehrotra (1988) studied biological activity of neem (*A. indica*) in laboratory against the chrysomelid, *Aulacophora foveicollis*, a major pest of cucurbits in India. Tests using leaves of musk melon (*Cucumis melo*) as feeding substrate revealed that the effective concentration for 50 percent antifeedant activity was 0.01 percent methanolic neem seed kernel extracts and 0.4 percent neem oil. No-choice feeding of adults for a period of 11 days on leaves of musk melon treated with 0.5-2.0 percent neem seed kernel extract led to nearly 50 percent mortality within 4-7 days; however, even at the highest concentration of 2.0 percent neem oil, there was no mortality by the end of 11 days. Neem oil as a residual film on petridishes had an LC50 of 0.7 percent.

Zehnder and Warthen (1988) studied the mortality effects and feeding inhibition of neem (*A. indica*) seed extract against colorado potato beetle, *Leptinotarsa decemlineata* in the laboratory at 25°C, 65 percent RH. Treatment dilutions were prepared from a viscous concentrate of neem seed extract that contained 2.72 mg azadirachtin/g extract. Adult mortality did not exceed 25 percent in any neem treatment, but 73 percent of larvae died within 72 hrs after feeding on foliage treated with 1.2 percent neem extract. The toxicity of neem extract to larvae was enhanced, and feeding of larvae and adults was inhibited to a greater extent by the additin of the synergist piperonyl butoxide at a ratio of 10:1 (piperonyl butoxide/azadirachtin). In studies in potato fields in Virgtinia, weekly sprays of neem extract at 0.8 and 1.2 percent significantly reduced the numbers of larvae and adults of *L. decemlilneata* 3 days after application.

Mostafa (1989) tested the efficacy of neem (*A. indica*) flower and fruit powder against *Trogoderma granarium*. Flowers and fruits of neem were sun-dried and powdered, and added to stored rice. The effect on artificially introduced *Tribolium castaneum* was assessed. In rice treated with the powdered flowers at 1 and 2 percent, adult mortality was 66 and 61 percent, respectively, after 7 days, compared to 22 percent in untreated rice. When the treated rice had been stored for 12 months, mortality of artificially introduced adults was 95 and 91 percent by 7 days after introduction, compared with 22 percent in untreated rice. When adults were introduced after 6 months of storage, 100 percent mortality had occurred in the rice treated at 4 percent by 7 days after introduction and by 9 days in rice treated at the other concentration.

Schmutterer (1989) studied insecticidal effects of methanolic extracts of leaves and bark of philippine neem tree, *Azadirachta integrifoliola* Merr. and compared with *A. indica* and *Melia azedarach* L. Crude extract of 3 insecticidal trees were tested at concentration of 50-800 ppm against the coccinellid *Epilachna varivestis* in the laboratory. In causing mortality and morphogenetic defects, leaf extracts of *A. integrifoliola* were superior to those from *M. azedarach*; extracts from leaves of *A. indica* were the least effective. The same sequence of effectiveness was found in cage tests as in petri dish bioassays, but higher concentrations were needed. Among bark extracts, concentrations of 200 and 400 ppm *Azadirachta* spp. were about equally effective, but these concentrations of *M. azedarach* had no effect.

Kossou (1989) evaluated different products of neem against curculionid, *Sitophilus zeamais*. Various components of the neem tree (*A. indica*) were tested for their insecticidal effects on the *S. zeamais* in stored maize. Seed kernel, leaf, flower and bark from the tree were used in powdered and extract (80% ethanol) form at a rate of 2.5 g and 2 ml, respectively, per 50 g of grain infested with 20 insects. Pirimiphosmethyl and ethanol were applied at 25 mg and 2 ml, respectively, for comparison. Ethanolic extracts showed a higher toxicity to *S. zeamais* than the powdered neem forms; after being confined on treated grain for 72 hrs, mortality was 32.6-79.0 percent for the powdered forms compared with 34.8-90.0 percent for the ethanolic extracts. The index of susceptibility ranged from 0.45 for the kernel to 3.87 for the bark in powdered forms of neem, and from zero for the kernel or leaf to 1.76 for bark in the ethanolic extracts. When leaf or kernel powder was used to protect grain over a 60-day storage period in aerated conditions, insect mortality was 90 percent after treatment with the kernel powder compared with 69.8 percent after treatment with the leaf powder.

Reddy *et al.* (1990) evaluated plant extracts for the control of brinjal leaf beetle, *Henosepilachna* [*Epilachna*] *vigintioctopunctata*. Field experiments were conducted during the kharif season of 1988 in India to evaluate the efficacy of petroleum ether extracts of *Eucalyptus globulus* (1%), *Calotropis gigantia*, *Pongamia glabra*, *Annona squamosa, Azadirachta indica* and *Lantana camara* for control of *H. vigintioctopunctata* sprayed on brinjal. At 24 hrs and 3 days after spraying, the most effective extracts were those obtained from *A. indica* and *A. squamosa*, reducing the number of pest larvae by 88.00 and 92.99, and 85.98 and 91.02 percent, respectively.

Jacob and Sheila (1990) tested the effectiveness of crude palm oil, refined palm oil, sesame oil, neem oil and coconut oil against *Callosobruchus chinensis* in green gram (*Vigna radiata*). All the oils caused more than 60 percent mortality of the bruchid after 3 days, sesame and neem oil at 1 ml/ 100 g being the most effective.

Ivbijaro (1990) tested the efficacy of neem seed oil against *Callosobruchus maculatus* F. Oils from the seeds of neem (*A. indica*) and the brown pepper (*Piper guineense*) afforded protection to cowpea seeds (*Vigna unguiculata*) from *C. maculatus* in the laboratory beyond 3 months; 24 hrs after surface treatment with piper seed oil (PSO) at 2 and 3 ml/kg, bruchid mortality reached 100 percent, but with neem extract mortality was 65-100 percent at 3-5 days. Oviposition averaged 15.5 eggs for PSO, 49.5 for neem and 60.75 for no treatment.

Agrawal (1990) studied ovicidal activity of some phytochemicals against *Myllocerus undecimpustulatus* Faust. In the laboratory, 5 extracts from *A. indica, Swietenia macrophylla* and *Calophyllum inophyllum* were tested for their ovicidal activity against day-old eggs of *M. undecimpustulatus*. The eggs were treated in a 1 percent solution of each extract. The 2 neem extracts caused 100 percent mortality, those of *C. inophyllum* and *S. macrophylla* caused 94.6 and 76 percent mortality, respectively.

However, azadirachtin, an isolates of neem seed showed no ovicidal activity against the freshly oviposited eggs of *Dysdercus koenigii* (Koul, 1984) and *Oncopeltus fasciatus* (Dorn, 1986). According to Agrawal (1988), the thionemone, a neem seed extract, however, showed significant ovicidal activity against eggs of *D. koenigii*. It is worth mentining that Schmutterer (1985) reported that under field conditions, neem seed extracts did not show any ovicidal activity in coleoptera and lepidoptera.

Patil *et al*. (1991) evaluated the effect of insecticides and neem extract against flea beetles on sorghum. In a field experiment in Maharashtra, India, in 1988-89, sorghum was treated 1 month after sowing with 250 and 500 g/ ha each of endosulfan, malathion, quinalphos, dimethoate, demeton-S-methyl, acephate and Neemark (neem extract). All treatments reduced the numbers of *Chaetocnema* sp. on the plants.

Naqvi *et al*. (1991) investigated the toxicity of NfD neutral fraction of winter neem leaves containing the tetranor-triterpenoids nimocinolide and isonimocinolide) against 1-week-old adults of the stored products pest *Sitophilus oryzae*. The LD50 was observed to be 0.0392 mg/cm^2 when the neem extract was appllied to filter paper. Inhibition of alkaline phosphatase was 13.17 percent and of acid phosphate 4.28 percent; protein metabolism was also inhibited.

Xie *et al.* (1991) tested effect of azadirachtin on the western corn root worm, *Diabrotica virgifera* Leconte in the laboratory. It was observed that azadirachtin was toxic to the larvae.

Seek *et al.* (1991) studied the effect of neem powders against *Callosobruchus maculatus*. Powders and aqueous solutions of an ecotype of *Azadirachta indica* from Senegal were tested against *C. maculatus* in stored cowpeas. Both neem leaves and neem seed powders, mixed with cowpeas at 3 percent w/w caused 85-90 percent mortality to adult bruchids in 72 hrs. Seed powders were more effecitve than leaf powders. With aqueous solutions, the best result was obtained with dry neem seeds in water, as compared with fresh neem seeds or fresh dry neem leaves. Dipping cowpeas in an aqueous solution of dry seeds (60 g/lt) significantly reduced fecundity and oviposition by a parental population of the bruchid.

Rao *et al.* (1992) studied the insecticidal properties of some plant extracts against *Henosepilachna vigintioctopunctata*. Petroleum ether extracts of leaves of 4 selected plants were tested in the laboratory for their antifeedant and insecticidal properties against *H. vigintioctopunctata* at concentrations of 0.2 and 0.5 percent applied to brinjal leaves. Mortality after 24 hrs at both concentrations of *Bougainvillea spectabilis* and *Parthenium hysterophorus*, and 0.5 percent concentration of *A. indica* was 100 percent.

Niber *et al.* (1992) studied toxicity of plant extracts against storage beetles. Eleven indigenous Ghanaian plant species were tested for their toxicity to 3 storage pest species (*Acanthoscelides obtectus, Prostephanus truncatus* and *Sitophilus oryzae*) under laboratory conditions. *P. truncatus* was reared on loose maize grains, *S. oryzae* on wheat grains and *A. obtectus* on cowpeas. The plant extracts evaluated were obtained from leaves and seeds of *Azadirachta indica*, leaves of *Chromolaena odorata* [*Eupartorium odoratum*], leaves and roots of *Cissampelos owariensis* [*C. pareira*], seeds of *Datura metel, D. stramonium* and *Ricinus communis*, leaves of *Erythrophleum suaveolens, Grewia carpinifolia, Sida acuta, Solanum nigra* and *Strophantus hispidus*. Crude ethanol extracts at 10 percent (w/v) were used as topical applications. The effect of the plant extracts was studied and the extracts of *A. indica. R. communis* and *S. nigrum* were observed most toxic to the 3 pest species.

Schmutterer and Doll (1993) observed that philippine neem tree *Azadirachta excelsa* (=*A. integrifoliola*) having good sourse of insecticidal properties with growth regulating properties against *Epilachna indica* and *E. varivestis*. Investigations were carreid out in the Philippines on the island of Palawan, and in Germany. Bioassays of leaf, bark and seed extracts of both species *E. indica* revealed greater bioactivity for *A. excelsa*, which contains azadirachtin as well as marrangin, a new compound closely related to azadirachtin, in its seed kernels. Marrangin was more toxic than azadirachtin to *E. varivestis*.

Azmi *et al*. (1993) observed toxic effect of neem formulation against *Sitophilus oryzae*. In laboratory studies, the toxicity of a compound containing 10 percent cyfluthrin (Solfac) and a neem formulation containing crude extract from ripe berries of *Azadirachta indica* against *S. oryzae* was determined. The tests were carreid out by releasing the curculionids on treated filter papers treated with different concentrations of the compounds. A mortality rate of 90 percent was obtained with 0.05 percent of cyfluthrin and a 1 percent concentration of the neem compound.

Dey and Sarup (1993) tested 8 vegetable oils (mustard, soyabean, coconut, neem, groundnut, cotton, sesame and castor) at 5 doses against adults of the stored products pest *S. oryzae* in 3 varieties of stored maize in India. Irrespective of the oils used and the maize varieties, weevil mortality was highest 1 day after treatment. Coconut and soyabean oils at 3.30 ml/kg grain were highly effective in the varieties Pusa Arun, AEB comp. (white) and VL-42. The average percentage increase in adult weevil mortality, reduction in population and avoidable loss due to damaged grains as a result of the application of oils was highest in Pusa Arun, being 98.35, 53.81 and 38.42 percent, respectively.

longiswaran and Venugopal (1994) conducted field studies with neem oil applied in combination with either monocrotophos, carbaryl or phosphamidon gave a level of control of a pre-harvest infestation of *Sitotroga cerealella* on rice similar to that obtained with either monocrotophos, carbaryl or phosphamidon applied alone, and at a reduced cost.

Naumann *et al*. (1994) tested systemic action of neem seed extracts on mountain pine beetle. A botanical insecticide derived from seeds of *Azadirachta indica* was tested for the "bait and kill" strategy for control of outbreaks of *Dendroctonus ponderosae* in British Columbia. Application of neem into frills at the base of beetle-attacked *Pinus contorta* var. *latifolia* trees reduced numbers of beetle larvae per unit area of bark, decreased densities of larvae per unit length of adult gallery, and increased larval mortality when applied at doses > 0.25 g a.i. (azadirachtin) per tree.

Saxena (1995) mixed various vegetable oils with organic insecticides and evaluated in 4 ratios against adults of *S. oryzae* in the laboratory, and the type of action (synergistic, antagonestic or additive) was determined. The mortality count of insects in three applications of each concentration was recorded and average mortality in each concentration was calculated by multiple regression analysis for determining the synergistic activity by mixing vegetable oils and chemical insecticides. The results indicated that neem with pp-DDT and lindane gave synergistic effect while with malathion and pyrethrum, it gave additive effect. However, with permethrin, antagonestic effect was obtained.

Pal *et al*. (1996) compared relative toxicity of certain plant extracts to khapra beetle, *Trogoderma granarium*. The insecticidal properties of alcohol extracts from the leaves of chandani (*Taberna montana* [*Tabernaemontana*] *coronaria*), curry (*Murraya paniculata*), asripel (*Lantana camara*), eucalyptus (*Eucalyptus* sp.), neem (*A. indica*), rhizomes of sweet flag (*Acorus calamus*) and seeds of imli (*Tamarindus indica*), and oil from neem kernels compared with the commonly used organophosphate insecticides malathion 50EC and chlorpyrifos-methyl 50EC were tested for their comparative toxicity against the fully gorwn larvae of *T. granarium*. The relative toxicity on the basis of LC50 in descending order was : malathion > chlorpyrifos-methyl > *A. indica* oil > *A. calamus* > *T. coronaria* > *T. inidca* > *A. indica* leaves > *L. camara* > *M. paniculata* > *Eucalyptus* sp.

Sridevei and Dhingra (1996) evaluated some non-toxic vegetable oils as synergists for many synthetic pyrethroids in mixed formulations against *Tribolium castaneum* Herbst. On the basis of joint action ratios, the type of action of insecticides was determined when three synthetic pyrethroids (deltamethrin, cypermethrin, fenvalerate) were used separately in mixed formulations with each of the four non-toxic vegetable oils (sesame oil, karanj [*Pongamia pinnata*] oil, neem [*A. indica*] oil, citronella [*Cymbopogon nardus*] oil and the well known standard synergist piperonyl butoxide (PBO) in four ratios (1:1, 1:2, 1:4, 1:8) and evaluated against a susceptible strain of *T. castaneum* adults by direct spray and film residue methods. Out of 30 combinations tested, additive action was obtained when the oils were used in mixed formulations with cypermethrin and fenvalerate. With the exception of neem oil by the direct spray method, the remaining three vegetable oils also showed additive action with deltamethrin. Antagonistic action, however, was produced by neem oil in the mixed formulations with deltamethrin. Only the well known synergist PBO synergised cypermethrin and fenvalerate.

Singh *et al*. (1996) studied the effect of neem saw dust on the mortality of *Callosobruchus chinensis*. Mixing dried neem (*A. indica*) sawdust at rates of 100, 200, 300 and 400 mg/50 g Vigna mungo seed provided effective protection against C. chinensis, 400 mg being the most effective rate.

Haque *et al*. (1996) tested the effect of neem oil on the food consumption and survival of *Epilachna dodecastigma* [*E. pusillanima*] Wied. In a laboratory study, 1st and 3rd instar larvae and adults of *E. dodecastigma* were exposed to 0.25, 0.50, 1.0 or 2.0 percent neem oil applied on brinjal leaf discs. All the 1st instar larvae were killed at all the concentrations before feeding. The feeding activity in 3rd instar larvae and adults decreased with increasing oil concentration. Highest larval mortality was recorded at the 2 higher oil concentrations.

Chandel *et al.* (1996) observed the insecticidal properties of some oii cakes against *Brahmina coriacea*, white grubs. Three oil cakes (neem (*A. indica*); mahua (*Madhuca longifolia*) and mustard (*Brassica* sp.) were evaluated for their insecticidal properties at 3 and 4 kg/m^2 in glass jars under laboratory conditions against larvae of *B. coriacea*. Neem cake gave the highest mortlaity (24.06-33.66%) of 1st and 2nd instar larvae after 15 days followed by mahua cake (17.16-19.56%). However, mustard cake had no effect.

Singh *et al.* (1996) conducted laboratory studies at 29.6±2°C to determine the effects of extracts of neem (*A. indica*), garlic, oranges, *Eucalyptus hybrida*, *Lantana camara*, *Vitex negundo* and *Ferula assafoetida* on *Rhyzopertha dominica* on wheat. Treatment with extracts of F. assa-foetida (1 and 2%), *A. indica* (2 and 3%) and *L. camara* (3%) were the most effective in terms of adult mortality and reduced grain damage.

Ramamurthy and Venugopal (1997) studied the effect of plant product and insecticides on the incidence of *Sitotroga cerealella* under field condition on sorghum variety CO-26. It was observed that neem oil (3%) significantly checked the damage and it was followed by neem oil (2%), malathion (0.1%), neem seed kernel extract (2, 3%) and endosulfan (0.04%) which were all on par. Two spraying may be given once at milky stage and the second at 10 days thereafter was significantly effective in controlling the pest.

Padmanaban *et al.* (1997) evaluated plant products and oil cake against the 2nd and 3rd instar larvae of white grub, *Leucopholis burmeisteri* in pot experiments under field conditions in Karnataka, India. In the first experiment, fresh leaves, dry leaves and dried leaf powder of *Vitex negundo* and some other plant products were evaluted against 2nd instar larvae. In the second expriment, oil cakes of karanj (*Pongamia pinnata*) neem (*A. indica*) and mahua (*Madhuca indica* [*M. longifolia*]) at the rate of 8.5, 19.0, 34.53 g/pot basis were tried against 3rd instar larvae. The experiment was laid out in 1993-94 with Vitoxyl (2%), Nimecidine (2%) and Achook (2%) and *V. nigundo*, fresh leaves, dry leaf and dry leaf powder (85 g/pot and a control. Treatment wise observations on the grub mortality were recorded after 3 days of application of oil cakes. The result indicated higher grub mortality with dry leaf powder of *V. negundo* followed by 2 percent Achook. These results indicated that the application of plant products could be incorporated in the IPM of arecanut white grub.

Emosairue and Ukeh (1997) did field trials with neem products for the control of okra beetle, *Podagrica* spp. in the south eastern Nigeria. The efficacy of 2 neem products (3% neem oil (NO) and 5% neem seed kernel extract (NSKE)) and a standard synthetic insecticide, lambda-cyhalothrin, were evaluated in the field during the early season (March-June) and late season (September-December) in 1994 for the control of *Podagrica uniformis* and *P. [Nisotra] sjostedti*, both major pests of okras. All treatmnets significanlty reduced the pest population (p=0.05) within lambda-cyhalothrin giving the best level of control followed by 3 percent NO and 5 percent NSKE. Both neem products showed promise as a substitute for synthetic insecticides for the control of both pest species. It was concluded that higher concentrations and closer spray regimes would probably improve their efficacy.

Das and Borah (1997) studied comparative efficacy of certian insecticides and neem formulations against *Sitophilus oryzae* Linn. and *Sitotroga cerealella* Oliv. in stored wheat seeds. Wheat seeds (var. Sonalika) were treated with three pesticides, each at three dosages, viz., BHC (HCH) 10 percent dust and malathion 5 percent dust both @ 2, 3 and 4 g/kg of seeds and neem kernel powder (*Melia azedarach*) at 10, 20 and 30 g/kg of seeds alone and coupled with the fungicide carbendazim 50 WP to study their comparative effectiveness against *S. oryzae* and *S. cerealella*. One set was kept only for fumigation with ethylene di-bromide (EDB). The highest dosage of malathion (4g/kg) and medium dosage of malathion (3 g/kg) coupled with carbendazim (4 g/kg) were the best treatments up to 2 months of storage. Whereas, malathion (4 g/kg) with carbendazim was the best treatment up to 4 months of storage followed by malathion alone (3 g/kg), malathion (2 g/kg) coupled with the carbendazim and neem kernel powder (20 g and 30 g/kg) with fungicide. However, neem kernel powder had to be remixed with seeds after 4 months of storage for lowering infestaion by insects and for good germination.

Marklowsky *et al.* (1997) did greenhouse and field experiments in Germany with Neem-Azal T/S. It was found that NeemAzal T/S at 1.00 percent was effective against the larvae of *Galerucella nymphaeae* on waterlilies (*Nymphaea* spp.) causing upto 70 percent mortality.

Basedow *et al.* (1997) did control trials by azadirachtin formulation (NeemAzal T/S), *Bacillus thuringiensis* and by combination of both. A combined field-laboratory experiment against *Leptinotarsa decemlineata* in potato near Frankfurt/Main, Germany, in 1995 showed, after the first treatment on 24 June (second treatment on 12 July), maximum efficacy against larvae of 71 percent (Neem Azal T/S, 0.5%), 55.4 percent (Novodor (*B. thuringiensis* subsp. *tenebrionis* 3 lt/ha) and 5 percent (combination of Neem Azal, 0.1%). The maximum feeding damage (mainly by larvae) on 24 July was 7 percent (Neem Azal), 20 percent (Novodor), 44 percent (combination) and 63 percent (untreated).

Baumgart *et al*. (1997) carried out a field trial at South Dakota in 1993 to investigate the effectiveness of the neem formulations Neem Azal-F, Align and *B. thuringiensis* var. *sandiego* (M-One) against colorado potato beetle *L. decemlineata* Say in organic potato production. A single application of Neem Azal-F at 50 g a.i./ha was sufficient to control larvae (82% effectiveness) and adults (91%). A higher dosage of active ingredient up to 100 g a.i./ha did not increase mortality. Align did not control the larval population, but was effecitve against the adults (94%). The *B. t*. preparation M-One demonstrated moderate control of both larvae (73% mortality) and adults (77% mortality). It is concluded from these findings that a single Neem Azal-F application provides appropriate and environmentally sound control of *L. decemlineata* which is low in cost and labour for organic potato growers.

Goltwald *et al*. (1997) did trial on the effectiveness of neem products against insect pest, *L. decemlineata*. Field trials carried out in Germany in 1993-95 showed that 3 Neem Azal formulations were effective against *L. decemlineata* on potato and *Oulema melanopus* on barley and wheat. The preparations could exceed 90 percent effectiveness depending on time of application.

Hummel annd Kleeberg (1997) got effetive results from Neem Azal formulations. Field, greenhouse and laboratory trials were carried out during 1996 in Germany to investigagte different formulations of Neem Azal; Neem Azal-T/S against *L. decemlineata, Macrosiphum rosae* and *Dysaphis plantaginea*, and Neem Azal-BD and Neem Azal-WF against *Anthrenus vorax* [*A. flavipes*] and *Tineola bisselliella*. Results indicated that all formulations were highly efficient.

Patel and Purohit (1998) tested the bioefficacy of various botanical materials and formulations viz., garlic bulb extract (3% suspension), mint leaves extract (3% suspension), NSKE (3% suspension), *Ardusa* leaves extract (3% suspension), *L. camara* leaves extract (3% suspension), Neemark (0.5%), Fieldguard (1%) alongwith the chemical insecticides viz. DDVP (0.03%) and endosulfan (0.075%) against Epilachna beetle, (*H. vigintioctopunctata*) under laboratory condition at Navsari, India. The treatment of DDVP was found superior in effectiveness followed by endosulfan against both the larvae and adults. Though the per cent mortality obtained in the treatments of various botanical materials was less than that of chemical insecticides, the neem-based insecticides viz. Fieldguard, Neemark and NSKE gave considerable percent of mortality.

Ji-Baozhong *et al.* (1998) tested the effects of neem extracts and quinalizidine alkaloids on the survival and reproduction of *Apriona germarii* Hop. When adult *A. germarii* were fed with twigs of *Broussonetia papyrifera* treated with a dose of 1.0 micron g/mm^2 thermospine or cytisine, thermospine showed no feeding deterrent effects and cytisine showed significant feeding deterrent effects. With the same dose, both alkaloids caused the longicorne beetle to reduce oviposition and egg hatching significantly, without causing their death. When *A. germarii* was fed with twigs of *B. papyrifera* treated with the doses of 8x10.2, 4x10.2, 2x10.2, 1x10.2 or 5x10.3 micron g/mm^2 of neem extracts (containing azadirachtin 10%), the first two doses caused 100 percent mortality, the dose of 2x10.2 micron g/mm^2 caused 80 percent mortality and complete inhibition of oviposition, and the last two doses caused 30 and zero percent mortalities, respectively, with considerable reduction in oviposition and egg hatching.

Barranco *et al.* (1998) crried out the experiment for evaluation of pesticides in two assays, the first with 7-day-old larvae and the second with 1-month-old larvae of *Rhynchophorus ferrugineus*. In both cases, the insecticides were incorporated into the semi-artificial diet which is used to rear this species in the laboratory. The insecticides tested were a nartural extract of neem tree seeds (azadirachtin at 3 cc/lt) and EXP60720 A (fipronil at 1500, 0.4, 0.2, 0.1, 0.05 and 0.01 ppm). It showed that for the younger larvae, the mortality reached 100 percent at doses greater than 0.1 ppm of fipronil. On the other hand, the mortality caused by azadirachtin was less than 50 percent. Mortality of 100 percent was only reached for 1-month-old larvae with doses of fipronil exceeding 0.2 ppm.

Cardet *et al.* (1998) tested the comparative efficacy of neem (*A. indica*) kernel oil, groundnut oil and a synthetic insecticide (K-Othrine [deltamethrin]) in protecting stocks of leguminous tree seeds against seed beetles. The following insect/seed combinations were used as models. *Caryedon acaciae* [*C. serratus*]/*Acacia nilitica, C. longispinosus/A. raddiana* and *C. serratus/Tamarindus indica*. Neem oil, when used at concentrations of 5-20 ml/kg of seeds, had pronounced adulticidal and ovicidal effects which were maintained for 5 months. No significant fall in seed viability was observed except in the case of *C. acaciae/A. nilitica*. This study concluded that it is possible to use neem oil, a product that can be obtained locally at low cost, to protect tree seed stocks efficiently in the Sahel.

Reddy and Singh (1998) tested neem products against pulse beetle, *Callosobruchus maculatus* F. Neem seed extracts and the pure compound present therein have been evaluated for various biological effects on insect and other pests. There is, however, perhaps no report on the evaluation of isolated neem seed volatiles against insect pests. Neem seed volatiles were isolated from neem seed oil and evaluated for their bioactivity against eggs and 2 and 15-day old larvae and adults of *C. maculatus*. All stages were found susceptible to different doses of volatiles. At 50-200 micron l, they caused 81-100 and 25.4-1— percent mortality of adults and larvae, respectively. Eggs, when exposed to 100-200 micron l, for 3 days failed to hatch. At a 50 micron l dose with a 5-day exposure period, the eggs that hatched failed to reach the adult stage and terminated as larvae. The adults were found to be most susceptible, followed by the larvae and eggs. The effects of the volatiles were time-dependent.

Azmi *et al.* (1998) copared the toxicity of a neem extract RB-a and a pyrethroid, Coopex (permethrin + bioallethrin) and effects on GPT (alanine aminotransferase) and GOT (aspartate amonitransferase) activities and oxygen consumption were studied in one-day-old adults of *Sitophilus oryzae* in the laboratory at 30±2°C and 70±5 percent RH. The LD50 of Coopex was 6.128 micron g/cm^2 whereas RB-a did not act as an acute contact poison. Even at a dose of 1257 micron g/cm^2, it caused 34 percent mortality of *S. oryzae*. The oxygen consumption following treatment with Coopex decreased less (from 0.0005191 to 0.0006589 with 6.12 micron g/cm^2), compared to RB-a (from 0.0005191 to 0.0002076 with 1257 micron g/cm^2). Coopex inhibited GPT activity by 62.77 percent and caused almost no inhibition of GOT activity in insects treated with the LD50. In the case of neem extract, it was noted that GPT was inhibited by about 57.47 percent and almost no effect on GOT was observed at a dose of 1257 micron g/cm^2.

Singh *et al.* (1999) tested the neem dust formulation against khapra beetle *Trogoderma granarium* Everts and cigarette beetle *Lasioderma sericorne* F. The dust was tested on 1st and 3rd instar larvae of *T. granarium* with concentrations 2.0, 1.0, 0.5, 0.25, 0.125, 0.06, 0.03 and 0.015 percent. The dust was mixed in 50 g of wheat seed and three replications were kept for each treatment. In case of *L. sericorne*, dust was evaluated against 1st instar larvae. The concentrations tested were 2.00, 1.00, 0.50, 0.25 and 0.0125 percent.

The results revealed that dust had no effect on the hatching of eggs of *T.*

granarium and caused 20-100 percent mortality of 1st instar hatched larvae as against 100 percent survival in control by 7th day. By 14th day, all the larvae died in all the concentrations. The dust also caused high activity against 3rd instar larvae. It was also observed that at higher concentration (0.25%) , the larval mortality, the pupal mortality and percent deformed adult was 80.0, 66.26 and 88.88 percent. However, at lower concentrations 0.06-0.15 percent, 6.6-23.33 percent larvae were survived even on the 55th day. In *L. sericorne*, the dust had no effect on hatching and it was found that at 1 percent concentration, all the 1st instar larvae died within a week. However, the lowest concentration (0.05%) had very little effect on the larvae.

Tripathy and Sahoo (2000) studied the efficacy of some neem formulations and neem seed kernel extract (NSKE) against *Callosobruchus chinensis* infesting blackgram. Four commercial neem formulations viz. Forture Aza, Nivar, Neemgold and Econeem, each at 0.25 and 0.5 percent concentrations were mixed with black gram seeds. After 24 hours of treatment adults were released but the percentage mortality after 72 h was high and it was not possible to get any adult from the treatment although very low oviposition was recorded. Neemgold recorded lower oviposition than other treatments whereas Forture Aza recorded the highest number of eggs.

The crude neem seed kernel extract was found to be equally effective in protecting the pulse seeds against bruchids attack with out deteriorating germinability of seeds.

2. Diptera

Srivastava *et al.* (1984) carried out field-plot tests at New Delhi, India to evaluate the effectiveness of 2 spray applications of 5 or 8 percent neem (*Aazdirachta indica*) kernel extract (NKE), 0.02 percent fenvalerate or 0.05 percent quinalphos at an interval of 15 days at pod formation for the control of *Melanagromyza obtusa* and *Heliothis armigera* on red gram (*Cajanus cajan*). The results showed that treatment with 8 percent NKE significantly reduced the damage caused by the 2 pests and was comparable to the treatments with fenvalerate and quinalphos. The fenvalerate treatment gave the highest grain yield, followed by the neem treatments and quinalphos.

Larew *et al.* (1985) did soil drenches of crude neem extract to chrysanthemum against larvae and pupae of *Liriomyza trifolii* Burgess. Three conentrations viz. 0.01, 0.1 and 0.4 percent were taken in this experiment. Five plants were used per treatment and in each pot, 150 ml of the solution was applied in the soil. The concentration at 0.4 percent causing significant mortalilty of pupae and late-instar larvae. However, 0.1 and 0.4 percent concentrations caused significantly lower adult emergence. The mortality percentage between pupal and adult stages with 0.1 and 0.4 percent neem was 89 and 100 percent, respectively, when longevity effect was observed for 3 to 21 days, 0.1 percent neem caused significant or complete mortality between the pupal and adult stages. It was also observed that a single drench at 0.4 percent concentration did not inhibit adult females from feeding or oviposition. The control of prepupal larvae at 0.4 percent was 84 and 95 percent in the two trials conducted. The mortality of prepupae of the agromyzid that were reared on untreated plants and placed on neem drenched soil was 89 percent.

Lindquist *et al.* (1986) evaluated neem products against leafminer *Liriomyza trifolii*. Rooted cuttings of cv. Iceberg were soaked for 2, 4 or 24 hrs in 0.1, 0.3, 1.0 or 3.0 percent Margosan-O (containing azadirachtin islated from *A. indica* seeds). Two and four weeks after treatment, plants were placed for 6 hrs in cages with *L. trifolii*. All Margosan-O treatments reduced plant growth. Placing cuttings in 3 percent Margosan-O for 2 or 4 hrs gave the best control of *L. trifolii* with minimal long-term plant growth effects.

Kondel *et al.* (1986) tested Margosan-O, a neem formulation of neem seed extract, as a systemic soil drench and a foliar spray against *Liriomyza trifolii* on chrysanthemum. The authors found that the systemic uptake of a 0.33 percent formulation significantly reduced the number of pupae and adults reared from treated plants, but did not reduce the number of mines. Residual activity of the formulation applied to the soil at 0.33 percent was evident for up to 21 days after treatment. No plant growth inhibtion occurred after 5 repeated drenchings with Margosan-O. When it was applied as a foliar spray, concentrations of 0.41, 0.84 and 1.25 percent significantly reduced the number of adults reared from treated plants and did not inhibit plant growth.

Larew (1988) observed the effect of neem seed extract toxin to control *L. trifolii*. A 1.0 percent solution of neem seed extract used against *L. trifolii* failed to increase the toxicity of untreated chrysanthemum foliage (*Dendranthema morifolium*) positioned above or below sprayed foliage. Untreated bean (*Phaseolus limensis* [*P. lunatus*]) leaves opposite to leaves painted with 0.4 percent extract were not as toxic to *L. trifolii* as painted leaves. Foliage on chrysanthemum cuttings that were root-dipped in 0.4 percent extract was toxic to leafminers but subsequent new foliage was not. Bean seeds soaked in 4 percent extract for 6 hrs germinated to seedlings which were not significantly more toxic to *L. trifolii* than untreated seedlings. In contrast, drenching the germination medium with 0.8 percent extract re sulted in seedlings from which no leafminers were reared. The results indicate that complete coverage is required for maximum efficacy of neem seed extract foliage sprays against *L. trifolii*.

Marion *et al.* (1990) observed systemic activity of neem extract against birch leafminer, *Fenusa pusilla*. Micro-injection of neem seed extract into boles of 7-yr-old paper birch (*Betula papyrifera*) in a nursery in Dansville, New York, USA, was evaluated for control of *F. pusilla*. Treatments at 1.5 and 3.0 percent concentrations caused significant decrease of foliar damage, and neem was found to be as effective as insecticide Bidrin [dicrotophos] and Meta-Systox-R [oxydemeton-methyl].

David *et al.* (1991) tested chemical insecticides and neem oil against blossom midge, *Contarinia* sp. The cecidomyiid is recorded on Jasminum sambac in Tamil Nadu, India, for the first time. Methyl parathion, dimethoate, endosulfan, monocrotophos, chlorpyrifos, cypermethrin, fenvalerate, fenpropathrin, BPMC (fenobucarb}, carbofuran and neem oil were evaluated for the control of this pest. Purple discoloration due to *Contarinia* sp. was lowest on bushes treated with 0.1 percent monocrotophos (2.62%). Neem oil and chlorpyrifos were as effective as monocrotophos one week after spraying. Neem oil caused paleness of leaves 2 weeks after treatment. Most of the insecticides tested, except monocrotophos, dimethoate and carbofuran, induced secondary infestations of *Tetranychus cinnabarinus*.

Singh *et al.* (1991) tested efficacy of plant extracts against mustard sawfly, *Athalia proxima* Klug. In laboratory studies, petroleum ether extracts of 10 different plant species at concentrations of 0.5, 1.0 and 1.5 percent were tested against the larvae of *A. proxima* [*A. lugens*]. At all the concentrations tested, extracts of *Saussurea leppa* were the most toxic followed by those of *Azadirachta indica* and *Cucurbita maxima*. Larval mortality was highest with the 1.5 percent extracts.

Azam (1991) determined the toxicity of neem (*A. indica*) oil (0.5, 0.75, 1.0, 1.25 and 1.5%) against the larvae of leafminer, *L. trifolii* in cucumber in the laboratory. Author found that the concentrations of 1.0-1.25 percent caused >80 percent mortality of larvae and pupae and the concentrations of <1 percent were less effective and 1.5 percent did not cause a significant increase in mortality.

Rao and Panwar (1992) conducted the field studies to determine the efficacy of insecticides/plant products to control infestations of shootfly, *Atherigona soccata* and *A. naquii* in maize at New Delhi, India. There was no significant difference between pre- or post-treatments using either carbofuran 3G, carbofuran 25STD, 3 percent neem kernel suspension, neem oil 20EC, Mohua oil 20EC (from *Madhuca indica* [*M. lognifolia*]) or phorate 10G (PEG 6000) on the average number of eggs laid/plant by either pest species. The average percentage dead hearts was least in plots treated with phorate 10G (8.82) and was highest in plots treated with neem oil (45.62).

Thomas *et al.* (1992) tested neem seed extracts as a control agent for spruce budworm. The effects on larvae of *Choristoneura fumiferana* produced by ingestion of Margosan-O (a neem formulation containing 0.3 azadirachtin) were investigated. Bioassays with the test material were conducted using various instars. The dose-response data on feeding reduction, developmental retardation and mortality suggested that Margosan-O has promise as a control agent for *C. fumiferana* in a pest management programme.

David *et al.* (1992) conducted the field experiment in Tamil Nadu, India, to study the effect of 10 insecticides and neem (*A. indica*) oil on a yet unidentified species of *Contarinia* which causes purple discoloration and drying of flower buds in *Jasminum sambac*. The lowest percentage of discoloured buds (2.62%) was observed on bushes treated with 0.1 percent monocrotophos, followed by bushes treated with synthetic pyrethroids, namely 0.012 percent cypermethrin (4.31% discoloured buds). Treatment with 2 percent neem oil mixed with 0.05 percent teepol before diluting with water (5.22% discoloured buds) was on a par with 0.05 percent chlorpyrifos treatment and compared favourably with the control (18.33% discoloured buds).

Samalo *et al.* (1993) tested the effects of neem oil, neem cake and Hind-O-Meal (containing neem oil seeds and *Melia azedarach* cake) on the incidence of major rice insect pests during the wet season of 1989 in Orissa, India. These treatments were compared with applications of monocrotophos (0.5 kg a.i./ha) 25, 40 and 55 days after transplanting. Application of 3 percent neem oil at 10, 25, 40 and 55 days after transplanting or substitution of a spray with topdressing with neem cake or Hind-O-Meal or a mixture of both significantly reduced populations of gall midge (*Orseolia oryzae*). The maximum grain yield (2707 kg/ha) was achieved with fortnightly sprays of monocrotphos, followed by 3 percent neem oil sprays and a combined treatment with neem cake and Hind-O-Meal.

Borkar and Urade (1995) conducted a study in Akola, India, on pigeonpea to evaluate the efficacy of insecticides and the number of applications for the suppression of pod borer complex of pigeonpea (*Cajanus cajan*). Applications of endosulfan (0.07%) followed by methyl parathion (0.05%) were effective, recording 5.10 and 5.24 percent pod incidence, respectively, in green pods. Parathion-methyl (0.05%) alternated with neem seed kernel extract (NSKE) was effective resulting in 10.23 percent *Melanagromyza obtusa* infestaion in dry pods. Similarly, in dry pods two spray schedules were effective in reducing incidence of the fly. The interaction of endosulfan (0.07%) with three sprays recorded minimum pod fly infestaion (8.70%) followed by two sprays of parathion-methyl alternated with NSKE (9.70%) infestation.

Rao and Pawar (1996) conducted field studies in maize fields in India to compare the relative efficacy of imidacloprid 70 WS at 3 concentrations (5, 7.5 and 10 g kg-1 seed) as a seed treatment, laboratory prepared phorate in a PEG 6000 matrix (PEG-Phorate 10) applied at 0.25 g m-1 row, carbofuran 5G (slow release) at 0.5 and 1 g m-1 row, a commercial formulation of phorate (Thimet 10G), carbofuran (Furadan 3G) at 1 g m-1 row as furrow treatments at the time of sowing, neem kernel dust 10 percent at 20 kg/ha, neem WDP 25 percent, neem oil 25EC and Nimbecidine R applied at 0.1 percent each as a foliar application, for the control of *Atherigona soccata* and *A. naquii* during the spring season. Oviposition, dead-heart formation, chlorophyll content of leaves, leaf area and plant height were studied. *A. soccata* and *A. naquii* did not discriminate between plants receiving different treatments while ovipositing. All the synthetic insectivides, including the seed treatment at higher concentrations, significantly reduced dead-heart formation. Phytotonic effects on plants expressed in terms of a higher chlorophyll content, larger leaf area, and higher plant height were recorded for plants receiving the neem formulations.

Malik *et al*. (1996) did field evaluation of some plant extracts and insecticides against linseed budfly, *Dasyneura lini*. Ethyl alcohol extracts of various parts of seven plants i.e., seeds of *Tamarindus indica*, leaves of *Lantana camara*, leaves of *Azadirachta indica*, rhizomes of sweet flag (*Acorus calamus*), leaves of *Eucalyptus globulus*, leaves of *Murraya paniculata* and leaves of *Tabernaemontana corconaria* each in 0.5 and 1.0 percent concentrations along with four insecticides i.e., phosphamidon 85 SL (0.03%), decamethrin (deltamethrin) 2.8EC (0.002%), chlorpyrifos 20EC (0.05%) and endosulfan 35EC (0.07%) were tested against *D. lini* under field conditions in India. Among the plant extracts, sweet flag (1.0%) received significantly lower (11.70) bud infestation at harvest with a higher yield of 12.30 g/ha followed by neem (1.0%) having 12.75 percent bud infestation and 11.80 g/ha yield.

Reddy *et al*. (1996) found Repelin as an effective botanical insecticide against Ber fruitfly, *Carpomyia vesuviana* Costa. A field trial was laid out in a RBD with three replications for evaluating insecticides and the Repelin against fruitfly. The treatments were Dichlorvos (0.05%), malathion (0.05%), carbaryl (0.15%), phosalone (0.07%), monocrotophos (0.05%), Repelin (2.00%) and a control. The fruitfly infestation after spraying was 13.2, 22.5, 19.7, 16.5, 8.1, 6.2 and 38.1, respectively considering the infestation in control (38.1%).

It is observed that 2 percent Repelin was the most effective followed by 0.05 percent monocrotophos. Repelin (2%) can be recommended during fruiting season as it has the additional advantage of ecological safety.

Krishnamoorthy and Srinivasan (1996) evaluated neem products and standard chemicals for the management of beanfly, *Ophiomyia phaseoli* (Tryon) in french beans. Field experiments were conducted with 6 treatments, namely 2, 4 and 5 percent NSKE; 1 percent neem oil; 1 percent Repelin; 0.07 percent endosulfan and control. Count of the larvae and pupae and the number of stem mines in 10 plants selected at random were taken. The best treatment was 0.07 percent endosulfan followed by 5 percent NSKE in first year trial while in second year, 5 percent NSKE was the best effective treatment. It was found that application of NSKE at the time of appearance of puncture marks in the unifoliate leaf stage and again at petiole mining stages was the most appropriate.

Pawar *et al*. (1996) carreid out field studies in Maharashtra, India, to evaluate the efficacy of NSKE (5%), cypermethrin 25EC (0.01%), endosulfan 35EC (0.07%), dimethoate 30EC (0.03%) dichlorvos 76EC (0.07%), deltamethrin 2.8EC (0.001%), malathion 50EC (0.05%) and phorate 10G (1 kg/ha) to control *Liriomyza trifolii* on tomatoes (cv. Dhanashree). The best pest control was by NSKE, and cypermethrin, endosulfan and dimethoate. The best cost-benefit ratio was found with NSKE (1:14), followed by cypermethrin (1:12), dimethoate and endosulfan (both 1:10).

Omara *et al*. (1997) studied the effects of aqueous neem seed kernel powder (ANSKPE) and NeemAzal-F (5% azadirachtin) on *Liriomyza congesta* and *Aphis craccivora* infesting faba bean in the field in Egypt. Spraying at a high concentration (100 ppm of NeemAzal-F and 4% of ANSKPE) had a slight effect on mean numbers of mines and larvae of *L. congesta* as feeding and oviposition were not completely inhibited, but were effective as an antifeedant. ANSKPE at 4 percent reduced the number of *A. craccivora* by 79.48 and 52.39 percent in 1993 and 1994, respectively, while the corresponding reductions for NeemAzal-F at 100 ppm were 87.07 and 56.56 percent.

Jayakumar and Uthamasamy (1997) tested the bioefficacy of some synthetic insecticides and botanicals against *L. trifolii*. Two seperate experiments were laid, each one for synthetic insecticides and botanicals in RBD with 7 treatments and 3 replications each. The number of alive larvae per leaf were recorded and then spraying was done on the leaves containing larvae. The mortality data was observed 1, 3, 6, 12 and 24 hrs after treatment. The treatments taken were phosalone (0.1%), quinalphos (0.08%), triazophos (0.07%), chlorpyrifos (0.07%), neem oil (0.2 and 0.3%) and control. The mortality count at 24 hrs was 96.6, 86.6, 90.0, 93.3, 73.3, 93.3 and 0.0, respectively. In the second experiment, seven treatments viz. 2 and 3 percent neem oil, 2.5 and 5 percent NSKE, 3 percent Leuppai oil, 3 percent Pongamia oil and control were taken with 3 replications. The data recorded at 24 hrs of the treatment revealed that larval mortality was 56.6, 93.3, 66.6, 90.0, 90.0, 50.0 and 0.0, respectively. It is, therefore, concluded that neem oil at 3 percent and NSKE at 5 percent were very effective in both the trials.

Patanaik (1997) tested 6 neem formulations against serpentine leaf miner *L. trifolii* and tobacco caterpillar *Spodoptera litura* on tomato in Orissa, India. Multineem was the most effecitive, reducing leaf infestation by 82.2 percent. Neemazol was the least effective, with a reduction in infestation of 73.1 percent. However, yields in treated plots were lower than those in untreated ones, suggesting possible adverse effects on fruit setting.

Scholz Dobelin (1997) found that the control of main insect pest in tomatoes (*L. trifolii*) was achieved by *Dacnusa sibirica, Opius pallipes* and *Diglyphus isaea* or the use of a neem seed extract.

Saikia and Dutta (1997) tested insecticides and plant products against *Dacus [Bactrocera] tau*. A 2-year field investigation carried out at Jorhat, Assam, India, revealed that fenvalerate at 0.02 percent with 1 percent molasses was the best of 15 treatments tested against *D. tau* on ridge gourd (*Luffa acutangula*). The plant products Multineem (a preparation from *A. indica*) and *Polygonum hydropiper* leaf extract at different doses were less effective than fenvalerate, cypermethrin and malathion. Multineem, however, proved to be superior to *P. hydropiper* in suppressing attack by the pest.

Shamshad *et al*. (1998) evaluated the bioefficacy of neem products and phosphamidon (0.05%) against linseed budfly *Dasineura lini* Barnes. Out of seven neem-based insecticides namely, Nimbecidine (2%), NSKE (3%), Achook (2%), Neem Gold (2%), Neemol (2%), NeemAzal (2%) and Rakshak (2%). None was superior to phosphamidon but NSKE and Nimbecidine were found better.

Ameta and Gupta (1998) evaluated botanicals against sorghum midge. The efficacy of *Pongamia pinnata* seed and leaf extract (5%), *Azadirachta indica* seed and leaf extract (5%), *Parthenium* whole plant extract (5%), *Ricinus communis* leaf extract (5%) and *Adhatoda vassica* leaf extract (5%) against sorghum midge, *Contarinia sorghicola* Coq. was determined. The results revealed that spraying of 5 percent leaf extract at 90 percent panicle emergence significantly reduced the percentage of chaffy panicles (devoid of grains) and increased the yield over control. The minimum percentage of chaffy panicles (5.26%) was recorded in 5 percent *Parthenium* whole plant extract. The plots treated with *A. indica* leaf extract gave optinium higher grain yield (67.49 q/ha) followed by *Parthenium* whole plant extract (63.80 q/ha).

Singh (1998) tested neem seed kernel extract (NSKE), extracted with different solvents for their IGR and oviposition deterrent effects on melonfly, *Bactrocera cucurbitae* and oriental fruitfly, *B. dorsalis*. The IGR experiments were conducted with three extracts viz. aqueous suspensions of neem seed kernel (NSKS) and deoiled neem seed kernel powder (DNSKPS), and deoiled neem seed kernel powder (DNSKP) dust at 0.3, 0.6, 1.25, 2.5 and 5.0 percent concentrations admixed with sand and kept for pupation. All the three extracts were significant. NSKS and DNSKP at 1.25 percent for *B. cucurbitae* and 2.5 percent for *B. dorsalis* cuased 100 percent pupal mortality and consequently no adult emergence was recorded at these and subsequent higher concentrations. DNSKPS gave similar results at 0.6 percent for *B. cucurbitae* and at 2.5 percent for *B. dorsalis*.

Ragumoorthi and Subba Rao (1998) tested neem formulations against maggots of moringa fruitfly *Gitona distigma* Meigon. Econeem (0.3%), Neem Gold (0.15%) and Rakshak (0.15%) were assayed against *G. distigma* by filter paper method. The effect of neem formulations were evaluated against the soil inhabiting puparia of this pest. Bioassay result showed that all the neem formulations tested were toxic to the matured maggots of moringa fruitfly. LC50 value of all the neem formulations ranged between 0.022917 - 0.034454 ppm of azadirachtin. Among the neem formulations, Nimbecidine 150 ppm recorded the highest puparial mortality (60.22%), lowest, adult emergence (39.78%), fecundity (40.50 eggs/female), hatchability (53.25%) and larval survival (77.25%), when applied as soil drench.

Bharpoda *et al*. (1998) tested bio-efficacy of 10 botanical and chemical insecticides against American serpentine leaf miner, L. trifolii Burgess and fruit borer, Helicoverpa armigera Hubner on tomato (variety Pusa Ruby). Two sprays of methyl-o-demeton (0.025%) and azadirachtin (0.0018%) + endosulfan (0.07%) effectively suppressed *L. trifolii* while one spray itself was found to be efffective against *H. armigera* with significantly higher fruit yield.

Das *et al*. (1998) tested neem derivatives against *Melanagromyza obtusa* and *H. armigera* in pigeonpea. Studies on neem products on pests of medium maturing pigionpea were conducted using Achook (0.5%), Nimbecidine (0.2%), NSKE (5%), Repelin 93EC (1%) with endosulfan (0.07%) and monocrotophos (0.04%) as standards along with an untreated control. The crop was sprayed thrice starting at 50 percent flowering stage and repeated at 10 days interval. The neem products were not as effective as chemical insecticides in controlling the borer infestation. However, they reduced the infestation and a significant increase in yield was recorded than the untreated plots.

Tomar (1998) tested efficacy of some insecticides against okra petiole maggot *Melanagromyza hibisci*. Field experiments were conducted at Bilaspur, Madhya Predesh, India to determine the efficacy of endosulfan, fenvalerate, Multineem, carbaryl, acephate and Dipel (*Bacillus thuringiensis* subsp. *kurstaki*) for the control of *M. hibisci*. Endosulfan gave the largest percentage reduction in leaf infestation, number of larvae, pupae and larvae + pupae, followed by carbaryl and fenvalerate, whereas, Dipel and Multineem were less effective.

Walter (1999) did field and laboratory trials for the Neemix (4.5% azadirachtin) in controlling citrus leaf miner and vegetable leaf miner, *Liriomyza trifolii* and *L. hedobrensis*. Neemix at the rate of 0.5 gal/acre gave 87 percent mortality against *L. trifolii*. Research indicated that Neemix in combination with insecticide increases the efficacy of both the material and possibly slow the development of tolerance to them.

3. Hemiptera, Homoptera, Orthoptera

Rao and Rao (1979) tested the effect of biocides on brown plant hopper (BPH) adult, *Nilaparvata lugens* (Stal) on rice. In tests in India in 1977, potted rice seedlings 40 days old were sprayed to run-off with Thuricide HP (which contains *Bacillus thuringiensis*) at 0.25-1 kg/100 in water, with an aqueous extract of roots or shoots of *Eclipta erecta* [*E. elba*] or with an aqueous extract of leaves of neem (*Azadirachta indica*). When adults of the BPH were caged on the seedlings, all those placed on seedlings treated with *B. thuringiensis* were dead within 24 hrs. Of the remaining treatments, only a hot-water neem-leaf extract gave 100 percent mortality after 48 hrs, but all the root, shoot and leaf extracts tested afforded some mortality.

Abraham and Ambika (1979) tested the effect of leaf and kernel extract of neem on *Dysdercus cingulatus* Fabricius. The results are reported of laboratory experiments carried out in Kerala, India, on the effects of topical applications of extracts of the leaves and seed kernels of neem on 3rd-, 4th- and 5th-instar nymphs of *D. cingulatus*, with particular reference to moulting and vitellogenesis. The kernel extract caused 80 percent mortality and the leaf extract 55 percent, but both extracts caused similar degeneration of the ovarioles of 6th-instar nymphs, with no vitellogenesis, and they also had similar effects on moulting, which resembled those of juvenile hormone. All surviving treated nymphs produced a 6th-instar, but nymphs treated in the 3rd-instar moulted went directly into a nymph that was of the 6th-instar type but smaller than the 6th-instar nymph. The results of this and previous experiments suggest that the leaves and kernels of neem contain a juvenile hormone mimic.

Tondon and Lal (1980) carried out two field experiments in 1975-77 in Lucknow, India, to determine the effectiveness of various insecticides applied in the soil against 1st-instar nymphs of the mango mealy bug, *Drosicha mangiferae* Green. Of the 9 compounds tested, a dust containing methyl parathion at 10 g a.i./tree and an emulsion containing 0.3 percent aldrin at 20 lt/tree were most effective against the pest. HCH (BHC), quinalphos and neem seed cake were ineffective.

Mariappan and Saxena (1983) tested the effect of custard apple (*Annona squamosa* L.) oil and neem oil on survival of *Nephotettix virescens* Dist. and on rice tungro virus (RTV) transmission. Crude neem oil and custard apple oil was obtained by methanolic extractive of decorticated seeds. These oils were tested at 5, 10, 20, 30 and 50 percent concentrations and a control was taken. The experiment was replicated five times using 440 viruliferous insects and 440 seedlings for each replication. It was observed that both the oils reduced insect survival and RTV transmission. The survival decreased progressively with increased concentrations of both oils. After two days at 30 percent concentration in the custard oil treatment, the survival was 1.5 percent of insects and at 50 percent none survived. In case of neem oil, only 2.5 percent survived at both concentrations on 3rd day.

Schauer (1984) obtained toxic effects of neem seed kernel extracts against *Acyrthosiphon pisium* Harris and *Aphis fabae* Scopoli.

Bhanotar and Srivastava (1984) tested effect of neem kernel suspension on the development of eggs of locust, *Schistocerca gregaria*. In laboratory studies in India, pairs of adults of *S. gregaria* in the gregarious phase were released over sterilized sand in tubes measuring 4 x 15 cm that had been treated with 25 ml of an aqueous suspension of neem kernels at 0.125, 0.25, 0.5 or 1.0 percent. Five oothecae were laid in the sand treated at 0.125 percent and 1 in each of the tubes containing sand treated at the higher concentrations. On average, only 22.6 hoppers/ootheca hatched in the tube containing the sand treated at 0.125 percent as compared with an average of 60 hoppers/ootheca in untreatd sand and none hatched in the other treatments. It is suggested that neem oil could be used in the field to reduce the hatch rate of this locust.

Koul (1984) tested the interaction of azadirachtin (prepared from the methanol extract of neem seed kernels (*A. indica*) with the reproductive behaviour of *Dysdercus koenigii* F. was studied in the laboratory in India. Various concentrations of azadirachtin were applied to adults topically or by abdominal injection. Mortality was almost 50 percent by the 7th day after injection of the extract at 1 micron g/insect. The fertility of the surviving insects was reduced considerably, especially in females. Embryogenesis was severely impaired by topical application, preliminary histochemical studies indicated damage to the trophocytes.

Mariappan and Saxena (1984) observed the effect of mixture of custard apple and neem oil on survival of *N. virescens* and on RTV transmission. Methanolic extracts of custard and neem decorticated seeds were obtained. Both these oils were mixed in the ratio of 1:1, 1:2, 1:4 (v/v) and 1 percent teepol was added as detergent. Three concentrations were taken viz. 5, 10 and 20 percent. In the experiment, 1600 insects and 3352 seedlings of rice were taken in the split plot design. The data were taken on the survival of insects and RTV infection. It was observed that the higher the oil concentration, the lower the insect survival. In conntrol, the survival rate varies between 97-100 percent even after 2 to 3 days of feeding. In case of 20 percent of both oils, or oil mixture was < 7 percent after 2 days. However, on 3rd day, the survival was zero. Sprays of mixture of seed oils ofcustard-apple and neem in 1:1, 1:2 and 1:4 proportions at 5, 10 or 20 percent concentrations were significantly more effective in reducing the survival of *N. virescens* on rice seedlings and its transmission of RTV than sprays of the individual oils.

The author opined that these oil mixtures against *N. virescens* should be less prone to problem of development of resistance. The mixtures of these two oils improve the potentiation of insecticidal action which is an additional value in the control of *N. virescens*. It is also observed that low cost, safety for the environment and easily availability to marginal farmers are some of the advantages of the use of these oils.

Deva Kumar *et al.* (1986) evaluated neem limonoids and azadirachtin against safflower aphid, *Dactynotus carthami* H.R.L. Hexane extracted neem oil, pure oil lipid associated limonoids (LM1) and polar limonoids (LM2) were obtained from the fresh neem seeds. Limonoidal residue from expeller grade neem oil and groudnut oil were evaluated in the laboratory. The adult viviparous females which were shining black were used for the test. Ten insects were placed in each petridish and sprayed directly. These insects were airdried and transferred to separate glass tubes. Mortality data were recorded after 24 hrs. It was observed that treatments having limonoids were highly aphicidal. The pure oil at 1 percent gave 40 percent mortality. The LC50 of hexane extracted oil, lipid soluble limonoids (LM) and expeller grade residual oil and azadirachtin-groundnut oil mixture (1:19) was 0.240, 0.114, 0.125 and 0.177, respectively. The deacetyl salannin + groundnut mixture (1:19) gave 40 percent kill at 0.5 percent concentration. The azadirachtin was thus found to be highly toxic to the aphid. The aphicidal action of azadirachtin and deacetyl salannin was reported by the authors for the first time.

Saxena (1986) studied the effect of neem oil and neem cake on green leaf hopper (GLH) and on RTV infection. In field trials in the Philippines, a mixture of custard apple in the ratio of 1:4 and neem seed oils were sprayed at 50 percent for the control of the GLH, *N. virescens* on rice and of RTV on the variety IR42, which is susceptible to the virus. A mixture of neem cake in the ratio of 3:10 (w/w) and urea was also applied at 60, 30 and 30 kg nitrogen/ha at the seedling stage, maximum tillering stage and panicle initiation, respectively. The mixture either alone or with neem cake and urea effectively controlled the vector and the disease.

Pandey *et al.* (1987) carried out a laboratory experiment to determine the effectiveness of extracts of certain plnats viz. *Lantana camara, Azadirachta indica, Cassia fistula, Ageratum convioidea, Ipomoea carnea* and *Thevetia neriifolia* [*T. peruviana*] against mustard aphid, *Lipaphis erysimi* Kalt. Three concentrations (0.5, 1.0 and 1.5%) of each extract was tested. leaves were dipped in desired concentration and then dried under fan and thereafter transferred to another petridish. Twenty healthy aphids were released in to each petridish. The mortality counts were taken after 24, 48 and 72 hrs of the release. It was observed that neem formulations at 0.5 percent gave 53.33, 60.00 and 60.00 percent mortality after 24, 48 and 72 hrs while at 1 percent, it was 63.33, 76.66 and 80.00 percent, respectively. Of these, 1.0 and 1.5 percent were found to be effective. Extracts of *A. indica, L. camara* and *I. carnea* proved effective against the aphid giving (80,66.6 and 66.6% mortality, respectively, at 1.5%.

Venkatetan *et al.* (1987) studied the efficacy of various neem products against *Aphis gossypii* on cotton cultivar MCU5 in India in 1985. The treatments were neem leaf extract (3%), neem kernel extract (3%), neem oil (3%), neem oil with sesame oil (3%) in a 5:1 ratio, neem oil with sesame oil (3%) in a 10:1 ratio and endosulfan (0.07%). The crop was sprayed on 15th day when it was 10 days old and again when the aphid population exceeded 15 percent infestation. The treatments were equally effective in reducing the aphid populations 1-3 days after treatment, but by 7 days after treatment, endosulfan was significantly more effective. Neem leaf extract reduced the aphid population more than the other neem products tested but the difference between the neem treatments were not significant.

Olaifa and Adenuga (1988) studied the effect of neem products in controlling acridid grass hopper, (*Zonocerus variegatus*) damage. The efficacy of 3 neem products (an emulsifiable cocentrate of neem oil, neem kernel water extract and neem leaf water extract) against the acridid on cassava in Nigeria was compared with that of 0.3 percent lindane, 0.006 percent cypermethrin and 0.24 percent monocrotophos. All the neem products protected cassava for varying periods. The efficacy of neem oil and neem kernel water extract was superior to that of the insecticides. The neem products recommended to control *Z. variegatus* on cassava during the dry season are an emulsifiable concentrate of neem oil at 0.5-2 percent to be applied every 8 days; an emulsifiable concentrate of neem oil at 3-4 percent to be applied every 10 days; neem kernel water extract at 7-10 percent to be applied every 12 days; or neem leaf water extract at 50 percent to be applied every 6 days.

Singh *et al*. (1988) evaluated the activity of neem seed kernel extract (NSKE) against mustard aphid, *Lipaphis erysimi*. The toxicity of aqueous, ethanol and hexane extracts obtained from neem (*A. indica*) seed kernels to the aphid was investigated in the laboratory. The hexane extract, which exhibited a much higher activity than the 2 other extracts, had an LC50 of 0.674 percent. When the hexane extract was partitioned with ethanol, the ethanol-soluble fraction had an LC50 of 0.328 percent, whereas the ethanol-insoluble part showed no activity even at 1 percent. Column chromatography of the ethanol-soluble fraction yielded 8 compounds : nimbin, epinimbin, desacetylnimbin, salannin, desacetylsalannin, azadirachtin and 2 unidentified compounds, a salannin derivative and a nonterpenoid. Of these, only 5 could be tested. Nimbin and epinimbin exhibited no toxicity at 0.3 percent, whereas salannin, the salannin derivative and the non-terpenoid gave LC50s of 0.055, 0.096 and 0.104 percent, respectively.

Singh *et al*. (1988) evaluated the efficacy of endosulfan 35EC, dimethoate, phosphamidon 85EC and neem seed powder against *Uroleucon carthami* on safflower in India. Two sprays of endosulfan at 800 ml or dimethoate at 750 ml at fortnightly intervals gave the most effective control.

Phadke *et al*. (1988) tested the efficacy of neem products, Neemark (0.5, 0.4%), water extract of neem (1%) and insecticides, endosulfan (0.1%) and fenvalerate (0.0125%) agaisnt *Bemisia tabaci* and *Heliothis* on cotton. Six sprays of all 5 treatments were given at intervals of 15 days. Neemark (0.4%) was as effective as endosulfan in controlling both pests, while Neemark (0.5%) was more effective than fenvalerate in controlling *B. tabaci*. Neemark was significantly superior to neem water extract against both pests. The yields were greatest in plots treated with Neemark (0.5%), fenvalerate and endosulfan. The yield data give definite conclusions that Neemark (0.5%) is effective in protecting wide variety of insects.

Ramaraju and Babu (1989) tested 9 plant derivatives for their inhibitory effect on nymphs of the BPH (*Nilaparvata lugens*) and the WBPH (*Sogatella furcifera*) on rice. Forty-day-old potted IR20 plants were sprayed with 1 percent oils of neem (*A. india*), pinnai (*Calophyllum inophyllum*), pungam (*Pongamia glabra* [*P. pinnata*]) and mahua (*Madhuca longifolia*), 2 percent neem seed kernel and 2 percent seed extracts of the other plants, and a 5 percent neem cake extract. All the plant derivatives tested were effective in reducing adult emergence of the pests. The greatest reduction in the case of *N. lugens* was obtained with 5 percent neem cake extract, followed by 2 percent pungam seed extract, and in the case of *S. furcifera* the greatest reduction was also obtained with the neem cake extract, followed by the 2 percent neem seed kernel.

Flint and Parks (1989) tested Margosan-O for control of *B. tabaci*, *Frankliniella* spp., *Bucculatrix thurberiella* and *Tetranychus* spp. in small field plots of cotton (*Gossypium hirsutum*) at Phoenix, Arizona. A significant reduction (60%) was observed in the numbers of immature whiteflies on cotton leaves after applications of aqueous sprays containing 160 ppm of azadirachtin (138.7 ml a.i./ha) but lower concentration, including the recommended rate of 20 ppm were ineffective. No phytotoxicity was observed in cotton plants after application.

Prabhakar *et al.* (1989) tested the toxicity of malathion, methyl parathion, monocrotophos, sulprofos, permethrin and neem seed extract to various life stages of insecticide resistant and susceptible populations of the aleyrodid cotton pest *B. tabaci* in the greenhouse. All insecticides were most active against 1st-instar nymphs with the lowest LC50s ranging from 0.001 to 0.014 mg/ml for the synthetic insecticides and 1.44 mg/ml for neem seed extract. Toxicities decreased between successive stages. Adults of the resistant strain were the most tolerant, regardless of the insecticide tested, with LC50s ranging from 2.824 to 25.50 mg/ml. First-instar nymphs were the least resistant to all the insecticides, with resistance increasing to moderate levels in pupae and to high levels in adults. Neem seed extract was the least toxic material to any of the stages; however, no resistance to neem was detected in any stage. The LC50 of neem in 1st-stage larvae, 2nd- and 3rd-instar larvae and pupa was 1.435, 2.360, 6.946 and 8.640, respectively for GH strain while for BR strain, it was 1.440, 1.609, 4.808 and 10.090, respectively. However, the LC50 values in five chemical insecticides were less indicating very high toxicity.

Saxena *et al*. (1989) tested neem seed bitters for the control of GLH (N. virescens) on rice in the field in Philippines during the wet season of 1987. Populations of GLH were significantly reduced on plants treated with neem seed bitters before transplanting, sprayed with 10,000 ppm neem seed bitters at 1- or 2-week intervals or sprayed with monocrotophos at 0.75 kg a.i./ ha. Populations of the predatory mirid *Cyrtorhinus lividipennis* and *Araneae* were lower in plots teated with monocrotophos than in those treated with neem. Tungro virus infection was lower in insecticide- and neem-treated plots than in untreated plots at 45 and 65 days after transplanting.

Naqvi *et al*. (1989) tested the effect of malathion and 3 formulations of neem extracts against whitefly, *Aleurobus [Aleyrobus] barodensis* on brinjal crop in Pakistan. All 4 insecticides were effective in reducing aleyrodid populations after 48 hrs, malathion being the most effective; after 96 hrs, the population reduction was lowest for malathion.

Kumar *et al*. (1989) tested kerosene and neem extracts in the laboratory and in the field in Karnataka against *Planococcus citri* and *P. lilacinus* on coffee. Kerosene was tested alone at 1.5 percent and with 0.025 percent a.i. methyl parathion. The combination of kerosene and methyl parathion was more effective than methyl parathion alone giving 99.96 and 96.7% mortality, respectively, on the 4th day, while kerosene alone was also effective giving 64.10% mortality. Effective control was also obtained with 3, 5 and 7 percent neem oil agianst *P. citri*.

Kumar *et al*. (1989) tested kerosene, alone or in combination with neem extract, in sprays in the laboratory against *Coccus viridis* on coffee. Both 1 percent kerosene and 4 percent neem-kerosene emulsion were considered effective, giving up to about 80-90 percent mortality in 13-19 days.

Jothi *et al*. (1990) tested oils and extracts of neem (*A. indica*), mahua (*Madhuca longifolia*) and pongamia (*Pongamia pinnata*) against *Toxoptera citricidus* on lime in the field in Karnataka, India. On the basis of the effectiveness and cost of the treatments, mahua and pongamia oils at 1 percent and neem and pongamia seed extracts at 2 percent were recommended for control of the pest, and should be appliled at emergence of the new flush, when infestation begins.

Jena *et al*. (1990) tested 12 insecticides for control of *Leptocorisa oratorius* on rice in greenhouse and field experiments in Orissa, India. Etofenprox, monocrotophos and oxydemeton-methyl (each at 0.5 kg a.i./ha) controlled the pest and were effective for the entire 'milky grain' period of the crops. BPMC (fenobucarb), methyl parathion, malathion, quinalphos and chlorpyrifos gave good knockdown control but were not persistent. Phosphamidon, diazinon, neem oil and phosalone did not control the pest.

Gujar and Mehrotra (1990) tested toxicity and morphogenetic effects of neem oil against *Dysdercus koenigi* F. The last-instar nymphs were used in the experiment. Neem oil was tested for its toxicity to the 0-24-hr-old last-instar nymphs. Ten nymphs of uniform size were released in treated petriplates and the experiment was replicated thrice. The mortality of nymphs was observed after 24 hrs. The morphogenetic effects of neem oil were investigated at various concentrations (0.1-1.0%). The morphogenetic effects at moulting and subsequently adult and the nymphal mortality were recorded daily. The results indicated that neem oil caused dose-dependent mortality in the last-instar nymphs. Most of the nymphs were found shrivelled and dehydrated at the time of death after 24 hrs of treatment.

Patil et al. (1990) studied the effect of some indigenous plant materials against *Aphis gossypii* and *Dactynotus carthami*. The cold alcoholic extracts of plant material viz. *Acorus calamus, Azadirachta indica, Butea frondosa, Calotropis procera, Datura alba, Ipomea carnea, Parthenium hysterophorus, Spilanthes acmella* and *Tephrosia purpurea* were taken for the study. Twenty five healthy aphids of equal size were kept in the petridish on which 1 ml of the extracts was sprayed. The petridish was dried under fan and then transferred to another petridish containing food material. Mortality data were taken after 24 hrs. The alcoholic extracts of P. hysterophorus appeared to be comparatively more toxic against both the test insects of *Butea frondosa*. However, the cold alcoholic plant extracts secreened in the present studies, exhibited 23 to 57 times less toxicity when compared with LC50 values of dimethoate as 0.0162 and 0.1930 against *A. gossypii* and *D. carthami*, respectively. The LC50 value of dimethoate was 0.0162 while that of *P. hysterophorus* was 0.2707 and 0.4761 of *A. indica* in case of *A. gossypii*. In case of *D. carthami*, the LC50 of dimethoate for *P. hysterophorus* and *A. indica* were 0.0177, 0.3864 and 0.5095, respectively.

Patil *et al*. (1990) carried out the field trials comparing the efficacy of four plant products viz. neem seed extracts, neem oil, nicotine sulphate and tobacco decant, an animal product viz. fish oil resin soap and a chemical insecticide viz. monocrotophos for three seasons against *Bemisia tabaci*. The authors found that neem products are less effective in controlling the pest.

Nandagopal *et al*. (1990) tested the effects of neem products i.e. neem oil, neem leaf extact (500 g leaves extracted with 500 ml of distilled water-DW) and neem decoction (500 g of leaves boiled in 500 ml of DW), on *Balclutha hortensis* were tested on dried groundnut leaves. Highest mortality (97.5%) of the cicadellid resulted from neem oil at 25 percent. The most cost-effective treatment was neem leaf extract at 5 percent with a resulting mortality of 83.4 percent.

Rao *et al.* (1990) tested a number of insecticides against cotton pests in the field in Andhra Pradesh, India, in 1986-87. They found that triazophos (0.1%), mineral oil (2%) and deltaphos (a mixture of 0.05% triazophos and 0.002% deltamethrin) were the most effective against *B. tabaci*. Natural products such as neem oil gave good results against *B. tabaci*, but their ineffectiveness agaisnt *H. armigera* resulted in lower yields of seed cotton.

Nandihalli *et al.* (1990) observed in the field studies carried out in Karnataka, India, in cotton cv. Laxmi that neem products (Neemguard and Neemark) at 3 ml/lt used in conjunction with monocrotophos (1 ml/lt) were more effective against *B. tabaci* than monocrotophos alone. Thus the combination gives synergistic effect.

Krishnaiah *et al.* (1990) carried out greenhouse studies to compare the efficacy of seed oils of pinnai (*Calophyllum inophyllum*) and mahua (*Madhuca latifolia*) with neem (*A. indica*) on BPH and GLH. Oil of pinnai was similar to neem on orientation and settling of BPH. Pinnai oil was superior to neem and mahua as antifeedant effects against BPH and growth disruptant effects against BPH and GLH. Among ready to use formulaitons of botanicals Repelin (RD-9) spray exhibited lower survival of BPH and GLH nymphs than Neemta 2100 and Neemguard. Efforts to utilize botanicals in combination with insecticides revealed that the mixture of neem oil and monocrotophos showed significantly higher toxicity than monocrotophos alone against BPH but not so against GLH. Combination of Repelin and insecticides viz. monocrotophos and ethofenprox at lower concentration recorded significant reduction in survival of BPH nymphs than insecticide treatments alone while such effect was observed with monocrotophos against GLH.

Nimbalkar *et al.* (1990) studied bioefficacy of some new insecticides including neem seed extract and neem oil against *B. tabaci*. It was found that neem seed extract (5%) and neem oil (0.5%) also registered considerable reduction of nymphal population. Neem oil at 0.5 percent and neem seed extract at 5 percent reduced whitefly population to the extent of 56.36 and 43.70 percent, respectively. Data taken after 48 hrs of treatment revealed that dimethoate (0.1%), monocrotophos (0.1%), phosalone (0.1%), fenvalerate (0.015%) and neem seed extract (5%) were equally effective rendering 91.40, 87.00, 86.90, 86.90 and 85.90 percent reduction in population, respectively. Neem oil (0.5%) also gave 75.20 percent reduction in population. At 7th day neem seed extract (5%) was equally effective with the fenvalerate (0.015%), monocrotophos (0.1%), phorate (1 kg a.i./ha at 10 cm depth), phosalone (0.1%) in reducing adult population of whitefly.

Haq *et al.* (1991) studied the efficacy of botanicals and organophosphate pesticides against plant hopper, *Sogotella furcifera*. The toxicity of nicotine (1250 ml/ha), neem oil (4000 ml/ha), moncocrotophos (1250 ml/ha) and methidathion (2000 ml/ha) to *S. furcifera* was evaluated in rice cv. KS-282 in Islamabad, Pakistan. After 72 hrs, monocrotophos gave the highest mortality (93.15%), followed by methidathion (89.16%), nicotine (61.63%) and neem oil (33.39%).

Rajasri *et al.* (1991) compared the efficacy of certain newer insecticides including neem products against chilli pest complex. Six synthetic insecticides, 4 neem preparations and one chitin inhibotor were evaluated as foliar sprays for the control of pests of chilli (*Capsicum annuum*) in the filed in Andhra Pradesh, India. Triazophos and phosalone (0.1%) caused up to 90 percent mortality of *Scirtothrips doraslis* and *Myzus persicae*, and >78 percent mortality of *Hemitarsonemus* [*Polyphagotarsonemus*] *latus*. Amitraz (0.1%) also gave 78 percent mortality of *P. latus*. Triazophos, Trebon (etofenprox 0.1%) and fenvalerate (0.01%) were effective against the noctuids *H. armigera* and *S. litura*. Neem oil, Neemguard, Repelin and Biosol gave significant control of the pests, but were less effective than the synthetic insecticides.

Facknath (1991) evaluated commercial neem fomulation Margosan-O for the control of *Dysdercus flavidus*. The efficacy of the neem-based insecticides, Margosan-O for control of *D. flavidus* was assessed in the laboratory. Concentration of azadirachtin greater than 150 ppm were toxic to eggs and nymphs; at lesser concentrations the insecticide had a growth regulating effect on both eggs and nymphs.

Price and Schuster (1991) studied the effects of abamectin, neem seed extract (from *A. indica*), pyrethrins + piperonyl butoxide and certain organic pyrethroid, carbamate, organophosphorus and chlorinated hydrocarbon insecticides against different life stages of *Bemisia tabaci* infesting poisettias, and their effects on its natural enemies, parasitoids belonging to the genus *Encarsia*, in the greenhouse. Permethrin applied alone and in combination with piperonyl butoxide, and fenpropathrin, fenvalerate, abamectin and pyrethrins formulated with piperonyl butoxide reduced the numbers of nymphs and adults of *B. tabaci* by the largest amount. Neem seed extract reduced the number of nymphs only, and endosulfan and lindane reduced the number of adults only.

Guddewar *et al.* (1991) compared ether and ethanol extracts of *Azadirachta indica* (neem) and Chandni, *Taborneemontana caronaria,* agaisnt *Dysdercus koenigii*. Ten nymphs of red cotton bug were placed in petridish and topically sprayed with the plant products. These were then dried and kept for recording mortality data. The data revealed that all the extracts were superior to control. The neem extracts i.e. ether and ethanol recorded 13.33-80.00 percent mortality and 13.33-70.00 percent mortality, respectively. Mortality increased with the increased doses. It was also observed that ether extraxt of both chandni and neem gave higher mortality than ethanol extracts. The ovicidal effect of these two products revealed that all the extracts were superior than control. At 1.5 percent concentration, the neem kernel ether extract, neem kernel ethanol extract, chandni ether extract and chandni ethanol extract gave the hatchability reduction viz. 63.33, 63.33, 63.33 and 70.00 percent, respectively.

Beevi *et al.* (1992) tested the efficacy of ten insecticides against eggs of mealy bug, *Maconellicoccus hirsutus* Green on *Morus alba* L. in the laboratory. It was observed that hatching was least in eggs treated with neem oil, followed by monocrotophos, demethon methyl and fish oil rosin soap + dichlorvos.

Wilps *et al.* (1992) studied the effects of neem oil and azadirachtin in the laboratory as well as in field cages on mortality of *Schistocerca gregaria* collected in the Tamesna desert (Niger) by topically applied or injection methods. Enriched neem oil sprayed on flying *S. gregaria* increased the mortality rate up to 75 percent within 10 days. Treatment with neem oil generally decreased flight performance substantially up to 70 percent.

Nasseh *et al.* (1992) investigated in cage experiments in the Tamesna desert, Niger, the effects of the growth inhibitors Alsystin (triflumuron) and Teflubenzuron/Nomolt (teflubenzuron) and extracts of *A. indica* and *Melia vokensii* on immature *S. gregaria*. The growth inhibitors and neem extract caused 100 percent mortality in 8-13 days, while the extract from *M. volkensii* caused 40 percent mortality. All the substances caused serious malformations in the developing adults.

Nandagopal and Soni (1992) determined the residual efficacy of monocrotophos, dimethoate, cypermethrin, phosphamidon and 2 concentrations of neem oil to *Balclutha hortensis* infesting groundnut in field trials conducted in Gujarat, India. Cypermethrin was the most persistent and toxic insecticide over 9 days causig >50 percent mortality, followed by dimethoate and phosphamidon. Neem oil and monocrotophos were the least persistent insecticides and caused >50 percent mortality only up to 24 hrs after application.

Lawand *et al.* (1992) compared Neemark (azadirachtin) 90EC with chemical insecticides namely dimethoate, DDVP, monocrotophos and endosulfan separately or in various combinations against the aleyrodid *Aleurolobus barodensis* on sugarcane in Maharashtra India. The botanical insecticide at 0.5 and 0.7 percent with removal of infested leaves was as effective as treatment with the chemical insecticides in reducing numbers of aleyrodid egg masses and nymphal populations.

Crocker (1992) evaluated neem extract for the control of the lygaeid chinch bug, *Blissus insularis* in St. Augustinegrass lawns (*Stenotaphrum secundatum*) in Texas during 1990. Azadirachtin (720 F at 0.042 and 0.084 kg a.i./ha) was compared with diazinon (5G at 8.5 kg a.i./ha) for the control of lygaeid. Diazinon was effective against adults and all instars of nymphs. None of the azadirachtin treatments were significantly different from the untreated control.

Dhawan *et al.* (1992) compared Navneem 95EC at 250, 500 and 750 ml/ha and Neemark at 1250 ml/ha (both extracts of *A. indica*) with dimethaote at 625 ml/ha and quinalphos at 2000 ml/ha for control of sucking pests and bollworms of cotton (*Gossypium hirsutum* var. F286) in plots. The crop was either sprayed once in July during the vegetative phase using Navneem, Neemark and dimethoate, or at 10-day intervals starting in August i.e. during the reproductive phase using Navneem, Neemark and quinalphos. All treatments significantly reduced jassid (*Amrasca biguttula* [*A. devastans*]) populations during the vegetative phase.

Jackai *et al.* (1992) tested the insecticidal activity of neem, *A. indica*, on major cowpea pests using different concentrations of aqueous extracts, and powders of both neem seed and kernel. The pests were larvae of the pyralid *Maruca testulalis* and the coreid *Clavigralla tomentosicollis*. Neem proved effective in acting as an insecticide and affected the rates of developmennt of both pests at concentrations as low as about 9 percent solution (w/v). Imaginal survival was found to derease with increasing neem concentration. Most *C. tomentosicollis* nymphs died within 9 days of infestaion.

Nimbalkar *et al.* (1993) studied the efficacy of phorate, endosulfan, phosphamidon, dimethoate, methyl demeton, monocrotphos, phosalone, acephate, fenvalerate, neem seed extract and neem oil for the control of *Bemisia tabaci* on cotton in the field in Maharashtra, India. Monocrotophos, fenvalerate and phorate applied at a depth of 10 cm at sowing were the most effective. Neem seed extract and neem oil gave considerable reductions of adult and nymphal populations.

Lowery *et al.* (1993) studied laboratory and field evaluation of neem for the control of aphids. Neem seed oil (NSO) (0.5, 1, 2%), neem seed extract (NSE) (1%) + pyrethrum, NSO (1%) + pyrethrum were tested againt strawberry aphid, *Fimbriaphis fimbriata* Richards and *Chaetosiphon fragaefolii, Myzus persicae, Nasonovia ribisnigri*. Laboratory and field trials with formulated NSO and NSE demonstrated that these materials are effective aphicides. NSO sprays to intact plants in the laboratory resulted in significant reductions in numbers of *M. persicae* on peppers (*Capsicum*) and rutabagas (swedes), *N. ribisnigri* on lettuce and *C. fragaefolii* on strawberry. NSO reduced aphid numbers in a dose-dependent manner, with estimated concentration for a 50 percent reduction in aphid populations ranging from 0.2 to 1.4 percent. Under field conditions in British Columbia, NSO and NSE treatments were as effective as the botanical insecticide pyrethrum (pyrethrins) for the control of aphids on pepper and strawberry, but ineffective for the control of aphids on lettuce. The effectiveness of neem apears to be influenced by host plant, aphid species and weather conditions.

Patel *et al*. (1993) tested Neemark (azadirachtin) against *Aleurolobus barodensis* on sugarcane in the field in Gujarat, India. Neemark (0.3%) had greater ovicidal action than demeton-O methyl (0.05%), fluvalinate (0.005%) and naled (0.1%) but was less effective than 0.1 and 0.05 percent triazophos and 0.005 percent fenpropathrin. Seven days after application, Neemark was more effective than naled, but less effective than triazophos and fenpropathrin.

Bhathal and Singh (1993) evaluated three neem products, AZT-VR-K, Neemguard and Neemark against aphid, *Lipaphis erysimi* by exposing nymphs to treated radish leaf surfaces. It was observed that 1 percent concentration of AZT-VR-K resulted in 98.3 percent mortality of nymphs. AZT-VR-K was also highly toxic to 1st- and 2nd-instar nymphs. Neemguard (2%) caused insignificant mortality and 20 percent of the adults formed were abnormal. Neemark at 0.25-1 percent, failed to cause any significant toxic and developmental effects.

Nasseh *et al*. (1993) studied the effect of neem products against *S. gregaria*. The effects of various neem products on nymphs of *S. gregaria* were tested in semi-field (cage) trials in the southern Tamesna desert (Republic of Niger). The neem products applied were from different sourcess and differed with regard to both formulaiton and concentration of active compounds. all products were tested either on locusts that hatched from eggs of laboratory colonies or on nymphs caught in the field. The products supplied by Trifolio GmbH, Lahnau, caused 100 percent mortality in both test populations when applied at 1 lt/ha, compared with mortality rates of 60-70 percent for the products from Giessen when applied at 10 lt/ha. The laboratory-reared nymphs proved to be more sensitive, i.e., these populations died 4-6 days earlier and exhibited higher rates of morphogenetic defects after application of sublethal dosages.

Naik and lingappa (1993) tested efficacy of some newer insecticides and neem oil against green peach aphid, *M. persicae* on tobacco. The suitabilities of methamidophos (0.04 or 0.05%), triazophos (0.04 or 0.06%), acephate (0.08%), neem oil (0.5%), endosulfan (0.04%) and phosphamidon (0.06%) for control of *M. persicae* were investigaged in a field of tobacco var. Anand 119 at Nipani, Karnataka, India. Methamidophos (0.04 & 0.05%) gave the best immediate control, followed by acephate. Both of these insecticides controlled the aphids for up to 20 days. yield increases of 85.63 percent and 76.6 percent were recorded in plots sprayed with methamidophos and acephate, respectively. However, neem oil had no effect on the pest.

Braman (1993) observed antifeedant and moulting disruption effects in male cricket, *Scapteriscus vicinus* treated once a week for 3 weeks with azadirachtin (0.3%). The highest concentrations of formulated product (7.1 lt/378.5 lt) resulted in 100 percent mortality by the 7th week after initial application, compared withh 15 percent mortality in the untreated control, and 85 and 75 percent mortality in the 4.7 and 2.3 lt/378.5 lt treatments, respectively.

Naqvi *et al.* (1993) studied toxicity of neem extract fractions against *Amritodus atkinsoni* in comparision with dimethoate in the field on mangoes in Pakistan. The LC50 of dimethoate was 0.14 percent, with decreased prophenoloxidase activity up to 52 percent and inhibited cholinesterase activity. The LC50 for RBA was 0.4 percent, with 15 and 19 percent prophenoloxidase and cholinesterase inhibition. The LC50 for URN was 1.4 percnet, with 16 and 70 percent prophenoloxidase and cholinesterase inhibition and activation, resepectively.

Singh and Singh (1993) tested the efficacy of biopesticides on eggs and adults of cowpea bug, *Chauliops fallax*. Laboratory experiments were conducted to investigate the efficacy of cartap hydrochloride (0.05%), derris extract (0.10%), nicotine extract (0.20%), neem oil (0.30 and 0.50%), neem seed water extract (3.0%), neem oil 2EC (0.02 and 0.01%) and monocrotophos (0.036%) in the control of eggs and adults of *C. fallax* on soyabean. Results show that all the insecticides demonstrated significantly greater mortality of eggs compared with the control, in particular cartap hydrochloride and neem oil (0.5%) resulted in 87.5 and 85 percent mortality, respectively. Mortality of adults was also significantly greater on treated leaves. Monocrotophos and cartap hydrochloride gave 100 percent martality at 24 hrs, while neem oil 2EC (0.2%) resulted in 100 percent mortality at 48 hrs after exposure of the test insects to the treated leaves. Nocotine, derris and neem seed water extracts were much less effective for both egg and adult mortality.

Murthy *et al.* (1994) evaluated the effect of a mixture of oils of *Azadirachta indica, Pongamia glabra* [*P. pinnata*], *Bassia latifolia* [*Madhuca longifolia*] at concentrations of 0.25, 0.5, 0.75, 1.0 and 1.25 percent and a new plant product Plexin (a soap oil formulation) for the control of sugarcane whitefly, *Aleurolobus barodensis* on sugarcane cv. CoC671. Toxic effect of this chemical on egg, nymph and adult mortality was recorded with the 1.25 percent mixture.

Sontakke *et al.* (1994) tested effect of neem oil and chemical insecticides alone and their mixture against the plant hopper, *Sogatella furcifera* (Horvath). In field studies conducted in Orissa, India. Applying neem oil and the insecticides (monocrotophos, chlorpyrifos, carbaryl and quinalphos) either alone or in combination reduced numbers of *S. furcifera* compared with untreated plots.

Stark and Rangus (1994) studied the lethal and sublethal effects of the neem insecticide formulation, Margosan-O on pea aphid, *Acyrthosiphon pisum*. Azadirachtin significantly reduced population increase of test insect in a concentration-dependent manner. At a concentration equivalent to 100 mg/lt of azadirachtin, population increase was 3.5 times less than in the control. Azadirachtin also reduced the longevity and fecundity of young adults of *A. pisum* exposed to treated faba beans. Mortality caused by azadirachtin stabilized 7 days after newborn *A. pisum* were exposed to treated faba beans and 10 days for adults. However, azadirachtin was slower-acting when applied topcially. Mortality in adults of *A. pisum* caused by topically applied azadirachtin stabilized 17 days after treatment with a resultant LD50 of 2.91 micron g azadirachtin/g.

Rajgopal *et al.* (1994) tested the efficacy of Limonool, a neem-based formulation against *Aphis craccivora* Koch in field bean. Twenty healthy nymphs were released on each petridish and spraying of solution was done. The treatments were Limonool (1, 2, 3 and 4 ml/lt), endosulfan + Limonool (1 ml + 1 ml/lt), monocrotophos (0.075 ml + 1 ml/lt), methyl parathion + Limonool (0.5 ml + 1 ml/lt) and control. The data revealed that Limonool (1, 2, 3 and 4 ml/lt) gave 73, 83, 86 and 100 percent mortality after 36 hrs. In case of mixture of chemical insecticide + Limonool, cent percent mortality was recorded after 12 hrs. It was further observed that mortality percentage increased gradually with an increase in concentration. The neem-based pesticide Limonool alone and in combination with chemical insecticide at low dosage exhibited a very good effect in the control of the aphid.

Roychoudhry and Jain (1994) did comparative evaluation of neem and bakayan oils with some common insecticides for the control of *B. tabaci*. and virus transmission. Whiteflies were raised on their hosts sunhemp (*Crotolaria juncea* L.) and soybean (*Glycine max* L.). Both neem oil obtained from *A. indica* and bakayan oil obtained from the seed of *Melia azadarach* L. were sprayed on host plants and the whiteflies. Besides, one to two days old eggs or 1st- or 3rd-instar nymphs on the leaves were also sprayed once. The eggs and nymphs which became dark and shrivelled were considered dead for mortality rating. Results indicated that neem oil (2%) has good effect on the eggs and nymphs after 24 hrs with 74.4 and 100 percent mortality, respectively. Whereas at 72 hrs, the egg mortality becomes 100 percent. In case of bakayan oil, the effect on eggs and nymphs were 41.77 and 100 percent mortality, respectively. On comparision with phosphamidon (0.05%), the mortality of eggs and nymphs at 24 hrs were 58.69 and 100 percent, respectively. In case of adult whiteflies in neem and bakayan oils at 2 percent at 24 hrs, the mortality was 82.57 and 100 percent for neem oil and 92.96 and 98.33 percent for bakayan oil, respectively. These sprays also reduced the virus transmission on the tested crops.

Durairaj and Venugopal (1994) compared the efficacy of phosalone (0.2%), Neem Azal-F (0.005, 0.0075, 0.01%), neem oil (3.0%), NSKE (5.0%), Neemark (0.1%) and a control. The results revealed that in phosalone, the adult population was 4.53/3 leaves, followed by neem oil (3%) when the population was 5.6/3 leaves. In Neem Azal-F, the concentrations 0.01, 0.0075 and 0.005 percent recorded 6.73, 6.46 and 6.53/3 leaves adult populaiton.

Tappetzhofen (1995) reported the effect of neem products on *B. tabaci* with different adjuvants. As neem seed is still quite expensive in the Dominican Republic, it was attempted to reduce the dosage by increasig the efficiency of the insecticide using the adjuvant Tersan mark (alcohol-9-ethylene-oxide). The effect of various aqueous neem seed extracts (25, 50 and 60 g seeds/lt, and oilcake at 15 g/lt) and formulated neem oil mixed with the adjuvant was tested in laboratory and field trials against *B. tabaci* on egg-plant. In laboratory trials, the effectiveness of all tested products except neem oil was increased by 24-48 percent.

Stark and Walter (1995) studied the efficacy of neem oil (NO) and neem oil components when mixed with commercial neem insecticides. A series of toxicity studies with *Acyrthosiphon pisum* and the neem extract formulations Margosan-O (MO), MO devoid of NO, Azatin, RH-9999, Azatin with NO (5%), RH-9999 with NO (5%) and NO (5%) were conducted. It was found that addition of NO from MO increased the efficacy of neem insecticides that did not contain the oil, while removal of NO from MO reduced ints efficacy by 62 percent. NO was also extracted with methanol. When methanol-extracted

NO was added to MO devoid of NO, its efficacy was still 30 percent lower than MO. Addition of canola oil gave a similar response. Six limonoids, nimbandiol, deacetylnimbin, 6-acetylnimbandiol, decaetylsalannin, nimbon and salannin, and 2 unidentified chemicals, believed to be limonoids, were identified in NO. The data indicated that NO and other oils increased the efficacy of neem inscticides, but polar components of NO also contributed to increase biological activity of neem insecticides.

Rai *et al.* (1996) conducted a field trial during in Gujarat, India, to test the effectivness of various insecticides against the sugarcane pest *A. barodensis*. One spray each of 10 insecticides was sprayed in a year coinciding with peak activity of the pest. The treatments were evaluated on the basis of incidence of eggs, nymphs and pupae in different treatments and of the insecticide cost/benefit ratio. Based on abundance of the pest, quinalphos (0.03%), phosphamidon (0.3%), acephate (0.05%), monocrotophos (0.05%), methyl-O-demeton (0.05%), endosulfan (0.075%), dimethoate (0.03%), triazophos (0.05%), DDVP (0.03%) and Neemark (0.3%) were comparable and the most effective of the compounds treated. However, based on the cost/benefit ratio, the trends of insecticide in descending order were DDVP (1:44.55), dimethoate (1:34.33), acephate (1:30.18), quinalphos (1:27.70), phosphamidon (1:26.21), endosulfan (1:25.06), triazophos (1:19.43), monocrotophos (1:16.10), Neemark (1:4.63) and methyl-O-demeton (1:4.13).

Raguraman and Rajasekaran (1996) tested the effect of neem products on insect pests of rice and the predatory spider. NO (3%) and NSKE (5%) high volume applications were superior to monocrotophos, neem cake extract (NCE) (10%) and neem coated urea (NCU) treatmnets in suppressing BPH (*Nilaparvata lugens*) populations 3 days after spraying rice in the kharif season. No and NSKE also effectively checked leaf folder, *Cnaphalocrocis medinalis*, up to 14 days after spraying and were on par with monocrotophos (0.04%). Ear head bug, *Leptocorisa* spp. populations remained very low in all the plots treated with neem products. There was a better recolonization of the predatory wolf spider, *Lycosa pseudoannnulata* in neem treatments. The yield level in these treatments was comparable to that of insecticide treatment. Yellow stem borer, *Scirpophaga incertulas* incidence was not reduced by neem products in the rabi season.

Sharma *et al.* (1996) observed that *Azadirachta indica* and *Annona squamosa* (castor apple) seeds contain terpenoids and acetogenins which possess the toxic and antifeedant properties. A trial was laid out on a number of insects. The toxic effect of these materials are effective against the spotted stem borer (*Chilo partellus*) oriental armyworm (*Mythimna separata*), head bugs (*Calocoris angustatus*), and the aphid (*Melanaphis sacchari* Zehnt.).

Neem extract is non-toxic to the parasites and predators of the sorghum midge.

Akbar *et al.* (1996) observed that comparative efficacy of synthetic insecticides (monocrotophos and methryl parathion and neem extracts-neemoil and Repelin) revealed that monocrotophos proved most effective against nymphs and adults of *Sogatella furcifera* at all post treatment intervals and in its overall effect. Maximum overall mortality (63.5 and 66.8%) was reocrded against nymphs of *S. furcifera*, respectively, on rice cvs. Basmati 370 and Basmati 385 by application of neem oil and Repelin.

Sontakke and Dash (1996) conducted field studies in Orissa, India to determine the efficacy of synthetic insecticides and neem formulations for the control of *Lipaphis erysimi* and *Crocidolomia binotalis* infesting mustard, and their nontarget effects on *Apis florea* visiting mustard flowers. All chemicals reduced the incidence of *L. erysimi* and *C. binotalis* compared with the untreated control. However, quinalphos followed by chlorpyrifos and endosulfan, were most effective against both pests and resulted in the highest seed yield. Nimbicide and azadirachtin gave a comparatively better level of control of both pests and a higher seed yield.

Ragumoorthi and Arumugam (1996) studied the field evaluation of insecticides and botanicals and effective time of application for the control of the guava pest, *Helopeltis antonii* in Tamil Nadu, India which showed that malathion (0.1%) followed by monocrotophos (0.04%), NO (2%), pinnai (*Calophyllum inophyllum*) (2%) and pongamia (*P. pinnata*) oil (2%) sprayed twice at the pea stage of fruits followed by another spray 15 days later, effectively reduced numbers of the pest and of corky growths on the fruits.

Dimetry *et al.* (1996) assayed the bioactivity of different formulations of neem seed extracts against *B. tabaci* in semi-field. Tomato plants (cv UC-97) were cultivated in pots and left to become naturally infested with *B. tabaci* in an open field and were sprayed with various concentrations of each extract. The high concentration of all the extracts tested exhibited activity. Also, the different treatments reduced the population density of the adults compared with the (untreated) control. The percentage reduction in the population reached its maximum level 1 hr after treatment with any tested extract.

Dange *et al.* (1996) tested the field efficacy of chemical insecticides and neem insecticides against safflower aphid, *Uroleucon sonchi* during rabi in India. The insecticides included dimethoate (0.03%), phosphamidon (0.02%), monocrotophos (0.04%), Neemark (0.5%), Indiara [diallyl disulfide and allyl propyl disulfide] (1%), Chinmix [beta-cypermethrin] (0.003%), fluvalnate (0.003%), lambda-cyhalothrin (0.003%), quinalphos (0.05%) and carbaryl

(0.1%). The lowest numbers of *U. sonchi* were found in plots treated with monocrotophos (1.26 aphids/2.5 cm shoot), followed by phosphamidon, lambda-cypermethin, quinalphos and dimethoate (1.78, 1.93, 2.04 and 2.32 aphids/2.5 cm shoot, resepctively). The highest grain yield was reocrded in carbaryl-treated plots (12.4 g/plant), followed by phosphamidon-treated plots (11.4 g/plant). However, Neemark has very little effect.

Katole *et al.* (1996) studied the efficacy of some oils and insecticides against citrus blackfly nymphs, *Aleurocanthus woglumi* on Nagpur mendarins, in Akola, India. Monocrotophos (0.025 and 0.05%), Phosalone (0.025 and 0.05%), neem oil, karanj oil and mahua oil (1% each) were tested. Pooled results revealed that monocrotophos (0.05%) was superior to all treatments 3, 7 and 14 days after treatment. Phosalone (0.05%) and moncrotophos (0.025%) were at par and superior to the rest of the treatments, followed by phosalone (0.025%). Neem oil (1%) was as effective as phosalone (0.025%) up to 7 days after treatment and superior to karanj and mahua oils.

Hohn *et al.* (1996) tested two plant extracts against insect pests of temperate fruits in Switzerland in 1994 and 1995. Extract fo *Quassia amara* appliled against larvae of the tenthredinid *Hoplocampa* sp. on peaches was almost completely effective in protecting the fruits. Against the same pest on apples, infestation was below 5 percent, compared with nearly 20 percent for untreated trees. Neem extract was highly effective against *Dysaphis plantaginea* and *Dysaphis* sp. on apples. In a glasshouse trial, extracts of both plants reached about 100 percent effectiveness agaisnt *Aphis pomi* in 10 days.

Debaraj *et al.* (1996) tested chemical insecticides and neem products against cabbage aphid *Brevicoryne brassicae*. In laboratory tests, monocrotophos, quinalphos and malathion were more effective than endosulfan, dichlorvos, carbaryl or the neem-based insecticides and Godrej Achook.

According to Hussain *et al.* (1996), on the basis of a field trial in Karnataka, India, neem oil and pongamia oil (both at 4%) are recommended for the control of *Planococcus citri* on guava; they caused 93.23 and 89.39 percent mortality of the pest, respectively.

Hussain *et al.* (1996) tested various plant oils against *Orthezia insignis* infesting the ornamental crop *Crossandra* sp. in Kranataka, India. Neem and pongamia oils (both at 4%), caused the greatest mortality and are recommended for controlling this pest on *Crossandra*. They were also more effective than quinalphos and monocrotophos.

Bhatnagar (1996) conducted an experiment during the winter season of

1993-94 and 1994-95 to evaluate the efficacy of 6 insecticides and a neem formulation on incidence of *Acyrthosiphon pisum* on 'Rachna' pea (*Pisum sativum*) in Madhya Pradesh, India. Monocrotophos (0.04%) and chlorpyrifos (0.05%) were most effective in controlling the pest causing 35.66 and 30.54% mortality, respectively. However, neem formulation was less effective.

Dash *et al.* (1996) evaluated the efficacy of neem derivatives alone and in combination with monocrotophos and chlorpyrifos against BPH (*Nilaparvata lugens*) and WBPH (*Sogatella furcifera*). The data showed that neem derivatives were very effective at suppressing both species of planthoppers. The maximum protection treatment schedule (MPTS), which comprised application of seedling root dip with chlorpyrifos (0.02%), carbofuran granules (1 kg a.i./ha) at 20 days after transplanting and spraying of monocrotophos (0.5 kg a.i./ha) at 45 and 70 DAT, was most effective against the planthoppers and recorded the greatest grain yield (4120 kg/ha) followed by a neem oil (1.5%) + monocrotophos (0.05%) mixture (3692 kg/ha). Neem oil (NO) in combination with either monocrotophos or chlorpyrifos performed significantly better in controlling BPH + WBPH compared with its application alone at a 3 percent concentration. Among the neem derivatives, NSKE (5%) and Repelin (1%) spray performed best against the planthoppers. Following the application of neem derivatives, there were greater populations of natural enemies, such as spiders and mirid bugs as compared with insecticide-treated plots. The greatest natural enemy populations were recorded with a NO (3%) spray.

Nanda *et al.* (1996) tested neem pesticides in insect pest management of rice. NO (2%) was used in the field proved moderately effective against the stem borers, *Scirpophaga incertulas* Walker and *Chilo suppressalis* Walker; gall midge, *Orseolia oryzae* Wood-Manson; leaffolder, *Cnaphalocrosis medinalis* Guenee; GLH, *Nephotettix virescens* Distant and *N. nigropictus nigropictus* Stal and WBPH, *Sogatella furcifera* Harvarth recording 5.4 to 7.3 percent deadhearts (DH) and 3.0 to 8.0 percent whiteheads (WH); 0.7 to 8.2 percent silvershoot (SS), 3.1 to 7.0 damaged leaves, 4.4 to 18 number of GLH/sweep/hill, respectively compared to 8.1 to 13.4 percent DH and 9.2 to 13.6 percent WH; 3.6 to 17.5 percent SS; 8.1 to 15.7 damaged leaves, 28.9 to 40.6 number of GLH/sweep and 8.6 number of WBPH/hill in untreated check plot. In other two protection schedules of the same experiment when NO and NSE (2%) at 20 and 70 DAT with an intermediate spray of monocrotophos (0.4 kg a.i./kg) at 40 DAT applied to Jaya rice provided better control compared to neem pesticide alone and registered 4.8 to 6.7 percent DH and 2.9 to 3.8 percent WH; 0.6 to 8.0 percent SS, 2.6 to 5.6 percent damaged leaves by leaffolder, 4.8 to 18.3 number of GLH/sweep and 1.8 to 2.6 number of WBPH/hill. However, the treatments of NO or

NSE (2%) alone as 3 sprays or 2 sprays of these two neem pesticides coupled with monocrotophos (0.4 kg a.i./ha) as intermediate spray at 40 DAT provided to a moderately resistant rice variety Lalat, exhibited relatively better control and maintained the magnitude of damage inflicted by the major insect-pests below the economic threshhold (ETL).

Two round foliar application of NO or NSE (2%) at 20 and 70 DAT in combination with monocrotophos spray (0.4 kg a.i./ha) at 40 DAT may be considered as an acceptable technology for IPM in rice.

Bhathal and Singh (1996) studied the toxic, developmental and reproductive effects of neem products against mustard aphid, *Lipaphis erysimi* (Kaltanbach). Two commercial neem products (Neemguard and Neemark) and two neem extracts (NO and AZT-VR-K) were evaluated for their biological activities against mustard aphid. Only 0.1 micron of each dose was topically applied on to the dorsum of 3rd- and 4th-instar nymphs and adults separately. There were three replications with 20 nymphs and 5 adults in each replication. The observations were taken every day for mortality, moulting and abnormalities. Commercial products and neem oil proved highly toxic to 3rd- and 4th-instar nymphs. On the other hand AZT-VR-K was more toxic to 4th-instar nymphs while in 3rd-instar treated nymphs considerable delayed mortality and abnormality in adults formed was observed.

Wilps *et al*. (1996) tested the effects of various neem formulations on mortality rate and morphogenetic defects upon *S. gregaria* larvae. The trials were conducted in southern and western Sahara desert (Republic of Niger). The products tested were enriched NO (with 0.2% AZT), pure NO (with 0.01% AZT), Azal-F (with 50% AZT), pure NO (with 0.04% AZT) and 0.04 percent 3-Tigloyl (azadirachtol + approximately 10% suspended matter). These products were applied with hand sprayers. It was observed that 1 lt/ha dose of Azal-F inflicted 100 percent mortality on both field and laboratory insects. The 100 percent mortality in laboratory insects was established on 9thday while in field insects, it took 13 days. The pure NO causes 100 percent mortality on 11thday in laboratory insects and on 15th day on field-reared insects. The laboratory-reared nymphs proved to be more sensitive i.e. died 4 to 6 days earlier and exhibited higher rates of morphogenetic defects after application of sublethal dosages.

Srivastava *et al*. (1996) evaluated 2 enriched NSKE against mango hoppers, *Idioscopus nitidulus* Walker and *Amritodus atkinsoni* Leth. Two neem seed enriched extracts (an oil-based and a cake-based) were isolated. The oil-based formulation contained a non-polar solvent and polysorbate as emulsifier while cake-based formulation had polar solvent and polysorbate as emulsifier.

Table 1. Effect of extracts on the development of nymphs of mango hopper, *I. nitidulus* (1989)

Treatment	Post-treatment observation days after								
	% Mortality nymphs			% Malformed emergence			% Adult		
	2	5	9	2	5	9	2	5	9
Oil-based extract (1%)	100								
Cake-based extract (0.2%)	46.6	60.0	66.6	0.0	2.2	2.2	6.6	15.5	22.2
Control	3.8	12.2	17.7	0.0	0.0	0.0	7.7	55.6	78.8

Two trials were conducted one in the laboratory and the other in the field. In laboratory, fresh mango panicles along with the nymphs of hopper were sprayed separately with oil-based (1%) and cake-based (0.2%) extracts. Three replications were made for each formulation and each replicate has 30 nymphs. The other laboratory test was conducted for evaluation of effect of the two neem extracts with same doses on settling and ovipositional response of adult of *A. atkinsoni* under choice condition. In the fiield trial, spraying with the above doses of extracts were done on the plants and the panicles were bagged after spraying. Population before and after spray was recorded. The data revealed that the treatment with oil-based concentrate resulted in 100 percent mortality of nymphs of *I. nitidulus* within 48 hrs, while cake-based concentrate caused 46.6 percent mortality in 48 hrs which increased to 66.6 percent on 9th day whereas 17.7 percent mortality was recorded in control (Table 1). Both the enriched extracts at the above concentrations, when tested for settling response of adults of *A. atkinsoni* under choice condition, only 1/4th adults settled on plants treated with extracts in comparison to control (Table 2). In the field trial with oil-based extract revealed 100 percent mortality of nymphs of *I. nitidulus* with cake-based extract, there was not much difference in the population of hoppers on control and treated panicles. Hormonal action of the extracts were also observed as a few hoppers with imperfect moulting were observed in bagged panicles. In a large scale field trial, oil-based extract (2%) and cake-based extract (0.3%) bringing down the population of mango hoppers i.e. nil and 2.7 hoppers per panicle, respectively as compared with 16.50 hoppers per panicle in control (Table 3).

Gopa Kumar *et al.* (1996) evaluated neem derivatives and *Lantana camara* against cardamom pests. Three neem-based insecticides viz. Neemine

Table 2. Effect of neem extracts on the settling and ovipositional response of adults of mango hopper, *Amritodus atkinsoni* (1989)

Treatment	Settling pattern (%)			Ovipositional activity after 24 hrs*
	10	20	40	
Oil-based extract (1%)	10.0	4.0	4.0	0.0
Control	22.0	16.0	16.0	0.0
Cake-based extract (0.2%)	7.5	12.5	13.0	0.0
Control	60.0	45.0	46.0	0.0

* All adults died by this time.

(0.025, 0.038, & 0.05% a.i.), Neem SSFF21 (0.004, 0.006 & 0.008% a.i.) and Margoside CK (0.04, 0.06% a.i.) and quinalphos (0.025%) as standard chemical and an absolute control were imposed in seven rounds per year in two different blocks. There were 12 plants per plot with three replications. Thrip *Sciothrips cardamomi* (Ramk). damage was assessed from harvested capsule. The other expeeriment was done under field where NO (0.05%) was compared with chemical insecticides against whitefly (*Dialeurodes cardamomi* David & Subr.). Data on impact of seven rounds of applications of neem derivatives on thrips damage was assessed. Treatment of quinalphos recorded the lowest thrip damage with all the other treatments more than 50 percent of the capsule recorced thrip damage. Data on impact of the different treat-

Table 3. Evaluation of neem extracts for control of mango hopper, I. nitidulus in the field (1990)

Treatment	Avg pre count population	Avg post count hopper population at indicated days									
		3	5	7	10	12	14	16	19	21	23
Oil-based	10.80	0.00	0.15	0.05	0.00	0.20	0.00	0.00	0.05	0.025	0.00
extract (2%)	(3.427)*	(1.000)	(0.070)	(1.024)	(1.000)	(1.086)	(1.000)	(1.000)	(1.024)	(1.104)	(1.000)
Cake-based	13.95	1.85	3.25	4.50	3.65	4.05	2.70	2.65	1.40	2.10	1.25
(0.3%)	(3.823)	(1.603)	(2.031)	(2.316)	(2.081)	(2.181)	(1.851)	(1.864)	(1.527)	(1.704)	(1.460)
Control	12.90	19.80	27.80	25.10	24.40	22.90	16.50	16.15	10.05	12.95	9.20
	(3.707)	(4.547)	(5.356)	(5.075)	(4.987)	(4.810)	(4.144)	(3.966)	(3.258)	(3.657)	(3.160)
CD at 5%	0.438	0.510	0.884	1.288	1.019	0.949	0.841	0.932	0.966	0.067	
SEM	0.157	0.184	0.318	0.464	0.367	0.342	0.303	0.336	0.348	0.219	

* Figures in parenthesis are transformed mean values

ments on mortality of whitefly nymphs indicated that with NO treatment, the nymphal population was drastically reduced to 12.2 percent which is highhly significant over that of control and some other insecticides. hence, NO (0.5%) is effective against whitefly nymphs.

Hapke *et al.* (1997) tested the efficacy of NeemAzal T/S for the control of aphids, *Dysaphis plantaginea* on sour cherry and *Phorodon humuli* on hops in 1995 under field trials in southwestern Germany. Application of 30 ppm NeemAzal T/S (3 lt/ha) was done thrice, one application in each treatment plot- took place at bud burst, just before and after flowering. Applications before and after flowering were the most promising treatments. On one organic farm, NeemAzal T/S was tested against *P. humuli*. For comparison another plot was treated withh Spruzit, a natural preparation of pyrethrum (pyrethrins). The neem plot was treated twice, and the Spruzit plot was treated only once. After the first treatment, the aphid colonies decreased greatly in both plots. In the Spruzit plot, the colonies were increasing again in the middle of July, while in the neem plot they stayed at a low level after the second treatment.

Vogt *et al.* (1997) carried out the field investigations at Dossenheim, Germany, on the effectiveness of NeemAzal T/S at 0.3 percent against *D. plantaginea* on apple and on its side-effects on *Chrysoperla carnea*. A single application of NeemAzal-T/S in April gave very good control of *D. plantaginea* for about 5-6 weeks. Yield losses caused by *D. plantaginea* were significantly lower in the neem-treated plot than in the untreated control plot. The side-effect test revealed that in the field, NeemAzal-T/S is harmless to larvae of *C. carnea*. The test was based on evaluation of the survival of the larvae, their further development and pupation and the fecundity and fertility of the resulting adults.

Bollhalder *et al.* (1997) observed in a laboratory test that NeemAzal-T/S, known to have a very strong toxic effect on the fertility of aphids such as *Dysaphis plantaginea* and *D. devecta*. They also found that it was effective against larvae and adults of *Myzus persicae*.

Prashad (1997) tested the efficacy of neem products and oxydemeton methyl against *Lipaphis erysimi* on rapeseed crop. In the field experiment, oxydemeton methyl (0.05 and 0.025%), Magocide (CK) (0.2, 0.1, 1.2 and 0.8%) were taken. Data were recorded before and after 1, 2 and 3 days of spraying. Oxydemeton methyl at both concentrations recorded very effective in checking the aphid population. The mean number of aphids at 0.05 percent was found 91-98 percent reduction in population while in 0.025 percent concentration, it was found 74-93 percent reduction. The neem formulations were comparatively less effective. The mean number of aphid/shoot varied from 19-38 giving 39 to 73 percent reduction. This is suggested that efficacy of neem products lasted only upto 3 days after which the population started increasing. In another trial, the efficacy of various treatments followed simillar trend.

Huwer *et al.* (1997) tested the effect of neem on the banana spotting bug, *Amblypelta lutescens* Distant in Australia on papaya, french beans (*Phaseolus vulgaris*) and banana in the laboratory and under ambient conditons. The formulations had no ovicidal effect. High nymphal mortality was observed under laboratory conditions, and was comparable with that given by endosulfan. NeemAzal was the most effective of the products, followed by Green Gold One and powdered neem seeds.

Schulz *et al.* (1997) tested the effects of different NeemAzal formulations (containing 1% azadirachtin-A [AZA-A], 51% plant oils) on apple aphid and *Aphis fabae*. NeemAzal formulations proved to be effective aphicides in laboratory and field trials in south-western Germany. NeemAzal sprays on broad beans (faba beans) and apple trees resulted in significant reductions in numbers of *A. fabae* and *D. plantaginea*, respectively. Results with *A. fabae* showed that an admixture of plant oil improves efficacy. To test the effect of various formulations and application dates, selected formulations were applied to apple trees on 4 fruit farms in different regions during the red bud stage at a concentration of 20-30 g AZA-A/ha. One treatment took place during full bloom. All treatments controlled *D. plantaginea* effectively and there were no significant differences between treatments. NeemAzal-T/S applied at a rate of 50 mg/kg AZA-A on potted apple trees resulted in high mortality of the offspring of *D. plantaginea* after 12 days.

Manger *et al.* (1997) observed in field trials in Netherlands that NeemAzal-T/S at 1.5-3.0 lt/ha had some effect on *Delia brassicae* [*D. radicum*], *Plutella xylostella, Mamestra brassicae, Pieris rapae* and *Brevicoryne brassicae* on Brussels sprouts, but could not compete with esfenvalerate (Sumicidin Super).

Hummel *et al.* (1997) tested effect of NeemAzal-T/S on pea aphid, *Acyrthosiphon pisum* in the laboratory. Laboratory trials showed that the transportation of active ingredient of neem kernels from the upper leaf surface into the leaf is a fast and effective process in order to reach the sucking sites of *A. pisum* for pest control purposes. With 0.5 percent NeemAzal-T/S, the activity lasted for about 2 weeks after treatment .

Kienzle *et al.* (1997) carried out field trials in organic apple orchards in Germany over 2 years (1994-95) which showed that NeemAzal-T/S (30 g AZA-A/ha) to be effective against *D. plantaginea*. It was also effective against

Operophtera brumata.

Dutta *et al*. (1997) tested the comparative bioefficacy of 6 insecticides as a foliar spray against rice hispa, *Dicladispa armigera*, in farmer's fields in Assam, India. Monocrotophos (0.10%) proved to be the most effective insecticide agaisnt the pest. Based on the larval mortality, the effectiveness in descending order was monocrotophos (0.10%), diflubenzuron (0.20%), quinalphos (0.075%), deltamethrin (0.003%), neem oil (1%) and phosphamidon (0.128%).

Henkel *et al*. (1997) did field trials in northern Germany with NeemAzal-T/S and Telmion (a formulated rapeseed oil product) which demonstrated a non-uniform efficiency of these preparations against *D. plantaginea* and *Rhopalosiphum insertum* on apple. During the red bud stage, one application took place with one of the following : NeemAzal-T/S (0.2%) + Telmion (2%); NeemAzal-T/S (0.15%) + Telmion (1.5%); Pirimor (0.05%); or water. The neem treatment was very effective against *D. plantaginea* giving 76-100 percent mortality, it was only 40-58 percent effective against *R. insertum* durig the first week. It did not cause great harm to beneficial insects.

Singh and Srivastava (1997) tested the effect of formulated neem products in controlling mango leaf hopper, *Idioscopus nitidulus* Walker. Two neem seed enriched extracts, an oil-based and a cake-based were isolated. Oil-based formulation contained non-polar solvent and polysorbate as emulsifier while cake-based formulation had a polar solvent and polysorbate as emulsifier. Oil-based (1.5%) and cake-based (0.5%) were taken along with control for such studies and replicated 6 times. The data were subjected to statistical analysis. Both the products gave significantly superior results over control and remained effective upto 14 days. Hence, both the extracts are recommended (Table 4).

In another trial, 4 concentrations of neem kernel an ether extract 2% obtained from IARI*, New Delhi and 3 prepared by CISH, Lucknow viz. 2 & 3% alcoholic, 3% hexane, 2 & 3% acetone extractives were tested against *I. nitidulus*. The data revealed that ehter extractive (3%, IARI product) was best followed by alcoholic extractive (3% CISH product) (Table 5) (Fig. 4.9, 4.10, 4.11).

Hiremath *et al*. (1997) tested methanol extracts from 84 samples of 49 Indian plant species in 30 families for insecticidal activities against *Nilaparvata lugens* using a topical application method. The activity varied between both plant species and tissue sampled. At a dose of 0.5 micron g/female, the following 11 extracts indicated significant insecticidal activity : *Adhatoda vasica* leaves (100% mortality), *Annona squamosa* seeds (100%), *Nerium*

Table 4 : Field efficacy of neem-based extracts on mango hoppers

Treatment/conc.%	Avg. % reduction in hopper population after indicated days				
	1	3	7	10	14
1. Hexane extract (oil-based)	79.56	89.14	70.41	85.28	52.94
1.5% (2 lt/20 lt of water)	(63.83)**	(71.84)	(58.07)	(67.67)	(46.69)
2. Alcoholic extract (neem-based)	84.82	89.78	85.86	87.51	90.32
0.5% (2.5 lt/20 lt of water)	(72.16)	(68.91)	(69.91)	(72.26)	(72.26)
3. Control	9.10	41.13	14.64	6.86	41.47
	(14.36)	(39.28)	(17.53)	(9.72)	(39.89)
CD at 5%	(8.08)	(9.61)	(13.62)	(12.79)	(8.15)

indicum [*oleander*] stems (100%), *Clerodendrum inerme* whole plants (90%), *Pongamia pinnata* seeds (95%), *Prosopis chilensis* stems (90%), *Vitex negundo* leaves (90%), *Azadirachta indica* seeds (89%) and stems (85%), *Aegle marmelos* leaves (88%) and *Madhuca indica* [*longifolia*] seed oil (88%). However, the extract of *A. squamosa* seeds alone exhibited significant activity (100% mortality) even at 0.25 micron g/female. As naturally occurring insecticides, these tropical plant-derived materials could be useful as an alternative for synthetic insecticides in management of field populations of *N. lugens*.

Kulat *et al*. (1997) conducted field trials in Nagpur, Maharashtra, India, during the kharif seasons of 1994-96, to determine the efficacy of 6 plant extracts and 2 insecticides for the control of *Aphis gossypii* and *Amrasca devastans* on okras. Aqueous leaf extracts of tobacco (2%), *Ipomoea carnea* (5%) and a seed extract of *A. indica* and *Pongamia glabra* (both at 5%) gave a smilar level of control compared to endosulfan (0.06%) and monocrotophos (0.05%).

Mourier (1997) investigated the effect of neem kernel water extract (NKWE) in different concentrations on *Phenacoccus manihoti* in the laboratory at 23°C and in a greenhouse on cassava. A choice-test showed that neem-treated cassava leaves were less attractive to 1st-instar nymphs than untreated leaves. Pseudococcids that started feeding on treated leaves died in the 2nd-instar. A greenhouse experiment showed that three applications of NKWE at weekly intervals protected cassava against early-instar nymphs.

Table 5 : Field evaluation of neem products against mango hopper, *Idioscupus nitidulus*

Treatment/conc.%	Avg. mean	Avg. % reduction in hopper population after indicated days			
Ether extract	10.87	95.50	98.67	98.03	100.00
(IARI product) 2.0%	(19.21)**	(76.52)	(84.58)	(83.59)	(90.00)
Alcoholic extract	10.07	45.60	89.60	82.17	95.03
(CISH product) 3.0%	(18.28)	(41.46)	(71.78)	(69.26)	(79.44)
Hexane extract	24.07	70.20	90.90	91.20	90.77
(CISH product) 3.0%	(29.25)	(57.49)	(73.25)	(73.04)	(75.46)
Alcoholic extract	11.67	34.43	89.00	74.97	84.33
(CISH product) 2.0%	(19.77)	(35.42)	(71.42)	(60.06)	(70.95)
Acetone extract	9.93	51.03	90.70	89.27	100.00
(CISH product) 3.0%	(18.28)	(45.73)	(72.60)	(71.74)	(90.00)
Acetone extract	8.47	8.33	76.63	75.37	83.53
(CISH product) 2.0%	(16.91)	(10.00)	(62.26)	(60.61)	(67.70)
Control —	16.87	32.80	28.37	61.77	69.27
—	(24.24)	(34.93)	(30.55)	(51.83)	(56.38)
CD at 5%	—	(21.55)	(17.14)	(12.51)	(17.95)

** Figures in paranthesis are angular transformed values.

* IARI = Indian Agricultural Research Institute, New Delhi, India.

^ CISH = Central Institute for Sutropical Horticulture, Lucknow, India.

Singh *et al*. (1997) tested ether and ethanol extracts of various parts of *Azadirachta indica* against *Dysdercus koenigii* on cotton under laboratory condition. Different parts of neem tree viz, leaves, epicarp, kernels and whole fruit were dried under shade and ground with the electrical grinder. Ten grams of the powders were extracted with ether and ethanol with the help of Soxhlet apparatus. Ten third instar (1-day-old nymphs) were released in the petridishes having treated seeds. The nymphs were allowed to feed on the treated seeds. Observations on mortality were recorded after 24, 48 and 72 hrs of treatment. Ethanol was the most suitable solvent for extracting more of the toxic component from leaves and epicarp, whereas ether was the most suitable for extracting the toxic component from fruits and epicarp. The toxicity of extracts was time and concentration dependent, and was greatest for ether extract of seed kernel. In ether extract (1.5%), the mortality was 70 percent with leaves extract, 70.3 percent in epicarp, 80.6 percent in kernel and 63.4 percent in fruit extracts. In case of ethanol extract (1.5%), the mortality was 66.7 percent in leaves extract, 63.2 percent in epicarp, 73.4 percent in kernel and 66.7 percent in fruit extracts.

Gupta *et al*. (1997) conducted laboratory studies to determine the effect of RD-9 Repellin on the growth and development of *D. koenigii*. At high concentrations, RD-9 Repelin prolonged the nymphal period and consequently the number of generations was reduced to almost half that of untreated populations. The emergence of adults was also affected, and those that did emerge did not oviposit eggs. Five concentrations (0.25, 0.5, 1, 2 and 3%) were tested agaisnt test instar. Mortality of test instars in different concentrations were noted after 48 and 72 hrs of treatment. It was observed that on 4th-instar nymphs with above cocentrations, the mortality was 20.8, 37.5, 72.0, 76.0 and 88.0 percent, respectively, while on 5th-instar, it was 43.1, 47.7, 86.3, 90.0 and 90.0 percent, respectively.

Gopal and Senguttuvan (1997) tested the effect of insecticides and neem products on cotton aphid in watermelon field in India during 1992 which showed that malathion gave the largest reduction (96.43%) in numbers of *Aphis gossypii* followed by endosulfan (95.60%), carbaryl (92.93%) and NSKE 10% (91.50%).

Wyss (1997) carried out field studies in organic apple orchards in Switzerland to investigate several strategies for the control of *D. plantaginea*. The product NeemAzal-T/S, also tested in combination with fungicides, gave good control of *D. plantaginea*. The predators *Adalia bipunctata* and *Episyrphus balteatus* were tested in a 4-step screening system for inundative releases during the hatching period in spring. Preliminary results of field cage experiments showed a significant effect on the number of aphids of *A. bipunctata* and *E. balteatus*.

Partridge and Borden (1997) evaluated a NSE enriched with azadirachtin in three experiments for control of *Elatobium abietinum* on potted spruce trees (*Picea sitchensis*) in a greenhouse. In experiment 1, foliar applications to the runoff point of 150 ppm azadirachtin applied three times (biweekly) held populations below 5.3 aphids/6 cm twig samples; populations on trees treated at 75 ppm, or with emulsifier only, reached 32.2 and 26.9 aphids/sample, respectively. With 5 weekly applications (experiment 2), suppression to <14 aphids/sample was achieved with both 150 and 75 ppm treatments;

populations on control trees peaked at 36.8 aphids/sample. The extract at 150 ppm in experiment 3 significantly reduced the number of aphids on trees that initially had high populations. The results indicate that neem has potential for operational use against aphids.

Chitra *et al*. (1997) tested efficacy of plant extracts such as Allitin (garlic extract), Neknool (neem extract), Neemguard, Monocrotophos (0.04%), endosulfan (0.07%) and sweet sop extract (*Annona* sp.), custard apple (*Annona squamosa*), mexican prickly poppy (*Argemone mexicana*) and margosa (*Azadirachta indica*) for the control of aphid, *Aphis gossypii* infesting cotton in field studies in Tirupati, Andhra Pradesh, India. Monocrotophos gave the best level of control followed by endosulfan. All plant extracts gave an overall efficacy of 88.81-90.06 percent.

Verghese (1997) studied the effects of 5 and 2.5 percent NSE and Econeem at 9 ppm on newly hatched and late 1st-instar (4-day-old) nymphs of *Maconellicoccus hirsutus* in the laboratory. After 24 and 48 hrs, mortality of early 1st-instar nymphs was greatest with 5 percent NSE. For late 1st-instar nymphs, mortality after 24 and 48 hrs was greatest with 9 ppm azadirachtin.

Schulz *et al*. (1997) conducted intensive field tests over three years in Europe demonstrated a high efficacy of NeemAzal-T/S (1% AZA-A) for control of *D. plantaginea*. A single application at a rate of 3 lt NeemAzal-T/S per hactare (30 g AZA-A/ha), preferably at the red bud stage of apples, prevented a population development of this aphid. The pre-bloom application also resulted in good control of some other fruit damaging insects. Side effects of NeemAzal-T/S on beneficial arthropods and other nontarget organisms were generally scarce or were tolerable. In semi-field trials with *Chryoperla carnea*, no negative effects of NeemAzal-T/S (3 lt/ha) could be observed, whereas in laboratory trials, development was disrupted. Preliminary results on honeybees indicate no acute toxicity on adults and no long term toxicity on larval development. Predatory mites were also unaffected under field conditions. With this information, it may be assumed that neem formulation NeemAzal-T/S can be used as an environmentally safe insecticide in apple orchards.

Balu *et al*. (1997) found that spraying with monocrotophos (0.075%) and dithane M45 (mancozeb) or NSE on teak trees (*Tectona grandis*) against leaf hopper (Cercopidae) along with association with the fungal leaf pathogen Phomopsis, gave satisfactory control in Kerala and Tamil Nadu, India.

Borad and Bhalani (1997) carried out laboratory and field tests in Gujarat with pigeon peas on the effectiveness of 18 insecticides against *Coccidohystrix insolita*. Quinalphos (0.05%) and methyl parathion (0.05%), dichlorvos (0.04%), triazophos (0.04%) monocrotophos (0.04%) and Neemark (0.5%) were highly effective against 3rd-instar nymphs of *C. insolita* under laboratory conditions, whereas quinalphos, triazophos and monocrotophos were most effective and gave above 98 percent pest mortality within 10 days of application under field condition.

Pasini *et al*. (1997) observed the efficacy of some natural or recently synthesized active ingredients against the flatid *Metcalfa pruinosa* on *Hibiscus syriacus*. The products tested were imidacloprid (Confidor 0.125a.i.); etofenprox (30% Trebon at 30ml/100 lt), malathion (50% Mastro 50EC at 125 g/100 lt), and NO (4%) at 300 ml/100 lt. The results showed that etofenprox had a sufficient and prolonged efficacy, contrasting with the high and brief activity of malathion. Neem oil was not active against the adults, but it is suggested that it could be active against the immature stages.

Krishnaiah *et al*. (1998) observed in greenhouse studies that neem formulation containing 10,000 ppm AZT were translocated when applied as a root dip to rice resultig in the mortality of BPH (*Nilaparvata lugens*) and GLH (*Nephotettix virescens*) feeding on leaves. Downward translocation was also demonstrated by spraying 100-500 ppm AZT on leaves. Mortality was observed in nymphs of WBPH (*S. furcifera*), *N. lugens* and *N. virescens* after confining them to unsprayed portions of rice seedlings.

Laprade and Cerdas (1998) studied management of the spiralling whitefly, *Aleurodicus dispersus* Russell in banana with insecticides and neem. Three insecticides [a mixture of buprofezin (Applaud WP 25) + thiocyclam (Evisect 50 PS), pyrethrins (Pyrenone 66EC), naphthalene oil (Triona LME 80), and one natural oil (neem oil CE 80)]; were tested for the control of *A. dispersus* in Costa Rica. Four applications (150 lt/ha) werre carried out with a backpack mist blower at 8-day interals. The insecticidal effect was evaluated in the eight leaf of eight newly flowered banana plants per treatment. The tank mixture Applaud + Evisect showed the best effect in reducing populations of the pest. However, neem oil was not so effective.

Ram and Verma (1998) conducted field experiments against major insect pests of cowpea viz., jassid, *Empoasca kerri* Purthi and the legume pod borer, *Moruca testulalis* (Geyer) during 1994 and 1995 with a view to evaluating the efficacy of neem products. The neem products tested were azadirachtin 0.03EC (1500 ml and 750 ml/ha). Monocrotophos was used as a check. These were sprayed on the foliage of pants at 20 and 35 days after sowing. Performance of all the neem products was statistically superior over control in minimizing the damage of jassids and pod borer, while among the neem products, performance of azadirachtin 0.03EC at 1500 ml/ha was found most effective with highest cost benfit ratio of 1:1.90, however, it was at par with NO [1500 ml/ha] and Repelin 100EC [1000 ml/ha]. The performance of neem product was slightly inferior in comaprison to monocrotophos 36EC at 1000 ml/ha but looking to the enviromental benefits, neem product can be used as substitute for insecticides,

Pasupathy *et al.* (1998) studied management of tea mosquito bug with varities and botanicals in guava. Guava has attained importance in arid zones as in the wasteland developement programme with special reference to acid, alkali and sodic soil conditions. The market value of this fruit is much affected by the attack of tea mosquto bug, *H antonii*. Among eight botanicals treid to manage the tea mosquito bug in Lucknow 46 variety, NSKE (5%) was significantly effective in reducing the infestation of tea mosquito bug. This was better than chemicals like endosulfan and malathion. Pongamia oil (3%) was the least effective.

Pasupathy *et al.* (1998) obseerved the significant differences among the different botanicals against the mango hopper. All the botanicals could effectively reduce the population of hoppers. NSKE (5%), NO formulation (developed by ADAC&RI, Trichy), karanj (pungam) seed kernel extract (5%) and NO (3%) reduced the hopper populations significantly and they were all on par. All these treatments were as effective as the chemical phosalone 35EC (0.05%). Palmarosa oil (3%), karanji oil (KO at 3%) and tobacco leaf extract (10%) were next to the above effective botanicals. NSKE (5%) recorded the lowest population of 0.4 hopper/inflorescence when compared to the population 23.8 hoppers in untreated check.

Akashe *et al.* (1998) conducted field trials with NSKEs in comparison with dimethoate to evaluate its effectiveness against aphid (*Uroleucon compositar* Theobald) on safflower. Treatments with one spray of NSKE (5%) followed by one spray of dimethoate 30EC (0.05%), two spray of NSKE (5%) and two Sprays of dimethoate (0.05%) were applied. The treatment with NSKE (5%) one week after first aphid infestation followed by dimethoate (0.05%) 15 days after first spray was found as effective as two spays of dimethoate to check the pest.

Singh (1998) studied the bioefficacy of some essential oils against *Lipaphis erysimi* Kalt. The aphicidal action of oils of *Azadirachta indica, Cymbopogan martinii, C. winterianous* and *Mentha spicata* was studied at variouus concentrations (0.5, 1.0 and 1.5%) for different periods of exposure under laboratory as well as field conditions. Significantly higher mortality of *L. erysimi* was recorded by all the essential and neem oils with 1.5 percent concentration after 72 and 96 hrs of application under laboratory and field studies, respectively. The highest mortality (88 and 98%) was observed by neem oil followed by *C. martinii* (73 and 89%), *C. winterianous* (66 and 85%) and *M. spicata* (56 and 78%) under laboratory and field environments, respectively. One of the reasons of higher aphicidal activity of NO as compared to the essential oil was the synergistic effect of limonooids present in neem oil. Therefore, it was concluded that essential and neem oils may be fitted in integrated management of *L. erysimi*.

Singh and Haridayesharya (1998) tested the insecticidal activity of neem and clove oil mixture against mustard aphid, *L. erysimi* infesting rapeseed, *Brassica compestris*. The mixture containing both the oils in ratio of 1:1 triturated with Triton X-100 and suspended in glass distilled water to prepare stock solution of 8 percent concentration. Required concentrations of the test solutions were prepared by its dilution with distilled water and sprayed on the potted plants infested with counted number of 3rd-instar nymphs and adults of aptero viviparous females. A concentration of 1.5 percent was found to be most effective as it caused 100 percent mortality in 24 hrs. Fecundity was also reduced significantly. The lower concentrations (1 and 0.5%) were also found effective but required longer period to cause complete mortality.

Soundararajan *et al*. (1998) evaluated bioefficacy of neem formulations against curry leaf psyllid, *Diaphorina citri* K. Neem derivatives are becoming popular in IPM due to the effectiveness and ecosafety, solvent free neem formulations viz., NO 60EC (C) and NOPO 60EC (C) which are effective against pests of rice, pulses, oilseeds and tea. Studies were carried out to find out the efficacy of these formulations against *C. citri* and the efficacy was compared with Nimbecidine (0.03%, 2 ml/lt) and endosulfan (0.07%). Results revealed that all neem formulations were effective in reducing psyllid populations. Endosulfan reocrded the highest reduction (70.30%) followed by NO (54.0%), Nimbecidine (49.0%) and NOPO (42.0%). Owing to the effectiveness, ecological safety and compatibility with other components, neem formulations can be used in IPM programmes.

Jyani *et al*. (1998) evaluated 9 insecticidal spray schedules against aphid, *Aphis gossypii* G., jassid, *Amrasca biguttula biguttula* I. and shoot and fruit borer, *Earias vittella* Fb. of bhendi during kharif 1994 to 1996. The spray schedules consisting of endosulfan (0.035%) followed by neem seed kernel (5%) suspension and endosulfan (0.035%) followed by Gronim (AZT) (0.4%). Gronim was found to be most effective against suckig pests and fruit borer of bhendi.

Thakur and Singh (1998) evaluated bioefficacy of three neem compounds viz., Achook, Niconeem and Neemark and some recommended insecticides like, chlorpyriphos, monocrotophos and trebon against jassids (A. bigutella) on okra. The results indicated that treating the month-old okra at 10 days interval with trebon (0.015%) followed by monocrotophos (0.04%) was the most effective in reducinng jassid population. All the neem compounds failed to provide effective control of jassids on okra. However, chlorpyriphos (0.05%) and Niconeem (4 lt/ha) were significantly at par with each other in checking jassid population.

Dimetry *et al*. (1998) tested bioactivity of some formulations of neem seed extracts against whitefly, *Bemisia tabaci* Genn. on tomato. Naturally infested tomato plants in an open field were sprayed with various concentrations of each extract. The high concentration of all the extracts tested exhibited activity. The percentage reduction in the population reached its maximum level 1 hr after treatment with any tested extract.

Singh (1998) studied aphicidal action of plant extracts against *L. erysimi*. Under an environmental friendly approach for control of *L. erysimi*, three concentrations (0.5, 1.0 and 1.5%) of the alcoholic extract of different parts of *Allium cepa* (bulb of garlic), *Allium sativa* (clove), *Azadirachta indica* (bark), *A. indica* (leaf), *Capsicum annuum* (fruits), *Curcuma longa* (roots), *Piper longum* (spike) and *Zingiber officinales* (rhizomes) were tested under laboratory as well as field conditions. The highest aphid mortality (80 and 98%) was obtained at 1.5 percent of *A. sativa* after 96 hrs of spray under both the environments. *A. indica* (bark and leaf) and *Z. sativa* were found at par with *A. sativa* in their bioefficacy under both the conditions while *C. longa* (leaf) and *P. longum* were equally effective as *A. sativa* in laboratory and field conditions, respectively. The efficacy of *A. cepa* and *C. annuum* was significantly higher than control. Therefore, it was conluded that these plant extracts may be used as biopesticides (plant origin) for ecofriendly control of *L. erysimi*.

Gupta *et al*. (1998) evaluated the efficacy of imidacloprid and some neem products as seed treatments against the pest complex (*Amrasca devastans, Earias* spp., *Pectinophora gossypiella* and *Helicoverpa armigera*) in New Delhi in 1993-94. Seed treatment with imidacloprid (0.5%) w/w followed by 5 sprays at 15-day intervals with synthetic insecticides (monocrotophos-500 g a.i./ha, deltamethrin-12.5 g a.i./ha, endosulfan-750 g a.i./ha, cypermethrin-60 g a.i./ha, triazophos-600 g a.i./ha) were effective in controlling cotton pests and increasing seed cotton yield. The neem products were not effective, but the build-up of *Bemisia tabaci* population was lower in the neem-treated plots.

Mahapatro and Nanda (1998) tested neem in the management of rice pest complex. Effectiveness of NO, and NSE as foliar sprays with 2 percent concentration at 20, 40 and 70 DAT or 2 sprays with these neem derivatives at 20 and 70 DAT alongwith 1 spray of monocrotophos at 0.4 kg a.i./ha at 40 DAT against the major rice pests like stem borers, leaffolders, whorl maggot and green leafhoppers was moderate. Superiority of protection schedules with the combined application of neem derivatives with monocrotophos over only neem derivatives was not clearly understood. Plots treated with neem derivatives as well as the same neem derivatives with monocrotophos gave grain yields of 25.35 to 27.50 q/ha and 24.16 to 28.88 q/ha, respectively.

Prashad *et al.* (1998) tested the efficacy of neem products against mustard aphid *L. erysimi* during 1996-97 crop season. Two products namely neem oil and Achook applied at much higher doses of 1400 and 2800 lt/ha did not prove effective beyond 2 days when sprayed on Pusa Kalyani variety of brown sarson (*Brassica compestris*) at recommended concentrations of 0.2 percent and 4 ml/lt of water, respectively. These 2 chemicals reduced the aphid population by 13.2 and 36.8 percent at 2 days after spraying. The population in the untreated control plot increased by 21.7 percent. In another trials, it was found that NO (0.5 and 1.0%) reduced the aphid population upto 69 percent upto 2 days and 32.5 percent at 10 days after spraying. Achook applied 1, 2 and 5 ml/lt of water reduced the aphid population upto 2 days, but after 10 days spraying aphid increased in all the treatments.

Tuncer and Aliniazee (1998) tested acute and chronic effects of neem on *Myzocallis coryli*, a major pest of hazelnuts worldwide. Adults and immatures of *M. coryli* were exposed to foliage treated with differrent concentrations of a botanical insecticide, Margosan-O. Both mature and immature stages were highly sensitive to this compound. Acute toxicity causing immediate mortality of young nymphs, particularly at higher rates (50 ppm or more), and chronic effects resultig in reduced offspring production, lengthening of nymphal development time and survival of the subsequent generation. The offspring production of adult aphids fed on treated leaves was reduced in a dosage-dependent manner, nearly 50 percent at 2.5 ppm to over 80 percent at 62.5 ppm. However, no nymphal survival was recorded at 25 ppm or higher doses.

Singh and Gupta (1998) tested the effect of neem soap on the aphidicidal activity of 4 plant-based synthetic compounds, viz., benzaldehyde, benzyl benzoate, emthyl benzoate and methyl salicylate in combination against mustard aphid, *L. erysimi*. These combinations were found to be superior in comparison to chemicals used without neem soap in reducing the aphid population. The bioefficacy of all the chemicals at 0.1 percent concentration when used alongwith 0.25 percent neem soap was almost similar to 0.5 percent concentration of synthetic chemical when used alone. This indicates that neem soap enhances the activity by nearly 5 times and acts as synergist to boost the aphicidal activity of these plant-based synthetics.

Senguttuvan (1998) carried out field trials in Tamil Nadu, India to determine the effectiveness of various botanicals against the cashew pest, *Helopeltis antonii*. NSKE (5%) was found to be equally as effective 0.05% endosulfan, 0.05% monocrotophos and 0.1% carbaryl durig flushing, fruiting and flowering, respectively.

Tabassum *et al.* (1998) dertermined the toxicity of fenpropathrin (Danitol) and neem formulation (RB-a + pipeeronyl butoxide + Triton X 100) on adults of *Alphitobius diaperinus*. in the laboratory at 30±2°C and 60±5% RH. A 117.8 micron g/cm^2 dose of neem formulation caused 70 percent mortality, while a 1.96 micron g/cm^2 dose of fenpropathrin caused 80 percent mortality. The LD50 of fenpropathrin was 0.2749 micron g/cm^2, and that of the neem formulation was 58.92.

Singh *et al.* (1998) conducted 2 field trials to observe the effect of neem derivatives on mango mealy bug, *Drosicha mangiferae* Green. Findings of the first trial showed that spraying of mango plants with cake-based neem derivative (1%) and oil-based neem derivative (2%) could cause 77.13 and 62.50 percent reduction in population of 2nd-instar nymphs, respectively, while in the second trial, 63.49 and 48.67 percent reduction in population of 3rd-instar nymphs was obtained with the spraying of above two derivatives, respectively. In the above treatments, percentage reduction in nymphal population was significantly higher as compared to control but significantly less as compared to monocrotophos (0.04%), the recommended insecticide.

Krishnamoorthy and Rajagopal (1998) tested the effect of insecticides + NO against california red scale, *Aonidiella aurantii* in a citrus orchard in Bangalore, Karnataka, India. The effect on the activity of 2 natural enemies, *Aphytis melinus* and *Chilocorus nigrita*, under field conditions was also assessed. Monocrotophos (0.08%), dimethoate (0.07%), chlorpyrifos (0.1%) and NO (2%) when applied once, gave only initial control of the scale. The population increased after 14 days of treatment. However, when applied twice at 10-day intervals, the scale was kept under control. All treatments were significantly superior to the control. All insecticides except NO were highly toxic to the natural enemies.

Dhaliwal *et al.* (1998) conducted field studies with 2 formulations of neem : Achook and Nimbecidine for the contnrol of insect pests on cabbage. The neem formulations were evaluated at 1, 2 and 4 kg/ha and compared to 0.5 kg a.i./ha of endosulfan used as the control. The mortality of insects was highest with 4 kg/ha of the neem formulations, 7 days after treatment. The mortality of *L. erysimi* in Achook, Nimbecidine and endosulfan was 95.77, 91.24 and 97.24 percent, respectively. Mortality of *Spodoptera litura* was 81.6, 70.66 and 85.66 percent, respectively, whereas with *Pieris brassicae*, it was 80.00, 73.33 and 86.66 percent. Endosulfan was the most effective against all the insect pests, followed by Achook and Nimbecidine. The neem formulations were safer to the parasitoids, *Microplitis similis* and *Apanteles* [*Cotesia*] *glomeratus*, which parasitized *S. litura* and *P. brassicae* larvae, respectively. The feeding efficiency of the coccinellid predator, *Coccinella septempunctata* on *L. erysimi*, treated with neem-based insecticides, was higher than for aphids treated with endosulfan.

Holmes and Hassan (1999) evaluated the contact and systemic action of neem seed extract (NSE) at 60, 120 and 180 ppm against green peach aphid, *Myzus persicae* Sulzer on miniature reoses. Pirimor, a commonly used contact actions aphicide, was also used for comparison. All neem concentrations gave poor to very slight control of *M. persicae* when applied as contact action foliar sprays, with Pirimor providing the greatest contact kill. Neem at 180 ppm, when applied as a soil drench possibly through existing irrigation systems, gave total aphid control within 24 hrs, apparently through systemic action. Aphid parasites and other beneficial insects did not appear to be affected by neem treatments, whereas Pirimor treatments reduced beneficial insects.

Singh and Korpradits Kul (1999) evaluated some plant extracts against aphids which is a key vector of chilly-virus complex. Alcoholic extracts of leaf, neem seed and garlic were used and the extracts were sprayed at a concentration of 5000 ppm on chilly seedlings. As a substitute for synthetic insecticides, the efficacy of neem (*Azadirachta indica*), garlic (*Allium sativum*) and tagak-tagak (*Rhinocanthus nasuta*) were compared with malathion on chilly, against aphids at different time intervals. The effect of malathion was the best 4 days after spraying, when it recorded almost no aphid on the plants. Neem extract (5000 ppm) generally recorded low aphid populations when compared with unsprayed control malathion and other plant extracts, especially at 8 and 12 days after spraying. Neem extract also recorded an average fewer plants infected with virus symptoms when compared with other treatments.

Walter (1999) observed that adjuvants, activators and synergists improve the activity of azadirachtin, a potent insecticide with exceptional environmental characteristics. However, it is also a very expensive agrochemical that will be limited in it's applications due to its high cost. One way to reduce the cost of application of azadirachtins is through the use of adjuvants, activators and synergists. The laboratory trials were conducted against whitefly, *Trialeurodes vaporariorum*. The tomato plants were sprayed with solution of Neemix (0.5 gal/100 gal) (5 ppm azadirachtin) containing various adjuvants. The result indicates that most of the adjuvants improved the performance of Neemix but the adjuvants containing vegetable and mineral oils gave the most significant improvement. Neemix + Soydix (0.025%), Neemix + Joint Venture (0.5%) and Neemix + Dynamic (0.25%) being the best treatments gave 92.11, 91.27 and 83.02 percent mortality as against 32.88 percent mortality with Neemix alone.

4. Hymenoptera

Marion *et al.* (1990) evaluated the effect of micro-injection of neem (*A. indica*) seed extract into holes of 7-yr-old paper birch (*Betula papyrifera*) in a nursery In Dansville, New York, USA, for control of the leafminer *Fenusa pusilla* at 1.5 and 3.0 percent concentrations. This treatment caused significant decreases of foliar damage, and neem was found to be as effective as injecticide B (Bidrin) [dicrotophos] and Meta-Systox-R [oxydemeton-methyl].

Lyons *et al.* (1996) did laboratory bioassays using systemically-treated red pine (*Pinus resinosa*) branches and potted white pine (*P. strobus* L.) trees indicated that neem seed extracts containing azadirachtin were efficacious in controlling *Acantholyda erythrocephala*. Significant larval mortality on these branches occurred in a dose-dependent manner at concentrations greater than or equal to 32 micron g a.i./200 ml of water. Population reduction reached 100 percent at 4000 micron g a.i./200 ml of water resulted in significant population reductions.

Helson *et al.* (1996) studied neem seed extracts (NSE) against forest pests. NSE containing azadirachtin has great potential for managing several forest insect pests. In laboratory activity spectrum studies, neem was very potent to 12 species of lepidopteran and sawfly larvae and competitive with standard insecticides. Sawfly larvae are particularly susceptible to NSE (e.g. LC50 - 0.04 g azadirachtin/ha, introduced pine sawfly, *Diprion similis* (Hartig).

Field trials were conducted with Neem EC (2% azadirachtin) on the pine false webworm, *Acantholyda erythrocephala* L. infesting red pine, *Pinus resinosa* Ait. At 50 g azadirachtin/ha (ca. 0.02 g/1.5-4 m tree), mistblower (Solo Port 423, 87 5 lt/ha) applications during egg hatch significantly reduced larval numbers and provided very good protection of 1 year old and current year foliage (85-89% and 98-100%, respe.) in both 1994 and 1995. In 1995, lower dosages of 12.5 and 25 g/ha also provided similar foliage protection. An ultra-low volume application of undiluted Neem EC at 2.5 lt/

ha = 50 g/ha with Micron ULVA + sprayer provided 53 and 72 percent protection of old and new foliage, respectively. At 10 lt/ha = 200 g/ha, excellent foliage protection (+90%) was achieved.

Helson *et al.* (1999) tested Azatin EC, a commercial neem formulation containing 3 percent azadirachtin, in the laboratory. The Azatin is sprayed on the leaves and then larvae were placed for feeding. The larvae were observed every after 3 days until death or pupation. The treatment caused 95 percent mortality of pine sawfly at 0.17 percent (*Neodiprion swainei*). Field trials were conducted on the pine false webworm of *A. erythrocephala* and pine sawfly, *Diprion similis*. Applications of both neem formulations on *A. erythrocephala* at 50 g azadirachtin/ha provided over 80 percent protection. In another trial, applications of 12.5 and 25 g azadirachtin/ha provided similar foliage protection to that of 50 g/ha. It was found that an ULV applications at this dose with a micron ULVA+ sprayer was less effective than a mistblower application. The field trial conducted with the application of individual trees at 0.02 g azadirachtin/tree resulted in 89 percent reductions in frass in 12 days after treatment. Larval population reductions on treated trees were 88 percent after 14 days of treatment. In another set of experiment, 3rd-instar larvae when exposed to 1-hr-old treated foliage for 7 days in laboratory on treated foliage, 100 percent mortality occurred and 0.77 g of frass was produced compared to 9 percent mortality and 14.15 g of frass in control.

5. Isoptera

Grace and Yates (1992) studied behavioral effects of a neem insecticides on *Coptotermes formosanus*. The effect of a commercial insecticide formulation (Margosan-O) containing 0.3 percent azadirachtin and 14 percent neem oil on orientation, tunnelling, and feeding behaviour of *C. formosanus* were studied in the laboratory at 29°C and 80 percent RH. In short-term orientation assays, termite workers did not avoid papers treated with a 1000 ppm solution of azadirachtin, although significant mortality occurred in the 100 ppm treatment. However, subsequent observatios indicated that termites avoided long-term contact with the treated sand. In 2 multiple-choice feeding assays, *C. fomosanus* workers fed significantly less on papers containing azadirachtin at _100 ppm.

Panda *et al.* (1998) evaluated nine insecticides against termite on sesamum in Orissa. The result indicated that lindane was the most toxic insecticide followed by neem (*A. indica*).

6. Lepidoptera

Fagoonee (1980) studied the integrated cabbage pest management. In preliminary field trial, neem extract sprays on cabbage gave significant control of both *Crocidolomia binotalis* Zell and *Plutella xylostella*.

Klocke and Kubo (1982) observed that limnoid from neem (*Azadirachta indica*) fruit was a potent toxicant at 10 ppm against *Spodoptera frugiperda* and *Heliothis zea*.

Ho-DT and Kibuka (1983) verified some of the reported insecticidal properties of neem (*A. indica*). Tests were carried out on rice in western Kenya using neem oil, neem cake and urea-neem cake (a blend of urea fertilizer, 46% nitrogen and neemcake i.e. 4:1). Neem cake and urea-neem cake were incorporated into the soil before transplanting and neem oil was sprayed on the rice plants 1, 30 and 55 days after transplanting. Neem oil at high concentrations gave better protection at the early vegetative growth stages than neem cake or urea-neem cake against the main pests (*Maliarpha separatella* Rag., *Sesamia calamistis* Hmps., *Diopsis macrophthalma* Dalm. [*thoracia* Westw.] and *D. apicalis* Dalm.). Plots treated with urea-neem cake had more deadhearts than those treated with neem cake, indicating that nitrogen attenuated the systemic effect of the neem cake. At harvest, the percentages of stems infested by *Maliarpha* was significantly lower in plots treated with neem cake and urea-neem cake than in untreated plots and those treated with neem oil, however, there was no significant difference in the number of whiteheads. The highest grain and straw yields were obtained from plots treated with neem cake and urea-neem cake.

Rao and Srivastava (1984) evaluated neem formulations against sorghum earhead worm. In tests in India on the effectiveness of neem oil to protect sorghum from damage by larvae of *Heliothis armigera*, earheads at the milky stage were immersed for 5 min in a 1 percent pure neem kernel suspension, a 1 percent suspension from a water dispersible powder or a 1 percent emulsion. When larvae 8 or 13 day old were confined on the treated sorghum, larval weight gain was much lower than on untreated sorghum, larval mortality was high and the percentage of larvae that pupated was very low. The pure kernel suspension was significantly more effective than the other formulations.

Singh and Azam (1986) carried out field observations in Andhra Pradesh, India, in 1984-85 on the seasonal occurrence and population dynamics of *Phyllocnistis citrella* on lime and 6 insecticidal treatments were tested against the pest. It remained active throughout the year, but the heaviest infestaion occurred during February and March. Neem cake extract at a rate of 1 kg/10 lt water was the most effective insecticidal treatment, causing 95.26 percent mortality after 3 days, followed by 0.05 percent dimethoate (94.25%) and 0.05 percent metasystox (demeton-s-methyl) (89.09%).

Prashad *et al.* (1986) carried out field trials in Andhra Pradesh, India, to evaluate the relative effectiveness of deltamethrin (0.0015%), fenvalerate (0.01%), dicarbasulf (thiocarb) (0.044%), chlorpyrifos (0.026%) and neem seed kernel suspension (NSKS) in water (2% in the nursery and 0.5% in the planted crop) against *Spodoptera litura* on F.C.V. tobacco both in the nursery and on the transplanted crop. All the insecticides and NSKS gave protection against damage by *S. litura*. In the nursery, deltamethrin gave maximum protection 2 days after spraying, while after 9 days thiodicarb was the most effective compound, followed by deltamethrin and fenvalerate which were equally effective and superior to NSKS and chlorpyrifos. Laboratory studies on the persistence of toxicity showed that thiodicarb was the most persistent, causing 100 percent mortality for up to 13 days, followed by deltamethrin (8 days), fenvalerate (7 days) and chlorpyrifos (3 days).

Parmar *et al.* (1987) compared the physico-chemical characteristics, shelf-life and relative efficacy of 3 neem formulations. A water-dispersible powder (WDP) (25% w/w) and a dust preparation (10% w/w) both based on crushed neem seed kernels and an emulsifiable concentrate (25% w/w) hased on neem seed oil, were evaluated in laboratory bioassays at 28°C and 75 percent RH with 2nd to 5th instar larvae of the arctiid *Spilosoma obliqua*. All 3 formulations controlled the larvae. The WDP and the emulsifiable concentrate (both at 1%) were compared with conventional insecticides in the field against the lycaenid *Euchrysops cnejus* on green gram (*Vigna radiata*) and cowpeas (*V. unguiculata*) in India. The neem formulations gave a level of control which was comparable to that of cypermethrin (0.007%), flucythrinate (0.0075%), fenvalerate (0.01%), quinalphos (0.05%), malathion (0.05%), neem oil-synergized malathion EC (oil 1%, malathion, 0.05%) and endosulfan (0.07%).

Kirsch *et al.* (1987) studied the efficacy of neem extracts in controlling major insect pests in tobacco and cabbage crop. Neem kernel aqueous extract and the enriched formulation extract AZT-VR-K were evaluated under field conditions in the Philippines for control of the noctuid *Heliothis* spp. on tobacco and *Plutella xylostella* on cabbage. The extracts were compared with commercially available insecticides and the insect growth regulator BAY SIR 14591. Applications of both extracts (at 0.15%) in tobacco, at spray intervals of 7 or 10 days, resulted in control of *Heliothis* spp, which was comparable to that achieved by methamidophos (0.09%). On cabbages, sufficient control of small larvae (1st to 2ndinstar) of *P. xylostella* was achieved only by the growth regulator. Neem applications were considered to be more effective than the microbial pesticide.

Fagoonee *et al.* (1987) found in field trials in Mauritius on Chinese cabbages, cabbages and cauliflowers that neem seed kernel extract (NSKE) was to be as effective as deltamethrin (Decis) against *P. xylostella* and *H.* [*Helicoverpa*] *armigera* on tomatoes. The use of the neem extract alternately with the insecticide gave the best protection against the latter pest.

Ram Prashad *et al.* (1987) carried out a field trial in Andhra Pradesh, India, to determine the effectiveness of a 2 percent NSKS, cypermethrin (0.0025%), fenvalerate (0.01%), methamidiphos (0.053%), phenthoate (0.067%), BPMC [fenobucarb] (0.09%), etrimfos (0.05%) and endosulfan (0.047%) against *S. litura* in tobacco nuseries. The number of leaves damaged in each bed was recorded 2 and 9 days after spraying and the percentage of leaves damaged was calculated out by counting the total number of leaves on 10 randomly selected seedlings in each bed. The mean seed kernel suspension was as effective as fenvalerate, cypermethrin or methamidophos. Phenthoate, fenobucarb, etrimfos and endosulfan were less effective.

Sombatsiri *et al.* (1987) compared the toxicity of an aqueous NSKE and an improved extract in bioassays using *S. litura* and *P. xylostella* and in field trials against these species on kale in Thailand. The improved extract was prepared by alcohol and petroleum ether in the extraction process. The LC50 of an aqueous extract against 4th instar larvae of *S. litura* was 7.3 percent while that of improved extract was 0.78 percent. For 2nd and 3rd instar larvae of *P. xylostella*, these values were 0.84 and 8.6 percent, respectively, for the aqueous extract and 0.46 and 4.6 percent for the other extract. When 0.1 percent piperonyl butoxide was added to the 0.67 percent improved extract, the mortality of 2nd instar larvae and pupae of *P. xylostella* rose from 51.1 to 93.3 percent. In field trials, the improved extract at a concentration of 33.3 percent gave the best result in reducing populations of *S. litura* and *P. xylostella* when compared with 1.67 percent improved extract or a 33.3 percent aqueous extract. There was no significant difference in the control achieved by the 33.3 percent improved extract and 0.025 percent cypermethrin.

Agrawal and Mall (1988) carried out laboratory experiments on the insecticidal and antifeedant activity of extracts of neem (fraction C and thionemone) and of seeds of *Calophyllum inophyllum* against 3rd instar larvae of *Diacrisia obliqua* at concentrations of 1, 2.5 and 5 percent by contact and oral application. There was a positive correlation between concentration and mortality. Better results were obtained with the contact method

(33.33-100% mortality) than with the feeding method (2.66-80%). None of the extracts showed antifeedant activity.

Kareem *et al.* (1988) evaluated the efficacy of neem seed bitters at 5000 ppm, 3 percent NSKE and monocrotophos (0.3 kg a.i./ha) for the control of insect pests on mungbean (*Vigna radiata*), in a rice-mungbean cropping system in the Philippines in January-March 1988. Damage by the pyralids *Sylepta* [*Sylepte*] sp., *Lamprosema* sp., *Etiella* sp. and *Mauca* sp. and the noctuid *Heliothis* sp. was significanlty reduced by all 3 treatments. Yield was greatest in plots treated with monocrotophos, but the cost-benefit ratio was greatest for neem seed bitters.

Bhat *et al.* (1988) tested seven insecticides i.e. folidol [parathion] (2%), carbaryl (10%) quinalphos (5%) and toxaphene [camphechlor] (20%) as dusts, and monocrotophos (0.05%), quinalphos (0.05%), phosalone (0.05%) and endosulfan (0.5%) and neem [*A. indica*] extract for the control of the noctuid *H.* [*Helicoverpa*] *armigera*, the pyralid *Maruca testulalis*, the lycaenids *Lampides boeticus* and *Euchrysops cnejus*, and the tortricid *Cydia* [*Leguminivora*] *ptychora* on green gram (*V. radiata*) in the field in India. Treatments with monocrotophos, quinalphos or neem seed extract (5%) were the most effective and resulted in 24.90, 34.99 and 36.66 percent incidence of the pests, 7 days after treatment.

Bhat *et al.* (1988).carried out field experiments in Karnataka, India, to determine the effectiveness of various insecticidal formulations and of a neem seed extract for the control of *Maruca testulalis* and *Cydia ptychora* on cowpea (*Vigna unguiculata*). Pest incidence was lowest in plots treated with monocrotophos, 250 ml/ha (29.97%) and grain yield was highest (5.75 g/ha), followed by neem seed extract at 25 kg/ha (42.34%), phosalone at 250 ml/ha (43.06%) and quinalphos at 250 ml/ha (44.83%), with grain yields of 4.75, 4.70 and 4.14 q/ha, respectively. Dust formulations of folidol (2%), carbaryl (10%), toxaphene (20%) and quinalphos (5%) each at a rate of 20 kg/ha were less effective against the pests.

Rwamputa and Schabel (1988) prepared aqueous extracts of neem (*A. indica*) seed kernel (0.5 and 1.0%) and of neem leaves (1.0%) from powdered kernel or leaf, and used to soak fresh leaves of khaya nyasica (host of *Heteronygmia dissimilis*), which were subsequently tested in the laboratory as feed for *H. dissimilis*. First, 3rd and last instars of *H. dissimilis* were used in the test which was carried out for 3 days, using fresh soaked leaves each day. The 1 percent kernel extracts was the most effective preparation, causing cumulative mortality of 100 percent to 1st instar larvae, and 80 percentt to 3rd and last instar larvae over 3 days. The 0.5 percent kernel extract was less effective, and the leaf extract least effective.

Gahukar (1988) investigated the efficacy of neem cake extract at 1 and 3 kg a.i./ha against insect pests of groundnut and compared with treatments of cypermethrin at 70 g a.i.and dimethoate with deltamethrin at 41 g a.i./ha in Senegal in 1986. The major insect pest was the noctuid *H. armigera*, but a number of coleopterans, notably *Anaemerus fuscus* and *Brachycerus granulatus*, were also observed. The neem extracts and treatment of dimethoate with deltamethrin gave satisfactory control of all the insect pests.

Thakur *et al*. (1988) studied the comparative efficacy of neem seed kernel and some insecticidal formulations against the gram (*Cicer arietinum*) pod borer, *H. armigera*. A field trial was carried out in Madhya Pradeah, India, to determine the effectiveness of a NSKE, a neem leaf extract and some widely used insecticides against the pest. On the basis of grain yield, endosulfan (0.07%) was the most effective treatment, followed by monocrotophos (0.04%) and the neem leaf extract (5%). On the basis of profitability, endosulfan was the most effective, followed by monocrotophos and the NSKE (5%). It is concluded that NSKE can be used in place of the highly toxic synthetic insecticides because of its safety to beneficial insects and its lower cost.

Degaonkar *et al*. (1988) tested five insecticides i.e. karanja oil (0.2%) + soap (1%), NSKE (5%) + soap (1%), neem leaf extract (5%), BHC (10% dust), endosulfan (0.07%) and untreated control against *H. armigera* and *Agromyza* [*Melanagromyza*] *obtusa*. They also inestigated the cost benefits on Arhar in Maharashtra, India. The data revealed that all the treatments were significantly superior over control in reducing the pod damage. The data of economics for the control of Arhar pod borer complex alongwith cost details were also evaluated and it was observed that best cost-benefit ratio was with BHC (HCH 10D), followed by endosulfan (0.07%) and neem leaf extract (5%).

Sardara and Kumar (1989) tested the efficacy of plant oils agaisnt leaf hoppers and shoot and fruit borer on okra. Field trials in Karnataka, India was conducted with okra cultivar Pusa Sawani. Neem oil and karanja oil (each at 0.5, 1, 2%) and garlic oil (at 0.25, 0.50 and 1%) were tested for first year and in the second year, castor oil (at 1%) and monocrotophos (at 0.05%) were added as new treatments. Among various oils evaluated, neem oil (2%) proved to be most effective in reducing the borer damage *Earias vittella*. In the second trial, neem oil (1%) was most effective. Neem oil was as effective as monocrotophos (0.05%) and can, therefore, be recommended for use in integrated control management of the okra pests.

Mikolajczak *et al*. (1989) tested the effect of Meliaceous seed extracts on growth survival of *Spodoptera frugiperda* (J.E. Smith). Hexane and ethanol extracts of seeds from 10 species of the family Meliaceae were incorporated into a synthetic diet at 10000, 2000, 400, 80 or 16 ppm and fed to neonate larvae of the noctuid S. frugiperda in no-choice tests in the laboratory at 26-28°C, 55-65% RH and continuous darkness. All the extracts produced significant mortality, reduced the larval growth rate and increased the time to pupation. The 2 highest doses of all the ethanol extracts caused 100 percent mortality before pupation, but the hexane extracts were less effective.

Ivbijaro and Balaji (1990) conducted field trials in Ibadan, Nigeria, to determine the effects of cypermethrin + dimethoate, the oil extract from ripe seeds of *Piper guineense* and the aqueous extract of ripe seeds of neem (*A. indica*) on the insect pests and yield of cowpea (*V. unguiculata*). The insecticides were applied at 30, 42 and 53 days after planting. Cypermethrin + dimethoate was consistently superior to all other treatments in substantially reducing the population densities of the thrips *Megalurothrips sjostedti* and the pyralid *Maruca testulalis*, which are serious pests on the flowers and buds, and in reducing the densities of 4 pod-sucking hemipterous insects (the coreids *Clavigralla shadabi* and *Riptortus dentipes*, and the pentatomids *Aspavia armigera* and *Nezara viridula*). Pest population densities were also reduced by foliar sprays with neem or *P. guineense* extracts. Seed yield after treatment with cypermethrin + dimethoate was 0.99 t/ha, more than double the yield from the other treatments. A marginal increase in seed yield compared with the untreated control was achieved by foliar spraying with *A. indica* extract.

Tanzubil and McCaffery (1990) observed that treatment of larvae of *Spodoptera exempta* with azadirachtin and aqueous neem seed extracts produced a range of adverse effects. High doses of up to 10 micron g/larva of azadirachtin resulted in 100 percent larval mortality, but this effect was delayed and prolonged.

Chari *et al.* (1990) studied the bio-efficacy of 9 neem formulations viz. Margosan-O (0.0018%), neemrich-II (0.1%), F6 (0.75%), Repelin (1%), Wellgro (2%), Neemark (0.01%), dichloroethane extract (DCE) (1%), alcoholic extract (1%), neem oil emulsion (NOE) (2%) and NSKS in water (2%) against *S. litura* in tobacco nerseries. These neem formulations differed significantly in protecting tobacco from the *S. litura* both in laboratory and nursery. In laboratory tests, Repelin, DCE and NSKS gave better protection than all other neem formulations. Margosan-O and neem rich-II were the next best treatments. Mortality of larvae was also observed and it was 66.7 and 33.3 percent for Margosan-O, 33.3 and 66.7 percent for Repelin in 1988 and 1989, respectively. In nursery tests, Repelin, DCE and NSKS gave significant protection to tobacco seedlings at 9 days after spraying. It is concluded that apart from NSKS (2%) in water, Repelin (1%) and DCE (1%) were most effective in protecting tobacco seedlings from *S. litura* damage followed by Margosan-O (0.0018%), neemrich-II (0.1%), alcoholic extract (1%) and Wellgro (2%).

Rao and Rao (1990) tested the efficacy of neem products Repelin, Neemark and Neemguard. Maximum protection check (with 0.1% quinalphos at floral initiation stage), acephate (0.1% at 50% flowering stage) and deltamethrin (0.002% at pod formation stage) and untreated check against *H. armigera*. The data revealed that Repelin was found effective when sprayed thrice at 1.5 percent concentration at floral initiation, peak flowering and pod maturity stages of pegionpea crop. The reduction of the pest in terms of eggs (1.3/5 twigs), larvae (5.3 larvae.5 plants) and bored pods (7.9%) with the mximum protection check comprising of three insecticides applied at 10 days intervals were significantly at par. The treatment Neemark and Neemguard were also at par with the maximum protection check, in case of untreated check.

Nimbalkar *et al.* (1990) tested the comparative efficacy of neem products i.e. neem seed extracts (5, 4, 3, 2, 1%), neem oil (0.5%), karanj seed extract (5, 4, 3, 2, 1%), Indiana (1%), endosulfan (0.06%) and untreated control against cotton bollworm. The data revealed that on the counts of cumulative percent infestation of bollworm in green bolls, none of the plant extracts showed insecticidal or antifeedant properties. In cut open method, none of the plant extracts exhibited marked differences in reducing bollworm infestation. In locule damage basis, all the treatments except Indiana (1%) recorded a reduction in the infestation. The results thus obtained will still be inadequate to prove the insecticidal nature of these plant extracts. The data on the basis of bad kapas revealed that 0.06 percent endosulfan and 4 percent neem seed extract registered lowest percentage of bad kapas. However, none of the other plant products has recorded lower percentage of bad kapas than control. Hence, these results also did not favour to establish the insecticidal properties of the plant extracts against cotton bollworm.

Sachan and Lal (1990) evaluated relative performance of NSKE (5%) + soap (1%), neem leaf extract (NLE) (5%) + soap (1%), fenvalerate (0.02%), cypermethrin (0.006%), endosulfan (0.07%), quinalphos (0.05%), endosulfan 4D (25 kg/ha), BHC 10D (25 kg/ha), quinalphos 1.5D (25 kg/ha), methyl parathion 2D (25 kg/ha) and untreated control against *Helicoverpa armigera*. From the data, it is observed that all the treatments including NSKE and NLE were singinficantly superior over control both in the pest control as well as in yield. The grain yield obtained from NSKE-treated plot (1350 kg/ha) was at par with NLE-treated plot (1368 kg/ha) and these were at par with quinalphos, endosulfan, BHC and methyl parathion treatments giving grain yields 1364, 1455, 1348, 1426 kg/ha, respectively. Thus, the neem products have good scope in the management of *H. armigera* on chickpea and can be fitted in the IPM schedule. However in case of pigeonpea, the NLE, neem oil, karanj oil and henge oil did not provide effective control of *H. armigera* as observed in chickpea.

Butani and Mittal (1990) carried out ivestigation on comparative efficacy of botanical insecticide, NSKS (3%) and other insecticides viz., permethrin (0.012%), decamethrin (0.00125%), fenvalerate (0.012%), cypermethrin (0.002%), monocrotophos (0.04%), endosulfan (0.07%), phosalone (0.07%), carbaryl (0.15%), phenthoate (0.05%), DDT (0.2%), quinalphos (0.05%), fenitrothion (0.03%), malathion (0.05%), chlorpyriphos (0.04%), methyl parathion (0.03%) and DDVP (0.05%) against pod borer showed that all the tested insecticides had significantly reduced the larval population over control. The yield was also increased significantly in all the insecticidal treatments over control. However, the insecticides malathionm, DDT were found as good as NSKS (3%) in rduction of larval population.

Singh *et al*. (1990) evaluated neem oil at various concentraions (1, 2, 3 and 4%) mixed with teepol (0.4%) against rice leaf folder (*Cnaphalcrocis medinalis* Guenee) on PR 106 and stem borer (*Scirpophaga incertulas* Walker) on Basmati 370, respectively. The other three treatments were teepol (0.4%), monocrotophos (0.05%) and untreated control. For stem borer, monocrotophos was the best treatment both in terms of reducing stem borer incidence (4.15% dead hearts and 3.57% white heads) and increase in yield (2027 kg/ha). Neem oil (4%) was at par with monocrotphos in respect of stem borer incidence (5.26% dead hearts and 5.99% white heads) but inferior in yield (1792 kg/ha). The untreated control recorded 7.18 percent dead hearts and 13.86 percent white heads with a yield of 1523 kg/ha.

Samuthiravelu and David (1990) carreid out a study to test the effectiveness of neem oil (0.1, 0.2, 0.3, 0.4 and 0.5% alone) and incombination with deltamethrin (0.09%) on cotton under field conditions in Tamil Nadu, India, during 1986-87. Deltamethrin was the most effective treatment for minimizing infestation *Earias vittella* (with a yield loss due to the pest of 121.33 kg/ha). The next best treatments were neem oil (0.4 and 0.5%) or neem oil (0.5%) with deltamethrin (0.09%) (which gave yield losses of 146.1, 143.66 and 124.80 kg/ha, respectively).

Sehgal and Ujagir (1990) studied the effect of synthetic pyrethroids, neem extracts and other insecticides for the control of *H. armigera*. Among the insecticides tested for 2 or more years, cypermethrin (36 and 45 g/ha), deltamethrin (12 g/ha), fenvalerate (120 g/ha), endosulfan (420 g/ha) and monocrotophos (24 g/ha) significantly and consistently reduced pod damage to <22.5 percent, giving grain yields of >1.7 t/ha. Less effective but still significanlty better than the control were quinalphos at 300 g, HCH at 2600 g and chlorpyrifos at 240 g.ha, and NSKE at 5 percent.

Meisner *et al.* (1990) reported the effect of Margosan-O, on the development of larvae of noctuid (*Earias insulana*) and their rate of penetration into cotton bolls in the laboratory by incorporating Margosan-O into semisynthetic diet, spraying cotton leaves and bolls, or standing cotton seedlings or twigs (with one boll/twig) in aqueous dilutions of the compound. After 11 days on Margosan-O-treated diets practically none of the larvae survived at any of the concentrations tested (0.02-0.2%), whereas on untreated diet 62 percent of the larvae survived. Survival rate and average weight of the larvae were much lower on leaves treated with 0.5 or 0.2 percent Margosan-O or exposed to its systemic effect at 0.5 or 0.25 percent than on untreated leaves. larval panetration into Margosan-O-sprayed both at concentration of 0.01-0.5 percent was about 30-45 percent of that into untreated bolls.

Gupta *et al.* (1990) tested bioefficacy and economics of certain insecticides i.e. monocrotophos, endosulfan, fenvalerate, decamethrin (deltamethrin) and cypermethrin and plant oils of neem and karanj against *H. armigera* on chickpea (*Cicer arietinum*) in the field in Madhya Pradesh, India. All treatments significantly reduced the larval population. The highest grain yield was obtained with endosulfan (0.07%), followed by monocrotophos (0.04%).

Samuthirvelu and David (1991) tested bioefficacy of neem oil and endosulfan against the fruit borer, *Earias vittella* on okra. Neem oil (at 0.1, 0.3 and 0.5%) was tested alone and with endosulfan (0.035 and 0.07%) for the control of *E. vttella* in Tamil Nadu, India. Applications of neem oil and endosulfan, alone and together, reduced damage and the maximum yield was obtained with ensosulfan (0.07%).

Salem (1991) evaluated neem seed oil (at 20, 40, 60, 80 and 100 ppm) as tuber protectant with sevin (carbaryl) 10D (at 1.0 kg/ton potatoes) against *Phthorimaea operculella* Zell on potatoes stored covered with a rice hill in a store in Menoufiya, Egypt, in summer 1990. Storage losses after 6 months in potatoes treated with 100 ppm neem oil and carbaryl were 25 and 10 percent, respectively. In laboratory experiments, 100 ppm neem oil and carbaryl caused 78.57 and 88.10 percent mortality of larvae, respectively, and both allowed only 10 percent successful pupation. All adults emerging from pupae of larvae treated with 100 ppm neem oil were deformed.

Jagginavar *et al.* (1991) tested the efficacy of 6 different insecticides in controlling *Maruca testulalis, Cydia* [*Leguminovora*] *ptychora* and *Lampides boeticus* in plots of cowpeas cv. C-152 (in Karnataka, India). Pod damage was lowest and yields greatest with a spray application of monocrotophos (at 0.6 lt/ha), followed by endosulfan spray (1.2 lt/ha), endosulfan dust (20 kg/ha), malathion spray (1.2 lt/ha) and neem oil spray (3 lt/ha), which were on a par with each other.

Rohitha *et al.* (1991) tested the activity of neem (*A. indica*), nutmeg (*Myristica fragrans*) and pine (*Pinus radiata*) oils against the tortricid *Epiphyas postvittana* (light brown apple moth) using 0.01, 0.05, 0.25 and 6.25 percent aqueous emulsions incorporated into artificial diets. All 3 oils were toxic to neonate larvae, with LC50s of 1.17, 1.47 and 0.44 percent for neem, nutmeg and pine oils, respectively.

Rovesti and Desco (1991) tested aqueous NSKEs against eggs and larvae of the apple pest *Leucoptera malilfoliella* in the laboratory, inconsistent results were found for kernels from different origins. An extract from a batch of kernels at 20 g/lt gave a 60 percent reduction of egg hatching rate, while no significant reduction was obtained when using kernels from another batch at 25 g/lt. In both experiments, the growth of hatched larvae was disrupted and no pupation occurred. A concentration as low as 1.25 g/lt of the more effective kernles caused 80 percent larval mortality. Complete kill was obtained when 1st to 3rd instar larvae were treated at 5 or 20 g/lt. Systemic activity was confirmed when the extract was applied to apple plants as a soil drench or sprayed on the lower leaves of the plants.

Jackai and Oyedivan (1991) tested neem (*A. inidca*) oil and different formulations of neem oil (with a hot water extract of *Tetrapleura tetraptera* fruits as an emulsifier) including neem oil emulsifiable concentrate (NOEC), neem oil slurry EC (NOSEC) and NOEC (5%) [all obtained from the seeds of the neem plant] against the pyralid *Maruca testulalis* on cowpea in the laboratory. NOEC at 5, 10 and 20 percent exhibited a high degree of insecticidal activity to 3rd instar larvae. All the treated flowers were protected from larval damage 2 days after treatment as compared with the 100 percent damage on untreated flowers and flowers treated with *T. tetraptera* extract alone. NOSEC and NOEC (5%) also exhibited some insecticidal activity which decreased with a decrease in concentration. NO and NOEC were superior to NOSEC.

Thakar *et al.* (1992) tested the relative efficacy of indigenous plant extract against *H. armigera*. Monocrotophos (0.04%) and aqueous extract (5 and 10%) and ether extract (2%) of neem leaves and kernels, extracts of

Ipomoea and Lantana, other neem preparations and BT (*Bacillus thruringiensis*) were compared in their effectiveness in reducing pod damage to gram (*Vigna*) by *H. armigera*. Aqueous neem extracts (10%) gave comparable damage reduction as in quinalphos (0.04%), and are recommended as an economically viable alternative.

Simwat *et al.* (1992) tested the efficacy of plant insecticides against *H. armigera*. Experiments were conducted in 1990-91, in which control of *H. armigera* on upland cotton (*Gossypium hirsutum*) and chickpea (*C. arietinum*) was tested using Neemark (0.05 and 0.75%), RD-9 Repelin (combined extracts of *A. indica, Pongamia glabra* and *Madhuca indica* at 1 and 2%), with quinalphos (0.2%) and cypermethrin (0.02%) as standards. The crop was sprayed once and then 10 third-instar larvae collected and reared on treated food. On cotton, pest mortality was significantly lower after Neemark and RD-9 Repelin treatments than after the standard treatments. On chickpea, both the botanical insecticides and the cypermethrin treatment were less effective than the quinalphos treatment.

Chandrakar *et al.* (1992) studied toxicity and cost benefit ratio of some insecticides against *Mythimna separata*. Monocrotophos, quinalphos, methyl parathion, phenthoate, endosulfan, cypermethrin, dichlorvos, malathion, chlorpyrifos, phosalone, phosphamidon, fenvalerate, BPMC (fenobucarb) and neem (*A. indica*) oil were evaluted against *M. separata* in fields of rice in Madhya Pradesh, India. Maximum larval mortality (almost 100%) within 24 hrs after treatment was obtained from a cypermethrin (0.05 kg a.i./ha) treated plot resulting in the largest yield (57.75 q/ha). Parathion methyl (0.6 kg a.i./ha) and chlorpyrifos (0.3 kg a.i./ha) were the 2nd and 3rd most effective treatments, respectively. Neem oil has less insecticidal property.

Karelina *et al.* (1992) tested the effectiveness of preparations based on azadirachtin against some pests of the order Lepidotera. The preparations NeemAzal and NeemAzal-S (from neem, *A. indica* seeds), caused 86-100 percent mortality when tested agaisnt early-instar larvae of *Mamestra brassicae, H. armigrera, Pieris rapae* and *Hyphantria cunea*.

Datkhile *et al.* (1992) evaluated 12 treatments (0.02% fluvalinate, 0.05% quinalphos, 0.07% endosulfan, 0.01% fenvalerate, 0.01% fenpropathrin, 0.0075% cypermethrin, 0.1% BHC, 5% neem seed extract, 0.2% Neemguard and NPV at 250 larval equivalents (LE)/ha) for the control of *H. armigera* on chickpea in Rahuri, India. Larval mortality was greatest with cypermethrin (0.01%), followed by cypermethrin (0.0075%) and fenvalerate (0.01%). The greatest yield was recorded with cypermethrin (both at 0.01 and 0.0075%) and fenvalerate (0.01%). The neem seed extract and the Neemguard, however, gave less larval mortality.

Meisner and Nemny (1992) investigated the effect of Margosan-O (Margosine-O) on the development of larvae of *Spodoptera littoralis* by feeding the larvae treated cotton leaves or by topical application. Seedlings of cotton were sprayed in the laboratory with varying concentrations of the insecticide and offered to larvae weighing 30-50 mg for 2 days. The insecticide was active down to a concentrations of 2 x 10-3% a.i. azadirachtin in the spray. Topical application showed that at 0.1 to 4 micron g a.i., azadirachtin caused mortality of larvae up to pupating ranging from zero to approx. 70 percent. The effective dose was 2 micron g/larva.

Dhawan *et al*. (1992) evaluated Neemrich 20EC for the management of insect pests of cotton. Neemrich at 1.25 and 1.5 lt/ha was compared with Neemark at 1.25 lt/ha and quinalphos 25EC at 2.0 lt/ha for the control of cotton bollworms (*Pectinophora gossypiella, Earias vittella, E. insulana* and *Helicoverpa armigera* in plots of upland cotton (*G. hirsutum*) in Punjab, India. Damage to cotton was significantly reduced by all treatments and was significantly less in Neemrich (1.5 lt/ha) and quinalphos treatments.

Chaudhry (1992) tested the effect of botanical pesticides against *Plecoptera reflexa* Guen, a shisham (*Dalbergia sissoo*) defoliator. Aqueous extracts of different parts of 4 plants (*A. indica, Dutura alba*, *Melia azedarach, Calotropis procera*) were tested for control of larvae of *P. reflexa* plantations in Pakistan. In the case of *A. indica* and *M. azedarach*, the seed powder was used for making a 10 percent water extract, and for *D. alba* and *C. procera* leaves, branches and stems were boiled in water to make 15 and 21 percent decoctions, respectively. All the test extracts gave significant mortality with the maximum by *D. alba* (90%), followed by *A. indica* (73.3%), *C. procera* (66.6%) and *M. azedarach* (60%). Control mortality was 13.3 percent.

Anwar *et al*. (1992) studied plant with insecticidal activity against 4 major pests in Pakistan. Acidified alcohol extracts from flowers and leaves of *Aesculus indica* caused up to 75 percent mortality of *Saccharicoccus sacchari*. Leaf extracts from *Schinus molle* and leaf and flower extracts from *A. indica* caused 60-70 percent mortality of larvae of *Scirpophaga incertulas*. *Azadirachta indica* leaf powder fed to H. armigera caused failure of pupation and adult emergence. Significant mortality (60-90%) was observed in adults of *Pyrilla perpusilla* treated with extracts from leaves of *A. indica, Dryopteris ramosa* and *S. molle*.

Venkateswarlu et al. (1992) tested the efficacy of neem oil alone and in combination with certain chemical insecticides on pod borer complex of black gram (*Vigna mungo*). Neem oil, alone and with either of 2 chemical insecticides, was tested in sprays against *S. litura, Maruca testulalis* and *H. armigera* on *V. mungo* in Andhra Pradesh, India. Neem oil (at 1%) in combination with endosulfan (0.07%) was the most effective in reducing infestation and increasing yields.

Ragumoorthi and Arumugam (1992) tested 5 chemical insecticides and 3 plant extracts against 2 pests on *Moringa oleifera* grown as a vegetable crop in Tamil Nadu, India. All the treatments agaisnt *Gitona* sp. caused significant reductions in the percentage of fruits infested and themean number of larvae per fruit, as compared with the untreated control. Treatment agaisnt *Noorda blitealis* took place during the early vagetative stage and flowering, and all treatments caused significant reductions in larval populations up to 21 days after treatment, as compared with the control. The best results against the 2 pests were obtained with 0.04 percent dichlorvos and fenthion and 1 percent neem cake extract and neem oil.

Patel *et al*. (1993) tested extracts of neem, *Annona spuamosa, Momordica charantia, Lantana camara, Pongamia glabra*, *Diospyros cordifolia* and garlic against *Amsacta moorei* in field cages in Gujarat, India. Five percent suspension sprays of neem seed and leaf, custard apple seed and leaf, bitter gourd seed, *Lantana* leaf, kanzi seed, dheki seed and garlic bulb have been found considerably effective against 2nd instar larvae of *A. moorei*. However, neem seed was found signnificantly superior to rest of the treatments by giving 92.22 percent mortality followed by neem leaf extract which gave 72.22 percent larval mortality. The custard seed extract gave 36.67 percent larval mortality. The rest of the extracts gave larval mortality in the range of 46.67 to 54.44 percent.

Shama and Bhatnagar (1993) studied the effect of non-edible oils on *Chilo partellus* in the laboratory. It was observed that the number of eggs deposited by the female were less in number on neem and karanj oil-treated strips as compared with castor and mahua oil. The survival of the larvae in neem treatment at 2 percent after 24, 48 and 72 hrs, was 81.66, 78.33 and 21.66 percent, respectively; at 6 percent, it was 61.66, 30.00 and 16.66 percent; at 8 percent, it was 56.66, 26.66 and 11.66 percent and at 10 percent, it was 55.00, 23.23 and 11.66 percent, respectively. Thus, it was observed that higher the oil concentration, the lower was the larval survival.

Robert *et al*. (1994) studied biological activity of NSKE and synthetic azadirachtin against larvae of *Plutella xylostella*. The activity of two neem extracts, AZT and NeemAzal (containing 30 and 3 mg azadirachtin ml^{-1}, respectively) and synthetic azadirachtin (AZ) against 2nd instar larvae (L2) of *P. xylostella* was examined using-dip bioassays. On Chinese cabbage, AZ was significantly ($P<0.05$) less toxic (3 to 4-fold; LC50 0.54 ug AZ ml^{-1}) than either neem extract against a laboratory strain of *P. xylostella* (FS). Larval mortality on 13 days was found to increase with increasing exposure time of *P. xylostella* (FS) larvae to AZT-treated chinese cabbage, although there was little difference in mortality between 48 and 120 hrs exposure. When AZT, NeemAzal and AZ wer applied at a dose (1 ug AZ ml^{-1}) which gave end-point mortalities between 50 and 90 percent (at day 13). AZT was also shown to have ovicidal activity against *P. xylostella* (Wellcome) at relatively high dose ranges (10-1000 ug AZ ml^{-1}) as well as some contact activity (FS strain) in topical bioassays. AZT showed little or no activity at rates up to 1000 ug AZ ml^{-1}. In medium-volume (MV, 200 lt ha^{-1}) and ULV (c. 1lt ha^{-1}) spray bioassays on Brussels sprout. AZT gave 16-92 percent and 88-100 percent mortality, respectively (Wellcome strain) at rates approximating to 1-20 g AZ ha^{-1}.

Rajgopal *et al.* (1994) tested the efficacy of limonool (neem-based pesticide) against web worm *Crocidolomia binotalis* Zell. infesting cabbagge. In the experiment, 11 treatments were given viz. limonool (1, 2, 3, 4ml/lt water), endosulfan(2ml/lt), mococrotophos (1.5 ml/lt), methyl parathion (1ml/lt), endosulfan + limonool (1ml + 1ml/lt), methyl parathion + limonool (0.5 ml + 1 ml/lt) and untreated control. The percentage mortality in all the treatments was significantly higher than that of untreated check. Cent per cent mortality of larvae was noticed in insecticides alone as well as in combination with limonool treatments after 12 hrs. The percentage mortality also increased with increased concentration and duration. Among the limonool concentrations, the maximum mortality of 100 percent was noticed in limonool (4 ml/lt) followed by 93 percent in limonool (3 ml/lt), 90 percent in limonool (2 ml/lt) and 86 percent in limonool (1 ml/lt) after 72 hrs. However, the lowest (10%) was recorded in untreated (water spray) check.

Sarode and Gabhane (1994) studied the performance of neem seed extract with reduced insecticidal dosage on the infestation of okra fruit borer. NSKE was evaluated along with reduced doses of 3 recommended insecticides (endosulfan, monocrotophos and malathion) against *E. vittella* on okra in Akola, Maharashtra, India. Among 14 treatments, application of ensosulfan (0.06%) was the most effective, followed by NSKE (5%) + 3/4 dose of endosulfan (0.045%), monocrotophos (0.05%), NSKE + 1/2 dose of endosulfan (0.03%), and NSKE + 3/4 dose of monocrotophos (0.037%) which recorded 14.83, 16.09, 18.13 and 18.16 percent fruit infestaion, respectively.

Dhawan and Simwat (1995) evaluated some botanical insecticides against American bollworm and tobacco caterpillar (*Heliothis armigera*) infesting cotton. Three botanical insecticides, prepared from neem (*A. indica*) -RD-9 Repelin (1 and 2%), Neemark (0.5 and 0.75%) and Neemrich 20EC (0.1 and

0.15%) were compared with quinalphos (0.2%) as standard, against young larvae of *H. armigera* in plots of *Gossypium hirsutum*. The insecticides were sprayed onto the plots, and larvae collected 30 minutes later and reared on treated food for 72 hrs. All 3 botanicals were also compared with quinalphos against *Spodoptera litura* in the laboratory. Repelin and Neemrich at the highest concentraions were more effective than Neemark in the crop spraying experiments (mortality 70, 70 and 66.7%, resep.), but not as effective as quinalphos (100%). All the insecticides caused 100 percent mortality of *S. litura*.

Bahtnagar (1995) tested the toxic effect of plant extracts on eggs of *Chilo zonellus*, a pest of maize. Several semi-solid crude plant metabolites were isolated by soxhlet-sequential extraction and bioassayed against 3 day old eggs of *Chilo partellus* under laboratory conditions. Petroleum ether fractions of *Annona squamosa* and *Azadirachta indica* were the most toxic and these completely inhibited hatching. The LC50 values of these extracts were 0.03545 and 0.03588, respectively.

Behera and Satapathy (1996) did screening of 7 indigenous plant extracts for their insecticidal properties against *S. litura*. Two treatments were given to 4th instar larvae by exposing them to castor (*Ricinus communis*) leaves dipped in the extract and by topical application with an atomizer. The concentrations used were 1.25, 2.5, 5.0 and 10.0 percent. NSKE was the most effective, causing 100 percent mortality 10 days after treatment. The other treatments caused 50 percent mortality or less.

Sahayaraj and Sekar (1996) tested the efficacy of plant extracts against tobacco caterpillar larvae in groundnut. The leaves of *Azadirachta indica* (neem), *Vitex negundo*, rind of *Citrus sinensis* (orange) and rhizome of *Zingiber officinale* (ginger), 10 g of each, were macerated in hot distilled water and applied to groundnut leaves to assess their efficay against larvae of *S. litura*. Larval mortality was higest after 96 hrs on leaves treated with extracts of orange (90%), followed by *V. negundo* (83%), *A. indica* (80%) and ginger (70%).

Shukla *et al.* (1996) studied the efficacy of some insecticides and plant products agaisnt *E. vittella* on okra. Field experiments were conducted to determine the efficacy of neem oil, Achook, fenvalerate, cypermethrin, karanj oil, dichlorvos and malathion for control of *E. vittella* infesting okras. Fenvalerate (0.005%) gave the highest level of control and highest level of control and highest healthy fruit yield (70.75 q/ha) and was also highly cost-effective with a cost-benefit ratio of 1:10.3. There was no statistical difference between crop yields following treatment with cypermethrinn (0.005%), malathion (0.05%), Achook (1%) and neem oil (1%) sprays (which produced 69.46, 64.43, 64.37 and 64.04 q/ha healthy fruits, respectively.

Leskovar and Boales (1996) tested azadirachtin in controlling lepidopteran pests and increasing marketability of cabbage. Azadirachtin (ATI) was evaluated for the control of cabbage looper (*Trichoplusia ni*), diamondback moth (*Plutella xylostella*) and silverleaf whitefly (*Bemisia argentifoli*) in cabbage cv. Green Cup grown in SW Texas. In autumn 1992, ATI was tested at 1, 22, 33 and 44 g a.i./ha. In 1993, ATI was evaluated at 33 g/ha and in combinnation with M-Pede (1%, v/v), an organic insecticide. Two commercial (Align and Neemix) and one expeerimental hydrogenated (LDF) ATI formulations were evaluated at 11 g/ha in 1994. The cabbage looper counts were reduced by ATI compared with the nontreated control. Insect mortality was similar for all ATI rates tested in 1992. In 1993, ATI at 33 g/ha + M-Pede reduced the number of cabbage looper and diamondback moth larvae. ATI efficacy against cabbage looper and diamondback moth was enhanced when crop oil polyol fatty acid esters with polyethoxylated derivatives was mixed with Align or LDF formulations in 1994. Using ATI on lepidopterous pests appears to be beneficial for integrated pest management strategies.

Ambethgar (1996) tested the efficacy of different neem products, i.e. neem cake (200 kg/ha basal application), neem cake (200 kg/ha basal application) + NSKE 5 percent (25 and 32 days after transplanting), NSKE 5 percent, neem leaf decoction (0.5 kg/lt of water), and neem oil 3 percent was compared with chlorpyrifos (0.5 kg a.i./ha) and quinalphos (0.4 a.i./ha) against *Cnaphalocrocis medinalis* in rice fields in Pondicherry, India. Chlorpyrifos proved to be the most effective against the pest, the leaf damage being 11.10 and 10.08 percent, after the first and second spraying, respectively. Neem cake + NSKE 5 percent was the next best treatment, followed by quinalphos, NSKE and neem oil. Both, the applications of neem leaf decoction and neem cake, proved least effective in reducing the percentage of damaged leaves due to the insect.

Chari *et el*. (1996) tested bioefficacy of neem pesticides, *Pongamia* oil and chemical insecticides against *S. litura*. Three recent neem formulations, NeemAzal F.I. containing 10000 ppm azadirachtin (AZT) at 50 and 30 ppm AZT, NeemAzal F.II. containing 50000 ppm at 50 and 25ppm AZT, and Fortune Aza containing 11000 ppm at 50, 30 and 15ppm AZT were tested. The above were evaluated in comparison with 2 percent aqueous NSKS and chlorpyrifos (0.05%) for their bio-efficacy against *S. litura* on FCV tobacco in the laboratory and nursery during 1994-95. Results indicated that NeemAzal FI and II at 50 ppm AZT concentration, NSKS (2%) and chlorpyrifos (0.05%) gave significant protection to tobacco seedlings from pest damage, followed by NeemAzal FII (25 ppm AZT) and NeemAzal FI (30 ppm AZT) Fortune Aza at 50 ppm AZT gave equal protection as chlorpyrifos (0.05%) 9 days after spraying.

Evaluation of neem formulations CTRI-I, CTRI-II, CTRI-III and DOR-I from seeds of neem, all at 50 ppm AZT concentrations, and DOR-II from seeds of *Melia azedarach* (1% extract) and 4 *Pongamia* formulations prepared from seeds of *Pongamia pinnata*. It is revealed that all neem formulations were significantly more effective against *S. litura* on tobacco than *Pongamia* formulations. Contrary to earlier findings, neem formulations from seeds of *A. indica* gave better protection than those from seeds of *M. azedarach*.

Breithaupt and Schmutterer (1996) tested NSKE to control Asian corn stem borer, *Ostrinia furnacalis* in Papua New Guinea. Aqueous NSKE from *Azadirachta siamensis* (from Thailand) and *A. indica* (from Papua New Guinea) have been used in field experiments to control *O. furnacalis*. The use of the imported neem kernels from *A. siamensis* showed that only an ANSKE with a high concentration of 50 g NSK-powder/lt of water had an effect on larval populations in the field, while a concentration of only 20 g NSK-powder/lt of water from locally grown neem trees applied between 50 and 60 days after sowing, observed to be sufficient to control later instars of *O. furnacalis* larvae in the field.

Kevin *et al*. (1996) tested efficacy of neem seed extracts for the control of eastern spruce budworm defoliator damage. Foliar and systemic applications of neem seed extract in a white spruce (*Picea glauca* (Moench Voss.), seed extract effectively controlled defoliation by larvae of eastern spruce budworm, *Choristoneura fumiferana* (Clem.) (Lepidoptera : Torticidae). Azatin EC was applied at vegetative budswell and vegetative budflush in 1995 using a backpack compressed air sprayer (75 ppm azadirachtin). Insect observation 4 weeks after treatment (4 branches/tree) were reduced to 82 percent and 81 percent at budswell and budflush, respectively. The incidence of shoot grazing (two 45 cm branches/tree) was reduced by 50 percent at both application times. At 60 ppm azadirachtin, the incidence of shoot damage was reduced to 43 percent (from 82 to 47%) and the proportion of grazed shoots severely damaged was reduced to 72 percent (from 74 to 21%).

Suhas *et al*. (1996) tested the effect of neem seed extract (NSE) on ber fruit borer, *Meridarchis scyrodes* Meyrick. The application of 5 percent NSE was as toxic as fenvalerate, decamethrin (deltamethrin) and cypermethrin in the control of *M. scyrodes*.

Patel and Patel (1996) tested ovicidal property of some indigenous plant extracts against *S. litura*. Leaf extracts of *Ailanthus excelsa* produced the highest egg mortality (48%) and was superior to *Azadirachta indica* extracts.

Kiarie and Saxena (1996) evaluated neem extracts against maize stalk borer, *Chilo partellus* Swinh in Coastal Kenya. Neem extract at 3, 6 and 12 percent was sprayed on maize (cv. Coast Composite) for the control of *C. partellus*. Dipterex (5% G at 10 kg/ha) served as insecticide check and unprotected control plots were maintained. Extracts were sprayed thrice at 3, 4 and 5 weeks after plant emergence on plants which had been infested at 5 larvae (1 to 2 day old) per plant one day before the first spray application. Neem reduced the leaf damage, tunnelling, and larval population. control efficacy increased with the concentration. Neem extract (12%) was at par with dipterex. Yield increases with 3, 6 and 12 percent neem extract were 17, 24 and 28 percent, resepectively as compared with 30 percent increase with insecticide.

Facknath (1996) found that the extracts of neem were found to be successful in breaking down the resistance developed by *P. xylostella* to some commonly used pyrethroids, namely deltamethrin and cypermethrin, and to have a synergistic effect with the organophosphate prothiofos, both in laboratory bioassays as well as in field trials. Comparable studies on (*Melia azederach*) extracts did not give significant results. Tamaron which produced 29.3 percent mortality alone killed 88 percent of the larvae when mixed with neem (*A. indica*) and 86.7 percent when mixed with *M. azederach*. Larvicidal effect of Cymbush was enhanced from 17.3 to 77.3 percent with neem and 73.3 percent with *Melia*.

Subba Rao and Mohan (1996) tested efficacy of NSKE on pigeonpea pod borer (PPB). A field trial was conducted in kharif 1994. The neem products namely NSKE (5%), Nimbecidine (0.2%) and Achook (0.5%) were evaluated along with endosulfan (0.07%) and mococrotophos (0.04%). NSKE was prepared by dissolving 12.5 kg of finely ground neem seed kernel in 250 lt of water. The solution was kept overnight and filtered and the volume made upto 250 lts. A total of 3 sprayings was given at an interval of 10 days starting from 50 percent flowering of the crop. The pod borers recorded during the experimental period were *H. armigera* and *E. atomosa*. The treatment Achook was significantly superior recording 22.95 percent of pod damage as against control (37.1%), it was on par with NSKE (25.45%), endosulfan (25.95%) and monocrotophos (29.75%). Considering the cost, spraying with NSKE can be recommended for the control of pob borers and it is also safe to environment.

Vyas *et al*. (1996) studied the effect of 4 plant extracts on three important noctuid pests. The data on percentage mortality for *E. vittella* in case of neem, annona, jatropha and mahua formulations were 58, 89, 68, 21; for *H. armigera* 100, 100, 60, 76 and for *S. litura* 28, 44, 11 and 5, respectively.

Shukla *et al*. (1996) evaluated neem products against rice, pigeonpea and chickpea insect pests. The field trials were conducted with neem oil (NO 3%), NSKE (5%) along with conventional insecticides monocrotophos, endosulfan and cypermethrin. The result indicated that there was significant variation in pod and grain damage by the pod borer complex, namely *Heliothis, Exelostis* and *Melanagromyza* of pigeonpea with the treatments. The lowest damage was in cypermethrin followed by endosulfan and monocrotophos. Among neem products, the yield was on par in NSKE (5%), Neemark (3%) and NO (3%) which was 26.29, 13.25 and 4.90 percent higher than the control. Thus neem products can be safely and very effectively utilized for the control of pod borer of pigeonpea. In case of chickpea, the pod borer, *H. armigera* population was very moderate in comparision to synthetic insecticides which gave control at the early stage of crop growth. The grain yield was 1940, 1874, 1808, 1775 kg/ha against the treatment cypermethrin, NSKE, Neemark and NO, respectively.

Dhaliwal *et al*. (1996) observed that NO (4%) proved effecitve agaisnt rice stem borer, *Scirpophaga incertulus* (Walker) resulting in 5.99 percent whiteheads as against 3.58 percent in mococrotophos (0.05%) and 13.86 percent in untreated control. NO (1, 2, 3, 4%) was also effective against rice leaf folder, *Cnaphalocrocis medinalis* (Guinee) (20.56-26.39% folded leaves) as compared to monocrotophos (20.71%) and untreated control (33.32%). Among neem-based insecticides, RD-9 Repelin (4 kg/ha) was most effective against stem borer with 10.60 percent whiteheads, followed by Neemrich-II and Neemrich-I (both 2 kg/ha) with 11.21 and 12.79 percent whiteheads as against 12.20 percent in phosphamidon (0.527 kg a.i./ha) and 18.06 percent in untreated control. All the treatments were at par with respect to yield, but proved significantly better than neem oil and control.

Sachan and Gujar (1996) studied the effect of neem products in the management of *H. armigera*. Large scale multi-location field trials have revealed that NSKE (5%) and NLE (5%) were found at par with the spraying of quinalphos (0.05%), but these treatments were slightly inferior to fenvalerate (0.2%), cypermethrin (0.006%) and endosulfan (0.07%) and endosulfan 4D. The NSKE and NLE provided significant and cost effective control of *H. armigera* compared with chemical insecticides in chickpea. Laboratory studies on commercial formulations of neem showed that Repelin, Neemguard, Neem 25EC and Neem 25WDP provided leaf area protection up to 75 to 90 percent from *Helicoverpa* damage in chickpea. Another commercial formulation, Neemark, proved quite effective against *H. armigera* in chickpea. In pigeonpea, 2 sprays of NSKE (5%) and a single spray of cypermethrin (0.01%) gave significantly better cost benefit ratio of (1:3.81) and endosulfan (0.07%) (1:3.39) applied alone, against *H. armigera*.

Akhauri *et al.* (1996) observed NO as possible biorational insecticide for the management of spotted pod borer, *Maruca testulalis* Geyer in pigeonpea. Field trials were done in RBD with early grown pigeonpea cv. UAAS-120 to compare the effectiveness of two indigenous plant products. It was observed that both NO and mahua oil (each at 2%) as spray were able to reduce the larval population. However, NO proved better than mahua oil spray in significantly reducing the mean pod damage (27.2 to 18.9%) while increasing the mean number of matured pods (67.4 to 85.8/plant) and grain yield (13.5 to 17.4 g/plant) in comparison to untreated check. NO, therefore, might be exploited more efficiently as a biorational insecticide in pod borer management in early pigeonpea cultivation.

Chari *et al.* (1996) studied a bioefficacy of some neem formulaitons against *S. litura*. The bioefficacy of neem formulations viz. Repelin A (RD-9 Repelin at 1%), karanjia (*P. pinnata*) and Mahua (*M. indica*) containing 300 ppm of azadirachtin in its natural form along with 500 ppm of pongamin and pongamol, respectively, Neemgold (0.003 and 0.0015% AZT), Vepacide-I and -II (0.25%), dichloroethane (DCE) extract of deoild neem cake (DOC) and Navnem (0.1%) was studied against *S. litura* on tobacco in comparison with aqueous NSKS (2%0, DCE extract of neem seed kernel (NSK) (1%) and Neemrich-II (0.1%). The results indicated that Repelin A, NSKS, Neemgold (0.003%), DCE of NSK and DCE of deoiled neem cake (DOC) (!%) effectively controlled *S. litura* as compared with Neemgold (0.0015%), Vepacide-I and -II (0.25%) and Neemrich-II (0.1%). Cent per cent mortality was obtained in the laboratory with DCE of NSK within 5 days, followed by Neemgold (0.003%) in 6 days, NSKS and Neemrich-II in 7 days, Vepacide-II in 8 days, DCE of DOC, Repelin A, Neemgold (0.0015%) and Vepacide-I in 9 days.

John et al. (1996) tested the efficacy of Turplex containing 3 percent AZT against black cutworm, *Agrotis ipsilon* larvae. Turplex was applied at the rate ranging from 25 to 125 g AZT a.i./ha and compared with chlorpyrifos. It was found that turplex gave excellent control of cutworm larvae even at 25 g a.i./ha rate. In general, Truplex has shown very effective performance against all the lepidopterous caterpillar complex. The AZT formulaiton Align was tested against *Trichoplusia ni*, the cabbage looper. Align 50 and 100 g a.i./ha and pyrethroid (Asana at 50 g a.i./ha) were sprayed and it was observed that Align even at 50 g a.i./ha rate sprayed twice showed excellent efficacy in controlling the pest in the field within 14 days.

Ali Niazee *et al.* (1997) evaluated neem insecticides both in laboratory and field against *Archips rosanus*. Margosan-O (0.3% AZT) was tested in Oregon against the tortricid *A. rosanus*. In laboratory tests, a 1 percent aqueous solution of neem insecticide produced 100 percent larval mortality within 48 hrs of treatment. At lower concentrations of 0.1 and 0.01 percent, 48-hr mortality rates were reduced. Prolonged deleterious effects of sublethal treatments of neem were observed even at the lowest concentration (0.001%) tested. Mortality in the 0.1 and 0.01 percent treatment groups reached 100 percent within 32-39 days after treament. Field data showed that 3 weeks after application, 1 percent Margosan-O was as effective as the organophosphate insecticide diazinon, and significantly more effective than the microbial insecticide Dipel (*Bacillus thuringiensis*).

Jaglan *et al.* (1997) evaluated Methanol and chloroform:methanol (9:1) extracts of neem seeds and green leaves against *H. armigera*. The chloroform:methanol (9:1) neem seed and leaf extracts showed better insecticidal properties than methanol extracts.

Hernandez and Vendramim (1997) studied bioactivity evaluation of aqueous extracts of 12 Meliaceae species on the development of *Spodoptera frugiperda*. The extract was incorporated into an artificial diet and offered to *S. frugiperda* during its entire larval stage. The variables evaluated were length and viability of larval and pupal stages and weight of pupae. Larval mortality above 80 percent was obtained with the extracts from fruits of *M. azedarach* and seeds of *A. indica* and *Cedrela odorata*.

Kulkarni and Joshi (1997) studied insecticidal action of neem and some other plant extracts against *Rhesala imparata*. The efficacy of methanolic extracts from seeds of *A. indica* and leaves of *Aloe vera*, *Lantana camara* var. *aculeata, Datura metel, Ipomoea carnea* subsp. *fistulosa* and *Annona squamosa*, was evaluated against larvae of *R. imparata* in the laboratory. Three concentrations viz. 1, 3 and 5 percent were taken in each case. It was observed that seed extracts of *A. indica, A. squamosa* and *L. camara* were statistically significant giving 84.35, 81.07 and 84.45 percent mortality after 72 hrs. The performance of neem products at 5 percent was also quite efficacious giving 68.69 percent mortality while at 1 percent, it gives only 27.62 percent mortality.

Kwanchai and Cholikamhaeng (1997) studied the effect of Thai neem extracts for the control of *Spodoptera exigua*.

Formulation-I : neem seed extract (NSE) - Kilograms of Thai neem seed were ground to powder by a grinding machine and put into the extracter tank. The powder was stirred in 100 lt of 95 percent methanol for 4 hrs then left to stand overnight and then stirred again for 1 hr. The solution was filtered and concentrated by vacuum evaporation. After removal of neem oil in the upper layer of the crude product, neem solution was determined for AZT content by HPLC. It contained 0.514 percent AZT w/v.

Formulation-II : neem fruit extract (NFE) - The extraction process was the same as the above but using dried neem fruit powder. The AZT contents of NFE was 0.09 percent w/v.

The efficacy test of neem extract formulations against the beet armyworm, *S. exigua* on asparagus was conducted in farmer's field and under laboratory conditions. NSE (0.5% AZT) and NFE (0.09% AZT) at the rate of 100, 150 ml/20 lt mixed formulation of NSE + NFE at the rate of 50 + 50 and 75 + 75 ml/20 lt were compared with chlorpyrifos at the rate of 30 ml/20 lt. First spray on asparagus was started when severe infestation occurred and continued once every 5 days for 5 times. The results showed that NSE (150 ml) gave the best control, whereas, NSE (100 ml) mixed formulation 50 + 50, 75 + 75 ml and chlorpyrifos was moderately effective, followed by NFE (100 and 150 ml) yielding low levels of control. In the laboratory, all treated neem extracts had no contact poison on eggs and larvae of the beet armyworm but NSE (100, 150 ml) mixed formulations NSE + NFE (50 + 50, 75 + 75 ml) showed high stomach poison activity except NFE (100, 150 ml). NSE (150 ml) and NSE + NFE (75 + 75 ml/20 lt) have shown good residual activity for 5 days.

Erbach *et al.* (1997) evaluated the effect of NeemAzal T, NeemAzal T/S and Delphin on grape berry moth, *Lobesia botrana* and the predatory mite, *Typhlodromous pyri*. Field tirals in Germany showed that NeemAzal T and NeemAzal T/S, and Delphin (*Bacillus thuringiensis* [subsp. *kurstaki*]) provide environmentally friendly alternatives for control of the grape pest *L. botrana*. The treatments caused 58-80 percent mortality of the pest, while there was no measurable difference in the population development of *T. pyri* between treated and untreated populations.

Flemming (1997) carried out the control measures against the pine pest *Thaumetopoea processionea* on a large scale in 1996. Efforts were concentrated on oaks (*Quercus* spp.) in the north-western region of the country Brandenburg, Germany. Control rates of nearly 100 percent were obtained with helicopters, but only 70 percent with ground techniques of sprayings. The pesticides tested were NeemAzal, Delfin and Foray 48 B (both based on *Bacillus thringiensis*). Foray 48 B applied by helocpters and Delfin applied on the ground were the most effective.

Jaglan *et al.* (1997) extracted seven different plant parts including kernels and leaves of neem (*A. indica*) in solvent by different methods of extraction, i.e. room temperature extraction method, hot extraction method and Soxhlet extraction method, which were tested at different concentrations against larvae of *H. armigera*. Of the three methods employed for extraction to compare their effectiveness, the room temperature extraction gave on average larval mortality 38.7% and hot extraction methods gave larval mortality 37.7%, which were statistically on a par. The Soxhlet extraction method was found to be slightly inferior to the other two methods.

Sanap and Pawar (1997) conducted field experiments in RBD replilcated thrice. Three sprays of each treatment were given from initiation of flowering at a fortnightly intervals. The observations on pod damage were recorded on 5 randomly sampled plants. The total number of healthy and damaged pods were observed. The treatments were Repelin, Nimbecidine, Achook, NSKE, NPV (250 LE/ha), endosulfan (0.07%) and untreated control. All the neem-based formulations recorded significantly lower pod damage over control. During 1994-95 season, the lowest damage was registered in the treatments with nimbecidine (0.2%), followed by Achook (0.5%) and Repelin (0.13%). However, neem formulations were found more efective than endosulfan, considering the average of 3 years trial. The lowest damage of 8.91 percent was found in nimbecidine (0.2%), followed by 9.22 percent in Achook (0.5%).

Rao and Dhingra (1997) studied synergistic activity of some vegetable oils when mixed with cypermethrin against different instars of *S. litura*. The non-toxic vegetable oils, karanj oil, neem oil, citronella oil and piperonyl butoxide (PBO) were separately used in mixed formulations with cypermethrin at various ratios (1:1, 1:2, 1:5, 1:10) and evaluated against 1, 4, 7 and 9 day-old larvae of *S. litura* by leaf dip and direct spray methods. With the direct spray method, the synergistic activity of neem oil in the formulation was pronounced against 4 and 9 day-old larvae. Also, karanj oil exhibited synergism against 1, 7 and 9 day-old larvae. However, citronella oil showed synergistic activity only against 7 day-old larvae.

Perez *et al*. (1997) Studied the effect of neem natural pesticides to control pests of cabbage, corn and cowpea. Three natural insecticides, obtained from crude extracts of neem seeds, were tested against pests in the field (*Plutella xylostella* on cabbage and *Spodoptera frugiperda* on corn) as well as in storage (*Acanthoscelides obtectus* onseed cowpea). The experiments were carried out in Cuba. The following neem tree products and doses were tested : Cubanim-t (dry powder) at 5 kg/ha/application as a 25 g/lt suspension; Oleonim-80 CE (water emulsion) at 1.5 lt/ha/applied as a 5 ml/lt dilution; Oleonim-50 CE at 3 lt/ha/applied as a 10 ml/lt dilution; as a commercial control, Karate (2.5 EC lambda-cyhalothrin), at 1.2 lt/ha/applied as a 4 ml/lt dilution. Cubanim-t achieved 80 percent efficiency against *P. xylostella* after 4 applications, statistically equal to the commercial insecticide (Karate), and decreased pest intensity from 0.95 to 0.02 larvae/plant by the 4th application. *S. frugiperda* on maize decreased from 0.65 to 0.07 larvae/plant after 3 sprays with Oleonim 50CE; from 1.18 to 0.06 with Oleonim 80CE; and from 1.50 to 0.10 with Cubanim-t.

Patel *et al.* (1998) tested field bio-efficacy of neem against leaf eating caterpillar, *S. litura* Fab. and *Achoea janata* Linn.on castor. The bio-efficacy of Gronim (0.3EC, azadirachtin-based) was tested at 0.0012 and 0.0018 percent and combination with endosulfan (0.07%) against the pests at Anand, India during 1995-96 and 1996-97. A mixture of azadirachtin (0.0018%) and endosulfan as well as monocrotophos (0.04%) alone gave cent per cent larval mortality of *S. litura* and *Achaoea janata* after 7 and 3 days of their spray, respectively. The higher (0.0018%) dose of azadirachtin was found more effective than it's lower (0.0012%) dose against *S. litura* and *A. janata* after 7 and 3 days of spray, respectively.

Singh (1998) evaluated some biopesticides against pea bod borer complex. The bioinsecticidal trial was laid out at Mathura, India during two consecutive seasons i.e. rabi, 1994-95 and rabi, 1995-96 to get safer and effective insecticides to control the pests. For this, Nimbicidine, Repelin, Achook, Suneem, Nethrin (all at 1.5%), Dipel-8L, Biobit, Biotox, Bioline (all at 0.15%) and endosulfan (0.07%) were tested alongwith control. It was evident from the results that all treatments were superior over control. Endosulfan proved to be the best (0.66% pod damage), followed by Nimbicidine (5%), Suneem (5%), RD-9 Repelin (5.66%) and Achook (6%).

Gupta *et al.* (1998) tested neem-based pesticides against different life stages of *Helicoverpa armigera*. Different life stages of American bollworm *H. armigera* starting from egg to adult were taken for testing the efficacy of neem-based pesticides (Godrej Achook, neem oil and neem seed powder-NSP) under laboratory conditions. The ovicidal action of the commerrcial formulation of Godrej Achook was not observed. However, dipping of larvae (3rd instar) with different concentrations of this product (0.1, 0.25, 0.5, 1.0, 1.5 and 2.0%) was found effective and there was intermediate formation of insect for moderate concentrations and higher mortality with higher concentrations.

Srivastava and Proksch (1998) studied bioactivity of seven plant extracts against *Spodoptera littoralis*. Methanol extract from the 7 plant species viz. *Ageratum conyzoides, A. houstonianum, Aglaia roxburghiana, A. odorata,*

A. *indica, Cedrela toona* and *Melia azaderach* were individually incorporated in the artificial diet at different concentrations viz. 1, 3, 5, 7 and 9 mg crude extract (dry wt) per standardised piece of artificial diet. A batch of thrity neonates of S. littoralis was released in each concentration and the observations were recorded on survival of larvae at 6 days after feeding on spiked diet, 10 days after feeding (6 days on spiked + 4 days on normal diet), and weight gained by the larvae at 6- and 10th-day after feeding. *A. conyzoides* was slightly more bioactive than *A. soustonianum* owing to slightly higher concentration of Preconcene II in it, as evidenced by high performance liquid chromatograms. Amongst Meliaceae, *A. odorata* leaves were most active followed by *M. azaderach* stem, leaves and fruits, *A. indica, A. rosburghiana* leaves and *C. toona* leaves. In the second observations recorded 10 days after feeding all the plant extracts exhibited very high level of insecticidal activity. At still higher doses survivorship was close to zero.

Mahadevan (1998) studied azadirachtin for the control of tomato fruit borer, *H. armigera*. A new neem-based insecticide 'Eco Neem' PJM 01/94 with an azadirachtin content of 0.3 percent was tested against the pest during the winter and summer seasons of 1994 in Tamil Nadu, India. A total of 2 rounds of spray was given at 15-day intervals. Damage was recorded on days 3, 7 and 15, after each spray. The results revealed that spraying 2 rounds of Eco Neem at 10 ppm (3.3%) was the best means of controlling the pest. During the winter, spraying of monocrotophos, combined with endosulfan spray, resulted in only 4.2 percent damage to the crops compared with 6.8 and 28.9 percent damage for Eco Neem (10 ppm)-treated and untreated crops, respectively.

Sahayaraj and Paulraj (1998) screened relative toxicity of some plant products against groundnut pest, *S. litura*. A laboratory study was carried out to evaluate the effect of water extracts of 4 plant leaves in insect pest control. Various concentrations, of the extracts (0.5, 1, 2, 4 and 6%) were used against last-instar larvae of the groundnut pest. *Calotropis gigantea* was found to be the most toxic plant product followed by *Vitex negundo, Azadirachta indica* and *Pongamia glabra*.

Nestor *et al.* (1998) studied the mortality of leafminer, *Phyllocnistis citrella* with mineral oil and neem products. The effect of mineral oil (Citrolina) and neem on the mortality of *P. citrella*, was determined in 3-year-old Persian lime (*Citrus aurantifolia*) orchards at Cuitlahuac, Veracruz, Mexico. The concentrations tested were : 2, 5 and 8 percent Citrolina, 30 percent neem and 50 g of ground neem seed/lt of water. The treatments were applied to shoots with different instar larvae. The most effective level of control was obtained with 5 percent Citrolina and 50 g neem/lt of water, with mortalities of 72.7 and 70.5 percent, respectively. Mortality was observed between 24 and 36 hrs following the application of Citrolina, and between 48 and 72 hrs for neem.

Dhawan *et al*. (1998) evaluated different biopesticides against *H. armigera*. The efficiency of commercially available formulations of *B. thuringiensis* subsp. *kurstaki* (Dipel, Bioasp and Biolep), neem (*A. indica*) (Nimbecidine, Neemgold and NeemAzal) and insect growth regulators (lufenuron, flufenoxuron and RH-2485 a diacylhydriazne) were tested against *H. armigera* larvae. All the 3 formulations of *B. thuringiensis* at the rate of 1.50 kg/lt/ha were as effective as quinalphos against the young larvae. Neem-based insecticides were significantly inferior to standards, quinalphos and chlorpyrifos against young and mature larvae except NeemAzal which was as effective as quinalphos against young larvae. In another experiment, Dipel, Nimbecidine and lufenuron were evaluated against *H. armigera* population collected from Ludhiana, Barnala and Bhatinda, Punjab (India) and the data showed variation in susceptibility to the pest. The Barnala population was comparatively more susceptible to all the 3 biopesticides. Similarly, the Bhatinda population was more susceptible to Nimbecidine and Dipel.

Dhembare *et al*. (1998) investigated the potential use of plant extracts as control agents of *M. scyrodes* on ber fruit (*Ziziphus mauritiana*). Extracts from 11 plant species were obtained using methanol as the solvent. Extracts were put on ber fruit in petri plates. Ten *M. scyrodes* larvae were released on to each plate and mortality was assessed after 12-48 hrs. Mortality was compared to that in a standard endosulfan treatment. Extracts from *Annona reticulata, Azadirachta indica* and *Ocimum sanctum* gave effective control, 64., 70.3 and 63.3 percent mortality after 48 hrs with 1 percent, respectively, compared to 73.3 percent mortality with endosulfan.

Das (1998) studied synergistic action of NSKE against pests of pigeonpea. During field experiments in Madhya Pradesh, India, in the kharif season of 1997-98, four insecticides and a NSKE alone or in combinations in half doses were tested against *Melanagromyza obtusa* and *H. armigera*. The lowest damage and highest grain yield were obtained using dimethoate (0.15%) + NSKE (5%).

Sahayaraj and Paulraj (1998) studied relative toxicity of some plant extracts to groundnut leaf miner, *Aproaerema modicella*. Larvae of this pest were exposed to neem (*A. indica*), *Pongamia glabra* and *Calotropis gigantea* extracts (0.5-6%) for 4 days. All the plnat extracts were toxic. The larval mortality was highest (63-75%) at the highest concentration.

Jaglan *et al.* (1998) evaluated methanol and chloroform : methanol (9:1) extracts of neem seeds and green leaves against *H. armigera* on chickpea, pigeonpea and cotton. Various concentrations of extracts were used to treat leaves and pods/bolls of chickpea, pigeonpea and cotton. Treated foods were offered to early stage (neonate) and advanced stage larvae of pest for 48 hrs, and then untreated food was offered for the rest of the life cycle. The chloroform : methanol (9:1) neem seed and leaf extracts showed better insecticidal properties than methanol extracts. However, neem seed extract in chloroform : methanol (9:1) was the most promising in causing adverse morphogenic effects on various biological parameters of *H. armigera*. Early stage larvae were more sensitive to the exposure of neem extracts than advanced stage larvae.

Singh and Pandey (1998) conducted a field trial using NDR 637 in kharif 1995 and Pusa Basmati in kharif 1996 to test the efficacy of seven neem products viz. Nimbecidine, Neemax, Neemgold, Econeem, NeemAzal, Fortune Aza and Achook along with Durmet a chemical check against yellow stem borer, *Scirpophaga incertulas* Walk. During both the years, all botanicals were at par in controlling the pest and in increasing the yield over control except Fortune Aza. Neemgold was found superior in producing higher yield and minimising incidence.

Pareek and Batta (1998) did laboratory and field evaluation of plant products against maize stem borer, *Chilo partellus* Swinhoe. Plant products viz. neem seed kernel, neem oil, Datura, papaya, Ghikanvar, Aak, Tulsi seed, mint, Henna, Ipomea, garlic, clove and Lantana were evaluated against the pest. Neem oil (1%) was found most effective giving highest mortality after 12, 24 and 48 hrs of treatment. The neem seed kernel (10%) and neem oil (0.5%) were at par. The effect of neem oil (0.5 and 1%), neem seed kernel (5 and 10%) and Ghikanvar (50%) was also studied on the development of the borer. All the plant products tested significanlty lowered down the survival of larvae, larval pupation, adult emergence, longevity and growth index as compared to control.

Patel *et al.* (1998) studied bioefficacy of some neem-based formulations against hairy caterpillar, *Selepa celtis* infested Amla. The neem-based formulations and plant materials NSKS (5%), neem leaf extract (NLE 5%), Parasmani (0.2%), Neemark (1%), Nimbicidine (0.2%), Repelin (1%) and Gronim (0.4%) were evaluated for their bioefficacy against *S. celtis* in Amla (Gujarat-1). NSKS was found more effective than all the neem-based formulations tested against *S. celtis*.

Sarode and Gabhane (1998) compared the efficacy of 13 insecticidal treatments viz. NSKE (5%), endosulfan (0.06%), monocrotophos (0.05%), malathion (0.05%), NSKE + 3/4th dose of endosulfan, NSKE + 3/4th dose of monocrotophos, NSKE + 3/4th dose of malathion, NSKE + 1/4th dose of endosulfan, NSKE + 1/4th dose of monocrotophos, NSKE + 1/4th dose of malathion and the control against *Earias vittella*. The data indicated that all the treatments were significantly superior over control. Endosulfan (0.06%) was the most effective treatment. However, the treatments NSKE (5%) + 3/4th dose of endosulfan (0.045%), monocrotophos (0.05%) and NSKE (5%) + 1/2 dose of endosulfan recorded 13.05, 14.50 and 15.28 percent fruit damage, respectively were observed promising.

Nelson and Logiswaran (1998) did three field trials to study the efficacy of insecticides against *Nephopteryx eugraphella* Ragonot. The insecticides tested were carbaryl (0.1%) phosalone (0.05%), moncrotophos (0.03%), quinalphos (0.05%), Dipel (0.05%), NSKE (5%) and untreated control. It was observed that on 14th day after treatment, carbaryl (0.1%) was the best treatment followed by Dipel (0.05%) and NSKE 5%) and phosphamidon (0.05%0 recording 11.13, 12.25, 12.25, 12.50 percent bud damage, respectively against control 21.88 percent. The carbaryl and NSKE controlled bud damage effectively and recorded high cost:benefit ratio.

Moorthy *et al*. (1998) studied the effect of NSKE (4%) sprayed thrice on different dates at an interval varying from 6-13 days on cabbage crop. Three applications of NSKE were adequate to reduced the population of *P. xylostella* of more than 40 per plant to negligible level. In the three trials conducted, the reduction in pest population was 95.97, 90.57 and 97.70 percent.

Pandey and Singh (1998) evaluated natural products as an insecticide against top borer, *Scirpophaga excerptalis* Walker in sugarcane. The treatments taken were 1, 2, and 3 percent extract of kaner fruits (*Nerium indicum*) and flowers; 1, 2 and 3 percent extract of *Eucalyptus rostrata* leaves; 1, 2 and 3 percent extract of neem (*Azadirachta indica*) leaves; 1, 2 and 3 percent extract madar (*Calotropis procera*) leaves and the untreated control. The data revealed that least incidence (6.35 and 6.17%) of top borer was recorded in 3 percent *E. rostrata* leaf extract compared to control (16.88 and 15.77%) in two years trial. This was followed by 2 percent neem leaf extract. Incidence of the borer was 10.62 and 7.02 percent, respectively in two trials. It is, therefore, proposed that these natural products can be exploited in integrated insect pest management to control this pest.

Yee and Toscano (1998) studied in laboratory the effect of synthetic and natural insecticides on lettuce leaf and consumption by larvae of *Spodoptera exigua*. Two synthetic and 4 natural inscticides were compared : Capture (bifenthrin), Lannate (methomyl), Success (spinosad), Xentari (*B. thuringiensis* subsp. *aizawai*), Kryocide (cryolite) and Trilogy (90% neem oil extract-NOE). Capture, Lannate, Success and xentari at 50 ppm a.i. caused 100 percent mortality within 1day after exposure and prevented any detectable feeding on leaf disks by neonate *S. exigua* larvae. Neither Kryocide nor Trilogy at 50 ppm reduced leaf consumtion by neonate, 3rd instar and 5th instar *S. exigua* compared with controls. Higher concentrations of Kryocide and Trilogy (1000 and 10,000 ppm) significantly reduced leaf consumption by 3rd-instar larvae.

Sahayaraj and Paulraj (1998) studied the effect of neem extract on *Amsacta albestriga*. When applied in the laboratory against 5th-instar larvea of A. albistriga, NLE caused 8.33, 13.33, 23.33, 36.66 and 85.71 percent mortality at 0.5, 1.0, 2.0, 4.0 and 6.0 percent, respectively. Mortality of pupae treated with the lower concentrations was 22.73 percent; the higher concentrations did not affect the pupae. Adult emergence was 30 percent in the control, decreasing to 14.11 percent when treated at 0.5 percent and 10.76 percent at 1 percent. Adults resulting from treated larvae laid no eggs.

Agrawal *et al.* (1998) tested some neem formulations viz. Neemarin, NeemAzal, Econeem, Achook and Nimbecidine at 1 percent concentration against silkworm, *Bombyx mori*, larvae under laboratory conditons. Mulberry, *Morus alba* leaves were dipped in the eperimental petridishes wherein the larvae were released for feeding. In control, only water-treated leaves were kept. The results showed that silkworm larvae started dying after 3rd-day of treatment and continued upto 5th and 7th day of treatment. Amongst the different neem products Neemarin proved better followed by NeemAzal and the actual cumulative mortality percentage was 55.83 and 52.47 percent, respectively. Achook proved to be thrid effective neem product which resulted in 45.82 percent cumulative mortality within 3-7 days of treatment. Nimbecidine was least effective formulation with over all mortality percentage of 27.47 percent. Econeem was slightly better to that of Nimbecidine and registered 29.15 percent mortality.

Prashad *et al.* (1998) studied action of plant extracts against *Amsacta moorei*. The rhyzome extract pf *Acorus calamus*, bulb extract of *Allium sativum* and the seed extracts of *Azadirachta indica, Annona squamosa, Croton tiglium* and *Pongamia glabra*, alongwith *A. indica* alone at its two concentrations of 1.5 and 3.0 percent were taken for this experiment.

The results revealed that all the plant extracts were proved significantly superior over control. The extract of *A. indica* when sprayed in combination with the extract of rhyzome extract of A. calamus offered 74.34 percent mortality of the larvae and proved significantly suprior to rest of the combinations, whereas *A. indica* alone at 3.0 percent concentration resulted only 61.18 percent mortality. The former results were closely followed by the combination of *A. indica* with the extract of *P. glabra* which resulted 58.015 percent mortality. The poorest performance was reocrded with joint action of *A. indica* and *A. sativum*, which caused 37.45 percent larval mortality. The findings, therefore, revealed *A. calamus* has synergistic effect when used in combination with *A. indica* extracts.

Vyas *et al.* (1999) tested the effects of plant extracts including neem product on growth and development of *Earias vitella, H. armigera* and *S. litura*. The effect of methanolic seed extracts of custard apple (*Annona squamosa*), wild castor (*Jatropha curcas*) and mahua (*Bassia latifolia*) *vis-a-vis* a commercial neem formulation currently being marketed in India as Achook. This formulation is in a water-soluble powder form containing azadirachtin, azadiradion and nimbocinol + epinimbocinol at 300, 500 and 2000 ppm, respectively. All the three plant extracts and neem formulation Achook were tested at 1 percent concentration against 4-day-old larvae. Larvae were allowed to feed on treated plant material for 24 hrs followed by subsequent transfer to control. Observations were recorded on the percentage larval mortality, time taken for pupation and adult emergence. The data were taken on percent mortality with neem formulation Achook, *A. jatropha* and *B. latifolia* extracts were 57.89, 89.47, 68.42 and 21.05 percent for *E. vittella*, 100, 100, 60, 66 percent for *H. armigera*, respectively with the above 4 extracts.

Gupta *et al.* (1999) conducted field experiments to evaluate the bioefficacy of neem products against cotton bollworms viz. *Earias* spp., *Pectinophora gossypiella* and *H. armigera* and their impact on whitefly, *Bemisia tabaci* (Genn.). Neem products were applied alone as well as in combinations with synthetic insecticides against the pests. Neem-based insecticides such as RD-9 Repelin, Nimbecidine, Achook and NeemAZ and bioinsecticides *Bt* were evaluated alone and in combinations of chemical insecticides against pest complex in cotton crop. The experiments were done in RBD with three replications. Observations were recorded on the incidence of bollworm on locule basis in green bolls and open bolls in randomly selected plants. The percent incidence of bollworms are subjected to angular transformations for statistical analysis. Data were also taken on whitefly infestation. The observation recorded revealed that the application of neem alone or in combination with *Bt* formulations or broad spectrum conventional insecticides failed to check the incidence of bollworms. However, neem in combiation with one spray of synthetic pyrethroids gave significant control of bollworms, as minimum incidence (10.8, 18.2%) and miximum yield of seed cotton (1970, 1819 kg/ha) were noted.

Chari *et al*. (1999) did laboratory and field experiments with neem (*A. indica*), pongamia (*P. pinnata*) and chinaberry (*M. azedarach*) extracts and commercial neem formulaitons against tobacco caterpillar, *S. litura*. Three neem formulations - NeemAzal FI containing 10,000 ppm azadirachtin (AZA) at 50 and 30 ppm, NeemAzal FII containing 50,000 ppm AZA, Fortune Aza conatining 11,000 and 1500 ppm AZA at 50, 30 and 15 ppm concentrations and neem (*A. indica*) extracts (CTRI-I, II and DOR-I) containing 15,000 ppm AZA, CTRI-III containing 80,000 ppm AZA, CTRI-IV containing 30,000 ppm AZA, all at 50 ppm concentrations, 1.5 percent oil and seed extract of chinaberry (DOR-II) at 0.005 percent a.i. and four pongamia seed extracts (1%) including pongamia oil (1.5%) were evaluated in comparison with aqueous NSKS (2%) and chlorpyrifos (0.05%) in the laboratory and nursery against *S. litura* during the 1994 and 1995. Results revealed that NeemAzal FI and II (50 ppm), NSKS (2%) and chlorpyrifos (0.05%) gave significant protection to tobacco seedlings from the *Spodoptera* damage followed by NeemAzal FII (25 ppm) and NeemAzal FI (30 ppm). Fortune Aza (50 ppm) gave equal protection as chlorpyrifos (0.05%) and NSKS (2%) up to 9 dyas after spraying. These results confirm the earlier findings that neem formulations at 30 ppm AZA concentrations could effectively protect the tobacco seedlings from the *Spodoptera* damage up to 9 days after spray. All the neem extracts were significantly more effective against *S. litura* on tobacco than pongamia treatments. Neem oil (1.5%) was found highly effective and on par with NSKS (2%), chlorpyrifos and CTRI-III neem extract.

7. Thysanoptera

Pillai and Ponniah (1988) did experiments to control rice thrips with neem products. Neem (*A. indica*) products (5 and 10% neem seed and neem cake extract and 2% neem oil) were evaluated for the control of *Stenchaetothrips biformis* on rice in Tamil Nadu, India in June-July, 1987. All the treatments significantly reduced the pest population and 2 percent neem oil was as effective as phosphamidon (100 EC at 2500 ml/ha) and fenthion (100 EC at 500 ml/ha) in controlling *S. biformis*.

Madhusudhan and Gopalan (1988) did experiments to control thrips *S. biformis* in rice nursery. Spray formulations of phosphamidon, monocrotophos, phosalone, chlorpyrifos, metasystox (demeton-S-methyl), endosulfan, dimethoate and neem oil were evaluated for the control of *S. biformis* in Tamil Nadu, India. All the insecticides gave significant control of the pest, with chlorpyrifos, dimethoate and monocrotophos being most effective. Neem oil significantly reduced the thrips population 48 and 72 hrs after application but not after 24 hrs.

Ivbijaro and Bolaji (1990) conducted field trials in Ibadan, Nigeria, to determine the effects of cypermethrin + dimethoate (Sherpa Plus), the oil extract from ripe seeds of *Piper guineense* and the aqueous extract of ripe seeds of neem (*A. indica*) on the insect pests and yield of cowpea (*Vigna unguiculata*). The insecticides were applied at 30, 42 and 53 days after planting. Cypermethrin + dimethoate was consistently superior to all other treatments in substantially reducing the population densities of the thrips *Megalurothrips sjostedti* and the pyralid *Maruca testulalis*, which are serious pests on the flowers and buds, and in reducing the densities of 4 pod-sucking hemipterous insects (the coreids *Clavigralla shadabi* and *Riptortus dentipes*, and the pentatomids *Aspavia armigera* and *Nezara viridula*). Pest population densities were also reduced by foliar sprays with neem or *P. guineense* extracts. Seed yield after treatment with cypermethrin + dimethoate was 0.99 lt/ha, more than double the yield from the other treatments. A marginal increase in seed yield compared with the untreated control was achieved by foliar spraying with *A. indica* extract.

Bhardwaj and Gupta (1992) observed in laboratory experiments, aqueous extracts of garlic leaves or cloves or neem (*A. indica*) bark or leaves had no effect on *Thrips tabaci* Lindemann while neem oil caused considerable mortality.

Chandramohan and Nanjan (1993) conducted field studies in India, dimethoate 0.006 percent applied at 20-day intervals was the most effective treatment for controlling infestations of *Thrips tabaci* on garlic, compared with acephate, furathiocarb, demeton-o-methyl and neem oil.

Eapen (1994) tested insecticides and neem oil against the thrips *Sciothrips cardamomi* R, an important pest of small cardamom causing about 30-80 percent crop loss. The field studies were conducted in Karnataka, India, during 1989-93 to determine the efficacy of 3 granular pesticides, alone and in combination with neem oil to control various soil-borne pests of cardamom. Among the pesticides, phorate applied at 5 and 2.5 g a.i./clump gave the highest reduction in damage by *S. cardamomi* (32.9 and 29.2%), followed by guinalphos (18.6 and 15.4%). Carbofuran was effective only at 5 g a.i./ clump. Neem oil had no efffect.

Ascher *et al*. (1996) conducted extensive experiments with *Thrips tabaci* and *Frankliniella occidentalis* (Pergande) (Thysanoptera:Thripidae). The insect growth and development effect and persistence to the industrial neem formulations listed above (applied as sprays or drenches to cotton seedlilngs) in pre- and post-infestation were compared in laboratory experiments and, in part, in glasshouse/laboratory and field/laboratory trials. All trials recorded high activity. Addition of plant-derived oils in combination with PBO improved the effect and extended the persistence of simple, 'home-made' neem seed kernel extracts.

Schmidt *et al.* (1997) carried out greenhouse experiments in eastern Germany in 1995 showed that NeemAzal T/S sprayed on Fuchsia at 0.3 percent was 91.8 percent effective against *Frankliniella occidentalis* and was not phytotoxic.

Saxena and Kidiavai (1997) studied neem seed extract (NSE) spray applications as low-cost inputs for management of the flower thrips in the cowpea. The control of larvae of *Megalurothrips sjostedti* on cowpea (*V. unguiculata*) was studied at Nairobi and Mbita, Kenya, during 1992-94. Applications of 2 or 3 percent NSE at 200 lt/ha with a knapsack sprayer, 38, 47 and 51 days after emergence of the crop or 5, 10 or 20 percent NSE sprayed at 10 lt/ha at 31, 39 and 49 days after emergence, often significantly reduced the number of larvae in flower 2 days after treatment. Also, fewer adults occurred in flowers in plots sprayed at 5, 10 or 20 percent NSE, 51 days after emergence. Grain yield was significantly higher in plots sprayed with 20 percent NSE than in untreated plots and was comparable to the grain yield obtained in plots sprayed three times with cypermethrin. Because of the low cost of NSE, the net gain was often greater when the crop was sprayed withh the extract than with cypermethrin.

References

Abraham, C.C. and Ambika, B. (1979). Effect of leaf and kernel extracts of neem on moulting and vitellogenesis in *Dysdercus cingulatus* Fabr. (Heteroptera : Pyrrhocoridae). *Current Science*, **48**(12) : 554-556.

Agrawal, I.L. (1988). Studies on the biological effects of some plant products on certain insects. **Ph.D. Thesis**, Univ. of Kanpur, Kanpur, India : pp. 186.

Agrawal, I.L. (1990). Ovicidal activity of some phytochemicals on *Myllocerus undecimpustulatus* Faust (Coleoptera : Curculionidae). *Indian Journal of Entomology*, **52**(1) : 35-38.

Agrawal, I.L. and Mall, S.B. (1988). Studies on the insecticidal and antifeedant activity of some plant extracts on Bihar hairy caterpillar, *Diacrisia obliqua* Walker (Lep., Arctiidae). *Journal of Applied Entomology*, **105**(5) : 529-532.

Agrawal, N.; Gupta, R.K.; Srivastava, J.P.; Mathur, Y.K. and Tripathi, R.A. (1998). Effect of some commercial neem formulations on silk worm reared on mulberry leaves. Abst. *ICPPMSA*, 11-13 Dece., CSAU A & T, Kanpur : 192.

Akashe, V.B.; Makar, P.V.; Karkeli, M.S.; Mallapur, C.P. and Nandkarni, P.G. (1998). Evaluation of neem seed kernel extract for control of safflower aphid; *Uroleucon compositae. National Symposium in Entomology in 21st Century*, April 30-May 2, 1998, Udaipur : 137.

Akbar, S.; Ahmed, S. and Hassan, M.U. (1996). Efficacy of neem products and insecticides against whitebacked plant hopper, *Sogatella furcifera* (Horv.). *Pakistan Journal of Zoology*, **28**(1) : 5-7.

Akhauri, R.K.; Sinha, M.M. and Yadav, R.P. (1996). Neem oil as possible biorational insecticide for the managemnet of spotted pod borer (*Maruca testulalis* Geyer) in early pigeon pea. *Neem and Environment*, **1** : 559-564, Ed. R. P. Singh *et al*.

Ali Niazee, M.T.; Amin Alhumeyri and Saeed, M. (1997). Laboratory and field evaluation of a neem insecticide against *Archips rosanus* L. (Lepidoptera : Tortricidae). *Canadian Entomologist*, **129**(1) : 27-33.

Ali, S.I.; Singh, O.P. and Misra, U.S. (1983). Effectiveness of plant oils against pulse beetle *Callosobruchus chinensis* Linn. *Indian Journal of Entomology*, **45**(1) : 6-9.

Ambethgar, V. (1996). Effectiveness of neem (*Azadirachta indica* A. Juss) products and insecticides against rice leaf folder, *Cnaphalocrocis medinalis* Guenee. *Journal of Entomological Research*, **20**(1) : 83-85.

Ameta, O.P. and Gupta, H.C.L. (1998). Evaluation of botanicals against sorghum midge, *Contarinia sorghicola* under field conditions. *Nat. Sem. in Ent. in 21st Century*, April 30-May 2, 1998, Udaipur : 131.

Anwar, T.; Jabbar, A.; Khalique, F.; Tahir, S. and Shakeel, M.A. (1992). Plants with insecticidal activities against four major insect pests in Pakistan. *Tropical Pest Management*, **38**(4) : 431-437..

Ascher, K.R.S.; Meisner, J. and Klein, M. (1996). Efficacy of commercial neem products against lepidopteran pests, leafhoppers and thrips, and improving home-made preparations by adjuvants. *Neem Newsletter*, **13**(3) : 33-34.

Azam, K.M. (1991). Toxicity of neem oil against leaf miner (*Liriomyza trifolii* Burgess) on cucumber. *Plant Protection Quarterly*, **6**(4) : 196-197.

Azmi, M.A.; Naqvi, S.N.H.; Akhtar, K.; Jahan, M.; Habib, R. and Tabassum, R. (1993). Toxicological studies of Solfac and neem formulation against *Sitophilus oryzae. National Academy Science Letters*, **16**(5-6) : 187-190.

Azmi, M.A.; Naqvi, S.N.H.; Khan, M.F.; Akhtar, K. and Khan, F.Y. (1998). Comparative toxicological studies of RB-a (neem extract) and Coopex (permethrin + bioallethrin) against *Sitophilus oryzae* with reference to their effects on oxygen consumption and GOT, GPT activity. *Turkish Journal of Zoology*, **22**(4) : 307-310.

Balu, A.; Rajarishi, R.; Deeparaj, B. and Durairaj, S. (1997). Curling and crinkling of teak leaves. *Indian Forester*, **123**(8) : 775-777.

Barranco, P.; Pena, J. de-la; ;Martin, M.M.; Cabello, T. and de-la Pena, J. (1998). Efficiency of chemical control of the new palm pest *Rhynchophorus ferrugineus* (Olivier, 1790) (Col.:Curculionidae). *Boletin de Sanidad Vegetal*, Plagas, **24**(2) : 301-306.

Basedow, T.; Peters, A.; Kleeberg, H. (ed.) and Zebitz, C.P.W. (1997). Control of colorado potato beetle (*Leptinotarsa decemlineata* Say) by an azadirachtin-formulation ('Neem-Azal T/S'), by *Bacillus thuringiensis tenebrionis* ('Novodor'), and by combinations of both : short-term and long-term effects. *Proceedings in 5th Workshop Wetzlar*, Germany, 22-25, January, 1996 : 59-65.

Baumgart, M.; Brocke, K. vom; Crow, M.H.; Vom Brocke, K.; Kleeberg, H. (ed.) and Zebitz, C.P.W. (1997). Control of the Colorado potato beetle (*Leptinotarsa decemlineata* Say) in organic gardening systems with *Bacillus thruingiensis* var. *san diego* (M-One) and new products from the neem tree (NeemAzal-F and Align TM). *Proceedings in 5th Workshop Wetzlar*, Germany, 22-25, January, 1996 : 67-73.

Beevi, N.D.; Janarthanan, R. and Natarajan, K. (1992). Efficacy of some insecticides against eggs of *Meconellicoccus hirsutus* Green on mulberry, *Morus alba* L. *Journal of Insect Science*, **5**(1) : 114.

Behera, U.K. and Satapathy, C.R. (1996). Screening indigenous plants for their insecticidal properties against *Spodoptera litura* Fab. *Insect Environment*, **2**(2) : 43-44.

Bhanotar, R.K. and Srivastava, Y.N. (1984). Effect of neem kernel suspension on the development of eggs of desert locust, *Schistocerca gregaria* (Forskal). *Neem Newsletter*, **1**(3) : 30.

Bhardwaj, B.S. and Gupta, R.P. (1992). Studies on the effect of plant extract in controlling onion thrips. *Newsletter*, Associated Agricultural Development Foundation, **12**(3) : 9-10.

Bharpoda, T.M.; Patel, J.J.; Jyani, D.B.; Matholia, N.G.; Patel, N.C.; Patel, G.P.; Patel, U.P. and Patel, J.R. (1998). Bio-efficacy of synthetic and botanical insecticides against pests of tomato. *Neem Newsletter*, **15**(2) : 21.

Bharpoda, T.M.; Patel, V.B.; Patel, J.J.; Matholia, N.G.; Patel, U.P.; Patel, G.P. and Patel, J.R. (1998). Field bio-efficacy of neem (gronim 0.3%EC) against leaf eating caterpillar on castor. *National Seminar in Entomology in 21st Century*, April 30-May 2, 1998, Udaipur : 137.

Bhat, N.S.; Manjunatha, M.; Raju, G.T.T.; Nagabhushana, G.G. and Deshpande, V.P. (1988). Chemical control of pod-borer complex of greengram (*Vigna radiata*). *Indian Journal of Agricultural Sciences*, **58**(11) : 864-866.

Bhat, N.S.; Raju, G.T.T.; Manjunatha, M.; Nagabhushana, G.G. and Deshpande, V.P. (1988). Chemical control of cowpea pod borers. *Indian Journal of Plant Protection*, **16**(2) : 197-200.

Bhathal, S.S. and Singh, D. (1993). Toxic, developmental and reproductive effects of AZT-VR-K and some neem products agaisnt mustard aphid, *Lipaphis erysimi* (Kaltenbach). *Neem and Environment*, **1** : 287-296.

Bhathal, S.S. and Singh, D. (1996). Toxic and developmental effects of some neem products against mustard aphid, *Lipaphis erysimi* (Kalt.) through leaf surface treatment. *Journal of Insect Science*, **6**(2) : 226-228.

Bhatnagar, A. (1995). Efficacy of indigenous plant extracts against the eggs of maize stem borer, *Chilo partellus* Swinhoe. *Annals of Plant Protection Sciences*, **3**(1) : 15-18.

Bhatnagar, A. (1996). Efficacy of economics of some insecticides and a neem formulation on incidence of pea aphid (*Acyrthosiphum pisum*) on pea, *Pisum sativum*. *Annals of Plant Protection Sciences*, **4**(2) : 131-133.

Bollhalder, F.; Zuber, M.; Kleeberg, H. (ed.) and Zebitz, C.P.W. (1997). NeemAzal-T/S against *Myzus persicae*. *Proc. 5th Workshop Wetzlar*, Germany, 22-25 January 1996 : 141-145.

Borad, P.K. and Bhalani, P.A. (1997). Bioefficacy of various insecticides against *Coccidohystrix insolita* (Green) on pigeonpea. *Gujarat Agricultural University Research Journal*, *22*(2) : 78-87.

Borah, D. and Saharia, D. (1982). Field evaluation of certain insecticides for the control of *Henosepilachna vigintioctopunctata* F. (Coleoptera : Coccinellidae). *Journal of Research*, Assam Agricultural University, **3**(2) : 224-226.

Borkar, S.L. and Urade, P.N. (1995). Comparative efficacy of insecticides in the management of pod fly (*Melanagromyza obtusa*) on pigeonpea (*Cajanus cajan*) L. *Bioved.*, **6**(1) : 91-96.

Braman, S.K. (1993). Azadirachtin affects growth and survival of immature tawny mole crickets (Orthoptera : Gryllotalpidae). *Florida Entomologist*, **76**(3) : 526-530.

Breithaupt, J. and Schmutterer, H (1996). Aqueous neem seed kernel extracts to control the Asian corn stem borer *Ostrinia furnacalis* in Papua New Guinea. *Neem Newletter*, **13**(3) : 36.

Butani, P.G. and Mittal, V.P. (1990). Comparative efficacy of botanical insecticide (neem seed kernel suspension) and other insecticides against gram pod borer (*Heliothis armigera* Hubner). *Proceedings Symposium Botanical Pesticides in IPM*, Rajahmundry : 276-281.

Cardet, C.; Kandji, T.; Delobel, A. and Danthu, P. (1998). Efficiency of neem and groundnut oils in protecting leguminous tree seeds against seed beetles in the Sahel. *Agroforestry Systems*, **40**(1) : 29-40.

Chandel, R.S.; Chander, R. and Verma, T.D. (1996). Evaluation of insecticidal properties of some oil cakes against *Brahmina coriacea* (Hope) white grubs. *Crop Research Hisar*, **11**(3) :391-393.

Chandrakar, H.K.; Shrivastava, S.K. and Kaushik, U.K. (1992). Toxicity and cost benefit study of some insecticides against *Mythimna separata* Walker. *Oryza*, **29**(2) : 167-168.

Chandramohan, N. and Nanjan, K. (1993). Field evaluation of insecticides against onion thrips, *Thrips tabaci* Lind., on garllic. *Madras Agricultural Journal*, **80**(1) : 53-54.

Chari, M.S.; Ramaprasad, G.; Sitaramaiah, S. and Murthy, P.S.N. (1990). Bio-efficacy of neem formulations against *Spodoptera litura* F. in tobacco nurseries. *Proceedings Symposium Botanical Pesticides in IPM*, Rajahmundry : 145-153.

Chari, M.S.; Ramaprasad, G.; Sitaramaiah, S. and Murthy, P.S.N. (1996). Bioefficacy of natural pesticides from neem and *Pongamia* against *Spodoptera litura* Fabricius (Lepidoptera : Noctuidae). *Neem Newsletter*, **13**(4) : 53-54.

Chari, M.S.; Ramaprasad, G.; Sitaramaiah, S. and Murthy, P.S.N. (1996). Bioefficacy of some neem formulations against *Spodoptera litura* F. *Neem and Environment*, **1** : 595-612.

Chari, M.S.; Ramaprasad, G.; Sitaramaiah, S. and Murthy, P.S.N. (1999). Laboratory and field evaluation of neem (*Azadirachta indica* A. Juss.), pongamia (*Pongamia pinnata* L.), chinaberry (*Melia azedarach* L.) extracts and commercial neem formulations against tobacco caterpillar, *Spodoptera litura* F. **Azadirachta indica A. Juss**. Ed. Singh and Saxena : 111-130.

Chaudhry, M.I. (1992). Efficacy of botanical pesticides against *Plecoptera reflexa* Guen. (Noctuidae, Lepidoptera), shisham defoliator. *Pakistan Journal of Forestry*, **42**(4) : 199-202.

Chitra, K.C.; Rao, S.J. and Rao, P.K. (1997). Efficacy of plant extracts for controlling cotton aphid (*Aphis gossypii*). *Indian Journal of Agricultural Sciences*, **67**(3) : 134-135.

Crocker, R.L. (1992). Evaluation of neem extract for control of southern chinch bug in St. Augustinegrass. *Progress Report*, Texas Agricultural Experiment Station, **4891-4921** : 85-87.

Dange, N.M.; Surywanshi, D.S. and Zanwar, P.R. (1996). Efficacy of insecticides agaisnt safflower aphid. *Journal of Maharashtra Agricultural Univerisities*, **21**(1) : 149-150.

Das, A. and Borah, B. (1997). comparative efficacy of certain insecticides agaisnt *Sitophilus oryzae* (Linn.) and *Sitotroga cerealella* (Oliv.) in stored wheat. *Journal of the Agricultural Science Society of North East India*, **10**(2) : 275-280.

Das, G.P. (1986). Pesticidal efficacy of some indigenous plant oils against the pulse beetle, *Callosobruchus chinensis* Linn. (Coleoptera : Bruchidae). *Bangladesh Journal of Zoology*, **14**(1) : 15-18.

Das, G.P. (1987). Efficacy of neem oil on the egg and grub mortality of *Callosobruchus chinensis* Linn. (Bruchidae : Coleoptera). *Tropical Grain Legume Bulletin*, **34** : 14-15.

Das, S.B. (1998). Synergistic action of neem seed kernel extract and insecticides against pests of pigeonpea. *Insect Environment*, **4**(3) : 83-84.

Das, S.B.; Kandalkar, V.S. and Verma, O.P. (1998). Use of neem derivatives against *Melanagromyza obtusa* (Malloch) and *Helicoverpa armigera* (Hubner) infesting medium maturing pigeonpea. *Neem Newsletter*, **15**(2) : 18.

Dash, A.N.; Senapati, B. and Mishra, P.R. (1996). Efficacy of neem derivatives in combination with synthetic insecticides against population of brown and white backed plant-hoppers and their natural enemies in rice. *Journal of Insect Science*, **9**(2) : 137-142.

Datkhile, R.V.; Pawar, S.A.; Mote, U.N.; Khaire, V.M. and Goel, S.C. (1992). Bioefficacy of different insecticides against gram pod borer *Heliothis armigera* Hubner on chickpea *Cicer arietinum* L. *Proceedings of the National Symposium on Growth, Development and control Technology of Insect Pests* : 156-160.

David, P.M.M.; Haniffa, M.D. and Natarajan, S. (1991). Evaluation of some insecticides and neem oil against Jasmine blossom midge, *Contarinia* sp. *Madras Agricultural Journal*, **78**(1-4) : 82-84.

David, P.M.M.; Haniffa, M.D. and Natarajan, S. (1992). Evaluation of some insecticides and neem oil against Jasmine blossom midge. *Madras Agricultural Journal*, **79**(1) : 15-17.

Debaraj, Y.; Singh, M.P. and Singh, T.K. (1996). Laboratory testing of some insecticides and neem products against cabbage aphid, *Brevicoryne brassicae* (L.). *Annals of Plant Protection Sciences*, **4**(2) : 174-175.

Degaonkar, A.M.; Kurtadikar, J.S. and Rekawar, P.V. (1988). Economical control measures against arhar pod borer complex in Marathwada region. *Pesticides*, **22**(8) : 15-16.

Devakumar, C; Saxena, V.s. and Mukerjee, S.K. (1986). Evaluation of neem (*Azadirachta indica* A. Juss) limonoids and azadirachtin against safflower aphid (*Dactynotus carthami* H.R.L.). *Indian Journal of Entomology*, **48**(4) : 467-470.

Dey, D. and Sarup, P. (1993). Feasibility of protecting maize varieties with vegetable oils to save losses in storage due to *Sitophilus oryzae* Linn. *Journal of Entomological Research*, **17**(1) : 1-15.

Dhaliwal, G.S.; Gill, R.S.; Dilawari, V.K.; Arora, R. (ed.); Randhawa, N.S. (ed.) and Dhawan, A.K. (1998). Management of insect pest complex of cabbage with neem based insecticides. *Proc. of International Conference on Ecological Agriculture : Towards Sustainable Development*, Volume **II**, Chandigarh, India, 15-17 November, 1997 : 306-314.

Dhaliwal, G.S.; Jaswant Singh and Dilawari, V.K. (1996). Potential of neem in insect pest management in rice. *Neem and Environment*, **1** : 425-432.

Dhawan, A.K. and Simwat, G.S. (1995). Evaluation of some botanical insecticides against American bollworm and tobacco caterpillar infesting cotton. *Allelopathy Journal*, **2**(2) : 213-214.

Dhawan, A.K.; Simwat, G.S.; Dhaliwal, G.S. (ed.); Arora, R. (ed.); Randhawa, N.S. (ed.) and Dhawan, A.K. (1998). Evaluation of different biopesticides against cotton bollworm, *Helicoverpa armigera* (Hubner). *Proceedings of International Conference on Ecological Agriculture : Towards Sustainable Development*, Chandigarh, India, 15-17 November, 1997 : 274-280.

Dhawan, A.K.; Simwat, G.S.; Tauro, P. (ed.) and Narwal, S.S. (ed.) (1992). Management of sucking pests of cotton with natural plant products. *Proc. First Nat. Sym. Allelopathy in Agroecosystems* (*Agriculture & Forestry*), Feb. 12-14, 1992 held at CCS Haryana Agricultural Univ., Hisar, India : 154-155.

Dhawan, A.K.; Simwat, G.S.; Tauro, P. (ed.) and Narwal, S.S. (ed.) (1992). Field evaluation of Neemrich 20EC for management of insect pests on cotton during reproductive phase. *Proceedings First National Symposium Allelopathy in Agroecosystems* (*Agriculture and Forestry*), Feb. 12-14, 1992, held at CCS Haryana Agric. Univ., Hisar, India : 152-153.

Dhawan, A.K.; Srimannarayana, G.S.; Simwat, G.S.; Nagaiah, K.; Tauro, P. (ed.) and Narwal, S.S. (ed.) (1992). Management of cotton pests on upland cotton *Gossypium hirsutum* with Navneem 95EC. *Proc. First Nat. Sym. Allelopathy in Agroecosystems* (*Agriculture & Forestry*), Feb. 12-14, 1992 held at CCS Haryana Agricultural Univ., Hisar, India : 156-157.

Dhembare, A.J.; Pondhe, G.M.; Mandakini Padol; Shobha Gulave; Padol, M. and Gulave, S. (1998). Efficacy of some plant extracts against ber fruit borer *Meridarchis scyrodes* (May.). *Advances in Plant Sciences*, **11**(1) : 299-301.

Dimetry, N.Z.; Gomaa, A.A.; Salem, A.A. and Moniem, A.S.H. Abd El (1998). Bioactivity of some formulations of neem seed extracts against the whitefly, *Bemisia tabaci* (Genn.). *Neem Newsletter*, **15**(3) : 38.

Durairaj, C. and Venugopal, M.S. (1994). Comparative evaluation of NeemAzal F and some other neem formulations against cotton whitefly, *Bemisia tabaci* (Gennadius) (Aleyrodidae : Hemiptera). *Neem Newsletter*, **11**(3) : 33.

Dutta, B.C.; Bairagi, P.P. and Bhattacharyya, B. (1997). Bioefficacy of certain insecticides against rice hispa, *Dicladispa armigera* (Oliv.). *Journal of the Agricultural Science Society of North East India*, **10**(2) : 202-205.

Eapen, S.J. (1994). Effect of three granular pesticides on damage by thrips (*Sciothrips cardamomi* R.) in small cardamom (*Elettaria cardamomum* Maton). *Journal of Entomological Research*, **18**(2) : 181-183.

Emosairue, S.O. and Ukeh, D.A. (1997). Field trial of neem products for the control of okra flea beetles (*Podagrica* spp.) in south eastern Nigeria. *Global Journal of Pure and Applied Sciences*, **3**(1) : 13-19.

Erbach, J.H.; Holst, H.; Kleeberg, H. (ed.) and Zebitz, C.P.W. (1997). Effects of NeemAzal-T, NeemAzal-T/S and Delfin on the grape berry moth (*Lobesia botrana*) and the predatory mite (*Typhlodromus pyri*). *Proceedings 5th Workshop Wetzlar*, Germany, 22-25 January, 1996 : 151-154.

Facknath, S (1996). Efficacy of neem and *Melia* in overcoming tolerance of *Plutella xylostella* to some commonly used synthetic insecticides. *Neem and Environment*, **1** : 505-516.

Facknath, S. (1991). Evaluation of a commercial neem formulation (Margosan-O) for the control of *Dysdercus flavidus*. *Revue Agricole et Sucriere de I'lle Maurice*, **70**(3) : 34-38.

Fagoonee, I. (1980). The life cycle, biionomics and control of the cabbage webworm *Crocidolomia binotalis* Zell. (Lepidoptera, Pyralidae). *Revue Agricole et Sucriere de I'lle Maurice*, **59**(2) : 57-62.

Fagoonee, I. and Lauge, G. (1981). Noxious effects of neem extracts on *Crocidolomia binotalis*. *Phytoparasitica*, **9**(2) : 111-118.

Fagoonee, I.; Schmutterer, R. (ed.) and Ascher, K.R.S. (1987). Use of neem in vegetable crop protection in Mauritius. *Proceedings of the 3rd International Neem Conference*, Nairobi, Kenya, 10-15 July, 1986 : 419-429.

Flemming, G. (1997). First experiences at control of *Thaumetopoea processionea* in the north-western region of the Land Brandenberg in the year 1996. *Gesunde Pflanzen*, **49**(2) : 54-57.

Flint, H.M. and Parks, N.J. (1989). Effect of azadirachtin from the neem (Azadirachta indica) tree on immature sweetpotato whitefly, *Bemisia tabaci*, (Homoptera : Aleyrodidae) and other selected pest species on cotton. *Journal of Agricultural Entomology*, **6**(4) : 211-215.

Gahukar, R.T. (1988). Use of neem (*Azadirachta indica*) extracts in the control of insect pests of groundnut. *Insect Science and Its Application*, **9**(5) : 639-640.

Gopakumar, B.; Kumaresen, D.; Chandersekar, S.S. and Varadarasan, S. (1996). Evaluation of neem derivatives and *Lantana camara* L. against cardamom pests. *Neem and Environment*, **1** : 589-594.

Gopal, S. and Senguttuvan, T. (1997). Effect of insecticides and neem products on cotton aphids in water melon. *Madras Agricultural Journal*, **84**(2) : 111-112.

Gottwald, R.; Kleeberg, H. (ed.) and Zebitz, C.P.W. (1997). A three-year trial on effectiveness of neem products against insect pests of field crops. *Proceedings in 5th Workshop Wetzlar*, Germany, 22-25, January, 1996 : 51-58.

Grace, J.K. and Yates, J.R. (1992). Behavioural effects of a neem insecticide on *Coptotermes formosanus* (Isoptera : Rhinotermitidae). *Tropical Pest Management*, **38**(2) : 176-180.

Guddewar, M.B.; Shukla, A. and Pandey, S. (1991). Assessment of insecticidal and ovicidal properties of *Tabernaemontana coronaria* against red cotton bug (*Deysdercus keonigii*). *Indian Journal of Entomology*, **53**(2) : 311-315.

Gujar, G.T. and Mehrotra, K.N. (1988). Biological activityy of neem against the red pumpkin beetle, *Aulacophora foveicollis*. *Phytoparasitica*, **16**(4) : 293-302.

Gujar, G.T. and Mehrotra, K.N. (1990). Toxicity and morphogenetic effects of neem oil on the red cotton bug, *Dysdercus koenigii* F. *Proc. Symp. Botanical Pesticides in IPM*, Rajahmundry : 217-224.

Gupta, G.P.; Katiyar, K.N. and Sharma, K. (1999). Neem in the management strategies of insect pest of cotton. **Azadirachta indica A. Juss**. Ed. Singh and Saxena : 176-190.

Gupta, G.P.; Kundu, S.K. and Sharma, K. (1997). Effect of RD-9 Repelin on growth and development of red cotton bug, *Dysdercus koenigii*. *Indian Journal of Entomology*, **59**(1) : 54-58.

Gupta, G.P.; Lal, R. and Roshan Lal (1998). Utilization of newer insecticides and neem in cotton pest management system. *Annals of Plant Protection Sciences*, **6**(2) : 155-160.

Gupta, G.P.; Mahapatro, G.K. and Chandra, A. (1998). Neem-based pesticides : Targeting different life stages of *Helicoverpa armigera*. *National Seminar in Entomology in 21st Century*, April 30-May 2, 1998, Udaipur : 123.

Gupta, M.P.; Parsai, S.K. and Gupta, D.P. (1990). Bioefficacy and economics of certain insecticides and vegetable oils against gram pod borer, *Heliothis armigera* Hubner on chickpea. *Indian Journal of Plant Protection*, **18**(2) : 207-211.

Hapke, C.; Schutz, C.; Kienzle, J.; Zebitz, C.P.W.; Kleeberg, H. (ed.) and Zebitz, C.P.W. (1997). NeemAzal T/S for the control of aphids on sour cherry and hop. *Proc. 5th Workshop Wetzlar*, Germany, 22-25, Jan., 1996 : 101-104.

Haq, E.; Mohsin, A. and Hashmi, A.A. (1991). Efficacy of botanical and organophosphate pesticides against white backed planthopper *Sogatella furcifera* (Horv.). *Proc. Eleventh Pakistan congress of Zoology*, **11** : 149-152.

Haque, M.A.; Parvin, N. and Ahmed, K.S. (1996). Effect of neem oil on the food consumption and survival of *Epilachna dodecastigma* (Wied.). *Bangladesh Journal of Entomology*, **6**(1-2) : 1-5.

Helson, B.; Barry Lyons and Peter de Groot (1996). Development of neem seed extract for forest insect pest management in Canada. *Neem Newsletter*, **13**(3) : 35-36.

Helson, B.; Barry Lyons and Peter de Groot (1999). Evaluation of Neem EC formulations containing azadirachtin for forest insect pest management in Canada. *Azadirachta indica A. Juss*, Ed. Singh and Saxena : 79-92.

Henkel, G.; Mester, E.; Heyne, P.; Kleeberg, H. (ed.) and Zebitz, C.P.W. (1997). The effect of NeemAzal-T/S on *Dysaphis plantaginea*, *Rhopalosiphum insertum* and beneficial insects. *Proc. 5th Workshop Wetzlar*, Germany, 22-25, Jan., 1996 : 93-99.

Hernandez, C.R. and Vendramim, J.D. (1997). Bioactivity evaluation of aqueous extracts of Meliaceae to *Spodoptera frugiperda* (J.E. Smith). *Revista de Agricultura Piracicaba*, **72**(3) : 305-318.

Hiremath, I.G.; Ahn-YoungJoon; Kim-SoonIl; Ahn-Y.J. and Kim, S.I. (1997). Insecticidal activity of Indian plant extracts against *Nilaparvata lugens* (Homoptera : Delphacidae). *Applied Entomologyy and Zoology*, **32**(1) : 159-166.

Ho-DT and Kibuka, J.G. (1983). Neem (*Azadirachta indica* A. Juss) products for control of rice stem borers. *International Rice Research Newsletter*, **8**(5) : 15-16.

Hohn, H.; Hopli, H.U. and Graf, B. (1996). Quassia and neem : exotic insecticides in fruit culture. *Obst und Weinbau*, **132**(3) : 62-63.

Holmes, M.S. and Hassan, E. (1999). The contact and systemic action of neem seed extract against green peach aphid *Myzus persicae* Sulzer (Hemiptera : Aphididae). ***Azadirachta indica*, A. Juss**. Ed. Singh & Saxena : 93-102.

Hummel, E. and Kleeberg, H. (1997). New results for the practical application of NeemAzal-formulations. *Mitteilungen der Deutschen gesellschaft fur Allgemeine und Angewandte Entomologie*, **11**(1-6) : 331-336.

Hummel, E.; Kleeberg, H. (ed.) and Zebitz, C.P.W. (1997). Effect of neem extract formulations NeemAzal-T/S on the pea aphid *Acyrthosiphon pisum* in the laboratory. *Proc. 5th Workshop Wetzlar*, Germany, 22-25, Jan., 1996 : 33-39.

Hussain, M.A.; Puttaswamy and Viraktamath, C.A. (1996). Effect of botanical oils on lantana bug, *Orthezia insignis* Browne infesting crossandra. *Insect Environment*, **2**(3) : 85-86.

Hussain, M.A.; Puttaswamy and Viraktamath, C.A. (1996). Management of citrus mealybug, *Planococcus citri* Risso on guava using botanical oils. *Insect Environment*, **2**(3) : 73-74.

Huwer, R.; Kleeberg, H. (ed.) and Zebitz, C.P.W. (ed.) (1997). Effect of neem on the 'banana-spotting bug' *Amblypelta lutescens lutescens* (Distant) (Hemiptera : Coreidae) in Australia. *Proc. 5th Workshop Wetzlar*, Germany, 22-25, Jan., 1996 : 213-222.

Ivbijaro, M.F. (1990). The efficacy of seed oils of *Azadirachta indica* A. Juss and *Piper guineense* Schum and Thonn on the control of *Callosobruchus maculatus* F. *Insect Science and its Application*, **11**(2) : 149-152.

Ivbijaro, M.F. and Balaji, O.O. (1990). Effects of cypermethrin + dimethoate and extracts of *Piper guineense* and *Azadirachta indica* on the pests and yield of cowpea, *Vigna unguiculata*. *Journal of Agricultural Science*, **115**(2) : 227-231.

Ivbijaro, M.F. and Balaji, O.O. (1990). Effects of cypermethrin + dimethoate and extracts of *Piper guineense* and *Azadirachta indica* on the pests and yield of cowpea (*Vigna unguiculata*. *Journal of Agricultural Science*, **115**(2) : 227-231.

Jackai, I.E.N. and Oyedivan, I.O. (1991). The potential of neem *Azadirachta indica* A. Juss. for controlling post-flowering pests of cowpea, *Vigna unguiculata* Walp-I. The pod borer, *Maruca testulalis*. *Insect Science and Its Application*, **12**(1-3) : 103-109.

Jackai, L.E.N.; Inang, E.E. and Nwobi, P. (1992). The potential for controlling post-flowering pests of cowpea, *Vigna unguiculata* Walp. using neem, *Azadirachta indica* A. Juss. *Tropical Pest Management*, **38**(1) : 56-60.

Jacob, S. and Sheila, M.K. (1990). Treatment of greengram seeds with oils against the infestation of the pulse beetle, *Callosobruchus chinensis* Linn. *Plant Protection Bulletin*, Faridabad, **42**(3-4) : 9-10.

Jagginavar, S.B.; Kulkarni, K.A. and Lingappa, S. (1991). A note on chemical control of cowpea (*Vigna unguiculata* (L.) Walp.) pod borer complex. *Karnataka Journal of agricultural Sciences*, **4**(1-2) : 56-58.

Jaglan, M.S.; Khokhar, K.S.; Malik, M.S. and Taya, J.S. (1997). Standardization of method for extraction of bioactive components from different plants for insecticidal property. *Indian Journal of Agricultural Research*, **31**(3) : 167-173.

Jaglan, M.S.; Khokhar, K.S.; Malik, M.S.; and Singh, R. (1997). Evaluation of neem (*Azadirachta indica* A. Juss) extracts against American bollworm, *Helicoverpa armigera* (Hubner). *Journal of Agricultural and Food chemistry*, **45**(8) : 3262-3268.

Jaglan, M.S.; Khokhar, K.S.; Malik, M.S.; and Singh, R. (1998). Evaluation of neem (*Azadirachta indica* A. Juss) extracts against American bollworm, *Helicoverpa armigera* (Hubner). *Neem Newsletter*, **15**(4) : 47.

Jaipal, S.; Zik Singh; Malik, O.P. and Singh, Z. (1984). Insecticidal activity of various neem leaf extracts against *Rhyzopertha dominica*, a stored grain pest. *Neem Newsletter*, **1**(4) : 35-36.

Jena, M.; Dani, R.C. and Rajamani, S. (1990). Effectiveness of insecticides against rice gundhi bug. *Oryza*, **27**(1) : 96-98.

Jeyakumar, P. and Uthamasamy, S. (1997). Bio-efficacy of some synthetic insecticides and botanicals against *Liriomyza trifolii*. *Indian Journal of Entomology*, **59**(4) : 347-350.

Ji, BaoZhong; Zhao BoGuang; Wu Ruqi; Ji, B.Z.; Zhao, B.G. and Wu, R.Q. (1998). Effects of neem extracts and quinolizidine alkaloids on the survival and reproduction in *Apriona germari* (Hope) (Coleoptera : Cerambycidae). *Journal of Nanjing Forestry University*, **22**(1) : 83-86.

John Immaraju; Timothy Wood; Nelson, R.; Ruggero W. and Wells S. (1996). Efficacy profile of a commercial neem insecticide. *Neem and Environment*, **2** : 897-908.

Jothi, B.D.; Verghese, A. and Tandon, P.L. (1990). Evaluation of different plant oils and extracts against citrus aphid, *Toxoptera citricidus* (Kirkaldy). *Indian Journal of Plant Protection*, **18**(2) : 251-254.

Jyani, D.B.; Patel, N.C.; Patel, I.J. and Patel, J.R. (1998). Use of azadirachtin based formulations in the strategy of pest management in bhendi. *Neem Newsletter*, **15**(2) : 19.

Kareem, A.A.; Saxena, R.C. and Palanginan, E.L. (1988). Effect of neem (*Azadirachta indica*) seed bitters (NSB) and neem seed kernel extract (NSKE) on pests of mungbean following rice. *Interantional Rice Research Newsletter*, **13**(6) : 41-42.

Karelina, T.N.; Kliiberg, G.; Pukhal Skaya, N.A. and Kovalev, B.G. (1992). Effectiveness of preparations based on azadirachtin against some pests of the order Lepidoptera. *Agrokhimiya*, **6** : 104-111.

Katole, S.R.; Mahajan, R.K. and Satpute, U.S. (1996). Efficacy of some non edible oils and insecticides against citrus blackfly nymphs. *PKV Research Journal*, **20**(1) : 25-27.

Kevin W. Wanner, Berry C. Kostyk and Blair V. Helson (1996). Efficacy of neem seed extract for control of eastern spruce budworm defoliation damage to white spruce seed orchards in Ontario, Canada. *Neem Newsletter*, **13**(3) : 39-40.

Kiarie, M. and Saxena, R.C. (1996). Exploratory evaluation of neem extract sprays on maize for control of stalk borers in Coastal Kenya. *Neem and Environment*, **1** : 571-574.

Kienzle, J.; Schulz, C.; Kleeberg, H. (ed.) and Zebitz, C.P.W. (1997). Two years of experience with the use of NeemAzal in organic fruit orchards. *Proceedings 5th Workshop Wetzlar*, Germany, 22-25 January, 1996 : 27-31.

Kirsch, K.; Schmutterer, R. (ed.) and Ascher, K.R.S. (1987). Studies of the efficacy of neem extracts in controlling major insect pests in tobacco and cabbage. *Proceedings of the 3rd International Neem Conference*, Nairobi, Kenya, 10-15 July, 1986 : 495-515.

Klocke, J.A. and Kubo, I. (1982). citrus limonoid by-products as insect control agents. *Entomologia Experimentalis et Applicata*, **32**(3) : 299-301.

Knodel, J.J.; Larew, H.G. and Webb, R.E. (1986). Margosan-O, a commercial formulation of neem seed extract, controls *Liriomyza trifolii* on chrysanthemums. *Journal of Agricultural Entomology*, **3**(3) : 249-254.

Kossou, D.K. (1989). Evaluation of different products of neem *Azadirachta indica* A. Juss for the control of *Sitophilus zeamais* Motsch on stored maize. *Insect Science and its Application*, **10**(3) : 365-372.

Koul, O. (1984). Azadirachtin : I - Interaction with the development of red cotton bugs. *Entomol. Exp. Appl.*, **36** : 85-88.

Koul, O. (1984). Azadirachtin. II. Interaction with the reproductive behaviour of red cotton bugs. *Zeitschrift fur Angewandte Entomologie*, **98**(2) : 221-223.

Krishna Moorthy, P.N. and Srinivasan, K. (1996). Evaluation of neem products and standard chemicals for the management of bean fly, *Ophiomyia phaseoli* (Tryon) in french beans. *Neem and Environment*, **1** : 433-438.

Krishnaiah, N.V.; Katti, G.; Krishnaiah, K.; Pasalu, I.C.; Kumar, K.M. and Lingaiah, T. (1998). Studies on systemic action of neem formulations in rice. *Pesticide Research Journal*, **10**(2) : 155-159.

Krishnamoorthy, A. and Rajagopal, D. (1998). Effect of insecticides on the California red scale, *Aonidiella aurantii* (Maskell) and its natural enemies. *Pest Management in Horticultural Ecosystems*, **4**(2) : 83-88.

Kulat, S.S.; Nimbalkar, S.A. and Hiwase, B.J. (1997). Ralative efficacy of some plant extracts against *Aphis gossypii* Glover and *Amrasca devastans* (Distant) on okra. *PKV Research Journal*, **21**(2) : 146-148.

Kulkarni, N. and Joshi, K.C. (1997). Insecticidal action of some plant extracts against Albizia defoliator, *Rhesala imparata* Walker (Lepidoptera : Noctuidae). *Entomon*, **22**(2) : 135-139.

Kumar, M.G.; Bhat, P.K. and Ramaiah, P.K. (1989). Potential role of kerosene and neem derivatives in integrated management of mealybugs on coffee. *Journal of Coffee Research*, **19**(1) : 17-29.

Kumar, M.G.; Bhat, P.K. and Vijayalakshmi, C.K. (1989). Screening of kerosene and neem extract against coffee green scale. *Journal of Coffee Research*, **19**(1) : 51-55.

Kwanchai, S. and Choeikamhaeng, A. (1997). Thai neem extracts from pilot plant (KU) for the control of beet armyworm (*Spodoptera exigua*) on Asparagus. *Proceedings of first International Symposium on Biopesticides* held at Naresuan University, Phitsanulok, Thailand, Oct. 27-31, 1996 : 165-168.

Laprade, S. and Cerdas, V.H. (1998). Managemet of the spiralling white fly (*Aleurodicus dispersus* Ressell, Homoptera : Aleyrodidae) in banana (Musa AAA) using insecticides and oil. *CORBANA*, **22**(48) : 89-94.

Larew, H.G. (1988). Limited occurrence of foliar-, root- and seed-applied neem seed extract toxin in untreated plant parts. *Journal of Economic Entomology*, **81**(2) : 593-598.

Larew, H.G.; Knodel, M.J.J.; Webb, R.E. and Warthen, J.D. (1985). *Liriomyza trifolii* (Burgess) (Diptera : Agromyzidae) control on chrysanthemum by neem seed extract applied to soil. *Journal of Economic Entomology*, **78**(1) : 80-84.

Lawand, P.C.; Patil, A.S.; Mali, M.S.; Borole, D.H.; Waghulde, J.K.; Chaudhary, Y.R. and Patil, A.D. (1992). Evaluation of Neemark, a botanical insecticide against sugarcane white flies. *SISSTA Sugar Journal*, **18**(1) : 21-27.

Leskovar, D.I. and Boales, A.K. (1996). Azadirachtin : potential use for controlling lepidopterous insects and increasing marketability of cabbage. *HortScience*, **31**(3) : 405-409.

Lindquist, R.K.; Casey, M.L. and Krueger, H.R. (1986). Evaluation of neem extract, a natural plant product, for leafminer control on chrysanthemum. *Ohio Florists' Association Bulletin*, **880** : 7-8.

Longiswaran, G. and Venugopal, M.S. (1994). Efficacy of neem oil, insecticides and their combinations on pre-harvest infestaion by rice angoumois grain moth, *Sitotroga cerealella* Oliv. *Persticide Research Journal*, **6**(1) : 98-99.

Lowery, D.T.; Isman, M.B. and Brard, N.L. (1993). Laboratory and field evaluation of neem for the control of aphids (Homoptera : Aphididae). *Journal of Economic Entomology*, **86**(3) : 864-870.

Lyons, D.B.; Helson, B.V.; Jones, G.C.; McFarlane, J.W. and Scarr, T. (1996). Systemic activity of neem seed extracts containing azadirachtin in pine foliage for control of the pine false webworm, *Acantholyda erythrocephala* (Hymenoptera : Pamphilidae). *Proceddings of the Entomological Society of Ontario*, **127** : 45-55.

Madhusudhan, V.V. and Gopalan, M. (1988). Chemical control of thrips *Stenchaetothrips biformis* in the rice nursery. *International Rice Research Newsletter*, **13**(3) : 42.

Mahadevan, N.R. (1998). Azadirachtin to control tomato fruit borer. *Insect Environment*, **3**(4) : 92.

Mahapatro, G. and Nanda, U. (1998). Role of neem in managing rice insect pest complex in kharif season. *National Seminar in Ent. in 21st Century*, April 30-May 2, 1998, Udaipur : 122.

Malik, Y.P.; Singh, S.V. and Pandey, N.D. (1996). Field evaluation of some plant extracts and insecticides against linseed bud fly, *Dasyneura lini*. *Indian Journal of Entomology*, **58**(1) : 32-35.

Manger, W.; Koeckhoven, J.; Kleeberg, H. (ed.) and Zebitz, C.P.W. (1997). Control of insects in Brussels sprouts with NeemAzal-T/S. *Proc. 5th Workshop Wetzlar*, Germany, 22-25 January, 1996 : 41-49.

Mariappan, V. and Saxena, R.C. (1983). Effect of mixtures of custard-apple oil and neem oil on survival of *Nephotettix virescens* (Homoptera : Cicadellidae) and on rice tungro virus transmission. *Journal of Economic Entomology*, **76**(3) : 573-576.

Mariappan, V. and Saxena, R.C. (1984). Effect of mixture of custard-apple oil and neem oil on survival of *Nephotettix virescens* (Homoptera : Cicadellidae) and on rice tungro virus transmission. *Journal of Economic Entomology*, **77**(2) : 519-521.

Marion, D.F.; Larew, H.G.; Knodel, J.J. and Natoli, W. (1990). Systemic activity of neem extract against the birch leafminer. *Journal of Arboriculture*, **16**(1) : 12-16.

Markowsky, E.; Kleeberg, H. (ed.) and Zebitz, C.P.W. (1997). The control of larvae of the waterlily leaf beetle *Galerucella nymphaeae* L. with the preparation NEEM AZAL T/S produced by the company TRIFOLIO-M GmbH. *Proceedings in 5th Workshop Wetzlar*, Germany, 22-25, January, 1996 : 123-127.

Meisner, J. and Nemny, N.E. (1992). The effect of Margosan-O on the development of the Egyptian cotton leafworm, *Spodoptera littoralis* (Boisd.) (Lep., Noctuidae). *Journal of Applied Entomology*, **113**(4) : 330-333.

Meisner, J.; Klein, M. and Keren, S. (1990). Effect of Margosan-O on the development of *Earias insulana*. *Phytoparsitica*, **18**(4) : 287-297.

Mikolajczak, K.L.; Zilowski, B.W. and Bartell, R.J. (1989). Effect of meliaceous seed extracts on growth and survival of *Spodoptera frugiperda* (J.E. Smith). *Journal of chemical Ecology*, **15**(1) : 121-128.

Moorthy, P.N.K.; Kumar, N.K.K.; Selvaraj, c. and Daniel, J.S. (1998). Neem seed kernel extract applications for diamondback moth management : transfer of technology for mechanised farming. *Pest Management in Horticultural Ecosystems*, **4**(2) : 128-130.

Mostafa, T.S. (1991). The efficiency of neem flower and fruit powders against *Trogoderma granarium* Everts adults infesting stored rice grains (Coleoptera : Dermestidae). *Bulletin of the Entomological Society of Egypt*, Economic Series, **17** : 93-99.

Mourier, M. (1997). Effects of neem (*Azadirachta indica*) kernel water extracts on cassava mealybug, *Phenacoccus manihoti* (Hom., Pseudococcidae). *Journal of Applied Entomology*, **121**(4) : 231-236.

Murthy, V.; Mala, S.R.; Vadivelu, M. and Solayappan, A.R. (1994). Evaluation of new plant product 'PLEXIN' for the control of sugarcane whiteflies, *Aleurolobus barodensis* (Hemiptera : Aleyrodidae). *Cooperative Sugar*, **25**(5-6) : 209-210.

Naik, L.K. and lingappa, S. (1993). Efficacy of some newer insecticides against green peach aphid, *Myzus persicae* (Sulzer) on beedi tobacco. *Indian Journal of Plant Protection*, **21**(1) : 55-58.

Nanda, U.; Mohapatra, G.K.; Sahu, A. and Mohapatra, S.C. (1996). Perspective of neem pesticides in insect-pest management of rice. *Neem Newsletter*, **13**(3) : 44.

Nandagopal, V. and Soni, V.C. (1992). Residual effect of some insecticides and neem-oil against jassid on groundnut. *Journal of Maharashtra Agricultural Universities*, **17**(3) : 420-422.

Nandagopal, V.; Jaroli, R.K.; Kumar, R. and Reddy, P.S. (1990). Neem products : possible insecticides on the groundnut jassid *Balclutha hortensis*. *International Arachis Newsletter*, **8** : 22.

Nandihalli, B.S.; Hugar, P. and Patil, B.V. (1990). Evaluation of neem and neem products against cotton whitefly, *Bemisia tabaci* (Gennadius). *Karnataka Journal of Agricultural Sciences*, **3**(1-2) : 58-61.

Naqvi, S.N.H.; Akhtar, K. and Azmi, M.A. (1991). Toxicity of NfD against *Sitophilus oryzae* L. exposed to impregnated filter paper and its effects on phosphatases and protein metabolites. *Acta Biologica Cracoviensia Series Zoologia*, **33** : 49-58.

Naqvi, S.N.H.; Nurulain, S.M.; Azmi, M.A. and Asdaque, T. (1989). Effect of neem fractions and malathion against whiteflies, *Aleurobus barodensis* on brinjal crop (*Solanum melongena*). *Sarhad Journal of Agriculture*, **56**(1-3) : 25-28.

Naqvi, S.N.H.; Raza, S.; Khan, M.F.; Ahmad, M. (ed.) and Shakoori, A.R. (1994). Toxicity of neem extract fractions against mango leaf hoppers *Amritodus atkinsoni* in comparison with dimethoate and their effect on some enzymes. *Proc. of Pakistan Congress of Zoology*, **Vol. 12** held under auspices of the Zoological Society of Pakistan Govt. College, Lahore, April, 1992 : 477-486.

Nasseh, O.; Krall, S.; Wilps, H.; Salissou, G.B. and Nasseh, O.M. (1992). The effects of growth inhibitors and of biotic insecticides on larvae of *Schistocerca gregaria* (Forskal). *Sahel PV Info.*, **45** : 5-19.

Nasseh, O.; Wilps, H.and Krall, S. (1993). Neem products - effective biopesticides for combatting the desert locust, *Schistocerca gregaria* (Forskal). *Zeitschrift fur Pflanzenkrankheiten und Pflanzenschutz*, **100**(6) : 611-621.

Naumann, K.; Rankin, L.J. and Isman, M.B. (1994). Systemic action of neem seed extract on mountain pine beetle (Coleoptera : Scolytidae) in lodgepole pine. *Journal of Economic Entomology*, **87**(6) : 1580-1585.

Nelson, S.J. and Logiswaran, G. (1998). Status of the sapota leaf webber and bud borer, *Nephopteryx eugraphella* Ragonot in Tamil Nadu. *Pest Management in Horticultural Ecosystems*, **4**(1) : 25-28.

Nestor, Bautista M.; Oscar, Morales G.; Carrillo, S.J.L.; Hiram, Bravo M.; Nestor, B.M.; Oscar, M.G. and Hiram, B.M. (1998). Mortality of *Phyllocnistis citrella* with mineral oil and neem. *Manejo Integrado de Plagas*, **50** : 29-33.

Niber, B.T.; Helenius, J.; Varis, A.L. and Tierto Niber, B. (1992). Toxicity of plnat extracts to three storage beetles (Coleoptera). *Journal of Applied Entomology*, **113**(2) : 202-208.

Nimbalkar, S.A.; Khodke, S.M.; Taley, Y.M. and Patil, K.J. (1990). Bioefficacy of some new insecticides including neem seed extract and neem oil for control of whitefly, *Bemisia tabaci* Genn. on cotton. *Proc. Symp. Botanical Pesticides in IPM*, Rajahmundry : 256-260.

Nimbalkar, S.A.; Khodke, S.M.; Taley, Y.M. and Patil, K.J. (1990). Bioefficacy of neem and karanj seed extracts along with other commercial plant products against bollworms on cotton. *Proceedings Symposium Botanical Pesticides in IPM*, Rajahmundry : 252-255.

Olaifa, J.I. and Adenuga, A.O. (1988). Neem products for protecting field cassava from grasshopper damage. *Insect Science and Its Application*, **9**(2) : 267-270.

Omara, S.M.; Kelany, I.M.; Kleeberg, H.; Kleeberg, H. (ed.); Zebitz, C.P.W. (1997). Effect of an aqueous neem seed kernel powder extract (ANSKPE) and NeemAzal-F on the *Liriomyza congesta* (Becker) and *Aphis craccivora* (Koch) infesting broad bean at Zagazig Region, Sharkia Governorate, Egypt. *Proceedings 5th Workshop Wetzlar*, Germany, 22-25 Jan. 1996 : 223-235.

Padmanaban, B.; Daniel, M. and Srimannarayana, G. (1997). Evaluation of plant products and oil cake against white grub *Leucopholis burmeisteri*. *Indian Journal of Entomology*, **59**(4) : 362-365.

Pal, R.K.; Tripathi, R.A.; Prasad, R. and Rameshwar Prasad (1996). Relative toxicity of certain plant extracts to khapra beetle, *Trogoderma granarium*. *Annals of Plant Protection Sciences*, **4**(1) : 35-37.

Panda, P.; Panigrahi, D. and Sasmal, A. (1998). Evaluation of some insecticides against termites on Sesamum in Orissa. *Annals of Plant Protection Sciences*, **6**(1) : 118-120.

Pandey, K.P. and Singh, S. (1998). Evaluation of natural plant product as an insecticide against top borer *Scirpophaga excerptalis* Wlk. in sugarcane. *Entomon*, **23**(3) : 235-237.

Pandey, N.D.; Singh, L.; Singh, Y.P. and Tripathi, R.A. (1988). Effect of certain plant extracts against *Lipaphis erysimi* Kalt. under laboratory conditions. *Indian Journal of Entomology*, **49**(2) : 238-242.

Pareek, R.A. and Batta, J.K. (1998). Laboratory and field evaluation of plant products against maize stem borer *Chilo partellus*. *National Seminar in Entomology in 21st Century*, April 30-May 2, 1998, Udaipur : 127.

Parmar, B.S. and Dutta, S. (1987). Evaluation of some neem oils as malathion synergists. *International Journal of Tropical Agriculture*, **5**(3-4) : 323-326.

Parmar, B.S.; Srivastava, K.P.; Schmutterer, H. (ed.) and Ascher, K.R.S. (1987). Development of some neem formulations and their evaluation for the control of *Spilosoma obliqua* in the laboratory and *Euchrysops cnejus* in the field. *Proceedings of the 3rd International Neem Conference*, Nairobi, Kenya, 10-15 July, 1986 : 205-215.

Partridge, M.J. and Borden, J.H. (1997). Evaluation of neem seed extract for control of the spruce aphid, *Elatobium abietinum* (Walker) (Homoptera : Aphididae). *Canadian Entomologist*, **129**(5) : 899-906.

Pasini, M.; Tosi, L. and Galbero, G. (1997). Control trials against adults of *Metcalfa pruinosa* (Say) with different active ingredients. *Informatore Agrario*, **53**(20) : 68-70.

Pasupathy, S.; Bahskaran, R.; Ayyamperumal, A. and Rajukkannu, K. (1998). Management of mango hoppers, *Idioscopus niveosparsus* using botanicals. *National Seminar in Ent. in 21st Century*, April 30-May 2, 1998, Udaipur : 126.

Pasupathy, S.; Bahskaran, R.; Ayyamperumal, A. and Rajukkannu, K. (1998). Management of tea mosquito bug with varieties and botanicals in guava. *National Seminar in Entomology in 21st Century*, April 30-May 2, 1998, Udaipur : 120.

Patanaik, H.P. (1997). Studies on neem formulations against serpentine leafminer, and tobacco caterpillar, on tomato. *Insect and Environment*, **3**(1) : 10.

Patel, C.B.; Jhala, R.C.; Patel, M.M. and Shah, A.H. (1993). Field bio-efficacy of Neemark in comparison to chemical insecticides against sugarcane whitefly, *Aleyrolobus barodensis* Maskell (Aleyrodidae : Homoptera). *Botanical Pesticides in Integrated Pest Management* : 343-350.

Patel, J.R.; Patel, J.I.; Mehta, D.M. and Shah, B.R. (1993). Integrated management of *Amsacta moorei* Butler with botanical insecticides. *Botanical Pesticides in Integrated Pest Management* : 327-331.

Patel, K.N. and Purohit, M.S. (1998). Evaluation of indigenous botanical materials against Epilachna beetle, *Henosepilachna vigintioctopunctata*. *National Seminar on Entomology in 21st Century*, April 30-May 2, 1998, Udaipur, pp. 136.

Patel, V.B. and Patel, J.R. (1996). Ovicidal property of some indigenous plant extracts against *Spodoptera litura* Fab. *Gujarat Agricultural University Research Journal*, **21**(2) : 137-140.

Patel, V.B.; Patel, N.C. and Patel, J.R. (1998). Bio-efficacy of some neem based formulations against hairy caterpillar, *Selepa celtis* infesting amla. *National Seminar in Entomology in 21st Century*, April 30-May 2, 1998, Udaipur : 127.

Patil, B.V.; Nandihalli, B.S.; Hugar, P.; Thimmanagouda, B.R. and Lingappa,

S. (1990). Efficacy of some plant and animal products against cotton pest complex. *Proc. Symp. Botanical Pesticides in IPM*, Rajahmundry : 183-189.

Patil, C.S.; Pawar, S.A.; Mote, U.N. and Khaire, V.M. (1991). Evaluation of insecticides against flea beetles on sorghum. *Annals of Applied Biology*, **118**(Supplement) : 22-23.

Patil, K.J.; Deshkar, M.M.; Rane, A.E. and Nimbalkar, S.A. (1990). Some indigenous plant materials against *Aphis gossypii* G. and *Dactynotus carthami* H.R.L. *Proc. Symp. Botanical Pesticides in IPM*, Rajahmundry : 238-244.

Pawar, D.B.; Lawande, K.E. and Warade, S.D. (1996). Control of *Liriomyza trifolii* on tomato. *Journal of Maharashtra Agricultural Universities*, **21** : 1.

Perez, G.; Padron, R.; Soto, R. and Bertsch, F. (1998). Effect of three neem natural pesticides on control of pests of field cabbage, field corn and stored cowpea seed. *Agronomia Costarricense*, **21**(2) : 259-266.

Phadke, A.D.; Khandal, V.S. and Rahalkar, S.R. (1988). Use of a neem [*Azadirachta indica*] product in insecticide resistance management (IRM) in cotton. *Pesticides Bombay*, **22**(4) : 36-37.

Pillai, M.A.K. and Ponniah, S. (1988). Neem for control of rice thrips. *International Rice Research Newsletter*, **13**(5) : 33-34.

Prabhakar, N.; Toscano, N.C. and Coudriet, D.L. (1989). Susceptibility of the immature and adult stages of the sweetpotato whitefly (Homoptera : Aleyrodidae) to selected insecticides. *Journal of Economic Entomology*, **82**(4) : 983-988.

Prasad, R.; Pal, R.K. and Mathur, Y.K. (1998). Joint action of certain plant extract against red hairy caterpillar on mungbean. Abst. *ICPPMSA*, 11-13 Dece., CSAU A & T, Kanpur : 162.

Prasad, S.K. (1998). Efficacy of neem products agaisnt mustard aphid, *Lipaphis erysimi* under field condition. *National Seminar in Entomology in 21st Century*, April 30-May 2, 1998, Udaipur : 135.

Prasad, S.K. (1998). Efficacy of some neem products vis-a-vis oxydemeton methyl against *Lipaphis erysimi* on rapeseed crop under field condition. *Indian Journal of Entomology*, **59**(2) : 147-150.

Prashad, R.G.; Joshi, B.G. and Rao, S.N. (1986). Ralative efficacy of some insecticides and neem seed kernel suspension against tobacco caterpillar, *Spodoptera litura* Fabricius on tobacco. *Indian Journal of Plant Protection*, **14**(2) : 69-74.

Price, J.F. and Schuster, D.J. (1991). Effects of natural and synthetic insecticides on sweetpotato whitefly, *Bemisia tabaci* (Homoptera : Aleyrodidae) and its hymenopterous parasitoids. *Florida Entomologist*, **74**(1) : 60-68.

Qadri, S.S.H. and Brahmanand Rao, B. (1979). Effect of combining some indigenous plant seed extracts against house-hold insects. *Pesticides*, **11**(12) : 21-23.

Radhakrishnan, S.; Rajamanickam, B. and Madhava Rao, S. (1983). Efficacy of certain insecticides against *Callosobruchus chinensis* (Lino.) on *Cajanus cajan* seeds in storage. *Pesticides*, **17**(11) : 19-20.

Ragumoorthi, K.N. and Arumugam, R. (1992). Control of moringa fruitfly *Gitona* sp., and leaf caterpillar *Noorda blitealis* with insecticides and botanicals. *Indian Journal of Plant Protection*, **20**(1) : 61-65.

Ragumoorthi, K.N. and Arumugam, R. (1996). Control of guava tea mosquito bug, *Helopeltis antonii* Sign. (Miridae : Hemiptera) with insecticides and botanicals. *Indian Journal of Plant Protection*, **24**(1-2) : 1-5.

Ragumoorthy, K.N. and Subba Rao, P.V. (1998). Dose-mortality response of neem formulations to maggots of moringa fruit fly *Gitona distigma* (Meigon) (Diptera : Drosophilidae) and their effect on puparia of moringa fruit fly as soil drench. Abst. *ICPPMSA*, 11-13 Dece., 1998, CSAU A & T, Kanpur (India) : 167-168.

Raguraman, S. and Rajasekaran, B. (1996). Effect of neem products on insect pests of rice and the predatory spider. *Madras Agricultural Journal*, **83**(8) : 510-515.

Rai, A.B.; Pastagia, J.J.; Patel, H.M.; Patel, M.B. and Patel, C.B. (1996). Evaluation of certain pesticides against sugarcane whitefly, *Aleurolobus barodensis* Mask. (Hemiptera : Aleyrodidae). *Cooperative Suage*, **27**(6) : 439-441.

Rajasri, M.; Reddy, C.P.V.; Krishnamurthy, M.M. and Prasad, V.D. (1991). Bioefficacy of certain newer insecticides including neem products against chilli pest complex. *Indian Cocoa, Arecanut and Spices Journal*, **15**(2) : 42-44.

Rajgopal, D.; Gowda, K. and Gowda, K.N.M. (1994). Efficacy of 'Limonool' (neem based pesticide) against web worm *Crocidolomia binotalis* Zell infesting cabbage. *Neem Newsletter*, **11**(2) : 19-20.

Rajgopal, D.; Gowda, K. and Thippaiah, M. (1994). Efficacy of 'Limonool' (neem based pesticide) against *Aphis craccivora* Koch in field bean (*Lablab purpureus*). *Neem Newsletter*, **11**(2) : 17-18.

Ram, C. and Verma, J.P. (1998). Bio-efficacy of different neem products against insect pests of cowpea. *National Seminar in Entomology in 21st Century*, April 30-May 2, 1998, Udaipur : 130.

Ramamurthy, R. and Venugopal, M.S. (1997). Effect of plant products on the incidence of *Sitotroga cerealella* under field condition. *Indian Journal of Entomology*, **59**(1) : 8-10.

Ramaraju, K. and Babu, P.C.S. (1989). Effect of plant derivatives on brown planthopper (BPH) and whitebacked planthopper (WBPH) nymph emergence on rice. *International Rice Research Newsletter*, **14**(5) : 30.

Ramprasad, G.; Sitaramaiah, S.; Joshi, B.G. and Rao, S.N. (1987). Relative efficacy of neem seed kernel suspension, some synthetic pyrethroids and insecticides against *Spodoptera litura* F. in tobacco nurseries. *Indian Journal of Plant Protection*, **15**(2) : 190-192.

Rao, D.V.S.; Thippeswamy, M.; Murty, P.S.S. and Siva Rao, D.V. (1984). Studies on field evaluation of insecticides on rainfed groundnut. *Pesticides*, **18**(12) : 35-38.

Rao, G.R. and Dhingra, S. (1997). Synergistic activity of some vegetable oils in mixed formulations with cypermethrin against different instars of *Spodoptera litura* (Fabricius). *Journal of Entomological Research*, **21**(2) : 153-160.

Rao, K.R. and Panwar, V.P.S. (1992). Efficacy of insecticides/plant products against shoot fly species (*Atherigona soccata* and *A. naquii*) in spring sown maize. *Pesticide Research Journal*, **4**(1) : 55-57.

Rao, K.R. and Panwar, V.P.S. (1996). Relative efficacy of some insecticides and neem products for the control of shoot flies (*Atherigona* spp.) in spring sown maize. *Pesticide Research Journal*, **8**(2) : 124-131.

Rao, K.T. and Rao, N.V. (1990). Repelin, a plant origin insecticide for controlling pod borer, *Helicoverpa armigera* of pigeonpea, *Cajanus cajan*. *Proceedings Symposium Botanical Pesticides in IPM*, Rajahmundry : 232-235.

Rao, N.V.; Reddy, A.S. and Reddy, P.S. (1990). Relative efficacy of some new insecticides on insect pests of cotton. *Indian Journal of Plant Protection*, **18**(1) : 53-58.

Rao, P.K.; Aleem, M.A.; Chitra, K.C. and Mani, A. (1990). Efficacy of some botanicals and ash against pulse beetle, *Callosobruchus chinensis* (Linnaeus). *Proc. Symp. Botanical Pesticides in IPM*, Rajahmundry : 282-287.

Rao, P.R.M. and Prakasha Rao, P.S. (1979). Effects of biocides on brown planthopper adults on rice. *International Rice Research Newsletter*, **4**(3) : 20.

Rao, R.V.S. and Srivastava, K.P. (1984). Evaluation of neem formulaitons against sorghum earhead worm. *Neem Newsletter*, **1**(4) : 37-38.

Rao, S.J.; Chitra, K.C.; Rao, P.K. and Reddy, K.S. (1992). Antifeedant and insecticidal properties of certain plant extracts against brinjal spotted leaf beetle, *Henosepilachna vigintioctopunctata* (Fab.). *Journal of Insect Science*, **5**(2) : 163-164.

Reddy, A.V. and Singh, R.P. (1998). Fumigant toxicity of neem (*Azadirachta indica* A. Juss.) seed oil volatiles against pulse beetle, *Callosobruchus maculatus* Fab. (Col., Bruchidae). *Journal of Applied Entomology*, **122**(9-10) : 607-611.

Reddy, P.V.R.; Chitra, K.C. and Rao, P.K. (1990). Field evaluation of certain plant extracts for the control of brinjal spotted leaf beetle, *Henosepilachna vigintioctopunctata*. *Journal of Insect Science*, **3**(2) : 194-195.

Reddy, V.V.S.; Ramesh Babu, T. and Narasimha Rao, B. (1996). Repelin - An effective botanical insecticide against Ber fruitfly, *Carpomyia vesuviana* Costa. *Neem and Environment*, **1** : 575-580.

Reed, D.K. and Reed, G.I. (1984). Control of vegetable insects with neem seed extracts. *Proceedings of the Indiana Academy of Science*, **94** : 335-339.

Robert, H.J. V. and Denis, J.W. (1994). Biological activity of neem seed kernel extracts and synthetic azadirachtin against larvae of *Plutella xylostella* L. *Neem Newsletter*, **11**(2) : 26-27.

Rohitha, B.H.; Pottinger, R.P.; Lauren, D.R. and Popay, A.J. (ed.) (1991). Activity of neem, nutmeg and pine oils against the lightbrown apple moth. *Proceedings of the Forty Fourth New Zealand Weed and Pest Control Conference* : 58-60.

Rovesti, L. and Desco, K.V. (1991). Effectiveness of neem seed kernel extract against *Leucoptera malifoliella* Costa (Lep., Lyonetiidae). *Journal of Applied Entomology*, **111**(3) : 231-236.

Roychoudhry, R. and Jain, R.K. (1994). Comparative evaluation of neem and bakayan oils with some common insecticides for control of whitefly (*Bemisia tabaci* Genn.) and virus transmission. *Neem Newsletter*, **11**(2) : 21-25.

Rwamputa, A.K. and Schabel, H.G. (1988). Effects of *Azadirachta indica* kernel and leaf extracts on the larvae of *Heteronygmia dissimilis*. *Journal of the Tanzania Association of Foresters*, **6** : 3-10.

Sachan, J.N. and Gururaj, Katti (1996). Neem in *Helicoverpa* [*Heliothis*] *armigera* management : pulses ecosystem. *Neem and Environment*, **1** : 459-468.

Sachan, J.N. and Lal, S.S. (1990). Role of botanical insecticides in *Helicoverpa armigera* management in pulses. *Proceedings Symposium Botanical Pesticides in IPM*, Rajahmundry : 261-269.

Sahayaraj, K. and Paulraj, M.G. (1998). Effect of neem leaf extract on *Amsacta albistriga* Walker. *Insect Environment*, **4**(2) : 42-43.

Sahayaraj, K. and Paulraj, M.G. (1998). Relative toxicity of some plant extracts to groudnut leaf miner, *Aproaerema modicella* Dev. *International Arachis Newletter*, **18** : 27-29.

Sahayaraj, K. and Paulraj, M.G. (1998). Screening the relative toxicity of some plant extracts to *Spodoptera litura* Fab. (Insecta : Lepidoptera : Noctuidae) of groundnut. *Fresenius Environmental Bulletin*, **7**(9-10) : 557-560.

Sahayaraj, K. and Sekar, R. (1996). Efficacy of plant extracts against tobacco caterpillar larvae in groundnut. *International Arachis Newsletter*, **16** : 38.

Saikia, D.K. and Dutta, S.K. (1997). Efficacy of some insecticides and plant products against fruit fly, *Dacus tau* (Walker) on ridge gourd, *Luffa acutangula* L. *Journal of the Agricultural Science Society of North East India*, **10**(1) : 132-135.

Salem, S.A. (1991). Evaluation of neem seed oil as tuber protectant against *Phthorimaea operculella* Zell. (Lepidoptera : Gelechiidae). *Annals of Agricultural Science*, Moshtohor, **29**(1) : 589-595.

Samalo, A.P.; Senapati, H.; Satpathy, C.R. and Jacob, T.J. (1993). Effect of neem derivatives on incidence of some major insect pests of wet season rice. *Botanical Pesticides in Integrated Pest Management* : 197-202.

Samuthiravelu, P and David, B.V. (1990). Bio-efficacy of neem oil and deltamethrin against spotted boll worm *Earias vittella* (Fab.) (Noctuidae : Lepidoptera) on cotton (MCU 5). *Madras Agricultural Journal*, **78**(7-8) : 294-298.

Samuthirvelu, P. and David, B.V. (1991). Bioefficacy of neem oil and endosulfan against the fruit borer *Earias vittella* (Fab.) (Noctuidae : Lepidoptera) on okra. *Madras Agricultural Journal*, **77**(1-4) : 77-78.

Sanap, M.M. and Pawar, V.M. (1997). Evaluation of neem based formulations against *Helicoverpa armigera* (hubner) infesting chickpea. *Neem Newsletter*, **14**(4) : 57-59.

Sardara, H.R. and Kumar, N.K.K. (1989). Effectiveness of plant oils against leaf hopper and shoot and fruit borer on okra. *Indian Journal of Entomology*, **51**(2) : 167-171.

Sarode, S.V. and Gabhane, A.T. (1994). Performance of neem seed kernel extract with reduced insecticidal dosage on the infestation of okra fruit borer, *Earias vittella* Fab. *Journal of Entomological Research*, **18**(4) : 327-330.

Sarode, S.V. and Gabhane, A.T. (1998). Effect of neem and insecticide combinations for okra fruit borer management. *Pest Management in*

Horticultural Ecosystems, **4**(1) : 54-56.

Saxena, P. (1995). Toxicity of mixtures formulated separately in four ratios with commonly used insecticides and non-toxic chemicals to the adults of *Sitophilus oryzae* Linn. *Indian Journal of Entomology*, **57**(3) : 301-304.

Saxena, R.C. (1986). Effect of custard apple oil, neem oil, and neem cake on green leafhopper (GLH) population and on tungro (RTV) infection. *International Rice Research Newsletter*, **11**(2) : 25.

Saxena, R.C. and Kidiavai, E.L. (1997). Neem seed extract spray applications as low-cost inputs for management of the flower thrips in the cowpea crop. *Phytoparasitica*, **25**(2) : 99-110.

Saxena, R.C.; Kareem, A.; Palanginan, E.L.; Malayba, M.T. and Kareem, A.A. (1989). Systemic and foliar applications of neem [*Azadirachta indica*] seed bitters (NSB) to control leafhopper (GLH) and rice tungro virus (RTV) disease. *International Rice Research Newsletter*, **14**(1) : 31.

Schauer, M. (1984). Effect of variously formulated neem seed extracts on *Acyrthosiphon pisum* and *Aphis fabae* in natural pesticides from the neem tree and other tropical plants. *Proc. 2nd Int. Neem Conf.*, Rauischhalzhausen, 1983, GTZ, Eschborn : 141-150.

Schmidt, R.; Kleeberg, H. (ed.). and Zebitz, C.P.W. (1997). NeemAzal-T/S against flower thrips (*Frankliniella occidentalis*) on greenhouse fuchsia. *Proceedings 5th Workshop Wetzlar*, Germany, 22-25 Jan., 1996 : 139-140.

Schmutterer, H. (1985). Which insect pests can be controlled by application of neem seed kernel extracts under field condition ? *Z. Angew. Ent.*, **100** : 468-475.

Schmutterer, H. (1989). First investigation on the insecticidal effects of methanolic extracts form the leaves and bark of the Philippine neem tree, *Azadirachta integrifoliola* Merr., and comparison of these effects with those of extracts from *Azadirachta indica* (A. Juss) and *Melia azedarach* L. *Journal of Applied Entomolgoy*, **108**(5) : 483-489.

Schmutterer, H. and Doll, M. (1993). The marrango or Philippine neem tree, *Azadirachta excelsa* (= *A. integrifoliola*) : a new source of insecticides with growth-regulating properties. *Phytoparasitica*, **21**(1) : 79-86.

Scholz Dobelin, H. (1997). Experiences with natural enemies. *Gemuse Munchen*, **33**(4) : 246-248.

Schulz, C.; Kienzle, J.; Hermann, P and Zebitz, C.P.W. (1997). NeemAzal-T/S - a new botanical insecticide for fruit growing. *Gesunde Pflanzen*, **49**(3) : 95-99.

Schulz, C.; Kienzle, J.; Kleeberg, H. (ed.) and Zebitz, C.P.W. (1997). Effects of different NeemAzal-formulations on apple aphids and *Aphis fabae* Scop. *Proc. 5th Workshop Wetzlar*, Germany, 22-25 Jan. 1996 : 81-92.

Seck, D.; Sidibe, B.; Haubruge, E. and Gaspar, C. (1991). Protection of stores of cowpea (*Vigna unguiculata* (L.) Walp) at farm level : the use of different formulations of neem (*Azadirachta indica* A. Juss) from Senegal. *Mededelingen van de faculteit Landbouwwetenschappen, Rijiksuniversiteit Gent.*, **56**(3b) : 1217-1224.

Sehgal, V.K. and Ujagir, R. (1990). Effect of synthetic pyrethroids, neem extracts [from *Azadirachta indica* seed kernels] and other insecticides for the control of pod damage by *Helicoverpa armigera* (Hubner) on chickpea and pod damage-yield relationship at Pantnagar in northern India. *Crop Protection*, **9**(1) : 29-32.

Senguttuvan, T. (1998). Effects of botanicals against tea mosquito bug in cashew. *Insect Environment*, **4**(3) : 78.

Shamshad Ali; Rajnish Kumar and Rizvi, S.M.A. (1998). Bioefficacy of neem products against linseed budfly, *Dasineura lini* Barnes. *Neem Newsletter*, **15**(2) : 18.

Sharma, H.C.; Sankaram, A.V.B. and Nwanze, K.F. (1996). Utilisation of natural pesticides derived from neem and custard apple for integrated pest management. *Neem Newsletter*, **13**(3) : 35.

Sharma, V.K. and Bhatnagar, A. (1993). Studies on the effects of non edible seed oils on *Chilo partellus* Swinhoe. *Botanical Pesticides in Integrated Pest Management* : 351-359.

Shukla, A.; Pathak, S.C. and Agrawal, R.K. (1996). Efficacy and economics of some insecticides and plant products against the infestation of okra shoot and fruit borer, *Earias vittella* Fab. *Crop Research Hisar*, **12**(3) : 367-373.

Shukla, B.C.; Kaushik, U.K. and Gupta, R. (1996). Evaluation of neem products against insect pests of rice and pulse crops. *Neem and Environment*, **1** : 497-504.

Simwat, G.S.; Dhawan, A.K.; Tauro, P. (ed.) and Narwal, S.S. (ed.) (1992). Use of plant insecticides against *Heliothis armigera* (Hubner) infesting cotton and chickpea. *Proceedings First National Symposium Allelopathy in Agroecosystems (Agriculture and Forestry)*, Feb. 12-14, 1992, held at CCS Haryana Agric. Univ., Hisar, India : 158-159.

Singh, A.K.; Khan, A.M.; Jain, M.K.; Chandel, B.S. and Pandey, U.K. (1987). Insecticidal properties of some oils against *Rhizopertha dominica* Fabr. (Bostrichidae : Coleoptera), a pest of stored grains. *Zeitschrift fur*

Angewandte Zoologie, **74**(4) : 411-415.

Singh, C.P. (1998). Aphidicidal action of plant extracts against *Lipaphis erysimi*. *National Seminar in Entomology in 21st Century*, April 30-May 2, 1998, Udaipur : 134.

Singh, C.P. (1998). Bioefficacy of some essential oils against *Lipaphis erysimi* in tarai of Uttar Pradesh. *Nat. Sym. in Entomology in 21st Century*, April 30-May 2, 1998, Udaipur : 134.

Singh, C.P. and Gupta, K.C. (1998). Synergistic effect of neem soap on the activity of plant based synthetics against mustard aphid *Lipaphis erysimi* (Kaltenbach.) (Hemiptera : Aphididae). Abs. *ICPPMSA*, 11-13 Dec., 1998 at CSAU A&T, Kanpur : 170-171.

Singh, H. and Korpraditskul, V. (1999). Evaluation of some plant extracts against aphids, key vectors of chilly-virus complex. ***Azadirachta indica*, A. Juss**, Ed. Singh & Saxena : 139-146.

Singh, H.; Mrig, K.K. and Mahla, J.C. (1996). Efficacy and persistence of plant products against lesser grain borer, *Rhyzopertha dominica* (F.) in wheat grain. *Annals of Biology*, Ludhiana, **12**(1) : 99-103.

Singh, J.; Sukhija, H.S. and Singh, P. (1990). Evaluation of neem oil against rice leaf folder and stem borer. *Proceedings Symposium Botanical Pesticides in IPM*, Rajahmundry : 288-290.

Singh, K. and Haridayesharya (1998). Insecticidal activity of neem and clove oil mixture against mustard aphid. Abs. *ICPPMSA*, 11-13 Dec., 1998, CSAU A&T, Kanpur : 142.

Singh, K.J. and Singh, O.P. (1993). Efficacy of biopesticides on eggs and adults of cowpea-pod bug (*Chauliops fallax*) infesting soybean (*Glycine max*). *Indian Journal of Agricultural Sciences*, **63**(11) : 756-758.

Singh, N.N.; Rai, D.K. and Rai, P.C. (1988). Judicious use of insecticides against safflower aphid, *Uroleucon carthami* (HRL.) on safflower. *Journal of Aphidology*, **2**(1-2) : 22-27.

Singh, R.A. and Pandey. V. (1998). Efficacy of botanical pesticides against rice yellow stem borer *Scirpophaga incertulas* (Wlk.). *Neem Newletter*, **15**(2) : 18.

Singh, R.P. and Srivastava, R.P. (1997). Effect of formulated neem products in controlling mango leaf hopper, *Idioscopus nitidulus* (Walker). *Proc. of First International Symposium on Biopesticides* held at Naresuan University, Phitsanulok, Thailand, Oct. 27-31, 1996 :133-136.

Singh, R.P.; Devakumar, C. and Dhingra, S. (1988). Activity of neem (*Azadirachta indica* A. Juss) seed kernel extracts against the mustard aphid, *Lipaphis erysimi*. *Phytoparasitica*, **16**(3) : 225-230.

Singh, R.P.; Doharey, K.L. and Padmaja Saxena (1999). Evaluation of neem dust formulation against khapra beetle, *Trogoderma granarium* Everts and cigarette beetle, *Lasioderma sericorne* F. *Azadirachta indica* A. Juss. Ed. Singh and Saxena : 76-78.

Singh, R.P.; Masarrat Haseeb and Srivastava, R.P. (1998). Effect of neem derivatives on mango mealy bug, *Drosicha mangiferae* (Green). Abs. *National Symposium on Mango Production and Export*, June 25-27, 1998 at Central Institute for Subtropical Horticulture, Lucknow : 59.

Singh, R.S. (1998). Evaluation of some bioinsecticides against pea pod borers. National Seminar in Entomology in 21st Century, April 30-May 2, 1998, Udaipur : 121.

Singh, S. (1998). Effectiveness of NSKE in the management of melon and oriental fruitflies. Abs. *ICPPMSA*, 11-13 December, 1998, CSAU A & T, Kanpur (India).

Singh, S.C.; Pandey, N.K. and Kumar, A. (1996). Effect of neem saw dust on mortality of the pulse beetle (*Callosobruchus chinensis* L.) on black gram (*Vigna mungo*). *Journal of Ecotoxicology and Environmental Monitoring*, **6**(1) : 69-71.

Singh, S.V.; Sanjeev Pandey; Guddewar, M.B.; Malik, Y.P. and Pandey, S. (1997). Response of Neem extractives agianst red cotton bug, *Dysdercus koenigii* on cotton seed. *Indian Journal of Entomology*, **59**(1) : 41-44.

Singh, T.V.K. and Azam, K.M. (1986). Seasonal occurrence, population dynamics and chemical control of citrus leafminer, *Phyllocnistis citrella* Stainton in Andhra Pradesh. *Indian Journal of Entomology*, **48**(1) : 38-42.

Singh, Y.P.; Pandey, S.; Guddewar, M.B.; Shukla, A. and Saini, M.L. (1991). Efficacy of some plant extracts against mustard sawfly (*Athalia proxima* Klug). *Plant Protection Bulletin Faridabad*, **43**(3-4) : 26-30.

Sombatsiri, K; Temboonkeat, K.; Schmutterer, H. (ed). and Ascher, K.R.S. (1987). Efficacy of an improved neem kernel extract in the control of *Spodoptera litura* and of *Plutella xylostella* under laboratory conditions and in field trials. *Proceedings of the 3rd International Neem Conference*, Nairobi, Kenya, 10-15 July, 1986 : 195-203.

Sontakke, B.K. and Dash, A.N. (1996). Studies on the effects of insecticides and neem formulation against major pests of mustard and their safety to honey bees. *Indian Journal of Environment and Toxicology*, **6**(2) : 87-88.

Sontakke, B.K.; Panda, S.K.; Satapathy, C.R. and Samal, T. (1994). Effect of neem oil, certain sprayable insecticides and their mixture on the incidence of white backed planthopper, *Sogatella furcifera* (Horvath) (Delphacidae : Homoptera). *Orissa Journal of Agricultural Research*, **7**(3-4) : 89-91.

Soundararajan, R.P.; Chitra, N.; Gunasekaran, K.; Anbalagan, G. and Kumar, K. (1998). Bioefficacy of solvent free neem formulations against curry leaf psyllid. Abs. *ICPPMSA*, 11-13 Dec., 1998 at CSAUA&T, Kanpur : 155.

Sridevi, D. and Dhingra, S. (1996). Ealuation of some non-toxic vegetable oils as synergists for different synthetic pyrethroids in mixed formulations against *Tribolium castaneum* (Herbst). *Journal of Entomological Research*, **20**(4) : 335-343.

Srivastava, K.P.; Agnihotri, N.P.; Gajbhiya, V.T. and Jain, H.K. (1984). Relative efficacy of fenvalerate, quinalphos and neem kernel extracts for the control of pod fly, *Melanagromyza obtusa* (Malloch) and pod borer, *Heliothis armigera* Hubner infesting red gram, *Cajanus cajan* (L.) Millsp. together with their residues. *Journal of Entomological Research*, **8**(1) : 1-4.

Srivastava, R.P. and Peter Proksch (1998). Bioactivity of seven plant extracts against egyptian cotton leafworm, *Spodoptera littoralis*. National Seminar in Entomology in 21st Century, April 30-May 2, 1998, Udaipur : 120.

Srivastava, R.P.; Singh, R.P.; Haseeb, M. and Abbas, S.R. (1996). Evaluation of enriched neem (*Azadirachta indica* A. Juss) seed extracts against mango hoppers, *Idioscopus nitidulus* Walker and *Amritodus atkinsoni* Leth. *Neem and Environment*, **1** : 581-588.

Stark, J.D. and Rangus, T.M. (1994). Lethal and sublethal effects of the neem insecticide formulation, 'Margosan-O', on the pea aphid. *Pesticide Science*, **41**(2) : 155-160.

Stark, J.D. and Walter, J.F. (1995). Neem oil and neem oil components affect the efficacy of commercial neem insecticides. *Journal of Agricultural and Food Chemistry*, **43**(2) : 507-512.

Subba Rao, P.V. and Mohan, S. (1996). Efficacy of neem seed kernel extract on pigeonpea pod borer. *Neem Newsletter*, **13**(1) : 1-2.

Suhas, Y.; Balikai, R.A.; Biradar, B.M.; Mokashi, A.N. and Yelshetty, S. (1996). Management of ber fruit borer, *Meridarchis scyrodes* Meyrick (Carposinidae : Lepidoptera). *Karnataka Journal of Agricultural Scienices*, **9**(3) : 529-532.

Tabassum, R.; Naqvi, S.N.H.; Jahan, M.; Nurulain, S.M.; Khan, M.F. and Azmi, M.A. (1998). Determination of the toxicities of fenpropathrin (pyrethroid) and neem formulation (RB-a + PBO + Tx-100) against *Alphitobius diaperinus* adults and their effects on transaminases. *Turkish Journal of Zoology*, **22**(4) : 319-322.

Tandon, P.L. and Lal, B. (1980). Control of mango mealy bug, *Drosicha mangiferae* Green (Margarodidae : Homoptera) by application of insecticides in soil. *Entomon*, *5*(1) : 67-69.

Tanzubil, P.B. and McCaffery, A.R. (1990). Effects of azadirachtin and aqueous neem seed extracts on survival, growth and development of the African armyworm, *Spodoptera exempta*. *Crop Protection*, **9**(5) : 383-386.

Tappertzhofen, S. (1995). On the effect of neem products on *Bemisia tabaci* Genn. on aubergines using differet adjuvants. *Anzeiger fur Schadlingskunde Pflanzenschutz, Umweltschutz*, **68**(6) : 137-139.

Thakar, A.V.; Ameta, O.P.; Tripathi, N.N.; Vyas, A.; Tauro, P. (ed.) and Narwal, S.S. (ed.) (1992). Relative efficacy of indigenous plant extract against gram pod borer, *Heliothis armigera* (Hub.). *Proceedings First National Symposium Allelopathy in Agroecosystems* (*Agric. & For.*), Feb. 12-14, 1992, held at CCS Haryana Agric. Univ., Hisar, India : 162.

Thakur, A.K. and Singh, A.K. (1998). Evaluation of neem formulations in comparison to some recommended insecticides against jassid on okra. *Neem Newsletter*, **15**(2) : 17.

Thakur, R.C.; Nema, K.K. and Kango, K.N. (1988). Comparative efficacy of neem seed kernel and some insecticidal formulations against the gram pod borer, *Heliothis armigera* (Hubner). *Legume Research*, **11**(3) : 114-116.

Thomas, A.W.; Strunz, G.M.; Chiasson, M. and Chan, T.H. (1992). Potential of Margosan-O, an azadirachtin-containing formulation from neem seed extract, as a control agent for spruce budworm, *Choristoneura fumiferana*. *Entomologia Experimentalis et Applicata*, **62**(1) : 37-46.

Tomar, R.K.S. (1998). Efficacy of some insecticides against okra petiole maggot *Melanagromyza hibisci*. *Indian Journal of Entomology*, **60**(1) : 22-24.

Tripathy, M.K. and Sahoo, P. (2000). Efficacy of some neem formulations and neem seed kernel extract (NSKE) against *Callosobruchus chinensis* (Linn.) infesting blackgram. *Neem Newsletter*, **17**(1) : 7-11.

Tuncer, C. and Aliniazee, M.T. (1998). Acute and chronic effects of neem on *Myzocallis coryli* (Homoptera : Aphididae). *International Journal of Pest Management*, **44**(2) : 53-58.

Venkataramireddy, P.; Chitra, K.C. and Rao, P.K. (1990). Efficacy of the plant extracts in the control of brinjal spotted leaf beetle, *Henosepilachna vigintioctopunctata* F. *Proc. Symp. Botanical Pesticides in IPM*, Rjajhmundry : 225-227.

Venkateswarlu, P.; Raghavaiah, G. and Nagalingam, B. (1992). Efficacy of neem oil alone and in combination with certain insecticides on pod borer complex of blackgram, *Vigna mungo* (L.) Hepper. *New Agriculturist*, **3**(2) : 147-152.

Venkatetan, S.; Balasubramanian, G.; Jayaraj, S. and Gopalan, M. (1987). Studies on the efficacy of neem products agaisnt the aphid *Aphis gossypii* Glover on cotton. *Madras Agricultural Journal*, **74**(4-5) : 255-257.

Verghese, A. (1997). Effect of neem on first instar crawlers of the grape mealybug, *Maconellicoccus hirsutus* Green. *Insect Environment*, **2**(4) : 121-122.

Verma, R.R. and Rachna, Joshi (1988). Prophylactic control measures of *Dyscerus fletchari* Marshall (Coleoptera : Curculionidae). *Progressive Horticulture*, **20**(3-4) : 355-356.

Vogt, H.; Handel, U.; Vinuela, E.; Kleeberg, H. (ed.); Zebitz, C.P.W. (1997). Field investigations on the efficacy of NeemAzal-T/S against *Dysaphis plantaginea* (Passerini) (Homoptera : Aphididae) and its effects on larvae of *Chrysoperla carnea* Stephens (Neuroptera : Chrysopidae). *Proceedings of 5th Workshop Wetzlar*, Germany, 22-25 Jan., 1996 : 105-114.

Vyas, B.N.; Ganesan, S.; Raman, K.; Godrej, N.B. and Mistry, K.B. (1999). Effects of three plant extracts and Achook : a commercial neem formulation on growth and development of three noctuid pests. **Azadirachta indica A. Juss**. Ed. Singh and Saxena : 103-110.

Vyas, B.N.; Raman, K.; Godrej, N.B. and Mistry, K.B. (1996). Effects of four plant-based products on growth and development of three important noctuid pests. *Neem Newletter*, **13**(2) : 15.

Walter, J.F. (1999). Adjuvants, activators and synergists. Do additives improve the activity of azadirachtin? ***Azadirachta indica*** **A. Juss**, Ed. Singh & Saxena : 47-56.

Walter, J.F. (1999). Neemix and Neemix 4.5 for the management of fruit and vegetable pests on a commercial scale. *Azadirachtin Indica A. Juss*, Ed. Singh and Saxena : 57-66.

Wilps, H.; Kirkilionis, E. and Muschenich, K. (1992). The effects of neem oil and azadirachtin on mortality, flight activity, and energy metabolism of *Schistocerca gregaria* Forskal - a comparison between laboratory and field locusts. *Comparative Biochemistry and Physiology C, Comparative Pharmacology and Toxicology*, **102**(1) : 67-71.

Wilps, H.; Nasseh, O. and Krall, S. (1996). The effects of various neem formulations on mortality rate and morphogenetic defects upon

Schistocerca gregaria (Forskal) larvae. *Neem and Environment*, **1** : 221-236.

Wyss, E. (1997). Different strategies to control the rosy apple aphid *Dysaphis plantaginea* in organic fruit growing. *Mittellungen-der-Deutschen-Gesellschaft-fur-Allgemeine-und-Angewandte Entomologie*, **11**(1-6) : 233-236.

Xie, Y.S.; Gagnon, D.; Arnason, J.T.; Philogene, B.J.R.; Lambert, J.D.H.; Kaminski, J.; Morand, P.; Timmins, G. and Werstiuk, N.H. (1991). Effects of azadirachtin on the western corn rootworm, *Diabrotica virgifera* (LeConte) (Coleoptera : Chrysomelidae). *Canadian Entomologist*, **123**(3) : 707-710.

Yee, W.L. and Toscano, N.C. (1998). Laboratory evaluation of synthetic and natural insecticides on beet armyworm (Lepidoptera : Noctuidae) damage and survival on lettuce. *Journal of Economic Entomology*, **91**(1) : 56-63.

Zehnder, G. and Warthen, J.D. (1988). Feeding inhibition and mortality effects of neem-seed extract on the colorado potato beetle (Coleoptera : Chrysomelidae). *Journal of Economic Entomology*, **81**(4) : 1040-1044.

Chapter 5

Antifeedant, deterrent, IGR and repellent effects of neem products on insects

From centuries, farmers in the Indian subcontinent and in other part of the world have been protecting their crops from damage of insects with natural repellent, antifeedant, deterrent and insect growth regulators (IGR) found in the fruits and leaves of neem tree (*Azadirachta indica*). These products even work as an inhibitor of nitrifying bacteria. Botanical pesticides are predominant and the most promising source of biological pesticides today is the neem. These neem products have been found effective against more than 200 agricultural and household pests. The feeding deterrents, repellents, antifeedants and IGR are chemicals that can protect plants from insect herbivore through regulation of feeding. Feeding deterrence can be caused by an effect of the chemical on the chemoreception and/or centres that regulate feeding and metabolism. Feeding deterrent primarily affects the feeding behaviour of insects but they can also be toxic if fed upon.

Azadirachtin isolated from neem (*A. Indica*) is one of the example of neem deterrent/ antifeedant/repellent/IGR. The activity of azadirachtin, a limonoid from the neem tree as a strong antifeedant, IGR and as an elicitor of reproductive effects is now well understood and documented although its biochemical effects at the cellular level are still unknown. Antifeedancy varies markedly between species with Lepidoptera being particularly sensitive to azadirachtin. The physiological effects on growth, moulting and reproduction are more consistent between species although cuticle or gut may provide barriers to bioavailability in some species. The mode of action of azadirachtin lies in (i) effects on deterrent and other chemoreceptors resulting in antifeedancy, (ii) effects on ecdysteroid (moulting hormone) and juvenile hormone tittres through a blockage of morphogenetic peptide hormone release and (iii) direct effects on most other tissues studied resulting in an overall loss of fitness of the insect. The complexity of the molecular structure of azadirachtin has precluded its synthesis for pesticide use although novel synthesis of the parent molecule is now almost complete and research into simpler mimetic substances is ongoing. Applied research has concentrated on a variety of natural formulations from neem seed kernels which contain azadirachtin together with several structurally related molecules.

Earliest recorded indication on neem acting as an insect repellent is perhaps by Mann and Burns in 1927 who have quoted an observation that during the locust cycle of 1926-27 the adult locusts did not feed on neem leaves (*Melia azadirachta*). The next report in the literature is by Astrakhantzev *et al.* (1936) who reported the efficacy of water and alcohol extracts of *M. azedarach* as aphicide against *Brevicoryne brassicae* L. The alkaline alcohol extracts prepared to contain 32 g fruit per 100 ml were the most effective producing 98.2 percent mortality after 48 hrs. This is first report indicating insecticidal activity of the neem extracts, though the concentration used was very high. Pruthi (1937) found that neem leaves mixed with grain or kept in 5 or 7 cm layer over the stored grain, protected the commodity from damage by storage pests.

Volkonsky (1937, a & b) reported that extracts of leaves of *M. azedarach* when sprayed on other plants gave protection against locust. During that period Fry et al. (1938) also reported efficacy of 8 percent air dried leaves of *M. azedarach* for the protection of cacao beans against *Ephestia cautella* Wlk.

Systematic studies on repellent and insecticidal properties of different parts of neem and its products started in the early 60's when Pradhan, Jotwani and Rai (1962) reported for the first time very high antifeeding properties of the neem seed kernel against desert locust, *Schistocerca gregaria*. These findings created great interest and a number of research workers throughout the world started working on this peculiar plant, they isolated and identified the active principles from the fruits of the two neem varieties viz. *Azadirachta indica* and *Melia azedarach*.

1. Coleoptera

In the earlier period work, only neem leaves were tested by different workers and that too mostly against the stored grain pests. In the later years, however, neem seed was also tested against a large number of field pests. It was for the first time, Jotwani and Sircar (1965) protected the wheat seed by mixing the powdered neem seed kernels at the rate of 1 to 2 parts per 100 parts of seed (w/w) against three major pests viz. *Sitophilus oryzae* Linn., *Rhizopertha dominica* F. And *Trogoderma granarium*. The seeds were protected for at least 269, 321 and 379 days, respectively. Jotwani and Sircar (1967) reported that seeds of mung, bengal gram, cowpea and pea can effectively be protected from damage of *Callosobruchus maculatus* Fabr. for about 237, 236 and 282 days, respectively by mixing the seed with crushed neem seed at the rate of 1 to 2 parts per 100 parts of seed (w/w). The neem kernel powder at 2 percent is also reported most effective as jowar protectant against S. oryzae (Deshpande, 1967).

Pandey *et al*. (1976) tested powders of sweetflag rhizomes caner drupes, adhatoda and sadabahar leaves, petroleum ether extract of garlic and onion and neem oil for their repellent property against *Callosobruchus chinensis* Linn. infesting gram seed. These products afford protection to stored pulses gram (*Cicer arietinum*). The results showed that 1 to 2 parts of powdered caner (*Thevetia neriifolia*) or rhizomes of sweet flag (*Acorus calamus*), 8 parts sadabahar leaf powder (*Ipomoea carnea*) and 1-3 parts of petroleum ether extract of garlic (*Allium sativum*) and neem oil (*A. indica*) per 100 parts of seed effectively protect gram seeds from damage by the bruchid for at least 135 days. The Adhatoda leaf powder (*Adhatoda vasica*) (8 parts per 100 parts of seed) and petroleum ether extract of onion (*Allium cepa*) (5 parts/100 parts of seed) did not afford protection against the attack of bruchid.

Ladd *et al*. (1978) tested solvent extracts from seeds of the neem tree (*A. indica* A. Juss). *Popillia japonica* Newman from feeding on sassafras leaves (*Sassafras albidum* Nutt) and soybeans leaves (*Glycin max* L.). In most of the untreated plants, the foliage was completely destroyed whereas leaves protected with neem extracts remained untouched. In the experiment in 1975, three extracts viz. hexane, ethyl ether and ethyl alcohol were prepared and used for the experiments. All the 3 extracts appeared to be outstanding deterrents to feeding. It was observed that at the lowest concentration (0.25%) there were slight indications of the removal of parenchymatous tissue from any of the treated surfaces. However, when beetles were offered only treated leaves (0.5 and 2.0%), there were only occasional small scars on the surfaces. Some beetles died rather than consume the treated foliage. The experiment conducted in 1976, with aqueous preparations made from the neutral portion of the ethyl ether extract (i.e. extract freed by acids and bases by successive washings with 5 percent potassium hydroxide solution and 5 percent hydrochloric acid solution). The experiment was conducted to evaluate the residual deterrence to beetles of a single application of 1 percent emulsion of the neutral fraction. The results indicated that the neem extract provided a high degree of repellency for several days. In the laboratory experiment, the neem treated leaflets collected 3 days after treatment remained undamaged, those collected after 12 days had only slight damage, and those tested 17 days after treatment showed moderate feeding. The detterence in both tests in 1977 in feeding on neem treated and untreated plants are striking. The tests with soybeans indicated that the untreated plants were completely eaten up by ninth day while there was no feeding in the treated plants. Ladd and Jacobson (1978) carried out laboratory and field studies in Ohio USA to evaluate the effectiveness of solvent extracts prepared from seeds of neem (*A. indica*) as feeding deterrents to *Popillia japonica* Newm. Initial laboratory studies, using beetles confined to sassafras leaves, aqueous preparations of extracts as low as 0.25 percent gave good foliage protection. Subsequent assays demonstrated that single spray applications of 1 percent extracts could protect soybean foliage form feeding by *P. Japonica* for up to 2 weeks. In small-scale field tests carried out in 1977, plantings of soybeans sprayed every 3-4 days with 1 percent solutions of neem extracts remained relatively undamaged by beetle feeding at population levels that resulted in the total defoliation of the untreated plants.

Annonymous (1980) reported powdered neem seed (*A. indica*) as a protectant against *Trogoderma granarium* in stored wheat grains.

Ahmad *et al.* (1980) dried and powdered leaves of tobacco, neem (*A. indica*), *Vitex negundo* and *Polygonum serrulatum* in Bangladesh as repellents against adults of *Sitophilus oryzae* L., attacking stored wheat grains under laboratory conditions. The most effective repellent, as observed from the relative numbers of infested grains and the total weight loss from wheat grains treated with the plant preparations, was V. Negundo, in wheat treated with powdered leaves of this plant, however, the mortality rates of *S. oryzae* observed 1-3 months after application were the lowest of all treatments.

According to Chin *et al.* (1980), *Cordia cyclindristachya*, which was introduced from the America into Malaysia and grown as a hedge plant since the 1950s, was recognised in 1977 to have become a noxious weed in some parts, and in 1977-78. *Schematiza cordiae* Barber was successfully released for its control. Since it is important to know whether this beetle is likely to attack other plants when *Cordia* becomes scarce, its biology and especially its food-plant specificity and feeding habits were studied in the field and in the laboratory. In the laboratory, its fecundity and life-span were not significantly different at 22 and 28°C. Starvation tests with larvae and adults showed S. cordiae to be very specific to *C. cyclindristachya*; it fed on 4 other unrelated plants, especially if these were treated with extracts of *C. cyclindristachya*, but it could not complete its development on any of them. The deterrent effect of extracts of non-food-plants (*Brassica rapa*, tobacco, *Nephelium lappaceum* and *Azadirachta indica*) applied to *Cordia* leaves was tested and *A. indica* produced the most effective in preventing feeding.

Schmutterer and Rembold (1980) carried out laboratory studies in the German Federal Republic, some fractions of neem seeds from West Africa were purified, and 4 pure fractions had strong to very strong growth disruption effects on *Epilachna varivestis* Muls., while a fifth had a lesser effect. After uptake of the neem seed ingredients by feeding on treated bean leaves, various more or less typical symptoms were observed in fourth-instar larvae, depending on the individual fraction, resulting in death or failure to moult. Only one fraction exhibited obviously repellent properties.

Ambika *et al.* (1981) carried out laboratory experiments in India to compare the effects of extracts of neem leaves (*A. indica*) and the juvenile hormone analogues hedroprene (ZR-512) and MV-678) [1-(8-methoxy-4,8-dimethylnonyl)-4-(1-methylethyl) benzene] on adults of *Callosopbruchus chinensis* L. When the insects were reared on seeds of green gram (*Vigna radiata*) treated with 0.25 m µ g hydroprene applied in a 5-ml spray, a significant reduction in fecundity and the number of progeny was observed; the analogue also caused the highest mortality of any of the treatments after 24 hrs (30.5%). The neem leaf extracts at 2.5 and 5.0 percent and MV-678 at 0.5 mµ g in a 5-ml spray were also effective in inhibiting progeny production.

Reed and Pierce (1981) carried out studies in the USA to test neem extracts for their repellency to the striped cucumber beetle (*Acalymma vittatum* F.) by cutting pieces from cantaloupe leaves, dipping them in the extract solution and placing them in a dish with untreated leaf pieces. Five fasting beetles were placed in the dish; some of the extract concentrations protected the leaves for 22 hrs. In another test, muskmelon seedlings were sprayed and placed in a cage with 24 beetles. At 0.1 percent azadirachtin gave good protection for at least 3 days and in some cases for up to 10 days. Application of 0.25 percent azadirachtin also protected the plants form bacterial wilt symptoms.

Ascher and Gsell (1981) carried out a laboratory study in the German Federal Republic, 4th-instar larvae of *E. varivestis* Muls. were fed for 3 days on beans (*Phaseolus vulgaris*) sprayed with low concentrations of a methanolic seed kernel extract of neem (*A. indica*). Treatment with solutions of 15 and 20-30 mg/lt resulted in the development of only 12 and 1 percent normal adults, respectively. AT 15 mg/lt, a high percentage of abnormal adults were found.

Reed *et al.* (1981) found that *Diabrotica undecimpunctata* Howardi Barber and *Acalymma vittatum* F. were important pests of melon in the Midwest of the USA both because of their feeding damage and transmission of bacterial wilt disease. Tests were carried out in the laboratory to evaluate the effectiveness of 350 natural products as antifeedants for the beetles. The most promising antifeedant compounds were extracts from neem seeds (*A. indica*) and tung, neriifolin (from *Thevetia thevetioides*) and naphthoquinones.

Keyserlingk *et al.* (1982) described that video equipment and a microcomputer can be used to facilitate laboratory assessment of behavioural responses of insects to volatile attractants or repellents. Laboratory experiments in Berlin showed that adults of *Scolytus scolytus* F. on a simple bioassay arena responded to volatile chemicals with behaviour patterns similar to those displayed in their natural environment on trees in response to the entrance holes of other beetles of the same species. Responses were recorded simultaneously with 2 video camerae and analysed with the slow-motion and single-frame facilities of the video recorder. In the Berlin tests, 3 monoterpene compounds from conifers and *Acorus calamus* oil and 2 non-toxic neem (*A. indica*) oils appeared to disturb or repel the beetles.

Pereira and Wohlgemuth (1982) tested the effectiveness of ground neem seeds and leaves (*A. indica*) obtained from Togo as a protectant of stored maize against attack by pests in the laboratory. Neem seed was found to be toxic to adults of *Sitophilus oryzae* L., *S. zeamais* Motsch., *Cryptolestes ferrugineus* Steph. and *Rhyzopertha dominica* F. but not to *Tribolium castaneun* Hbst. Neem seed and leaves effectively reduced the progeny production of all the pests screened and the adult emergence of *S. cerealella* and *Ephestia cautella* Wlk. The ground leaves and seeds acted by disrupting larval development or by reducing adult fecundity. Neem seeds are the effective in protecting stored maize against *S. zeamais* and *R. dominica* but not against *S. oryzae*. However, when stored for a long period under warm and humid conditions, the grain could spoil due to growth of the neem-seed-borne fungus, *Aspergillus* sp.

Devi and Mohandas (1982) studied the relative efficacy of 11 antifeedants and deterrents in protecting stored paddy pests from infestation by *R. dominica, S. cerealella* and common pests occurring in godown. The mean percentage of infestation and the mean germination of paddy grain mixed with various antifeedants and deterrents was calculated and it was found that neem extract (1.0 and 5.0%) and coconut oil (1.0%) gave good protection to the grains from *R. dominica* upto six months during storage. Neem oil applied at 1.0 percent concentration gave relatively good protection and was on par with neem extract (0.5%) during third month. The neem extract at 1.0 percent was found to be the best among the treatments in protecting the grains from *S. cerealella* and neem extract at 0.5 percent also occupied relatively top rank. The authors, therefore, opined that considering the relative cost, effect on germination and effectiveness in protecting grains, neem extract at 0.5 percent can be considered as the best treatment for protecting stored paddy from insect infestation.

Lange and Schmutterer (1982) carried out a laboratory study in the German Federal Republic showed that the synergist piprotal (Tropital) increased the growth-disrupting effect of a methanolic neem kernel extract (ratio 1:2) in 4th-instar larvae of *Epilachna varivestis* about 6-fold. Sesamex (Sesoxane) also increased the effectiveness of the extract, but not significantly so. Piperonyl butoxide and 1,1'-oxybis[2,3,3,3-tetrachloropropane] (S-421) had neither a synergistic or antagonistic effect. In 1st-instar larvae of *E. varivestis*, piprotal also increased the effectiveness of the extract but to a lesser extent than in 4th-instar larvae. In *P. xylostella*, sesamex and piprotal improved the effectiveness of the neem kernel extract about 3-fold; the pupal mortality rate was considerably higher than in untreated insects.

Luca (1982) studied various attractants, antifeedants, deterrents, repellents and lethal (obtained from various plants) and used as non-specific alternatives to chemical treatment for protecting stored products against bruchid damage. In his experiment, he has used an antifeedant obtained from the seeds of *A. indica*. He concluded that these oils which both acted as physical barriers to the pests and killed them. These oils did not prevent seed germination and were harmless to vertebrate consumers.

Meisner and Mitchell (1982) studied the effects of various extracts of seeds of neem (*A. indica*) and of the triterpenoid azadirachtin on the feeding of adults of *Phyllotreta striolata* F. on radish in the laboratory in Canada. A 0.4 percent neem kernel extract had a strong inhibitory effect on feeding, whereas aqueous, methanol and hexane extracts were less active. Neem oil was ineffective. Azadirachtin had a moderate inhibitory effect on feeding only at 0.1 percent. No systemic activity was observed when an aqueous extract of neem kernel was incorporated into soil containing radish seedlings. Neem kernel extractive in a concentration of 0.4 percent strongly suppressed feeding, whereas aqueous, methanolic and hexane extracts showed less activity. Neem oil was ineffective. Zhang and Zhao (1983) studied in China on the protection of stored rice, it was found that when root-bark powders of *Cetastrus angulatus* and *Tripterygium wilfordii* were mixed with the rice at 0.5 percent, or oil of neem (*A. indica*) was mixed in at 5 ml/kg, the population growth of weevils (*S. oryzae* and *S. zeamais*) was inhibited by about 90 percent. The root-bark powder of *C. angulatus* had no contact or fumigant action but inhibited the development of the eggs and young larvae. Volatile essences from *Litsea cubeba*, *Cymbopogan caesius, Citrus poonensis* and orange used at 0.5 ml/petri dish killed adult weevils in 3 days and completely suppressed population build-up.

Pereira (1983) conducted studies in which 5 traditionally extracted oils and 1 industrially extracted oil were evaluated as protectants against attacks by *C. maculatus* on stored cowpeas (*Vigna unguiculata*) and *Bambara groundnuts* (*Vigna* [*Voandzeia*] *subterranea*). On cowpea, only neem (*A. indica*) kernel oil reduced oviposition (at 8 ml oil/kg seed). On Bambara groundnuts, neem kernel oil, palm oil and groundnut oil (industrially extracted) reduced oviposition at 3 ml/kg, and karite oil (extracted from the fruits of the tree

Butyrospermum parkii) was effective at 5 ml/kg. The ovicidal action was higher on Bambara groundnuts than on cowpeas for all the oils. Only neem oil showed larvicidal properties. Even after storage for 90 days, the cowpea testa retained most of its toxic activity. Good protection of cowpea was afforded for 3 months by neem oil at 8 ml/kg and of Bambara groundnuts for 6 months at 5 ml/kg.

Jilani and Su (1983) evaluated the repellency of parts of 3 plants that are common in Pakistan, to adults of the stored-grain pests *T. castaneum*, *Sitophilus granarius* L. And *R. dominica* in laboratory studies. Turmeric was the most effective as a plant powder against *S. granarius* and *R. dominica* and as a solvent extract against *T. castaneum*. During the experiment repellency studies by powdered plant material and plant extracts prepared by acetone, ethanol and petroleum ether were made by using three plants namely Rhizomes of *Curcuma longa* L. (turmeric), leaves of *Azadirachta indica* A. Juss (neem) and leaves of *Trigonella foenum graecum* L. (fenugreek). It was concluded that turmeric powder which had the lowest insect population and the highest population in its control was the most effective repellent among the three plant material used against red flour beetles, it was observed that all extracts indicated some degree of repellency at one or both application rates. However, most of the extracts had a lower value in repellency (40.1-60.0%). However, none of the extracts produced average repellency between 80.1-100 percent. Again turmeric extract was found to be most effective among all and petroleum ether extract was best. However, neem and fenugreek produced the average repellencies between 40.1 and 60.0 percent. Earlier the extracts of neem leaves and seeds were studied by Jilani and Molik (1973) as repellent and feeding inhibitors against various stored grain insects viz. *Tribolium castaneum, Trogoderma granarium* and *R. dominica*. The present studies by Jilani and Su (1983) are in conformity with the observation by Jilani and Malik (1973).

Ivbizaro (1983) observed that maize grains mixed with dry ground seeds of neem (*A. indica*) at 0.5, 1.0 and 2.0 g/20 g maize were protected for 6 months in storage form damage by *S. oryzae*. At the highest rate, neem seed prevented oviposition, at the lowest rate it markedly reduced oviposition, and at all the rates tested it completely halted post-embryonic development. Mortality among the parent weevils reached 100 percent within 5 days. Adult weevils placed on maize that had been treated 4 months previously with 2.5 g neem/20 g maize failed to oviposit, and 76 percent of them died within 10 days; when adults were placed on maize treated 4 months previously with 1 g neem seed/20 g maize, progeny emergence was delayed for 6 months and very few were produced. Neem seed had no adverse effects on the viability of the maize grains and no fungus or shrinkage was observed. Again the author in 1983 studied in Nigeria that the addition of 1-3 g dry ground neem seed to 20 g stored cowpeas (*V. unguiculata*) protected the seeds from damage by *C. maculata* for more than 4 months. The neem seed did not affect adult survival, but it significantly reduced the fecundity of the females, did not allow the emergence of any F1 adults in the tests with 2 or 3 g/20 g cowpeas, and prolonged the development period for the reduced numbers of larvae that hatched in the test with 1 g/20 g seeds. More than 99 percent of the untreated cowpeas had adult exit holes, but no damage was visible after treatment with 2 or 3 g neem seed. The texture, colour and general appearance of the treated seeds were normal, and 82 percent of them were viable (Ivbizaro, 1983).

Sowunmi and Akinnusi (1983) added powdered kernel of neem seed (*A. indica*) to stored seeds of cowpea in the laboratory in Nigeria which were then kept in open glass jars at ambient temperatures for 8 months. Treatment at 0.5 parts per 100 parts of seed was effective in controlling *V. maculatus* F. for upto 4 months, but thereafter considerable damage occurred. Treatment at 1.0 and 2.0 parts/100 parts of seed was effective for 8 months; the number of holed seeds per 500 grains averaged 32 and 18, in 1 and 2 parts per 100 parts of seed, respectively, as compared with 445 for no treatment.

Malik *et al*. (1984) tested 7 plant species for their repellent activity against *T. castaneum* and antifeedant activity against *R. dominica*. The best repellent activity was found in the rhizomes of *Saussurea lappa* and antifeedant activity in the leaves of *Chenopodium ambrosioides* and in azadirachtin isolated from neem (*A. indica*).

Bowry *et al*. (1984) evaluated certain oil seed cake powder as grain protectant against *S. oryzae* Linn., fire oil seed cake powders viz. neem (*A. indica*), linseed (*Linum usitatesimum*), castor (*Ricinus communis*), Mahua (*Bossia latifolia*) and mustard (*Brassica juncea*) were tested. To explore the repellent effect, the powders of oil seed cakes were thoroughly mixed with maize seeds at 1.0, 2.5, 5.0 and 7.5 parts/100 parts of seed (w/w). It is concluded that the oviposition of S. oryzae in maize seeds treated with neem cake powder in various ratio laid significantly minimum number of eggs (24-62) than any of the cake powder. It was earlier reported that the neem kernel powder at concentrations 1 and 2 percent were found to be most effective in reducing the rate of oviposition of *R. dominica* and *S. cerealella* (Savitri, 1975). It was also concluded by Bowry et al. (1984) that the maize seeds when treated with powdered neem cake in various ratios gave the minimum (7.86%) grain damage in comparison to all other treatments.

According to Ladd *et al*. (1984), topical applications of azadirachtin, a purified extract from seeds of *A. indica*, to larvae of *Popillia japonica*, completely disrupted subsequent normal development to the adult stage. Prepupae and newly formed pupae were somewhat less susceptible, but pupae 3 days old and above were not affected. Azadirachtin (purity 90+%) was isolated and concentrated from kernel of *A. indica* by acetone washes, open column adsorption and reversed phased liquid chromatography, as well as by reversed phase, high performance liquid chromatography. The azadirachtin was dissolved in ethyl alcohol and then topically applied on to the latero-ventral thoracic regions of immature form (3-5 micron lt/insect). In another trial, azadirachtin was applied to prepupae and pupae (72 hr-old) in amounts that ranged from 1 to 4 and 0.25 to 4 micron lt/insect, respectively. Another trial was conducted to estimate the dose response relationship of larvae and most susceptible stage. It was observed that all prepupae, regardless of azadirachtin dose, died or were unable to emerge as normal adults. The IGR activity shown by azadirachtin against wider variety of immature stages were studied and none were able to complete development but high degree of IGR activity continued until pupae were 24 hrs old. Topical application of 1.6 μg azadirachtin per larva resulted in death or deformity of all individual. Calculated LD50s and LD90s of topically applied azadirachtin were 0.1 and 4.0 μg/larva, respectively. Azadirachtin also increased the duration of the immature stages. Larvae treated with 1.6 μg of azadirachtin, for example, had significantly longer larval periods than did untreated larvae. Length of prepupal and pupal stages were also increased by applications of azadirachtin.

Singh and Singh (1985) conducted studies to explore the possibility of utilizing deoiled neem kernel powder against Khapra beetle (*Trogoderma granarium* Eversts). Powdered neem seed kernel was extracted seven times with hexane to remove the oil contents and then used for experiments. Three groups of ten 1st-instar or 3rd-instar larvae were added to 10 g wheat flour mixed with 100 mg of deoiled powdered neem seed kernel dissolved in 20 ml acetone. None of the 30 1st-instar larvae developed beyond the 2nd-instar and all died after the 2nd week, while 10 of 20 control larvae emerged as adults. Of the thirty 3rd-instar larvae, 5 pupated in the 4th week and one deformed adult emerged in the 5th week, compared with the control in which all 20 had emerged as normal adults by the end of the 3rd week. Further development of larvae introduced at 3rd-instar was arrested. The growth-retarding effect and death of the larvae is due to the biologically active compound, azadirachtin, present in deoiled kernel. It is suggested that these findings offer a cheap and effective method for controlling *T. granarium* in stored wheat.

Rajasekaran *et al.* (1985) determined the effectiveness of treatments with extracts of karanja (*Pongamia glabra*) or neem (*A. indica*) for the control of *S. oryzae* and *Callosobruchus chinensis* on sorghum or green-gram (*Vigna radiata*) grain in laboratory tests in India. Coating sorghum grain with karanja extract at 0.4 percent v/v or with neem extract at 1.0 percent w/w gave complete protection from S. oryzae. Coating green-gram grain with the 2 extracts at 0.6 percent v/v and 0.8 percent w/w, respectively, gave significant protection from *C. chinensis*.

Sujatha and Punnaiah (1985) observed that stored seed of green-gram (*V. radiata* L.) could be effectively protected from the pulse beetle, *C. chinensis* L. When the seed was treated with 0.125, 0.25 and 0.75 percent concentrations of the oils of sesamum (*Sesamum indicum* L.), capal (cotton seed oil), oil palm (*Elaeis quineensis* Jacq.) and neem (*A. indica*), groundnut (*Arachis hypogaea* L.) and coconut (*Cocos nucifera* L.). The observations revealed that longevity of the beetle was more in all the treatments than in control and at 0.125 and 0.25 percent except in capal 0.25 percent. All the oil treatments with 0.5 and 0.75 percent concentrations resulted in less longevity than in control. It was observed that the number of eggs laid decreased as the concentration increased. The adult emergence was very low in all the oils at 0.125 percent. However, no adult emerged at 0.25, 0.5 and 0.75 percent. At 0.125 percent concentration, the least infestation was noticed (3.2%) in capal followed by neem oil (5.91%). There appeared to be no adverse effect of these oils on the viability of seeds.

Pandey *et al.* (1986) tested five plant extracts viz. leaves and twigs of *A. indica*; leaves and flowers *Lantana camera* L.; leaves, flowers and buds of *Ageratum conyzoides*; leaves, twigs and young branches of *Theretia merifolia* A. Juss and leaves of *Ipomea carnea* Jacq. against *C. chinensis* Linn. Three concentrations viz. 0.5, 1.0 and 1.5 percent was taken in each of the extracts. The authors compared the plant extracts with the treatment of absolute control and benzene. It is evident from the data that all the plant extracts at the three doses repelled the beetles from 98.83 to 100 percent i.e. those beetles did not enter in to the sacklets containing the treated seed of green-gram, while the absolute control was about 79 percent. Benzene, however, proved to have a potent repellent effect keeping 99 percent beetles away from sacklets. It was also observed that plant extracts a higher dose (1.5 parts/100 parts of seed by w/w) and benzene alone proved very effective in reducing the egg laying. The maximum percentage of seed containing eggs was 0.97 percent only. The observation further revealed that maximum percentage of weight loss was in absolute control (37.78%), while *A. indica*, *L. camera* and *A. conyzoides* gave better performance in terms of weight loss. No adverse effects on seed germination were observed.

Tanzubil and Schmutterer (1986) carried out experiments in Ghana in 1985-96 to evaluate the effectiveness of several locally available materials for the protection of stored cowpeas from the bruchid *C. maculatus*. The materials tested were : neem fruit dust (2, 5 and 10%), neem leaf dust (5 and 10%), mahogany bark dust (5%), rice husk ash (10%), groundnut oil (4ml/kg seed), wood ash (10%), neem seed kernel oil (2 and 5 ml/kg of seed) and pirimiphos-methyl (Actellic) (15 ppm). Groundnut oil and neem fruit dust at 10 percent by weight of seed, protected stored seed for at least 4 months, while neem leaf dust and pirimiphos-methyl were effective for 3 months. In a 2nd experiment, neem seed kernel oil at 5 ml/kg seed gave effective control.

Singh and Kotaria (1986) studied the effect of deoiled neem kernel powder (DNKP) on the development of *T. granarium* in wheat grain. The deoiling of NKP was done by hexane extraction method. DNKP thus obtained was passed through 200 mesh sieve. Various concentrations viz. 0.06, 0.125, 0.25, 0.5, 1.0 and 2.0 percent were mixed with wheat seeds (w/w). Observation on the hatchability of eggs and larval development were taken on every 3rd day after removal of adults. Eggs hatched in all the treatments while none of the hatched larvae could develop further and died as 1st instar larvae by 11th day.

Gupta and Karim (1986) treated samples of seeds of grasspea (*Lathyrus sativus*), chickpea (*Cicer arietinum*) and lentils with neem seed oil at 4 ml/100 g seeds and exposed to infestation with 3 pairs of adults of *C. chinensis* remained uninfested after storage for 5 months. Untreated seeds suffered 77.33 to 91.00 percent infestation.

According to Ketkar *et al.* (1986), edible and non-edible oils were evaluated against the bruchid stored product pests *C. maculatus* on cowpea and *C. chinensis* on green-gram (*Phaseolus radiatus*). Non-edible oils from neem (*A. indica*), castor (*R. communis*), *P. glabra* [*P. pinnata*], *Calophyllum inophyllum*, a neem extract and 9 edible oils were evaluated at 0.25, 0.5, 0.75 and 1.0 percent w/w on cowpea. Neem oil was the best surface protectant giving complete protection at 0.75 and 1.0 percent even after 150 days, while *P. pinnata* did so at 1.0 percent. The growth index of *C. maculatus* was lowest (0.19%) with neem oil and this treatment had the greatest ovicidal effect. In the case of *C. chinensis* on *P. radiata*, neem *Shorea robusta, Schleichera trijuga, P. pinnata* and coconut oils gave protection for up to 90 days at 0.5 percent.

Ketkar *et al.* (1986) observed that the effect of several neem extract formulations on the chrysomelid *Ootheca bennigseni* was tested in the greenhouse and field on beans *Phaseolus vulgaris* in Tanzania in 1984 and 1985.

Neem seed kernel extracts of 2 and 4 percent had a deterrent effect when applied to leaves but none of the formulations gave complete protection. In greenhouse studies, 1 percent aqueous leaf and NSKEs were highly active as antifeedants.

Singh and Vir (1987) tested the emulsions of oils of neem, *Pongamia glabra, Brassica juncea juncea* (Indian mustard) and *R. cummunis* as grain protectants against the bruchid *C. maculatus* in stored cowpeas (*V. unguiculata*). The results showed that insect growth was most reduced in grains treated with neem oil.

Schmutterer (1987) conducted a long-term experiment in the German Federal Republic. He offered 20 females of *Leptinotarsa decemlineata* to potato plants treated with a simple aqueous extract of 30 g pulverized neem seeds/lt or 2.5 ml/lt of an enriched neem seed extract formulated as a 20 percent emulsifiable concentrate (AZT-VR-K) for 24 hrs and were kept in pairs with untreated males and an adequate supply of fresh untreated potato shoots throughout the reproductive period (over 3 months). The overall oviposition and hatching rates were reduced by 87 and 89 percent, respectively, for the aqueous extract and by 86 and 91 percent for AZT-VR-K relative to an untreated variant. Some females laid no eggs. Two in each variant had apparently refused the treated plants and were responsible for most of the eggs. In 2 short-term feeding experiments females ready to oviposit received only leaf pieces treated with the 2 neem formulations. The experiments confirmed the refusal of some females to feed, especially at a temperature of 22°C as opposed to 27°C, and indicated a tendency for the repellent effect of the extracts. No clear relationship was observed between feeding rate and reduction in fecundity. In combined field and laboratory experiments, either males and females were kept for 6 days in large, fine-netting bags in the field on potato plants previously treated with AZT-VR-K at 0.625, 1.25 or 2.5 ml/lt and then kept in the laboratory, or they were kept in the laboratory and fed for 2 days on shoots of potato plants taken immediately or at 2-day intervals after treatment of the plants in the field with AZT-VR-K at 2.5 ml and thereafter on untreated food. In the first experiment oviposition and hatching were reduced by 86 and 92 percent with 2.5 ml AZT-VR-K over 30 days in the laboratory relative to an untreated control, while in the second the reductions were 87 and 95 percent after 14 days for freshly contaminated material and declined sharply with residue age.

Mohiuddin *et al.* (1987) tested 17 oils from plants available in Pakistan in the laboratory for their repellency to the tenebrionid *T. castaneum*. Seed oil of Intsia bijuga compared favourably with neem (*A. indica*), *Ocimum basilicum, Tagetes erecta, Momordica charantia*, celery and garlic were less repellent than *I. bijuga* oil having 60.1-80% repellency, but more repellent than oils from *Cuminum cyminum*, bottle gourd (*Lagenaria siceraria*) or Indian mustard (*B. juncea*) having 40.1-60% repellency.

Schmutterer and Ascher (1987) observed in his laboratory and greenhouse experiments that NSKE greatly reduced the fecundity of the chrysomelid *L. decemlineata*. Females which consumed potato leaves treated with an enriched formulated extract (2.5 ml of 20% EC) or an aqueous extract of 30 g seed kernels/lt laid only a few eggs, and some females were completely sterile. The aqueous extract had a slightly stronger antifeedant effect than the enriched extract.

Echendu *et al*. (1988) treated coated dry ginger (*Zingiber officinale*) root powder, pulverized dried neem (*A. indica*), fruit or cashew nut oil to cowpeas and tested in the laboratory for control of *C. maculatus*. All reduced infestation, number of eggs laid and adult emergence.

Maheswaran and Ganesalingam (1988) studied the effect of air containing 13 percent volatile substance from crushed seeds of *A. indica* on the reproductive biology of newly emerged adults of *T. castaneum* on rice grains. The volatile substances reduced the number of eggs laid by females by 27.5 percent, prolonged the development period from egg to larva by 2.5 days and reduced the number of larvae emerging from eggs by 22.8 percent. Newly hatched to 12-day-old larvae exposed to the volatile substances failed to develop further, whereas treated larvae more than 13 days old successfully completed their development into adults.

Jilani *et al*. (1988) tested the repellency of turmeric oil (100, 500 or 1000 ppm), sweet flag (*Acorus calamus*) or neem (*A. indica*) and in Margosine-O. The repellency increased with increasing concentration of the oils and Margosine-O. In second choice test, filter paper strips treated with turmeric oil or *A. calamus* oil at 200, 400 or 800 µg/cm2 repelled insects during the first 2 weeks and thereafter, repellency decreased more rapidly than with neem oil or Margosine-O. Adults which were fed on wheat flour that had been treated with the test materials at 200 ppm produced fewer and underweight larvae, pupae and adults.

Ramachandran *et al*. (1988) found neem extract incorporated in wheat flour favoured survival, growth and development of *T. castaneum* at doses of <1 ppm but adversely affected these parameters at higher doses.

Gupta *et al*. (1988) evaluated the effectiveness of non-edible oils of neem, *P. glabra* [*P. pinnata*], *Bassia latifolia* [*Madhuca longifolia*], *Salvadora persica, Calophyllum inophyllym, Butea frondosa* [*B. monosperma*] and

Vateria indica against insect pests of wheat grain in the laboratory at the rates of 2.5 and 5 ml/kg wheat seeds. All the oils resulted in lower infestation and higher yields than in an untreated control. The effect of seeds were observed after 6 months of treatment. It was observed that all the treatments were significantly superior over control under both the doses, but effectiveness of the treatments at 2.5 ml/kg of seed was low. Neem oil used at 5 ml/kg of seed was found most effective. The effect of seeds were observed again after 12 months of treatment and seed treated with oils at 5 ml/kg has shown more effectiveness and here again neem oil was found to be the most effective in comparison to other plant extracts. The germination was not affected by the treatments.

Doharey and Singh (1989) tested NSKEs against chafer beetles. Ethanolic and hexane extracts were formulated by the authors and tested against the beetle, water and ethanolic extract offered complete protection of leaves against all the four species i.e. *Apogonia blanchardi* Retsema, *Maladera insannabiles* Brenske, *Adoretus bicolor* Brenske and *Adoretus* sp. Whereas hexane extract was not so effective and beetles were found to nibble the leaves. Hexane extracts also cause phytotoxicity and brittled the leaves. It was also observed that both water and ethanolic extracts inhibited the feeding of the *A. blanchardi* beetles at 2.5 and 1.0 percent concentrations. The authors emphasized that the NSKEs are cheap, easily obtainable and environmentally safe material for managing the above beetles.

Mukherjee and Ramachandran (1989) studied the effects of azadirachtin on the feeding and development of *T. castaneum* Herbst. Azadirachtin incorporated into wheat flour at up to 10 ppm reduced growth and survival of the larvae of the stored product pest *T. castaneum* at 1 ppm and above. No antifeedant effect against adults or larvae was observed, but fecundity was reduced. Topical application to eggs or larvae at 1, 2 and 5 mg did not have any adverse effects. Application to pupae less than 6 hr-old reduced the emergence of normal adults.

Mishra *et al.* (1989) studied the effect of neem oil on the ovipositional behaviour of *Henosepilachna* [*Epilachna*] *sparsa*. In a laboratory study, females of the coccinellid *H. aparsa* were fed with brinjal (aubergine) leaves treated with 0.025 and 0.05 percent neem oil to determine its effect on ovipositional behaviour. The preoviposition period of females fed leaves treated with 0.05 percent neem oil was 21.09 percent longer than that of insects fed untreated leaves. Females fed leaves treated with either concentrations of neem oil had a shorter oviposition period, a reduced number of egg laying a smaller number of egg masses/female and a smaller number of eggs/female in comparison to the untreated controls.

Kossan (1989) evaluated different products of neem (*A. indica*) against *Sitophilus zeamais* Motsch on stored maize. Various components of the neem tree were tested for their insecticidal effects on the curculionid *S. zeamais* in stored maize. Seed kernel, leaf flower and bark from the tree were used in powdered and ethanol extract and used at a rate of 2.5 g and 2 ml, respectively, per 50 g of grain infested with 20 insects. Pirimiphos-methyl and ethanol were applied at 25 mg and 2 ml, respectively, for comparision. Ethanolic extracts showed a higher toxicity deterrent effect to *S. zeamais* than the powdered neem forms; after being confined on treated grain for 72 hrs, mortality was 32.6-79.0 percent for the powdered forms compared with 34.8-90.0 percent for the ethanolic extracts. Makanjuola (1989) evaluated extracts of neem for the control of store product pests. Laboratory investigations on the activity of neem leaf and seed extracts in water or methylated spirit on *C. maculatus*, the curculionids *S. oryzae* and *S. zeamais* and the apionid *Cylas puncticollis* were made at 27±2°C and 65±5% RH. Field trials were conducted to determine their effectiveness as protectants of cowpeas (*V. unguiculata*) and maize in storage at 28±4°C and 82±7% RH for 5 months. The effectiveness of neem was affected by differences in insect behaviour. The extracts were more active as suppressants of *C. maculatus* than of *S. oryzae* and there was no effect on *C. puncticollis*. All of the extracts tested resulted in a significant reduction in oviposition, egg hatch and adult emergence of *C. maculatus* and in adult emergence of *S. oryzae*. Field trials showed that the extracts generally gave good protection to cowpea against *C. maculatus* for 5 months, but only moderate protection to maize against *S. zeamais*. Seed extracts were more effective than leaf extracts.

Lobbinah and Appiah (1989) observed the effect of neem products on stored maize weevil, *S. zeamais*. Various neem (*A. indica*) products were compared with copra oil, palm kernel oil and 0.25 percent diazinon dust for protection of stored maize against the curculionid *S. zeamais*. Maize kernels treated with neem oil or ash were damaged less than untreated kernels. Although copra and palm kernel oils reduced attack by *S. zeamais*, these edible oils were not as effective as neem oil. Neem wood ash was, however, less effective than 0.25 percent diazinon dust. Neem derivatives inhibited oviposition and deterred feeding.

Karel (1989) studied the effect of neem extract against foliar beetle *O. bennigseni* Weise. Seven treatments were tested viz. Aqueous neem leaf extract (4%), aqueous NSKE (2%), aqueous NSKE (4%), neem seed kernel dust (2%), alcoholic neem kernel extract (2%), rogor (dimethoate 40 EC, 600 g a.i./ha) and control. Treatments' effects were assessed by counting insects and observing foliar damage by the beetle. These various formulations of neem (*A. indica*) extracts were evaluated under field and greenhouse conditions in Tanzania to examine the effects of *O. bennigseni*, a serious pest of common beans (*P. vulgaris*) and other legumes in many parts of Africa. At concentration of 2.0 and 4.0 percent, neem formulations from seed kernels deterred *O. bennigseni* from feeding on bean leaves. None of the neem formulations protected bean plots completely. However, formulations of NSKE resulted in lower foliar damage and less incidence of *O. bennigseni* than other formulations. In most instances, untreated leaves were severely damaged, whereas leaves protected with neem extracts were not attacked or were only slightly damaged. In greenhouse studies, 1.0 percent aqueous leaf and seed kernel extracts were highly active as an antifeedant against *O. bennigseni*, which consumed little or none of the leaves in free-choice experiments, bean seedlings treated with 1.0 and 2.0 percent NSKEs were well protected. Because of their effectiveness, safety for the environment, lower cost, availability, and easy preparation, neem extracts have great potential in protecting beans against foliar beetles.

Babu *et al*. (1989) studied the effect of pre-storage treatment of mungbean (*Vigna radiata*) variety PS-16 with neem (*A. indica*), karanja, mustard, groundnut and castor oils, each at 2.5, 5.0 and 10.0 ml/kg seed, on infestation by the bruchid *C. chinensis*. Treatments with karanja oil (5 and 10 ml/kg) and castor oil (10 ml/kg) effectively reduced oviposition by the bruchid under conditions of artificial infestation, while maintaining a high level of germination for over 18 months of storage under ambient conditions. After 24 months of storage, the yield of neem-oil-treated seed was significantly lower.

Chowdhary and Pathak (1989) found that oils of neem, groundnut, sesame and soyabean were highly effective even at a low concentration (0.25 ml/100 g seed).

Ivbizaro (1990) observed that oils from the seeds of (*A. indica*) and the brown pepper (*Piper guineense*) afforded protection to cowpea seeds (*V. unguiculata*) from *C. maculatus* in the laboratory beyond 3 months; 24 hrs after surface treatment with Piper seed oil (PSO) at 2 and 3 ml/kg, bruchid mortality reached 100 percent, but with neem extract mortality was 65-100 percent at 3-5 days. Oviposition averaged 15.5 eggs for PSO, 49.5 for neem and 60.75 for no treatment. The corresponding reproductive efficiencies were nil, 1.7-2 percent and high. PSO prevented the emergence of bruchid offspring and neem prolonged development. Neem reduced seed damage by 94. 67 percent.

Jeyarajan and Babu (1990) studied the efficacy of certain azadirachtin rich neem seed fractions on *Epilachna* beetle (*Henosepilachna vigintioctapunctata*). Neem seed fractions were dissolved in acetone and then brinjal leaves were dipped in this solution and then placed in moistened filter paper in a petridish. Five fourth instar larvae were released on to these leaves and feeding area of the leaves was measured after 24 hrs. Similarly adults were also released. It was observed that Nem-75, 1000 ppm was the best antifeedant for grubs in which only 370.33 mm sq. leaf area was fed while in case of NK-100, the fed area was 371.67 and in case of Nemidin, it was 435.00 mm sq. In case of adults, NK-100 (1000 ppm) was most effective having feeding area 388.67 mm sq., the next best was Nemidin (1000 ppm) and V-74 (1000 ppm) where the fed area were 655.0 and 639.33 mm sq., respectively.

Pascual *et al*. (1990) observed that when azadirachtin was injected into newly moulted pupae of *Tenebrio molitor* at doses ranging from 0.01 to 10 μg/insect. It induced a delay of the development and an inhibition of apolysis and acdysis, which was dose dependent. Low doses of azadirachtin suppressed adult ecdysis and high doses also prevented apolysis and secretion of adult cuticle. The use of azadirachtine as a potential component for integrated control of stored food pests is emphasized by the author

Jilani and Saxena (1990) tested repellency of oils of turmeric (*Curcuma longa* L.), sweetflag (*Acorus calamus* L.) and neem (*A. indica* A. Juss) and the neem based insecticides Margosan-O against lesser grain borer (*Rhyzopertha dominica* F.) for 8 weeks. The tests were carried out by the paper strip method. This paper is dipped in acetone solution of known concentrations of the test materials to get deposits at 200, 400 and 800 μg/cm sq.). Acetone treated strips served as control. The authors observed that all the test materials strongly repelled *R. dominica* showing more than 50 repelling upto 8 weeks. Turmeric oil and weetflag oil were significantly more repellent during the first 2 weeks than neem oil and Margosan-O, but thereafter their repellency decreased more rapidly than that of neem oil or Margosan-O. *R. dominica* adults made significantly fewer and smaller feeding punctures in filter paper discs treated with the test materials at 100, 500 or 1000 μg/cm sq. than in control discs.

Buiyah and Quiniones (1990) tested leaf powder and whole dried leaves of lagundi (*Vitex negundo*) against *S. zeamais* on maize grains. Whole dried leaves and leaf powder were applied at 1 and 5 percent by weight, while neem seed oil and actellic (pirimiphos-methyl) were applied at 5 ml/kg grain and 6 ppm, respectively. After 6 months, grain treated with leaf powder had the highest infestation rate 90.22 and 90.15 for 1 and 5% leaf powder, respectively. Grains treated with pirimiphos-methyl and neem seed oil had the lowest infestaion 14.91 and 31.75%, respectively.

Mishra *et al.* (1990) studied the effect of neem oil on the growth and development of *Henosepilachna sparsa* Hbst. Feeding brinjal (aubergine) leaves treated with 0.025 and 0.05 percent neem oil to *B. sparsa* increased the duration of life stages in the subsequent generation. Neem oil also reduced the weight of treated insects compared with the untreated control.

According to Fauzia *et al.* (1991), the most important group of insects that injure ethnological and natural science specimens and other types of museum exhibits are dermestid beetles. The study demonstrated efficacy of deoiled neem seeds and custard apple (*Annona squamosa*) seeds as a means for protecting insects in storage *A. vorax*, which is a very common pest of woollen material. When larvae were exposed on the test piece of woollen material, treated with 0.05 and 0.1 g of neem seed and custard apple seed powder, antifeedant effects irrespective of the doses applied were apparent.

According to Tanzubil (1991), under laboratory conditions, neem (*A. indica*) products can protect stored cowpeas from damage caused by artificial infestations of the bruchid *C. chinensis* for 4 months. As a result, farmers are being encouraged to use neem since it is easy to prepare, apply and costs less than currently available insecticides. In field trials carried out in Ghana, aqueous extracts of neem seeds and leaf extracts reduced the incidence of *Megalurothrips sjostedti* infesting cowpeas and increased yields significantly. Extracts also had an adverse effect on the biology of *Maruca testulalis*. These findings have generated interest in a neem-based pest management system for cowpeas, and possibly on other crops.

Kaethner (1991) observed that seed kernel extract (500 and 1000 ppm), and seed oil (3%) of neem (*A. indica*) reduced the fecundity of females of *Melolontha hippocastani* in combined laboratory and field trials. The results were confirmed in a field trial in an oak forest near Karlsruhe, Germany, but immigration of untreated beetles into treated plots prevented a clear estimation of the efficacy of the neem treatment. The fecundity of females of *M. melolontha* was reduced after neem treatment in the laboratory. The food intake of the beetles was reduced for several days and the ability to fly decreased or was lost.

Durairaj *et al.* (1991) studied the effect of 1 and 2 percent water extracts of neem seed and cake, *Pongamia glabra* [*P. pinnata*] seed, *Calophyllum inophyllum* seed and *Bassia latifolia* [*Madhuca longifolia*] seed on feeding and mortality of the chrysomelid *Scelodonta strigicollis* in the laboratory at 25°C. Neem seed extract (2%) had the greatest antifeedant activity followed by *P. pinnata* seed extract.

Khaire *et al.* (1992) applied sunflower, castor, mustard, safflower, palm, groudnut, sesame, neem karanja and maize oil, each at 5, 7.5 and 10 ml/kg to protect pigion peas (*Cajanus cajan*) from attack by *C. chinensis*. Effects on progeny emergence, loss in grain weight and seed germination up to 100 days after treatment were investigated. Adult emergence was prevented by karanja oil at 0.75 and 1.0 percent, and neem oil at all concentrations for up to 100 days. None of the oils affected seed germination. Neem, palm and karanja oils were cheaper to use in comparision to the other oils.

Kaethner (1992) studied fitness reduction and mortality effects of neem based pesticides on the colorado potato beetle (*L. decemlineata* Say). Adutls and larvae of *L. decemlineata* were treated in the laboratory with extracts and oil from seed kernels of *A. indica*. Beetles exhibited a decreased feeding rate and reduction of fecundity and vitality. Morphogenetic defects were observed in the F1 progerny of treated females. The number of eggs masses produced by females which fed on neem-treated potato leaves were smaller. Trials in small scale potato plots (cv. Christa) near Eberstadt, Germany, confirmed that reduction of fecundity in treated beetles. The number of adults in the F1 progeny was reduced to 1 percent of the control group. Larvae showed a greatly increased mortality when treated directly with neem oil in the laboratory or fed on oil-treated foliage. This result was confirmed in small plot field trials especially when neem extract and neem extract + neem oil were used. These neem products were highly efficient against all instars of the pest.

Mohiuddin *et al*. (1993) testd 12 vegetable oils for their toxic and/or repellent effect against *T. castaneum* and *R. dominica* for a period of 8 weeks. The greatest repellent activity (80-100%) was found in oil of *A. indica*. Oils of *Trigonella foenumgraecum* and *Myristica fragrans* had a repellency of 60.1-80 percent and oils of *Piper nigrum, Nigella sativa* and *Curcuma longa* had a repellency of 40.1-60 percent. Oils of *M. fragrans, A. indica* and *P. nigrum* exhibited the greatest toxicity at 0.25 percent by surface treatment of wheat grains.

Susha and Karnavar (1993) found azadirachtin to reduce the vitellogenic count of *T. granarium* when topically applied to pupae at 0.5 or 1.0 µg/pupa. It is concluded that the hormonal control of oocyte development was disrupted by the compound.

Khaire *et al*. (1993) tested 7 plant products for their effectiveness as protectants of grains of Bengal gram (chickpea) against *C. chinensis*. The lowest number of eggs of the bruchid was laid on chickpeas treated with 1 percent of *A. indica* and 2 percent extract of *P. glabra*.

Patel *et al*. (1993) studied the effect of powdered neem seed kernel for the control of lesser grain borer, *R. dominica* in wheat (*Triticum aestivum*). The efficacy of powdered neem seed keernel at 1, 2, 3, 4 and 5 percent in the control of *R. dominica* in stored wheat was assessed in the laboratory. The 5 percnet treatment was most effective; at this dose, percentage damage was reduced to 2.55, 3.15 and 7.13 percent, compared with control percentage damage of 6.57, 13.6 and 24.71 percent after 32, 64 and 96 days, respectively.

Jood *et al.* (1993) evaluated some plant products against *T. granarium* Everts in stored maize and their effects on nutritional composition and organoleptic characteristics of kernels. Oil and powder of leaves and kernels of *A. indica*, powders of lemon leaves, garlic bulbs and leaves of *Mentha spicata* were used in stored maize kernels at 1 and 2 percent against larvae of *T. granarium*. Powdered kernels and oil of *A. indica* gave complete protection to grains for 6 months, whereas substantial insect infestation (7-19%) was noticed after 3 months in the other treatments. Moisture, ash, fibre, fat, protein and carbohydrate contents of treated kernels remained unaffected after 1 month of storage. After 6 months, changes in nutritional composition were proportional to insect damage.

Akhtar *et al.* (1993) studied the determination of toxicity, fecundity and emergence inhibiting effects of neem factor against *Callosobruchus analis* L. Toxicity, fecundity and emergence inhibition were observed for *C. analis*. The LD50 (neem extract) was 0.0169 mg/cm sq, in the filter paper treatment method and 3.6 mg/g in the grain treatment method. Fecundity inhibition was 36.20 and 76.21 percent, respectively. Inhibition of adult emergence was about 100 and 79.65 percent, respectively.

Jacob and Sheila (1993) did exhibited with indigenous plant products for protection of stored rice form the lesser grain borer *R. dominica* Fabr. Leaf powders of plant namely datura (*Datura alba* Nees), oak (*Calotropis procera* Br.), crofen weed (*Chromolaena oderatum* Linn.) and neem (*A. indica* A. Juss) at two doses i.e. 2.5 and 5.0 percent were used to see the effect on adult emeregence. The leaves were dried up in shade, ground timely and sieved through a 60 ml jar. This was then mixed thoroughly with the leaf powders at two doses and were shaken in electric shaker for 2 hrs. Each treatment was replicated thrice. It was observed that minimum number of adults were recorded in rice grain treated with *Chromolaena* at 5 percent followed by *Calotropis* at two doses. However, all the treatments significantly reduced the number of adults emerging form grains. Maximum number of adults emerged in control treatment.

Ahmad *et al.* (1993) conducted laboratory experiments to ascertain the efficacy of oils of neem (*A. indica*), linseed (*L. usitatissimum*), safflower (*Carthamus tinctorius*), sunflower (*Helianthus annuus*) and sesame (*S. indicum*) against *C. chinensis* on mungbean seeds (*V. radiata*). Four concentrations viz. 0.5, 1.0, 2.0 and 4 percent, of each oil were tested. All the concentrations of each oil significantly reduced oviposition and adult emergence of the pest. Only 1.33, 1.11, 10.00, 1.67 and 0.78 eggs and 0.00, 0.44, 3.11, 0.22 and 0.44 adults were recorded on mungbean seeds treated with 4 percent oils of neem, linseed, safflower, sunflower and sesame, respectively, compared to 31.67 eggs and 13.78 adults in the control. Neem oil at 4 percent completely protected the pulse seed from attack, while only 2.00, 11.27, 2.08 and 1.96 percent of seed loss were recorded at the same concentrations of linseed, safflower, sunflower and sesame oil, respectively, compared to 56.27 percent in the control. The effectiveness of oils increased with the increase of the concentrations in all cases. Among the oils tested, neem oil was the best protectant of pulse seeds against *C. chinensis*.

Jood *et al.* (1993) studied the effect of neem oil and powders of neem leaf and neem kernel, lemon leaf, garlic bulb and *Mentha spicata* leaf in wheat grains at 1 and 2 percent against larvae of *T. granarium*. Neem products completely prevented damage by an introduced larval population. In other treatments, grain damage was noticeable after 3 months, although significantly less than in untreated grains. Thereafter, damage increased progressively and reached 45 percent (lemon), 46 percent (*M. spicata*) and 72 percent (garlic) after 6 months. Nutritional composition, viz. Moisture, ash, fibre, fat, protein and carbohydrates of treated grains after 1 month remained unaffected. After 6 months, changes in the level of nutritional compounds were proportional to insect damage. Neem products did not produce an adverse effect on nutritional composition. Chapatis made from flour of treated grains after 1 month did not show adverse effects on colour, appearance and texture but neem oil, neem kernel and garlic powders produced unacceptable taste and aroma. After 6 months of treatment, taste, aroma and overall acceptability of cahpatis were either affected by insect damage or by plant products except neem leaf treated grains, which were organoleptically acceptable.

Mochizuki (1993) observed in laboratory studied that the length of feeding scars produced by *Lissorhoptrus oryzophilus* [*Oryzophagus oryzae*] decreased on rice leaves treated with a neem solution above 2.5 percent. Leaves treated with a >10 percent neem solution darkened, indicating a phytotoxic effect. The concentration (4.4%) of neem solution estimated to produce a 50 percent reduction in feeding.

Kuntal (1993) observed development of feeding materials viz. khejri

(*Prospis* sp.), ber (*Ziziphus mauritana*) and neem (*A. indica*) on *Holotrichia consanguinea*. Khejri was the best diet and neem the worst for development of the scarabaeid.

Doll and Schmutterer (1993) tested crude methanolic and aqueous extracts and oil from the seeds of *Azadirachta excelsa* for the first time on insects and compared with those of *A. indica* (neem). Five concentraions were applied in bioassays using the coccinellid pest *Epilachna varivestis* as test insect. Extracts were superior to those of neem in causing mortality and morphogenetic defects. Marrango extracts also reduced the lifespan and fecundity of adult females. The most obvious differences were obtained with oil, and marrango oil was 5 times as active as neem oil.

Oroumchi and Lorra (1993) applied aqueous extract of neem (*A. indica*) seed kernels and leaves, and chinaberry (*Melia azedarach*) leaves to alfalfa (lucerne) leaves in the laboratory caused high mortality and strong growth-disturbing effects in the larval stages of *Hypera postica*. Most larvae died prior to or during moulting. Also, an antifeedant effect was observed when lucerne leaves treated with a 5 percent aqueous NSKE were observed to 4th-instar larvae under choice conditions. Momentary dipping of 4th-instar larvae in the same extract reduced the survival rate to 15 percent. In a cage-test under field conditions in Iran, 2.5 and 5 percent neem seed kernel suspensions applied 4 times at weekly intervals to naturally infested lucerne plants completely interrupted the larval development of the pest and increased yield.

Palaniswamy and Wise (1994) tested the effects of neem-based products on the number and feeding activity of a crucifer beetle *Phyllotreta cruciferae* (Goeze) on canola. In laboratory and field studies conducted in the USA, 3 neem-based formulations were tested to determine their repellent and feeding deterrent effects on *P. cruciferae* on canola (rape). At concentration ranging from 0.05 to 12.0 percent a.i., Safer's Neem Insecticide (SNI) was more effective than RD-9 Repellin (a moxture of extracts of *Pongamia glabra, Madhuca indica, Sesamum indicum* and *Ricinus communis* or Margosan-O (azadirachtin) in decreasing plant damage through a higher mortality and repellency. RD-9 Repellin showed repellency for 1-2 days at 12 percent, and for 6 hrs at 1.2 and 6 percent. However, the repellency of Margosan-O was limited to 6 hrs at a concentraion of 6 percent. The decrease in feeding injury in neem-treated rape seedlings was attributable largely to the repellent effects of neem products. In field studies, SNI significantly reduced the recolonization of rape seedlings for 1-2 days. It was concluded that the short residual activity of neem formulations limited their effectiveness as a seedling protectant for *P. cruciferae* on rape.

Salom *et al.* (1994) studied laboratory evaluation of biologically based compounds as antifeedants for the control of pales weevil, *Hylobius pales* (Herbst). Twenty plant-produced compounds or mixtures, and one insect-produced semi-chemical were evaluated as potential antifeedants. Initially, a choice laboratory feeding bioassay was conducted to screen compounds and identify antifeedant activity. This was followed by a no-choice dose-response bioassay to further evaluate the most active compounds from the choice test. When applied to twigs of *Pinus strobus*, borneol, bornyl acetate, cucurbitacin, myrcene, neem extract, (S)-carvone, (R)-carvone and limonin were the most effective antifeedants for both of males and females of *H. pales* after 24 hrs.

Prokopy and Powers (1995) studied the influence of NSKE on oviposition and mortality of beetle, *Contotrachelus nenuphar* and adult fly, *Rhagoletis pomonella* adults. In the laboratory and in a commercial apple orchard in Massachusetts, the neem seed extract Margosan-O was evaluated at a rate of 1 lt/100 lt water for potential oviposition-deterring and mortality effects against adults of *C. nenuphar* and *R. pomonella*. It was observed that neem had no effect.

Xie *et al.* (1995) studied repellency and toxicity of azadirachtin and neem concentrates to three stored grain beetles. The repellency and toxicity of azadirachtin (98% azadirachtin) and 3 neem extracts (48, 23 and 7% AZA) to 3 stored-product insects, *Cryptolestes ferrugineus, Sitophilus oryzae* and *Tribolium castaneum*, were infestigated in the laboratory. Each test material repelled all 3 species in a standard repellency test. Significant negative correlations were found between insect settling response and extract concentration. *T. Castaneum* was more sensitive to the repellent action of neem than the other 2 species. The test materials were also toxic to the 3 pest species, with *C. ferrugineus* being the most susceptible. The numbers of F1 adults of all 3 species in almost all treatments were significantly reduced compared with controls. This reduction was significantly dose-dependent. It was confirmed that azadirachtin was largely responsible for both repellent and toxic actions of neem on stored-product insects. However, the neem extracts are slightly more active than pure azadirachtin.

Miah *et al.* (1996) studied lentil seeds as a protectant against *C. chinensis*, powdered leaves of *Amoora ruhituka*, *Azadirachta indica* and *Vitex negundo* and oils of *A. indica*, linseed (*L. usitatissimum*) and sesame (*A. indicum*) under laboratory conditions. Uninfested healthy lentil seeds were placed in petri dishes on top of perforated paper discs covering powdered leavess or oil mixed with powdered soil and exposed to 3 pairs of newly-emerged adults fo the bruchids. *V. negundo* leaf powder significantly ($P<0.05$) reduced oviposition, adult bruchid emergence and the weight loss of lentil seeds. The powdered leaves of *A. rohituka* and *A. indica* had no significant effect on these aspects. The other oils were marginally effective at some doses and *A. indica* oil at 2.0 ml/2 g soil reduced oviposition more than the smaller doses and the sesame oil treatments.

Mohapatra *et al.* (1996) studied comparative bioassay of different solvent extracts of neem seed kernels against *S. oryzae* L. In rice. Solvent (water, acetone, methanol, ethanol and hexane) extracts of neem (*A. indica*) seed kernels were evaluated against *A. oryzae* adults at different concentrations (0.001, 0.01, 0.1, 0.5 and 1.0%). Only aqueous and alcoholic extracts were effective. Alcohol extracts were superior to aqueous extracts providing 100 percent protection to rice grains for 6 weeks at a concentration of 1.0 percent.

Jood *et al.* (1996) evaluated some plant products against *T. granarium* Everts in sorghum. The efficacy of neem (*A. indica*), citrus (*Citrus limon*), garlic (*Allium sativum*) and podina (*Mentha spicata*) powders and neem oil at 1 and 2 percent level in sorghum grains was evaluated against larvae of *T. granarium*. Neem kernel powder and neem oil completely prevented damage by an introduced larval population. In other treatments, grain damage was noticeable after 3 months, although significantly less than in untreated grains. Thereafter, the percentage of damaged kernels increased progressively and reached as high as 29 percent (neem leaf), 44 percent (citrus), 56 percent (podina) and 71 percent (garlic) compared to 85 percent in the control after 6 months. Nutritional composition, i.e. moisture, ash, fibre, fat, protein and carbohydrates of treated grains remained unaffected after 1 month of storage. After 6 months, changes in nutritional composition were proportional to insect damage.

Ogunwolu and Odunlami (1996) evaluated the reproductive suppression properties of root bark powder of *Zanthoxylum zanthoxyloides* against the bruchid *C. maculatus* in comparision with neem (*A. indica*) seed powder (NSP) and pirimiphos-methyl (PM). At application rates ranging from 0.125 to 3 g/20 g cowpea seed, *Z. zanthoxyloides* was effective as oviposition suppressant due to high contact toxicity to the bruchid adults, but significantly more effective than NSP.

Chiranjeevi and Sudhakar (1996) studied the effect of plant products on the fecundity, adult emergence and development of *C. chinensis* in blackgram. Powders from neem (*A. indica*), sweetflag (*Acorus calamus*), apamarga (*Achyranthus aspera*), kesarachettu (*Crinum defixum*), and lantana (*Lantana camara*); and cow dung, *Acacia arabica* [*nilitica*] wood, neem wood and *Casuarina indica* wood ashes were tested for their effect on egg laying, development period and adult emergence of *C. chinensis* infesting *Vigna mungo*. The application of 0.5-2.0 neem seed powder completely prevented the development of the pest. Cow dung ash and *A. calamus* powder/100 parts of seed (w/w) and reduced the development at higher concentrations in other treatments i.e. in 1.5 and 2.0 parts/100 parts. The egg laying capacity of the pest and total number of adults emerging decreased with increase in concentrations of the powders and ashes.

Schnetter *et al*. (1996) sprayed Neem Azal-T/S against *Melolontha melolontha* in orchards in Germany in 1995. This azadirachtin preparation causes a long-lasting feeding inhibition after intake with food. Neem Azal-T/S reduced egg producing to less than 3 percent. It is as efficient as the synthetic insecticide Rubitox (phosalone) in the control of the pest population. The side-effects against non-target organisms were clearly lower for Neem Azal-T/S than for Rubitox.

Rouf *et al*. (1996) tested the leaf powder of neem (*A. indica*), nishinda (*Vitex negundo*) or biskatali (*Polygonum hydropiper*), and combinations of these materials against *C. chinensis* on lentil seeds. Four gram biskatali leaf powder/50 g lentil seeds was most effective in reducing oviposition and adult emergence, and also reduced pest damage and seed weight loss. The combination of neem + biskatali leaf powder ranked second followed by neem leaf powder alone. At low doses (1-2 g) these 3 plant materials applied either alone or in combination were less effective.

Rhode *et al*. (1996) tested a number of insecticidal treatments against the forest pest *Melolontha hippocastani* in southern Germany, in 1994. Neem extracts showed no direct toxicity, but resulted in behavioural changes; feeding stopped after 2-3 days, flight activity was reduced, and larvae did not seek shelter from rain. The extracts also resulted in loss of fertility. A garlic extract and the entomogenous fungus *Beauveria brongniartii* were ineffective. Phosalone (Rubitox) showed lethal contact toxicity in 3-4 hrs and the effect lasted for around 1 week. Decis (deltamethrin) showed lethal contact toxicity and repellency, which lasted about 10 days.

Singh *et al*. (1996) conducted laboratory studies at 29.6 + 2°C to determine the effects of extracts of neem (*A. indica*), garlic, oranges, *Eucalyptus hybrida* [*E. tereticornis*], *Lantana camara*, *V. negundo*, asafoetida (*Ferula assa-foetida*) and 'pudinhara' against *Rhyzopertha dominica*. Treatment with *A. indica* (2.0 and 3.0%) and *F. assa-foetida* (1.0 and 2.0%) resulted in lower fecundity per female and adult emergence. Extracts of 'pudinhara' (2.0 and 4.0%) were the least effective.

Ignatowicz and Wesolowska (1996) studied the repellency of powdered plant material of Indian neem and other products against stored grain pests. Using an olfactometric method, the repellency of several plant powders against stored product pests (*C. chinensis, Sitophilus granarius* and *S. oryzae*) was confirmed and compared. Powdered seed kernels of neem (*A. indica*) were more effective repellents of the stored product pests than the powders of dry leaves and seed shells. Repellency of neem products increased with their concentration in the product, and 5.0 percent concentration, by weight, was the most effective. A repellent effect was also identified in powdered twigs of *Ledum palustre* and *Acorus calamus* against the stored product pests. The lowest concentration (0.5%) of these powders in wheat or lentil seeds was highly repellent. These powders were more active than the neem tree products.

Chandel *et al*. (1996) reported that non-edible seed oils of Margosa, Indian beech and mowra were found as feeding deterrents to the apple defoliating beetle *Brahmina coriacea* at least for 2 days.

Kiarie and Saxena (1996) observed various neem derivatives against maize weevil, *S. zeamais*. The maize grain was mixed with 0.6, 1.25, 2.5 or 5.0 percent powder of dried neem leaves, seed, kernel or neem cake. The growth and development of first instar larvae were completerly arrested in maize grain applied with neem oil at 0.02 percent.

Bhuiyan *et al*. (1996) tested the efficacy of neem oil as an antifeedant and growth inhibitor to the beetle *Epilachna dodescastigma* Weidmann. The neem oil concentration of 1.5 percent emulsified was used. It was observed that food consumption by the test insect was significantly reduced and growth inhibition indicated by mortality was 100 percent in the 1st-3rd instar. A few grubs with dark spots and low body weight pupated at lower concentration but failed to emerge as adults. Some emerged as deformed ones with rippled hind wings. Oviposition was significantly reduced. The study indicated that neem oil served as a potential antifeedant and growth inhibitor of *E. dodescastigma* beetle.

Singh *et al*. (1996) evaluated neem seed kernel powder (NSKP), deoiled neem kernel powder (DNKP), neem leaf powder (NLF), hexane extract of seed kernel (HE), ethanol soluble part of hexane extract (EtoSHE) were evaluated as protectant of greengram (*V. radiata*) against pulse beetle, *C. maculauts*. NSKP (100 mesh) when admixed at the rate of 0.25 parts/100 parts of the greegram gave 100 percent protection against the beetle. There was considerable loss of activity of NSKP after 6 months of treatment as 1 part of NSKP/100 parts of greegram was needed to get complete protection. DNKP and NLF failed to offer satisfactory results but DNKP completely arrested growth and development of the beetle. However, HE and EtoSHE gave complete protection at 1 percent concentration. The DNLP, however, was found highly effective against cigarette beetle, *Lasioderma sericorne*. The IGR activity of the DNKP was noticed when eggs of the beetle were released in the 0.06 parts of DNKP/100 parts of the flour. None of the eggs reached to adult stage whereas in control all the 30 eggs released reached to adult stage.

Schmutterer and Sen (1996) observed growth disrupting effect of some neem-based products on mexican bean beetle, *Epilachna varivestis* (Mulsant). Experiments were taken up to test the comparative efficacy of neem products viz. Repelin, Neemark, Neempest, Jawan, Neemguard and Margosan-O. The observation revealed that a majority of the products were effective as they reduced leaf feeding and larval weight considerably. These products cause disturbance in growth, moulting and pupation. The larvae treated with Neemark and Margosan-O died without reaching the adult stage. In case of Repelin, 50 percent of the adults were deformed.

Bhuiyan (1996) tested the efficacy of crude oil and extracts of neem and other indigenous plants for the control of *Dicladispa armigera* Olivier. Crude oils, methanol and acetone extracts of seeds of *A. indica* (neem), *Melia azedarach, Aphanamixis polystachya* Parker (pithrai), *Annona squamosa, A. reticulata* Linn. (Bulluck's heart) and *Polyalthia longifolia* Benth & Hook (devdaroo) were evaluated against rice hispa, *D. armigera* (Coleoptera : Chrysomilidae), a serious pest of rice especially in Bangladesh. The active phyto-chemicals present in these plants acts as IGR activity which altered feeding, oviposition and growth of the insect. Feeding and oviposition by adult rice hispa were significantly reduced on treated young rice seedlings with 1 percent methanol and acetone extracts of NSK (neem), PSK (pitharai), CSK (custard apple) and BSK (bulluck's heart) and 2-3 percent extracts of CHK (chinaberry) and DSK (devdaroo). It was also observed that 1-2 percent crude oil of neem were more persistent than the extracts alone. However, it was also noted that 1 percent acetone extracts of NSK demonstrated better IGR activity than the methanol extracts.

Singh and Gour (1996) observed the effectivity of certain plant extracts against larvae of *Mythimna separata* Walker and adults of *C. chinensis* Linn. Seeds of neem (*A. indica*), custard apple and *Cleome viscosa* Linn., pods of *Tephrosia villosa* Pers., debarked stem of *Milletia recemosa* and rhizomes of turmeric were extracted in petroleum ether and ethanol, and the extracts were tested against larvae of oriental armyworm, *M. separata* and adults of pulse beetle *C. chinensis*. Nine such extracts were preserved at 5°C. Leaf discs of sorghum leaves (4.5 cm dia.) washed and after drying dipped in 0.1 percent concentration of the extracts for five minutes. These discs were airdried and then 40 third instar larvae of *M. separata* were released in the centre. After 2 hrs, unconsumed area was determined. It was observed that neem petroleum ether extract 0.1 percent was found to be repellent while at 0.06 percent, it showed antifeedant activity. Simillar results were obtained with ethanol extract when 0.1 percent petroleum ether extract was applied to 6th instar larvae, larval period was slightly extended but adults failed to emerge from pupae. Pupal mortality was observed at as low as 0.02 percent. Other plant extracts were less effective. Potter's tower studies showed petroleum ether extract of neem to be more toxic (LC50 : 0.0165) than the ethanol extract (LC50 : 0.142). LC50 value of DDVP was 0.0155. custard apple seed petroleum extract (LC50 : 0.8288) was close to neem seed petroleum extract.

In case of *C. chinensis*, 20 grams of pigeonpea was treated with 6 concentrations of neem seed extracts of petroleum ether and 40 adults (2 day-old) of *C. chinensis* were introduced. The experiment was conducted in randomised block design and replicated thrice. The data revealed that 12 hrs after the release, the insect population decreased with increase in concentrations. Redgram grains mixed with petroleum ether extract of neem at 0.06 percent and 0.4 percent ethanol extract kept the adult pulse beetle, *C. chinensis* away. Growth index on treated grain was observed to be low. Toxicity to adult pulse beetle with these extracts was very low. These preliminary studies may open up avenues for neem utilization in the pest management of *M. separata* and *C. chinensis*.

Yadav (1996) studied, olfactory, oviposition and developmental response of four species of pulse beetle to four neem oils. Fresh neem kernels were obtained, decorticated and dried at 40oC for 24 hrs, powdered and passed through a 40 mesh sieve. Hundred grams of powder was added to 50 ml of petroleum ether at 40-60°C and stirred thoroughly. After 24 hrs, solvent was decanted through filter paper. The solvent was allowed to evaporate and the oil left was kept in referigerator. Subsequently 1 percent stock solution of oil was prepared in carbon-tetrachloride and further diluted to treat 100 g seeds of chickpea, cowpea, soybean and french bean. Seven doses varying between 20 and 500 ppm were tested. The observations were taken on percent olfactory attraction of adults, oviposition, hatch and emergence of adult. The results indicated that 200-500 ppm concentrations were ovicidal and *Zabrotes subfasciatus* Boheman was the most susceptible bruchid on french bean. *Callosobruchus maculatus* Fab. And *C. analis* Fab. were least susceptible. Larval development was also affected at 200-500 ppm concentrations and adult emergence was arrested. Adults with deformed wing and prepupal abnormal bahaviour were also observed. The results have shown absence of olfactory response, dose dependent ovicidal and larval toxicity and juvenile hormone activity.

Otto *et al*. (1996) reported sterile effect of neem preparation Neem Azal-F on young females of *Leptinotarsa decemlineata*.

Rajapakse and Senanayake (1997) studied in Sri Lanka the effectiveness of 7 vegetable oils agtainst *C. chinensis* in pigeonpea *Cajanus cajan* L. One kg of pigeonpea was mixed with 7 oils at three concentrations viz. 0.4, 0.6 and 0.8 parts/100 parts of seed (w/w). Fifteen grams of treated seeds at three concentrations were kept in jar and mixed by mechanical shaker. The experiment was conducted at 30±3°C and 75±6 percent RH. Observations were taken at 1, 15, 30 and 45 days after treatment. The 7 plant oils tested were *Cymbopogon citratus, Bassia* [*Madhuka*] *longifolia*, *Ricinus communis, Cocos nucifera, Arachis hypogaea, Glycine max* and *Azadirachta indica*. All the oils significantly affected oviposition at all 3 concentrations tested at one day after treatment. *C. citratus, R. communis* and *B. longifolia* at 0.8 percent concentration were found to reduce oviposition significantly compared to the control. All the oils tested did not appear to affect seed germination significantly. However, the most effective concentration was 0.6 (w/w) for *G. max* and 0.8 percent concentration recorded on *A. indica*.

Suss *et al*. (1997) studied repellent activity of *A. indica* extracts on food stuff insects. The repellent effect of a neem formulation extracted from the seeds of *A. indica* and containing 3 percent azadirachtin (31.7 g/lt) was evaluated against adults of *S. oryzae, R. dominica, T. castaneum* and *Lasioderma serricorne*. Concentrations of 0.5, 1.0 and 1.5 percent sprayed on corrugated fibreboard in a 4-way olfactometer. It was observed that 1 percent neem concentration repelled 76.1 percent of S. oryzae and 66.3 percent of *R. dominica*. The addition of piperonyl butoxide (0.1%) increased the repellency to *R. dominica* by 94.4 percent.

Hein and Hummel (1997) did bioassay for the determination of feeding activity and mortality of *Epilachna varivestis*. Bean leaves (*Pahseolus*), treated with azadirachtin A, were fed to 4th-instar larvae, and weight changes and prolongation of larval development were recorded. A disadvantage of this method was the natural variation of bean leaves over the seasons, influencing feeding activity and the degree of efficiency.

Pandey and Singh (1997) studied the effect of neem bark powder on infestaion of pulse beetle *C. chinensis* in stored chickpea. The experiment was conducted with dried neem bark powder in four concentraions viz. 0.2, 0.4, 0.6 and 0.8 percent (w/w) along with untreated control. The seeds were mixed with powder and shaken vigorously. Twenty tow pairs of *C. chinensis* were introduced in each jar and kept at 28±2°C and 70 percent RH. The observation revealed that neem bark powder was an effective pesticide. The population growth of the beetle was significantly low and mortality was high as compared to control during summer season. The neem bark powder thus reduced seed damage and insect population growth.

Deka and Hazarika (1997) observed neem (*A. indica*) seed oil (NSO) acted as a potential antifeedant against adults of the chrysomelid, *Dicladispa armigera*. Under laboratory conditions, daily adult consumptin of fresh rice leaves was 1.05±0.08 mg which was reduced by 50 percent when leaves were treated with 6.46 percent NSO. The gain in adult body weight per day was significantly reduced. Similarly, adverse effects of NSO were also observed on the consumption index (CI), growth rate (GR) and approximate digestibility (AD), efficiency of conversion of ingested food (ECI), and efficiency of conversion of digested food (ECD).

Chinwada and Giga (1997) studied the effectiveness of wood ash, sand, finger millet husks and ground neem (*A. indica*) leaves as protectants of stored beans (*P. vulgaris*) against *Acanthoscelides obtectus* and *Zabrotes subfasciatus*. In the laboratory, ash mixes significantly reduced the emergence of F1 adults and seed damage levels at application rates of 1:4, 1:2 or 1:1 v/v. The other three protectants gave satisfactory protection only at 1:1 v/v.

Mansour *et al.* (1997) evaluated Neem Azal-S, soybean and sunflower oils in the laboratory as protectants of seeds of mung bean (*V. radiata*) against infestaion by *C. chinensis*. Among the oils used, Neem Azal-S at 0.5 percent could protect the seeds from serious damage up to one year. Neem Azal-S at 0.5 percent could be used most effectively when the stored mung beans are to be used for seeding.

Raguraman and Singh (1997) screened oils obatined from *A. indica* seed kernels, *Cedrus deodara* wood and their combination (1:1) each at 3, 2 and 1 percent concentrations, for activity against *C. chinensis*. *C. deodara* exhibited the highest fumigant potential, followed by the combination. Topical application at 0.5 micron/insect at 3, 2 and 1 percent showed negligible mortality after 24 and 48 hrs of treatment. Insects treated to the combination quickly recovered from the anaesthetic effects when compared with insects in the *A. indica* and control treatments. In an ovipositional behavioural experiment, *C. deodara*-treated *Cicer arietinum* seeds had mean numbers of 134.3, 114 and 112.3 eggs/5 females while *A. indica*-treated seeds had 98.3, 118.7 and 57 eggs/5 females at 3, 2 and 1 percent concentrations, respectively.

Rajapakse *et al*. (1998) studied the effect of dried leaf powders of *Piper nigrum, Annona reticulata, Azadirachta indica* and *Capsicum annuam* and dried peel of *Citrus limon* on oviposition, adult emergence and adult mortality of *Callosobruchus maculatus* on cowpea seeds, in the laboratory at 27-30°C and 60-80 percent RH. *A. indica* gave the highest reduction in oviposition of *C. maculatus* (37.5%), followed by *A. reticulata* (39%). *A. indica* caused a significant reduction in adult emergence (20.3%).

Rahim (1998) evaluated the residual activity of an ethanolic neem (*A. indica*) kernel extract, containing azadirachtin, in the laboratory against *R. dominica* on wheat stored for up to 48 weeks. All treatments in both fresh and residual studies did not have any effect on mortality of the parents. On freshly treated grain, the application rate of 5 mg/kg of azadirachtin was effective in inhibiting F1 progeny production by >98 percent against three test strains. The minimum application rate for >95 percent reduction of F1 progeny of the multi-resistant strain throughout the 48-week storage period was 25 mg/kg, with complete prevention estimated to occur with 75 mg/kg. The effectiveness of azadirachtin was not synergised by piperonyl butoxide.

Murugesan *et al*. (1998) studied the efficacy of a neem derivative, Fortune-Aza (0.15%) for the control of *Bruchidius* sp. and *Caryedon serratus* both seed pests of *Acacia nilotica* and *Tamarindus indica*. Different concentrations of neem formulations viz. 15, 25, 75 and 100 ppm were prepared with distilled water. These concentrations were sprayed on 100 g seeds in glass jar. Pupae and adults were exposed to the treated seeds. In the seed treatments, 75 and 100 ppm antifeedancy and insect growth regulation effects were observed on *Bruchidius* sp. and *C. serratus*. In lower concentrations, very little or no antifeedant activtiy was observed. The adult emergence, mortality and oviposition of both pest species dicreased following exposure. Seeds treated with higher dosages of Fortune-Aza gave 100 percent protection against attack by bruchid beetles. Besides, it was observed that Fortune-Aza interrupts insect reproduction in the form of antiovipositinal effect. The adverse effect was noted on fertility rate of the concentrations of 75 and 100 ppm, as very few females deposited their eggs on treated seeds in comparision to untreated seeds.

Kraus *et al*. (1998) tested methanol extracts of a number of plants, including *Ailanthus altissima, Acanthospermum hispidum* and *Hertiera littoralis*, and seed cake of *Azadirachta indica* (neem cake), for various biological activities. Antifeedant activities towards *Epilachna varivestis* were observed with *A. altissima, A. hispidum, H. littoralis* and neem cake extracts, and insect growth regulating activities were found in *A. altissima* and neem cake extracts.

Murugan *et al.* (1998) observed in Tamil Nadu, India, the interactive effect of neem products on the control of pulse beetle, *Callosobruchus maculatus* Fab. The insect was cultured in laboratory on cowpea at 27±1°C, 10:14L:D and 65-70 percent RH. The extracts of neem seed kernel (NSKE), neem leaf powder (NLP), and neem gum extract (NGE) and neem oil (NO) used in the experiment. The experiment was conducted on cowpea seeds. The repellent action of neem extract, oil and powder against *C. maculatus* was assessed in a choice bioassay test.

It was observed that the neem products considerably, protected the cowpea seed against *C. maculatus* (Table-1). It is observed that there is a repellency effect and mortality of adults after the treatment of NSKE and NLP. The neem oil exhibited moderate effect on repellency and mortality. The neem products considerably, affected *C. maculatus* progeny at 2 percent NSKE and NGE.

Table-1 : Effect of different neem treatments against *Callosobruchus maculatus* in cowpea

Treatment	Repellency (%)	Damage (%)	Weight gain (%)	Mortality (%)	No. of F1 adults progeny
Control	00	32.0	100	00	20
Neem oil (2%)	55	12.8	24	79	4
NSKP (2 g)	91	9.7	35	61	6
NLP (2 g)	81	11.5	64	41	14
NGP (2 g)	51	10.8	78	31	3
NSKE (2%)	94	9.6	20	82	7
NLE (2%)	87	11.2	47	51	13
NGE (2%)	81	8.1	60	43	10
NSKP (0.25 g) + NLP (1.0 g)	95	7.1	18	83	1
NSKP (0.25 g) + NGP (1.0 g)	83	8.3	28	71	0
NSKE (0.25%) + NLE (1%)	98	8.7	11	91	2
NSKE (0.25%) NGE (1%)	89	7.2	26	78	1

NSKP = Neem seed kernel powder, NLP = Neem leaf powder,
NGP = Neem gum powder, NSKE = Neem seed kernel extract,
NLE = Neem leaf extract, NGE = Neem gum extract
[Table obtained from Murugan *et al.* (1998)]

References

Ahmed, A.; Sultana, P. and Ahmed, A. (1980). Comparative efficacy of some indigenous plant materials as repellents against *Sitophilus oryzae* Linn. *Bangladesh Journal of Agricultural research*, **5**(2) : 31-35.

Ahmed, K.S.; Haque, M.A. and Islam, B.N. (1993). Efficacy of edible and non-edible oils against the pulse beetle, *Callosobruchus chinensis* (Coleoptera : Bruchidae). *Bangladesh Journal of Entomology*, **3**(2) : 1-7.

Akhtar, K.; Naqvi, S.N.H. and Azmi, M.A. (1993). Determination of toxicity, fecundity and emergence inhibiting effects of neem factor (NfQ) against *Callosobruchus analis* L. *Zeitschrift fur Angewandte Zoologie*, **80**(2) : 141-154.

Ambikam, B.; Abraham, C.C. and Nalinakumari, T. (1981). Effect of neem leaf extract and two JH analogues on the development of *Callosobruchus chinensis* Linn. (Coleoptera : Bruchidae). *Agricultural Research Journal of Kerala*, **19**(1) : 72-75.

Ascher, K.R.S. and Gsell, R. (1981). The effect of neem seed kernel extract on *Epilachna varivestis* Muls. larvae. *Zeitschrfit fur Pflanzenkrankheiten und Pflanzenschutz*, **88**(12) : 764-767.

Astrakhantzeu, P.I.; Belugin, N.O.; Bogolyubov, N.V. and Borozdina, K.I. (1936). Chemico-toxicological investigation of *Melia azedarach*. In summary of the *Scientific Research Work of the Institute of Plant Protection for the Year 1935* (in Russian) Roy 800, 596 pp. *Lenin Acad. Agric. Sci*. (RAE (A) : 25-159, 1937).

Babu, T.R.; Reddy, V.S. and Hussaini, S.H. (1989). Effect of edible and non-edible oils on the development of the pulse beetle (*Callosobruchus chinensis* L.) and on viability and yield of mungbean (*Vigna radiata* (L.) Wilczek). *Tropical Science*, **29**(3) : 215-220.

Bhuiyan, N. Islam (1996). Efficacy of crude oils and extracts of neem and other indigenous plants for control of *Dicladispa armigera* Olivier [Coleoptera : Chrysomelidae]. *Neem Newsletter*, **13**(4) : 55-56.

Bhuiyan, N. Islam; Haque, M.A. and Ahmed, S. (1996). Neem oil : A potential antifeedant and growth inhibitor of the beetle *Epilachna dodecastigma* (Wiedmann) [Coleoptera : Coccinellidae]. *Neem Newsletter*, **15**(3) : 46-47.

Bowry, S.K.; Pandey, N.D. and Tripathi, R.A. (1986). Evaluation of certian oilseed cake powders as grain protectant against *Sitophilus oryzae* Linnaeus. *Indian Journal of Entomology*, **46**(2) : 196-200.

Buiyah, M.I.M. and Quiniones, A.C. (1990). Use of leaves of lagundi (*Vitex negundo* Lin.) as corn seed protectant against the corn weevil, *Sitophilus* zeamais Motesch. *Bangladesh Journal of Zoology*, **18**(1) : 127-129.

Chandel, R.S.; Chander, R. and Gupta, P.R. (1996). Non-edible oils as feeding deterrent to apple defoliating beetle (*Brahmina coriacea*). *Neem Newsletter*, **13**(1) : 4.

Chin, S.L.; Sudderuddin, K.I. and Khoo, S.G. (1980). Studies on the biology, host specificity, feeding behaviour and toxicity of the coleopteran *Schematiza cordiae* Barb. *Malaysian Applied Biology*, **9**(1) : 59-66.

Chinwada, P. and Giga, D.P. (1997). Traditional seed protectants for the control of bean bruchids. *Tropical Science*, **37**(2) : 80-84.

Chiranjeevi, C. and Sudhakar, T.R. (1996). Effect of indigenous plant materials on the fecundity, adult emergence and development of pulse beetle, *Callosobruchus chinensis* (L.) in blackgram. *Journal of Research APAU*, **24**(3-4) : 57-61.

Choudhary, B.S. and Pathak, S.C. (1989). Efficacy of organic materials for the control of *Callosobruchus chinensis* Linn. *Indian Journal of Plant Protection*, **17**(1) : 47-51.

Cobbinah, J.R. and Appiah Kwarteng, J. (1989). Effects of some neem products on stored maize weevil, *Sitophilus zeamais* Motsch. *Insect Science and Its Application*, **10**(1) : 89-92.

Das, G.P. and Karim, M.A. (1986). Effectiveness of neem seed kernel oil as surface protectant against the pulse beetle, *Callosobruchus chinensis* Linn. (Bruchidae : Coleoptera). *Tropical Grain Legume Bulletin*, **33** : 30-33.

Deka, N. and Hazarika, L.K. (1997). Effects of neem seed oil on food consumption and utilization by rice hispa. *Pesticide Research Journal*, **9**(1) : 113-116.

Deshpande, A.D. (1967). Neem as a seed protectant against storage pests. **M.Sc. Thesis**, Post Graduate School, I.A.R.I., New Delhi.

Devi, D.A. and Mohandas, N. (1982). Relative efficacy of some antifeedants and deterrents against insect pests of stored paddy. *Entomon*, **7**(3) : 261-264.

Doharey, K.L. and Singh, R.P. (1990). Evaluation of neem (*Azadirachta indica* A. Juss) seed kernel extracts against chafer beetles. *Indian Journal of Entomology*, 1989, publ. 1990, *51*(2) : 217-220.

Doll, M. and Schmutterer, H. (1993). Comparision of the effects of extracts and of oil from seed kernels of *Azadirachta indica excelsa* (Jack) (=*A. integrifoliola* Merr.) and *Azadirachta indica* (A. Juss.) on the Mexican bean beetle *Epilachna varivestis* (Muls.). *Mitteilungen der Deutschen Gesellschaft fur Allgemeine und Angewandte Entomologie*, **8**(4-6) : 775-780.

Durairaj, C.; Ramaraju, K. and Sundarababu, P.C. (1991). Effect of plant derivatives on feeding and mortality of grapevine flea beetle, *Scelodonta strigicollis* L. (Eumolpidae : Coleoptera). *Madras Agricultural Journal*, **78**(1-4) : 14-16.

Echendu, T.N.C.; Njoku, B.O. (ed.); Oti, E. (ed.); Odurukwe, S.O. and Ene, L.S.O. (ed.) (1988?). Preliminary investigations into the use of ginger, neem and cashew nut shell liquid (CNSL) to reduce damage caused to stored cowpeas by *Callosobruchus maculatus*. *Proceedings of the First National Ginger Workshop*, Umudike, Nigeria, 17-21 October, 1988 : 124-128.

Fauzia, Shaheen; Shashi, D.; Shaheen, F.; Dhawan, S.; and Agrawal, O.P. (ed.) (1991). Natural plant products : a safe preventive measure against *Anthrenus vorax* (Copleoptera : Dermestidae). *Proceedings of the International Conference*, 20-25 Feb., 1989, Lucknow, India : 186-192.

Gupta, G.P.; Mahapatro, G.K.; and Ajanta, Chandra (1998). Neem seed powder : targeting the quiescent stages of *Helicoverpa armigera* Hubner. *Annals of Plant Protection Sciences*, **6**(2) : 170-173.

Gupta, H.C.; Verma, J.P.; Bereth, S.S. and Methur, B.N. (1988). Evaluation of some non-edible oils as grain protectant in wheat and their subsequent effect on germination. Indian *Journal of Entomology*, **50**(2) : 147-150.

Hein, D.F. and Hummel, H.E. (1997). Feeding activity and degree of efficiency of the neem compound azadirachtin A : a common bioassay with *Epilachna varivestis* Muls. compared with the wafer test with *Spodoptera littoralis* Boisd. *Mitteilungen der Deutschen Gesellschaft fur Allgemeine und Angewandte Entomologie*, **11**(1-6) : 385-390.

Ignatowicz, S. and Wesolowska, B. (1996). Repellency of powdered plant material of the Indian neem tree, the Labrador tea, and the sweet-flag, to some stored product pests. *Polskie Pismo Entomologiczne*, **65**(1-2) : 61-67.

Ivbijaro, M.F. (1990). The efficacy of seed oils of *Azadirachta indica* A. Juss and *Piper guineense* Schum and Thonn on the control of *Callosobruchus maculatus* F. *Insect Science and Its Application*, **11**(2) : 149-152.

Ivbijaro, M.F. (1983). Preservation of cowpea, *Vigna unguiculata* (L.) Walp, with the neem seed, *Azadirachta indica* A. Juss. *Protection Ecology*, **5**(2) : 177-182.

Ivbijaro, M.F. (1983). Toxicity of neem seed, *Azadirachta indica* A. Juss, to *Sitophilus oryzae* (L.) in stored maize. *Protection Ecology*, **5**(4) : 353-357.

Jacob, S. and Sheila, M.K. (1993). A note on the protection of stored rice from the lesser grain borer, *Rhizopertha dominica* Fabr. by indigenous plant products. *Indian Journal of Entomology*, **55**(3) : 337-339.

Jeyarajan, S. and Babu, P.C.S. (1990). Efficacy of certain azadirachtin rich neem seed fractions on brinjal epilachna beetle, *Henosepilachna vigintioctopunctata* (Coleoptera : Coccinellidae). *South Indian Horticulture*, **38**(1) : 46-48.

Jilani, G. and Malik, M.M. (1973). Studies on neem plant as repellent against stored grain insects. *Pakistan Journal of Scientific and Industrial Research*, **16** : 251-254.

Jilani, G. and Saxena, R.C. (1990). Repellent and feeding deterrent effects of tumeric oil, sweetflag oil, neem oil, and a neem-based insecticide against lesser grain borer (Coleoptera : Bostrychidae). *Journal of Economic Entomology*, **83**(2) : 629-634.

Jilani, G. and Su, H.C.F. (1983). Laboratory studies on several plant materials as insect repellants for protection of cereal grains. *Journal of Economic Entomology*, **76**(1) : 154-157.

Jilani, G.; Saxena, R.C. and Rueda, B.P. (1988). Repellent and growth-inhibiting effects of turmeric oil, sweetflag oil, neem oil, and "Margosan-O" on red flour beetle (Coleoptera : Tenebrionidae). *Journal of Economic Entomology*, **81**(4) : 1226-1230.

Jood, S.; Kapoor, A.C. and Singh, R. (1993). Evaluation of some plant products against *Trogoderma granarium* Everts in stored maize and their effects on nutritional composition and organoleptic characteristics of kernels. *Journal of Agricultural and Food Chemistry*, **41**(10) : 1644-1648.

Jood, S.; Kapoor, A.C. and Singh, R. (1993). Evaluation of some plant products against *Trogoderma granarium* Everts in stored wheat and their effects on nutritional composition and organoleptic characteristics of treated grains. *Journal of Pest Management*, **39**(1) : 93-98.

Jood, S.; Kapoor, A.C.; and Singh, R.; (1996). Evaluation of some plant products against *Trogoderma granarium* Everts in stored sorghum and their effects on nutritional composition and organoleptic characteristics. *Journal of Stored Products Research*, **32**(4) : 345-352.

Jotwani, M.G. and Sircar, P. (1965). Neem seed as a protectant against stored grain pests infesting wheat seed. *Indian Journal of Entomology*, **27**(2) : 160-164.

Jotwani, M.G. and Sircar, P. (1967). Neem seed as a protectant against bruchid *Callosobruchus maculatus* (F.) infesting some leguminous seeds. *Indina Journal of Entomology*, **29**(1) : 21-24.

Jotwani, M.G. and Srivastava, K.P. (1981). Neem - insecticide of the future.

I - As protectant against stored grain pests. *Pesticides*, **15**(10) : 19-23.

Kaethner, M. (1992). Fitness reduction and mortality effects of neem-based pesticides on the colorado potato beetle, *Leptinotarsa decemlineata* Say (Col., Chrysomelidae). *Journal of Applied Entomology*, **113**(5) : 456-465.

Kaethner, M. (1991). Potential of neem seed kernel products for the control of the cockchafers *Melolontha hippocastani* F. and *M. melolontha* L. (Col., Scarabaeidae). *Journal of Applied Entomology*, **112**(4) : 345-352.

Kahare, S.N.; Kahare, N.P.; Harinkhere, J.P.; Kandalkar, V.S. and Thakur, S.K. (1993). Exploration of herbal products as grain protectants against *Callosobruchus chinensis*. *Journal of Soils and Crops*, **3**(1) : 33-36.

Karel, A.K. (1989). Response of *Ootheca bennigseni* (Coleoptera : Chrysomelidae) to extracts from neem. *Journal of Economic Entomology*, **82**(6) : 1799-1803.

karel, A.K.; Schmutterer, H. (ed.) and Ascher, K.R.S. (1987). Response of *Ootheca bennigseni* (Coleoptera : Chrysomelidae) to neem extracts. *Proceedings of the 3rd International Neem Conference*, Nairobi, Kenya, 10-15 July, 1986 : 393-403.

Ketkar, C.M.; Schmutterer, R. (ed.) and Ascher, K.R.S. (1987). Use of tree-derived non-edible oils as surface protectants for stored legumes against *Callosobruchus maculatus* and *C. chinensis*. *Proceedings of the 3rd International Neem Conference*, Nairobi, Kenya, 10-15 July, 1986 : 535-542.

Keyserlingk, H. von (1982). The measurement of attractant or repellent effects of chemicals on walking insects. *Proceedings in XXXIV International Symposium on Crop Protection*, Mededelingen van de Faculteit Landbouwwetenschappen, Rijksuniversiteit Gent., **47**(2) : 547-555.

Khaire, V.M.; Kachare, B.V. and Mote, U.N. (1992). Efficacy of different vegetable oils as grain protectants against pulse beetle, *Callosobruchus chinensis* L. in increasing storability of pigeonpea. *Journal of Stored Products Research*, **28**(3) : 153-156.

Khan, M.A. (1982). Repellency of chemical compounds to stored product insect pests. A review of literature. *Mitteilungen aus der Biologischen Bundesanstalt fur Land und Forstwirtschaft*, **206** : pp. 33.

Kiarie, M. Kega and Saxena, R.C. (1996). Neem derivatives for management of *Sitophilus zeamais* Motschulsky [Coleoptera : Curculionidae]. *Neem Newsletter*, **13**(3) : 39.

Kossou, D.K. (1989). Evaluation of different products of neem *Azadirachta indica* A. Juss for the control of *Sitophilus zeamais* Motsch on stored maize. *Insect Science and Its Application*, **10**(3) : 365-372.

Kraus, W.; Koll, M. Weber; Maile, R.; Wunder, T.; Vogler, B. and Atta ur Rahman (1998). Biologically active constituents of tropical and subtropical plants. *Neem Newsletter*, **15**(4) : 56.

Kuntal, J.S. (1993). Postembryonic development and various growth indices of white grub beetle, *Holotrichia consanguinea* Blanch on kherhri, ber and neem. *Uttar Pradesh Journal of Zoology*, **13**(1) : 33-36.

Ladd, T.L. Jr. and Jacobson, M. (1978). Neem seed extracts as feeding deterrents to Japanese beetles. *Proceedings of the North Central Branch of the Entomological Society of America*, **33** : 41.

Ladd, T.L. Jr.; Jacobson, M. and Buriff, C.R. (1978). Japanese beetles : extracts from neem tree seeds as feeding deterrents. *Journal of Economic Entomology*, **71**(5) : 810-813.

Ladd, T.L. Jr.; Warthen, J.D. Jr. and Klein, M.G. (1984). Japanese beetle (Coleoptera : Scarabaeidae) : the effects of azadirachtin on the growth and development of the immature forms. *Journal of Economic Entomology*, **77**(4) : 903-905.

Lange, W. and Schmutterer, H. (1982). Experiments with synergists to improve the effect of metamorphosis disruption of a crude methanolic extract of the neem tree (*Azadirachta indica*). *Zeitschrift fur Pflanzenkrankheiten und Pflanzenschutz*, **89**(5) : 258-265.

Luca, Y. de (1982). Products of vegetable origin that can be used against bruchids (Col.) (Attractants, antifeedants, deterrents, repellents, lethal). *Frustula Entomologica*, **2** : 19-29.

Maheswaran, P. and Ganesalingam, V.K. (1988). Effect of volatile substances from *Azadirachta indica* (neem) seeds on the reproductive biology of *Tribolium castaneum* (Coleoptera : Tenebrionidae). *Vingnanam Journal of Science*, **3**(1-2) : 20-26.

Makanjuola, W.A. (1989). Evaluation of extracts of neem (*Azadirachta indica* A. Juss) for the control of some stored product pests. *Journal of Stored Products Research*, **25**(4) : 231-237.

Malik, M.M.; and Mujtaba, Naqvi, S.H. (1984). Screening of some indigenous plants as repellents or antifeedants for stored grain insects. *Journal of Stored Products Research*, **20**(1) : 41-44.

Mansour, M.H.; Kleeberg, H. (ed.) and Zebitz, C.P.W. (Editors) (1997). The effectiveness of plant oils as protectants of mung bean *Vigna rediata* against *Callosobruchus chinensis* infestation. *Proceedings of 5th Workshop Wetzlar*, Germany, 22-25 June, 1996 : 189-200.

Meisner, J. and Mitchell, B.K. (1982). Phagodeterrent effect of neem extracts and azadirachtin on flea beetles, *Phyllotreta striolata* (F.). *Zeitschrift fur Pflanzenkrankheiten und Pflanzenschutz*, **89**(8-9) : 463-467.

Miah, M.R.U.; Rahman, N.H.; Sufia, B.; Islam, B.N.; and Sutradhar, G.N.C. (1996). Application of leaf powders and oils as a protectant of lentil seeds against *Callosobruchus chinensis* Linn. *Bangladesh Journal of Scientific and Industrial Research*, **31**(3) : 137-142.

Mishra, B.K.; Mishra, P.R. and Patnaik, N.C. (1990). Effect of neem oil on the growth and development of epilachna beetle, *Henosepilachna sparsa* (Hbst.). *Orissa Journal of Agricultural Research*, **2**(3-4) : 169-172.

Mishra, B.K.; Mishra, P.R.; Patnaik, N.C. and Mohapatra, H. (1989). Effect of neem oil on the ovipositional behaviour of epilachna beetle. *Orissa Journal of Agricultural Research*, **2**(1) : 70-71.

Mochizuki, A. (1993). Antifeedant activity of neem oil to the rice water weevils, *Lissorhoptrus oryzophilus* Kuschel (Coleoptera : Curculionidae). *Applied Entomology and Zoology*, **28**(2) : 254-256.

Mohapatra, S.; Sawarkar, S.K. and Patnaik, H.P. (1996). Comparative bioassay of different solvent extracts of neem seed kernels against *Sitophilus oryzae* L. in rice. *Current Agricultural Research*, **9**(1-2) : 89-93.

Mohiuddin, S.; Qureshi, R.A.; Khan, A.; Nasir, M.K.A.; Khatri, L.M. and Qureshi, S.A. (1987). Laboratory investigtions on the repellency of some plant oils to red flour beetle, *Tribolium castaneum* Herbst. *Pakistan Journal of Scientific and Industrial Research*, **30**(10) : 754-756.

Mohiuddin, S.; Qureshi, R.A.; Ahmed, Z.; Qureshi, S.A.; Jamil, K.; Jyothi, K.N. and Prasuna, A.L. (1993). Laboratory evaluation of some vegetable oils as protectants of stored products. *Pakistan Journal of Scientific and Industrial Research*, **36**(9) : 377-379.

Mukherjee, S.N. and Ramachandran, R. (1989). Effects of azadirachtin (from *Azadirachta indica*) on the feeding, growth and development of *Tribolium castaneum* (Herbst) (Col., Tenebrionidae). *Journal of Applied Entomology*, **107**(2) : 145-149.

Murugan, K.; Senthilkumar, V.; Jeyabalan, D.; Babu, R.; Senthil Nathan, S. and Sivaramakrishnan, S. (1998). Interactive effect of neem products on the control of pulse beetle, *Callosobruchus maculatus* Fab. *Neem Newsletter*, **15**(4) : 41-44.

Murugesan, S.; Balu, A.; Durairaj, S.; Pankajam, S. and Sunitha, B. (1998). Insecticidal performance of a neem product in control of two major seed pests of forestry tree species. *Entomon*, **23**(3) : 197-201.

Ogunwolu, E.O. and Odunlami, A.T. (1996). Suppression of seed bruchid (*Callosobruchus maculatus* F.) development and damage on cowpea (*Vigna unguiculata* (L.) (Walp.) with *Zanthoxylum zanthoxyloides* (Lam.) Waterm. (Rutacease) root bark powder when compared to neem seed powder and pirimiphos-methyl. *Crop Protection*, **15**(7) : 603-607.

Oroumchi, S. and Lorra, C. (1993). Investigation on the effects of aqueous extracts of neem and chinaberry on development and mortality of the alfalfa weevil, *Hypera postica* Gyllenh (Col., Curculionidae). *Journal of Applied Entomology*, **116**(4) : 345-351.

Otto, D.; Kleeberg, H. (ed.) and Zebitz, C.P.W. (1997). Inhibition of vitellogenin and sex pheromone production in females of the Colorado potato beetle, sterilized by neem ingredients (azadirachtins). *Proceedings of 5th Workshop Wetzlar*, 22-25 June, Germany, 1996 : 263-267.

Palaniswamy, P. and Wise, I. (1994). Effects of neem-based products on the number and feeding activity of a crucifer flea beetle, *Phyllotreta cruciferae* (Goeze) (Coleoptera : Chrysomelidae), on canola. *Journal of Agricultural Entomology*, **11**(1) : 49-60.

Pandey, N.D.; Mathur, K.K.; Pandey, S. and Tripathi, R.A. (1986). Effects of some plant extracts against pulse beetle, *Callosobruchus chinensis* Linnaeus. *Indian Journal of Entomology*, **48**(1) : 85-90.

Pandey, N.D.; Singh, S.R. and Tewari, G.C. (1977). Use of some plant powders, oils and extracts as protectants against pulse beetle, *Callosobruchus chinensis* Linn. *Indian Journal of Entomology*, **38**(2) : 110-113.

Pandey, N.K. and Singh, S.C. (1997). Effect of neem bark powder on infestation of pulse beetle *Callosobruchus chinensis* in stored chickpea. *Indian Journal of Entomology*, **59**(2) : 161-163.

Pascual, N.; Marco, M.P. and Belles, X. (1990). Azadirachtin induced imaginal moult deficiencies in *Tenebrio molitor* L. (Copleoptera : Tenebrionidae). *Journal of Stored Products Research*, **26**(1) : 53-57.

Patel, K.P.; Valand, V.M. and Patel, S.N. (1993). Powder of neem-seed kernel for control of lesser grainborer (*Rhizopertha dominica*) in wheat (*Triticum aestivum*). *Indian Journal of Agricultural Sciences*, **63**(11) : 754-755.

Pereira, J. and Wohlgemuth, R. (1982). Neem (*Azadirachta indica* A. Juss) of West African origin as a protectant of stored maize. *Zeitschrift fur Angewandte Entomologie*, **94**(2) : 208-214.

Pereira, J. (1983). The effectiveness of six vegetable oils as protectants of cowpeas and Bambara groundnuts against infestation by *Callosobruchus maculatus* (F.) (Coleoptera : Bruchidae). *Journal of Stored Products Research*, **19**(2) : 57-62.

Pradhan, S.; Jotwani, M.G. and Rai, B.K. (1962). The neem seed deterrent to locusts. *Indian Farming*, **12**(8) : 7 & 11.

Prokopy, R.J. and Powers, P.J. (1995). Influence of neem seed extract on oviposition and mortality of *Conotrachelus nenuphar* (Col., Curculionidae) and *Rhagoletis pomonella* (Dip., Tephritidae) adults. *Journal of Applied Entomology*, **119**(1) : 63-65.

Pruthi, H.S. (1937). Report of the imperial entomologist. *Sci. Rep. Agric. Inst.*, New Delhi, 1935-36 : 123-127.

Raguraman, S. and Singh, D. (1997). Biopotentials of *Azadirachta indica* and *Cedrus deodara* oils on *Callosobruchus chinensis*. *International Journal of Pharmacognosy*, **35**(5) : 344-348.

Rahim, M. (1998). Biological activity of azadirachtin-enriched neem kernel extracts against *Rhyzopertha dominica* (F.) (Coleoptera : Bostrichidae) in stored wheat. *Journal of Stored Products Research*, **34**(2-3) : 123-128.

Rajapakse, R.; Senanayake, S.G.J.N. and Ratnasekera, D. (1998). Effect of five botanicals on oviposition, adult emergence and mortality of *Callosobruchus maculatus* Fabr. (Coleoptera : Bruchidae) infesting cowpea, *Vigna unguiculata* L. Walp.). *Journal of Entomological Research*, **22**(2) : 117-122.

Rajapakse, R.H.S. and Senanayake, S.G.J.N. (1997). Effectiveness of seven vegetable oils against *Callosobruchus chinensis* L. in pigeon pea *Cajanus cajan* L. *Entomon*, **22**(3-4) : 179-183.

Rajasekaran, B.; Kumaraswami, T.; Regupathy, A (ed.) and Jayaraj, S. (ed.) (1985). Control of storage insects with seed protectants of plant origin. *Behavioural and Physiological Approaches in Pest Management* : 15-17.

Ramachandran, R.; Mukherjee, S.N. and Sharma, R.N. (1988). Hormetic effect of azadirachtin on *Tribolium castaneum* (Herbst) (Coleoptera : Tenebrionidae). *Indian Journal of Experimental Biology*, **269**(11) : 913-914.

Reed, D.K. and Pierce, R. (1981). Natural products repel cucumber beetle. *Agricultural Research*, USA, **30**(2) : 12.

Reed, D.K.; Jacobson, M.; Warthen, J.D. Jr.; Uebel, E.C.; Tromley, N.J.; Jurd, L. and Freedman, B. (1981). Cucumber beetle antifeedants : laboratory screening of natural products. *Technical Bulletin, Science and Eduction Administration*, United States Dept. of Agri., **1641** : pp. 13.

Rohde, M.; Bressem, U.; Bornholdt, G. and Brenner, U. (1996). Studies on the control of the forest chafer in southern Hessen in 1994. *Forschungsberichte Hessische Landesanstalt fur Forsteinrichtung, Waldforschung und Waldokologie*, **22** : pp. 122.

Rouf, F.M.A.; Sardar, M.A. and Ahmed, K.S. (1996). Individual and combined effects of some plant materials for protection of lentil seeds against pulse beetle, *Callosobruchus chinensis* L. *Bangladesh Journal of Entomology*, **6**(1-2) : 13-21.

Salom, S.M.; Carlson, J.A.; Ang, B.N.; Grosman, D.M. and Day, E.R. (1994). Laboratory evaluation of biologically-based compounds as antifeedants for the pales weevil, *Hylobius pales* (Herbst) (Coleoptera : Curculionidae). *Journal of Entomological Science*, **29**(3) : 407-419.

Savitri, P. (1975). Study on the efficacy of neem kernel powder and neem oil as a mixture with grain and also by impregnation on paddy bags in the control of certain important storage pests of paddy. *M.Sc. Thesis*, College of Agriculture, Bapatla (A.P.).

Schmutterer, H. and Rambold, H. (1980). Effects of some pure fractions from seeds of *Azadirachta indica* on feeding activity and metamorphosis of *Epilachna varivestis* (Col. Coccinellidae). *Zeitschrift fur Angewandte Entomologie*, **89**(2) : 179-180.

Schmutterer, H. and Sen, R. (1996). Growth-disrupting effect of some neem-based products on Mexican bean beetle *Epilachna varivestis* Mulsant (Coleoptera : Coccinellidae). *Neem Newsletter*, **13**(2) : 13.

Schmutterer, H. (ed.) and Ascher, K.R.S. (1987). Fecundity-reducing and sterilizing effects of neem seed kernel extracts in the Colorado potato beetle, *Leptinotarsa decemlineata*. *Proceedings of the 3rd International Neem Conference*, Nairobi, Kenya, 10-15 July, 1986 : 351-360.

Schmutterer. H. (1987). Experiments on reduction in fecundity and sterlization of the Colorado potato beetle, *Leptinotarsa decemlineata* by application of neem seed extracts. *Mitteilungen der Deutschen Gesellschaft fur Allgemeine und Angewandte Entomologie*, **5**(1-4) : 152-161.

Schnetter, W.; Mittermuller, R. and Froschle, M. (1996). Control of the cockchafer *Melolontha* in the Kraichgau with NeemAzal-T/S. *Bulletin OILB-SROP*, **19**(2) : 95-99.

Singh, H.; Mrig, K.K. and Mahla, J.C. (1996). Effect of different plant products on the fecundity and emergence of lesser grain borer, *Rhyzopertha dominica* (F.) in wheat grains. *Annals of Biology* Ludhiana, **12**(1) : 96-98.

Singh, M.P. and Vir, S. (1987). Effects of some natural products on oviposition and development of (*Callosobruchus maculatus* Fabr.) on cowpea. *Transactions of Indian Society of Desert Technology*, **12**(2) : 143-147.

Singh, R.P. and Kataria, P.K. (1986). Deoiled neem kernel powder as protectant of wheat seeds against *Trogoderma granarium* Everts. *Indian Journal of Entomology*, **48**(1) : 119-120.

Singh, R.P. and Singh, S. (1985). Evaluation of deoiled neem (*Azadirachta indica* A. Juss) seed kernel against *Trogoderma granarium* Eversts. *Current Science*, **54**(18) : 950-951.

Singh, R.P.; Saxena, P. and Doharey, K.L. (1996). Evaluation of neem seed kernel and its derivative against three important insect pests of stored products. *Neem Newsletter*, **13**(3) : 42-43.

Singh, T.G.M. and Gour, T.B. (1996). Effectivity of certain plant extracts against larvae of *Mythimna separata* Wlk. and adults of *Callosobruchus chinensis* Linn. *Neem and Environment*, **1** : 349-356.

Sowunmi, O.E. and Akinnusi, O.A. (1983). Studies on the use of neem kernel in the control of stored cowpea beetle (*Callosobruchus maculatus* F.). *Tropical Grain Legume Bulletin*, **27** : 28-31.

Sujatha, A. and Punnaiah, K.C. (1985). Effect of coating stored seed of greengram with vegetable oils on the development of pulse beetle. *Indian Journal of Agricultural Sciences*, **55**(7) : 475-477.

Susha, C. and Karnavar, G.K. (1993). Effect of azadirachtin on vitellogenic oocyte development in *Trogoderma granarium* Everts (Coleoptera : Dermestidae). *Indian Journal of Experimental Biology*, **31**(2) : 188-190.

Suss, L.; Locatelli, D.P. and Cavalieri, M. (1997). Evaluation of repellent activity of *Azadirachta indica* A. Juss extracts on foodstuff insects. *Tecnica Molitoria*, **48**(10) : 1105-1112.

Tanzubil, P.B. (1991). Control of some insect pests of cowpea (*Vigna unguiculata*) with neem (*Azadirachta indica* A. Juss.) in northern Ghana. *Tropical Pest Management*, **37**(3) : 216-217.

Tanzubil, P.B.; Schmutterer, R. (ed.) and Ascher, K.R.S. (1987). The use of neem products in controlling the cowpea weevil, *Callosobruchus maculatus. Proceedings of the 3rd International Neem Conference*, Nairobi, Kenya, 10-15 July, 1986 : 517-523.

Volkonsky, M. (1937a). [Surla action acridifuge des extraits de ſeuilles de *Melia azedarach*.]. *Archs. Inst. Pasteur*, Alger., **15**(3) : 427-492.

Volkonsky, M. (1937b). [Sur un procede' nouveae de protection des cultures centre les cicridiens.]. *C.R. Soz.* Biol., Paris, **125** : 417-418.

Xie, Y.S.; Fields, P.G. and Isman, M.B. (1995). Repellency and toxicity of azadirachtin and neem concentrations to three stored-product beetles. *Journal of Economic Entomology*, **88**(4) : 1024-1031.

Yadav, T.D. (1996). Olfactory, oviposition and development response of four species of pulse beetles to four neem oil treated legumes. *Neem and Environment*, **1** : 343-348.

Zhang, S. and Zhao, S.H. (1983). Experiments on some substances from plants for the control of rice weevils. *Journal of Grain Storage Liangshi Chucang*, **1** : 1-8.

2. Diptera

Steffens and Schmutterer (1982) studied the effect of crude methanolic neem seed kernel extract on metamorphosis and quality of adults of the Mediterranean fruitfly, *Ceratitis capitata* Wied. A laboratory study carried out in the German Federal Republic showed that *C. capitata* was highly sensitive to the effects of a crude methanolic neem seed kernel extract (*A. indica*) incorporated in an artificial larval diet. The treatment resulted in a prolonged larval period and a high rate of pupal mortality in the surviving insects. Pupal size decreased in a dose-dependent manner with increases in neem extract concentration, as did the percentage of adults unable to fly. A multifactorial test showed that even very low concentrations of the extract could significantly affect the quality of the adults resulting from treated larvae.

Singh and Srivastava (1983) studied the ovipositional deterrent effect of alcoholic extract of neem seed oil against *Dacus cucurbitae*. A laboratory study carried out in India showed that an alcoholic extract of neem seed oil (obtained from *A. indica*) completely deterred oviposition by *D. cucurbitae* on bitter gourds (*Momordica charantia*) at 5 percent, and was highly effective at 2.5 percent. The extract was also effective in inhibiting oviposition by the pest on guava, although the concentration required was much higher (20%).

Webb *et al*. (1983) studied the repellency and toxicity of foliar applications of aqueous solutions of seed extracts of neem (*A. indica*) to adults of *Liriomyza sativae* Blanch. and *L. trifolii* (Burgess), 2 important pests of ornamental and vegetable field crops in southern USA and of greenhouse crops in North America, were evaluated in studies in Maryland and Ohio in 1979-80. The primary leaves of Henderson bush lime bean (*Phaseolus lunatus*) were used in all the tests except one, when chrysanthemum leaves were used. Significantly fewer eggs were laid by *L. trifolii* females in neem-treated foliage than in water-treated foliage; this was not the case for *L. sativae*. However, larval mortality of both species shortly after hatch was high (100% for *L. trifolii* and 98.2% for *L. sativae*). Neem residues 2-6 hrs-old on leaves repelled *L. sativae* adults, thus reducing oviposition, but residues 22 hrs-old were not appreciably repellent. Neem applied as 0.1 percent solution on leaves was active against *L. sativae* eggs 1-3 day-old as well as against larvae in all 3 instars. Against *L. sativae* eggs and larvae, neem was highly effective as 0.1 and 0.05 percent solutions giving 91-100% mortality and moderately effective as a 0.025 percent solution giving 49-92% mortality. Neem applied to leaves was effective for up to 7 days in killing larvae hatching from eggs, but its effectiveness declined steadily from the initial day after application (80% mortality) to the 7th day (44% mortality). Adults caged for 25 hrs with foliage freshly dipped in 0.1 percent neem extract showed no adverse effects. Similarly, 0.1 percent neem extract applied to chrysanthemum leaves showed that *L. trifolii* adults were slightly inhibited from making ovipositor punctures in leaves. Leaf-mining larvae failed to develop.

Meisner et al. (1986) sprayed an aqueous neem seed kernel extract on maize seedlings at 1, 0.5, 0.25, 0.1 and 0.05 percent and assayed against larvae of the pyralid *Ostrinia nubilalis*. No pupation occurred following treatment of plants with concentrations of 1.0, 0.5 and 0.25 percent. Good control was obtained when the maize seedlings were exposed to larvae 4 and 6 days after soil treatment. When azadirachtin was applied to seedlings at 0.005, 0.006, 0.0075, 0.01, 0.025 and 0.05 percent, all concentrations except the lowest reduced the average weight of treated larvae. No pupation occurred with larvae fed on leaves treated with 0.006-0.05 percent azadirachtin. The residual effect of 0.025 percent azadirachtin was studied and pupation rates were 0, 11 and 16 percent when plants were infested 2, 4 and 6 days, respectively, after treatment with 0.025 percent azadirachtin. Neem extracts were applied to bean seedlings (*Phaseolus vulgaris*) and the effect on *L. trifolii* was assessed. When both sides of leaves were painted, 1 percent ethanol and methanol extracts had the highest activity. Following spray applications before exposure to ovipositing females, aqueous, methanolic and ethanolic extracts were active. When plants were sprayed immediately after the appearance of larval mines, only the 1 percent methanolic extract showed high activity. A 95.4 percent reduction in pupal yield was achieved when plants were exposed to ovipositing females 5 days after soil treatment with 1 percent aqueous neem extract.

Areekul *et al.* (1988) studied the repellency effect of 130 plant extracts from 110 species to *Dacus dorsalis* was assessed in the laboratory on the basis of the rate of oviposition by 15-day-old females on cups containing 20 percent citrus juice coated outside with a mist spray of one gram equivalent of each plant extract per cup. The extracts which exhibited high repellency were obtained from *Azadirachta indica* var. *siamensis* (neem), *Bixa orellana, Citrus hystrix, Cucumis melo* (melon), *Cymbopogon citratus, Hedychium coccineum* var. *angustifolium, Heliotropium indicum, Homalomena, Ocimum gratissimum, Ricinus communis, Ternstoemia japonica* [*T. gymnanthera*] and 2 unidentified species. Moderate repellency was exhibited by 19 extracts, low repellency by 17 and 80 were ineffective.

Sanderson *et al.* (1989) carried out a trial in the USA indicated that an in-transit treatment of chrysanthemum cuttings with neem extracts (as Margosan-O) was effective in controlling *Liriomyza trifolii*. Treatment lasted long enough to disrupt the life-cycle and reduced the chances of reinfestation from subsequent generations of the agromyzid.

Wilps (1989) studied the extract of neem tree on flight activity, food uptake, reproduction and carbohydrate metabolism in the *Phormia terraenovae*. Adults of the calliphorid *P. terraenovae* were fed at 25°C and RH 70 percent with various concentrations of aqueous and methanolic extracts of neem seed kernel (NSKE) between the 2nd and 4th-day after eclosion. The methanolic extracts were enriched 50-60 percent with azadirachtin. The intake of either extract shortened lifespan by 4 days, and also reduced food intake, egg deposition and caused a complete loss of flying ability. Measurements of abdominal haemolymph concentrations of the low molecular weight carbohydrates and glycogen showed that NSKE-fed flies lost the ability to alter carbohydrate concentrations and mobilize glycogen.

Stark *et al*. (1990) studied the effects of azadirachtin on metamorphosis, longevity and reproduction of three tephritid fruitfly species. The effects of azadirachtin on the of *Ceratitis capitata, Dacus [Bactrocera] dorsalis* and *D. [Bactrocera] cucurbitae* exposed as late 3rd-instar larvae and pupae to treated sand were determined. Formation of puparia was not affected by azadirachtin at the concentrations tested. Adult emergence was completely inhibited at concentrations of 14 ppm for *C. capitata* and *D. dorsalis* or at 10 ppm for *B. cucurbitae*. *B. cucurbitae* was significantly more susceptible to azadirachtin than *C. capitata* and *B. dorsalis* at the LC50. Even though adult emergence was inhibited, approximately 95 percent of treated puparia contained living adults. Adults that emerged from treatments appeared normal but significantly more died compared with controls. Ten days after emergence, 75 percent of *B. dorsalis* and 64 percent of *C. capitata* exposed as larvae and pupae to 4.66 ppm azadirachtin had died. Approximately 24 percent of *B. cucurbitae* exposed to 2.77 ppm azadirachtin died within 10 days after emergence. Azadirachtin had no significant effect on the number of eggs laid by adult *B. dorsalis* and *C. capitata* that had survived larval-pupal treatments. *B. cucurbitae* laid significantly more eggs than adults that had survived larval-pupal treatments with 1.85 ppm. Azadirachtin had no significant effect on egg hatch or the growth and development of F1 progeny.

Parkman and Peinkowski (1990) studied sublethal effects of neem seed extracts on adult of *L. trifolii*. Fecundity and longevity were determined for adults of *L. trifolii* that had survived treatments with neem seed extract. Soil drenches of 1.0 and 2.0 ppm azadirachtin applied to infested chrysanthemums (*Dendranthema morifolium*) significantly reduced female fecundity. Longevity of males was significantly reduced by 2.0 ppm drench treatments. Fecuncity of females treated with 1.0 ppm of azadirachtin peaked later and decreased more rapidly than that of untreated females, whereas overall fecundity of females treated with 2.0 ppm azadirachtin exhibited no consistent pattern of peak and declilne.

Mote and Jadhav (1993) conducted field studies in Maharashtra, India, to determine the efficacy of extracts of *Azadirachta indica, Argemone mexicana* and *Calotropis gigantea* with that of carbofuran for the control of *Atherigona soccata* infesting sorghum. The results suggested that all the plant extracts were effective in reducing levels of the muscid.

Dimetry *et al.* (1995) evaluated two NSKE against *L. trifolii* Burg. Laboratory evaluation of NeemAzal-S and Margosan-O was carried out against the adults and larvae of *L. trifolii*. The feeding deterrent activity of both compounds was significant against the adults particularly at high concentrations, and lasted for 5 days ater treatment. Both formulations also deterred the females from laying eggs and the percentae of oviposition deterrent index (ODI) reached 80.7 and 52.6 for NeemAzal-S and Margosan-O (2%). respectively. Post-infestation treatment of bean seedlilngs containing 3rd instar larvae indicated that the toxic action of the tested extracts was fairly limited. However, different percentages of malformed pupae were obtained with 2 percent NeemAzal-S, where all the resulting progeny were female.

Smirle *et al.* (1996) studied the effects of neem oil on feeding behaviour and development of the pear sawfly, *Caliroa cerasi*. Neem oil deterred feeding by larvae of the tenthredinid *C. cerasi*, both in choice and in no-choice cherry leaf disk bioassays. Deterrence was greater in the choice tests, with 50 percent inhibition occurring at 0.49 percent aqueous neem oil compared with 1.11 percent in the no-choice tests. Antifeedant activity towards larvae was slilghtly less than has been observed for lepidoptera, but was higher than deterrence to other insects such as aphids. Topical application of neem oil to larvae resulted in reduced feeding, increased mortality, and a trend towards slower development. Most larval mortality after neem treatment resulted from incomplete subsequent moults.

Ranganath *et al.* (1997) studied a number of botanical and chemical insecticides against *Bactrocera cucurbitae* on cucumber (*Cucumis sativus*) and ridge gourd (*Luffa acutangula*) in South Andaman, India. Neem oil at 1.2 percent was the most effective treatment in reducing damage to cucumber (neem percentage damage 6.2%, as compared with 39.0% in the control), while neem cake at 4.0 percent and DDVP (dichlorvos) at 0.2 percent were the most effective against the pest on ridge gourd, reducing damage to 9.1-9.5 percent as compared with 32.9 percent in the control.

Kostyk and Wanner (1997) applied a neem seed extract containing 10, 30 or 90 ppm azadirachtin to black spruce (*Picea mariana*) trees at two stages of reproductive development to control insect damage to the seed cones in seed orchards at Dryden, Ontario, Canada. Applied after pollination, neem had no effect on damage by the spruce cone maggot (*Strobilomyia appalachensis*) but did provide some protection against defoliation by the spruce budworm (*Choristoneura fumiferana*). Neem had no effect on the number of spruce budworm larvae for the first 9 days of the experiment, but after 23 days, populations declined on trees with 30 and 90 ppm azadirachtin compared with control trees. The application of neem seed extract for controlling spruce budworm damage appears promising, but more information is needed on earlier applications in the field to determine if this product could be used operationally without causing substantial flower abortion.

Singh and Singh (1998) studied the oviposition deterrent by the use of NSKEs and azadirachtin against *Bactrocera cucurbitae* and *B. dorsalis*. Neem (*A. indica*) seed kernel extracts, (i.e. seed kernel aqueous suspension, ethanolic extract of seed kernel, hexane extract of seed kernel and acetone extract of deoiled seed kernel powder) at 1.25-20 percent, and pure azadirachtin at 1.25-10 ppm, were evaluated as oviposition deterrents to *B. cucurbitae* and *B. dorsalis* on pumpkin and guava, respectively, in the laboratory at 27°C and 65 percent RH. Neem seed kernel aqueous suspension at 5 percent in choice tests, and at all concentrations (_1.25%) in no-choice tests significantly deterred oviposition in both species. Similarly, the ethanolic extract was significantly active at all the concentrations tested for both species in choice and no-choice tests. However, with neem oil and its ethanolic extract sensitivities of the two species differed considerably. Both extracts deterred oviposition by *B. cucurbitae* at all the concentrations tested under both choice and no-choice test conditions. However, with *B. dorsalis*, neem oil was significantly deterrent only at 20 percent in both tests, and at 5 and 20 percent for oil ethanolic extract in choice and no-choice tests, respectively. The acetone extract of deoiled kernel powder significantly deterred oviposition by both species at all concentrations tested. Azadirachtin failed to deter oviposition in either species.

Venkateswarlu and Singh (1998) tested six different non-edible oils against oriental fruitfly, *B. dorsalis* Hendel and compared their efficacy with three insecticides in mango orchard (cv. Dashehari). Among the oils tested, neem oil (1.0%) exhibited highest reduction in infestation (68.89%) of *B. dorsalis* followed by mahua oil 1.0 percent (66.00%), karanj oil 1.0 percent (64.47%), menthol oil 1.0 percent (64.29%), Eucalyptus oil 1.0 percent (57.60%). However, there was not much difference between these treatments. All the oils exhibited same response like that of monocrotophos 0.05 percent (68.89%) and quinalphos 0.05 percent (69.21%). But, decamethrin 0.002 percent showed highest percent reduction of infestation (80.43%) and was significantly superior than all the treatments tested. Both oils and insecticides were superior over control in reducing the larvae of *B. dorsalis*. Highest per cent reduction of larvae both in physiologically matured fruits as well as fully ripened dropped fruits was found in decamethrin 0.002 percent whereas, lowest percent reduction of larvae was found in menthol oil.

Ragumoorthi *et al.* (1998) tested neem and other plant products against moringa fruitfly, *Gitona distigma* (Meigon). Five neem (*A. indica*) formulations : Nimbecidine (0.03%), NeemAzal (1.0%), Neem Gold (0.15%), Econeem (0.3%) and Rakshak (0.15%) were evaluated against *G. distigma*. Effectiveness of these formulations was measured in terms of ovicidal action, adult emergence, fecundity, hatchability, maggot survival and behavioural response. Among the neem formulations, Nembecidine was effective with high antioviposition effect (15.33 eggs/female at 150 ppm concentraion and 54.67 eggs/female in control), reduced longevity of adult (11 days at 150 ppm and 23.33 days in control), reduced adult emergence (71.25% treatment, 85.50% in control), reduced fecundity (50.25 eggs/female against 62.75 eggs/female in control), reduced hatchability (57.25% against 69.75% in control) and reduced maggot survival (81.25% against 89.0% in control). Even after 7 days, only 22.6 percent of adult flies orientated on the Nimbecidine treated fruits and there was no orientation of adult flies till the end of two days after treatment. Aqueous and methanolic plant extracts from leaves of *Vitex negundo*, shoot of *Tribulus terrestris*, shoot of *Catharanthus ruseus*, seeds of *Melia azedarach*, leaves of *Azadirachta indica*, seed kernel of *A. indica* (NSKE), leaves of *Tagetes erecta*, and leaves of *Prosopis juliflora* were assessed for their reaction to *G. distigma*. Among them, methanolic plant extracts of *V. vegundo, M. azedarach* and NSKE were more effective with high feeding deterrency (43.75-62.50%), low fecundity (51.90-74.26% redution), low hatchability (46.75-51.25%), maggot mortality (18.50-24.75%), puparium mortality (24.75-31.25%) and high adult malformation (7.75-18.25%). Nimbecidine 0.03 percent at 150 ppm as foliar spray during 50 percent fruit set and soil application of NSKE at the rate of 2 lt/tree incorporated in the IPM programme for *G. distigma* has yielded higher numbers of moringa (*Moringa pterygosperma*) fruits.

Ragumoorthi and Rao (1998) tested 5 neem formulations, Nimbecidine (0.03%), NeemAzal (1.0%), Neem Gold (0.15%), Eco Neem (0.35%) and Rakshak (0.15%) against moringa fruit fly *G. distigma*. Effectiveness of these formulations were measured in terms of antioviposition effect, ovicidal action, adult emergence, fecundity, hatchability and maggots surviving to pupariation and orientational response. Among the neem formulations, Nimbecidine was effective. Antioviposition effect (15.33 and 54.67%), reduced longevity of adult (1.0 and 23.33 days), moderate ovicidal action (7.75

and 0.75% mortality), reduced adult emergence (71.25 and 85.50%), reduced fecundity (50.25 and 62.75 eggs/female), reduced hatchability (57.255 and 69.75%) and reduced maggot survival (81.25 and 89.00%) were observed in Nimbecidine at 150 ppm and untreated control, respectively. Even after 7 days, only 22.6 percent of adult flies oriented on the Nimbecidine treated fruits and there was no orientation of adult flies till the end of 2 days after treatment.

Vanranden and Roitberg (1998) studied the effect of neem products on oviposition deterrence survival, hehaviour and reproduction of adult cherry fruitfly, *Rhagoletis indifferens* Curran. An emulsifiable concentrate containing 1.75 percent and another containing 2.2 percent azadirachtin were used. Cherries infested *R. indifferens* were picked from infested trees and the dropped mature larvae to become puparia were stored at 4°C. For oviposition deterrence flies were collected every 3 days from cages and in groups 5-50 (4 female : 1 male). The studies on oviposition deterrence under field conditions were conducted and population level was calculated.

With pherocon-AM factory baited traps placed in the density of 20 traps/ha. Experiments were conducted at early season and late season density level. To study adult foraging behaviour, 12 flies for each treatment were observed for 15 mts or until fly left a branch section. The adult fertility and survival studies were done in laboratory. Four flies (3 female : 1 male) were placed in cages within 24 hrs of eclosion. The insecticide was tested for its potential effects on oviposition, survival, behaviour and reproduction. Using adult *R. indifferens*, evidence was found for slight oviposition deterrence in the field, but this effect was not duplicated in the laboratory. No behavioural changes were observed in adult fliles exposed to the insecticide under field conditions. A strong negative effect on the development of eggs and decreased survival of adults were observed when flies ingested the insecticide extract with food and water immediately after eclosion. Total suppression of egg development was obtained in adults exposed continuously to concentrations of the insecticide >0.5 percent. By delaying the onset of exposure to the insecticide, egg development was increased, and was not significantly different from controls. Applied as a foliar spray at the appropriate time within a cherry orchard, this neem-based insecticide may prevent adult female *R. indifferens* from maturing viable eggs and decrease adult survival.

References

Areekul, S.; Sinchaisri, P. and Tigvatan Anon, S. (1988). Effects of Thai plant extracts on the Oriental fruit fly. II. Repellency test. Kasetsart Journal, *Natural Sciences*, **22**(1) : 56-61.

Dimetry, N.A.; Barakat, A.A.; Abdalla, E.F.; El-Metwally, H.E.; El-Salam, A.M.E.A. (1995). Evaluation of two neem seed kernel extracts against *Liriomyza trifolii* (Burg.) (Dipt. Agromyzidae). *Anzeiger fur Schadlingskunde, Pflanzenschutz*, Umweltschutz, **68**(2) : 39-41.

Kostyk, B.C. and Wanner, K.W. (1997). Control of insect damage to black spruce seed cones with neem. *Northern Journal of Applied Forestry*, **14**(1) : 40-43.

Meisner, J.; Melamed Madjar, V.; Yathom, S.; Ascher, K.R.S.; Schmutterer, H. (ed.) and Ascher, K.R.S. (1987). The influence of neem on the European corn borer (*Ostrinia nubilalis*) and the serpentine leafminer (*Liriomyza trifolii*). *Proceedings of the 3rd International Neem Conference*, Nairobi, Kenya, 10-15 July, 1986 : 461-477.

Mote, U.N. and Jadhav, S.S. (1993). Effects of indigenous plant products on shoot fly damage in sorghum. *Journal of Maharashtra Agricultural Universities*, **18**(1) : 135.

Parkman, P. and Peinkowski, R.L. (1990). Sublethal effects of neem seed extracts on adults of *Liriomyza trifolii* (Diptera : Agromyzidae). *Journal of Economic Entomology*, **83**(4) : 1246-1249.

Ragumoorthi, K.N. and Rao, P.V.S. (1998). Management of Moringa fruitfly using neem products. *ICPPMSA*, 11-13 December, 1998, CSAU A&T, Kanpur (India) : 152-153.

Ragumoorthi, K.N.; Rao, P.V.S.; Dhaliwal, G.S. (ed.); Arora, R. (ed.); Randhawa, N.S. (ed.) and Dhawan, A.K. (1998). Neem products and plant extracts for managing Moringa fruit fly, *Gitona distigma* (Meigon). *Proceedings of International Conference on Ecological Agriculture : Towards Sustainable Development*, Chandigarh, India, 15-17 November, 1997 : 250-261.

Rambold, H.; Sharma, G.K.; Czoppelt, C. and Schmutterer, H. (1980). Evidence of growth disruption in insects without feeding inhibition by neem seed fractions. *Zeitschrift fur Pflanzenkrankheiten und Pflanzenschutz*, **87**(5-6) : 290-297.

Ranganath, H.R.; Suryanarayana, M.A. and Veenakumari, K. (1997). Management of melon fly (*Bactrocera* (*Zeugodacus*) *cucurbitae* Coquillett) in cucurbits in South Andaman. *Insect Environment*, **3**(2) : 32-33.

Sanderson, K.C.; Oetting, R.D. and Smith, D.A. (1989). In-transit neem insecticide treatment of rooted chrysanthemum cuttings controls leafminer. *HortScience*, **24**(5) : 856.

Shivendra Singh, Singh, R.P. (1998). Neem (*Azadirachta indica*) seed kernel extracts and azadirachtin as oviposition deterrents against the melon fly (*Bactrocera cucurbitae*) and the oriental fruit fly (*Bactrocera dorsalis*).

Phytoparasitica, **26**(3) : 191-197.

Singh, R.P. and Srivastava, B.G. (1985). Alcohol extract of neem (*Azadirachta indica* A. Juss) seed oil as oviposition deterrent for *Dacus cucurbitae* (Coq.). *Indian Journal of Entomology*, **45**(4) : 497-498.

Smirle, M.J. and Wei ShuGe (1996). Effects of neem oil on feeding behaviour and development of the pear sawfly, *Caliroa cerasi*. *Entomologia Experimentalis et Applicata*, **80**(2) : 403-407.

Stark, J.D.; Vargas, R.I. and Tahlman, R.K. (1990). Azadirachtin : effects on metamorphosis, lengevity, and reproduction of three tephritid fruit fly species (Diptera : Tephritidae). *Journal of Economic Entomology*, **83**(6) : 2168-2174.

Steffens, R.J. and Schmutterer, H. (1982). The effect of a crude methanolic neem (*Azadirachta indica*) seed kernel extract on metamorphosis and quality of adults of the Mediterranean fly, *Ceratitis capitata* Wied. (Diptera, Tephritidae). *Zeitschrift fur Angewandte Entomologie*, **94**(1) : 98-103.

Van Randen, E.J. and Roitberg, B.D. (1998). Effect of a neem (*Azadirachta indica*)-based insecticide on oviposition deterrence, survival, behaviour and reproduction of adult western cherry fruit fly (Diptera : Tephritidae). *Journal of Economic Entomology*, **91**(1) : 123-131.

Venkateswarlu, P. and Singh, G. (1998). Efficacy of non-edible oils on oriental fruit fly, *Bactrocera dorsalis* in mango. *National Seminar on Entomology in 21st Century*, April, 30-May, 2, 1998, Udaipur : 124

Webb, R.E.; Hinebaugh, M.A.; Lindquist, R.K. and Jacobson, M. (1983). Evaluation of aqueous solution of neem seed extract against *Liriomyza sativae* and *L. trifolii* (Diptera : Agromyzidae). *Journal of Economic Entomology*, **76**(2) : 357-362.

Wilps, H. (1989). The effect of extracts from the neem tree *Azadirachta indica* on flight activity, food uptake, reproduction and carbohydrate metabolism in the dipteran *Phormia terraenovae* (Diptera, Muscidae). *Zoologische Jahrbucher Abteilung fur Allgemeine Zoologie und Physiologie der Tiere*, **93**(2) : 271-282.

3. Hemiptera, Homoptera and Orthoptera

According to Pradhan and Jotwani (1971), the systematic work on neem products was initiated in the division of Entomology, Indian Agricultural Research Institute, New Delhi, India during the early sixties. In the present experiment, two of the isolates labelled as total bitters of neem (crude) and Nembidin 'T' (Pharmaceutical quality) were tested against the adults of *S. gregaria*, *Locusta migratoria* Linn., *Paecelocerus pictus* (Fab.) and

Aulacophora foveicallis (Lucas) and the final instar larvae of *Euproctis lunata* (Walker) and *Prodenia litura* (Fab.) The trials indicated varying degree of gustatory repellency of the two products against these insect pests. The antifeeding property as indicated by the amount of treated food eaten was more pronounced against the two species of locusts and the crude products was found to be superior to the refined product. In case of locusts, NSKS showed greater degree of repellency against the locusts whereas no feeding was observed in cabbage leaves treated with 0.001 percent suspension. However, with 0.0005 percent, there was 44.4 percent damage against 100 percent feeding on untreated leaves. The absolute antifeedant properties of NSKS at 0.01 percent was observed against *E. lunata*. However, higher concentrations of the suspension were found to significantly reduce feeding of *P. litura* larvae and *Acrida exaltata* Walker adults.

Griffiths *et al*. (1978) studied the influence of repellent materials and aphid extracts on settling behaviour and larviposition of *Myzus persicae* Sulz. When incorporated into an artificial diet or painted on the surface of the membrane containing the diet, solutions of the insect repellent GD 880 (2-methyl-2-(octylamino)-I-propanol) killed appids, whilst extracts of neem seeds (*A. indica*) or extracts of the bodies of *M. persicae* deterred settling and larviposition of apterous adults. The deterrent effect occurred even when the painted membrane was covered by a similar but untreated membrane. When applied in culture solutions to the roots of young kale plants, GD 880 and extracts of neem influenced aphids on the foliage only at concentrations that visibly affected plant growth, but neem extract (and to a lesser extent GD 880) decreased aphid colonisation when painted on kale leaves of non-damaging concentrations.

Jacobson *et al*. (1978) studied chemistry and biological activity of insect feeding deterrents from certain weed and crop plants. Concentrations of 0.01-1.0 percent of a hexane extract of neem seed (*A. indica*) and several of its chromatographic fractions significantly deterred feeding by *Aonidiella aurantii* (Mask.), *A. citrina* (Coq.), *Planococcus citri* (Risso), *Panonychus citri* (McG.) and *Aleurothrixus floccosus* (Mask.). An ether extract of the hexane-extracted seed reduced egg deposition by *Carpophilus hemipterus* L.,the ether and ethanol extracts repelled adults and larvae of this insect, and the hexane extract repelled only adults. A concentration of 1 percent of a pentane extract of the nuts of tung (*Aleurites fordii*) considerably deterred feeding by adults of *Anthonomus grandis* (Boh.); a concentration of 0.1 percent of an ether extract of the defatted nuts reduced feeding by *Acatymma vittalum* (F.) and *Conotrachelus nenuphar* (Hbst.), and a concentration of 0.2 percentt reduced feeding by *Diabrotica undecimpunctata* (Howardi Barber). Also, these extracts, and ethanolic and aqueous extracts as well deterred feeding by larvae of *Cydia* [*Laspeyresia*] *pomonella* (L.) and *Argyrotainia velutinana* (Wlk.). Several Polar and non-polar components of neem seeds and tung nuts appear to be responsible for the antifeedant properties of these plant materials.

Balasubramanian (1979) conducted pest management studies for rice brown plant hoppers in Coimbatore, India, on the control of *Nilaparvata lugens* (Stal) include the following : the identification of a highly resistant rice variety (Nira); light-trap studies; the establishment of various biotypes of the pest; the testing of several insecticides, of which FMC 35001, bendiocarb (Garvox), acephate (Orthene) and BPMC [2-(1methylpropyl)phenyl methylcarbamate] appear promising; the use of neem oil and neem seed kernel extract as antifeedants, which has been found to reduce numbers of *N. lugens* considerably.

Naraynan *et al.* (1980). studied phagodeterrency of six fractions of neem oil against *Schistocerca gregaria* Forsk. The extract 1 represents the total bitters extracted with cold aqueous alcohol. Extract 2 the residue left after water extraction of the total bitters. Extract 3, the petroleum ether soluble portion of extract 2. Extract 4, the residue left after the petroleum extraction of extract 2. Extract 5 is the total material got on petroleum ether extraction of deoiled neem cake. Extract 6 is the total acetone extract of the neem leaves. Sexually immature adult desert locusts were starved for 24 hrs and then released on treated cabbage leaf pieces. The data were taken after 24 hrs, and the unfed area left was measured. The results indicated that the total bitters extracted with cold aqueous alcohol and the total acetone extract of the neem leaves had a strong antifeedant effect, affording 99.28 percent protection in the case of the former and 98.57 percent in the case of the latter when used at a 1 percent concentration to treat cabbage leaves.

Jotwani and Srivastava (1981) tested the effecitveness of neem (*A. indica* and *Melia azedarach*) as a source of insecticide and antifeedant, showing the neem plant parts or products used, with their concentration or application rate; the plant or other commodity to be protected, the target arthropod pest species (largely locusts and lepidopterous larvae). It is concluded that the main action of neem, at least against arthropods growing plants, is as an antifeedant, since little direct mortality was observed. It is also pointed out that neem is effective not only against biting and chewing pests but also against sucking Hemiptera such as aphids and cicadellids. Some evidence on its inhibition of oviposition in some pests and its disruption of growth or moulting in others suggest the possibility that neem may cause sterility and interfere with growth and development.

Goel *et al.* (1983) observed the effect of neem formulations to locusts. It is estimated that 13 kg kernels could be obtained from 44 kg dried fruit, providig sufficient to spray 13 ha of crops. It is suggested that with the coming of the Rajasthan canal to the heart of the Thar Desert, an industry to grow neem and prepare antifeedants for the control of locusts could be established there. A 1 percent concentration is sufficient to act as an antifeedant against locusts.

Chiu *et al.* (1983) did experiments with neem formulations as antifeedant in brown plant hopper (*Nilaparvata lugens* Stal.). Lobaratory and greenhouse studies in China in 1981 showed that the seed oil of neem (*A. indica*) and petroleum ether extracts of the seed kernels of *Melia toosendan* and *M. azedarach* had strong antifeedant effects on nymphs of the rice pest *N. lugens*. The median antifeedant concentrations (AFC50) of the 3 compounds for 2nd and 3rd instar nymphs were 0.518, 0.519 and 1.15 percent, respectively. The compounds also exhibited moderate to slight systemic properties in rice plants and contact insecticidal activity against the plant hopper.

Saxena and Khan (1984) studied the effect of neem formulations on the survival and ability to transmit virus by brown plant hoppers. In studies in the Philippines, the duration of survival of *N. lugens* decreased markedly on rice seedlings that had been previously sprayed with neem oil emulsified in water with 1.6 percent liquid detergent and applied at 3, 6, 12, 25 and 50 percent. The effectiveness of the treatment increased with the concentration. The ability of the plant hopper to transmit persistent grassy stunt virus and ragged stunt virus was very markedly reduced on seedlings treated with neem. Viruliferous plant hoppers completely failed to transmit the viruses to plants sprayed with 50 percent neem oil, though transmission was still possible on unsprayed plants.

Koul (1984) studied the effect of azadirachtin on the development of red cotton bug, *Dysdercus koenigii*. The phytochemical evoked various specific and non-specific effects during the course of development. Prolonged development, wing deformities, unplasticization of wing lobes, development of wingless adults and larval mortality were the charactersitic features.

Coudriet *et al.* (1985) conducted bioassays against the immature stages of *Bemisia tabaci* with 0.2 and 2.0 percent aqueous solutions of neem-seed extract in laboratory and greenhouse studies. Cotton (*Gossypium hirsutum*) foliage treated with neem (*A. indica*) seed extract resulted in a reduction of egg viability and oviposition, prolonged larval periods and larval mortality. Although the mode of action of the extract is unknown, it was observed that adults were repelled, thus reducing oviposition. The failure of treated larvae to complete ecdysis would indicate an antiecdysteroid activity.

Chiu (1985) carried out laboratory and field experiments in China to determine the effectiveness of extracts of *Melia toosendan, M. azedarach* and *Azadirachta indica* (neem) against arthropod pests of agricultural crops. Neem seed oil or petroleum ether extracts of the seed kernels of *M. toosendan* and *M. azedarach* at 6 percent showed strong antifeedant effects on *N. lugens* and strong oviposition-deterring effecs on *Orseolia oryzae*, as did methanol extracts of *Melia*; 1 percent of the methanol extract of *M. toosendan* repelled 96.2 percent of females in the laboratory in choice tests. With the methanol extract of the seed kernels of *M. azedarach*, inhibition of feeding by 5th-instar larvae of *Spodoptera litura* was about 100 times stronger in choice than in no-choice tests. Preliminary field experiments showed that spraying with 500-800 mg/lt of toosendanin was effective in controlling 5th-instar larvae of *Pieris rapae*. A spray containing 0.5 percent emulsified oil from the seed kernels of neem. *M. toosendan* and *M. azedarach* was highly effecitve against the citrus pest *Panonychus citri* and comparatively safe to the predacious *Amblyseis newsami*.

Singh *et al.* (1986) studied the antifeedant and repellent qualities of neem against aphis, *Brevicoryne brassicae* L. on cauliflower and cabbage. The antifeedant and repellent effects of different neem extracts to the aphid *B. brassicae* was studied in pot and field studies in Himachal Pradesh, India. Three or 4 sprays of 0.1, 0.2 or 0.4 percent neem kernel suspension, 0.2, 0.4 or 0.8 percent oil emulsion or 0.5, 1.0 or 2.0 percent leaf water extract were applied at 15-day intervals. The different treatments exhibited a strong repellent and antifeedant effect, and significanlty reduced the aphid population on the potted plants compared with the control, except for the treatment of 0.5 percent leaf water extract on cabbage. In the field, all the treatments significantly reduced the population of aphids on both crops compared with the control.

Saxena *et al.* (1987) observed antifeedant effect of neem seed oil against rice pests. Laboratory and field studies which have confirmed the insect antifeedant properties of neem seed oil against the rice pests, *Cnaphalocrocis medinalis, Mythimna separata, Spodoptera mauritia acronyctoides* [*S. mauritia*], *Nilaparvata lugens, Nephotettix virescens* and *Sogatella furcifera*.

Singh *et al.* (1987) compared the insect antifeedant activity of neem seed kernels from 9 Indian ecotypes. Bioassays using the acridid *Schistocerca gregaria* showed large differences in antifeedant activity and extract yields from different regions. Antifeedant activity, extract yield and oil content could not be correlated with environmental conditions, although kernels obtained from humid areas yielded more oil and trees from areas near the desert possessed higher antifeedant activity than those growing in coastal areas.

The qualitative and quantitative values of extracts of seed coats and fallen leaves were compared to those of the seed kernel. Water and ethanolic extracts of the seed coat inhibited acridid feeding at concentrations of 0.012 and 0.048 percent, respectively. Chloroform extracts of the seed kernel, seed coat and fallen leaves had the highest activity with concentrations of 0.0007, 0.002 and 0.1 percent, respectively, being required to inhibit feeding.

Singh *et al*. (1987) studied the effect of neem (*A. indica*) kernel suspension on the oviposition behaviour of the acridid *S. gragaria* in the laboratory. Neem kernel suspensions at 0.1, 0.5 and 1.0 percent were mixed or sprayed on sand and offered for oviposition or applied directly to the abdomen of the locusts. When offered a choice of media for egg-laying, significantly fewer egg-pods were laid in sand either mixed or sprayed with all 3 concentrations of neem kernel suspension. The reduction in oviposition was dose-dependent. When locusts were not offered a choice of media for oviposition, they made more false borings than the control group and eggs were dropped indiscriminately on the floor of the cages. The application of neem kernel suspension onto the abdomen of the locusts did not affect oviposition on untreated sand.

Singh (1987) studied the effect of neem kernel extract mixed with insecticides on the oviposition and mortality of the *S. gregaria*. No eggs were laid in oviposition tubes treated with such mixtures. Mortality of locusts treated topically with 1 percent neem kernel extract in combination with fenitrothion or dichlorvos at an equivalent of 50 g a.i./ha or with chlorpyrifos or quinalphos at 150 g a.i./ha was 85.0, 82.5, 85.0 and 82.5 percent, respectively, as compared with 2.5 percent of untreated locusts.

Singh *et al*. (1987) studied the effect of neem (*A. indica*) kernel suspension on the hatchability of eggs of the acridid *S. gragaria*. The hatch rate of eggs laid in sand mixed or sprayed with 0.1 (9-14.9%), 0.5 oer 1.0 percent neem kernel suspension was significantly, lower than in the control. The lowest number of hoppers emerged from the eggs in sand treated with the 1 percent suspension.

Kabir and Mia (1987) studied the effectiveness of some indigenous materials as repellent against mustard aphid. Six indigenous materials were tested against *Lipaphis pseudobrassicae* [*L. erysimi*] on mustard in Bangladesh in 1983-85. Extracts of neem, garlilc, onions, and tobacco; straw ash, and a mixture of soap with kerosene reduced infestation and increased yields.

Olaifa *et al*. (1987) reported antifeedant and insecticidal effects of neem and other plant extracts against grass hopper, *Zonocerus variegatus*. The effects of extracts of *Azadirachta indica, Piper guineense, Petiveria alliacea, Ocimum gratissimum, Tetrapleura tetraptera* and *Nicotiana tabacum* on the acridid *Z. variegatus* in screenhouse experiments. Seedlings of cowpea (*Vigna unguiculata* subsp. *unguiculata*) and jack bean (*Canavalia ensiformis*) were treated with plant extracts, singly and in various combinations. The treated plants were used to feed the acridid. Extracts of *A. indica, P. guineense* and *P. alliacea* had antifeedant peroperties, resulting in 0, 16.7 and 25 percent defoliation, respectively. Following treatment with extracts of the remaining 3 plants, defoliation was greater than 50 percent. Extracts of *A. indica* and *P. alliacea* acted synergistically through the prolongation of residual activity. Combination of *A. indica* and *P. guineense* produced significantly higher mortality.

Dreyer (1987) studied the effect of neem aqueous extracts and neem oil on the pests of egg plant (*Solanum melongena*) in Togo. Aubergines planted out in small plots in Togo were treated weekly with an aqueous extract of 25 or 50 g pulverized neem seed kernels/lt or with 5 or 10 lt neem oil/ha (ULV) and examined for damage or infestaion, or both, by various pests and for effects on plant-reproductive phases and fruit yield in comparison with those in untreated plots. The aqueous extract gave good reduction in infestation and damage, especially at the higher concentration, in the case of *Scrobipalpa ergasima, Empoasca* [*Jacobiella*] *facialis* (larvae and adults), *Phycita melongenae, Selepa docilis* (larvae) and grasshoppers, including *Acanthacris ruficornis*, while neem oil was generally less effective, although often considerably better than no treatment. Both products were ineffecitve against *Tetranychus cinnabarinus*. Plant development was much improved and the yield doubled by the aqueous extract (with little difference between concentrations), neem oil being somewhat inferior, although it increased the yield significantly at 10 lt/ha.

Kareem *et al*. (1988) evaluated the effectiveness of rice seed treatment with crude neem seed kernel extract (NSKE) or neem cake against the brown planthopper (*N. lugens*) and the green leafhopper (*Nephotettix virescens* Distant) on young seedlilngs together with the effect of neem treatment on seed germination and seedling vigour. fewer nymphs of *N. lugens* reached the adult stage on TN1 seedlings treated with NSKE than on untreated ones, but only the treatment with 10 percent neem extract was significantly more effective. Only treatment with 10 percent NSKE reduced nymphal emergence of *N. virescens*. Nymphal emergence on IR42 was not significantly affected by the extracts, probably because it is moderately resistant to *N. virescens*. Seed germination, seedling root length, shoot length and chlorophyll content were not affected by neem treatment; soaking seed in NSKE increased seedling vigour, with significanty higher seedling root and shoot growth indices than those from untreated seed. Neem cake improves these parameters to a lesser extent.

Singh *et al*. (1988) tested the effect of neem cake on the oviposition of desert locust *S. gregaria*. The oviposition response of females of the acridid *S. gregaria* to sand mixed with neem cake (at rates of 90, 120 and 150 mg neem/kg sand) was investigated in the laboratory. When given a choice, females avoided the sand mixed with neem cake, but when there was no choice, there was no significant difference in the numbers of eggs laid in sand with neem or without neem.

Singh *et al*. (1988) observed the effect of neem kernel suspension (NKS) on the hatchlings of acridid, *S. gregaria*. The body weight and colour pattern of 1st-instar nymphs of the acridid, treated with 0.1, 0.5 or 1.0 percent NKS, were determined in the laboratory at 32.5°C. Hatchlings produced from media treated with 1 percent NKS were significantlyy lighter than those from untreated media. the number of green, hatchilings increased with increase in concentration of NKS. There was also a significant reduction in the weight of hatchlings with increase in concentration. It is concluded that the survival chances of hatchlings from treated media are reduced due to their smaller size and lighter colour.

Hague and Islam (1988) tested the effects of methanolic extracts of seeds of neem (*A. indica*) sand chinaberry (*M. azedarach*) on the orientatioon, feeding behaviour and growth of the cicadellid *Nephotettix nigropictus* in the laboratory. The extracts were shown to be strong repellents and antifeedants. The nymphal period was prolonged and mortality was increased in insects feeding on rice seedlings treated with the extracts. The effectiveness of the extracts increased with increase in concentration.

Kareem *et al*. (1989) compared the effect of neem seed and leaf bitters on oviposition and development of the rice pests, *Nephotettix virescens* and *Nilaparvata lugens* in the field cage and laboratory experiments in the Philippines. In the field, susceptible TN1 rice plants were sprayed with the extracts at 500, 2500 and 5000 ppm and each infested with 5 pairs of newly emerged *N. lugens* or *N. virescens*. In the laboratory, seedlings were dipped for 12 hrs in 100, 500 or 2500 ppm solutions and each transferred to a test tube where they were infested with a 1st-instar nymph of either species. The treatments significantly reduced oviposition by both pests, and adversely affected nymphal emergence and development. Fewer eggs were laid on plants sprayed with neem seed bitters at 5000 ppm. No nymphs of *N. virescens* developed to adults on seedlings treated with neem seed bitters or neem leaf bitters extracted from dried leaves at 500 ppm while 16 percent of nymphs developed to adults treated with neem leaf bitters extracted from fresh leaves at 500 ppm. Development on seedlings treated with neem leaf bitters (dried or fresh leaves) at 2500 ppm was completely suppressed. Development of *N. lugens* was reduced to 20 percent with neem leaf bitters extracted from dried leaves at 2500 ppm and to 39 percent with neem leaf bitters extracted from fresh leaves. No nymphs of *N. lugens* developed seedlings treated with neem seed bitters at 2500 ppm.

Gawande and Barkhade (1989) studied the effect of synthetic juvenile hormone analogues and juvenile hormone mimicking in some vidarbha plants on the pyrrhocrid, *Dysdercus cingulatus* Fab. In a laboratory study, the extracts of 12 plants from the Vidarbha region, India, and 2 synthetic juvenille hormone analogues were tested against the pyrrhocorid. Extracts from *Tectona grandis, Melia azadirachta* [*Azadirachta indica*], *Murraya koenigii, Citrus aurantifolia* (limes) and *Calotropis gigantea* produced a juvenile hormone-like effect, i.e. crumpling of wings.

Saxena *et al.* (1989) studied the effect of neem oil on courtship signals and mating behaviour of brown plant hopper females. Experiments were carried out to determine whether courtship signals and mating behaviour of the rice pest *N. lugens* were adversely affected by treatment with neem oil applied topically at 1, 2.5 or 5 μg/female, or applied at 3, 6 or 9 percent emulsified with Teepol to plants on which females were transferred 1 hr later. Both types of treatments disrupted normal courtship signal production and mating behaviour of females. Up to 25 percent of treated females did not emit signals and failed to mate, and up to 58 percent emitted abnormal signals that could not be recognized by mature males. The premating period was extended with treated females, and males often failed to copulate with treated females because the latter continuously moved their abdomens.

Songkittisuntorn (1989) tested the efficacy of neem oil and neem extracted substances on the rice leaf hopper, *N. virescens*. Neem oil was extracted from neem seeds, and further extraction was performed by water and other solvents. The repellent action of neem oil at 10 percent and upwards to the rice pest was found to increase with increasing oil concentration. Neem extracted by distillation was more effective than neem oil against female cicadellids. Of 3 extracts obtained by distillation (N1, N1M and N2, N1 was the best, followed by N2. Neem oil at 7 percent and upwards reduced population of 3rd-instar nymphs by >50 percent. Adding 0.1 percent piperonyl butoxide as a synergist to 5 percent neem oil increased mortality by 42.5 percent. Neem oil was better than neem extracted by distillation in reducing nymphal production. N1 was better than N1M, N2 and pure water extracts, respectively. N1 extract at 1.67 percent and 25 percent neem oil reduced female cicadellid survival to <50 percent after 6 days.

Ascher *et al.* (1989) tested the antifeedant effect of NSKE and azadirachtin on nymphs of acridid arthropod, *Eyprepocnemis plorans*. The antifeedant effect of azadirachtin and a methanolic NSKE on 4th-instar nymphs of the acridid was investigated using individuals collected in Sardinia, Italy. Nymphs were offered either saccharose-impregnated filter paper discs or leaves of broad bean (*Vicia faba*) treated with azadirachtin at 10-3-10-9 percent or with NSKE at 10-2-10-6 percent. The amount of substrate consumed was determined by weighing the filter paper or by measuring the leaf area. On filter paper, both compounds were higly active down to 10-4 percent. In the leaf treatment, however, azadirachtin was more active than the NSKE; phagodeterrence was 100 percent at 10-4 peercent, respectively. The NSKE was more active than the commercial preparation Neemark.

Kareem *et al.* (1989) tested neem as seed treatment for rice before sowing and effects of two homopterous insects and seedling vigour. Compared with treated seedlings, fewer 1st-instar nymphs of *N. virescens* reached the adult stage on rice seedllings (cv. TN1) raised from seeds treated before sowing with 2.5 percent neem kernel extract or with 2 percent neem cake. Likewise, fewer nymphs of *N. lugens* became aduts in treatments with neem exxtracts. Neem cake treatments were not effective. Neem treatments affected neither seed germination nor root length, shoot length or chlorophyll content of the rice cultivars IR36 and IR42. Seedlings from seeds treated with neem were more vigorous, and had higher root and shoot growth indexes and dry weights than those that germinated from the control.

Ramaraju and Babu (1989) tested the effect of plant derivatives on brown plant hopper (*Nilaparvata lugens*) (BPH) and white backed plant hopper (*Sogatella furcifera*) (WBPH) nymph emergence on rice. Nine plant derivatives were tested for their inhibitory effect on nymphs of two pests. Forty-day-old potted IR20 plants were sprayed with 1 percent oils of neem (*A. indica*), pinnai (*Calophyllum inophyllum*), pungam (*Pongamia glabra* [*P. pinnata*]) and illupai (*Madhuca longifolia*), 2 percent of neem seed kernel and 2 percent seed extracts of the other plants, and a 5 percent neem cake extract. All the plant derivatives tested were effective in reducing adult emergence of the pests. The greatest reduction in the case of *N. lugens* was obtained with 5 percent neem cake extract, followed by 2 percent pungam seed extract, and in the case of *S. furcifera* the greatest reduction was also obtained with the neem cake extract, followed by the 2 percent neem seed kernel.

Schmutterer and Freres (1990) studied the influence of neem seed oil on metamorphosis, colour and behaviour of the desert locust, *S. gragaria* and the african migratory locust, *Locusta migratoria migratorioides* (Rand F.). The application of low doses of cold-pressed neem seed oil and of AZT-VR-K-enriched neem seed oil on 1st-instar nymphs of *S. gregaria* and 1st- to 3rd-instar nymphs of *L. migratoria migratorioides* resulted in moulting disturbances and in morphogenetic defects of the wings, legs and antennae. Furthermore, the colour and behaviour of the locusts was influenced by spraying 1st and 2nd instar nymphs. Group-treated nymphs often developed into intermediate or green and brown forms similar to the solitary phase, whereas untreated nymphs showed the characteristic colour pattern of the gregarious phase. Treated nymphs were much less active than untreated ones.

Krishna (1990) tested plant volatiles and reproductive biology of *Dysdercus koenigii*. Exposure of 5th instar nymphs or adults of *D. koenigii* to eucalyptus oil odour had a considerable adverse effect on its reproductive performance, egg yield and hatch being most affected when treatment was restricted to females in mating pairs. The biochemical profiles of treated females were also affected. Exposure of both sexes of the insect to neem oil odour during the adult stage also resulted in a marked reduction in egg hatch.

Isman *et al.* (1990) tested insecticidal and antifeedant properties of neem oil and their relationship to azadirachtin content. The variation in azadirachtin content of 12 commercially obtained samples of neem oil (extracted from *Azadirachta indica*) was determined in the laboratory using liquid chromatography. The oils were bioassayed for larval growth inhibition and antifeedant activity against 5th instar larvae of the noctuid *Peridroma saucia*. Moulting disrupting activity (by topical application) was assessed against newly moulted 5th-instar larvae of the lygaeid *Oncopeltus fasciatus*. Among the 12 neem oil samples tested, the azadirachtin concentration ranged from <50 (limit of detection) to over 4000 ppm. The bioactivity of the oils, as measured by the EC50, was correlated with the growth inhibition (r = -0.92), antifeedant activity (r = -0.85) and moulting disrupting activity (r = -0.95). Between 72 and 90 percent of the variation in bioactivity of the oils was accounted for by the variation in the azadirachtin content. It is concluded that the determination of azadirachtin concentration may be a useful quality control criterion for neem oil as a precursor for insecticide production.

Natrajan and Sundaramurthy (1990) tested effect of neem oil on cotton whitefly, *Bemisia tabaci*. The effect of 0.5 and 1.0 percent neem oil and 0.08 percent monocrotophos on nymphs of *B. tabaci* on cotton was studied. Growth and development of nymphs were suppressed by treatment with neem oil but unaffected by monocrotophos. Only 14.3 and 13.0 percent of nymphs reached the adult stage following treatment with 0.5 and 1.0 percent neem oil, respectively, while 84.3 percent of those treated with monocrotophos became adults.

Rembold and Raychaudhuri (1991) reported a rich source of azadirachtins found in the seed kernel of neem (*A. indica*). These compounds disrupt the ecdysteroid-induced moulting process of insects. Their effect on the endocrine system of 2 insect species, *Locusta migratoria* and *Rhodnius prolixus*, was observed using a tritiated derivative. The labelled compound accumulated in the corpus cardiacum resulting in a reduced turnover of neurosecretory proteins.

Schmutterer (1991) tested contact effect of neem seed oil on nymphs of the gregarious phase of desert locust. In laboratory experiments, all 5 nymphal instars of the gregarious phase of *S. gregaria* were sprayed with seed oil from *Azadirachta indica*. The doses ranged from 2.5 to 10 lt/ha. The azadirachtin content of the neem oil was 0.09 percent. The treated locusts showed high mortality which often occurred during moults, prolonged nymphal development and disturbance of metamorphosis. The latter resulted in morphogenetic defects of eyes, antennae, legs and wings. There was also a clear tendency of treated nymphs to solitarization, indicated by the appearance of green haemolymph, and brown and green intermediate or solitary forms with little black pigment. A change in behaviour of the treated nymphs was also observed : the nymphs became sluggish and walking, jumping and climbing ability was reduced.

Krishnaiah and Kalode (1991) tested the efficacy of neem oil against rice insect pests. In greenhouse studies conducted during 1986-87, neem oil spray adversely affected the growth of 1st-instar nymphs of the delphacid *Nilaparvata lugens* on rice plants but its residual effect was relatively poor. At a concentration of 12 percent a.i., neem oil spray acted as an oviposition deterrent to *N. lugens*, but egg production was not significantly altered even at a concentration of 50 percent a.i. neem oil spray at 1 percent affected the growth of 1st-instar nymphs of *Nephotettix virescens*, and a concentration of 25 percent a.i. was required to adversely affect egg production. In field studies conducted in rice plots during the kharif sesason of 1986 in Andhra Pradesh, India, the toxicity of monocrotophos was significantly increased when mixed with neem oil against *N. lugens* but not against *N. virescens*. Also, the efficacy of neem oil against the ephydrid *Hydrellia philippina*, the chrysomelid *Dicladispa armigera*, the pyralid *Scirpophaga incertulas* and the cecidomyid *Orseolia oryzae* was poor.

Bhathal *et al.* (1991) observed that in laboratory experiments, both neem (*A. indica*) extracts and an extract of *Ageratum conyzoides* considerably reduced the hatchability of eggs of *Dysdercus koenigii*. In some cases, the eggs hatched but the larvae died within 24-48 hrs.

Olaifa *et al*. (1991) observed antifeedant and growth regulating effects of neem products on the grasshopper, *Zonocerus variegatus* L. In pot experiments, aqueous extracts from neem kernels at concentration of 5-20 percent were sprayed onto cowpeas and protected them from attack of *Z. variegatus* for 7 days. Extracts from neem leaves were less effective. Freshly prepared neem extracts were more effective than those that had been stored for 1 week at ambient temperature and relative humidity in the laboratory. Neem oil increased the rate of heartbeat of adults of *Z. variegatus*. Neem oil caused abnormal nymph-adult moult in *Z. variegatus*.

Das and Pande (1991) tested insecticides and neem leaf extract against *Dialeurodes pallida* Singh. Ten insecticides were tested in sprays at label recommendations against the immature stages and adults of *D. pallida* on *Piper betle*. Dichlorvos (350 g a.i./ha) was the best treatment, with an effective residual life of 3 days; 2 percent neem extract gave significant repellent action against adults for up to 3 days.

Radcliffi *et al*. (1991) strudied antifeedant effect of neem kernel extracts on *Kraussaria angulifera*, a sahelian grasshopper. NKE was tested in Niger against *K. angulifera*, a univoltine species of economic importance on grain sorghum and pearl millet. In field cage experiments using sorghum seedlings, aqueous NKE prepared was as effective as an ethanol extract. For 72 hrs post-treatment, the feeding injury sustained by neem-sprayed seedlings was not significantly different from that sustained by seedlings sprayed with a mixture of 3 percent cypermethrin and 4 percent dimethoate. Neem appeared to be more effecitve against nymphs than against adults. In feeding-choice experiments with nymphs, neem-sprayed seedlings sustained significantly less feeding injury than did non-sprayed seedlings for 144 hrs post-treament.

Freisewinkel and Schmutterer (1991) observed that the topical application of neem extract at 0.25-1.0 ml/m2 to the 5 nymphal instars of the gregarious phase of *Locusta migratoria migratorioides* led to increased mortality during moults, prolonged development and reduced fitness. Morphogenetic effects were observed on the legs, wings and antennae. A reduction in weight corresponded to reduced feeding activity. Colour changes and supernumerary moults suggested tendencies towards solitarization. It was observed that the earlier the nymphs treated, and the higher the amounts applied, the more distinct the effects.

Meisner *et al*. (1992) tested the effect of Margosan-O (neem extract) on the development of the leaf hopper, *Asymmetrasca decedens*. Laboratory tests of the effect of Margosan-O on *A. decedens* on cotton seedlings indicated that foliar and systemic treatments with 2 percent neem extracts were sufficient to prevent most individuals developing into adults. Settling and feeding were also found inhibited.

West *et al*. (1992) studied the influence of azadirachtin on the feeding behaviour of cereal aphids and slugs. The probing behaviour of *Rhopalosiphum padi* and *Sitobion avenae* and the feeding behaviour of several slug species (*Deroceras reticulatum, Arion distinctus, Agriolimax caruanae* [*Deroceras panomitanum*] and *Maximus* sp.) were assessed on seedlings of winter barley (*Hordeum marinum*) treated with different concentrations of azadirachtin. Settling behaviour of both aphid species was strongly biased towards the untreated seedlings or those treated with low concentration of azadirachtin. Concentration of <500 ppm were effective with topical application and probing activity was reduced for at least 4 days after application. Feeding behaviour of the slug species, as seen by the amuont of leaf eaten compared with the controls, was not affected by the presence of azadirachtin at those concentrations which deterred aphids from feeding.

Langewald and Schmutterer (1992) tested the effects of neem oil treatment on the phase status of desert locust *S. gregaria*. The effects of neem oil on S. gregaria was studied in a greenhouse at 28°C and in field cages in Benin. Neem oil with an azadirachtin content of <0.05 percent was applied in the greenhouse and of 0.2 and 0.4 percent in the field. The ability of adults to aggregate was reduced following treatment of 1st- and 2nd-instar nymphs. In field cages, treatment of 3rd-instar nymphs resulted in increased damage to the antennae of 5th-instar nymphs and adults. The proportions of yellow 4th-instar hoppers and green 5th-instar hoppers were greater in treated than in untreated populations. Positive correlations between the green (solitary phase) hoppers and antennal damage were observed for 4th-instar nymphs. In the 5th-instar a strong positive correlation was found between green coloration and damaged antennae.

Mishra and Singh (1992) tested antifeedant activity of neem formulations against desert locust *S. gregaria*. Extraction of seed kernel, seed coat and fallen leaves were powdered in domestic grinder and separately extracted from water and ethanol. Sexually mature adult locust (7 day-old) were starved for 16 hrs and then released on treated leaves with water and ethanol extracts of seed kernel, seed coat and fallen leaves were taken at 0.048, 0.012, 0.003, 0.0007, 0.00017, 0.00004 percent. In each case, percentage feeding was worked out. Among all the treatments, seed kernel was considered as the richest source of terpenoids. It was observed that both aqueous and ethanolic extracts of seed kernel are highly effective in deterring the feeding of locust at 0.003 percent concentration. The seed coat has very high antifeedant activity as only about 0.024 percent concentration of each extract inhibited the feeding. Aqueous and ethanolic extracts of fallen leaves exhibited poor antifeedant activity as 1.6 percent of each extract was required to protect completely against desert locust. The antifeedant activity of chloroform extracts of the 3 parts was also assessed and against seed kernel exhibited very high antifeedant activity at 0.0007 percent, this was followed by seed coat and leaves where 0.002 and 0.1 percent concentrations, respectively are required to set the same results as of neem seed kernel. It is also observed that chloroform extract has exhibited much higher activity as it inhibited feeding at 0.06 percent whereas 1 percent each of seed coat and fallen leaves was needed to stop the feeding by locust.

Venkateswarlu *et al.* (1993) studied the effect of neem oil on growth and development of mustard aphid, *Lipaphis erysimi*. In laboratory experiments, the effects of neem (*A. indica*) oil (0.10, 0.25, 0.50, 0.75, 1.00, 1.25 and 1.50%) on growth and development of *L. erysimi* was studied. At concentrations of 1.00, 1.25 and 1.50 percent, all the nymphs reared on treated Indian mustard leaves died before reaching the adult stage. At the lower concentration, nymphal survival, fecundity and growth index of the aphid decreased and developmental period increased.

Serra and Schmutterer (1993) observed the effect of neem extracts against *Bemisia tabaci* Genn. In field trials in the Dominican Republic, 3 formulations of neem extract (aqueous extracts from crushed neem seeds (including shells, at 40 or 50 g/lt) or neem cake (15 g/lt), fomulated neem-seed oil (1-1.5%)) and a commercial insecticidal soap showed a clear reduction of the numbers of puparia of the aleyrodid *B. tabaci* on the lower surface fo the leaves of tomato plants, as a consequence of pupal mortality. All neem products also caused reduced oviposition due to a repellent effect on adults; such effects lasted 2 weeks. Tomato plants were sensitive to weekly applications of neem extracts at high dosages. Lower dosages and frequencies are recommended during early developmental stages of the plants. However, yields were increased compared with those from untreated plots and plots treated with conventional combinations when alternating weekly applications of water extracts of neem seed (40 g/lt) or neem-seed cake (15 g/lt) with selective insecticides.

Baitha *et al.* (1993) studied the effectiveness of neem products against rice hispa, *Dicladispa armigera* Oliv. and *Hieroglyphus nigrorepletus* Bol. Different neem products : neem cake (150 kg/ha basal + neem seed kernel extract 5%); NSKE (5%); and neem oil (3%) were compared with monocrotophos 36 WSC (0.4 kg a.i./ha) and carbofuran 3G (0.5 kg a.i./ha) against the chrysomelid *D. armigera* and the acridid *H. nigrorepletus* on rice in Bihar, India, in 1990 and 1991. Carbofuran proved to be the most effective against D. armigera and showed the least leaf damage (9.48%) after the 1st application. Monocrotophos was the most effective against *H. nigrorepletus*, leaf damage being 11.08 and 12.10 percent after the neem 1st and 2nd sprayings, respectively. Neem cake + NSKE (5%) was the next best treatment, followed by NSKE and neem oil, which were equally effective against both pests.

Grahn (1993) studied different feeding behaviour of the two sexes of *Chrotogonus homalodemus* (Blanch) and the repellency of neem products. The feeding habits of adult males and females of *C. homalodemus* on cabbage leaf disks treated with a NSKE were studied. Males consumed less untreated leaf disks than females. In a choice bioassay, *C. homalodemus* did not feed on untreated cabbage following previous exposure to treated cabbage leaves, suggesting that the repellency effect of the extract lasted longer than the exposure time.

Schmutterer (1993) studied the effect of extracts from the seed kernel of *Azadirachta indica* on nymphs of locusts and grasshoppers was studied in the laboratory, and in field trials at Madagascar against *Nomadacris septemfasciata* on maize. There was a very strong feeding deterrent effect of neem oil on *S. gregaria, N. septemfasciata* and *Zonocerus variegatus*. However, *Locusta migratoria migratorioides* was repelled from food treated with neem oil to a much lesser extent. Morphogenetic defects on antennae, legs and wings were often observed on nymphs and adults of *S. gregaria*, and to a lesser extent on *Z. variegatus* and *L. migratoria migratorioides*. Slight effects occurred in *N. septemfasciata*. Solitarization effects on gregarious nymphs, e.g. green colour, mainly occurred in *S. gregaria*, along with a reduction of fitness. The highest mortality rate after topical application was recorded in *S. gregaria* and *Z. variegatus*, and after intake of neem-contaminated food in *L. migratoria migratorioides* and *Z. variegatus*. It is concluded that neem oil and other products (aqueous seed kernel extracts, neem seed powder) can be used for controlling some important locust and grasshopper pests in fields.

Nicol (1993) studied the Phago deterrency of neem products and their effects after oral in take on the *S. gregaria*. The antifeedant and oral effects of neem extracts on *S. gregaria* were evaluated in the laboratory. Nymphs and adults were strongly deterred by wheat seedlings on which neem seed extracts or neem oil had been, sprayed. The antifeedant response was related to the amount of neem product and the concentration of azadirachtin. The systemic action of NSKE had an antifeedant effect at concentraion of 15 g/lt. Third-instar nymphs were more deterred than 5th-instar nymphs. At low concentration, some of the contaminated food was ingested. Nymphs with treated food had prolonged development and resulted in images with reduced body length and weight. There was no effect on mortality.

Wilps *et al.* (1993) did fied-cage and laboratory studies on the effects of neem extracts on *S. gregaria*. Neem oil enriched with methanolic NSKE applied to resting adults of the acridid at 1 ml/m^2 did not increase mortality but reduced flight activity to more than 50 percent as compared with no treatment. The same products applied at the same volume to flying locusts increased mortality up to 65 percent within 4 days, whereas mortality for untreated locusts and those treated with palm oil remained below 10 percent. Locusts treated with neem extracts were unable to sustain flight for more than 30 minutes.

Nicol (1993) did studies on the contact effects of neem oil on the nymphs of *S. gregaria*. Third-instar nymphs of S. gregaria were kept in cages that had been sprayed with neem seed oil as ultralow-volume applications at 0.25-2.00 ml/m^2. Treatment caused increased mortality, delayed nymphal development and morphogenetic defects in the antennae, eyes, legs and wings; adults derived from treated nymphs were smaller than those from untreated nymphs. Tarsal contact of 3rd instar nymphs with neem oil for a relatively short time (5-135 min) led to greater effects (mortality and reduced nymphal development) than spraying with the same dose (0.5 ml/m^2). Keeping nymphs in treated cages had the strongest effects. When 3rd-instar nymphs were treated with neem oil on 12 different spots, application to the intersegmental membranes caused the greatest mortality. Application to strongly sclerotized spots caused mainly morphogenetic effects after ecdysis, whereas treatment of the extermities led mainly to local effects.

Freisewinkel (1993) studied oral and topical effects of neem oil on nymphs of african migratory locust, *L. migratoria migratorioides* (R. and F.). Nymphs were treated with neem extracts by spraying the oil at 0.25-1.00 ml/m^2, applying to the abdomen at 25 and 100 nl, or adminstering orally on maize leaves. Oral adminstration caused the highest mortality. Treatments caused prolonged nymphal development and reduced weight increases.

Nasseh *et al.* (1993) did field-cage experiments on the repellent effects of enriched neem oil on adults of *S. gregaria*. Enriched neem oil was applied at 10 lt/ha to one of 2 plants of *Schouwia thebaica* and 120 adult locusts were released into the cages containing them. The locusts consumed all of the leaves of the untreated plant in 10 days. When 120 locusts were confined with only a treated plant, more than 80 percent of them died within 10 days. On the 11th day, the surviving adults began feeding on the treated plant.

Agrawal (1993) tested 4 plant extracts, thionemone (*Azadirachta indica* preparation), heartwood extract (*A. indica*), extract of *Swietenia macrophylla* seed, and *Calophyllum inophyllum*, at 1.0, 2.5 and 5.0 percent against 3 different stages of the egg (1-, 3- and 5-day old) of *Dysdercus koenigii* by dipping. The adults and nymphs of *D. koenigii* were collected and acclamatized for one generation in the laboratory. The 1-, 3- and 5-day old eggs were taken for the experiment. Forty eggs of each age were dipped in the two concentrations for 10 seconds and then air dried. Number of eggs hatched was counted and mortality was calculated. It was observed that with 1-day-old eggs, the mortality increased (90.66%) with increased concentration of thionemone while 3- and 5-day-old eggs showed decreased mortality (50.0%) at the highest concentration (5%). The 5-day-old eggs were found to be least susceptible i.e. 18.66 percent.

Sauphanor *et al*. (1993) studied commercial formulations of neem extract against *Myzus persicae* and *Cydia molesta*. A commercial formulation of neem extract (Neem Azal F) was tested in the laboratory and in peach orchards in France on insect pests and beneficial insects. In the laboratory, mortality of *M. persicae* and *C. molesta* was observed, but this effect was insufficient to control field populations. At 50 ppm, neem extract increased the mortality of *Forficula auricularia* and reduced feeding in the laboratory. Field populations were also reduced. At this concentration, oviposition by coccinellids in the orchard was not affected but development of larvae of *Harmonia axyridis* was interrupted. Emergence of syrphid larvae in treated plots was reduced by 63 percent compared with those in untreated plots.

Nisbet *et al*. (1993) studied antifeedant effects of azadirachtin on the peach-potato aphid *M. persicae*. The electrical penetration graph (EPG) method was used to analyse the feeding behaviour of *M. persicae* on seedlings of *Nicotiana clevelandii* treated systemically with azadirachtin. A. preliminary experiment showed that the effects of tethering aphids for EPG recording were minimal. The percentage of the 9-h recording period devoted to non-penetration activities, and to stylet pathway patterns, increased as the azadirachtin concentration in the root treatment increased. The number of probes initiated and the numbers of sieve tube penetrations also increased with increased azadirachtin concentration. The mean time elapsing between the initiation of the first probe to reach a sieve element and contact with this tissue was not significantly altered by azadirachtin treatment. However, azadirachtin treatment significantly reduced the percentage of probes that reached sieve elements, and increased non-penetration activity before and after the first period of ingestion from the sieve elements. The percentage of the recording period spent in the EPG pattern associated with sieve tube penetration was significantly reduced by an azadirachtin concentration of 300 ppm, and the duration of each individual penetration was significantly reduced by an azadirachtin concentration of 100 ppm. When the total EPG was split into 3-h periods, significant interactions were seen between time period and azadirachtin concentrations for the duration of non-penetration, pathway, and sieve tube penetration patterns.

Saxena *et al.* (1993) observed in a leaf disc choice test bioassays in the laboratory at 17°C demonstrated that formulated neem seed oil was equally deterrent to 1st- and 3rd-instar nymphs and adults of *Chaetosiphon fragaefolii*. Concentrations of neem seed oil resulting in 50 percent feeding deterrence were approximately 1.1 percent. The rapid disruption of aphid feeding (<1 h) was not related to the presence of the limonoid azadirachtin, and deterrence probably resulted from the combined activity of several compounds. Activity to *C. fragaefolii* disappeared within 12-24 hrs following application to strawberries in the greenhouse. Neem seed oil was also deterrent to *Acyrthosiphon pisum* and *Chaetosiphon* spp., but was ineffective against *Fimbriaphis fimbriata*. *Nasonovia ribisnigri* and *Myzus persicae*. The antifeedant properties of neem did not appear to contribute significantly to the control of aphids and the viruses they transmit.

Lowery and Isman (1993) studied antifeedant activity of neem extracts to strawberry aphid *C. fragaefolii*. The emission of courtship signals as well as mating behaviour of females of *N. lugens* were impaired when individually treated with 1, 2.5 or 5 µg of neem oil, or when caged on rice plants sprayed with emulsified 3, 6 or 9 percent neem oil. A few treated females failed to produce mating calls; others emitted signals unrecognizable to males. Some treated females produced 'normal' signals. However, the pulse repetition frequency increased and the duration of each call decreased at higher doses or higher concentrations of neem oil. Even when females mated, the neem oil treatment prolonged the premating period; the mating duration was unaffected.

Lowery and Isman (1994) studied IGR effects of neem extract on aphids. Seed oil from neem (*A. indica*) applied to leaf discs at a concentration of 1 percent resulted in 94-100 percent mortality of 2nd-instar nymphs of *Nasonovia ribis-nigri* [*N. ribisnigri*] and *M. persicae* after 9 days. The equivalent amount of pure azadirachtin (_40 ppm) was as effective as the seed oil. The survival of adult aphids was unaffected by either treatment, but the survival of offspring from treated adults of *M. persicae* and *N. ribisnigri* was reduced significantly. The LC50s of azadirachtin for 2nd-instar nuymphs of 9 species of aphids ranged from 2.4 ppm for *M. persicae* on pepper to 635.0 ppm for *C. fragaefolii* on strawberry. For *M. persicae*, the growth regulating effect of azadirachttin was influenced by the food plant and the instar that was treated.

Puri *et al*. (1994) studied detergents and plant derived oils for the control of sweet potato white fly on cotton. In trials in Maharashtra, India, in September-November, 1991, 4 common detergent powders marketed in India (Nirma, Rin, Surf and Wheel) applied to single cotton leaves at concentrations of 0.25, 0.5 and 1.0 percent reduced the number of adults at *Bemisia tabaci* by 90-91 percent and nymphs by 87-90 percent. Treatments of 0.5 and 1.0 percent Nirma and Surf, as well as 1.0 percent Wheel, applied with a foot-operated sprayer to cotton, reduced the number of nymphs to an extent equal to that of a 0.05 percent triazophos spray. Treatments of cottonseed oil and neem oil at 0.5 and 1.0 percent with a foot-operated sprayer were equal in effectiveness to a 0.05 percent fenpropathrin spray against both adults and nymphs. The time required for spraying and the amount of spray needed were compared using a standard wand ((89 cm) with the nozzle at a 30 degree angle to the wand used both for 'over the top' spraying and for spraying the undersides of the leaves, and a modified wand which was shortened to 36 cm with a nozzle at right angles to the wand. The short wand was easier to use for spraying the undersides of leaves of mature cotton plants and took no more spray material and only slightly longer than 'over the top' spraying. Results also confirmed that the undersurfaces of the leaves must be thoroughly wet with spray liquid to obtain effective aleyrodid control.

Kearney *et al*. (1994) studied antifeedant effects of neem products against *S. gregaria*. Callus and micropropagated shoots were initiated from leaf explants of the neem tree (*A. indica*). A variety of whole plant and *in vitro* cell cultures from neem seedlilngs of Ghanian origin were tested for insect antifeedant compounds using *S. gregaria*. Feeding suppression occurred when whole extracts of seed, leaf, callus, suspension and shoot cultures were tested in no-choice feeding bioassays. Controls of sucrose, carrot callus and the plant growth medium showed no feeding deterrence. Azadirachtin was quantified from a seed extract by HPLC but was not detected in any of the other extracts. Antifeedancy was determined during batch growth of a suspension culture which had been in culture for 5 months; results indicated that antifeedants were still being formed and that levels increased after maximum biomass was attained.

Langewald *et al*. (1995) tested repellent effects of 3 different neem (*A. indica*) formulaitons on the red locust, *Nomadacris septemfasciata* Serv. in maize fields in the south west of Madagascar. A simple water extract from neem seed kernels (200 and 600 lt/ha), an enriched neem ULV-formulation (2 lt/ha) and a dust, prepared on the basis of neem seed cake (5 kg/ha), was applied. All 4 treatments provided protection from attack for at least 5 days. The water extract was significantly active up to 7 days and the neem-ULV-formulation up to 9 days after treatment. The application of neem seed kernel water extract could be a simple means for farmers to protect their fields against *N. septemfasciata*.

Dash and Senapati (1995) tested efficacy of neem derivatives against green leaf hoppers of rice, *Nephotettix virescens* (Dist.). The effeictiveness of NSKE and neem cake extract (NCE) when used as a rice seedling root-dip (SRD) to control infestation of *N. virescens* was studied under greenhouse conditions. There was a marked reduction in survival of the nymphs and adult emergence, a prolongation in the duration of the nymphal period and low growth indices in treated rice seedlings. NCE (10%) + Urea (1%) was the most effective treatment resulting in the lowest adult emergence (13.34%) and growth index (0.76%), followed by NSKE (5%) + Urea (1%) which resulted in 20.0 percent adult emergence and a growth index of 1.14 compared to 85 percent and 5.49 in the untreated control. However, a chlorpyrifos treatment at (0.02%) was superior to all neem products, reducing adult emergence to 5.0 percent with a growth index of 0.28. The adverse effect of neem products on N. virescens was more pronounced at higher cocentrations than at a lower concentrations. AT an NSKE (2.5%) + Urea (1%) and NCE (5%) + Urea (1%), nymphs developing into adults and the growth indices were 43.33 and 28.34 percent, 2.55 and 1.62 percent respectively.

Nandagopal and Rajgopal (1996) observed feeding site preference for sucking pests on groundnut and the effect of neem products. A labratory study with groundnut seedlings indicated that *Balclutha hortensis* [*B. incisa*] depended mainly on phloem (61.9%) for its sap requirement, while *Creontiades pallidus* and *Aphis craccivora* were depended entirely on phloem and xylem for its requirement, repectively. It is observed that the neem (*A. indica*) extracts had antifeedant effect on *B. incisa*.

Krishna Kumari *et al*. (1996) studied the impact of nimbecidine on feeding and enzyme activity of red pumpkin bug, *Aspongopis janus*. Nimbecidine, a neem-based pesticide, was tested for its antifeedant activity against *A. janus*. The insects fed Nimbecidine-sprayed leaves (1ml/gm leaf) showed reduction in feeding. Control insects showed normal feeding.

Matthew and Hassan (1996) tested the contact, systemic and repellent action of neem seed extract against *Myzus persicae*. Neem seed extract was tested at 60, 120 and 180 ppm. Pirimor, a commonly used contact action aphicide, was also used as a comparision against the neem treatments in the contact action trial. All neem concentrations gave poor to very slight control of *M. persicae* when applied as contact action foliar sprays, with Pirimor providing the greatest contact kill. However, neem displayed strong repellent effects for all concentrations tested, and total aphid control within 24 hrs through systemic action at 180 ppm, when applied as a soil drench. Aphid parasites and other beneficials insects did not appear affected by neem treatments, whereas Pirimor® treatments reduced beneficial insect numbers. It is observed that the neem still has a place in the control of aphids in situations such as organic crop production, or in crops where resistance to other chemicals by aphids has resulted.

Mordul *et al*. (1996) tested neem metabolites with insect antifeedancy and IGR properties produced in tissue culture of neem. The presence of antifeedant compounds has been measured using choice and no-choice bioassays for the desert locust, *S. gregaria*, with 2-4-day-old 5th-instar nymphs, and with azadirachtin as standard. A bioassay for insect growth regulatory effects has been established using one-day-old 5th-instar nymphs of the milkweed bug, *Oncopeltus fasciatus* Dallas. Isman *et al*. (1990) observed that Azadirachtin and some unknown compounds have been shown to be present in both callus and suspension cultures. Antifeedant products are present at relatively constant levels in callus cultures maintained over a four-year period. The azadirachtin contents of seed kernels from ripe fruits of identified Sri Lankan trees from different climatological zones have been compared with azadirachtin yields from tissues cultures derived from the same trees. The azadirachtin content of seeds from individual trees was variable. There was greater variability in the azadirachtin content of callus derived from leaves of those trees.

Bottenberg and Singh (1996) applied aqueous leaf extracts of neem at 5 and 10 percent (w/v) twice and four times during cowpea flowering on two cultivars (IT86D-715 and IT87D-941-1) in Kano, Nigeria, in 1993-94 wet season. Extracts were applied with a broom, an indigenous farmers' practice. Flower infestation by *Maruca testulalis* [*M. vitrata*] and *Megalurothrips sjostedti* was not reduced by neem application except in 1994 in the 10 percent neem treatment applied four times. Cv. 71 had consistently less pod damage than cv. 941. On average, neem applications reduced *M. testulalis* pod damage by 12 percent in cv. 715 and 16 percent in cv. 941 compared to the untreated control by chemical insecticide, it was 25% and 39% in cv. 715 and 941, respectively. Yield improvement by neem was variable. Significantly greater yield in comparision with the untreated control was obtained only with cv. 715, applied twice at 5 percent in 1994. Neem can be effective in reducing *M. testulalis* damage particluarly in combination with host plant resistance.

Wilps (1996) studied the effect of neem products against locust, *S. gregaria*. The neem products tested contained 300 to 1000 ppm azadirachtin as active ingredient and were designed as oily formulations. Ethanolic *Melia* extracts were formulated in palm oil, corn oil or pure neem oil without active ingredients of the neem seeds. The dosage applied varied between 300 and 1000 ppm/ha for the neem extracts; *Melia* extracts were used to 1000 ppm/ha. In field cage trials, dose dependent mortality rates to 100 percent were obtained. Besides this, malformations were observed, mostly indicated by deformed wings, legs, eyes, loss of tarsi, and incomplete moulting steps. In some cases, *Melia* extracts suppressed the adult moult completerly and the larvae died as overaged nymphs.

Dimetry *et al.* (1996) evaluated certain neem extracts against the tomato bug, *Cyrtopeltis tenuis* (Reut.). Different commercial formulations of neem seed extracts were tested against the pest under laboratory conditions. The results obtained indicated that in a choice chamber, the different formulations tested were phagodeterrent for *C. tenuis* nymphs at high concentrations. Also, the neem extracts tested had pesticidal properties agaisnt *C. tenuis* nymphs. The toxicity of the products decreased as the concentrations decreased. They had an apparent delayed effect on the resulting adults as their longevities significantly shortened and the resulting offspring was greatly reduced.

Baumgart (1996) tested the effects of neem oil on the grasshopper, *Zonocerus variegatus* L. in field condition. Pure neem oil was extracted with a hot press procedure at 80°C which had 0.013 percent azadirachtin. Two thousand 1st-instar hoppers were released in a net cage having teak plantations. The neem oil was applied in two application rates with 5 and 10 lt/ha ULV. Each treatment was replicated thrice. The data were recorded on migration, attraction to different plants and the feeding behaviour of hoppers. Morphogenetic defects on antennae and legs were also recorded. The data revealed that after 8 days of treatment, the behavioural studies were made and it was also found that in treated plots, the nymphs had a south orientation. In the treated plots, no feeding took place at the upper part of the vegetation. however, it is further observed that the neem oil treatments in both application rates cuased an inhibition of feeding between 10 and 40 percent compared to check. Both application rates of neem oil in the field affected the antennae and legs of nymphs effectively.

Krishnaiah and Kalode (1996) tested the bioefficacy of neem products against insect pests in rice. Black mirid bug (BMP), *Tytthus parviceps* (Reuter) reared on plants oviposited by brown plant hopper (BPH), *Nilaparvata lugens* from egg to adult stage. Three-day-old adults of BMB was used for the study.

Neem oil (50% EC) and crude neem oil were observed 221 and 102 times, respectively safer than malathion to BMB as compared to GLH. Resistance development studies revealed that after 8 generations, the LC50 of neem oil to GLH was 2.929 percent as compared to 1.308 percent in the beginning, while the corresponding values for the BPH were 1.54 percent and 2.242 percent, respectively. A field study on BPH resurgence for three seasons showed that BPH population in plots treated with neem oil (1 and 4%) was similar to untreated control while plots sprayed with deltamethrin (0.005%) recorded higher population and caused complete hopper burn. Among 12 botanical formulations, Repelin RD 9, Neemark, and Margoside CK at 0.5 and 2.0 percent exhibited signifcantly better growth inhibition against nymphs of BPH and GLH. Repelin at 0.5 percent significantly enhanced the toxicity of synthetic insecticides like ethofenprox and monocrotophos against BPH nymphs.

Sandhu *et al*. (1996) studied neem products on growth and development of white backed plant hopper (WBPH), *Sogatella furcifera* (Horvath). Five neem-based insecticides : Neemark (0.5, 0.25, 0.125%), RD 9 Repelin (1.0, 0.5, 0.25%), Neemrich I (0.4, 0.2, 0.1%), Neemrich II (0.1, 0.05, 0.025%), and neemoil (5.0, 2.0, 1.0%) were evaluated for the olfactory response and effect on fecundity and growth of WBPH. The repellence was studied in a horizontal cylindrical chamber open at both end for one hr. Twenty newly emerged nymphs of WBPH were released in chamber containing plant material. For determining the effect on fecundity, 4 pairs of newly emerged WBPH were caged on treated and control plants. Unhatched eggs were determined by staining the leaf portions where eggs were laid and were counted. Fecundity was worked out from fertility and the number of unhatched eggs.

All the treatments repelled the insect, with maximum repulsion (0.76) caused by Neemrich I 0.4 percent followed by Neemark 0.5 percent (-0.74), RD 9 Repelin 1.0 percent (-0.72), Neemrich II 0.1 percent (-0.68), and neem oil 5 percent (-0.44). The repellent effect decreased with decrease in concentration. Fecundity of the insect was minimum in case of RD 9 Repelin 1.0 percent (40.0 eggs/female) followed by Neemrich II 0.1 percent (46.0 eggs), Neemrich I 0.4 percent (46.66 eggs), RD 9 Repelin 0.5 percent (52.0 eggs), neemoil 1 percent (52.33 eggs), and Neemark 0.5 percent (55.66 eggs/female) as against 141.0 eggs/female in control. Fecundity increased with decrease in concentration of insecticides. The number of nymphs attaining adult stage was minimum in case of RD 9 Repelin 1 percent (28.0) followed by Neemrich II 0.1 percent (30.0), neemoil 1 percent (32.0), Neemrich I 0.4 percent (32.0), Neemrich I 0.2 percent (42.0), and Neemark 0.5 percent (42.0) as compared to 98.0 in control. Thus, RD 9 Repelin 1 percent was the most effective in reducing fecundity and growth of *S. furcifera*, while Neemrich I 0.4 percent caused maximum repulsion.

Nelson *et al*. (1996) studied antifeedant and growth inhibiting effects of azadirachtin-rich neem fractions on rice WBPH, *S. furcifera* and two lepidopteran pests. Three gravid females of WBPH were caged on IR 20 plants of rice treated with azadirachtin fractions in three replications. Observations of nymph emergence and the number of unhatched eggs were determined by dissecting out the tillers. Observations on the malformations and moulting disturbances were also made. Thirteen azadirachtin-rich neem fractions-Neemol, Ne-pet-et, NS 55/GSN/OU, NS 56.GSN/OU, NS 57/GSN/OU, NS 58/GSN/OU, Vemidin, V-75, NK-72, NK-100, Neem-75 and Nemidin-(all at 1000 ppm) were evaluated for their effect in comparision with neem cake extract (5%) on WBPH. Oviposition was highly deterred by the application of NS 58/GSN/OU, in which 89.13 percent reduction in egg laying was recorded compared to control. Feeding rate measured in terms of honeydew excretion indicated that NK-100 was the best antifeedant (69.34% reduced feeding rate) followed by NS 58/GSN/OU (69.24%), and both the treatments were on par. With reference to nymphal development, only 40.00 percent nymphs became adults in Vemidin treatment. Moulting disturbances were observed in NS 58.GSN/OU, NK-100, and Nemidin treatment. NS 58/GSN/OU was proved to posses high antifeedant and growth inhibiting effects against WBPH.

Studies with seven azadirachtin-rich fractions of neem on *Spodoptera litura* Fb. and *Helicoverpa armigera* Hb. revealed the antifeedant and growth ratardant effects of these fractions. In *S. litura*, Ne-pet-et showed higher antifeedant effect : 66.68 percent reduction in leaf area feeding was noticed compared to control. Nemidin inhibited the larvae from reaching the adult stage to the tune of 68.18 percent and it was on par with Nemidin, Ne-pet-et, Vemidin, and NK-100. In *H. armigera*, 91.56 percent feeding inhibition was recorded in Nem-75 treatment followed by Vemidin (87.69%) and NK-100 (75.74%). Moulting disturbances were noticed in Ne-pet-et, Vemidin, Nem-75 and Nemidin treatments.

Raguraman *et al*. (1996) tested the effects of neem extracts on BPH and its symbionts. In one experiment, twenty-day-old IR 20 rice plants were sprayed with neem oil (NO), neem oil + custard apple oil (CAO), neem seed kernel extract (NSKE) and neem cake extract (NCE). Neem seed bitters (NSB), a lyophilized extract of neem seed kernel by water was also used. In this experiment, the symbionts were counted from the one millimetre squares of the haemocytometer. The second experment was conducted to know the effect of neem extracts on honeydew excretion and its free aminoacids. Two sets of 30-day-old IR 20 plants having 8 tillers/ pot and one tiller/ pot in each treatment, were sprayed with neem products in four replications. The number of survivors were counted on the fourth and eighth day after release. Different extracts also affect the yield like symbionts (YLS). The YLS population was significantly lower in BPH that fed on neem-treated rice plants than on untreated ones. The decrease may be due either to the direct effects on the compounds present in the extracts on the YLS or through BPH fat bodies, which store YLS. It was also observed that honeydew excretion and total free aminoacids in honeydew were greatly reduced as compared to control. The weight of BPH was low in 5 percent NSKE treated plants (0.93 mg) while it was 1.88 mg in control. The length and breadth of BPH was shorter as compared to control. Significantly higher mortality occurred in neem treated plants than on control, mainly because of the growth disrupting effects of neem extracts. In case of neem treatment, they remained as nymph while in control they reached upto adult stage.

It is evident that these symbionts supply many essential nutrients which in turn control the rate of reproduction, adult longevity, size and weight etc. Thus these useful aspects of the insect and its dependence on endosymbionts may be exploited by using neem products. By this, we might use specific antibiotics or metabolic inhibitors or growth disruptors in the control of such insects which possess symbionts dependency.

Dimetry and Hawary (1997) studied the synergistic effect of some additives on the biological activity and toxicity of neem-based formulations against cowpea aphid, *Aphis craccivora* Koch. The effects of various concentrations of Neem Azal-T and Neem Azal-T/S (both containing extracts of *Azadirachta indica*) on the biology and percentage mortality of adult of *A. craccivora* were studied under laboratory conditions. Both extracts had an aphicidal effect against adults and significantly decreased fecundity and longevity. The addition of synergists such as ethyl oleate, sesame oil and dimethyl sulphoxide (DMSO) improved aphicidal activity. The activity of Neem Azal-T/S only was enhanced with suitable additives as they adversely affected fecundity and longevity in comparision to the extract alone. Adding $CaCO_3$ to the extracts had no clear effect on either neem seed extract formulation.

Koul *et al.* (1997) observed antifeedant activity of neem seed extracts and azadirachtin to cabbage aphid, *Brevicoryne brassicae* (L.). Using cabbage leaf disk choice bioassays, formulated neem seed extracts were highly deterrent to *B. brassicae*. The concentrations producing 50 percent feeding deterrence was approximately 0.83 percent for 2nd-instar larvae, irrespective of bioassay duration. The disruption of aphid feeding was related to the presence of azadirachtin concentration in the extract and the effective threshold concentration of azadirachtin was approximately 10.7 ppm. However, for a rapid antifeedant effect (within 6 hrs), 200 ppm of azadirachtin was required, which persisted for nearly 5 days.

Weissling *et al*. (1997) reported oviposition and feeding reduction in pear psylla (Homoptera) by the foliar application of many products. It was observed that summerform oviposition rates were reduced significantly by application of several mineral oils including neem and garlic barrier to pear foliage in no-choice tests, and by all oils, garlic barrier, and M-Pede in choice tests. Summerform feeding rates were reduced by application of mineral oils. These data suggest that 'summer type' oils and other compounds may have utility in *C. pyricola* management programms by decreasing feeding and oviposition.

Raguramann and Rajasekaran (1997) studied behavioural and physiological effects of neem on rice BPH, *N. lugens*. Greenhouse studies were conducted to determine the effect of neem oil and neem seed bitters applied at different concentrations as either high volume, low volume or ultra low volume sprays to the BPH. All neem products affected the orientation, probing and feeding time, food ingested and assimilated, and growth and development of *N. lugens*.

Deka *et al*. (1998) tested 8 aqueous plant extracts (*Azadirachta indica, Clerodendron inerme, Pongamia pinnata, Melia azadirach* [*M. azedirach*], *Polygonum orientale, Lantana camara, Adhatoda vasica* and *Cassia tora*) at all the concentrations tested (2-10%) exhibited ovicidal action against *Helopeltis theivora*. The ovicidal action was more promising with an increase in the concentration of extracts. The hatchability of normal eggs (70.83±5.35%) was significantly decreased to 36.67±2.35, 39.00±2.88, 42.50±3.81, 45.00±2.88 and 48.17±2.35 percent at 10, 8, 6, 4 and 2 percent aqueous extracts of *A. indica*, respectively.

Opender (1998) studied the effect of neem extracts on fertility and fecundity of *B. brassicae*, a cabbage aphid. Exposing adults of *B. brassicae* to neem seed extract and azadirachtin reduced fertility and fecundity at 30-60 ppm. The effect was linear and concentration dependent. Embryonic mortality just before parturition and failure of ecdysis appeared to be the major causes of reproductive failure.

Deka (1998) studied the effect of aqueous plant extracts on longevity and fecundity of tea mosquito bug, *Helopeltis theivora*. Aqueous extracts of 8 plants (*Azadirachta indica, Clerodendron* [*Clerodendrum*] *inerme, Melila azadirach* [*M. azedarach*], *Pongamia pinnata, Polygonum* [*Polygonatum*] *orientale, Lantana camara, Adhatoda vasica* and *Cassia tora*) in different concentrations (10, 8, 6, 4 and 2%) were screened in vitro under laboratory conditions (28±4°C and 80-95% RH) against *H. theivora* for adult longevity and fecundity. Increasing concentrations of all the plant extracts influenced inhibition of adult longevity and fecundity. Of all the plant species, *A. indica* offered maximum inhibition of adult logevity as well as fecundity of *H. theivora* followed by *C. inerme*. The resultant F1 adults obtained from adults fed with treated shoot extracts and reared on untreated shoots also exhibited inhibition of longevity and fecundity but to a lesser extent than the fresh adults.

Rup *et al*. (1998) studied the influence of neem seed extract on the development of *Lipaphis erysimi* (Kalt.). A study was carried out to investigate the effects of neem extract on the development of *L. erysimi*, a major pest of *Brassica* crops in India. The experiments were performed with 8000, 1600, 320, 64 and 12.8 ppm concentrations of NSKE in methanol, using leaf surface and topical treatments. The neem seed extract prolonged the developmental duration and decreased the percentage of emergence with increasing concentration in 1st- and 2nd-instar nymphs. Moreover, it increased the alatae production at lower concentrations along with a high percentage of morphological aberrations. Body weight of the adults showed a significant decrease. Longevity and fecundity of the adults decreased only with topical treatment.

Shah and Pandey (1998) tested the bio-efficacy of neem-based formulations for grass hopper management in Sehima, Heteropogon dominated grasscovers. Natural products are considered environmentally safer for pest management. In the present study, four commercially available neem formulations were evaluated for their efficacy for grasshopper management. The neem formulations used were Suneem (0.07%), Bioneem (0.35%), Jawan (0.20%) and Achook (0.20%) alongwith endosulfan (0.07%) as a check. The results indicated that all the neem formulations were signficantly superior in minimizing the pest damage there by increasing the biomass as compared to untreated control. However, Achook provided maximum protection followed by Suneem, Jawan and Bioneem. In Achook treated plots, an increase of 24.2 percentt (1.55 t/ha) higher green forage yield was recorded over control.

Dimetry and Haway (1998) studied the effect of various concentrations of Neem Azal-F (a commercial product of neem kernel extract with 5% azadirachtin content) on adults and first nymphal instars of *Aphis craccivora*. The extract had a deterrent and an antifeedant effect which hindered larviposition of the adults and decreased the reproductive period as well as the longevity of the adults. Also, the extract had an aphicidal effect against aphid, hindered development of the nymphs and elongated the duration of the nymphal stage.

References

Agrawal, I.L. (1993). Ovicidal activity of four plant extracts on *Dysdercus koenigii* Fabr. (Heteroptera : Pyrrhocoridae). *Indian Journal of Entomology*, **55**(4) : 440-443.

Ascher, K.R.S.; Streloke, M.; Schmidt, G.H. and Warthen J.D.Jr. (1989). The antifeedant effect of neem (*Azadirachta indica*) seed kernel extract and azadirachtin on nymphs of *Eyprepocnemis plorans*. *Phytoparasitica*, **17**(3) : 167-174.

Baitha, A.; Hameed, S.F. and Singh, R. (1993). Effectiveness of neem (*Azadirachta indica* A. Juss) products against rice hispa, *Dicladispa armigera* Oliv. and *Hieroglyphus nigrorepletus* Bol. *Journal of Entomological Research*, **17**(2) : 149-152.

Balasubramanian, M. (1979). Pest management studies for rice borwn planthopper in Tamil Nadu Agricultural Univesity. *Recent trends in rice brown plant hopper control. Colloquium on rice brown plant hopper*, 24 June 1979 : 9-10.

Bhathal, S.S.; Darshan Singh; Dhillon, R.S.; Nayyar, K. and Singh, D. (1991). Ovicidal effects of neem oil and plant extract of *Ageratum conyzoides* Linn. on *Dysdercus koenigii* Fab. *Journal of Insect Science*, **4**(2) : 185-186.

Bottenberg, H. and Singh, B.B. (1996). Effect of neem leaf extract applied using the 'broom' method, on cowpea pests and yield. *International Journal of Pest Management*, **42**(3) : 207-209.

Brajendra Singh; Singh, A.P. and Singh, B. (1987). Neem kernel suspension as an oviposition deterrent to the desert locust (*Schistocerca gregaria* Forsk.). *Plant Protection Bulletin*, India, **39**(1-2) : 9-16.

Chiu, S.F. (1985). Recent research findings on Meliaceae and other promising botanical insecticides in China. *Zeitschrift fur Pflanzenkrankheiten und Pflanzenschutz*, **92**(3) : 310-319.

Chiu, S.F.; Huang, B.Q. and Hu My (1983). Experiments on the use fo seed oils of some meliaceous plants as antifeedant in brown planthopper control. *Acta Entomologia Sinica*, **26**(1) : 1-9.

Coudriet, D.L.; Prabhaker, N. and Meyerdirk, D.E. (1985). Sweetpotato whitefly (Homoptera : Aleyrodidae) : effects of neem-seed extract on oviposition and immature stages. *Environmental Entomology*, **14**(6) : 776-779.

Das, M.L. and Pandey, Y.D. (1991). Evaluation of insecticides and leaf extract of *Azadirachta indica* L. against *Dialeurodes pallida* Singh. *Uttar Pradesh Journal of Zoology*, **11**(1) : 25-30.

Dash, A.N. and Senapati, B. (1995). Efficacy of neem derivatives as rice seedling root-tip against green leaf hopper, *Nephotettix virescens* (Distant) under greenhouse conditions. *Journal of Entomological Research*, **19**(1) : 33-38.

Deka, M.K.; Handique, R.; Singh, K. and Hazarika, L.K. (1998). Effect of aqueous plant extracts on longevity and fecundity of tea mosquito bug. *Crop Research Hisar*, **16**(1) : 102-105.

Deka, M.K.; Singh, K. and Handique, R. (1998). Bioefficacy of aqueous plant extracts on viability of eggs and subsequent development of tea mosquito bug. *Crop Research Hisar*, **16**(2) : 260-264.

Dimetry, N.Z. and El Hawary, F.M.A. (1997). Synergistic effect of additives on the biological activity and toxicity of neem-based formulations against the cowpea aphid, *Aphis craccivora* Koch. *Insect Science and Its Applications*, **17**(3-4) : 395-399.

Dimetry, N.Z. and Hawary, F.M.A.El (1998). Neem Azal-F as an inhibitor growth and reproduction in the cowpea aphid *Aphis craccivora* Koch. *Neem Newsletter*, **15**(3) : 40.

Dimetry, N.Z.; Gomaa, A.A.; Salem, A.A.; El Moniem, A.S.H.A.; Kleeberg, H. (ed.) and Zebitz, C.P.W. (1997). Evaluation of certain neem extracts against the tomato bug *Cyrtopeltis tenuis* (Reut.). *Proceedings 5th Workshop Wetzlar*, Germany, 22-25 Jan., 1996 : 201-211.

Dreyer, M. (1987). Results of field experiments on the effect of neem (*Azadirachta indica*) aqueous extracts and neem oil on the pests of eggplant (*Solanum melongena*) in Togo. *Mitteilungen der Deutschen Gesellschaft fur Allgemeine und Angewandte Entomologie*, **5**(1-4) : 144-151.

Freisewinkel, D.C. (1993). Oral and topical effects of seed oil from the neem tree, *Azadirachta indica* (A. Juss.) on nymphs of the African migratory locust *Locusta migratoria migratorioides* (R. & F.). *Mitteilungen der Deutschen Gesellschaft fur Allgemeine und Angewandte Entomologie*, **8**(4-6) : 785-789.

Freisewinkel, D.C. and Schmutterer, H. (1991). Contact action of neem oil on the African migratory locust, *Locusta migratoria migratorioides* (R. & F.). *Zeitschrift fur Angewandte Zoologie*, **78**(2) : 189-203.

Gawande, R.B. and Barkhade, U.P. (1989). Effect of synthetic juvenile hormone analogues and juvenile hormone mimicking substances in some Vidarbha plants on *Dysdercus cingulatus* Fab. *PKV Research Journal* **13**(2) : 173-175.

Goel, S.C.; Srivastava, Y.N.; Bhanotar, R.K. and Goel, S.C. (ed.) (1983). Neem trees in relation to locusts and Rajasthan Canal. *Insect Ecology and Resource Management* : 262-266.

Grahn, B. von (1993). Different feeding behaviour of the two sexes of *Chrotogonus homalodemus* (Blanch.) (Orthopt., Acrididae) and the repellency of an aqueous neem seed kernel extract (NSKE). *Anzeiger fur Schadlingskunde Pflanzenschutz*, Umweltschutz, **66**(3) : 46-48.

Griffiths, D.C.; Greenway, A.R. and Lloyd, S.L. (1978). The infulence of repellent materials and aphid extracts on setling behaviour and larviposition of *Myzus persicae*. *Bulletin of Entomological Research*, **68**(4) : 613-619.

Haque, M.A. and Islam, B.N. (1988). Effects of methanolic neem and chinaberry seed extracts on rice green leafhopper (*Nephotettix nigropictus* Stal.). *Bangladesh Journal of Agriculture*, **13**(1) : 53-57.

Isman, M.B.; Koul, O.; Luczynski, A. and Kaminski, J. (1990). Insecticidal and antifeedant bioactivities of neem oils and their relationship to azadirachtin content. *Journal of Agricultural and Food Chemistry*, **38**(6) : 1408-1411.

Jacobson, M.; Reed, D.K.; Crystal, M.M.; Moreno, D.S.; Soderstrom, E.L.; Chapman, R.F. (ed.) and Bernnays, E.A. (ed.) (1978). Chemistry and biological activity of insect feeding deterrents from certain weed and crop plants. *Entomologia Experimentalis et Applicata*, **24** : 448-457.

Jotwani, M.G. and Srivastava, K.P. (1981). Neem : insecticide of the future, II - Protection against field pests. *Pesticide*, **15**(11) : 40-47.

Kabir, K.H. and Mia, M.D. (1987). Effectiveness of some indigenous materials as repellent against the mustard aphid. *Bangladesh Journal of Zoology*, **15**(1) : 87-88.

Kareem, A.A.; Saxena, R.C.; Boncodin, M.E.M. and Malayba, M.T. (1989). Effect of neem seed and leaf bitters on oviposition and development of green leafhopper (GLH) and brown planthopper (BPH). *International Rice Research Newsletter*, **14**(6) : 26-27.

Kareem, A.A.; Saxena, R.C.; Boncodin, M.E.M.; Krishnasamy, V. and Seshu, D.V. (1988). Effect of neem seed treatment on rice seedlings vigor and survival of brown planthopper and green leafhopper (GLH). *International Rice Research Newsletter*, **13**(1) : 27-28.

Kareem, A.A.; Saxena, R.C.; Boncodin, M.E.M.; Krishnasamy, V. and Seshu, D.V. (1989). Neem as seed treatment for rice before sowing : effects on two homopterous insects and seedlings vigor. *Journal of Economic Entomology*, **82**(4) : 1219-1223.

Kearney, M.L.; Allan, E.J.; Hooker, J.E. and Mordue, A.J. (1994). Antifeedant effects of in vitro culture extracts of the neem tree, *Azadirachta indica* against the desert locust (*Schistocerca gregaria* (Forskal)). *Plant Cell*

Tissue and Organ Culture, **37**(1) : 67-71.

Kirpal Singh; Sharma, P.L. and Singh, K. (1986). Studies on the antifeedant and repellent qualities on neem (*Azadirachta indica*) against aphid (*Brevicoryne brassicae* L.) on cauliflower and cabbage. *Research and Development Reporter*, **3**(1) : 33-35.

Koul, O. (1984). Azadirachtin : I-interaction with the development of red cotton bugs. *Entomologia Experimentalis et Applicata*, **36**(1) : 85-88.

Koul, O.; Shankar, J.S. and Mehta, N. (1997). Antifeedant activity of neem seed extracts and azadirachtin to cabbage aphid, *Brevicoryne brassicae* (L.). *Indian Journal of Experimental Biology*, **35**(9) : 994-997.

Krishna Kumari, R.; Ramalingan, N. and Sabbanayagam, S. (1996). Impact of Nimbecidine on feeding and enzyme activity of *Aspongopus janus* (Fabricius) (Hemiptera : Pentatomidae). *Neem Newsletter*, **13**(2) : 18.

Krishna, S.S. (1990). Plant volatiles and reproductive biology of *Dysdercus koenigii* (F.) (Pyrrhocoridae : Hemiptera). *Phytophaga*, **3**(1) : 23-36.

Krishnaiah, N.V. and Kalode, M.B. (1991). Efficacy of neem oil against rice insect pests under greenhouse and field conditions. *Indian Journal of Plant Protection*, **19**(1) : 11-16.

Krishnaiah, N.V. and Kalode, M.B. (1996). Environmental impact and bioefficacy of neem products against insect pests in rice. *Neem and Environment*, **1** : 413-424, Ed. R. P. Singh *et al*.

Langewald, J. and Schmutterer, H. (1992). Effects of neem oil treatment on the phase status of desert locust, *Schistocerca gregaria*. *Proceedings of a Workshop on Biological Control of Locusts and Grasshoppers* held at the International Institute of Tropical Agriculture, Cotonou, Republic of Benin, 29 April-1 May, 1991 : 142-154.

Langewald, J.; Scherer, R. and Schmutterer, H. (1995). Repellent effects of different products of the neem tree on the red locust, *Nomadacris septemfasciata* Serv., in maize fields in the southwestern parts of Madagascar. *Anzeiger fur Schadingskunde, Pflanzenschutz*, Umweltschutz, **68**(3) : 55-57.

Lowery, D.T. and Isman, M.B. (1993). Antifeedant activity of extracts from neem, *Azadirachta indica*, to strawberry aphid, *Chaetosiphen fragaefolii*. *Journal of Chemical Ecology*, **19**(8) : 1761-1773.

Lowery, D.T. and Isman, M.B. (1994). Insect growth regulating effects of neem extract and azadirachtin on aphids. *Entomologia Experimentalis et Applicata*, **72**(1) : 77-84.

Martin, B. (1996). Effects of neem oil on the Grasshopper *Zonocerus variegatus* L. in the field. *Neem and Environment*, **1** : 405-412, Ed. R. P. Singh *et al*.

Matthew Scott Holmes and Hassan, E. (1996). The contact, systemic and repellent action of neem seed extract against green peach aphid, *Myzus persicae* Sulzer (Hemiptera : Aphididae). *Neem Newsletter*, **13**(3) : 34-35.

Meisner, J.; Klein, M. and Ben Moshe, E. (1992). Effect of Margosan-O on the development of the leafhopper, *Asymmetrasca decedens*. *Phytoparasitica*, **20**(1) : 15-23.

Mishra, P.K. and Singh, R.P. (1992). Antifeedant efficacy of neem (*Azadirachta indica* A. Juss) seed kernel, seed coat and fallen leaves extracts against desert locust, *Schistocerca gregaria*. *Indian Journal of Entomology*, **54**(1) : 89-96.

Mordue (Luntz), A.J.; Eeswara, J.P.; Wickramanada, I.R.; Johnson, S.; Morgan, E.D.; Stuchbury, T. and Allan, E.J. (1996). Neem metabolites with insect antifeedancy and growth regulatory properties are produced in tissue cultures. *Neem Newsletter*, **13**(3) : 29-30.

Nandagopal, V. and Rajgopal, K. (1996). Feeding site preference for sucking pest on groundnut and the effect of neem products on the *Balclutha hortensis* Lindb. *Karnataka Journal of Agricultural Sciences*, **9**(2) : 345-348.

Naraynan, C.R.; Singh, R.P. and sawaikar, D.D. (1980). Phagodeterrency of various fractions of neem oil against *Schistocerca gregaria* Forsk. *Indian Journal of Entomology*, **42**(3) : 469-472.

Nasseh, O.M.; Freres, T. and Krall, S. (1993). Repellent effect of *Azadirachta indica* A. Juss. on adults of *Schistocerca gregaria* Forskal - investigations in the Republic of Niger. *Mitteilungen der Deutschen Gesellschaft fur Allgemeine und Angewandte Entomologie*, **8**(4-6) : 835-838.

Natarajan, K. and Sundaramurthy, V.T. (1990). Effect of neem oil on cotton whitefly (*Bemisia tabaci*). *Indian Journal of Agricultural Sciences*, **60**(4) : 290-291.

Nelson, S.J.; Sundarababu, P.C.; Rajvel, D.S.; Sreemannarayana, G. and Geethanjali, Y. (1996). Antifeedant and growth-inhibiting effects of azadirachtin-rich neem fractions on rice white-backed planthopper *Sogaellla furcifera* Horvath, *Spodoptera litura* Fb. *and Helicoverpa armigera* Hb. *Neem and Environment*, **1** : 243-252, Ed. R. P. Singh, *et al*.

Nicol, C.M.Y. (1993). Further studies on the contact effects of seed oil from the neem tree, *Azadirachta indica* (A. Juss), on nymphs of the gregarious phase of the desert locust *Schistocerca gregaria* (Forsk.). *Mitteilungen der Deutschen Gesellschaft fur Allgemeine und Angewandte Entomologi*, **8**(4-6) : 791-793.

Nicol, C.M.Y. (1993). Phagodeterrency of neem products and their effects after oral intake on the desert locust, *Schistocerca gregaria*. *Mededelingen van de Faculteit Landbouwwetenschappen, Universiteit Gent*, **58**(2B) : 583-591.

Nicol, C.M.Y. and Schmutterer, H. (1991). Contact effects of seed oil from the neem tree, *Azadirachta indica* (A. Juss.), on nymphs of the gregarious phase of desert locust, *Schistocerca gregaria* (Forsk.). *Journal of Applied Entomology*, **111**(2) : 197-205.

Nisbet, A.J.; Woodford, J.A.T.; Strang, R.H.C. and Connolly, J.D. (1993). Systemic antifeedant effects of azadirachtin on the peach-potato aphid *Myzus persicae*. *Entomologia Experimentalis et Applicata*, **68**(1) : 87-98.

Olaifa, J.I.; Adedokun, T.A. and Adenuga, A.O. (1991). Antifeedant and growth-regulating effects of neem products on the variegated grasshopper, *Zonocerus variegatus* L.(Orthoptera : Pyrgomorphidae). *Journal of African Zoology*, **105**(2) : 157-162.

Olaifa, J.I.; Akingbohungbe, A.E.; Schmutterer, H. (ed.) and Ascher, K.R.S. (1987). Antifeedant and insecticidal effects of extracts of *Azadirachtin indica, Petiveria alliacea* and *Piper guineense* on the variegated grasshopper, *Zonocerus variegatus*. *Proceedings of the 3rd International Neem conference*, Nairobi, Kenya, 10-15 July, 1986 : 405-418.

Opender, K. (1998). Effect of neem extracts and azadirachtin on fertility and fecundity of cabbage aphid, *Brevicoryne brassicae* (L.). *Pesticide Research Journal*, **10**(2) : 258-261.

Pradhan, S. and Jotwani, M.G. (1971). Repeated confirmation of our discovery of antifeedant property of neem kernel. *Entomologist's Newsletter*, **1**(11) : 75-77.

Puri, S.N.; Bhosle, B.B.; Ilyas, M.; Butler, G.D. Jr. and Henneberry, T.J. (1994). Detergents and plant-derived oils for control of the sweetpotato whitefly on cotton. *Crop Protection*, **13**(1) : 45-48.

Radcliffe, E.B.; Dunkel, F.V.; Strzok, P.P. and Adam, S. (1991). Antifeedant effect of neem, *Azadirachta indica*, kernel extracts on *Kraussaria angulifera*, a Sahelian grasshopper. *Tropical Agriculture*, **68**(1) : 95-101.

Raguraman, S. and Rajasekaran, B. (1997). Behavioural and physiological effects of neem on rice brown planthopper, *Nilaparvata lugens*. *Madras Agricultural Journal*, **84**(6) : 326-328.

Raguraman, S.; Jayaraj, S. and Saxena, R.C. (1996). Effect of neem extracts on brown planthopper, *Nilaparvata lugens* and its symbionts. *Neem and Environment*, **1** : 375-380, Ed. R. P. Singh *et al*.

Ramaraju, K. and Babu, P.C.S. (1989). Effect of plant derivatives on brown planthopper (BPH) and white backed planthopper (WBPH) nymph emergence on rice. *International Rice Research Newsletter*, **14**(5) : 30.

Rambold, H. and Raychaudhuri, S.P. (1991). The azadirachtins - highly active insect growth inhibitors. *Recent advances in medicinal aromatic and spice crops* (**Volume I**). *International Conference* held on 28-31 January, 1980 at New Delhi, India : 31-37.

Rup, P.J.; Kalra, P.K.; Sohal, S.K.; Grover, I.S. (ed.) and Thukral, A.K. (ed.) (1998). Influence of neem seed kernel extract on the development of *Lipaphis erysimi* (Kalt.). *Environment and Development* : 205-211.

Sandhu, S.S.; Dhaliwal, G.S. and Dilawari, V.K. (1996). Effect of neem-based insecticides on growth and development of white-backed planthopper, *Sogatella furcifera* (Horvath). *Neem and Environment*, **1** : 281-286, Ed. R. P. Singh *et al.*

Sauphanor, B.; Lenfant, C.; Chen, X.; Sureau, F.; Eisenlohr, U. and Domange, A.L. (1993). Possibilities for utilization and secondary activity of neem (*Azadirachta indica*) extract in a peach orchard. *Bulletin OILB SROP*, **16**(4) : 10-13.

Saxena, R.C. and Khan, Z.R. (1984). Survival of brown planthopper and its ability to transmit grassy stunt and ragged stunt diseases on neem oil-treated rice seedlings. *Neem Newsletter*, **1**(3) : 25-27.

Saxena, R.C.; Greenhalgh, R. (ed.) and Roberts, T.R. (1987). Neem seed oil - a potential antifeedant against insect pests of rice. *Proceedings of the 6th International Congress of Pesticide Chemistry* held in Ottawa, Canada, 10-15 Aug., 1986 : 139-144.

Saxena, R.C.; Zhang, Z.T. and Boncodin, M.E.M. (1989). Effect of neem oil on courtship signals and mating behaviour of brown planthopper (BPH) females. *International Rice Research Newsletter*, **14**(6) : 28-29.

Saxena, R.C.; Zhang, Z.T. and Boncodin, M.E.M. (1993). Neem oil affects courtship signals and mating behaviour of brown planthopper *Nilaparvata lugens* (Stal.) (Hom., Delphacidae) females. *Journal of Applied Entomology*, **116**(2) : 127-132.

Schmutterer, H. and Freres, T. (1990). Influence of neem-seed oil on metamorphosis, colour and behaviour of the desert locust *Schistocerca gregaria* (Forsk.) and the African migratory locust *Locusta migratoria migratorioides* (R. & F.). *Zeitschrift fur Pflanzenkrankheiten und Pflanzenschutz*, **97**(4) : 431-438.

Schmutterer, H.; Baumgart, M.; Freisewinkel, D.; Langenwald, J. and Nicol, C.M.Y. (1993). The effects of neem oil and other neem products on nymphs and resulting adults of *Schistocerca gregaria, Nomadacris septemfasciata, Locusta migratoria migratorioides* and *Zonocerus variegatus*. *Journal of Applied Entomology*, **116**(2) : 178-186.

Serra, C.A. and Schmutterer, H. (1993). Control of the sweetpotato whitefly, *Bemisia tabaci* Genn. with neem extracts in tomato fields in the Dominican Republic. *Mitteilungen der Deutschen Gesellschaft fur Allgemeine und Angewandte Entomologie*, **8**(4-6) : 795-801.

Shah, N.K. and Pandey, K.C. (1998). Bio-efficacy of neem based formulations for grasshopper management in Sehima-*Heteropogon* dominated grasscovers. *National Seminar in Entomology of 21st Century*, April 30-May 2, 1998, Udaipur : 136.

Singh, Brijendra (1987). Effect of neem kernel extract mixed with insecticides on the oviposition and mortality of the desert locust (*Schistocerca gregaria* Forsk.). *Indian Journal of Plant Protection*, **15**(2) : 184-187.

Singh, Brijendra; Singh, A.P. and Singh, B. (1987). Effect of neem kernel suspension on the hatchability of eggs of the desert locust (*Schistocerca gregaria* Forsk.). *Journal of Advanced Zoology*, **8**(1) : 52-54.

Singh, Brijendra; Singh, A.P. and Singh, B. (1988). Effect of neem kernel suspension of the hatchlings of the desert locust (*Schistocerca gregaria* Forsk.). *Plant Protection Bulletin*, Faridabad, **40**(1) : 14-17.

Singh, Brijendra; Singh, A.P. and Singh, B. (1988). Effect of neem (*Azadirachta indica*) cake on the oviposition of the desert (*Schistocerca gregaria* Forsk.). *Plant Protection Bulletin*, Faridabad, **40**(1) : 33-34.

Singh, R.P.; Schmutterer, H. (ed.) and Ascher, K.R.S. (1987). comparison of antifeedant efficacy and extract yields from different parts and ecotypes of neem (*Azadirachta indica* A. Juss.) trees. *Proceedings of the 3rd International Neem Conference*, Nairobi, Kenya, 10-15 July, 1986 : 185-194.

Songkittisuntorn, U. (1989). The efficacy of neem oil and neem-extracted substances on the rice leafhopper, *Nephotettix virescens* (Distant). *Journal of the National Research council of Thailand*, **21**(2) : 37-58.

Venkateswarlu, P.; Sharma, C.M. and Singh, R.S.J. (1993). Effect of neem oil on growth and development of mustard aphid, *Lipaphis erysimi* Kalt (Homoptera : Aphididae). *Bioved*, **4**(2) : 139-142.

Weissling, T.J.; Lewis, T.M.; McDonough, L.M. and Horton, D.R. (1997). Reduction in pear psylla (Homoptera : Psyllidae) oviposition and feeding by foliar application of various materials. *Canadian Entomologist*, **129**(4) : 637-643.

West, A.J.; Mordue, A.J. and Luntz, A.J. (1992). The influence of azadirachtin on the feeding behaviour of cereal aphids and slugs. *Entomologia Experimentalis et Applicata*, **62**(1) : 75-79.

Wilps, H. (1996). Potential of phytopesticides in control of the desert locust, *Schistocerca gregaria* Forskal (Orthoptera). *Neem Newsletter*, **13**(4) : 57-58.

Wilps, H.; Nasseh, O. and Krall, S. (1993). Neem products - natural insecticides against *Schistocerca gregaria*. *Mitteilungen der Deutschen Gesellschaft fur Allgemeine und Angewandte Entomologie*, **8**(4-6) : 827-833.

4. Hymenoptera

Rambold *et al*. (1980) fed freshly moulted 4th instar larvae of *Epilachna varivestis* (Coleoptera) with bean leaves treated with two partially purified neem (*A. indica*) seed fractions, and weight gain was determined after 24 and 48 hrs as an indication of antifeedant activity. Larvae were then transferred to untreated leaves and observed until adult emergence. One fraction was also tested by a single topical application on spinning larvae of *Ephestia kuehniella* (Lepidoptera) and 3rd-instar larae of worker *Apis mellifera*. Both fractions caused severe growth disruptions, without acute toxicity, but only the less pure fraction showed a slight antifeedant effect. The fractions interfered with larval and pupal development causing malformations, high pupal mortality was observed. In honeybee larvae, one-tenth of the dose are effective and cuased disturbance in larval developmenet and reduced the degree of caste determination and survival.

Schmidt *et al*. (1987) studied the sterilizing effect of neem on ants. The sterilizing effect of neem extracts on queens and workers of *Formica polyctena* were tested by feeding and fumigation in the laboratory at 24°C. Fumigation with neem extract at 0.01 g/0.5 ml, 0.005 g/0.25 ml and 0.00125 g/ 0.0625 ml increased egg production. In feeding experiments, no toxic effect was observed but the egg laying capacity of the queens was reduced at the higher levels of neem extracts (0.25 and 0.25%). At the lowest neem concentration (0.0625%), there was a stimulatory effect on egg production. The sterilizing effect was reversible, as after feeding on untreated food for at least 10 days the queens began to oviposit again.

Feldhege and Schmutterer (1993) studied the side effects of neem product on *Encarsia formosa*, a parasitoid of whitefly. The influence of Margosan-O (azadirachtin) on the longevity, emergence and capacity of parasitism of *E. formosa* on whitefly, *Trialeurodes vaporariorum*, was studied. The results showed that application of 10 ppm azadirachtin was relatively non-toxic, whereas 20 ppm led to a slight but significant reduction in the fitness of *E. formosa*.

Ashok Kumar *et al*. (1998) studied the effect of neem leaf extract on ants on lemon plants. To control the ants, 10 and 15 percent neem leaf extracts were sprayed daily on plants up to 10 days on two plants on each concentration. It was observed that on 3rd day of spraying, the red and the other types of ants started leaving the treated plants. Some of them died and some migrated to other places. The 15 percent fresh green leaf extract was found to be more effective than 10 percent fresh neem leaf extract. The droplets of the extract that fell on the ground during spraying, repelled the ants due to which the ants were not observed near or on the lemon plants for a long period.

References

Ashok Kumar Singh; Maurya, B.R. and Ram, H. (1998). Neem leaf extract controls ants on lemon plant. *Neem Newsletter*, **15**(3) : 37.

Feldhege, M. and Schmutterer, H. (1993). Investigations on side-effects of Margosan-O on *Encarsia formosa* Gah. (Hym., Aphelinidae), parasitoid of the greenhouse whitefly, *Trialeurodes vaporariorum* Westw. (Hom., Aleyrodidae). *Journal of Applied Entomology*, **115**(1) : 37-42.

Rambold, H. and Czoppelt, C. (1981). Evaluation of insect growth regulators from *Azadirachta indica* (neem) using rearing tests on honeybee larvae. *Mitteilungen der Deutschen Gesellschft fur Allgemeine und Angewandte Entomologie*, **3**(1-3) : 196-198.

Rambold, H.; Sharma, G.K.; Czoppelt, C. and Schmutterer, H. (1982). Azadirachtin : a potent insect growth regulator of plant origin. *Zeitschrift fur Angewandte Entomologie*, **93**(1) : 12-17.

Schmidt, G.H.; Pesel, E.; Schmutterer, H. (ed.) and Ascher, K.R.S. (1987). Studies of the sterilizing effect of neem extract in ants. *Proceedings of the 3rd International Neem Conference*, Nairobi, Kenya, 10-15 July 1986 : 361-373.

5. Isoptera

Gold *et al*. (1989) studied mulching effects on termite scarification of drying groundnut pods. In a preliminary trial with 4 mulches to protect drying groundnut pods against scarification by *Odontotermes* spp. and *Microtermes obesi*, sunhemp mulches increased termite damage, while neem (*Azadirachta indica*) cake and *Ipomoea fistulosa* mulches seemed to act as repellent barries between the soil and the pods. *Celosia argentea* mulches afforded intermediate protection. Damage was less in pods placed on top of mulches than mixed in them.

Serit *et al*. (1992) evaluated the limonoids obacunone, nomilin and limonin (purified from the seeds of *Citrus natsudaidai*) and azadirachtin using a no-choice bioassay method as antifeedants of *Reticulitermes speratus*. Obacunone showed the highest degree of antifeedant potency (PC95=1133 ppm w/w), followed by nomilin (PC95=4475 ppm), azadirachtin (PC95=65293 ppm) and limonin (with PC95 estimates beyond the bioassay limits, the least potent compound). None of the 4 limonoids exhibited any evidence of acute toxicity towards *R. speratus*, although termites that fed on obacunone and nomilin at concentration higher than their PC95 estimates tended to die faster than those in the unfed (control) treatment. Obacunone and nomilin also caused a drastic reduction in consumption at concentration higher than 510 and 1360 ppm, respectively, but similar effects were not observed for azadirachtin and limonin. It was suggested that both obacunone and nomilin had an absolute antifeedant activity towards *R. speratus*.

Ishida *et al*. (1992) tested methanol extract of neem oil, purified to isolate 11 limonoid compounds varying in potency, with antifeedant activity against termites. Deacetylgedunin was the most active (95% activity at 113.7 µg/30-mg paper disc). Only 3 limonoids showed no activity in these bioassays. Structures and activities were compared and it appeared that the furan ring, alphabeta-unsaturated ketone and hydroxyl group each played an important role in determining antifeedant potency.

Serit *et al*. (1992) found in a no-choice bioassay that neem oil deterred feeding by *Reticulitermes speeratus*. A methanol extract of the oil was 4-fold more active than the whole oil. Eleven main limonoids, accounting for 81.5 percent of the activity, were purfied form the active chromatographic fractions of the methanol extract. No acute toxicity was found, although *R. speratus* that were given large doses tended to die faster than unfed ones.

Lopez *et al*. (1998) carried out several bioassays in order to evaluate the antifeedant activity of extracts from *Daphne gnidium*, *Anagyris foetida*, *Chrysanthemum coronarium*, *Ajuga multiflora*, *Ajuga linearifolia*, and *Azadirachta indica* on *Kalotermes flavicollis*, one of the most serious pests in Sherry vineyards. Ethanol extracts from leaves of *D. gnidium* and acetone extracts from leaves and flowers of *C. coronarium* added to the artificial diet showed antifeedant activity which was statistically significant at doses equal or higher than 20 µg/cm^2. In addition, results obtained at a dose of 40 µg/cm^2 of the extract obtained with a hydrochloric acid solution from *A. foetida* were significant. The neem extract, at doses equal or higher than 7 µg/cm^2, exhibited strong antifeedant and toxic effects at a dose of 28 µg/cm^2.

Sharma *et al*. (1998) tested the neem seed kernel extracts (NSKE) prepared by ethanolic extraction of kernels, concentration of the ethanolic extract, preparation of a fat-free methanol extract from the concentrate, and esterification of the acids obtained by hydrolysis of the fat-free extract. The efficacy of the ester fraction (NSEEF) was tested for the protection of the perishable timber of mango (*Mangifera indica*) in the laboratory against the test termite, *Microcerotermes beesoni*, by dipping test pieces of the wood in 0.00625-0.025 percent extract concentrations. NSEEF at 0.025 percent gave the best protection. In terms of the prevention of weight loss, which was 12.5 percent, compared with 32.5 percent in untreated controls.

References

Gold, C.S.; Wightman, J.A. and Pimbert, M. (1989). Mulching effects on termite scarification of drying groundnut pods. *International Arachis Newsletter*, **6** : 22-23.

Ishida, M.; Serit, M.; Nakata, K.; Juneja, L.R.; Kim, M. and Takahashi, S. (1992). Several antifeedants from neem oil, *Azadirachta indica* A. Juss., against *Reticulitermes speratus* Kolbe (Isoptera : Rhinotermitidae). *Bioscience Biotechnology and Biochemistry*, **56**(11) : 1835-1838.

Lopez, M.A.; Ocete, R.; Chi, D.; Darvas, B. and Coll, J. (1998). Antifeedant effects of various botanical extracts on *Kalotermes flavicollis*. *Boletin de Sanidad Vegetal*, Plagas, **24**(1) : 11-22.

Serit, M.; Ishida, M.; Hagiwara, N.; Kim, M.; Yamamoto, T. and Takahashi, S. (1992). Meliaceae and Rutaceae limonoids as termite antifeedants evaluated using *Reticulitermes speratus* Kolbe (Isoptera : Rhinotermitidae). *Journal of Chemical Ecology*, **18**(4) : 593-603.

Serit, M.; Ishida, M.; Nakata, K.; Kim, M. and Takahashi, S. (1992). Antifeeding potency of neem (*Azadirachta indica*) extractives and limonoids against termite (*Reticulitermes speratus*). *Journal of Pesticide Science*, **17**(4) : 267-273.

Sharma, P.; Ayyar, K.S.; Bhandari, R.S.; Rana, S.S. and Joshi, M.C. (1998). Efficacy of neem seed extracts in the protection of *Mangifera indica* wood against *Microcerotermes beesoni* Snyder (Isoptera : Termitidae) in laboratory. *Annals of Forestry*, **6**(1) : 89-94.

6. Lepidoptera

Chadha (1977) did four field tests on sesame in 0.2-0.8 percent suspension of powdered seeds of neem (*Azadirachta indica*) applied in sprays at 250 lt/ha 6 times at weekly intervals acted as a feeding inhibitor against larvae of *Antigastra catalaunalis* (Dup.), reducing infestation and increasing yields. The results compared favourably with those obtained when carbaryl and a mixture of BHC with DDT were used.

Mong-Ting-Tan *et al*. (1978) tested the antifeedant properties of water and alcohol extracts of different parts of the neem tree (*A. indica*), applied as soil or foliar sprays against the larvae of *Plutella xylostella* (L.) on turnip (*Brassica rapa*). None of the extracts gave complete protection to the plants against attack by the larvae, but the extract of the seed was the most effective, followed in order of descending effectiveness by the extracts of leaves, bark or twigs, and wood. Studies on the persistence of neem extacts and the effect of soil composition on its uptake by the plants showed the active principle to be effective for up to 10 days in treated plants. The rates of uptake by the plants in different soils were almost alike. The ability of adult females of *P. xylostella* to recognise oviposition sites was not affected by the presence of the extract. Neem seed extract applied to eggs by dipping and topically to larvae and pupae adversely affected hatching and larval and pupal development, and most of the treated insects surviving to the adult stage had deformed wings.

Joshi *et al*. (1978) tested the relative efficacy of water suspensions of neem kernel at concentrations (1 to 5%), fentin acetate (0.075%) and fentin hydroxide (0.075%) as antifeedants against the tobacco caterpillar, *Spodoptera litura* Fabricius in field-plot tests in India. The results showed that the treatment with fentin hydroxide, fentin acetate and 2 percent neem kernel suspension gave effective protection from larval attack.

Joshi and Sitaramaiah (1979) carried out the laboratory and nursery tests in India in 1977-78 which showed that a 2 percent aqueous suspension of the seed kernel of neem (*A. indica* [*Melia azadirachta*]) was an effective repellent and oviposition deterrent for adults of S. litura. The mean number of eggs laid on treated tobacco seedlings were 373 as compared with 1715 on untreated seedlings. Because adult females were repelled by the treated plants, they laid more eggs on cloth and glass surfaces in galss-jar experiments. Repellency was even more pronounced when moths were released in treated and untreated nursery beds enclosed in nylon-netting cages, the mean number of eggs laid on treated seedlings was 749, as compared with 19 713 in untreated control beds.

Meisner *et al*. (1981) studied the effect of azadirachtin and salannin, 2 triterpenoids isolated from the seeds of neem (*A. indica*), on the feeding response of larvae of *Spodoptera littoralis* (Boisd.) and *Earias insulana* (Boisd.) Lamellae of styropor (foamed polystyrene) were painted on both sides with a mixture of 5 percent sucrose with different concentraions of either azadirachtin or salannin dissolved in methanol and water (3:7). Azadirachtin strongly suppressed feeding by *S. littoralis* larvae, even at 0.001 percent, whereas salannin showed some antifeedant activity at 0.005 percent and above. Larvae of *E. insulana* were deterred from feeding on azadirachtin-treated lamellae, even at 0.005 percent, whereas salannin was effective only at 0.01 percent and above. Azadirachtin applied to cotton leaves deterred larvae of *S. littoralis* from feeding at all concentrations between 0.001 and 0.02 percent.

Kubo and Klocke (1982) carried out a laboratory study in Japan to determine the effecitveness of azadirachtin, isolated from neem (*A. indica*), in disrupting the behaviour and development of several lepidopterous pests. The compound was found to have a potent antifeedant effect on *Heliothis zea* (Boddie) and *Spodoptera frugiperda* (J.E. Smith), particularly on the second species. It also inhibited larval ecdysis in these 2 species and also in *Pectinophora gossypiella* (Saund.) and *H. virescnes* (F.) Concentrations of azadirachtin lower than that inhibiting ecdysis (1 ppm) inhibited the growth rate of feeding larvae by more than 50 percent.

Saxena (1982) carried out a laboratory test in India to evaluate the effectiveness of various concentrations of an extract from the neem seed kernel against *Amsacta moorei*, a serious pest of dew gram (*Vigna aconitifolia*). A stock solution was obtained by making a thin paste of seed powder (50 gm) diluted to 500 ml of distilled water. Three concentrations viz. 0.1, 0.5 and 1.0 percent were taken for the trial along with a control. The percentage of protection over the control was calculated on the basis of the mean dry weight of the leaves, consumed and also the mean gain in larval weight. It was observed that initial protection provided by the treatment ranged from 95.1 to 98.4 percent. No significant difference was observed within the concentrations . The weights of the larvae decreased in all the treatments indicated the antifeedant activity. No mortality was observed which indicates that it has no toxic effect. The residual effect of the extract was influenced by concentrations. The higher concentrations 0.5 and 1.0 percnt remaining highly effective up to 7 days. The extract possesses systemic effect but it took 72-96 hr to translocate in to plant system.

Schmutterer *et al.* (1983) studied morphogenetic effects of 4 partially purified fractions of neem (*A. indica*) and 2 methanolic seed extracts on larvae of rice pests, *Mythimna separata* (Walker) and *Cnaphalocrocis medinalis* (Guenee) in the laboratory. Larvae that fed for 24 hrs on rice leaf-cuts dipped in various solutins of the fractions and extracts exhibited pronounced developmental abnormalities and high mortality rates in the succeeding larval instars and in the pupal and adult stages. A fraction not containing azadirachtin showed the strongest growth-disrupting effect on early 6th-instar larvae of *M. separata*.

Joshi *et al.* (1984) observed neem seed kernel solution as an antifeedant against *S. litura*. In field tests on tobacco at Kateru, India, sprays containging

0.5, 0.75 or 0.1 percent of a suspension of seed kernels of neem (*A. indica*) protected the plants from attack by *S. litura* for 7 days and had no adverse effects on the yield or chemical and physical quality. On the basis of a smoke quality assessment and an organoleptic evaluation, the 0.5 percent spray is recommended.

Chiu and Zhang (1984) determined in choice and no-choice tests in China with 5th-instar larvae of *S. litura*, using the leaf-disc method, the median antifeedant concentraions (AFC50s) of petroleum ether (non-polar) extracts and methanol (polar) extracts of the seed kernels of *Melia azedarach* and *M. toosendan*. The antifeedant properties were found in the polar rather than the non-polar extracts. In the no-choice tests, more than 97.5 percent protection was afforded by the 2 methanol extracts used at 1-2 percent, by neem seed oil at 3 percent and by toosendanin (the toxic principle from *M. toosendan*) at 0.1 percent.

Redfern *et al*. (1984) formulated ethanol extracts of fresh seeds of neem (*A. indica*) as aqueous suspensions and de-fatted ground neem kernels as non-wettable dusts for greenhouse tests against larvae of *Spodoptera frugiperda* in the USA. First-instar larvae were fed on neem-treated sweet maize leaves. The antifeedant activity of 0.2 and 0.4 percent aqueous solutions of the neem seed extract was maintained for at least 21 days. Tests with formulated dusts showed moderate activity against the larvae.

Chari and Muralidharan (1985) observed neem as feeding deterrent of caster semilooper, *Achaea janata* linn., a pest of castor (*Ricinus cummunis*). The antifeedant activity of neem (*A. indica*) kernel suspension at 2. 3 and 5 percent and neem leaf suspension at 5 and 10 percent on 4th-instar larvae of *A. janata*, was investigated in the laboratory. After 24 hrs, none of the larvae had touched the castor leaves treated with either neem suspension. Larvae moved to treated castor leaves after 48 and 72 hrs when the food in the untreated control was exhausted, but excreta was only produced at the lowest concentraion of the neem leaf extract. In the untreated control, the leaves were completely devoured. It is suggested that neem seed suspension at 2 percent or neem leaf extract at 10 percent would be an effective and economical treatment for the control of the pest.

Sheike *et al*. (1985) tested ovipositional and repellent action of some vegetable oils/extracts against potato tuber moth. Of 7 vegetable oils or extracts tested against adults of the gelechiid, *Phthorimaea operculella* on potato tubers in the laboratory, 0.05 and 0.1 percent neem oil (*A. indica*), o.1 percent oils or extract of mahua (*Madhuca logifolia*) and karanj. *Pongamia glabra* afforded 91.66 to 100 percent oviposition deterrence. Varma and

Singh (1985) studied antifeedant effects of some plant extracts against *Amsacta moorei* (Butler). The antifeedant activity of 0.1 percent aqueous extract of the leaves of sacred basil (*Ocimum sanctum*) or 0.1 percent water emulsions of seed oils of neem. Pongamia (*Pongamia pinnata*) or bold-seeded mustard (*Brassica juncea*) against 4th-instar larvae of *Amsacta moorei*, an important pest of rainy-season crops in arid and semi-arid regions, was evaluated in laboratory tests in India. The emulsions of neem, pongam and mustard oils were prepared with the help of Triton X emulsifier. For each treatment, there were 4 replications each having five-fourth instar larvae kept in petriplates having food material. The left over unconsumed portion of the leaves after 24 hrs was meausred. It was observed that neem oil and pongam oil at 0.1 percent were excellent antifeedants. In case of neem treatment, the larvae fed only 10.57 percent of the leaves while in pongam oil, it fed only 16.5 percent leaves.

Rajasekaran *et al*. (1985) observed that addition of sesame oil to an aqueous extract of neem seed kernel at 6 percent (v/v) enhanced the antifeedant activity against the larvae of *Spodoptera litura*.

Lidert *et al*. (1985) studied that azadirachtin from *Azadirachta indica* and from the bark and stem wood of *Nymania capensis* exhibited strong antifeedant activity against tobacco budworm (*Heliothis virescens*), southern armywork (*Spodoptera eridania*) and Mexican bean beetle (*Epilachna varivestis*).

Hongo and Karel (1986) studied the effects of extracts of neem seed kernels, neem (*A. indica*) leaves, and fruits of pepper (*Capsicum annuum*) on the incidence of insect pests of common bean (*Phaseolus vulgaris*) and the damage caused by them in the field in Tanzania. Aqueous extracts from neem seed kernels and pepper fruits caused general deterrent effects, including an antifeedant effect against several important pests such as *Ootheca bennigseni* and larvae of *Marucca testulalis* and *Heliothis armigera*. Plots treated with these 2 extracts showed less damage and a significantly lower incidence of the pests than untreaed plots. These findings indicate the potential of these safe, indigenous and readily available extracts in the protection of *P. vulgaris* in Tanzania.

Kubo *et al*. (1986) observed from the fresh fruits of *Azadirachta indica* a new limonoid, deacetylazadirachtinol. In addition, 4 previously elucidated limonoids, including azadirachtin, salannin, 6-O-acetylnimbandiol and 3-desacetylsalannin were isolated from the same source. Deacetylazadirachtinol was found to be as potent as azadirachtin in the inhibition of insect ecdysis when fed in artificial diet to larvae of the noctuid *Heliothis virescens*. The other isolated limonoids were much less active against *H. virescens*.

Sitaramaiah et al. (1986) conducted the nursery tests in Andhra Pradesh, India, which showed that aqueous suspensions of diflubenzuron at 0.007, 0.009, 0.011 and 0.013 percent, a 2 percent neem seed kernel suspension in water and a combination of 1 percent neem seed kernel suspension plus 0.0035 percent diflubenzuron protected tobacco seedlings from damage by the noctuid *S. litura* for up to 9 days after spraying. Weekly treatments in a bulk nursery with 0.009 percent diflubenzuron and 2 percent neem seed kernel suspension commencing 4 weeks after germination and continued for 7 weeks, offered significant protection and significantly reduced *S. litura* larval population compared with an unprotected nursery.

Prabhakar *et al*. (1986) did laboratory evaluation of neem seed extract against the cabbage looper, *Trichoplusia ni* and armyworm, *Spodoptera exigua*. Neem seed extract was added in an artificial diet at 0.02, 0.2 and 2.0 percent (w/v) concentrations. The material for each concentration was blended for 5 minutes to ensure uniform dispersion of the extract in the diet. In each treatment, 10 larvae were taken and replicated five times for each stage. In case of the *T. ni* and *S. exigua*, 1st-, 2nd- and 3rd-instar larvae did not develop beyond the instar treated at 0.2 and 2.0 percent concentraions and caused prolonged development and induced mortality in all larval instars. Instars of *S. exigua* exposed to neem seed kernel extract failed to pupate and appeared to be more susceptible to the extract than were the larvae of *T. ni*. It was observed that several 5th-instar larvae developed to prepupal stage on 2 and 5 days at both 0.2 and 0.02 percent concentrations, but on 8 days, all died. This mortality of both the noctuids was most pronounced during ecdysis, indicating similar to that of other insect growth regulators.

Meisner *et al*. (1986) assayed methanolic solutions of azadirachtin, extracted from seeds of *Azadirachta indica* at various concentrations on maize (*Zea mays*) seedlings and the fresh deposits against larvae of *Ostrinia nubilalis*. At all concentrations tested, there was a significant difference in larval weight between treatmnets and the untreted controls. Practically no pupation occurred at 0.006-0.05 percent.

Joshi *et al*. (1986) observed the antifeedant and oviposition deterrent effects with the use of neem extract against the noctuid, *S. litura*.

Raffa (1987) studied the influence of the tetranortriterpenoid azadirachtin, a derivative of neem which caused greater inhibition of *Spodoptera frugiperda*. A two way choice test was conducted to determine the preference of *S. frugiperda* colony for either cotton or lima bean foliage. Forty insects were used in each test. The amount of lima bean and cotton leaf foliage was compared for each unit test by paired-t-analysis. The observation revealed that azadirachtin was very effective in reducing *S. frugiperda* feeding on treated cotton. Feeding deterrence was more pronounced down to 100 ppm. Azadirachtin was far less effective in deterring *S. frugiperda* feeding on lima beans. Feeding deterrence on cotton at 50-100 ppm was equivalent to beans at 100-1000 ppm. The compound inhibits larval growth and provides good crop protection with in 7 days. Feeding by surviving larvae was substantially reduced, relative to control units after 4 days but this was due to sickness rather than behavioral influence. The low level deterrence by azadirachtin on lima beans, therefore, does not necessarily means that this chemical could not provide good protection to this crop. The authors opined that such chemicals which exhibit more than one mode of action are most suitable for crop protection.

Pathak and Krishna (1987) studied the influence of the adours of certain botanical components or an organic solvent dichlormethane during breeding on the reproductive efficiency of *Earias vittella* F. It was observed that the rate of egg hatch in *E. vittella* declined sharply when mating pairs were continuously exposed to the odours of neem (*A. indica*), *Ocimum basilicum* or *Eucalyptus rostrata* [*E. camaldulensis*] in the laboratory. A similar effect was observed when adults were kept prior to mating in environments with vapours of eucalyptus oil or dichlormethane (methylene chloride).

Dreyer *et al.* (1987) carried out laboratory and field tests in Togo to demonstrate the efficacy of simple neem products (aqueous extracts of neem kernel powder, neem oil and ground neem kernels) for the control of arthropod pests on vegetables and field crops. Good control of the gelechiid *Scrobipalpa ergasima* on gboma (*Solanum aethiopicum*), the cicadellid *Jacobiella facialis* and several lepidopteran on aubergines, and the *Plutella xylostella* and the pyralid *Hellula undalis* on cabbages was achieved with weekly applications of 25 or 50 g neem kernel powder/lt. No activity was found against *Tetranychus cinnabarinus* on aubergines. Whorl treatment of maize plants with pure neem kernel powder or a 50:50 mixture of it with sawdust reduced populations of the noctuid *Sesamia calamistis* to almost zero. Aqueous extracts of neem cake and neem kernel powder were equally active in inhibiting the development of 3rd-instar larvae of the noctuid *Selepa docilis* in laboratory tests.

Singh *et al.* (1987) evaluated antifeedant and repellent qualities of various neem formulations against *Pieris brassicae* L. on cabbage and cauliflower. Seed kernel suspension (at 0.1, 0.2 and 0.4%), oil emulsion (at 0.2, 0.4 and 0.8%) and leaf water extract (0.5, 1.0 and 2.0%) of neem (*A. indica*) were evaluated for their repellent effect against 2nd- and 3rd-instar larvae of *P. brassicae* on cabbages and cauliflowers in pot and field trials in Himachal

Pradesh, India. All 3 concentrations of seed kernel suspension and oil emulsion and 2 percent leaf water extract had a significant antifeedant effect, reducing the mean leaf area consumed by 40-50 percent.

Koul (1987) studied the antifeedant and growth inhibitory effects of calamus oil and neem oil on *Spodoptera litura*. Distillates of the essential oil of the rhizomes of Indian calamus (*Acorus calamus*) and seed oil of neem (*A. indica*) were evaluated in the laboratory as emulsified foliage sprays against larvae of the noctuid *S. litura*. A dose-response relationship of both oils was established using food acceptance, feeding ratio, weight gain and larval development as parameters in choice and no-choice tests. At concentraions of 0.5 and 1.0 percent, calamus oil was effective in both tests inducing a significant reduction in feeding and inhibition of growth in early 3rd-instar larvae. Neem oil had such effects only at 2 percent, particularly in no-choice tests.

Umeh (1988) studied the effect of methanolic extracts of neem seed kernel on feeding, development and survival of *Spilosoma maculatus* Walker. Leaves of cowpea (*Vigna unguiculata*) that had been treated with high concentrations (5 and 10%) of extracts of neem (*A. indica*) seed kernel were found to be highly repellent and phagodeterrent to 2nd- and 4th-instar larvae of *S. maculatus*. Only the high concentration remained completely effective for 48 hrs. No acute toxicity was observed, but in the long run, all the 2nd-instar larvae that fed on treated leaves died. concentrations of 0.1 and 1.0 percent resulted in 90 and 10 percent survival of 4th-instar larvae, respectively.

Sinha and Mehrotra (1988) tested the efficacy of diflubenzuron and neem oil against *Heliothis [Helicoverpa] armigera*. Diflubenzuron (at 187.5, 375.0, 562.5 and 750.0 g a.i./ha) and neem oil (at 375.0, 562.0, 750.0 and 3750.0 ml/ha) were tested for the control of the noctuid *H. armigera* on chickpea (*Cicer arietinum*) in the field in Haryana, India. The first application was given at flowering stage when the insect population was 2.10 larvae/m row length. Observations on larval population and infestation on pods were recorded on 3, 13 and 23 days after second spraying. Seed yield was also recorded. Diflubenzuron reduced the larval population as well as borer incidence. The larval population was 3.49-4.58 larvae/3 m row length in treated plots by diflubenzuron which was observed significantly low. Neem oil remained at par with control. However, the results of neem oil do not corroborate the findings of Lal *et al.* (1986) perhaps, due to low dosage used.

Venkateswarlu *et al.* (1988) tested the efficacy of neem oil on certain behavioural aspects of *Spodoptera litrua* Fabricius. Neem (*A. indica*) oil at 8 and 16 percent exhibited complete repellent and antifeedant activity with larvae of *S. litura* on urd bean leavs (*Vigna mungo*). At 0.5-4 percent, repellency and antifeedannt activity increased with concentration. The numbers of eggs laid by females on treated leaves decreased with concentration. At 0.5 and 1 percent, neem oil lost its antifeedant property after 5 days, as compared with 6 days at 2 percent.

Shaw *et al.* (1988) tested the influence of different host plants on ovipoistional preference, larval survival, growth potential and fecundity of Bihar hariy caterpillar, *Diacrisia obliqua* Walker. The biology of the arctiid *D.* [*Spilosoma*] *obliqua* was studied on *Dolichos lablab* [*Lablab purpureus*], *Sesamum indicum* (Sesame), *Phaseolus mungo* [*Vigna mungo*], *Cynodon dactylon, Azadirachta indica* (neem), sweet potatoes, groundnuts, okra, maize and rice in the laboratory. Sweet potato and *P. mungo* were the most favourable food plants, resulting in the shortest larval period, and the greatest larval survival and size. Groundnut was the poorest food plant with only 33.3 percent larval survival. First- to 3rd-instar larvae did not survive on leaves of okra, maize, rice, *C. dactylon* or neem.

Bhandari *et al.* (1988) tested the effect of neem (*A. indica*) seed extracts on poplar *Pygaera cupreata* Butler. A fat-free methanol extract of neem seed and nine chromatographic fractions derived from it were tested against *P.* [*Clostera*] *cupreata* for their antifeedant activity. Preliminary studies indicated antifeedant activity in the whole methanol extract, and in 2 of the fractions (2 and 10). These three active extracts were further tested following the method of Nakanishi in which pairs of leaves are used, with leaf brushed on both sides with the test solution i.e. 1000 ppm extractive. Five replications of ten insects each were used for the test. The whole methanol extract and fraction 2 were found to have significant antifeedant activity compared with fraction 10 and untreated controls.

Shimizu (1988) studied the effect of neem extract on spermiogenesis in the noctuid *Mamestra brassicae* with diapausing adults in the laboratory. Even in the presence of both 20-hydroxyecdysone [ecdysterone] and ruptured testes, spermatocysts degenerated in Grace's medium containing 3 ppm of neem extract while at lower concentrations of neem extract (0.3 ppm), spermiogenesis was induced in 63.6 percent of cultures.

Cobbinoh and Osei-Owusu (1989) studied the effect of vrious neem formulation against insect pests of egg plant, okra and cowpea. The efficacy of various neem formulations (neem emulsion, aqueous-methanol extract of defatted neem cake and defatted neem cake applied as a dust) against some major pests of aubergine, okra and cowpea (*Vigna unguiculata*) was investigated in the field in China. Various concentrations of neem emulsion reduced the incidence of the noctuid *Selepa docilis*, the tingid *Urentus* sp. and the acridid *Zonocerus variegatus* on aubergine, however, some phytotoxicity was observed on plants treated with 10 or 20 percentt emulsion at the pre-flowering stage but did not recur in the subsequent spraying sessions. Damage cuased to okra by the pyralid *Sylepta* [*Syllepte*] *derogata* and the chrysomelid *Podagrica* [*Nisotra*] *sjostedti* was lower in plots treated with 5, 10 and 20 percent aqueous-methanol extracts of defatted neem cake. On cowpea, defatted neem cake applied as a dust not only reduced the incidence of the pyralid *Maruca testulalis* but also significantly increased pod yield.

Venkateswarlu *et al.* (1989) tested the effect of neem oil on growth and development of certain lepidopterous pests. In laboratory experiments, different conentrations of neem oil were sprayed on urdbean plants (*V. mungo*) in pots and tested against the noctuid *S. litura* and the arctiid *Pericallia ricini*. The developmental periods of both lepidopteran pests were extended when concentration of 0.5-4 percent were used. At 8 and 16 percent neem oil concentrations, larvae of both insects died before reaching the adult stage.

Melamed *et al.* (1989) tested the effecs of neem on the cornborer, *Sesamia nonagrioides* Lefebvre. An aqueous neem (*A. indica*) seed kernel extract was sprayed at different concentration on maize seedlings. The fresh residues were allowed to dry and assayed against larvae of *S. nonagrioides*. Larval survival, larval weight and percentage pupation served as criteria of activity. No pupation occurred with 1 and 0.5 percent neem extract, and only 10 percent of the larvae pupated at 0.1 percent. The residual effect of the 1 percent aqueous extract was also investigated 1, 3 and 7 days after treatment. Larval weight was significantly lower on 1- and 3-day-old residues than on untreated plants, with only 10 percent pupation occurring.

Rwamputa *et al.* (1989) studied the efficacy of crude aqueous neem against *Heteronygmia dissimilis*. An outbreak of H. dissimilis at Morogoro, Tanzania, prompted an investigation of, among other aspects, potential controls for this insect by locally available means. Boiled and unboiled crude aqueous extracts of the neem tree (*A. indica*), obtained from pulverized foliage (1% concn.) and seed kernel (0.5 and 1%), were applied to foliage of *K. nyasica*, and offered to first, third and last instar larvae of *H. dissimilis* in the laboratory. Feeding response was monitored by way of leaf areas consumed. Neem kernel extracts at 1 percent concentration afforded complete protection against defoliation by all instars. The 0.5 percent seed kernel and 1 percent leaf extracts offered only partial protection, i.e. about two-thirds and one-third, respectively, with decreases of efficiency against later instars. Boiling of extracts increased foliar protection slightly. Mortality of larvae was attributed to starvation.

Murthy *et al.* (1989) evaluated the relative effectiveness of 7 neem fractions against 3rd-instar larvae of *S. litura* on Lanka and FCV tobacco in the laboratory, and compared with that of a 2 percent neem seed kernel suspension. The results showed that 4 to 7 fractions were as effective as the 2 percent suspension in protecting leaves from larval damage. Fractionation of neem extracts by partition with different solvents enhanced antifeedant activity.

Ramachandran *et al.* (1989) observed the effects of food deprivation and concentration of azadirachtin on *Achaea janata* and *S. litura* on the leaves of castor. The relative efficacy of azadirachtin against the noctuids was investigated in the laboratory. Freshly moulted 4th-instar larvae were presented with weighed castor (*Ricinus communis*) leaf discs cut from young and mature leaves which had been dipped in azadirachtin solution of various concentrations. The larvae were allowed to feed on these leaves at 28±2°C for 24 hrs. Food consumption, faecal pellet production and weight gain of *A. janata* were found to be influenced by the level of food deprivation, age of leaf and concentration of azadirachtin (0.1-10.0 ppm). Differences in food consumption due to starvation were affected by the age of the leaf and concnetration of azadirachtin. Differences in consumption of young and mature foliage were affected by the latter factor alone, while it was also responsible for the differences in weight gain of *A. janata* on young and mature leaves. In *S. litura*, starvation of the larvae and concentration of azadirachtin on the leaf (10-50 ppm) independently and in combination affected weight gain. Age of the leaf did not affect any of the parameters. The percentage protection afforded against *A. janata* by azadirachtin was dependent on concentration. However, even at 50 ppm, azadirachtin did not offer significant protection against *S. litura*.

Ayyangar and Rao (1989) studied the effect of neem extract as larval repellent and ovipositional deterrent in *S. litura*. For the experiment, 25 g neem kernels were ground and extracted with 1.2 lt of methanol. The methanol soluble material was extracted several times with 0.75 lt hexane and the two extracts were partitioned, separated and concentrated. Effect of methanol and hexane extracts on larval hebaviour was studied by olfactometer. The number of larvae entered in each arm was recorded at the end of 30 mts from which percentage repellency was calculated. Ovipositional deterrency was also tested by offering neem extract filter paper for egg laying. The observations revealed that both the extracts had considerable repellency to different larval stages and methanol extracts exhibited greater repellency at 0.02 percent as compared with hexane extract at 0.46 percent. Percentage repellency increased with successive instars at the same concentrations from

63 percent in the 3rd-instar to 92 percent in the 6th-instar. The increased repellency shown by the successive instars may be attributed to their becoming more discrimatory to stimuli with fuller development of sensillae with successive moults. The ovipositional deterrency of hexane extract (0.018%) lasted only for 4 days while in methanol extract (0.01%) lasted for more than 5 days. It is, therefore, observed that both the extracts show ovipositional deterrency to *S. litura* but methanol extract was definitely superior to hexane extract.

Krishnaiah and Kalode (1990) studied the efficacy of selected botanicals against rice insect pests. In greenhouse studies in Andhra Pradesh, India, oils of *Bassia latifolia* [*Madhuca longifolia*] and *Calophyllum inophyllum* were highly toxic to nymphs of the brown planthopper (*Nilaparvata lugens*) on rice. *Hydnocarpus wightiana* oil was less effective. Oils of *C. inophyllum* and neem (*A. indica*) affected orientation and settling by the pest. *C. inophyllum* had a greater antifeedant effect than *A. indica* or *M. longifolia*. Field studies showed that soil incorporation of neem cake at 150 kg/ha followed after 10 days by a 3 percent spray of neem oil, was effective against the leaffolder (*Cnaphalocrocis medinalis*). However, 3 percentt sprays of neem oil, neem seed kernel suspension, *M. longifolia* oil and *C. inophyllum* oil did not show consistent effectiveness against *Scirpophaga incertulas, Orseolia oryzae* or *Hydrellia philippina*.

Koshiya and Ghelani (1990) studied the antifeedant activity of deterrent plant derivatives against *S. litura* on groundnut crop. All the plant materials, either leaves or seeds, were ground individually in grinder. The powder was mixed with acetone and kept for 24 hrs. The mixture was filtered through Buckner funnel. The acetone extracts were diluted to 5, 10 and 15 percent concentrations in soap water. The leaf discs were sprayed with the extracts and after drying under fan, transferred in to plastic tube. The larvae were allowed to feed on the treated discs for 12 hrs. The observations revealed that percent feeding was decreased with increased in the concentrations of antifeedant. The increased dose offered better protection to groundnut leaves against this pest. All the seven plant extracts of 5 percent concentration proved significantly superior to untreated control in providing protection against the 3rd-instar larvae. At 10 percent concentration of the leaf and seed extracts of neem were equally effective and gave 67.5 percent and 78.85 percent perotection of groundnut leaves, respectively. It was also observed that the neem extracts did not differ significantly form karanj seed extract.

Jeyarajan *et al*. (1990) tested 7 azadirachtin rich fractions namely Nemol, Ne-pet-et, NS58/GSN.OU, Vemidin, NK100, Nem75 and Nemidin for their antifeedant and growth retardant properties. Each fraction was dissolved in acetone and made up with water for required concentration. The leaves of castor were sprayed with the extracts and then after drying, placed in a tube with 5th-instar larvae. The observations revealed that in Ne-pet-et, the area fed and weight of faecal pellets produced were 793.00 mm2 and 518.33 mg, respectively and it was significantly lower than other treatments. Other treatments which were next in order were NK100, NS58/GSN/OU and Nemidin in which 65.88, 59.33 and 56.74 percent reduction in feeding was observed, respectively.

The development of *S. litura* after the larvae were fed with treated leaves for 24 hrs showed the strong growth retardant properties of fractions. In Nemidin and NK100 treatments, only 53.33 percent larvae became pupae and it was on par with Ne-pet-et, NS58/GSN/OU, Vemidin and Nemal treatments. In case of NS58/GSN/OU treatment, 23.33 percent larvae became adult whereas in Nemidin (26.67%), Ne-pet-et (30.00%), Vemidin (30.00%), Nk100 (30.00%). In acetone and control treatments, 63.33 and 73.33 percent larvae became adults, respectively. Malformation symptom was observed in adult in Ne-pet-et and Nemidin treatments.

Blaney *et al.* (1990) studied the antifeedant activity of azadirachtin-derivatives and related limonoids in choice and no-choice bioassays against 24- to 36-h-old final-instar larvae of the noctuids *Spodoptera littoralis, S. frugiperda, Heliothis virescens* and *H.* [*Helicoverpa*] *armigera* in the laboratory. The choice bioassay showed that the feeding behaviour of *S. littoralis* was affected by more of the compounds than that of either *S. frugiperda* or *H. virescens*, with *H. armigera* being the least affected. Azadirachtin and dihydroazadirachtin were the most potent of the 40 compounds tested. Hydrogenation of the C-22,23 double bond did not decrease antifeedant activity and the nature of the substitutes at C-1, C-3 and C-11 were important. Molecules with bulky substitutes at either C-22 or C-23 were usually ineffective antifeedants as were compounds lacking an epoxide. Compounds recorded as active antifeedants in the choice bioassay were not always as active in the no-choice test.

Tanzubil and Mc-Caffery (1990) tested the efficacy of azadirachtin on feeding and reproduction in *Spodoptera exempta*, an african armyworm. The effects of azadirachtin on the nutritional and reproductive physiology of the noctuid were studied in the laboratory. Azadirchtin (at 0.01, 0.1 and 1.0 µg/larva) retarded oocyte development and maturation in females obtained from treated final-instar larvae. Decreases in fecundity also resluted from treatment but were not accompanied by corresponding decreases in fertility. Topical treatment of larvae with azadirachtin reduced food intake and the efficiency of utilization of ingested and digested food. Growth rate was impaired, especially at doses >1 µg/larva.

Siddiqui et al. (1990) tested the effect of different neem products on mass rearing of the maize stalk borer, *Chilo partellus* in artificial diet. Varibale concentrations (0.01-0.2%) of each of 3 neem products (leaf powder, seed kernel and oil) were incorporated into an artificial diet for the pyralid *C. partellus* based on grain powder of green gram (*Vigna radiata*). All the products affected development except 0.01 percent neem oil, which, however, affected only fecundity and egg viability adversely. In these diets, there was either no larval establishment or poor larval survival.

Isman *et al*. (1990) determined the variation in azadirachtin content of 12 commercially obtained samples of neem oil in the laboratory using liquid chromatography. The oils were bioassayed for larval growth inhibition (incorporated in an artificial diet) and antifeedant activity against 5th-instar larvae of the noctuid *Peridroma saucia*. Moulting disrupting activity by topical application was assessed against newly moulted 5th-instar larvae of the lygaeid *Oncopeltus fasciatus*. Among the 12 neem oil samples tested, the azadirachtin concentrations ranged from <50 to over 4000 ppm. The bioactivity of the oils, as measured by the EC50, was correlated with the azadirachtin concentration for growth inhibition (r = -0.92), antifeedant activity (r = -0.85) and moulting disrupting activity (r = -0.95). Between 72 and 90 percent of the variation in bioactivity of the oils was accounted for by the variation in the azadirachtin content. It is therefore concluded that the determination of azadirachtin concentration may be a useful quality control criterion for neem oil as a precursor for insecticide production.

Osman and Port (1990) studied systemic action of neem seed substances against *Pieris brassicae*. The ability of neem seed powder and aqueous preparations applied to the soil to protect cabbage plants from damage by larvae of *P. brassicae* was tested. In a no-choice test, treatment with neem seed powder at 20, 40 or 80 g/kg soil resulted in a significant reduction in damage compared with untreated plants. Similarly, in choice tests, plants treated with 1.5, 3 or 6 g neem powder/kg soil or with 2.5, 5 or 10 percent neem aqueous solution were less damaged than untreated plants.

Sehgal and Ujagir (1990) compared the efficacy of synthetic pyrethroids, neem extracts and other insecticides against *Helicoverpa armigera* Hubner. The chickpea (*Cicer arietimum*) crop at Pantnagar in Uttar Pradesh, India, consistently suffers extensive pod damage due to *H. armigera* varying from 42.6 to 90.0 percent, with an average of 65.5 percent in unprotected fields. Grain yields were 0.02-1.36 t/ha, average 0.7 t/ha, during the 'rabi' winter seasons of 1986-87. Among the insecticides tested for 2 or more years, cypermethrin at 36 and 45 g, deltamethrin at 12 g, fenvalerate at 120 g, endosulfan at 420 g and monocrotophos at 24 g/ha significantly and consistently reduced pod damage to <22.5 percent, giving grain yields of >1.7 t/ha. Less effective but still significantly better than the control were quinalphos at 300 g, HCH at 2600 g and chlorpyrifos at 240 g/ha, and neem seed kernel extract at 5 percent. Mean grain yields were significantly negatively correlated with percentage pod damage.

Devaraj Urs and Srilatha (1990) studied the antifeedant and repellent properties of certain plant extracts against *Corcyra cephalonica* St. The antifeeant, repellency activities of neem, mustard, datura, eucalyptus and lemongrass was investigated. The tests were made on 3rd-instar larvae of *C. cephalonica* with 4 replication. The observations revealed that among all extracts, neem was found to be the best antifeedant against the test larvae as indicated by less number of holes on the treated paper, at 3.5 percent concentration at which the mean number of holes were 1.25, 3.5, 5.50, 6.75 at intervals of 24, 48, 72, 96 hrs, respectively. Datura was least effective antifeedanat among all products tested. It was also observed that eucalyptus was the most effective, giving 100 percent repellency against the test insect after 10 mts of treatment at the highest concentration followed by lemongrass, mustard, neem and datura. At lower concentrations (0.5, 1.0 and 1.5%), the percent repellency falls. The observations after 1 hr showed decreased percent repellency at different concentrations.

Bandara and Kudagamage (1991) studied the laboratory and field trials on cabbage in 1989 in Sri Lanka that neem oil, chlorfluazuron and triflumuron were promising against lepidoptera when *Plutella xylostella* and *Plusia [Chrysodexis] eriosoma* predominated.

Pathak and Krishna (1991) observed the postembryonic development and adult emergence of *C. cephalonica* were advesely affected in varying degrees when individuals were reared for the first 2 weeks of larval life or for a similar duration from the 16th day of their lives in an environment of eucalyptus oil volatiles. This was, however, not so in the progeny of parents exposed to these volatiles for only 5 min. Exposure to neem oil volatiles (extracted from the seeds of *A. indica*) during larval stages for similar periods failed to produce such adverse effects. A marked decline in the reproductive potential, in terms of egg output and egg hatchability, was observed when larvae were reared for the first 15 days in the presence of eucalyptus oil volatiles or when the parents were exposed for 5 min to such an environment during adult life.

Isman *et al.* (1991) reviewed the research aimed at developing a neem (*A. indica*)-based insecticide for use in Canada. The potent antifeedant action of azadirachtin in laboratory bioassays against the pyralid *Ostrinia nubilalis* and the noctuid *Peridroma saucia* and its moult-disrupting action against the acridid *Melanoplus sanguinipes* and the lygaeid *Oncopeltus fasciatus* are documented. Both the antifeedant and insect growth regulatory activities of various samples of neem seed oil were correlated with azadirachtin concentration.

Rovesti and Deseo (1991) tested the aqueous NSKEs against eggs and larvae of the apple pest *Leucoptera malifoliella* in the laboratory, inconsistent results were found for kernels from different origins. An extract from a batch of kernels at 20 g/lt gave a 60 percent reduction of egg hatching rate, while no significant reduction was obtained when using kernels from another batch at 25 g/lt. In both experiments, the growth of hatched larvae was disrupted and no pupation occurred. A concentration as low as 1.25 g/lt of the more effective kernels caused 80 percent larval mortality. Complete kill was obtained when 1st- to 3rd-instar larvae were treated at 5 or 20 g/lt. Systemic activity was confirmed when the extract was applied to apple plants as a soil drench or sprayed on the lower leaves of the plants.

Salem (1991) tested neem seed oil and insecticide-treatment programmes (chlorpyrifors + diflubenzuron (DC 702), cypermethrin (ploytrin), primicid (pirimiphos-ethyl) + IKI (chlorfluazuron) [Teper] and profenofos (kurakuron) in 1989, pirimiphos-ethyl + chlorfluazuron, profenofos, thiodicarb (Larvin) and fenvalerate (Sumicidin) in 1990 in field trials against *Pectinophora gossypiella* and *Earias insulana* on cotton. Yields following treatment with 4 sprays of 150 ppm neem oil and insecticides were 8.5 $\pm$ 0.61 and 7.7 $\pm$ 0.63 kentar, respectively, in 1989 and 6.29 $\pm$ 0.53 and 5.83 $\pm$ 0.29 kentar, respectively, in 1990.

Dreyer *et al.* (1991) briefly presented the results of trials carried out using water extracts of neem (*A. indica*) seeds against important pests. Control of many arthropods was good, with the exception of spider mites and thrips. Satisfying levels of control were achieved for lepidopterans (*Plutella xylostella* on cabbage, *Diaphania hyalinata* on cucumbers, fruit borers on tomatoes), polyphagous hemipterans (*Aphis gossypii, Bemisia tabaci*) and heteropterans (*Corythaica planaris* on eggplant). The use of neem extracts appeared to be acceptable to small farmers following a pilot programme, and is probably appropriate for horticultural crop production under irrigation. A major restriction to the increase in neem-based insecticide use is the lack of supplies from local neem trees. Economic costs of chemical and neem insecticides are similar, neem requiring more labour in preparation, but the seeds costing less to purchase.

Durairaj *et al.* (1991) studied the effect of three plant products on feeding of sapota leaf webber, *Nephopteryx eugraphella*. In a laboratory experiment, 4th-instar larvae of *N. eugraphella* were provided with sapota (sapodilla) leaaves treated with neem oil, *A. indica* seed-kernel extract, *A. indica* cake extract, pinnai (*Calophyllum inophyllum*) oil or illupai (*Brassia latifolia* [*Madhuca longifolia*]) oil. The *A. indica* seed-kernel extract (2%) caused the largest reduction in the area of the leaves eaten and in the weight of faecal pellets, with the 1 percent causing the next largest reduction.

Dubey *et al*. (1991) evaluated antifeeding properties of indigenous medicinal plants against *Heliothis* [*Helicoverpa*] *armigera*. Nine plant extracts were compared with 2 chemical insecticides in sprays against the noctuid *H. armigera* on chickpeas in Madhya Pradesh, India. Among the plant extracts, neem seed kernel extract and neem rind extract, due to their antifeedant action, provided maximum protection resulting in higher yields. Monocrotophos (0.04%) and a mixture of 5 percent BHC (HCH) with 5 percent DDT dust at 25 kg/ha were also effective.

Ahmad and Bhattacharya (1991) tested growth inhibitory effect of some plants for *Spilosoma obliqua* Walker. The leaves of neem (*A. indica*), fern (*Dipalazium esculantum*), Parthenium (*Parthenium hysterophorus*), bougainvillea (*Bougainvillea spectibilis*), ocimum (*Ocimum sanctum*) and sweet neem (*Murraya koenigii*) were dried in sun and again in oven at 45°C for 72 hrs. To obtain extract 50 g leaf powder was placed in conical flask and petroleum ether was added until leaf material was dipped completely. The flask was shaken several times and stock solution was prepared by filtration process. The decorticated bengal gram diet was used for testing the growth inhibitory effect of 6 plant leaves. An attempt was made to replace some proportion of basic commodity with cellulose. The larvae survival recorded at two days interval indicated no adverse effect due to addition of cellulose in the basic diet. It was observed that by adding 25 percent cellulose, slight larval period prolongation takes place while in 50 percent addition of cellulose, the larval period prolongation increased further. It was also observed that addition of 25 percent equivalent neem showed a gradual decrease in the population and 22 days after feeding, none of the larvae could survive to complete the development. If 50 g of leaf powder was added, the complete larval mortalilty was recorded in 12 days. Almost same trend was reported on other plant leaf powders. Sweet neem is generally observed avoiding *S. obliqua* when leaf powder of sweet neem was added to diet, it was observed that 86.7 and 80.0 percent larvae could not survive on 26th and 28th day after feeding. It was found that greater reduction in larval survival was seen in diet compounded with neem extracts of ethyl acetate (25 and 50%) on this diet complete larval mortality was recorded on 8 days after feeding with ethyl alcohal extract, a gradual decrease in larval survival was noticed on 8 days. The petroleum ether extract and hot water extract gave larval growth upto 16 and 18 days after feeding, respectively. On these diets, the larvae remained small. Several reports available to show that bitter principles are present in *A. indica* but only triterpenoids are responsible for phagodeterrancy and growth regulatory activity.

Ayyanger and Rao (1991) studied the effects of azadirachtin on development and reproduction of *Spodoptera litura*. The injection of neem extract at 1 µg body weight to 6th-instar larvae of the noctuid *S. litura* reduced fecundity, hatchability, pupation and adult emergence by 39.51, 53.77, 46.62 and 41.92 percent, respectively.

Sharma (1992) studied growth inhibitory activity of azadirachtin, a triterpenoid neem extract, on *Corcyra cephalanica*. Azadirachtin was found in laboratory experiments to inhibit further development of last-instar larvae of *C. cephalanica* in a dose-dependent manner. At 10 µg/larva, 55 percent remained in the larval stage. Larval to pupal and pupal to adult moulting disruption was ascribed to interference with the morphogenetic hormone pool size.

Sojitra and Patel (1992) tested the effect of plant extracts on ovipositional behaviour of spotted bollworm (*Earias vittella*) infesting okra (*Abelmoschus esculentus*). The experiment was conducted in complete randomized block design with three treatments. There were 15 treatments. The result indicates that Repelin (1%) and Neemark (0.1%) significantly reduced the oviposition by *E. vitella* compared with all other treatments. It is also observed that the plant sprayed with 5 percent paste suspension of neem seed kernel had significant low egg-laying than 5 percent leaf extract of neem, peppermint, sacred basil and ardusa.

Sharma *et al*. (1992) tested the effects of some plant substances as antifeedant against insect pests of jute (*Corchorus capsularis*). Three insect pests of jute viz. the stem weevil (*Apion corchori*), the semilooper (*Anomis sabulifera*) and the bihar hairy caterpillar (*Spilosoma obliqua*) were higjhly destructive pests of jute in Bihar. Laboratory reared 3rd-instar larvae of the last 2 of these pests were used in laboratory for antifeedant tests of crude aqueous extracts of leaves of neem (*A. indica*), lantana (*Lantana camara*), sanhund (*Euphorbia royleana*) and harsingar (*Nyctanthes arbortrytis* [*arbortrytis*]), and of whole plants of chirata (*Swertia chirata*). Jute leaves dipped in the extracts were fed to the larvae. All the extracts showed antifeedant activity, the most effective being those of chirata and sanhund, followed by neem, lantana and harsingar. Except for chirata, extracts at the higehr concentration of 100 percent were more effective than 50 percent extracts.

Zhang *et al*. (1992) analysed the azadirachtin content of seeds of *Azadirachta indica* trees planted in Wannin Country, Hainan Province, South China. Air-dried seeds were found to contain up to 0.48 percent azadirachtin. A methanol extract contained 2.94 percent azadirachtin. A fraction of the methanol extract (F3-5) with 0.883 percent azadirachtin was effective against 4th- to 5th-instar larvae of the pyralid *Ostrinia furnacalis*, when given at a concentration of 113 ppm mixed with synthetic diet. All larvae developed malformations and died or became 'permanent larvae' when treated with 500 ppm. After eating or coming into contact with leaves treated with 2 percent F3-5, all 4th-instar larvae of *Plutella xylostella* were inhibited in their growth and development, and then became malformed or died. Other fractions of the methanol extract containing no azadirachtin showed similar activity against these pests.

Loke *et al*. (1992) studied the occurrence, distribution and fruiting cycle of neem (*A. indica*) in Malaysia. The contact toxicity of neem oil (1.25, 2.5, 5.0, 10.0, and 40.0%) was evaluated against 2nd- and 3rd-instar larvae of *P. xylostella*. Significant mortality of larvae of both instars was observed at 10 percent concentration. Physiological and growth disruptive effects such as growth retardation, delayed adult emergence and abnormal adults were observed with sublethal concentration. Pupae were not affected by treatment with neem oil at 1.25, 2.5, 5.0 and 10 percent. However, adult males and females treated with 2.5 percent and higher concentrations of neem oil had significantly higher mortalities than untreaed adults. Using a leaf disc technique, it was shown that concentration of neem oil at 1.25 percent deterred feeding of larvae of *P. xylostella*. Neem oil at >1.25 percent significantly reduced egg hatch and at 20 percent, all treated eggs failed to hatch.

Singh *et al*. (1992) tested the effects of neem (*A. indica*) seed oil and two neem products (Neemark and Neemguard) on freshly moulted 5th-instar larvae of the tobacco caterpillar, *Spodoptera litura*, a polyphagous pest of field crops and vegetables in India. Tests were carried out by using leaf discs of *Ricinus communis* dipped into the different preparations as food. Feeding of *S. litura* was not deterred at any concentrations (1.25, 2.5 and 5.0%) of neem product tested but there was considerable larval mortality in all the neem treatments, with no larvae surviving at any concentration of Neemguard tested. Pupae emerging from neemark treatments were smaller than normal. Pupae emerging from neem oil treatments were of normal size but adult emergence was greatly reduced.

Jagannadh and Nair (1992) studied the azadirachtin induced effects on larval-pupal transformation of *Spodoptera mauritia*. Penultimate (5th-) and last (6th-) instar larvae of *S. mauritia* of various ages were injected with 0.5,

1 or 2 μg azadirachtin and the effects on moulting and larval-pupal transformation were analysed. Higher doses (1 and 2 μg) of azadirachtin induced a prolongation of the 5th-larval instar treated on day 0 and day 1. The resulting 6th-instar larvae failed to pupate. Sixth-instar larvae injected with 0.5, 1 or 2 μg azdirachtin showed prolongation of the 6th-larval instar. Azadirachtin treatments completely prevented normal pupation in 'day 0' and 'day 1' larvae even though the percentage pupation increased in treated larvae of other age groups. Injection of 2 μg azadirachtin prevented normal pupation in larvae of all age groups. Injection of 4 μg ecdysterone into 6th-instar larvae pre-treated with 1 μg azadirachtin on day promoted normal pupation in the majority of insects.

Maredia *et al.* (1992) studied the effect of neem on maize pests. Laboratory studies were conducted at 25°C and 60-80 percent RH to evaluate the efficacy of neem (*A. indica*) against the noctuids *Helicoverpa zea* and *Spodoptera frugiperda*, the pyralids *Diatraea saccharalis* and *D. grandiosella*, the curculionid *Sitophilus zeamais* and the cucujid *Prostephanus truncatus*. Neem seed powder at rates of 5 and 10 g and neem seed oil at the rate of 1 ml/kg of meridic diets, were incorporated into the diet of the first 4 pests. Maize grain was treated with 5 and 10 ml neem seed oil/kg of grain and infested with adults of *S. zeamais* and *P. truncatus*. The presence of neem seed powder and neem seed oil in the diets significantly reduced the survival and growth of *H. zea, S. frugiperda*, *D. saccharalis* and *S. grandiosella*. Neem oil also had significant effects on the survival of *S. zeamais*. While in case of *P. truncatus*, the effects of neem oil were not profound.

Kakoty *et al.* (1993) evaluated the antifeedant properties of a few commercial neem formulations against bunch and psychid caterpillar. Various commercial neem formulations viz. Green Mark, Neem Plus and Godrej Achook were applied to tea leaves and fed to *Andraca bipunctata* and *Clania* [*Eumeta*] *cramerii* in the laboratory showed promising antifeedant properties, with 62-100 percent of the leaf area of treated leaves remaining uneaten. It is concluded that there is the possibility of using neem formulations for the control of insect pests of tea in the near future.

Facknath (1993) assessed two neem formulations for the control of economically important insect pests in Mauritius to evaluate 2 commercial formulations prepared from the biologically active extracts of the neem (*A. indica*) tree. The test insects selected were lepidopterous pests of cruciferous crops, namely *Plutella xylostella* and *Crocidolimia binotalis*, and mosquitoes (a mixture of *Anopheles* and *Culex*). The laboratory bioassays showed the strong antifeedant action of Neemark and Neempest on the lepidopteran pests and their larvicidal action on mosquito larvae. Field testing on cabbages demonstrated the efficacy of both formulations in suppressing insect damage. At a concentration of 9 ml/lt, they were as effective as the recommended insecticide (1 ml/lt selecron [profenofos]) in protecting cabbages from insect damage.

Sinha (1993) studied field efficacy of neem products and insect growth regulators against *Helicoverpa armigera* Hubner. Diflubenzuron, neem (*A. indica*) oil and neem seed kernel extract were tested for control of *H. armigera* infesting chickpea in Karnal, Haryana, India. Diflubenzuron (0.05%), neem oil and kernel extract (5%) sprayed at an interval of 10 days gave a 2-fold reduction in infestation compared with the untreated control. Increase in seed yield was 39 and 47 percent with diflubenzuron and endosulfan, respectively. During the 1989-90 and 1990-91 seasons, some new neem products and chitin biosynthesis inhibitors were evaluated along with endosulfan. A large infestation was recorded (12.5% in 1989 and 37% in 1990-91). Flufenoxuron (0.0125 or 0.025%) reduced infestation by 45-50 percent and 77-78 percent in the 1989-90 and 1990-91 seasons, respectively. Neem emulsion and neem wettable powder gave 40-60 percent control. Neem kernel extract at 5.0 percent gave 40 percent reduction in infestation and was comparable to endosulfan at 0.07 percent. However, Neemguard was ineffective. There were no significant differences in the seed yield in plots treated with neem emulsion (0.125%), neem kernel extract (5%), flufenoxuron, and endosulfan.

Osman (1993) applied methanol extracts of neem (*A. indica*) seeds topically to larvae of *Pieris brassicae* disrupted subsequent development and decreased growth rates. Spinning larvae and 1-day-old pupae were less susceptible. Five-day-old pupae were unaffected.

Przybyszewski (1993) studied neem seed and leaf extracts to control the pests of cabbage. Three concentrations of simple, aqueous neem extracts from leaves (2.5, 5 and 10%) and seeds (1, 2 and 3%) were used against lepidopterous pests of cabbage in Mauritania. Extracts were made and applied with only locally available equipment. In a randomized split-block design replicated 4 times, 2 blocks separated differences between starting applications with younger or older transplants. The results showed lower pest damage levels in seed extract treatments beginning after the 3rd application. After the 6th application, all 3 seed extract treatments showed signifiantly lower pest levels, pest densities and percentage heads damaged by lepidopterous pests; however, these values were not significantly lower in the case of leaf extract treatments. All 3 concnetrations of seed extracts worked equally well. Separat control plots showed lower densities of lepitopterous than integrated control plots. The younger transplant block showed consistently lower densities of lepidoptera than the older transplant block.

Chari *et al*. (1993) tested bioefficacy of some neem extract formulations viz. 0.0018% Margosan-O, 0.1% Neem rich II, 0.75% F6, 1% Repelin, 2% Wellgro, 0.01% Neemark, 1% dichloroethane extract, 1% alcohal extract, 2% neem oil emulsion and 2% neem seed kernel suspension against the larvae of *S. litura* on tobacco in the laboratory and nursery in India. Repelin, dichloroethane extract and neem seed kernel suspension gave the best protection in laboratory tests. In nursery tests, these 3 formulations gave significant protection to tobacco seedlings 9 days after spraying.

Joshi *et al*. (1993) evaluated neem, mahua (*Madhuca longifolia*) and pongamia oils (2 and 4%) and neem and pongamia seed extracts (2%) against *Phyllocnistis citrella* on limes in Karnataka, India. All the treatments reduced the pest population, with neem seed extract being the most effective.

Devraj and Srilatha (1993) determined the antifeedant and repellent properties of extracts of neem, mustard, Datura, Eucalyptus and lemon grass (*Cymbopogon* sp.) against the pyralid *Corcyra cephalonica*. Neem extract had the greatest antifeedant activity followed by Eucalyptus, lemon grass, mustard and Datura extracts, resectively. Eucalyptus extract was the most repellent, followed by lemon grass, mustard, neem and Datura extracts.

Jayarajan *et al*. (1993) studied antifeedant and morphogenetic effects of azadirachtin against *S. litura*. Studies on the effects of azadirachtin-rich fractions on larvae of S. litura revealed that Ne-pet-et has strong antifeedant properties. Larval mortality due to malformation was observed with Vemidin, Nem-75 and Nemidin, and adult malformation with Ne-pet-et and Nemidin treatments.

Murthy *et al*. (1993) tested effect of neem extracts for the control of *S. litura* on tobacco in the laboratory and nursery in India. Dichloroethane extract was more effective than neem seed kernel suspension. Laboratory studies indicated that dichloroethane extract and neem cake could prevent damage by *S. litura*.

Koshiya and Ghelani (1993) determined the antifeedant activity of various plant extracts against 3rd-instar larvae of *S. litura* on groundnut leaves in the laboratory at 26°C. Neem leaf and seed extracts and extracts of seeds of *Pongamia glabra* [*pinnata*] were highly effective, giving 75.5, 88.96 and 66.4 percent protection, respectively at 15 percent concentration.

Rao *et al*. (1993) studied the effects of botanicals on certain behavioural responses and on the growth inhibition of *S. litura*. Biosol, Neemark, Repelin and neem oil were tested at 0.5-3.0 percent against larvae of *S. litura* in the laboratory. Repellency, antifeedant activity and development period increased with increase in concentration of all 4 biopesticides. Adult emergence, growth, survival, larval and pupal weight, number of eggs laid and hatchability of eggs decreased with increase in concentration. Neem oil had the greatest effects, followed by Neemark, Biosol and Repelin.

Hossain *et al*. (1994) evaluated the effectiveness of various indigenous materials and insecticides for their potential to control *Phthorimaea operculella* in storage. The study consisted of 12 treatments utilizing various indigenous materials including dry sand, ash, rice bran, sawdust, rice straw, eucalyptus, lantana, neem, water pepper, 2 chemicals and a control. All treatments except rice straw were effective in controlling the pest in storage for one month. After two months storage, the infestaion of potato tubers by the pest was 100 percent in the control while in treatments with ash, rice bran and neem, it was zero to negligible. After 3 months storage, no infestaion was observed in rice bran-treated potatoes and this was consistent with neem-treated potatoes, although the latter had greater infestation.

Bhathal *et al*. (1994) studied feeding deterrency of Neemark against 3rd-instar larvae of *Spilosoma obliqua*. The feeding deterrent activity of Neemark having 80% extract of *Azadirachta indica* water miscible formulation containing azadirachtin was evaluated against 3rd-instar larvae of *S.* [*Spilarctia*] *obliqua* by the leaf-disc dipping method. The damage within 48 hrs of feeding was significantly reduced over the untreated control. The antifeedant activity over control ranged between 6.7 percent at the lowest concentration (0.31%) and 86 percent at the highest concentration (5%).

Durairaj and Venugopal (1994) studied antifeedant effect of neem products against *S. litura*. Neem products viz. Neem Azal F (0.00125-0.01%), Neemark (1%), neem oil (3%) were tested against *S. litura* on fresh cotton leaves in laboratory. Five pre-starved 3rd-instar larvae were introduced in petridish containing treated leaves. After 48 hrs. the leaf area fed was measured. The result indicates that the lowest leaf area fed (8.6 cm^2) and minimum faecal pellet weight (29.9 mg) were recorded in Neem Azal F at 0.01 percent indicating the maximum antifeedant activity while in control the leaf area fed was 29.5 cm^2 with maximum faecal pellet weight (210.6 mg). In neem oil (3%), the fed area was 14.8 cm^2 with faecal pellet weight 45.9 mg. In case of Neemark (1%), the area fed was 12.7 cm^2 with a faecal pellet weight 58.0 mg. In the artificial medium after 24 hrs, the lowest faecal pellet weight (14.5 mg) was recorded in Neem Azal F (0.01%) indicating maximum antifeedant effect.

Sengutterval *et al*. (1995) studied effect of plant products on rice moth. A range of plant products was tested for efficacy in protecting stored groundnuts against *Corcyra cephalonica*. They were evaluated on the basis of percent dry matter loss, percent damaged kernals and pods, and the number of adults emerging from samples infested with larvae or exposed to oviposition. All the plant products and edible oils afforded some protection. However, *Vitex nigundo* leaf powder, neem leaf powder and neem oil were most effective followed by neem kernel powder.

Haumann and Isman (1995) tested neem products as ovipositional deterrents to noctuid moths. Applications of 3 concentrations of oil-free neem seed extracts (*A. indica*) to cabbage plants in cages did not deter oviposition by individuals of 3 species of noctuid moths, *Trichoplusia ni, Peridroma saucia* and *Spodoptera litura*. The concentrations used corresponded to 10, 50 and 100 ppm of the main active ingredient, azadirachtin. The total number of eggs laid per female, female longevity, and median day of oviposition were not affected. Sprays of the neem oil-based insecticide Margosan-OR, and a 1 percent aqueous emulsion of a refined neem seed oil similarly had no effect on any of the parameters studied. However, a 1 percent crude oil emulsion significantly reduced the proportion of eggs laid by *S. litura* on treated plants. It is suggested that literature reports of significant neem-based oviposition deterrence to *S. litura* are the result of compounds that are removed by higher levels of processing and thus not likely to be found in most commercial neem seed formulations.

Gerard and Ruf (1995) studied the neem extracts on the survival and feeding of larvae of four keratinophagous insects. Six concentrations of a neem (*A. indica*) seed extract ranging from 0.1-2.6 mg azadirachtin/g woollen cloth were tested against late-instar larvae of *Anthrenocerus australis, Tineola bisselliella, Tinea dubiella* and *Hofmannophila pseudospretella*. Feeding damage decreased with increasing rates for all species. A significant rate response was found for *T. bisselliella* and *T. dubiella* larval survival after 2 weeks, but there was no effect on *A. australis* and *H. pseudospretella*, apart from a rate response in larval moulting in the latter. When the bioassay was repeated using the 1st-instar of *H. pseudospretella*, survival increased with increasing rates of azadirachtin.

Rao *et al*. (1996) carried out a detailed study on the effects of neem (*A. indica*) on *S. litura*. Hexane and methanol extracts of neem seed kernels proved to be larval repellents and ovipositional deterrents. It was observed by the authors that methanolic extract at 0.02 percent concentration repelled 63.9 percent 3rd-instar larvae. The effect of this extract increased to 70.6, 81.9 and 92.1 percent in 4th-, 5th- and 6th-instar larvae, respectively. The methanolic extract was superior to 0.046 percent hexane extract. The two neem extracts were also observed as oviposition deterrent. The deterrence of hexane extract (0.02%) lasted for 4 days and the mathanolic extract (0.01%) continued upto 5 days. Injection of a single dose of azadirachtin (1 µg/g body weight) into 6th-instar larvae affected food consumption and utilization, mid-gut enzymes, haemolymph constituents viz. haemocytes, proteins, trehalose, uric acid, nin-hydrin positive substances and cations, secretions of the median neurosecretory cells of the brain, corpora allata size, and ecdysteroid titres. It reduced nucleic acid and protein concentrations of the larval fat body, reduced pupal weight, and disrupted ovarian development. Azadirachtin reduced the haemocyte count significantly with the maximum reduction for 2 days after treatment. The total haemocyte count is dependent on ecdysone titres, and in this context, the drastic reduction may be attributed to the interference of azadirachtin with the hormone system. The nin-hydrin positive substance concentration was higher in the treated larvae, suggesting greater breakdown of protein. It was also found out that the decreased intensity of neurosecretory material in control and the increased in neem treated larvae at 72 hrs suggest that more and more accumulation of neurosecretory material in the treated larvae. The azadirachtin was reported to reduce the size of *Corpora allata* volume resulting in reducing the juvenile hormone (JH). It was also observed that exogenous application of methoprene (20 µ/larva) and injection of ecdysterone (4 µg/g) 24 hrs after azadirachtin injection into 6th-instar larvae were not able to overcome azadirachtin effects on ecdysteroid titres and pupation. Adults emerged from azadirachtin-trearted larvae showed poor ovarian weight and reduced ovarian proteins and vitellogenin leading to reduced fecundity, hatchability, pupation, and adult emergence.

Sombatsiri (1996) tested Thai neem seed extracts (*A. indica* var. *siamensis*) for the control of insect pests on Chinese kale, namely *Plutella xylostella, Spodoptera litura*, and *Hellula undalis*. The experiment was carried out on Chinese kale field in randomized complete block design with five treatments and three replications. Various neem formulations were sprayed on the plots right on the 10th day after sowing and continued once every five days until seven days before harvest. It was observed from the data that all neem extracts 5%,MNSE, 0.25 and 0.5% MNPE could control *P. xylostella, S. litura* and *H. undalis* larvae in low level. Methanol extract showed better control of *H. undalis* larvae than *Bacillus thuringiensis* and insecticide. The neem extracts appeared to have withstood P. xylostella and S. litura and even showed better resistance to *H. undalis*. Methanolic neem extract at 5 percent concentration probably was the best among the neem formulations in controlling insects and resulted in greater marketable yield.

Jhansi and Singh (1996) studied the identification of effective and inexpensive neem seed kernel extracts (NSKE) for the management of *Helicoverpa armigera* Hubner. Six NSKEs viz. aqueous (AE), ethanolic (EtoHE), hexane (HE), ethanol soluble part of hexane extract (EtoHSHE), ethanol insoluble part of hexane extract (EtoHIHE), chloroform extract of deoiled kernel (CHEDK) and combination of CHEDK + HE (1:1) and CHEDK + EtoHSHE (1:1) were screened for their effects on egg hatching, oviposition, feeding and growth and development. All extracts at 2 percent concentration under choice condition significantly reduced egg laying by the moths on treated chickpea. The maximum deterrence was caused by CHEDK (90.4%). The order of effectiveness was : CHEDK + EtoHSHE > CHEDK > EtoHE > EtoHSHE > CHEDK + HE > HE + EtoHIHE > AE.

The percentage of egg hatched in different treatment ranged from 27.0 to 55.0 as against 61.0 to 66.0 in controls. EtoHSHE and CHEDK caused maximum inhibition in egg hatching. At 4 percent egg hatchability was found to range from 24.0 to 39.0 percent whereas it was 60.0 to 64.0 percent in controls. The order of activity was : EtoHSHE > CHEDK > CHEDK + EtoHE = EtoHE > AE > HE > HE + CHEDK > EtoHIHE. The percentage of feeding by 5, 10 and 15-day-old larvae on chickpea treated with 4 percent concentration of different extracts ranged from 6.5 to 28.0 (controls 29.0 to 33.5), 5.0 to 30.0 (controls 38.0 to 44.0) and 10.0 to 34.0 (controls 44.0 to 51.0), respectively. None of the extract could completely inhibit feeding by larvae. The efficacy of the extracts was as follows : CHEDK > EtoHSHE + CHEDK > EtoHE > AE > EtoHSHE > CHEDK + HE > EtoHIHE.

Five-day-old larvae when sprayed with 4, 2, 1 and 0.5 percent concentration of different extracts produced varied effects in different treatment. Significantly higher larval mortality (36.6 to 100%) was recorded in different treatments of the extracts in comparision with control mortality (6.6 to 13.3%). CHEDK, CHEDK + EtoHSHE and CHEDK + HE caused 100 percent mortality at 4 percent concentration. Extended larval period (20 to 45 days) were observed in these treatments and EtoHE. The percentage of pupal mortality in different treatments ranged from 6.1 to 33.4 as against 7.4 to 8.1 in controls. The highest pupal mortality was found in aqueous extract. EtoHIHE recorded highest percentage of adult emergence (30.3 to 56.7) while CHEDK the least (0.0 to 5.9) in different treatments. The adult emergence in control ranged from 79.3 to 85.4 percent. The overall effectiveness of different extracts was as follows : CHEDK > CHEDK + EtoHSHE > CHEDK + HE > EtoHE > EtoHSHE > AE > HE > EtoHIHE.

Mathur and Nigam (1996) studied the insecticide, antifeeding and juvenilizing effects of neem oil against *C. cephalonica* and *Epilachna vigintioctopunctata* F. The antifeedant effects of neem oil was mixed in 20 gram of wheat grains in each concentration. There were three replications and ten larvae were released in each replicate including control. After five days of treatment, the grains left in the tubes were weighed and the amount of consumed food was estimated. The antifeedant effect of the beetle pest was observed by dipping the brinjal leaves in various concentrations of the extract. The observation was recorded after 48 hrs of treatment. The juvenomimetic test in early instar larvae were noted with different stages of growth.

The results showed that the phytophagous insect, *E. vigintioctopunctata* was more susceptible than the stored product insect (*C. cephalonica*) in respect of the insecticidal value of neem oil in terms of LC50 values. The data for the antifeeding properties of neem oil clearly indicate that higher concentrations of neem oil, i.e. 4 and 5 percent offered the best results giving 100 percent antifeeding effects against *C. cephalonica*. Higher concentration of 5 percent neem oil when used against *E. vigintioctopunctata* gave maximum (89.16%) leaf area protection over control. However, lower concentration (4%) resulted in 16.24 percent protection over control. Neem oil has also elicited different juvenomimetic effects against both the test insects, infesting wheat grain and brinjal, respectively. The most spectacular morphogenetic effects appeared were 1-p-intermediates, deformed pupae and adults of the treated insects. Percentage inhibition of adult emergence was used as the criterion for evaluation of juvenomimetic efficacy of neem oil against the test insects.

Raman *et al.* (1996) developed neem formulations by successive treatment of industrial-grade neem oil to remove the major part of triglycerides and free fatty acids. This formulation was further processed to obtain a water soluble powder. Formulations were also extracted by solvents and the neem triterpene constituents were concentrated to yield a product. The final formulation is prepared by adding the above two formulations in the ratio of 95 : 5. The formulation Achook has azadirachtin (300 ppm) azadirachtin (500 ppm), nimbocinol and epinimbocinol. Bioassay was carried out on the larvae of *Helicoverpa armigera* and *Earias vitella* for antifeedant studies. It was revealed that Achook indicates antifeedant and growth disruptive effects on the larvae of two lepidopterous pests. It was also observed that at 1 percent concentration and above, 50 percent more larvae feeding inhibition was noticed in *E. vitella* and at 0.5 percent for *H. armigera*. The antifeedant effects and growth inhibition effects varied according to the concentration suggesting a dose-dependent response. Achook thus affects, larvae, pupae survival and adult emergence. It was observed that affected pupae either remained unhatched or often produced adults with deformed wings which normally do not survive. Field evaluation of the said neem formulation against the pests of major crops such as cotton, okra, brinjal, cabbage, jute, and tea were undertaken at different locations in the country. Results of the studies indicated significant reduction in pest population and associated crop damage, leading to increase in crop yields.

Saxena and Onyango (1996) studied the growth and development of maize stalk borer, *Busseola fusca* Fuller larvae on a semi-synthetic diet containing 10, 100, 500 or 1000 parts per billion (ppb) of 'NEEM bitters concentrate' (NBC) with 10 percent azadirachtin. Larval growth was significantly reduced and the development period prolonged on diet containing NBC at ≥ 100 ppb; at ≥ 500 ppb larval growth and development was totally inhibited. The fecundity of females developing on diet with NBC at 100 ppb significantly reduced if mated with normal males or with males reared on NBC-treated diet. Also the hatchability of eggs was reduced when females were reared on the diet containing NBC.

Bhiyan and Hossain (1996) studied the growth inhibiting effects of neem and sesame oils on jute hairy caterpillar, *Spilosoma* [*Diacrisia*] *obliqua* Walker. First- to 4th-instar larvae were fed on treated jute leaves with either 1-5 percent neem oil or 0.1-0.5 percent sesame oil alone or in combination of both. The results indicated serious growth inhibiting effects on the developing larvae depending upon the concentrations of the oils. Such effects were extended larval period (6-7 instars), reduction in larval weight and size, blackish spots on the larval bodies, hairless larvae, sluggish movement, slow feeding accompanied by high mortality in each moulting stage and development of abnormal pupae. It is found that most larvae fed jute leaves treated with 1-3 percent neem oil could survive to last instar with hairless and discoloured bodies and elongated larval period upto 20-28 days. Some larvae failed to complete the larval period or died before pupation. Very few of them pupated usually with abnormally low weight, or shrinked and cracked heads and bodies. At highher concentrations (4-5%) almost all larvae failed to survive to pupation and showed abnormalities in head and body with often transparent liquid oozing from the larval bodies.

Neem oil combined with sesame oil showed synergistic effects indicating higher mortality at early and late instars and occasionally very small number of larvae survived till pupation. Sesame oil alone did not show any remarkable growth inhibiting effect on jute hairy caterpillar. All these facts showed potential growth inhibitor in jute hairy caterpillar and can be integrated with appropriate cultural and biological pest control tactics.

Sharma *et al*. (1996) studied the oviposition deterrent and antifeedant activity of 2 formulations containing Neemrich I + oil of *Salvadora oleoides* and Neemrich I + neem extract (Plantmix I and Plantmix II, resp.) against

Phthorimaea operculella. Both formulations showed greater activity than their individual constituents. The oviposition deterrent activity of Plantmix I and II continued for 30 and 18 days, respectively, compared to 12 days for the individual constituents at the lowest concentration tested. Plantmix I had the highest antifeedant activity while Plantmix II was inactive. Nimbin was taken as a standard compound and did not have significant oviposition deterrent activity. It was concluded that Plantmix I is a considerable improvement to Neemrich I for preventing infestations of *P. operculella* on stored potatoes.

Desai and Parikh (1996) observed that neem-oil besed formulations (Nimbitor) shows excellent bio-effectiveness against insect pests of okra, brinjal (fruit and shoot borers), chickpea (pod borer), rice pest complex and cotton pest complex at concentration of 1-2 percent spray by reducing the crop damage and protecting the shoot, fruit, leaves, grains and cotton boll. Nimbitor helped to obtain increased yields and healthy produce. Nimbitor is compatible with synthetic insecticides and showed synergistic effect when used in combination. Laboratory studies have shown that Nimbitor has very good repellent activity against *Helicoverpa armigera* on castor leaves and *Spilosoma obliqua* on sunflower leaves.

Chitra and Rao (1996) studied the effect of certain plant extracts on the consumption and utilization of food by *Spodoptera litura*. The order of efficiency of plant extracts with respect to reducing food consumption, faeces excertion, weight gain of *S. litura* was neem seed kernel extract (NSK) (5%), *Nerium odorum* (5%) and *Argemone mexicana* (1%). Food consumption, faeces and weight gain increased with the advancement of larval age. Consumption index (CI), growth rate (GR) and approximate digestibility (AD) showed a negative relationship with larval age while efficiency of conversion of ingested food (ECI) and efficiency of conversion of digested food (ECD) to body substance were found to fluctuate between increase and decrease. In all the instars, CI and AD values were found to be high in NSK (5%) while these were low in the case of *A. mexicana* (1%). A reverse trend was, however, observed in case of GR, ECI and ECD.

Singh *et al.* (1996) tested an aqueous extracts of neem seed kernel at 0.5 percent for the control of *Papilio demoleus*, a pest of the medicinal tree *Aegle marmelos* in Karnataka, India. Spraying seedlings twice with an interval of 8 days was effective in providing protection against the pest. Although low mortality (23.3%) was reported, the extract had strong antifeedant and repellent effects and was effective as a moulting inhibitor.

Bomford and Isman (1996) studied desensitization of 5th-instar larvae of

S. litura to azadirachtin and neem. The deterrence of azadirachtin, in its pure form and as a consitituent of neem seed extract, to 5th-instar of pest. was measured using cabbage (*Brassica oleracea* var. *capitata*) leaf disc assays. Paired-choice assays, in which larvae could choose between feeding on a treated (1.3 μg azadirachtin/cm^2 leaf area) or an untreated leaf disc for 2 hrs, were conducted at 24 hrs intervals throughout the 5th-instar. In additon, no-choice assays, in which larvae could feed on only one leaf disc (10 μg azadirachtin/cm^2 leaf area) for 1.5 hrs, were conducted consecutively over a 6 hrs period at the beginning of the 5th-instar. After repeated exposures, larvae became desensitzed to pure azadirachtin in both choice and no-choice tests, but did not desensitize to neem containing the same absolute amount of azadirachtin in choice tests. Hunger was responsible for approximately one-third of the desensitization response in the no-choice tests. Sensitivity to azadirachtin was independent of age within the 5th-instar.

Lowery et al. (1996) studied the effects of neem on the growth and development of the oblique banded leaf roller *Choristoneura rosaceana*. Neem (*A. indica*) seed oil (NSO) added to meridic diet at 0.016 percent concentration reduced pupation and prevented adult eclosion. At a rate of 0.0016 percent, NSO reduced the fitness, resulting in longer developmental times, lower adult eclosion rates, and reduced egg production compared with controls. The neem insecticide Margosan-O produced comparable results based on concentration of the most biologically active constituent, azadirachtin. Pupation was completely inhibited at concentration of 0.25 and 1 percent for larvae exposed in the 5th- or 6th-instar, respectively; rates as low as 0.016 percent reduced pupal weights and adult eclosion rates. For larvae transferred to treated diet in the 5th-instar, physical abnormalities in the wings of adults occurred at a rate of 0.004 percent NSO and increased with increasing treatment rates. NSO at concentration as high as 2.0 percent was not antifeedant to the larvae, based on 24 and 48 hrs choice test bioassays.

Koul *et al*. (1996) studied the effect of neem allelochemicals on nutritional physiology of larvae of *S. litura*. The neem allelochemicals, azadirachtin, salannin, nimbinene annd nimbin were administered to 1st-, 4th-, 5th- and 6th-instar larvae of the noctuid, orally in an artificial diet, topically or via injection in the laboratory at 26°C, 65-75 percent RH and LD 16:8. Nutritional analysis revealed strong antifeedant and growth regulatory effects of azadiracthin which were independent of each other. While salannin and nimbinene induced concentration-dependent feeding deterrence only; nimbin was inactive to the 1000 ppm level. Salannin and nimbinene, however, did not interfere with the trypsin activity of the gut. These results and those from nutritional studies suggest that salannin and nimbinene have no toxicity mediated effects on *S. litura* larvae and antifeedant activity is a result of the effects on deterrent and other chemoreceptors.

Alrubeai *et al*. (1997) tested the efficacy of some plant extracts and chemicals insecticides on *Phyllocnistis citrella*. Spraying citrus plants with neem (*A. indica*) extracts or Sibahbah seed powder repelled *P. citrella* adults. The insecticidal effect on larvae was 16 percent for neem and 90 percent for Sibahbah.

Mayabini (1997) compared the efficacy of neem products against rice leaf folder, *Cnaphalocrosis medinalis* guenee. The efficacy of neem bark decoction, neem-based chemicals and neem derivatives (neem oil, leaf extracts and leaf decoction) against *C. medinalis* was studied in the net house and laboratory. All were applied as foliar sprays to pot-grown rice plants. Leaf samples were collected and laced in petridishes with ten-fourth instar larvae of the pest. Leaf area fed by the larvae was recorded after 48 hrs. Data were also recorded on pupation rates. Neem bark decoction appeared to be a very effective botanical for controlling the rate of feeding and reducing the rate of pupation.

Kulkarni *et al*. (1997) studied feeding inhibition properties of some plant extracts against Sessoo defoliator, *Plecoptera reflexa*, a major pest of *Dalbergia sissoo* in India. The antifeedant properties of 4 plant extracts were tested using a bioassay technique in the laboratory against *P. reflexa*. The extracts were prepared from leaves of *Aloe vera* [*A. barbadensis*], *Ipomoea carnea* and *Lantana camara* and methanol extract from seeds of *Azadirachta indica*. All the extracts showed some antifeedant activity, but extracts of seeds of *A. indica* (neem) and leaves of *I. carnea*, both at a concentration of 0.5 percent were very effective, giving 64.37 and 70.44 percent leaf protection, respectively.

Shukla *et al*. (1997) studied the effects of some plant odours in the breeding environment on fecundity and hatching of okra shoot and fruit borer, *Earias vittella* Fab. under laboratory conditions. Adult pairs of males and females of *E. vittella*, a pest of cotton and okra, were released in breeding chambers in different sets, containing the odours of the leaves of neem, *Azadirachta indica*; tulsi, *Ocimum basilicum*; eucalyptus, *Eucalyptus rostrata*; lantana, *Lantana camara*; bulbs of garlic, *Allium sativum* and one control set with no odour. Adult longevity did not differ significantly amonng the treatments. All the treatments significantly reduced the egg output as compared to the control. Similarly, all the odours significantly reduced egg hatching compared to the control. The lowest number of eggs (128) and hatching (68.15%) were recorded with *Azadirachta* leaves odour.

Sannaveerappanavar and Viraktamath (1997) studied the IGR activity of aqueous neem seed kernel extract (NSKE) against 3rd- and 4th-instar larvae of diamond moth, *Plutella xylostella* L. in the laboratotry. The NSKE (2-4%) caused complete mortality within 5-11 days. The majority of 3rd-instar larvae died without moulting and all larvae failed to pupate. At lower concentrations of NSKE (0.5-1%), lower mortality was observed, though several surviving larvae either failed to pupate or died in the pupal stage. At all concentrations of NSKE, there was a prolongation of the larval period, and death occurred at the time of moulting.

Kumar *et al.* (1997) observed the antifeedant and growth inhibiting effects of neem leaf exudate on *S. litura*. Young reddish neem leaves were collected and soaked in cold acetone. Various concentrations viz. 1.0, 2.5, 5.0 and 10.0 percent were taken for the study. Antifeedant activity of leaf exudate was assayed by using leaf disc no-choice test. The larval mortality was recorded at different concentrations after 48 hrs. The exudate caused tremendous antifeedancy, ovipositional deterrency and larval mortality in *S. litura* in a dose-dependent manner. At 10 percent, treatment pronounced antifeedancy (94.7%), ovipositional deterrency (91.5%) and mortality (87.9%) were observed. Extended larval and pupal durations and decreased oviposition period, adult longevity and fecundity were evident in *S. litura* after the treatment.

Gujar (1997) studied the biological effects of azadirachtin and plumbagin on *Helicoverpa armigera*. Azadirachtin from the neem tree (*A. indica*) and plumbagin, 2-methyl 5-hydroxy, 1,4-naphthoquinone from the medicinal plant *Plumbago* spp., were investigated for their biological effects on the pest. The 2nd-instar larvae (5.31 ± 1.55 mg body weight) were topically treated with different doses of azadirachtin formulation dissolved in acetone. The insects were then allowed to feed. Individually an artificial diet and observations were taken after 24 hrs. In another trial, a sublethal dose of 1 µg of azadirachtin was topically treated.

Azadirachtin sample SPIC SSF 22 was effective against 2nd-instar larvae. Treatment of 2nd-instar larvae with a sublethal dose of 1 µg/insect led to cosideratble growth inhibition for a period of time after which the insects died. Similarly, treatment of final instar larvae with different doses of azadirachtin led to a proportionate inhibition of larval growth, subsequently also affecting pupation. Plumbagin was not as effective as azadirachtin.

Aerts and mordue (1997) studied feeding deterrency and toxicity of neem triterpenoids against several pests. The biosynthesis of azdirachtin starts with a steroid precursor and proceeds via two levels of structural compexity, i.e., furan ring formation and C-ring opening, after which further modifications yield azadirachtin. Antifeedant and toxic properties of azadirachtin against insects were compared with those of various compounds from the two lower levels of structural complexity. To larvae of *Spodoptera litteralis*, several compounds from the lower structural levels showed strong antifeedant activity, comparable to that of azadirachtin. Nymphs of *Schistocerca gregaria* in antifeedant tests were more sensitive to azadirachtin than to the other compounds tested.

Katti and Saxena (1997) tested some neem formulations for antifeedant effect against semilooper, *Autographa nigrisigna* Walker in chickpea. The products used were Repelin (1%), Neemguard (2%), neem EC (0.15%0, neem 25 WDP (0.15%) and neem oil (0.6%). The antifeedant effect was worked out based on percentage pod feeding. The results presented revealed that all the neem treatments showed significant antifeedant effect in comparision to control, after 24 hrs of feeding by *A. nigrisigna*. Repelin, neemguard and neem EC treatments showed nil percent and pod feeding and were superior to neem WDP (13.3%) and neem oil (15.0%) treatmets which were on par with each other. The control treatment showed 21.7 percent feeding by *A. nigrisigna*. After 48 hrs of feeding by *A. nigrisigna* in repelin treatmnet while neemguard and neem EC treatments showed 1.7 and 3.3 percent pod feeding, respectively. These treatments were however on par with one another and superior to neem WDP and neem oil treatments which did not show any further increase in pod feeding by *A. nigrisigna*.

Behaera and Satapathy (1998) tested eight indigenous plant extracts at concentrations of 10.0, 5.0, 2.5 and 1.25 percent to assess their effects on growth and development of *Spodoptera litura*. Indian laurel (*Laurelia* sp.) and neem (*A. indica*) applied topically, and karanj (*Pongamia pinnata*) applied by the leaf dip method, prolonged the duration of the larval period significantly at all concentrations. Sweet flag (*Acorus calamus*) induced the highest percentage abnormalities due to morphogenetic effects both by leaf dip (38.6%) and topical (37.3%) application, followed by neem applied by leaf dip (33.3%), Indina laurel (33.3%) and karanj (33.3%), both applied topically.

Kumar and Babu (1998) studied the ovipositional deterrent effects of two commercial neem formulations, NeemAzal-T/S (1% azadirachtin) and Neem Azal-F (5% azadirachtin) against *Leucinodes orbonalis* showed that Neem Azal-F @ 1 ml/lt to be the most effective treatment. Repellency was found to be dose dependent. Investigations on their ovicidal effects against the eggs of *L. orbonalis* and *Henosepilachna vigintioctopunctata* [*Epilachna vigintioctopunctata*] in comparision to endosulfan, revealed the superiority of endosulfan in both cases.

Kulkarni (1998) conducted a study to test the feeding deterrence properties of methanolic leaf extracts of *Cassia tora, Ipomoea carnea* subsp. *fistulosa, Annona squamosa* and seed extract of neem (*A. indica*) against larvae of the poplar pest *Clostera cupreata* under laboratory conditions. Feeding was recorded on treated and untreated discs (control); in choice conditions (treated and untreated discs); and on leaf discs treated with each extract kept separately. The results indicated that leaf extracts of *A. squamosa* and seed extracts of neem were the most effective deterrent, and that *I. carnea* subsp. *fistulosa* was the least effective deterrent.

Datta and Saxena (1998) compared antifeedant activity of azadirachtin-A and parthenin against *S. litura* by short term dual choice and no-choice tests. Concentration dependent antifeedancy was noticed only in dual choice tests. Azadirachtin-A was considerably more effective than parthenin in causing antifeedancy.

Bajpai and Sehgal (1998) studied growth and development of neonate (24 hrs old) larvae of *Helicoverpa armigera* on semisynthetic diets treated with neem, karanj and tobacco formulations in the laboratory conditions. High larval mortality, >50%, was observed in the diets treated with benzene extracts (0.15 and 0.2%), ethyl acetate and butanol extracts (0.05 to 0.2%), methanol and chloroform extracts (0.025 to 0.2%), NSKWE (3 to 10%), Green mark and neem guard (0.4 to 0.8%), neem oil and karanj oil (2 to 6%) and nicotine sulphate (0.3 to 0.4%). Longer developmental period, with> 25 days, was observed with treatments of benzene extracts (0.025 to 0.15%) and karanj oil (1 to 4%). Strong antifeedant, high IGR and maximum toxicity was observed, with success index < 0.3, in diets treated with benzene extract (0.15%), NSKWE (3%), Green mark and neem guard (0.2%), neem oil (1%) and karanj oil (4%) in neonate larvae.

Bajpai and Sehgal (1998) studied the ovipositional deterrent effects of neem, karanj and tobacco formulations against the female moths of *H. armigera*. The chickpea plants treated with NSKE (chloroform extract, 0.15 and 0.2%; methanol extract, 0.1 to 0.2%; water extract, 7 and 8%); neem oil (6%) and nicotine sulphate (0.3 to 0.4%) strongly inhibited the oviposition as females laid significantly less number of eggs on treated plants compared to untreated plants. The commercial neem products, Green mark, neem guard and Achook from 0.1 to 0.8 percent and karanj oil 0.5 to 6 percent could not deter the female moths effetively.

Ghorpande and Thakur (1998) conducted field experiments during kharif seasons of 1992-93 to determine the relative efficacy of different plant products against sesamum leaf roller and capsule borer, *Antigastra catalaunalis*

Dup. All the 7 plant products viz. neem leaves and kernel extract, tobacco decoction, neem oil, karanj extract, mahua oil, thumba oil, margocide 80EC and 20 EC included in trials were found to be effective against *A. catalaunalis*. However, thumba oil (20 ml/lt), neem oil (10 ml/lt) and margocide 20EC (5 ml/lt) gave the best results both in respects of low degree of pest infestaion and higher seed yields.

Kulkarni and Joshi (1998) tested the antifeedant property of methanolic extracts of the leaves of four plant species and the seeds of *Azadirachta indica* (neem) against the maharukha webworm, *Atteva fabriciella* (a pest of Ailanthus). The extracts were compared with amrutguard (neem-based commercial product). The seed extracts of *A. indica* as well as leaf extracts of *Annona squamosa* and *Lantana camara* var. *aculeata* were very effective in protecting foliage. Absolute antifeedancy was exhibited by the leaf extract of *A. squamosa* and amrutguard at 2 percent as well as seed extract of neem at 3 percent.

Ocete *et al.* (1998) tested the antifeedant effects fo neem extracts against *Ocnogyna baetica* Ramb. The antifeedant activity of doses of 14.33 $\mu g/cm^2$ and 7.16 $\mu g/cm^2$ of an extract of neem were evaluated against larvae fo the 4th-instar of *O. baetica* using leaf discs of *Beta vulgaris* as food. Results showed that there was a statistically significant difference between the average consumption of dry weight of food/larva in the case of the disks treated with the higher dose compared with the control.

Rathore and Yadava (1998) studied non-preference of different plant extracts against the bihar hairy caterpillar, *Diacrisia obliqua* Walker, a serious pest of pulses crops in India. Late stage larvae are voracious feeders and difficult to control with insecticides. To study the non-feeding preference, the leaf extracts of neem (*A. indica*), Kaner (*Thevesia nerifolia*) and bulbs of garlic (*Allium sativus*) were tested against 4th-instar larvae. The respective plant parts were ground in a mixie without adding water and filtered. This was considerred as 100 percent solution. The filtrate was used in 100, 1, 0.5 and 0.25 percent concentrations. The castor leaves treatd with above concentrations were provided to the larvae to feed for 4 hrs at the rate of 3 larvae/petridish of 10 cm and replicated thrice. The test larvae were starved for 1.5 hrs before release for feeding. The leaf area in each treatment was measured before and after the experiment with the help of leaf area eaten was determined.

The extract of all the three plants significantly inhibted the feeding to the tune of 46.2, 32.2 and 48.9 percent in neem, kaner and garlic, repectively across the concentrations. The most effective concentration inhibiting the feeding was 100 percent but there was no significant difference between 100 percent and 1 percent. The reduciton at 100 percent concentration was 62.1, 28.1 and 71.6 percent in neem, kaner and garlic, respectively. There was no significant prefernce in consumption between the leaves treated with neem and garlilc. Neem is already known to be endowed with antifeeding property against many insect species.

Govindachari *et al.* (1998) studied the antifeedant and insect growth regulating activities of salanin, nimbin, and 6-deacetylnimbin, in comparision with azadirachtin-A on *Spodoptera litura*, *Pericallia ricini* and *Oxya fuscovittata*. Salanin deterred feeding, delayed moulting by increasing larval duration, caused larval and pupal mortalities, and decreased pupal weights into the two lepidoptera. Salannin also caused moult delays and nymphal mortalities in *O. fuscovittata*.

Aerts and Mordul (1998) studied feeding deterrence of neem. The neem tree (*A. indica*) produces a plethora of triterpenoids. One of these is azadirachtin, a complex compound with powerful insecticidal properties. Antifeedant and toxic properties of azadirachtin against insects were compared with those of various compounds from the two lower levels of structural complexity. To larvae of *Spodoptera littoralis*, several compounds from the lower structural levels showed strong antifeedant activity, comparable to that of azadirachtin. Nymphs of *Schistocerca gregaria* in antifeedant tests were more sensitive to azadirachtin than to the other compounds tested. To *S. littoralis, S. gregaria* and *Oncopeltus fasciatus*, azadirachtin was highly toxic, exerting severe growth and moult disrupting effects. In contrast, compounds from the lower structural levels than azadirachtin showed virtually no toxicity to these three insects. Thus, although several simpler triterpenoid strutures other than azadirachtin have insect antifeedant activity, the combination of strong antifeedant and toxic properties of azadirachtin may provide the more important chemical protection of the neem tree.

Srivastava and Mathur (1998) studied insect growth regulator effect of neem formulations against sesamum pod borer, *A. catalaunalis*. Six formulations viz. Bioneem, nemazol, Econeem, Achook, Nemata and Nimbicidine at 1 percent and 0.5 percent dosages of each were tested against last instar larvae of *A. catalaunalis*. Fresh and healthy leaves of sesamum were dipped in desired dosages of neem formulations and dried under shade. The treated leaves were provided to larvae for feeding. In control, the larvae were fed on leaves dipped in distilled water only. The results demonstrated that all the neem formulations manifest three types of morphological deformities viz. larval-pupal intermediate (L-P intermediates), larvae form pupae and deformed adults. Amongst the compounds tested Neem Azal proved to be the best followed by Bioneem. Insect growth regulatory (IGR) effectiveness was assessed on the basis of percentage inhibition of adult emergence. As high as cent percent inhibition of adult emergence was recorded in Neem Azal (1%) and it was closely followed by Bioneem 1 percent (90%). Higher dose (1%) gave better performance in all the formulations evaluated as compared to lower dose (0.5%). The sequence of effectiveness may be expressed as Neem Azal > Bioneem > Achook > Nimbicidine > Neemata > Econeem. In control, no IGR and mortality effects were noticed. The present findings suggest that Neem Azal and Bioneem may effectively be used in the management of *A. catalaunalis* infesting sesamum.

Jhala *et al*. (1998) studied antifeedant and larvicidal efficacy of plant extracts viz. ardusa (*Allanthus excelsa* Roxburgh) leaves extract (AALE), ardusia (*Adhatoda asica* L.) leaves extract (AILE), castor (*Ricinus communis* L.) leaves extract (CLE), castor seed extract (CSE), damra (*Ocimum acilicum* L.) leaves extract (DLEO), pudina (*Mentha piperita* L.) leaves extract (FLE), naffatiya (*Ipomea cornea* L.) leaves extract (NFLE), neem (*A. indica*) leaf extract (NLE), neem kernel extract (NKE) and tulsi (*Ocimum sanctum* L.) leaf extract (TLE) in the laboratory against cotton leaf roller, *Sylepta derogata* Fabricius. Based on percent reduction in leaf area consumption over control after three days of feeding by the larvae, the order of these plant extracts for their antifeedant efficacy was : NLE (86.45%) > NKE (82.72%) > NFLE (63.87%) > CLE (59.50%) > CSE (58.78%) > AALE (57.35%) > DLE (54.05%) > TLE (52.54%) > AILE (51.97%) > FLE (29.10%). While the order of these botanicals based on percent larval mortality after 4 days of feeding by the larvae on treated leaves was : NKE (66.71%) > NLE (43.28%) > AALE (39.96%) > DLE = CLE (33.23%) > FLE (29.97%) > NFLE (23.15%) > CSE (19.98%) > AILE = TLE (13.00%).

Babu (1998) studied the combined activity of neem and china berry extracts on feeding and oviposition of *S. litura* Fab. Antifeedant activity of NSKE, neem bark extract (NBE), chinaberry seed kernel extract (CSKE) and chinaberry bark extract (CBE) were assayed using a leaf disc choice test. The index of feeding deterrency was calculated as (C-T)/(C+T)x100, where C is the consumption of control discs and T is the consumption of treated leaf discs.

Oviposition deterrent bioassay was conducted and the deterrent activities of neem and chinaberry extracts were compared by calculating oviposition deterrence index (ODI), i.e. ODI = 100 (Control - Treated) / (Control + Treated). All data were subjected to analysis of variance and means were separated by using Duncan's multiple range test. In the present study 5th-instar larvae were strongly deterred from feeding on leaf discs treated with

NSKE followed by CBE and CSKE. Whereas in combined treatments of NSKE + CBE and NSKE + CSKE, 100 percent feeding deterrency was observed (Table 1). The present study clearly indicated that neem seed extracts alone had higher antifeedant effect against *S. litura* larvae than chinaberry extracts. The 100 percent antifeedant activity as expressed by combined treatment of NSKE + CBE and NSKE + CSKE, may be due to a synergistic interaction amongst the active compounds. The number of eggs laid on treated plants, decreased in a dose dependent manner. The 100 percent ovipositional deterrncy was seen in combined treatment NSKE + CBE. The NSKE when conbined with CBE and CSKE caused significant reduction in feeding and oviposition of *S. litura*.

Table-1 : Effect of neem and chinaberry extracts on feeding and ovipositional deterrency of *S. litura*

Treatment	Concentration (percent)	Feeding deterrency index (%)	Ovipositional deterrency index (%)
NSKE	1	46	37
	2	65	52
	4	89	76
CSKE	1	35	25
	2	49	44
	4	76	70
NBE	1	23	14
	2	37	23
	4	62	43
CBE	1	38	25
	2	56	41
	4	82	64
NSKE+CSKE	1	61	53
	2	82	75
	4	100	98
NSKE+NBE	1	52	40
2	69	58	
4	93	81	
NSKE+CBE	1	68	58
2	90	84	
4		100	100

(Within a column, means followed by a common letter(s) are not significantly differentat the 5% level by DMRT (Table obtained from Babu, 1998)).

Gupta *et al*. (1998) studied the impact of sunlight on the antifeedant effect of neem oil against *Helicoverpa armigera*. The experiment was laid out in a factorial completely randomised design with two treatments at three level viz. crude neem oil (1,3,5%) and various exposure time to sunlight (0,3,6 hrs) and was tested against *H. armigera* in laboratory (temperature 27±1°C, RH 65±5 percent and scoto : photo period 16:8).

On a clear sunny day, castor leaves on plants itself were treated with different concentrations of crude neem oil (1,3,5%) in morning hours. These treated leaves were plucked at different intervals after exposure to varied degree of sunlight (0,3,6 hrs). The intensity of sunlight was recorded regularly at half hourly intervals. These treated leaves were fed to freshly moulted 3rd-instar larvae of *H. armigera* for 48 hrs and thereafter fed with untreated castor leaves for 48-96 hrs. The leaf area of treated and untreated leaves before and after feeding was recorded with leaf area meter (LI-COR;LI-3000). Results show that the antifeedant effect of neem oil (1,3 and 3,5%) against test insect were at par (20.92, 18.03 and 18.03, 16.21 cm^2 leaf area/larva, resp.). However, 13.70, 18.15 and 23.31 cm^2 leaf area was consumed by larva when neem oil was exposed to sunlight for 0,3 and 6 hrs, respectively. The antifeedant effect of all the concentrations of neem oil at 0 hr exposure to sunlight were at par (14.83, 12.40 and 13.87 cm^2 leaf area/larva), whereas the effect of neem oil was reduced significantly after 3 and 6 hrs exposure to average sunlight intesity of 63,983 and 64,758 lux, respectively. No significant difference was observed in feeding efficiency of test insect on untreated castor leaves. It has been also observed that the test insect consumed larger area when fed on untreated leaves (av. 31.82 cm^2) during post exposure period than that of treated leaves (av. 18.38 cm^2). It shows that sunlight plays a crucial role and affects the stability of neem oil which inturn reduced its antifeedant effect against the test insect, *H. armigera*.

Patel and Patel (1998) studied oviposition-deterrent effect of botanical material on tobacco leaf eating caterpillar, *S. litura*. Pot plants of *Ricinus communis* were sprayed with leaf extracts (2 and 4%) of neem (*A. indica*), *Ailanthus excelsa* and *Nerium indicum*, and NSKE (2 and 4%) to investigate its deterring effects on oviposition of *S. litura*. All the plant extracts proved effective in reducing the oviposition by *S. litura*. However, among all Grownim (0.2 and 0.4%) proved highly deterrent to oviposition, Neemark (2%) and seed kernel suspension of neem (4%) also proved equally effective as Grownim but higher than the rest of the botanicals. Suspensions of neem leaf extract, leaves extract of yellow flower and leaves extract of *Obandor ardusa* proved equally deterrent to its oviposition. Suspension of neem seed kernel (2%) was found more effective than *O. ardusa* leaves (2%). It was also observed that all the concentrarions at lower (2%) and higher (4%) remained equally ineffective.

Chanda and Chakravorty (1998) studied the effect of neem oil on the development of rice moth, *Corcyra cephalonica*. Larvae were collected from the stock culture maintained in the laboratory. The antifeedant azadirachtin (0.03%) was treated as nimbecidine. The doses taken were 0.25, 0.50, 0.75 and 1.00 ml per 20 g seeds. The observation revealed normal, deformed and non-emerged adults. The deformed and non-emerged adults had crumpled or twisted wings. It was also observed that last instar larvae, pupae and adults from different treatments and control varied significantly with respect to their length, body weight and wing span. Thus the Nimbecidine (0.03%) is an effetive growth and development inhibiting chemical when taken with food by newly hatched to mature larvae of *C. cephalonica* (table-2).

Table-2 : Data on growth, duration of development and emerged morphs due to feeding in different doses of neem-food by the larvae of C. cephalonica (initial population of first instar larva (n=150)

Treatment (ml of Nimbecidine per 20 g of food)	Ultimate resultant morphs (% in parentheses)						Mortality %	Duration Mean±SE in days (range)
	Apparently normal adult		Deformed adult		Non-emerged adult			
	Male	Female	Male	Female	Male	Female		
0.25	32	38	8	3	4	—	65	44.22±8.81
	(21.33)	(25.33)	(5.33)	(2.00)	(2.67)	—	43.33	(35-72)
0.50	31	16	5	3	—	—	95	48.47±9.25
	(20.67)	(10.67)	(3.33)	(2.00)	—	—	63.33	(38-82)
0.75	22	10	3	3	—	—	112	54.39±11.52
	(14.67)	(6.67)	(2.00)	(2.00)	—	—	74.67	(45-90)
1.00	20	12	3	2	—	—	113	55.05±12.65
	(13.33)	(8.00)	(2.00)	(1.33)	—	—	75.33	(38-91)
Control	48	61	—	—	—	—	41	41.01±5.17
	(32.00)	(40.67)	—	—	—	—	27.33	(32.52)

(From Chanda and Chakravorty, 1998)

Sahayaraj (1998) studied the antifeedant effect of some plant extracts on the asian armyworm, *Spodoptera litura*, a pest in groundnuts. Plant extract of neem (*A. indica*), oranges (*Citrus sinensis*), *Vitex nigundo* and ginger (*Zingiber officinale*) were evaluated for their antifeedant and growth inhibitory activities against last instar larvae of *S. litura*. for evaluation of antifeedant effects of botanicals on food consumtion and assimilation in the laboratory, groundnut shoots were dipped in different extracts for 15 mts. Ten final instar larvae were placed in plastic trays and replicated 6 times. Deterrent effect were found in all plant extracts. The strongest deterrent effect was found in *V. negundo* and was indicated by very low food consumption and digestibility, faecal pellets production, and reduced body weight. The percentage of deformity in pupal as well as adult stages was the highest in *Z. officinale*, followed by *A. indica* and *C. sinensis*. All plant extracts showed antifeedant action and growth inhibiton.

References

Aerts, R.J. and Mordue, A.J. (1997). Feeding deterrence and toxicity of neem triterpenoids. *Journal of Chemical Ecology*, **23**(9) : 2117-2132.

Aerts, R.J. and Mordue, A.J. (1998). Feeding deterrence and toxicity of neem triterpenoids. *Neem Newsletter*, **15**(4) : pp. 45.

Ahmed, S.R. and Bhattacharya, A.K. (1991). Growth inhibitory effect of some plants for *Spilosoma obliqua* Walker. *Indian Journal of Entomology*, **53**(3) : 453-474.

Alrubeai, H.; Jborry, I.; Hafied, A.; Hasson, B. and Jabbo, N. (1997). Efficacy of some plant extracts and chemical insecticides on *Phyllocnistis citrella*. *Arab and Near East Plant Protection Newletter*, **25** : pp. 28.

Ayyangar, G.S.G. and Rao, P.J. (1990). Neem (*Azadirachta indica* A. Juss) extracts as larval repellents and ovipositional deterrents to *Spodoptera litura* (Fabr.). *Indian Journal of Entomology*, **51**(2) : 121-124.

Ayyangar, G.S.G. and Rao, P.J. (1993). Carry over effects of azadirachtin on development and reproduction of *Spodoptera litura* (Fabr.). *Annals of Entomology*, **9**(2) : 55-57.

Babu, R. (1998). Combined activity of neem and Chinaberry extracts on feeding and oviposition of *Spodoptera litura* Fab. *Neem Newsletter*, **15**(3) : 25-27.

Bajpai, N.K. and Sehgal, V.K. (1998). Effect of botanicals on oviposition behaviour of *Helicoverpa armigera*. *Proceedings of National Seminar in Entomology in 21st Century* held at Udaipur, April 30-May 2, 1998: pp. 137.

Bajpai, N.K. and Sehgal, V.K. (1998). Growth and development of *Helicoverpa armigera* Neonate larvae on diets treated with neem, karanj and tobacco formulations. *Proceedings of National Seminar in Entomology in 21st Century* held at Udaipur, April 30-May 2, 1998 : pp. 136.

Bandara, K.A.N.P. and Kudagamage, C. (1991). Potential of chitin inhibitors and botanical insecticides for the control of cabbage caterpillars. *Tropical Agriculturist*, **145** : 125-127.

Behera, U.K. and Satapathy, C.R. (1997). Effect of indigenous plant extracts on growth and development of *Spodoptera litura* Fb. *Environment and Ecology*, **15**(1) : 12-16.

Bhandari, R.S.; Jia Lal; Ayyar, K.S. and Pratap Singh (1988). Effect of neem seed extractives on poplar defoliator *Pygaera cupreata* Butler (Lepidoptera : Notodontidae) in laboratory. *Indian Forester*, **114**(11) : 790-795.

Bhathal, S.S.; Darshan Singh and Singh, D. (1994). Feeding deterrency of Neemark against third instar larvae of hairy caterpillar, *Spilosoma obliqua* Walker. *Journal of Entomological Research*, **18**(4) : 387-389.

Bhiyan, N.I. and Hossain, M.M. (1996). Neem and sesame oils : Potential growth inhibitors of *Spilosoma* (*Diacrisia*) *obliqua* Walker [Lepidoptera : Arctiidae]. *Neem Newsletter*, **13**(4) : 49-50.

Blaney, W.M.; Simmonds, M.S.J.; Ley, S.V.; Anderson, J.C. and Toogood, P.L. (1990). Antifeedant effects of azadirachtin and structurally related compounds on lepidopterous larvae. *Entomologia Expermentalis et Applicata*, **55**(2) : 149-160.

Bomford, M.K. and Isman, M.B. (1996). Desensitization of fifth instar *Spodoptera litura* to azadirachtin and neem. *Entomologia Experimentalis et Applicata*, **81**(3) : 307-313.

Chadha, S.S. (1977). Use of neem (*Azadirachta indica* A. Juss.) seed as a feeding inhibitor against *Antigastra catalaunalis* Dupon. (Lepidoptera : Pyralidae) a sesame (*Sesamum indicum* L.) pest in Nigeria. *East African Agricultural And Forestry Journal*, **42**(3) : 257-262.

Chandra, S. and Chakravorty, S. (1998). Food with neem oil affects life and development of rice moth, *Corcyra cephalonica* (Stainton) (Lepidoptera : Pyralidae). *Entomon*, **23**(2) : 153-156,

Chari, M.S. and Muralidharan, C.M. (1985). Neem (*Azadirachta indica* Linn.) as feeding deterrent of castor semilooper (*Achaea janata* Linn.). *Journal of Entomological Research*, **9**(2) : 243-245.

Chari, M.S.; Ramaprasad, G.; Sitaramaiah, S. and Murthy, P.S.N. (1993). Bio-efficacy of neem formulations against *Spodoptera litura* F. in tobacco nurseries. *Botanical Pesticides in Integrated Pest Management* : 145-153.

Chitra, K.C. and Ramakoteswara Rao, S. (1996). Effect of certain plant extracts on the consumption and utilisation of food by *Spodoptera litura* (Fab.). *Journal of Insect Science*, **9**(1) : 55-58.

Chiu, S.F. and Zhang, Y.G. (1984). Effects of some plant materials of Meliaceae on fifth instar larvae of *Spodoptera litura* as feeding inhibitors. *Neem Newsletter*, **1**(3) : 23-24.

Cobbinah, J.R. and Osei Owusu, K. (1988). Effects of neem seed extracts on insect pests of eggplant, okra and cowpea. *Insect Science and its Application*, **9**(5) : 601-607.

Datta, S. and Saxena, D.B. (1998). Parthenin and azadirachtin-A as antifeedants against *Spodoptera litura* (Fab.). *Neem Newsletter*, **15**(1) : 7.

Desai, A.C. and Parikh, K.M. (1996). Development of neem-based insecticide formulation "nimbitor" and its bioeffectiveness against vegetable and agriculture crops and ectoparasites of cattle. *Neem Newsletter*, **13**(3) : 38.

Devaraj Urs, K.C. and Srilatha, G.M. (1990). Antifeedant and repellent properties of certain plant extracts against the rice moth, *Corcyra cephalonica* St. *Proceeding Symposium on Botanical Pesticides in IPM* held at Rajahmundry : 159-165.

Devaraj Urs, K.C. and Srilatha, G.M. (1993). Antifeedant and repellent properties of certain plant extracts against the rice moth, *Corcyra cephalonica* St. *Botanical Pesticides in Integrated Pest Management* : 159-165.

Dreyer, M.; Pavis, C. (ed.) and Kermarrec, A. (1990). Neem : a promising natural insecticide for small scale vegetable producers in the Dominican Republic. *Rencontres Caraibes en lutte biologique Guadeloupe*, 5-7 November, 1990, Cristobal, Aptdo 21064, Santo Domingo, Dominican Republic.

Dreyer, M.; Schmutterer, R. (ed.) and Ascher, K.R.S. (1987). Field and laboratory trials with simple neem products as protectants against pests of vegetables and field crops in Togo. *Proceddings of the 3rd International Neem conference*, Nairobi, Kenya, 10-15 July, 1986 : 431-447.

Dubey, O.P.; Odak, S.C. and Gargava, V.P. (1991). Evaluation of anti-feeding properties of indigenous medicinal plants against the larvae of *Heliothis armigera* (Hubner). *Journal of Entomological Research*, **15**(3) : 208-211.

Durairaj, C. and Venugopal, M.S. (1994). Antifeedant effect of Neem-Azal F against cotton leaf worm *Spodoptera litura* (Fb.), (Noctuidae : Lepidoptera). *Neem Newsletter*, **11**(3) : 34-35.

Durairaj, C.; Asokan, G. and Venugopal, M.S. (1991). Effect of three plant products on feeding of sapota leaf-webber (*Nephopteryx eugraphella*)

(Lepidoptera : Phycitidae). *Indian Journal of Agricultural Sciences*, **61**(3) : 233-234.

Facknath, S. (1993). The assessment of two neem formulations for the control of some economically important insect pests. *Revue Agricole et Sucriere de l'Ile Maurice*, **72**(1-2) : 44-49.

Gerard, P.J. and Ruf, L.D. (1995). Effect of a neem (iAzadirachta indica A. Juss, Meliaceae) extract on survival and feeding of larvae of four keratinophagous insects. *Journal of Stored Products Research*, **31**(2) : 111-116.

Ghorpade, S.A. and Thakur, S.B. (1998). Management of sesamum leaf roller and capsu borer *Antigastra catalaunalis* with different plant products. Abstract in *Nantional Seminar in Entomology in 21st Century* held at Udaipur, April 30-May 2, 1998 : pp. 132.

Govindachari, R.; Narasimhan, N.S.; Suresh, G.; Partho, P.D.; Geetha, Gopalakrishnan and Gopalakrishnan, G. (1998). Insect antifeedant and growth regulating activities of salannin and other C-seco limonoids from neem oil in relation to azadirachtin. *Neem Newsletter*, **15**(3) : 38.

Gujar, G.T. (1997). Biological effects of azadirachtin and plumbagin on *Helicoverpa armigera*. *Indian Journal of Entomology*, **59**(4) : 415-422.

Gupta, G.P.; Jeyakumar, P. and Chandra, A. (1998). Impact of sunlight on the antifeedant effect of neem oil against *Helicoverpa armigera*. Abs. in *ICPPMSA*, 11-13 Dec., 1998, held at CSAU & T, Kanpur : pp. 142-143.

Hongo, H. and Karel, A.K. (1986). Effect of plant extracts on insect pests of common beans. *Journal of Applied Entomology*, **102**(2) : 164-169.

Hossain, S.M.Z.; Das, G.P. and Alam, M.Z. (1994). Use of various indigenous materials and insecticides in controlling potato tuber moth in storage. *Bulletin of the Institute of Tropical Agriculture*, Kyushu University, **17** : 79-84.

Isman, M.B.; Koul, O.; Arnason, J.T.; Stewart, J. and Salloum, G.S. (1991). Developing a neem-based insecticide for Canada. *Memoirs of the Entomological Society of Canada*, **159** : 39-47.

Isman, M.B.; Koul, O.; Luczynski, A. and Kaminski, J. (1990). Insecticidal and antifeedant bioactivities of neem oils and their relationship to azadirachtin content. *Journal of Agricultural and Food Chemistry*, **38**(6) : 1406-1411.

Jagannadh, V. and Nair, V.S.K. (1992). Azadirachtin-induced effects on larval-pupal transformation of *Spodoptera mauritia*. *Physiological Entomology*, **17**(1) : 56-61.

Jeyarajan, S.; Babu, P.C.S.; Srimannarayana, G. and Geethanjali, Y. (1993). Antifeedant and morphogenetic effects of azadirachtin rich fractions on *Spodoptera litura* F. *Botanical Pesticides in Integrated Pest Management* : 376-380.

Jeyarajan, S.; Sundra Babu, P.C.; Srimannarayana, G. and Geethanjali, Y. (1990). Antifeedant and morphogenetic effects of azadirachtin rich fractions on *Spodoptera litura* F. *Proc. Symp. on Botanical Pesticides in IPM* held at Rajahmundry : 376-380.

Jhala, R.C.; Darji, P.B. and Patel, N.C. (1998). Antifieedant and larvicidal efficacy of certain plant extracts against cotton leaf roller, *Sylepta derogata* Fabricius. *Neem Newsletter*, **15**(2) : 20.

Jhansi, B.R. and Singh, R.P. (1996). Identification of effective and inexpensive neem (*Azadirachta indica* A. Juss) seed kernel extract for management of *Helicoverpa armigera* (Hubner). *Neem and Environment*, **1** : 381-404 Edited by R. P. Singh *et al*..

Joshi, B.D.; Tandon, P.L. and Verghese, A. (1993). Evaluation of different plant oils and extracts against citrus leaf miner, *Phyllocnistis citrella* Stainton (Lepidoptera : Phyllocnistidae). *Botanical Pesticides in Integrated Pest Management* : 340-342.

Joshi, B.G. and Sitaramaiah, S. (1979). Neem kernel as an ovipositional repellent for *Spodoptera litura* (F.) moths. *Phytoparasitica*, **7**(3) : 199-202.

Joshi, B.G.; Ramaprasad, G. and Rao, S.N. (1984). Neem seed kernel suspension as an antifeedant for *Spodoptera litura* in a planted flue-cured Virginia tobacco crop. *Phytoparasitica*, **12**(1) : 3-12.

Joshi, B.G.; Ramaprasad, G. and Satyanarayana, S.V.V. (1978). Relative efficacy of neem kernel, fentin acetate and fentin hydroxide as antifeedants against tobacco caterpillar, *Spodoptera litura* Fabricius, in the nursery. *Indian Journal of Agricultural Sciences*, **48**(1) : 19-22.

Joshi, B.G.; Schmutterer, R. (ed.) and Ascher, K.R.S. (1987). Use of neem products in tobacco in India. *Proceedings of the 3rd International Neem Conference,* Nairobi, Kenya, 10-15 Jyly, 1986 : 479-494.

Kakoty, N.N.; Rahman, M.A.; Sarmah, M. and Singh, K. (1993). Evaluation of antifeedant property of a few commercial neem formulations against bunch and psychid caterpillars. *Two and a Bud*, **40**(1) : 18-20.

Katti, G. and Saxena, H. (1997). Laboratory testing of some neem formultaions for antifeedant effect against semilooper, *Autographa nigrisigna* Walker in chickpea. *Neem Newletter*, **14**(4) : 55-56.

Kaul, P.; Shankar, J.S. and Kapil, R.S. (1996). The effect of neem alletochemicals on nutritional physiology of larval *Spodoptera litura*. *Entomologia Experimentalis et Applicata*, **79**(1) : 43-50.

Koshiya, D.J. and Ghelani, A.B. (1990). Antifeedant activity of different plant derivatives against *Spodoptera litura* (Fab.) on groundnut. *Proc. Symp. on Botanical Pesticides in IPM* held at Rajahmundry : 270-275.

Koshiya, D.J. and Ghelani, A.B. (1993). Antifeedant activity of different plant derivatives against *Spodoptera litura* (Fab.) on groundnut. *Botanical Pesticides in Integrated Pest Management* : 270-275.

Koul, O. (1987). Antifeedant and growth inhibitory effects of calamus oil and neem oil on *Spodoptera litura* under laboratory conditions. *Phytoparasitica*, **15**(3) : 169-180.

Krishnaiah, N.V. and Kalode, M.B. (1990). Efficacy of selected botanicals against rice insect pests under green house and field conditions. *Indian Journal of Plant Protection*, **18**(2) : 197-205.

Kubo, I. and Klocke, J.A. (1982). Azadirachtin, insect ecdysis inhibitor. *Agricultural and Biological Chemistry*, **46**(7) : 1951-1953.

Kubo, I.; Matsumoto, A. and Matsumoto, T. (1986). New insect ecdysis inhibitory limonoid deacetylazadirachtinol isolated from *Azadirachta indica* (Meliaceae) oil. *Tetrahedron*, **42**(2) : 489-496.

Kulkarni, N. (1998). Feeding deterrence of some plant extracts against poplar defoliator, *Clostera cupreata* Butler (Lepidoptera : Notodontidae). *Insect Environment*, **3**(4) : 98.

Kulkarni, N. and Joshi, K.C. (1998). Antifeedant property of some botanical extracts aginst maharukha webworm, *Atteva fabriciella* (Lepidoptera : Yponomeutidae). *Journal of Tropical Forest Products*, **4**(1) : 11-16.

Kulkarni, N.; Joshi, K.C. and Shamila, Kalia (1997). Feeding inhibition property of some plant extracts against sissoo defoliator, *Plecoptera reflexa* (Lepidoptera : Noctuidae). *Indian Journal of Forestry*, **20**(4) : 390-394.

Kumar, N.S.; Jeyabalan, D. and Murugan, K. (1997). Antifeedant and growth inhibiting effect of neem leaf exudate on *Spodoptera litura*. *Indian Journal of Entomology*, **59**(2) : 151-154.

Kumar, S.P. and Babu, P.C.S. (1998). Ovipositional deterrent and ovicidal effects of NeemAzal against brinjal shoot and fruit borer and leaf beetle. *Shashpa*, **5**(2) : 193-196.

Lal, S.S.; Yadav, C.P. and Sachan, J.N. (1986). Strategies for the development of an integrated approach to control gram pod borer, *Heliothis armigera* (Hubn.) infesting chickpea. *Pesticides*, **20** : 39-50.

Lidert, Z.; Taylor, D.A.H. and Thirugnanam, M. (1985). Insect antifeedant activity of four prieurianin-type limonoids. *Journal of Natural Products*, **48**(5) : 843-845.

Loke, J.H.; Heng, C.K.; Rejab, A.; Basirun, N.; Mardi, H.C.A.; Ooi, P.A.C.

(ed.); Lim, G.S. (ed.) and Teng, P.S. (1992). Studies on neem (*Azadirachta indica* A. Juss) in Malaysia. *Proceedings of the 3rd International conference on Plant Protection in the Tropics*, Genting Highlands, Malaysia, 20-23 March, 1990, **No. 2** : 103-107.

Lowery, D.T.; Bellerose, S.; Smirle, M.J.; Vincent, C. and Pilon, J.G. (1996). Effects of neem on the growth and development of the obliquebanded leafroller, *Choristoneura rosaceana*. *Entomologia Experimentalis et Applicata*, **79**(2) : 203-209.

Maredia, K.M.; Segura, O.L. and Mihm, J.A. (1992). Effects of neem, *Azadirachta indica*, on six species of maize insect pests. *Tropical Pest Management*, **38**(2) : 190-195.

Mathur, Y.K. and Nigam, S. (1996). Insecticide, antifeeding and juvenilizing effects of neem (*Azadirachta indica* A. Juss.) oil against *Corcyra cephalonica* Staint. and *Epilachna vigintioctopunctata* F. *Neem and Environment*, **1** : 335-342.

Mayabini, Jena (1997). Efficacy of some neem products against rice leaf folder, *Cnaphalocrocis medinalis* Guenee. *Indian Journal of Plant Protection*, **25**(2) : 160-161.

Meisner, J.; Ascher, K.R.S.; Aly, R. and Warthen, J.D.Jr. (1981). Response of *Spodoptera littoralis* (Boisd.) and *Earias insulana* (Boisd.) larvae to azadirachtin and salannin. *Phytoparasitica*, **9**(1) : 27-32.

Meisner, J.; Melamed Madjar, V.; Tam, S. and Ascher, K.R.S. (1986). The effect of azadirachtin on larvae of the European corn borer, *Ostrinia nubilalis*. *Zeitschrift fur Pflanzenkrankheiten und Pflanzenschutz*, **93**(6) : 585-589.

Melamed Madjar, V.; Meisner, J. and Ascher, K.R.S. (1989). Effects of neem on the corn borer, *Sesamia nonagrioides* Lefebvre (Lepidoptera : Noctuidae). *Zeitschrift fur Pflanzenkrankheiten und Pflanzenschutz*, **96**(5) : 521-525.

Mong Ting Tan; Sudderuddin, K.I. and Tan Mong Ting (1978). Effects of a neem tree (*Azadirachta indica*) extracts on diamond-back moth (*Plutella xylostella* L.). *Malaysian Applied Biology*, **7**(1) : 1-9.

Murthy, P.S.N.; Gamaprasad, G.; Sitaramaiah, S.; Rao, B.V.K.; Prabhu, S.R. and Rao, S.N. (1989). Studies on the enrichment of neem fractions and their relative efficacy against tobacco caterpillar (*Spodoptera litura* F.) on Lanka and FCV tobacco. *Tobacco Research*, **15**(1) : 65-70.

Murthy, P.S.N.; Sitaramaiah, S.; Ramaprasad, G.; Rao, S.N.; Rao, B.V.K. and Prabhu, S.R. (1993). Neem extracts for the control of *Spodoptera litura* F. on tobacco. *Botanical Pesticides in Integrated Pest management* : 306-317.

Naumann, K. and Isman, M.B. (1995). Evaluation of neem *Azadirachta indica* seed extracts and oils as oviposition deterrents to noctuid moths. *Entomologia Experimetnalis et Applicata*, **76**(2) : 115-120.

Ocete, R. and Tio R. del (1998). Evaluation of the antifeedant effects of a commercial neem extract (*Azadirachta indica* Juss, Meliaceae) on *Ocnogyna baetica* Ramb. (Lepidoptera, Arctiidae). *Boletin de Sanidad Vegetal*, Plagas, **24**(1) :5-10.

Osman, M.Z. (1993). Effects of neem seed extract on growth and development of larvae of *Pieris brassicae* L. (Lep., Pieridae). *Journal of Applied Entomology*, **115**(3) : 254-258.

Osman, M.Z. and Port, G.R. (1990). Systemic action of neem seed substances against *Pieris brassicae*. *Entomologia Experimentalis et Applicata*, **54**(3) : 297-300.

Patel, K.B. and Patel, J.R. (1998). Oviposition-deterrent effect of botanical materials on tobacco leaf-eating caterpillar (*Spodoptera litura*). *Indian journal of Agricultural Sciences*, **68**(1) : 48-49.

Pathak, P.H. and Krishna, S.S. (1987). The influence of the odours of certain botanical components or an organic solvent dichloromethane during breeding on the reproductive efficiency of *Earias vittella* (F.) (Lepidoptera : Noctuidae). *Uttar Pradesh Journal of Zoology*, **7**(2) : 175-179.

Pathak, P.H. and Krishna, S.S. (1991). Postembryonic development and reproduction in *Corcyra cephalonica* (Stainton) (Lepidoptera : Pyralidae) on exposure to eucalyptus and neem oil volatiles. *Journal of Chemical Ecology*, **17**(12) : 2553-2558.

Prabhaker, N.; Coudriet, D.L.; Kishaba, A.N. and Meyerdirk, D.E. (1986). Laboratory evaluation of neem-seed extract against larvae of the cabbage looper and beet armyworm (Lepidoptera : Noctuidae). *Journal of Economic Entomology*, **79**(1) : 39-41.

Przybyszewski, J. (1993). The use of neem seed and leaf extracts to control lepidopteran pests of cabbage in Kaedi, Mauritania, West Africa. *Tropical Science*, **33**(3) : 217-225.

Raffa, K.F. (1987). Influence of host plant on deterrence by azadirachtin of feeding by fall armyworm larvae (Lepidoptera : Noctuidae). *Journal of Economic Entomology*, **80**(2) : 384-387.

Rajasekaran, B.; Kumaraswami, T.; Regupathy, A. (ed.) and Jayaraj, S. (ed.) (1985). Studies on increasing the efficacy on neem seed kernel extract. *Behavioural and Physiological Approaches in Pest Management* : 29-30.

Raman, K.; Ganesan, S.; Vyas, B.N.; Godrej, N.B. and Mistry, K.B. (1996). Development of a neem formulation and its evaluation for control of crop pests. *Neem and Environment*, **2** : 865-874.

Ramchandran, R.; Mukherjee, S.N. and Sharma, R.N. (1989). Effects of food deprivation and concentration of azadirachtin (from *Azadirachta indica*) on the performance of *Achaea janata* and *Spodoptera litura* on young and mature leaves of *Ricinus communis*. *Entomologia Experimentalis et Applicata*, **51**(1) : 29-35.

Rao, G.R.; Raghavaiah, G. and Nagalingam, B. (1993). Effect of botanicals on certain behavioural responses and on the growth inhibition of *Spodoptera litura* F. *Botanical Pesticides in Integrated Pest Management* : 175-182.

Rao, P.J.; Gupta, S.; Raj, D.M. and Kranthi, K.R. (1996). Neem effects on *Spodoptera litura* (Fabr.) : A holistic study. *Neem and Environment*, **1** : 357-374.

Rathore, Y.S. and Yadava, C.P. (1998). Studies on non preference of different plant extracts against bihar hairy caterpillar, *Diacrisia obliqua*. *Proceedings of Nantional Seminar in Entomology in 21st Century* held at Udaipur, April 30-May 2, 1998 : pp. 130.

Redfern, R.E.; Warthen, J.D.Jr.; Jacobson, M. and Stokes, J.B. (1984). Antifeeding potency of neem formulations. *Journal of Environmental Science and Health, A Environmental Science and Engineering*, **A19**(4) : 477-481.

Rovesti, L. and Deseo, K.V. (1991). Effectiveness of neem seed kernel extract against *Leucoptera malifoliella* Costa (Lep., Lyonetiidae). *Journal of Applied Entomology*, **111**(3) : 231-236.

Rwamputa, A.K.; Schabel, H.G.; Alfaro, R.I. (ed.) and Glover, S.G. (ed.) (1989). Effects of crude aqueous neem extracts on defoliation of *Khaya nyasica* by *Heteronygmia dissimilis* (Lepidoptera : Lymantriidae) in East Africa. *Insects Affecting Reforestation Biology and Damage* : 245-250.

Sahayaraj, K. (1997). Antifeedant effect of some plant extracts on the Asian armyworm, *Spodoptera litura* (Faricious). *Current Science*, **74**(6) : 523-525.

Salem, S.A. (1991). Response of cotton bolworms, *Pectinophora gossypiella* Saund. and *Earias insulana* Boisd. to see neem kernel pure oil. *Annals of Agricultural Science*, Moshtohor, **29**(1) : 597-607.

Sannaveerappanavar, V.T. and Viraktamath, C.A. (1997). The IGR activity of aqueous neem seed kernel extract against diamondback moth, *Plutella xylostella* L. larvae. *Mysore Journal of Agricultural Sciences*, **31**(3) : 241-243.

Saxena, R.C. (1982). Note on the use of neem kernel for the protection of dew-gram against *Amsacta moorei* Butler. *Indian Journal of Agricultural Sciences*, **52**(1) : 51-52.

Saxena, R.C. and Francis, O. O. (1996). Growth, development and reproduction of *Busseola fusca* (Fuller) [Lepidoptera : Noctuidae] on semi-synthetic diet containing 'neem bitters concentrate'. *Neem Newletter*, **13**(2) : 18.

Schmutterer, H.; Saxena, R.C. and Heyde, J von der (1983). Morphogenetic effects of some partially-purified fractions and methanolic extracts of neem seeds on *Mythimna separata* (Walker) and *Cnaphalocrocis medinalis* (Guenee). *Zeitschrift fur Angewandte Entomologie*, **95**(3) : 230-237.

Sehgal, V.K. and Ujagir, R. (1990). Effect of synthetic pyrethroids, neem extracts [from *Azadirachta indica* seed kernels] and other insecticides for the control of pod damage by *Helicoverpa armigera* (Hubner) on chickpea and pod damage-yield relationship at Pantnagar in northern India. *Crop Protection*, **9**(1) : 29-32.

Senguttuvan, T.; Kareem, A.A. and Rajendran, R. (1995). Effects of plant products and edible oils against rice moth *Corcyra cephalonica* Stainton in stored groundnuts. *Journal of Stored Products Research*, **31**(3) : 207-210.

Sharma, G.K. (1992). Growth-inhibiting activity of azadirachtin on *Corcyra cephalonica*. *Phytoparasitica*, **20**(1) : 47-50.

Sharma, I.N.S.; Singh, A.K.; Singh, S.P.; Tauro, P. (ed.) and Narwal, S.S. (ed.) (1992). Allelopathic potential of some plant substances as anti-feedant against insect pests of jute. *Proceedings First National Symposium on Allelopathy in Agroecosystems (Agriculture and Forestry)*, Feb. 12-14, 1992, held at CCS, Haryana Agric. Univ., Hisar : 166-168.

Sharma, R.N.; Tare, V.. and Deshpande, S.G. (1996). Improved plant based formulations for preventing *Phthorimaea operculella* damage on stored potatoes-Plantmix I and Plantmix II. *Pesticide Research Journal*, **10**(2) : 214-216.

Shaw, S.P.; Singh, B.B.P. and Parsad, D. (1988). Influence of different host plants on ovipositional preference, larval survival, growth potential and fecundity of Bihar hairy caterpillar, *Diacrisia obliqua* Walker. *Annals of Agricultural Research*, **9**(2) : 215-223.

Shelke, S.S.; Jadhav, L.D. and Salunkhe, G.N. (1985). Ovipositional and adult repellent action of some vegetable oils/extracts against potato tuber moth. *Journal of Maharashtra Agricultural Universities*, **10**(3) : 284-286.

Shimizu, T. (1988). Suppressive effects of azadirachtin on spermiogenesis of the diapausing cabbage armyworm, *Mamestra brassicae* in vitro. *Entomologia Experimentalis et Applicata*, **46**(2) : 197-199.

Shukla, A.; Pathak, S.C. and Agarwal, R.K. (1997). Effect of some plant odours in the breeding environment on fecundity and hatching of okra shoot and fruit borer, *Earias vittella* Fab. under laboratory conditions. *Crop Research Hisar*, **13**(1) : 157-161.

Siddiqui, K.H.; Sarup, P. and Marwaha, K.K. (1990). Effect of different neem products on mass rearing of the maize stalk borer, *Chilo partellus* (Swinhoe) in artificial diet. *Journal of Entomological Research*, **14**(1) : 8-11.

Singh, C.P. and Kapur, S. (1998). Antifeedant effect of certain plant extracts on *Spilosoma obliqua* Walker (Lepidoptera : Arctiidae). Abs. in *ICPPMSA*, 11-13 Dec., 1998, CSAU & T, Kanpur : pp. 170.

Singh, D.; Bhathal, S.S.; Tauro, P. (ed.) and Narwal, S.S. (ed.) (1992). Survival, growth and developmental effects of some neem based products on *Spodoptera litura* (Fabricius). *Proceedings First National Symposium on Allelopathy in Agroecosystems (Agriculture and Forestry)*, Feb. 12-14, 1992, held at CCS Haryana Agric. Univ., Hisar, India : 163-165.

Singh, K.; Sharma, P.L. and Singh, K. (1987). Evaluation of antifeedant and repellent qualities of various neem (*Azadirachta indica*) formulations against *Pieris brassicae* L. larvae on cabbage and cauliflower. *Research and Development Reporter*, **4**(1) : 76-78.

Singh, R.; Mahalakshmi, R.; Vijayachandran, S.N. and Singh, R. (1996). The lime butterfly : potential pest of *Aegle marmelos* Corr. in nurseery and its management with neem. *Insect Environment*, **2**(3) : 70.

Sinha, S.N. (1993). Control of *Helicoverpa armigera* Hb. infesting chickpea : field efficacy of neem products and insect growth regulators. *Indian Journal of Plant Protection*, **21**(1) : 80-84.

Sinha, S.N. and Mehrotra, K.N. (1998). Diflubenzuron and neem (*Azadirachta indica*) oil in control of *Heliothis armigera* infesting chickpea (*Cicer arietinum*). *Indian Journal of Agricultural Sciences*, **58**(3) : 238-239.

Sitaramaiah, S.; Ramaprasad, G. and Joshi, B.G. (1986). Efficacy of diflubenzuron and neem seed kernel suspension against the tobacco caterpillar, *Spodoptera litura*, in tobacco nurseries. *Phytoparasitica*, **14**(4) : 265-271.

Sojitra, I.R. and Patel, J.R. (1992). Effect of plant extracts [including *Azadirachta indica, Ricinus communis* and *Pongamia pinnata*] on oviopositional behaviour of spotted bollworm (*Earias vittella*) infesting okra (*Hibiscus esculentus*) [*Abelmoschus esculentus*]. *Indian Journal of Agricultural Sciences*, **62**(12) : 848-849.

Sombatsiri, Kwanchai (1996). Effects of the extracted substances from Thai neem seed, *Azadirachta indica* var. *siamensis*, on the insect pests of chinese kale. *Neem and Environment*, **1** : 565-570.

Srivastava, J.P. and Mathur, Y.K. (1998). Insect grwoth regulatory effect of some neem based formulations against sesamum pod borer. Abs. in *ICPPMSA*, 11-13 Dec., 1998, CSAU & T, Kanpur : pp. 228.

Tanzubil, P.B. and McCaffery, A.R. (1990). The effects of azadirachtin on feeding and reproduction in the African armyworm (*Spodoptera exempta* (Walker) and their implications for field control. *Proceedings : Integrated Pest Management In Tropical and Subtropical Cropping Systems*, **3** : 717-727.

Umeh, E.D.N. (1988). The effect of methanolic extracts of neem seed kernel on feeding, development and survival of *Spodoptera maculatus* Wlk. (Lep., Arctiidae). *Journal of Applied Entomology*, **106**(5) : 494-499.

Venkateswarlu, P.; Raghavaiah, G. and Nagalingam, B. (1988). Effect of neem oil on certain behavioural aspects of *Spodoptera litura* Fabricius in urdbean. *Indian Journal of Pulses Research*, **1**(2) : 118-123.

Venkateswarlu, P.; Raghavaiah, G. and Nagalingam, B. (1988). Effect of neem oil on growth and development of certain lepidopterous pests. *Indian Journal of Pulses Research*, **2**(1) : 85-87.

Verma, S.K. and Singh, M.P. (1985). Antifeedant effects of some plant extracts on *Amsacta moorei* Butler. *Indian Journal of Agricultural Sciences*, **55**(4) : 298-299.

Zhang, Y.G.; Zhang, X. and Zhao, S.H. (1992). Preliminary studies on azadirachtin content and insecticidal activity of the seeds from the introduced neem trees *Azadirachta indica* A. Juss planted in China. *Journal of south China Agricultural University*, **13**(1) : 14-19.

Chapter 6 Neem products in controlling plant diseases

During the last few years, the agricultural strategy has acquired a new dimension. The use of high yielding varieties and hybrid has become a significant aspect of new technology which is being accompanied by a series of related changes in agricultural practices such as massive application of fertilizers and plentiful irrigation. This has given rise to a drastic change in complexion of plant diseases. Major diseases have developed from minor ones which are previously of little or no importance.

A variety of chemicals including fungicides, fungistats, antibiotics, bactericidal and homeopathic drugs have been evaluated for their efficacy both as protectant and the therapeutant against various diseases. Currently, pesticides registrations are lost due to concerns about their safety and environmental impacts. Further pesticides are high energy inputs and are also costly.

Toxins from diseases inducing microorganisms are major factors in the development of a number of destructive plant diseases. Three classic cases often cited are the huge losses in oat production in North America from 1946 to 1948, rice production in India (Bengal famine) during 1942-43 and maize production in South America during 1970-71, caused by the fungus, *Helminthosporium* sp. In all cases, toxins were the major factors in the destructive processes and epidemic.

The botanical pesticides such as neem would be a potential method of disease control and increase the crop productivity. The neem product inhibit mycelial growth reduced leaf spot, rust, mildews, rot disease and mould. Additives are able to prolong the persistence of the neem products on the vegetation. Neem products have inhibited germination and growth of pathogens. Lesions formed by the pathogens on plant were observed restricted by pre-inoculation spray of neem extracts.

The research work conducted on the control of plant diseases with neem products are being described as follows.

1. Fungal diseases

Khan *et al.* (1973) observed that spraying of neem cake against *Rhizoctonia solani, Fusarium oxysporum* f.sp. *lycopersici* in tomato field produce suppression symptoms after 3 days. This clearly suggests that fungitoxic breakdown products were released at the time of neem cake breakdown. Khan et al. (1974a) also observed that in egg plant fields, addition of neem cake suppressed the growth of *R. solani, Fusarium* sp. and *Colletotrichum atramentarium* infection. Khan *et al.* (1974b) observed that *in vitro*, evaluation of aqueous extracts of neem oil cake showed inhibitory effect on *R. solani, F. oxysporum* f.sp. *lycopersici, Helminthosporium nodulosum, Alternaria tenuis* and *Curvularia tuberculata*. It was also observed that the inhibitory effect was fungistatic rather than fungicidal.

According to Singh *et al.* (1980), growth of *Fusarium oxysporum* f.sp. *ciceri, Rhizoctonia solani, Sclerotium rolfsii* and *Sclerotinia sclerotiorum* was inhibited in liquid medium by extracts of leaf, trunk bark, fruit pulp and oil of *Azadirachta indica*. Higher concentrations of oil inhibited seed germination. Oil-treated seeds sown in soil infested with the pathogens, singly or mixed, produced disease free seedlings whereas all the seedlings from untreated seeds exhibited dissease symptoms. The fruit pulp suppressed the formation of sclerotia by *R. solani*. A possible role of *A indica* extracts and oil in the control of *Cicer arietinum* diseases in the field is suggested. Extracts of *A. indica* leaf, trunk bark, fruit pulp and, in particular, seed oil inhibited fungal growth in 4 pathogens. Sulphur was the main constituent of neem oil and alkaloids, resins, glycosides and fatty acids were also present.

Nagarajan and Reddy (1980) observed that 1 percent neem seed suspension as a spray drench of tobacco seed beds to control *Phytophthora aphanidermatum* and *P. panasifica* var. *nicotianae* had little effect in comparison to fungicide. Singh *et al.* (1981) tested the effect of some plant extracts and their oils on conidia of *Erysiphe polygoni* DC. Garlic oil and neem oil at 100-5000 ppm and ginger rhizome extract 100-1000 ppm on germination and germtube growth of the conidia *in vitro*. It was observed that neem oil showed inhibitory effect. Bhowmick and Vardhan (1981) reported that aqueous leaf extract has also been found to inhibit at varying degrees to the foliar pathogen *Sarcocladium oryzae* and *F. oxysporum* f.sp. *cepae*.

Singh and Singh (1982) tested the effect of oil cake amended soil atmosphere on pigeonpea wilt pathogen (*Fusarium udum*). The atmosphere of autoclaved and unautoclaved soil amended with oil cake of *A. indica* and *Ricinus communis* was stimulatory for radial growth of *F. udum* when exposed 1 or 2 week after amendment. As the period of decomposition increased, the stimulatory effect declined, especially in unautoclaved soil. In autoclaved soil with a high dosage of *A. indica* cake, exposure of the fungus 3 week after amendment caused suppression of its growth. The atmosphere on the surface of the amended soil was more stimulatory than that in the soil. The surface atmosphere of *R. communis* amended soil was less stimulatory during the first 2 week and caused inhibition when tested 3 week after amendment.

Singh and Singh (1983) observed inhibition of pigeonpea wilt pathogen (*F. udum*) by ether distillate of margosa cake at 0.1 to 5.0 percent concentrations in soil amendment.

Pandy *et al.* (1983) reported the control of Pestalotia fruit rot of guava dry leaf extracts of two medicinal plants. Leaf extracts of *A. indica* and *Ocimum sanctum* inhibited germination of *Pestalotia psidii* spores *in vitro* and on guava fruits dipped in these extracts before or after inoculation. *O. sanctum* extracts are recommended as they do not affect fruit flavour.

Vir and Sharma (1985) observed that concentrations of 2.5 and 5 percent, the product prepared from *Azadirachta indica* inhibited radial growth of *F. moniliforme* [*Gibberella fujikuroi*] by 57.8-66.7 percent, that of *Aspergillus niger* by 77.8 percent, that of *Drechslera restratum* [*Setosphaeria restrata*] by 83.3-88.9 percent, and that of *Macrophomina phaseolina* by 61.1-75.6 percent. At 10 percent, the oil gave 100 percent inhibition of all the test pathogens. Khune *et al.* (1985) observed that bordeaux mixture, Bavistin (carbendazim) and Benlate (benomyl) completely inhibited growth of *Capnodium citri* and *Chaetothyrium* sp. isolates collected from infected citrus leaves at Akola, Nagpur and Katol. Dithane Z-78 (zineb), Adulsa + neem + tobacco decoction, Cuman L (ziram), neem leaf decoction, Dithane M-45 (mancozeb), Calixin (tridemorph) and Aureofungin consistently gave >75 percent inhibition.

Dwivedi and Chaube (1985) tested the effect of oil cakes on survival of *M. phaseolina* in soil. Of 5 different oil cakes tested, the best reduction in the number of recoverable propagules was given by Margosa (*A. indica*) and cotton. Mycelium was more susceptible than sclerotia to decomposition by oil cakes. The average number of colonies from cotton stem pieces colonized by the fungus in soil decreased with increasing dosages of oil cakes and duration of incubation. The time required to decrease recovery of *M. phaseolina* by 50 percent was 4 weeks.

Vir and Sharma (1985) observed neem product inhibited growth of *Alternaria alternata* by 61.1 percent at a concentration of 1 percent and 100 percent at 10 percent concentration.

Sadanandan and Iyer (1986) conducted trials for two years by amending the plots with neem cake, pongamia (*Pongamia pinnata*) cake and compared it with farm yard manure and the untreated control plots. The incidence of rhizome rot (*Pythium* sp.) on ginger (*Zingiber officinalis*) was 4.7, 8.5, 14.5 and 6.7 percent, respectively. The yield data were 7.2, 6.5, 6.6 and 5.4 tons/ha. Gunasekaran *et al*. (1986) observed that application of neem cake at 10 kg/palm per year increased the total population of fungi in the rhizosphere and also inhibited the growth of *Ganoderma lucidum*, the suspected pathogen of Thanjavur wilt. Treatment with neem cake alone gave similar results to those obtained with a combined application of Aureofungin-sol (stem injection) + soil drenching with bordeaux mixture (1%) + neem cake, and it was considerably cheaper. Dwivedi and Dubey (1986) studied the effects of volatile and non-volatile fractions (VF and NVF) of two medicinal plants on the germination of *M. phaseolina*. Deleterious effects of VF of hydrodistillates of *Azadirachta indica* and *Eucalyptus globulus* on germination of these sclerotia were more poronounced than NVF. A notable decrease in sclerotial germination due to VF and NVF of neem oil and neem oil-cake extract was observed after 24 hrs incubation. Extracts of both the plants showed fungicidal effects against the pathogen, whereas neem leaf distillate had no such effect.

Jeyarajan *et al*. (1987) studied the effect of neem and other plant products in the management of plant disease in India. Addition of neem cake to soil reduced preemergence and postemergence mortality of cotton seedlings caused by *Rhizoctonia solani*. Populations of fungi, bacteria and actinomycetes increasd in unamended soils. Neem cake gave similar reductions in root rot of soyabeans caused by *M. phaseolina* as soil drenching with quintozene. It was also effective in reducing wilt of coconut (*Ganoderma lucidum*), crossandra flowers (*F. solani*) and Piper betle (*Phytophthora capsici*). Thakore *et al*. (1987) did experiment by soil amendment with oil cakes in ginger field for the control of rhizome rot. Ginger rhizome rot, caused by *Pythium aphanidermatum* and *F. solani*, was reduced by amendments of cakes made from *Azadirachta indica*, *Calophyllum inophyllum* or *Pongamia glabra* [*P. pinnata*]. The maximum yield increase was given by cakes made from *P. glabra* follwed by *Hibiscus sabdariffa* and *Brassica campestris*.

Bhatnagar and Mccormick (1988) observed inhibitory effect of neem leaf extracts on aflatoxin synthesis in *Aspergillus parasiticus*. The extracts were prepared by blending 50 g (wet wt) of fresh leaves in 1 lt of 10 mM potassium phosphate (pH 7.0) or by boiling the leaves in the buffer. Extracts were added to fungal growth media at 1, 5, 10, 20 and 50 percent (v/v) concentrations prior to inoculation. The formulations did not affect fungal growth but assentially blocked (>98%) aflatoxin biosynthesis at concentration >10 percent (v/v). The inhibitory effect was diminished (60-70% inhibition) in heated leaf extracts. Volatile components of the extracts were analysed using capillary GC/MS; the major volatile component was 3-methyl 2-buten-1-ol. Howerver, volatiles from blended leaf extracts did not affect either aflatoxin synthesis or fungal growth. The neem-mediated inhibition appeared to involve regulation of secondary metabolism, because once secondary biosynthesis was initiated the inhibitory effect of the neem leaf constituents was lost.

Roy *et al.* (1988) studied the association of mycoflora with drug plants and aflatoxin producing potentiality of *Aspergillus flavus*. *Rauwolfia serpentina* roots, ginger rhizomes, *Emblica officinalis* [*Phyllanthus emblica*] fruit, seeds of *Azadirachta indica, Datura metel, Abrus precatorium, Holarrhena antidysenterica* [*H. pubescens*] and *Strychnos nux-vomica* were collected from India. *Aspergillus flavus, A. niger, A. candidus, A. luchuensis, A. nidulans, F. moniliforme* and *Penicillium citrinum* were frequently found whilst *Alternaria alternata, Chaetomium* sp., *Penicillium* sp. and *Rhizopus stolonifer* were less common. Of 50 *A. flavus* isolates from the medicinal plants, 21 were toxigenic. None of the plant samples were free from storage fungi. Among the toxigenic strains of *A. flavus*, 12 isolates had potentiality to produce aflatoxin B only, 7 had both B_1 and B_2 and 2 produced B_1, B_2 and G_1. The growth of *A. flavus* under storage might not only deteriorate the quantity of drug plants but also contaminate by elaborating aflatoxins.

Lakshmanan and Mohan (1989) observed antifungal properties of some plant extracts against collar rot of *Phaseolus aureus*. Water extracts of garlic, *Bougainvillea spectabilis* and *Azadirachta indica* significantly inhibited mycelial growth and sclerotial germination of *Thanatephorus cucumeris in vitro*. Under greenhouse conditions, all the plant extracts significantly reduced *P. aureus* [*Vigna radiata*] seedling mortality caused by the pathogen.

Singh *et al.* (1989) studied the fungitoxicity of some plant oils against *Sclerotium* [*Corticium*] *rolfsii* Sacc. and soil mycoflora. Of 6 plant oils tested against *S. rolfsii* and 10 soil fungi, the leaf oil of neem was the most effective, followed by that from *Eucalyptus globulus* and *Ocimum canum* [*O. americanum*]. The oil of *O. canum* possessed a broad fungitoxic spectrum, while oil of *Citrus* spp. and *Pinus roxburghii* had a narrow range. Banerjee *et al.* (1989) studied sensitivity of three sclerotial rice pathogens to plant oils. Purified sclerotia of *S. rolfsii, S. hydrophyllum* and *Rhizoctonia* [*Ceratobasidium*] *oryzae-sativae* were soaked in 5 percent plant oil emulsion and allowed to germinate on sterilized wheat grains in petri plates. Oils from *Azadirachta indica* and *Cymbopogon nardus* strongly inhibited germination of all 3 fungi. *Pongamia glabra* was only effective against *C. rolfsii* and *C. oryzae-sativae*. Results were similar in tests where oils were added to solid media and diameter of fungal colonies were measured.

Ghewande (1989) studied the effect of plant products on the management of foliar diseases of groundnut (*Arachis hypogoea*). In Junagadh, Gujarat, GAUG 1 groundnuts were sown and treated with aqueous leaf extracts of *A. indica, Lawsonia inermis, Triadax procumbens* and *P. pinnata* and the fungicides NCP-75 (0.3%) and carbendazim (0.05%) + mancozeb (0.2%). The incidence of *Phaeoisariopsis personata* [*Mycosphaerella berkeleyi*] and *Puccinia arachidis* was recorded 15 days before harvest. Carbendazim + mancozeb gave maximum control of *P. arachidis* and *M. berkeleyi*. *A. indica* and *L. inermis* extracts were effective in controlling both diseases. Yields increased by 15.42-39.84 percent due to treatment and the maximum increase was with chemical treatment.

Viswanathan and Kandiannan (1990) studied the effect of urea applied with neem cake on disease intensity and insect population in rice fields. Applications of neem cake + urea significantly reduced brown leaf spot (*Cochliobolus miyabeanus*) intensity as well as planthopper populations. Neem cake combined with pesticide treatment gave higher yields than the other treatments.

Lakshmanan *et al*. (1990) further studied the antifungal properties of some plant extracts against *T. cucumeris*, the causal agent of collar rot disease of *P. aureus*. Aqueous extracts of *Allium sativum, Azadirachta indica, Bougainvilea spectabilis, Cassia auriculata, Cascabola thevetia* [*Thevetia* sp.], *Jatropha villosa, Lantana camara, Piper betle, Pithecellobium dulce* and *Prosopis cineraria* were mixed with sterilized potato dextrose agar (PDA) medium to give 5 and 10 percent concentrations. Sterilized distilled water was used as a control. The plates were inoculated in the centre with 6 mm agar plugs taken from 2-day-old culture of T. cucumeris, grown on PDA, and incubated at 28 ± 3°C. A sclerotial germination assay, using uniform sized sclerotia recovered from 2-month-old dried PDA cultures, was carried out using the same plant extracts. *A. sativum, B. spectabilis* and *A. indica* were the most effective with regard to inhibiting mycelial growth and sclerotial germination.

Mishra and Tewari (1990) tested the toxicity of ethanolic extract of 3 botanicals against 5 fungal pathogens of rice. The results are presented with extracts from leaves of *Calotropics procera*, *Azadirachta Indica* and *Datura stramonium* against *Pyricularia oryzae, Rhizoctonia solani, Curvularia lunata* [*Cochliobolus lunatus*], *F. moniliforme* [*G. fujikuroi*] and *Aspergillus niger*. All the test extracts possessed toxic principles. Bhaskaran *et al.* (1990) did experiments to control *Ganoderma* wilt disease of groundnut. *G. lucidum* was consistently isolated from the roots of diseased coconut palms showing symptoms of Thanjavur wilt and caused root rottig when inoculated into healthy palms. Disease severity increased with increase in soil temperature and was less at higher soil moisture. The incorporation of organic manures, especially neem cake into the soil, and irrigation during the summer, reduced disease severity. Soil drenching with Bordeaux mixture (1%) and root application of aureofungin (2 g) + copper sulfate (1 g) in 100 ml water, 3 times at quarterly intervals (Jul., Oct. and Jan.) were very effective in management of the disease.

Ganapathy annd Narayanasamy (1990) tested the plant products against major disseases of groundnut. In laboratory and field tests, neem oil from *A. indica* and leaf extract from *Nerium odorum* reduced the incidence of late leaf spot (*Phaeoisariopsis personata* [*Mycosphaerella berkeleyi*]), rust (*Puccinia arachidis*) and ring mosaic (tomato spotted wilt virus). In addition, neem oil increased pod yield by 62.3 percent.

Zeringue and Bhatnagar (1990) observed inhibition of aflatoxin production in *A. flavus* infected cotton bolls after treatment with neem leaf extracts. In separate treatments, a spore suspension of *A. flavus* (control), an aqueous leaf extract of neem + a spore suspension of *A. flavus* or an aqueous neem leaf extract followed by an *A. flavus* spore suspension were injected 48 hrs later onto the surfaces of developing cotton bolls. Thirteen days after treatment, the seeds from the locules were harvested and fungal growth and aflatoxin production determined. Fungal growth was unaffected by the treatments but the seeds from locules receiving both neem leaf extracts and *A. flavus* simultaneously exhibited 16 percent inhibition of aflatoxin production, while the seeds in locules receiving *A. flavus* spores 48 hrs after neem extract was added exhibited >98 percent inhibition in aflatoxin production. It is concluded that neem leaf extracts contain an aflatoxin inhibiting factor, however, for maximal effect, translocation from the fibrous locule surface to the seed prior to the fungal inoculation may be required.

Bansal and Sobti (1990) tested the efficacy of plant products for the control of *Aspergillus niger* Van Tieghem and *A. flavus* Link. They cause crown rot under field condition and the aflatoxin causes heavy losses in storage. The seeds of groundnut were soaked in salt solution (2%), aqueous leaf extract of margosa (*Azadirachta indica* L.) (2%) and mehendi (*Lawsonia inermis* L.) (2%), thiram (0.03%), dithane M-45 (0.2%) and a control. Observation on *A. niger* and *A. flavus* incidence were reocrded after 7 days of incubation of 28±1°C. It was observe that *A. niger* incidence was reduced significantly over check (24%) by turmeric powder, neem extract, thiram and dithane M-45. The incidence with neem extract was least i.e. 2.66 percent. In case of *A. flavus*, all the treatments were at par with check (9%) except thiram which significantly reduced the incidence (1.03%). However, neem extract was most effective and significantly increase the seed germination followed by thiram and dithane M-45. It is revealed that neem seed extract was at par with thiram and mancozeb in reducing the incidence of both the species of *Aspergillus* and also increased the seed germination *in vitro*. Thus neem contains antifungal substances and is cheaper and easily available for use.

Pan (1990) observed that plant oil when mixed with fungicides shows synergistic effect and also antagonestic effects. In a test of germination and growth inhibition with *Drechslera oryzae* [*Cochliobolus miyabeanus*], the addition of neem and pongamia oils to fungicide Captan increased the fungicidal activity and showed synergistic effect. In contrary to this mixing neem and pongamia oils with carbendazim gave antagonestic effects. Garcia and Garcia (1990) conducted laboratory experiments to evaluate neem derivatives for their efficacy against *A. flavus* and *A. parasiticus*. The neem derivatives did not significantly inhibit growth, sporulation or spore germination in *A. flavus* and *A. parasiticus*. Aflatoxin formation by *A. flavus* and *A. parasiticus* treated with neem derivatives and the untreated controls did not vary markedly.

Chourasia and Roy (1991) studied the effect of temperature, relative humidity and light on aflatoxin B_1 production in neem and *Datura* seeds. The highest level of aflatoxin-B_1 production in both seeds (0.55-1.12 μg/g in *Datura* seeds and 0.47-1.15 μg/g in *Azadirachta* seeds) was reocrded at 30°C, 96 percent RH and total darkness after 3 weeks of incubation. The growth of *A. flavus* and aflatoxin production was inhibited at 33-55 percent RH and 10°C. Darkness favoured the synthesis of aflatoxin-B_1 whereas sunlight inhibited it completely.

Srivastava and Singh (1991) tested organic amendmennt against *Fusarium solani* and *Meloidogyne incognita* on *Phaseolus vulgaris*. Neem or mustard cake was applied at 5 percent to *P. vulgaris* inoculated with *M. incognita* and *F. solani* in a pot treatment. Both cakes reduced damage by the pathogens with neem cake being more effective than mustard. Doshi and Thakore (1991) observed the effect of oil cakes on the development of *Opium poppy* mildew under field condition. Oil cake amendments prepared from *Pongamia glabra* and *Azadirachta indica* were the most effective in increasing yields of latex and seed in *Papaver somniferum* grown under conditions culture, and in controlling both systemic and nonsystemic infection by *Peronospora arborescens*. Beneficial residual effects of the treatments extended into the following season.

Iyer and Williamson (1991) tested the efficacy of some plant extracts which inhibit protease activity of *Trichophyton* species. Protease was detected in all 3 species of *Trichophyton* (*T. rubrum*, *T. mentagrophytes*, *T. violaceum*) grown in Sabouraud's liquid medium. Extracts (100%) of *Allium sativum*, *Ocimum sanctum*, *Catharanthus roseus* and *Azadirachta indica* totally inhibited the protease activity. *Opuntia elatior* was found to be ineffective.

Rovesti *et al*. (1992) tested the effect of neem kernel extract on some phytopathogenic fungi. In greenhouse trials, aqueous neem kernel extract was as effective as sulfur against *Sphaerotheca fuliginea* on courgettes, *Erysiphe graminis* f.sp. *tritici* on wheat and *E.* graminis f.sp. *hordei* on barley when applied both before and after inoculation of the host with the pathogen. although not as effective as mancozeb, the neem extract also gave significant control of *Puccinia recondita* f.sp. *tritici* on wheat but was ineffective against *Phytophthora infestans* on tomato, *Cercospora beticola* on beet and *Septoria aplicola* on celery.

Rao and Ratnasudhakar (1992) tested the effect of rhizome powder of *Acorus calamus* and leaf powder of *Azadirachta indica* alone or in combination at 1 and 2 percent was investigated on fungal and bacterial population associated with rice during storage. The paddy variety BPT 2685 having the initial moisture content of about 13.6 percent was used for the experiment. The plant powder was mixed in paddy lots. The treatments were *Acorus* (1 & 2%), neem (1 & 2%), *Acorus* + neem (1% each), *Acorus* + neem (2% each) and a control. The observations were recorded on fungi and bacteria. It was observed that rhizome powder of *A. calamus* is very effective treatment followed by neem leaf powder (1 and 2%) treatment which minimized the fungus flora significantly at 60 days. The combination of *A. calamus* and *A. indica* followed by the leaf powder of *A. indica* were found least effective in minimising the incidennce of fungi. However, *A. calamus* (1%) and *A. calamus* + *A. indica* (1 and 2%) singnificantly reduced the bacterial number at 60 days. Mixing these plant powders with rice had no adverse effect on quality of cooked rice.

Hashmi *et al*. (1992) conducted studies on the seed mycoflora of lentil and its control. A total of 21 fungal species were isolated from 4 lentil cultivars in Pakistan using the blotter technique and of these 8 were pathogenic to various crops. The main species isolated were *F. moniliforme* [*Gibberella zeae*], *F. oxysporum* and *F. semitectum* [*F. pallidoroseum*]. Seed treatment with 80 ppm neem bitter extracts controlled the seedborne mycoflora. Achimu and Schlosser (1992) studied the effect of neem seed extracts against downy mildew (*Plasmopora viticola*) of grapevine. The underside of leaves of grape cv. Muller-Thurgau, susceptible to *P. viticola*, were sprayed with a fine mist of *A. indica* seed extracts or 1 of 3 commercial neem products 2 days before spray inoculation with a sporangial suspension of the pathogen. After 7 days incubation, the leaf area covered by the sporulating fungus was assessed. In tests performed both with detached leaves and in plants, all treatments gave a significant reduction in disease severity with an efficacy of >90 percent. When neemseed extracts were applied 4 days after inoculation, the efficacy was still >80 percent, indicating a curative effect. The antifungal properties of neem products were shown to be due to inhibition of the indirect germination of sporangia.

Lukade (1992) tested the effect of organic soil amendments on root rot incidence of safflower. Laboratory experiments were conducted to evaluate the effects of farmyard manure, press-mud cake, sunnhemp, subabul (*Leucaena leucocephala*) wheat straw, paddy straw, groundnut cake, neem cake and cotton seed cake on root rot of safflower variety Bhima, caused by *Rhizoctonia bataticola* [*Macrophomina phaseolina*]. Each organic amendment was incorporated at 2 percent by DW into soil that had been artificially inoculated with *M. phaseolina*. Seedling mortality was reduced by incorporation of both wheat and paddy straw. However, population size of *M. phaseolina* was smallest in soils amended with farmyard manure and press-mud cake.

Khan and Shah (1992) reported antifungal activity of leaf extract of neem on seed mycoflora of wheat. The antifungal activity of *A. indica* leaf extract on wheat seeds was investigated. Treatment caused a marked reduction in seed mycoflora as well as an enhancement of seed germination.

Ali *et al*. (1992) reported the effect of neem products as mould inhibitors against post harvest fruit rotting fungi on tomato *in vitro*. Neem oil, leaf extract and pericarp dust (from *A. indica*), togther with Tecto-60 [thiabendazole] and boric acid, were tested against isolates of *Penicillium italicum, Alternaria alternata* and *Aspergillus niger* from rotted fruits. Only neem oil was as effective as thiabendazole in checking growth of these fungi.

Moline and Locke (1993) reported the effect of neem seed oil and other fungicides for controlling post harvest apple decay. The antifungal properties of a hydrophobic seed extract from *A. indica* were tested against the postharvest apple pathogens; *Botrytis cinerea, Penicillium expansum* and *Glomerella cingulata*. The antifungal activity of neem seed oil was also compared with that of $CaCl_2$. A 2 percent aqueous emulsion of the clarified neem seed oil was moderately fungicidal to *B. cinerea* and *G. cingulata* in inoculated fruit, but had little activity against *P. expansum*. Ethylene production was reduced by 80 percent in fruit dipped in 2 percent neem seed oil compared with wounded, inoculated controls. Neem seed oil was as effetive as $CaCl_2$, but the effects of the 2 antifungal agents in combination were not additive.

Kazmi *et al.* (1993) evaluated some plant extracts having antifungal properties. Hexane extracts of *A. indica* seed, turmeric and *Valeriana officinalis* rhizomes and seed oil of mustard and *Anethum graveolens* were inhibitory to growth of *Alternaria alternata, Aspergillus flavus, A. fumigatus* and *A. wentii*, fungi commonly causing spoilage of stored grain. *A. graveolens* oil was the most effective, giving 100 percent inhibition at 0.025 percent concentration.

Sarramangala *et al.* (1993) evaluated plant extracts to control fungal pathogens of mulberry. Leaf extracts from *Azadirachta indica, Calotropis gigantea, Catharanthus roseus, Eucalyptus* sp., *Parthenium hysterophorus* and *Pongamia pinnata* were tested against *Cerotelium fici* and *Cercospora moricola*, pathogens of mulberry leaf rust and leaf spot, respectively. *In vitro*, effect of leaf extracts on *C. fici*, and conidial germiation of *C. moricola* was studied. Besides, a field experiment was conducted with K2 mulberry variety, naturally infested with leaf spot disease. Aqueous leaf extract (25% each) of 6 plants were sprayed on different plots. The extract was sprayed twice at an interval of 8 days. The data revealed that significant reduction in uredospore germination of *C. fici* was observed in all extracts. Maximum inhibition of spore germination was observed in leaf extract of *A. indica* (91.2%), while *P. pinnata* was least effective with 71.8 percent inhibition. The percentage of inhibition in gemination of *C. moricola* was 82.8 percent with *A. indica* and least effective was *P. pinnata* (57.2%). In foliar spray in the field, these extracts showed the same order of effectiveness as *in vitro*.

Narasimhan *et al.* (1993) tested the efficacy of plant products in managing sheath rot of rice. In field trials during the wet seasons of 1990-92, NSKE (5%), applied as foliar sprays at booting and 10 days later, gave control of *Sarocladium oryzae* and improvement of yield comparable to that achieved with 0.1 percent carbendazim.

Singh *et al.* (1993) tested leaf extracts of medicinal plants against diseases of banana. Treatment of infested banana fruits with aqueous leaf extracts of some medicinal plants (*Calotropis procera, Vitex negundo, Lantana camara, Azadirachta indica, Ficus religiosa, Ocimum sanctum, Thuja orientalis, Argemone mexicana, Achyranthes aspera, Datura fastuosa* and *Ricinus communis*) gave good control of disease (*Botryodiplodia theobromae,*

Fusarium oxysporum, Helminthosporium spiciferum [*Cochliobolus spicifer*], *Curvularia lunata* [*Cochliobolus lunatus*], *Aspergillus flavus* and *Trichothecium roseum*) development. Of the 11 leaf extracts, those of *A. indica* and *O. sanctum* were most effective, resulting in the minimum percentage loss in fruit weight. The appearance of the first disease symptom was also delayed in treated fruits as compared with untreated ones.

Meena and Mariappan (1993) tested the effect of plant products on seed borne mycoflora of sorghum. In *in-vitro* tests, leaf extract of *Azadirachta indica, Mentha arvesis, Aegle marmelos Catharanthus roseus, Lantana camara, Pongamia pinnata, Vitex negundo* and *Nerium odorum* [*N. oleander*], and flower extract of *C. roseus* inhibited mycelial growth and spore germination of seedborne mycoflora of sorghum including *Alternaria tenuis* [*A. alternata*], *A. flavus, C. lunata, F. moniliforme* and *Rhizopus stolonifer*. The neem extracts and those of *C. reseus* and *L. camara* were more effective than the other plant extracts tested.

Ghewande *et al*. (1993) conducted field experiments for the integrated control of early leaf spot (*Cercospora arachidicola* [*Mycosphaerella arachidis*]) and late leaf spot (*Phaeoisariopsis personata* [*M. berkeleyi*]) on groundnut during the rainy seasons of 1988-90. A mixture of carbendazim (0.05%) + mancozeb (0.2%) at 40, 55 and 70 days after sowing (DAS), intercropping groundnut with pigeonpea in combination with 2 sprays with fungicides at 55 and 70 DAS and intercropping with pigeonpea conbined with 1 spray each of neem leaf extract, fungicides and cell-free culture filtrate of *Penicillium islandicum* at 40, 55 and 70 DAS, respectively, were equally effective in controlling both diseases. The latter 2 treatments and intercropping groundnut with pigeonpea and spraying separately with neem leaf extract and *P. islandicum* inoculum at 55 and 70 DAS, gave significantly higher net monetrary returns (9569-11561 Rs/ha) with cost-benefit ratios ranging from 1:2.99 to 1:3.63 for cultivar Girnar 1 and 1:2.67 to 1:3.48 for cultivar JL 24.

Nair *et al*. (1993) conducted field trials at Panniyur, Kerala, India, during 1987-90, to investigate the effects of soil amendments and Bordeaux mixture on the control of *P. capsici* on *Piper nigrum*. The incidence of disease was reduced by all the treatments (Bordeaux mixture pasting, spraying and drenching (BMPSD), BMPSD with neem seed cake 2 kg/vine, BMPSD with lime 1 kg/vine, and BMPSD with neem seed cake 2 kg/vine + lime 1 kg/vine) compared with the control.

Ganguly (1994) studied fungitoxic effect of certain plant extracts against rice blast and brown spot pathogen. Water extracts of leaves of *Vinca rosea* [*Catharanthus roseus*], *Lantana camara, Ocimum tenuiflorum* [*O. sanc-*

tum], *Solanum melongena, Azadirachta indica, Polyalthia longifolia, Aegle marmelos* and *Datura metel* showed antifungal activity against *Pyricularia oryzae* [*Magnaporthe grisea*] and *Helminthosporium oryzae* [*Cochliobolus miyabeanus*] *in vitro*. Extract of *C. roseus* showed maximum inhibition of mycelial growth and spore germination of both fungi, followed by those of *P. longifolia* and *A. indica*.

Chandrasekar *et al*. (1994) tested the effect of foliar spray of potash and neem seed extract on Tikka leaf spot of groundnut casued by *Phaeoisariopsis personata* [*M. berkeleyi*] and *Cercospora arachidicola* [*Mycosphaerella arachidis*]. Twelve treatments were taken in the experiment. The results indicated that soil application of 55 kg K_2O/ha + K_2O (1%) foliar sporay + NSKE (2%) foliar spray reduced the disease significantly over the other treatments followed by 55 kg K_2O/ha + K_2O (0.5%) foliar spray + NSKE (2%) foliar spray. The pod and haulm yields were maximum in the recommended treatments with 1818.70 and 3174.00 kg/ha, respectively.

Ekpo and Banjoko (1994) tested the efficacy of Fernasan D (FD) (25% w/w thiram and 20% w/w lindane) and 4 wood ashes prepared from *Azadirachta indica, Vernonia amygdalina, Gliricidia sepium* and *Cassia siamea* against seedborne fungi of maize. The effects of the seed treatments on seed germination, pre-emergence mortality and seedling blight were also investigated. *A. flavus, A. niger, F. semitectum* [*F. pallidoroseum*], *F. moniliforme* [*G. fujikuroi*], *Curvularia pallescens* [*Cochliobolus pallescens*], *C. lunata* [*C. lunatus*], *Drechslera maydis* [*Cochliobolus heterostrophus*], *Penicillium* spp. and *Rhizopus* spp. were isolated from the seed. Seed treatment with thiram + lindane resulted in a reduction in percentage incidence of *A. flavus* and complete control of all the other isolated seedborne fungi. The various ash treatments gave complete control of *C. heterostrophus, C. pallescens* and *C. lunatus* and a reduction in percentage incidence of *Aspergillus, Gibberella, Fusarium* and *Penicillium* spp. Mean percentage incidence of *Rhizopus* spp. was increased in *A. indica*-treated seeds. Overall, the disease control efficacy of the seed treatments was in the order FD > *V. amygdalina* > *G. sepium* > *C. siamea* > *A. indica*.

Lehmann and Ibenthal (1995) purified extracts of neem by chromatography and tested against various sugarbeet pathogens *in vitro*. At least 5 bands with fungicidal potential were identified by bioautography using *Cladosporium cucumerinum*. Greenhouse and field experiments suggested that neem has potential as a practical treatment.

Sajid *et al*. (1995) studied comparative efficacy of neem products and Baytan against leaf rust of wheat. In the laboratory, neem oil and Baytan

25WP (triadimenol) completely inhibited germination of *Puccinia recondita* unrediospores. In the field, neem oil (4%) checked leaf rust on wheat after 4 applications but Baytan (0.1%) gave excellent rust control and the best improvement in yield.

Haque *et al.* (1995) Observed *Bradyrhizobium* sp. and *Paecilomyces lilacinus* eithor alone or mixed with cotton cake or neem cake significantly redused infection of *Macrophomina phaseolina, Rhizoctonia solani* and *Fusarium* spp.on roots of *Vigna radiata*. Mustrad cake was inhibitory to seed germination. The highest number of nodules/plant were produced where *Bradyrhizobium* sp. or *P. lilacinus* were used. The combined use of *Bradyrhizobium* sp. or *P. lilacinus* with a low amount of neem cake or cotton cake is suggested for the control of root rot disease of *V. radiatia*.

Nwosu and Okafor (1995) tested the antifungal activities of extracts of 10 medicinal plants collected from south-eastern parts of Nigeria against 7 pathogenic fungi using the broth dilution and agar plate methods. All the extracts at 1:10 dilution inhibited the growth of *B. haptosporus* and *B. ranarum* but did not inhibit that of *Aspergillus fumigatus, Geotrichum candidum* and *Candida albicans*. The *Jatropha curcas, Mitracarpus villosus, Azadirachta indica* and *Gongronema latifolium* failed to do so at 1:10 dilution.

Sobti *et al* (1995) did experiments with fungicidal compounds and plant extract against pathogens of pea nut, *Arachis hypogea*. Six treatments viz mother extract of *Azadirachta Indica* (2%), *Polyalthia longifolia* (5%), *Ocimum gratissium* (0.5%), TMTD (3%), carbendazium (0.1%) and a control. Observations were taken on the myclial inhibition, germination, healthy seedlings, rotted seed and seedling. It was observed that maximum mycelial inhibition of all the three pathogens, namely *Aspergillus flavus, A. niger* and *Rhizoctonia bataticola* were recorded with TMTD and carbendazim, followed by *P. longifolia*. The percentage mycelium inhibition was 99.8, 99.3 and 95.8 in *R. bataticola* and 97.0, 95.6 and 84.9 in *A. niger*. Neem, however, was not so effective as it gave mycelial inhibition as 34.1, 63.1 and 24.5 percent only in case of above three pathogens. However, maximum germination (91.1%) was observed in mother extract of *A. indica* and this extract of *A. indica* gave minimum germination (30.6%) rotted seeds and seedlings.

Anandaraj and Leela (1996) studied the toxic effect of some plant extracts on *Phytophthora capsici*, the foot rot pathogen of black pepper. Aqueous leaf extracts from *A. indica, Chromolaena odorata, Lantana camera, Piper colubrinum* and *Strychnos nux-vomica* were tested against vegetative and reproductive phases of *P. capsici, P. colubrinum* was resistant to *P. capsici*. Mycelial growth, sprorangial production, zoospore production and release, and zoospore germination were completely inhibited by *C. odorata* extracts at 2 percent. Similar results occurred with *A. indica* extracts although mycelial growth was inhibited by 75.5 percent. It is concluded that the water soluble toxic extracts of *A. indica* and *C. odorata* are ideal for developing into fungicides.

Louis *et al.* (1996) tested *in vitro* the screening of some medicinal plant extracts viz. *A. indica, Basella alba, Ocimum sanctum* and *Phyllanthus fraternus* against *P. capsici*, foot rot of black pepper. All extracts inhibited the fungus, with *A. indica* being the most effective.

Nevill and Safranyik (1996) tested the effects of neem seed extract on the growth of fungal associates of the mountain pine beetle. The fungicidal activity of neem seed extract, active ingredient azadirachtin, against *Ophiostoma clavigerum* and *O. ips*, the fungal associates of *Dendroctonus ponderosae* was evaluated. When added to malt extract agar a concentration of 100 ppm signficantly reduced the growth of both fungal associates, and concentrations above 250 ppm killed both fungi.

Rai (1996) observed a serious leaf-spot disease of mangoes (*Mangifera indica*) over the past 10 years in the Satpura plateau of India. Based on symptoms and cultural characteristics, the pathogen was identified as *Pestalotiopsis mangiferae*, not previously reported from the Satpura plateua of India. Screening of 17 medicinal plants against the pathogen revealed that 14 were antimycotic while 3 plants, *Argemone mexicana, Caesalpinia bonducella* and *Cassia fistula* accelerated the growth of the pathogen. The maximum antifungal activity was shown by *Eucalyptus globulus* (88%) and *Catharanthus roseus* (88%) followed by *Ocimum sanctum* (85.50%), *Azdirachta indica* (84.66%), *Ricinus communis* (75%) and *Lawsonia inermis* (74.33%), while the minimum activity was exhibited by *Jatropha curcas* (10%).

Rajappan *et al.* (1996) observed the management of three major rice diseases by neem formulations. Two new neem oil-based formulations [NO 60 EC(A), NO 60 EC (C)] and one neem oil-pungam oil-based formulation [NO-PO 60 EC (C)] were developed. Three dust formulations with neem derivatives like NSK 20 D, NO 10 D and NC 20 D and two dust formulations with pungam derivatives (PC 20 D and PO 10 D) were developed. Among the EC formulations, NO 60 EC (A) was tested for its efficacy against three major fungal pathogens : *Sarocladuim oryzae* (blast), and *Helminthosporium oryzae* (brown spot) in vitro condition at 9000, 4500 and 2250 ppm concentrations. All three concentrations effectively inhibited all the test fungi. Under field conditions, NO 60 EC (A) at 30 ml/lt, NO 60 EC (C) 30 ml/lt were registered as a disease percent (DP) of 23, 22 and 25 with a corresponding disease intensity (DI) of 3.2, 3.0 and 3.5 as against 40 DP and 5.6 DI in control. None of the dust formulations was effective in controlling the disease and comparable with the control.

Damayanti *et al.* (1996) tested the effect of plant extracts and fungicides on the pine apple fruit rotting fungus, *Ceratocystis paradoxa*. Antifungal activities of extracts of 16 plants were tested agaist *C. paradoxa* which causes soft rot of pineapples. *Xanthium strumarium* extract was the most effective followed by garlic extract. The effectiveness of other extracts against *C. paradoxa* occurred in the decreasing order; *Meriandra bengalensis, Mentha piperita, Curcuma longa, Phlogacanthus thyrsiflorus, Toona ciliata, Vitex negundo, Azadirachta indica, Eupatorium birmanicum, Ocimum sanctum* and *Leucas aspera*.

Govindachari *et al.* (1996) studied the antifungal activity of the cold expeller neem oil and the fractions derived through solvent partitioning, against *Drechslera oryzae* Subram. and Jain, *Fasarium oxysporum* Schlecht. ex Fr. and *Alternaria tenuis* Nees, indicated that the active fraction is a mixture of tetranortriterpenoids. Direct preparative High Performance Liquid Chromatography (HPLC) and subsequent bioassay of the semi-pure fractions showed that the active fractions contained major compounds such as 6-deacetylnimbin, azadiradione, nimbin, salannin and epoxy azadiradione. Fractions containing the azadirachtins did not show antifungal activity. Purification of the major tetranortriterpenoids, by preparative HPLC or crystalisation resulted in drastic reduction in anti-fungal activity. Mixtures of major tetranortriterpenoids affect fungal inhibitions.

Singh and Prithiviraj (1996) observed the efficacy of azadirachtin, a natural product isolated from neem (*A. indica*), against pea (*Pisum sativum* L.) powdery mildew (*Erysiphe pisi D.C.*) in a growth chamber. Azadirachtin significantly retarded several growth parameters of the pathogen, including multiple germ tube formation, number of haustoria, primary and secondary germ tube branches and colony size on pea leaves. Pre- inoculation treatment with azadirachtin was more effective in controlling disease development as compared to post- inoculation application. The effect of compound on the disease development was correlated with pheny alanine ammonia lyase (PAL) activity in pea leaves following treatment with azadirachtin. The activity increased significantly in the azadirachtin-treated leaves of pea which was dose-dependent. Statistical analysis of the data revealed signficant role of azadirachtin as resistance inducer in pea leaves against *E. pisi*.

Srivastava and Hooda (1996) tested the efficacy of neem cake as a binder of urea and azadirachtin individually or as component(s) of integrated management for control of blast (*Pyricularia oryzae*) of rice during 1992 and 1993. Neem coated urea (NCU) showed a decrease in disease and corresponding increase in yield due to slow and gradual utilisation of nitrogen. A dose of NCU (60 kg N/ha) was optimum from point of economical gain in disease supperession and yield. Among three azadirachtin formulations (Neemguard, Achook and Neemark) and fungicides (tricyclazole, carbendazim, edifenphos and hexaconazole), the formulation Neemguard appeared next best to tricyclazole and hexaconazole in terms of disease control. However, highest cost-benefit ratio was reocrded in case of hexaconazole while in Neemguard, it was approximately half of hexaconazole. Integration of Neemguard with tricyclazole revealed that two sprays of neem product and one spray of fungicide can be used effectively instead of 3 sprays of tricyclazole which is uneconomical. Integration of neem products with non-chemical means also appears effective and most economical as reflected by application of 60 kg NCU along with burnt rice husk at 400 g/ha and Neemguard spray at 0.5 percent.

Singh and Korpraditskul (1996) tested efficacy of some plant extracts on the control of *Colletotrichum capsici* (Syd.), the anthracnose disease of chilly crop. The authors studied the use of crude plant extracts as an alternative to commercial fungicide in the control of chilly anthracnose. The efficacy of neem, garlic and tagak (*Rhinocanthus nasuta*) (tagak 5000 ppm on green chilly) were comparable with the fungicide carbendazim (at 100 ppm). Neem extract was found to minimise the 'ripe chilly fruit rot'.

Gupta *et al*. (1996) studied plant extracts to control *Fusarium* diseases of mulberry. Leaf extracts of *Azadirachta indica, Calotropis gigantea, Eucalyptus* sp., *Parthenium hysterophorus* and *Pongamia pinnata* were evaluated for their antifungal activity against *Fusarium pallidoroseum* and *F. moniliforme* var. *intermedium* causing leaf blights, and *F. oxysporum* causing leaf spot diseases in mulberries. Leaf extracts (1:5) of *P. pinnata* were highly fungitoxic to *F. pallidoroseum* and *F. moniliforme* var. *intermedium* inhibiting mycelial growth on plates by 78.2 and 84.3 percent, respectively; whereas *C. gigantea* and *A. indica* were most effective against *F. oxysporum* inhibiting its mycelial growth by 78.5 and 73.2 percent, respectively. Under greenhouse conditions, aqueous leaf extracts (25%) of *P. pinnata* reduced the incidence of leaf blight caused by *F. pallidoroseum* and *F. moniliforme* var. *intermedium* by 63.3 and 67.1 percent, respectively; whereas in the case of leaf spot caused by *F. oxysporum, C. gigantea* and *A. indica* were both effective in reducing leaf spot incidence by 60.2 and 57.2 percent, respectively.

Aulerio *et al.* (1996) tested the efficacy of some natural plant extrac against some fungal pathogens of serious diseases of medicinal plants in vitr and in field studies in Casola Valsenio, Emilia-Romagna, Italy. The produc tested were micronized powder of garlic and *Equisetum arvense*, neem c and propolis. The fungal pathogens used in the laboratory tests wer *Rhizoctonia solani, Pythium ultimum* var. *ultimum* and *Fusarium solan* Only the aqueous extract of garlic (10,000 ppm of micronized powder) re duced the development of all the pathogens tested. The field studies wer conducted in 1994-95 in experimental nurseries of *Mentha piperita* an *Artemisia dracunculus* already infected by rust. Neem oil gave consistent good protection against rusts, *Puccinia menthae* and *P. dracunculina*. whic are the serious pathogens of *Mentha piperita* and *Artemisia dracunculus*.

Karthikeyan and Karunanithi (1996) studied the influence of organ amendments on the intensity of Fusarium wilt of banana. Panama diseas caused by *F. oxysporum* f.sp. *cubense* is the most important disease limitir banana cultivation. Field experiments were conducted with the banana var ety Rasthali. The treatments viz. neem cake, mahua cake (250 kg/ha each poultry manure and pleurotus decomposed coirpith (25 tons/ha each) we applied with inoculum of the fungus. The data were recorded on perce disease incidence and yield. The observation revealed that all the treatmen had a significant effect in disease control. However, neem cake (250 kg/h was the most effective treatment for the control of *F. oxysporum* f.sp. *cubens* in field experiments in Tamil Nadu, India, recording 1.7 percent diseas incidence as against 29.0 percent in control. Higher yield of 14.8 ton/ha wa recorded in neem cake treatment as against 9.2 ton/ha in control.

Panneerselvam and Saravanamuthu (1996) studied the saprophytic su vival of *F. moniliforme* [*G. fujikuroi*] in soil amended with oil cakes of groundn and neem by burying precolonized rice stem pieces. The neem and groundn oilcakes were ground and mixed with the soil to give amended concentratio of 0.5, 1, 2 and 3 percent. Unamended soil was used as the control. (*fujikuroi* was suppressed in the soil amended by both groundnut and neer but percent suppression was greater with neem cake at all concentraions. In second study, investigating the colonization by *G. fujikuroi* of sterilized substrat in amended soils, percent colonization was 76, 84, 90 and 97 in soil amend with 0.5, 1.2 and 3 percent groundnut oil cake but there was no recovery G. fujikuroi in neem-amended soils.

Khan *et al.* (1996) studied the comparative efficacy of plant extrac with fungicides in controlling the seed mycoflora of chickpeas. Seeds we subjected to a standard blotter method to isolate the mycoflora. Extracts fro *Azadirachta indica, Calotropis procera, Cuscuta reflexa, Euphorb* ulcherrima and *Solanum nigrum* were compared with benlate (benomyl), alixin (tridemorph), thiram, thiovit (sulfur) and sulfex as seed treatments. ungal species from 10 genera were recorded from the chickpea seeds. The ıost commonly isolated fungus was *Rhizopus nigricans* [*R. stolonifer*]. *olletotrichum dematium, F. oxysporum* and *M. phaseolina* were also isoited. All fungicides were effective in minimizing the occurrence of fungi. The ıost effective was banlate followed by calixin. The plant extracts also reuced the number of fungi and increased the percent germination.

Panda *et al*. (1996) studied the efficacy of homeopathic drugs (Arsenicum lbum [arsenic oxide], sulfur, Thuja, Teucrium and Baptisia) and leaf extracts om *Polyalthia longifolia, Aegle marmelos, Azadirachta indica, Cathaıus roseus, Ocimum sanctum* and *Allamanda cathartica* for the control of homopsis blight (*Phomopsis vexans*) and fruit rot in brinjal. Thuja and eucrium (each at 2%) were as effective as the fungicide bavistin (carbendazim) ı inhibiting the growth of *P. vexans*. However, leaf extracts of *A. indica* also ad no effect.

Fabry *et al*. (1996) investigated the extracts of the traditionally used ıedicinal plants, *Entada abyssinica* (stem bark), *Terminalia spinosa* (young ranches), *Harrisonia abyssinica* (roots), *Ximenia caffra* (roots), *Azadirachta ıdica* (stem bark), *Zanha africana* (stem bark) and *Spilanthes mauritiana ıcmella*] (roots and flowers) for fungistatic and fungicidal activity against *andida* spp. (*C. albicans, C. tropicalis, C. parapsilosis, C.* [*Torulopsis*} *labrata, C. guilliermondii, C. krusei*) and *Aspergillus* spp. by a microlitre erial dilution technique. *E. abyssinica, T. spinosa, X. caffra, A. indica* and *. africana* showed activity against various *Candida* spp. The minimum inibitory concentrations ranged from 0.006 to >8 mg/ml and the minimum ıngicidal concentration from 0.06 to >8 mg/ml.

Redondo *et al*. (1996) tested the extracts from figue (*Furcraea* sp.) and *zadirachta indica, Bacillus* sp., 3 *Trichoderma* sp. strains (T1C, T2C, '3C) and 10 chemicals (Banlate [benomyl], Plantvax [oxycarboxin], Fitora VP-78, Kocide 101 [copper hydroxide], Dithane M-45 [mancozeb], Ridomil 1Z-58 [mancozeb + metalaxyl], Antracol [propineb], Vitavax [carboxin], reolina and Limpido) in the laboratory for the control of *C. scoparium* on oybean. Of the fungicides, only benomyl inhibited normal developmennt of *. scoparium* (21% of fungal growth). The plant extracts inhibited fungal rowth by >20 percent with *A. indica* more effective.

Jagannathan and Sivaprakasam (1996) determined the ability of neem erivatives (neem oil and neem seed kernel extract), leaf extracts from *Vitex egundo, Acacia leucocephala* and *Polyalthia longifolia*, carbendazim, monocrotophos to control sheath rot (*Sarocladium oryzae*) of rice. All trea ments significantly reduced sheath rot compared with the control. Plants treate with NSKE, *V. nigundo, A. leucocephala, P. longifolia* and carbendazim ha higher yields than the control.

Gupta *et al.* (1996) studied plant extracts to control *Fusarium* disease of mulberry. Leaf extracts of *Azadirachta indica, Calotropis gigantea, Euc lyptus* sp., *Parthenium hysterophorus* and *Pongamia pinnata* were evalu ated for their antifungal activity against *F. pallidoroseum* and *F. moniliform* var *intermedium* causing leaf blight, and *F. oxysporum* causing leaf spc diseases in mulberries. Leaf extracts (1:5) of *P. pinnata* were highly fung toxic to *F. pallidoroseum* and *F. moniliforme* var. *intermedium* inhibitin mycelial growth on plates by 78.2 and 84.3 percent, respectively; wherea *C. gigantea* and *A. indica* were most effective against *F. oxysporum* inhibi ing its mycelial growth by 78.5 and 73.2 percent, respectively. Under greer house conditions, aqueous leaf extract (25%) of *P. pinnata* reduced the inc dence of leaf blight caused by *F. pallidoroseum* and *F. moniliforme* va *intermedium* by 63.6 and 67.1 percent, respecively; whereas in the case c leaf spot, *C. gigantea* and *A. indica* were both effective in reducing leaf spc caused by *F. oxysporum* by 60.2 and 57.2, respectively.

D' Aulerio and Zambonelli (1997) presented a list of the fungal disease of plants found in an epidemiological survey conducted in Emilia-Romagna Italy. Almost all species of the plants surveyed were affected with fungi ex cept *Leonurus cardiaca, Thymus vulgaris* and *Valeriana officinalis*. Th fungi affected the epigeal parts of the plants. Wild species were less affecte than cultivated species, where overcrowding favoured infections. Control tr als were carried out in the laboratory and field in 1994-95 using natura products; essential oils of thyme (*T. vulgaris*), lavender (*Lavandula angustifolia* and mint (*Mentha piperita*); neem oil; micronized powder of garlic (*Alliur sativum*) and *Equisetum propolis*. In the laboratory, essential oil of thym was active against all fungal species tested (*Phythium ultimum*, *Fusariur solani* and *Colletotrichum lindemuthianum*) from a minmum concentratio of 50 ppm. Mint and lavender oils were only effective at higher concentratio (400-800 ppm). In the field, neem oil gave protection against rusts on mir and tarragon (*Artemisia dracunculus*) in both years. Studies are being carrie out to identify additives that can prolong the persistence of the natural sub stances on the vegetation.

Singh and Prithiviraj (1997) studied the activity of NeemAzal against pe powdery mildew in detached leaf and intact plant experiments. Growth of *E pisi* was significantly retarded and a hypersensitive reaction was induced i the host.

Ushamaline *et al.* (1997) studied the effect of biocontrol agents and plant products on *Macrophomina phaseolina* (causal agent of charcoal rot of cowpea) and other soil microorganisms. Soil application of neem cake (150 kg/ha) and FYM (10 ton/ha) reduced charcoal rot incidence significantly and increased crop yield. Both *Trichoderma viride* and *T. harzianum* recorded a root rot incidence of 17.0 and 17.6 percent, respectively, compared with 38.3 percent in the control. The lesser charcoal rot incidence of 0.93, 1.24, 3.39, 4.01, 3.70 and 1.54 percent at 20, 30, 40, 50, 60 and 70 days was observed in neem cake-amended plots. However, neem cake amended plot recorded higher grain yield of 502.0 kg/ha as against control 388.2 kg/ha. The population of *M. phaseolina* was low both at 20 and 60 days in plots amended with neem cake. It is thus observed that the charcoal rot incidence and the population of *M. phaseolina* was significantly less in soil amended with neem cake. This may be due to the fact that antagonistic activity of the microflora is enhanced, besides stimulating germination of lysis of sclerotia of *M. phaseolina*.

Philip annd Sharma (1997) tested the fungitoxic effects of leaf and oil cake extracts of neem and karanj (*Pongamia pinnata*) on the mycelial growth, spore production and spore germination of *Fusarium solani* and *F. oxysporum* causing root rot disease in mulberries. Both extracts had inhibited germination and growth of both pathogens. The neem extracts were the most effective against both fungi.

Zaman *et al.* (1997) tested seed borne fungi of mustard and their control with indigenous plant extracts. Seven fungal genera namely *Alternaria, Fusarium, Aspergillus, Penicillium, Rhizopus, Chaetomium* and *Curvularia* were associated with mustard seeds. Percentage incidence of different fungi varied with location of collection and duration of storage period. Four plant extracts tested were effective in decreasing the prevalence of seedborne fungi. However, garlic and neem leaf were superior among the extracts followed by ginger and onion bulb. All the extracts gave highest control when used in crude form and their efficacy declined with increasing dilution.

Pani and Patra (1997) utilized plant extracts for the control of *Sclerotium rolfsii. In vitro* and *in vivo* studies to determine the effect of extracts of *Azadirachta indica, Psidium guajava, Lantana camara, Sopindus trifoliat* [*S. trifoliatus*], *Rauwolfia* [*Rauvolfia*] *serpentina*, *Lawsonia inermis, Agle marmelos, Cynodon dactylon, Tamarindus indica, Eichhornia crassipes, Adhatoda vasica, Moringa oleifera, Pongamia glabra* and *Tagetes erecta* on the mycelial growth of *V. volvacea* and *Sclerotium* [*Corticium*] *rolfsii*. The leaf extract of *T. indica*, followed by the seed extract of *S. trifoliat* and the root extract of *M. oleifera* supppressed the growth of *S. rolfsii*. Paddy straw inoculated with *S. rolfsii* and treated with *T. indica* leaf extract resulted in the highest sporophore yield followed by *S. trifoliat* seed extract and *M. oleifera* root extract. However, neem extract had very little effect.

Pasini *et al.* (1997) studied the control of rose powdery mildew. Weekly sprays of the mycoparasite *Ampelomyces quisqualis* (AQ 10 Biofungicide), of the mineral salt KH^2PO_4, of potassium salts of fatty acids (MYX F 4020 and MYX F 403.02-050), of a canlola oil (Synertrol), of wine vinegar, or neem extract (FU-3 Trifolio M) and a petroleum oil (JMS Stylet-oil) provided good control of *Sphaerotheca pannosa* var. *rosae* on roses, being as effective as dodemorph, in 2 separate experiments carried out under greenhouse conditions.

Das *et al.* (1997) evaluated some plant extracts against *Sclerotium ralfasii* causing stem rot in tuberose. Extacts from *Adhatoda vasica, Moringa oleifera, Azadirachta indica, Lawsonia inermis, Sapindus trifoliatus, Rauvolifia serpentina, Pongamia glabra* [*P.pinnata*] and *Lantana camara* were tested in the laboratory against *S. rolfsii* [*Corticum rolfsii*], the causal of stem rot of tuberose (*Polianthes tuberosa*). Root of *M. oleifera* and seed extract of *S. trifoliatus* at 1o percent showed maximum antifungal acivity, causing 96.6 and 82.7 percent growth inhibition of *C. rolfsii*, respectively. However, neem extract did not show any effect.

Kurucheve *et al.* (1997) screened 13 plants for fungitoxicity against *Rhizoctonia Solani in vitro*, the causal agent of sheath of paddy rice. Fresh leaf sample of *Allium cepa, Azadirachta indica, Caesalpinia pulcherrima, Eucalyptus globulus, Calotropis gigantea, Ipomoea carnia* [*carnea*], *Lawsonia inermis, Ocimum sanctum* [*tenuiflorum*], *Parthenium hysterophorus, Piper betel* [*betle*], *Pongamia glabra* [*pinnata*], *Prosopis juliflora* and *Thevetia peruviana* were collected and cold and hot water extracts were prepared. The maximum inhibition of mycelial growth was observed with the cold water extract of *P. juliflora* followed by *T. peruviana*. With hot water extraction, *T. peruviana* was the most fungitoxic followed by *P. juliflora*. Complete inhibition of sclerotial production was recorded with the cold water extracts of *C. pulcherrima, E. globulus* and *L. inermis*. The fungitoxicity of *C. pulcherrima* and *E. glubulus* was not lost even after air drying or autoclaving.

Ram (1997) studied the *in vitro* fungitoxicity of leaf extracts of *Cassia tora, Azadirachta idica, Anisomeles ovata, Aegle marmelos, Adhatoda vasica, Mentha arvensis, Dalbergia sissoo, Tinospora cordifolia, Pongamia pinnata, Cyperus rotundus, Ocimum adscendens* and *O. sanctum*, a resin extract of *Ferula foetida* and bulb extracts of *Allium sativum* and *A. cepa* against *Alternaria brassicae*. The leaf extract of *O. sanctum* was found to be most effective and completely inhibited spore germination of *A. brassicae*, the causal agent of *Alternaria* blight of mustard at 10,000 ppm. *A. sativa* and *A. cepa* were next in efficacy and inhibited spore germination by 40 percent at 10,000 ppm. However, neem formulation was less effective.

Valf and Steinhauer (1997) studied fungicidal activity of neem leaf extracts. To evaluate the fungicidal effect of neem leaves, a methanolic extract of ground dry leaves was prepared. After evaporation of the solvent, the residue was redissolved in a small amount of methanol and added to an agar medium. Growth of the test fungi *Phytophthora palmivora, Pythium debaryanum, Drechslera avenae* [*Pyrenophora chaetomioides*] and *Fusarium culmorum*, was inhibited by 10-65 percent, depending on the fungus sensitivity and the origin of the leaves (India or Dominican Republic). Powdery mildew of cucumber (*Sphaerotheca fuliginea*) was not inhibited by the extracts, the growth of the fungus was even promoted. Black spot of roses (*Diplocarpon rosae*) was influenced by one extract (origin Dominican Republic) in a detached leaf assay. The size and number of acervuli, the formation of yellow halos and the number of conidia/mm^2 were reduced.

Osnaya *et al*. (1997) studied the effect of alternative plant protection materials on black spot of roses. Preventive treatments with 0.5 and 1 percent solutions of $NaHCO_3$, K_2HPO_4 as well as formulated neem and rapeseed oils were tested singly and in combination for their efficacy against *Diplocarpon rosae*, the causal agent of black spot of roses. Detached leaflets of *Rosa canina* cv. *pollmeriana* were inoculated 24 hrs after the applications with 10 micron l drops containing 2000 conidia. After 48 and 72 hrs conidial germination was assessed, followed by an evaluation of halo formation, size of necrotic lesions, quantity of acervuli and number of conidia/mm^2 and conidia/lesion, 12 days after inoculation. The most effective treatments were combinations of formulated rapeseed oil (TELMION) with $NaHCO_3$, K_2HPO_4, resulting in an efficacy >90 percent. However, neem extract was less effective.

Fiume (1997) found that powdery mildew caused by *Oidiopsis* [*Leveillula*] *taurica* is responsible for infection of sweet pepper (*Capsicum annuum*) plants cultivated in greenhouse. In greenhouse conditions of southern Italy, the disease can cause high levels of damage even in late autumn and winter at high RH. In order to control the disease, trials were carried out to test the effectiveness of 2 new fungicides. The research included the following treatments : (i) untreated control, (ii) bupirimate, (iii) tetraconazole, (iv) neem oil extract, (v) bupirimate with neem oil and (vi) tetraconazole with neem oil. The experimental design was a randomized complete block, replicated thrice. Phytopathological, biological and physiological observations wer made at 10 day-intervals, 10 days after the first treatment. The percentage/plant of old and new leaves with powdery mildew was recorded. The most effective treatments were tetraconazole alone or combined with neem oil. The McKinney index ranged from 3 to 11 percent and from 3 to 7 percent, respectively, and there was no increase in diseased leaf number. These plants had the highest number of leaves (up to 400 leaves/plant), grew the highest and had the greatest number of lateral shoots.

Khilare and Gangawane (1997) used plant extracts in the management of thiophanate methyl resistant *Penicillium digitatum* causing green mold of Mosambi (*Citrus sinensis* L.). Twelve ioslates of *P. digitatum*, collected from infected Mosambi fruits in different orchards in Maharashtra, India, were screened for their sensitivity to thiophanate-methyl. The sensitivity tests were carried out at different concentrations i.e. 500-3500 micron g/ml. Aqueous extracts of plants were prepared by grinding 10 g fresh leaves with 10 ml distilled water and the extract was considered as 100 percent. Maximum inhibitory concentration (MIC) ranged from 2921 micron g/ml to 3977 micron g/ml on agar plates. Mosambi fruits treated with thiophanage-methyl alone and in combiation with aqueous extracts from 25 medicinal plants were inoculated with the highly resistance mutant EMS-PD-7 (resistant factor 7). Percentage control efficacy (PCE) was calculated. In general all the plant extracts were found to be individually effective. The PCE ranged from 22.4 to 72.4. *Azadirachta indica* was found to be highly effective. Most of the plant extracts increased PCE over thiophanate-methyl alone. Thiophanate-methyl combined with *A. indica, Calotropis procera, Terminalia chebula, Curcuma longa, Ocimum sanctum, Catbaranthus roseus* and *Zingiber officinale* extracts were the most effective, with PCEs of >70.

Coventry *et al.* (1997) observed the effect of neem based products on bacterial and fungal growth. The neem extract preparations NeemAzal-F and NeemAzal-T/S, and their active neem ingredient NeemAzal pulver, were screened *in vitro* for the presence of antibacterial and antifungal activity. NeemAzal-F and a 1 percent NeemAzal pulver solution (3000 ppm azadirachtin) inhibited the growth of *Bacillus cereus, B. mycoides, B. thuringiensis, B. subtilis, Erwinia carotovora* subsp. *carotovora* and *Pseudomonas syringae* pv. *phaseolicola* [*P. savastanoi* pv. *phaseolicola*] (Psp). These bacteria, with the exception of Psp, were very sensitive to NeemAzal-F but showed no sensitivity to NeemAzal-T/S. A 1 percent aqueous solution of NeemAzal-F

(500 ppm azadirachtin) completely inhibited colony radial growth of all fungus, *Gaeumannomyces graminis*. A NeemAzal pulver solution with the same concentration of active ingredient also reduced the colony radial growth rate significantly compared with the control. The results reveal the presence of antibacterial and antifungal activity in the neem extract.

Singh *et al*. (1997) evaluated the plant extracts of *Catharanthus roseus, A. indica, Pongamia pinnata, Tagetes erecta*, onion bulb and garlic extracts at 1, 2, 3, and 4 percent concentrations for radial growth and spore germination of *Colletotrichum capsici*, incitant of die-back of chilli (*Capsicum*). Extracts of garlic bulb at 3 percent concentration completely inhibited the growth and spore germination of *C. capsici* whereas, 4 percent extracts of onion bulbs, *A. indica leaf, P. glabra* leaf and *T. erecta* gave complete inhibition of fungal mycelial growth and spore germination.

Ramanujam *et al*. (1997) conducted a field trial in stem bleeding (a soilborne disease caused by *Thielaviopsis paradoxa* [*Ceratocystis paradoxa*])-affected gardens in India using either carbendazim or tridemorph, *Gliocladium virens*, a sealant (coal tar) and 2 organic amendments (neem [*A. inidca*] cake and farm yard manure) in various combinations. Observations after 4 years of treatment showed the lowest disease index (2.62) and the highest nut yield (59.4 nuts/palm annually) in the palms treated with tridemorph (4%) along with coal tar sealing and soil application of *G.* virens + neem cake + FYM.

Srivastava *et al*. (1997) investigated the fungicidal properteis of neem seed kernel powder (NSKP), karanj (*Pongamia pinata*) seed kernel powder (KSKP), neem leaf powder (NLP) and neem seed oil (NSO) against the mycoflora of neem seeds. NSKP, KSKP and NLP were used with both dry and wet seeds. The seeds were pelleted/treated with the plant preparations for 5 minutes before planting and incubating. Fungal growth was less from all the treated seeds than from control untreated seeds. NSO was the most effective treatment. NSKP, KSKP and NLP were more effective used with wet seeds than dry seeds. Neem seeds treated with NSO and NLP (on wet seeds) inhibited the growth of *Aspergillus flavus, Penicillium* sp. and *Mucor* sp. completely.

Suresh *et al*. (1997) observed antifungal fractions and compounds from uncrushed green leaves of neem. Using groundnut rust desease (*Puccinia arachidis*) as a bioassay system, 2 limonoids from the neem tree (*A. indica*) which evinced antifungal activity, were isolated by extraction, solvent partitioning reduced disease intensity considerably. The polar extract and impure HPLC fractions were more effective than the pure compounds in reducing the pustule numbers and, consequently, the disease severity.

Naik and Shivanna (1997) carried out a study on the control of mildew caused by *Oidium* sp. on polybag seedlings of *Acacia auriculiformis* in the nursery, in karnataka, India during kharif 1996. Treatments tested were : Calixin (tridemorph) and Bavistin (carbendazim), both at 0.1 percent; wettable sulfur at 0.3 percent; and Nimbicidin and *Clerodendron* leaf extract, both at 0.5 percent. Callixin was the most effective treatment, reducing disease incidence by 57.0 percent, followed by Bavistin (40.1%) and Nimbicidin (39.4%). The Clerodendron leaf extract was least effective (18.6%).

Shivpuri *et al*. (1997) tested fungicidal properties of 10 plant species extracts against pathogenic fungi. Ethanol extracts of 10 plant species (*Allium cepa, A. sativum, Azadirachta indica, Calotropis procera, Datura stramonium, Ocimum sanctum* [*O. tenuiflorum*], *Polyalthia longifolia, Tagetes erecta, Vinca rosea* [*Catharanthus roseus*] and *Withania somnifera*) were assessed against 5 pathogenic fungi (*Alternaria brassicola, Colletotrichum capsici, Fusarium oxysporum, Rhizoctonia solani* and *Sclerotinia sclerotiorum*) when tested under laboratory conditions at 2 concentrations (500 and 1000 micron g/ml). The leaf extracts of *A. indica, O. stramonium, O. sanctum, P. longifolia* and *V. rosea* were found more fungitoxic than other extracts. Their efficacy was more pronounced at 1000 micron g/ml as evidenced from the low radial growth.

Srivastava and Lal (1997) studied the biofungicidal properties of leaf extracts of some plants. An *in vitro* test indicated fungicidal properties in aqueous leaf extractsof *Calotropis procera, Azadirachta indica, Lantana camara* and *Ocimum basilicum* against *Curvularia tuberculata* and *Alternaria alternata* isolate 'A' obtained from fruits of pear and *A. alternata* isolate 'B' from fruit of pomegranate (*Punica granatum*). The observations revealed 64 to 85 percent control of fruit rot of pears and pomegranate.

Govindachari *et al*. (1998) did work on the identification of antifungal compounds from the seeds of neem. Evaluation of the activity of neem oil from *A. indica*, and the fractions derived through solvent partitioning, against *Drechslera oryzae* [*Cochliobolus miyabeanus*], *Fusarium oxysporum* and *Alternaria tenuis* [*A. alternata*], showed that the active antifungal fraction was a mixture of tetranortriterpenoids. The triterpenoid mixture derived from 90 percent methanol (MeOH) extraction of neem oil was investigated for antifungal activity against 13 phytopathogenic fungi. Direct preparative HPLC of the active fractions, and subsequent bioassay of the semi-pure fractions, suggested that the active fractions contained major compounds such as 6-deacetylnimbin, azadiradione, nimbin, salannin and epoxy-azadiradione. Pure azadiradione, nimbin, salannin and epoxy-azadiradione did not show antifungal activity. However, when these terpenoids were mixed, they showed antifungal activity, suggesting possbile additive/synergistic effects.

Hooda annd Srivastava (1998) studied the impact of neem coated urea (NCU) and potash on the incidence of rice blast (*Magnaporthe grisea*) was investigated in Hisar, India. All 3 levels of NCU used (30, 60, 90 kg N/ha) were effective in reducing the disease. However, for neck and node blast incidence, NCU at 30 kg N/ha had no effect compared with controls. NCU at 60 and 90 kg N/ha significantly reduced disease compared with controls. The highest cost:benefit ratio was recorded for NCU at 60 kg N/ha, followed by 90 kg N/ha. The 3 rates of potash used (15, 30 and 45 K_2O/ha) had no effect on the incidence of rice blast or yields.

Narasimhan *et al.* (1998) evaluated neem oil (NO) and pungam oil (PO) based emulsifiable concentrate (EC) formulations, viz., NO 60EC (acetic acid), NO 60EC (citric acid) and NO + PO 60EC (citric acid), which had been developed for their efficacy against sheath rot (*Sarocladium oryzae*) of rice. All 3 formulations effectively inhibited the mycelial growth of the causal pathogen under *in vitro* conditions. There was no significant difference between efficacy of the freshly prepared and stored formulations in arresting the growth of *S. oryzae*; efficacy was maintained even after 9 months of storage. These formulations effectively controlled rice sheath rot and led to increased yield.

Hundekar *et al.* (1998) observed that of the organic amendments tested in pot experiments, neem, cotton, groundnut and safflower cakes reduced the inoculum levels of *Macrophomina phaseolina* and *Fusarium moniliforme* [*Gibberella fujikuroi*] in the soil, and reduced the incidence of stalk rot of sorghum. Except for with neem cake, the populations of beneficial bacteria, fungi annd actinomycetes were higher in amended soils than in non-amended soils.

Enikuomehin *et al.* (1998) screened ash samples, from organs of 9 tropical plants for their abilities of inhibit mycelial growth and sclerotial germination of a Nigerian isolate of *C. rolfsii* [*Corticium rolfsii*] on agar and in the soil. Of the 11 samples tested, 10 showed some activity against mycelial growth of *C. rolfsii in vitro*. Ash samples from 9 sources protected seeds against pre-emergence rot. Ash from *M. indica* leaf, *V. amygdalina* leaf and *Azadirachta indica* leaf, protected seedlings against post-emergence infection.

Ehteshamal Haque *et al.* (1998) observed soil amendment with neem seed cake, cotton seed cake, *Datura fastuosa* and *Steochospermum marginatum* significantly reduced *Fusarium solani* infection of sunflowers in

Pakistan. Neem seed cake and cotton seed cake were also effective against *Macrophomina phaseolina* and *Rhizoctonia solani*, while *S. marginatum* and *D. fastuosa* were effective againnst *R. solani* infection of sunflower roots. *P. aeruginosa* significantly controlled root rot disease of sunflower caused by *M. phaseolina, R. solani* and *F. solani. P. aeruginosa* combined with neem seed cake or S. marginatum produced greater fresh weight of shoots and plant height, respectively, compared with separate applications.

Hundekar *et al*. (1998) tested *in vitro* evaluation of organic amendments against stalk rot of sorghum. Groundnut, neem, cotton, sunflower cake, wheat straw and farm yard maure were used and found effective.

Prashad *et al*. (1998) studied survival of sclerotia of *R. solani* in rice soils amended with oil cakes and green leaf manures. The effect of oil cakes (neem cake, groundnut cake, sesame cake) and 3 green leaf manure viz. *Calotropis* sp., *Crotalaria* sp. and *Phaseolus* sp. on sclerotial survival of *R. solani* (causing rice sheath blight) was studied in 3 rice land soils in Bapatla, Andhra Pradesh, India. Neem cake and *Calotropis* were the most effective in affecting the survival of sclerotia in the soil.

Pramanick and Phookan (1998) tested effect of 10 plant extracts in the management of sheath rot (*Sarocladium oryzae*) of rice inhibited mycelial growth *in vitro*. An aqueous extract of *Ocimum sanctum* was the most effective followed by *Eucalyptus citriodora* and *Azadirachta indica*. Lesions formed by the pathogen on rice plants (cv. IR-36) were also significantly restricted by pre-inoculation spray with these three plant extracts.

Fajardo *et al*. (1998) conducted a study to evaluate the efficacy of selected inducing agents in reducing the incidence serverity of green mould caused by *Penicillium digitatum* in mature oranges (cv. Valencia). The inducing agents tested were Margosan-O (a neem product), Aspire (biological post-harvest disease controlling agent) consisting of a water dispersible granule containing an antagonistic yeast *Candida oleophila* and chitosan (prepared from crab shells). These agents and chitosan + Aspire reduced fruit decay by 38, 41, 42 and 44 percent, respectively.

Rizvi *et al*. (1998) tested the effect of botanical pestcides in the germination of rust spores of lucerne. Several botanicals mainly containg azadirachtin as active ingredient have been tried for the bioassay against lucerne rust caused by *Uromyces striatus* Schroter. These botanicals behaved differently in inhibiting the uredospore germination in different dilutions and combination. The results suggests that botanicals pesticides would be a potential method of disease control and increase the crop productivity.

Chauhan and Singh (1998) tested *in vitro* effect of some neem products on mycelial growth of *Phytophthora drechsleri* f.sp. *cajani*. Neem products viz., Neemgold, Neemta, Repelin, Achook and Nimbicidine were evaluated against mycelial growth of *P. drechsleri* f.sp. *cajani*, the causal organism of Phytophthora blight of pigeonpea. Of these, Achook was most effective and completely checked the radial growth of the fungus in 14 and 21 days at 24 and 48 micron g/ml concentrations, respectively. Replin was also effective at 48 micron g/ml and inhibited the growth of mycelium, till 24 days. The effect of Achook at 24 micron g/ml was fungistatic. However, the concentration of 48 micron g/ml of Achook and Replin was fungicidal.

Jeylakshmi *et al*. (1998) tested neem oil against chilli fruit rot caused by *Colletotrichum capsici* (Syd.). Butler and Bisby, the causal organism of fruit rot of chillies under *in vitro* and pot culture. Investigations were carried out to assess the efficacy of different parts of neem viz., neem oil (3%), neem leaf (10%), NSKE (5%) and neem cake (10%) against *C. capsici*. Of these neem oil (3%) recorded the maximum inhibition of mycelial growth (67.52%) and spore germination (67.11%). In pot culture, two sprays of neem oil (3%) at 105 and 120 days after sowing recorded the maximum reduction (56.41%) of fruit rot intensity. However, carbendazim (0.1%) showed the maximum fungicidal effect.

D'-Aulerio *et al*. (1998) carried out tests in an experimental nursery in Casola Valsenio, Ravenna, Italy, during 1996-97 to investigate the efficacy of natural substances, adminstered alone or together with an additive (containing a mixture of plant and mineral oils with biodegradable surfactants), against *Puccinia menthae* and *P. dracunculina* on mint (*Mentha piperita*) and tarragon (*Artemisia dracunculus*). The natural products tested were micronized powder of garlic, neem oil, thyme (*Thymus vulgaris*) oil, lavender oil, mint oil and mustard (*Brassica nigra*) oil; these were compared with the chemical standard penconazole (Topaz 10% a.i. at 0.8 g/lt water). The additive notably increased the efficacy of the natural products. In both years of the study, all of products tested with the additive gave effective control of the rusts. In 1996, neem oil at 5 mg/lt was as effective as penconazole. Mustard oil, tested for the first time in this study, showed notable fungicidal activity at reduced doses, but was phytotoxic at higher doses.

Jeyalakshmi and Seetharaman (1998) did tests on biological control of fruit rot and die-back of chilli with plant products and antagonistic microorganism. Of the plant products and antagonists tested in the field in Tamil Nadu, India, palmarosa (*Cymbopogon martinii*) oil was the most effective for reducing fruit rot and die-back of *Capsicum* caused by *C. capsici*, followed by

Ocimum sanctum leaf extract, neem oil and *Saccharomyces cerevisiae*.

Lokhande *et al*. (1998) tested effect of different fungicides and neem products for the control of leaf spot of groundnut. The results of field trials conducted at Yavatmal, Maharashtra, India, on groundnut variety TAG-24 revealed that a combined spray of carbendazim + tridemorph was significantly superior to the other fungicide treatments viz. mancozeb, carbendazim and tridemorph alone in respect of lowest percentage disease incidence of leaf spot (19.48%), caused by *Mycosphaerella arachidis* and highest pod yield (20 q/ha). The neem products (neem seed extract and neem oil) were less effective than the fungicides but the treatments were significantly superior in comparison with the control.

Singh and Korpraditskul (1999) compared the efficacy of neem, garlic and tagak-tagak (*Rhinocanthus nasuta*) at 5000 ppm on green chilly with the fungicide carbendazim at 100 ppm. Green chilly fruits in 100 numbers were cleaned and sterillized and were dipped in *Colletotrichum capsici* spore suspension for 2 min and air-dried. Treated fruits were incubated over night. The chilly fruits were dipped for a 5 seconds in each plant extract and carbendazim. The results indicated that crude plant extracts of neem, garlic and tagak-tagak showed a significant effect ($P<0.01$) in the expression of spots on chilly when compared to control. The garlic extract was best. Neem extract minimized the ripe chilly rot. However, Korpraditskul *et al*. (1991) used these extracts successfully to control anthracnose of mango fruit (*C. gloeosporioides*).

Rettinassababady and Ramadoss (2000) studied the effect of neem products in the control of powdery mildew of rice fallow blackgram (*Erysiphe polygoni* DC). Five plant extracts/products namely neem seed kernel extract (NSKE 5%), neem oil (3%), neem cake extract (10%), *Prospis* leaf extract (10%) and *Ipomea* leaf extract (10%), carbendazim (0.1%) and mancozeb (0.2%) were secreened for antifungal properties. Among all NSKE at 5 percent was found to be superior in controlling the disease.

2. Neem and Mycorrhizal group of fungus

Habte *et al*. (1993) studied the effect of neem to mycorrhizal colonization. Thirty two-day-old seedlings of neem (*Azadirachta indica*) were transplanted into pots containing a subsurface Oxisol uninoculated or inoculated with *Glomus aggregatum* at three target soil solution P cocentrations. While no evidence of vesicular-arbuscular mycorrhizal fungal (VAMF) colonization was noted in uninoculated soil, neem roots were colonized in excess of 80 percent in inoculated soil, and soil. P did not influence VAMF colonization significantly. VAMF colonization led to significant increases in tissue P and dry matter accumulation at the first two increments of soil P, but not at the highest concentration. Dry matter yield of mycorrhizal neem at soil P concentration of 0.02 mg/lt was comparable to nonmycorrhizal growth of neem at soil P cocentration of 0.2 mg/lt. Based on these data neem was classified as a highly mycorrhizal-dependent species.

Phavaphut Anon *et al.* (1996) studied growth and nutrient uptake of neem seedlings in response to inoculation with the vesicular-arbuscular mycorrhizal (VAM) fungi *Glomus intraradices*, and the addition of 2 levels of phosphorus (0.65 and 1.30 mM P). Extensive VAM colonizatin in neem roots occurred at both P levels, and growth of VAM plants was similar at both P levels, while the growth of non-colonized (non-VAM) plants increased with increasing P supply. At the low P level, VAM plants had greater leaf area, shoot, root and leaf dry weight, and a greater root:shoot ratio than non-VAM plants. Low P, VAM plants had a greater leaf area and root dry weight and a comparable leaf and shoot dry weight and root:shoot ratio, when compared with high P, non-VAM plants. VAM inoculation also altered dry mass partitioning to root systems, resulting in greater root length and dry weight of suberized roots in VAM plants. The length of non-suberized roots increased with increasing P supply regardless of VAM inoculation. VAM inoculation increased leaf tissue concentration of N, P, S and B. It is concluded that there are benefits in VAM inoculation for growth enhancement and nutrient uptake of neem tree seedlings under low phosphorus conditions.

Srinivasan *et al.* (1996) assessed the effect of effulent irrigation on VAM status of 3 crop species (*Capsicum annum, Pennisetum typhoides* [*P. glaucum*], *Sorghum bicolor*), 4 tree species (*Azadirachta indica, Carica papaya, Musa paradisiaca, Psidium guajava*), and 7 weed species (*Ageratum conyzoides, Calotropis procera, Croton sparsiflorus, Euphorbia birta, Parthenium hysterophorus, Sidia acuta, Tridax procumbens*). The incidence of VAM infection was greater for weeds than for the crop and tree species. Effluent irrigation markedly altered the nutrient status of the soil and significantly increasing the root infection in *S. bicolor* and *P. guajava* whilst significantly reducing spore numbers in all plant species except *A. indica*.

Wan *et al.* (1997) determined the disappearance of azadirachtin-A (AZA-A) in the growth substrates of two biological systems in which one contained a gellan gum-based minimal medium (M-medium), and the other was vermiculite. AZA-A degradation in these media following pseudo-first order kinetics. The half-life of technical AZA-A was 44.4 days in thhe M-medium system, which contained the mycorrhizal fungus *Glomus intraradices* growing on Ri T-DNA transformed roots of carrots. In the vermiculite system, which contained the same spcies of symbiotic fungus growing on the roots of leeks, the half-life ranged from 13.2 to 46.2 days for technical and formulated AZA-A, respectively. There was evidence to suggest the occurrence of unidentified AZA-A conversion products in the treated and aged substrates of the two systems, as well as the presence of AZA-B residues in products formulated with natural neem extracts.

Wan and Rahe (1998) assessed the impact of the insecticide azadirachtin, neem extracts, and their products, Eneem 3G and Neemix R 4.5E, on the symbiotic VAM fungus *G. intraradices* by short-term subacute toxicity tests using root inducing transferred DNA (Ri T-DNA) transformed roots of carrots (*Daucus carota*) as the host in the minimal medium system. The concentration causing 50 percent inhibition of growth and development was determined, and effects of these test materials on carrot roots and colonization by *G. intraradices* were also evaluated. The inhinitory effect of azadirachtin and neem compounds was compared with that of several reference pesticides (benomyl, chlorothalonil, copper sulfate, dimethoate, glyphosate, and its metabolite aminomethyl-phosphonic acid). Azadirachtin, neem extracts, Eneem and Neemix, including their carriers and slovents, had low short-term toxicity to *G. intraradices* when compared with refernce pesticides. Neither *D. carota* nor the VAM colonization was affected by these compounds. The potential ecotoxicological impact of azadirachtin on the terrestrial environment is evaluated.

3. Bacterial diseases

Chaurasia and Jain (1978) studied the antibacterial activities of essential oils. Essential oils, obtained from seeds of *Anethum graveolens, Azadirachta indica* and *Cinnamomum zeylanicum* and from leaves of *Pongamia pinnata*, were evaluated *in vitro* for their antimicrobial activity against eleven bacteria, potentially pathogenic to plants and animals. All the oils were effective against 10 of the organisms but none was effective against *Shigella* spp.

Pierce (1981) observed that neem product AZ at 0.25 percent controls bacterial wilt caused by *Pseudomonas solanacearum* (E.F. Smith).

Jani *et al.* (1987) studied the effect of fatty acids from oil seeds as antibacterial agent. Fatty oils extracted from the seeds of 6 species (*Cassia tora, Wrightia tomentosa, Viburnum coriaceum* [*V. cylindricum*], *Michelia champaca, Pongamia pinnata* and *Azadirachta indica*) were tested *in vitro* against 5 species of bacteria, viz. *Bacillus pumilus, B. subtilis, Salmonella typhosa* [*S. typhi*], *S. paratyphi, Micrococcus pyogenes* var. *albus*, and *Staphylococcus aureus*. With 2 exceptions, all the oils showed antibacterial activity against all the bacteria. The exceptions were that oil from *P. pinnata* was not effective against *B. pumilus*, and that oil from *C. tora* was not effective against *M. pyogens*. Ali and Sanougi (1989) tested sensitivity of different aerobic bacterial species to azadirachtin. The azadirachtin obtained from leaves of neem did not inhibit growth of 4 organisms tested.

Pandey and Prashad (1992) evaluated *in vitro* certain preparations of neem kernel against peper bacterial leaf spot incited by *Xanthomonas compestris* (Cobb.) Dowson pv. *vesicatoria*. Neem oil as 5 percent EC and neem kernel at 4 percent were found bactericidal.

Vasishtha *et al*. (1992) observed that the bitter compound obtained from neem oil contained antibacterial activity. A 3-stage counter-current pretreatment (8 hrs for each stage) of neem seed kernels using aqueous ethanol (85%) gave negligibles of bitter (0.2%), and odoriferous compounds. Raw neem seed kernels contained about 5.5 percent bitter compounds and these bitter compounds when separated from the seed oil could be used as antibacterial preparations, insect repellents and medicines.

Maharishi (1993) observed encouraging results in controlling bacterial leaf spot of chilli caused by *Xanthomonas campestris* pv *vesicatona* with 2 percent of aqueous leaf extract of neem. Eswaramurthy *et al*. (1993) found that to control bacterial blight of rice, *X. campestris* pv *oryzae*, 2 percent spray of NSKE is effective.

Patil and Ghoderao (1997) did laboratory studies and found that plant extracts from *A. indica* and *Ipomea carnea* were effective at reducing the incidence and intensity of bacterial blight, *X. campestris* cv. *malvacearum*.

Hulloli *et al*. (1998) studied the management of bacterial blight of cotton induced by *X. axonopodis* pv *malvacearum* with the use of neem-based formulation. The MIC of the neem-based formulations plantolyte and agricare against *X. axonopodis* pv. *malvacearum* was 1 ug/ml. This was lower than the MIC of aminoglycosides, streptomycin and kanamycin (2.5 μg/ml). When mixed at 0.5 μg/ml with antibiotics/fungicides, these neem formulations reduced their MIC and completely inhibited the development of drug resistant mutants, suggesting that they acted not only synergistically with the antibiotics/fungicides but also reduced their hazardous effects. The neem formulations controlled cotyledonary infection alone or in combination with antibiotics at very low doses. It is concluded that neem-based formulations could prove useful for management of bacterial blight of cotton.

4. Neem Mixture with biopesticide in controlling diseases

Rajan *et al*. (1991) studied management of root rot disease of fenugreek. A field trial was laid out with 9 treatments using CO-1 variety in randomised block design. The treatments were seed dressing with carbendazim (0.3%), seed pelting with *Trichoderma vridis* (local), seed pelting with *Trichoderma viridis* (exotic), seed pelting with *Trichoderma harzianum*, FYM (1 ton/ha), neem cake (1 ton/ha), poultry manure (1 ton/ha), carbendazim drenching (0.1%) and control. The data revealed that the root rot incidence was 9.8, 6.7, 8.8, 10.7, 4.5, 3.1, 5.5, 1.3 and in control 13.1, respectively during 1988. In 1989, it was 14.5, 10.2, 10.9, 21.1, 9.0, 5.5, 11.9, 3.8 and in control 27.6, respectively. However in 1990, it was 32.3, 14.3, 14.5, 18.3, 17.9, 6.3, 24.2, 4.3 and in control 43.4, respcetively. The mean yield was 342 kg/ha, 482, 381, 386, 503, 680, 443, 594 and in control 313 kg/ha, respectively. Thus carbendazim (0.1%) is the best treatment for the control of root rot caused by *Rhizoctonia solani* followed by application of neem cake. These two treatments were statistically on par.

Jamadar and Desai (1996) carried out a field expriment on the management of charcoal rot (caused by *M. phaseolina*) of sorghum by soil amendments in Karnataka, India. The combined treatment of straw mulch + neem cake + seed treatment with *Trichoderma viride* recorded 7.4 percent charcoal rot incidence, 18.0 percent lodging, 0.76 nodes infection and 45.0 percent senescence compared with the untreated control which had 33.0 percent charcoal rot, 52.3 percent lodging, 2.1 nodes infection and 75 percent senescence. Significantly higher grain yield was obtained from the combined treatment (1913 kg/ha) compared with the untreated control (1412 kg/ha). The next best treatments were straw mulch + neem cake, straw mulch + seed treatment with *T. viride* and straw mulch alone. However, management of charcoal rot by straw mulch alone is economical and effective.

Karthikeyan (1996) tested the ability of organic amedments (neem cake, mahua cake, castor cake, farm yard manure [FYM], sheep manure and poultry manure), the biocontrol agnet (*Trichoderma harzianum*) and fungicides (carbendazim and thiram) to control seed and collar rot (*A. flavus* and *A. niger*) on goundnuts. The treatments were neem cake, mahua cake and castor cake (250 kg/ha each), FYM (25 ton/ha), sheep manure (500 kg/ha), poultry manure (500 kg/ha), soil application of Trichoderma viridis (500 g inoculum/sq m), carbendazim (2 g/kg of seed), thiram (4 g/kg of seed) and control. From the data, the results revealed that lowest disease incidence of 2.3 percent was observed in the treatment with carbendazim. Among the organic amendments, neem cake (250 kg/ha) and FYM (25 ton/ha) were highly effective in reducing the disease. The disease incidence in neem cake was 2.61 percent in 1994 and 2.35 percent in 1995 as against in control 9.15 in 1994 and 10.40 percent in 1995. The mean yield was 1075 kg.ha in 1994 as against control 508.3 kg/ha and 1150 kg/ha in 1995 as against control 525.7 kg/ha.

Alice *et al*. (1996) studied the effect of neem cake in controlling root rot of jasmine. Among 11 antagonists tested against *M. phaseolina, T. harzianum* isolate 2, *T. viride* native and *T. viride* commercial were effective. Pot culture experiments revealed that coating jasmine cuttings with *T. viride* along with neem cake application were effective in controlling root rot. This gave equal control to drenching plants with carbendazim (0.1%).

Sharma and Basandrai (1997) investigated the effect of biocontrol agents (*Trichoderma harzianum* [*viride*] and *Gliocladium virens*), fungicides (carbendazim, triadimefon and triacimenol) and crude leaf extracts (*Azadirachta indica, Lantana camara* and *Cannabis sativa*) on sclerotial viability of *Sclerotinia sclerotiorum* isolated from pea (*Pisum sativum*), chickpea (*Cicer arietinum*), cauliflower (*Brassica oleracea* var. *botrytis*) and cabbage (*B. oleracea* var. *capitata*). Fresh sclerotia of uniform size were dipped in each test solution for 1 hr before incubation at 22°C on potato dextrose agar. Sclerotial germination was recorded after 7 days. All the solutions were prepared by 25 g fresh leaves of each plant and macerated in 100 ml distilled water and centrifuged at 3500 rpm for 20 min. The effective treatments were *T. harzianum* (4 mil. spores/ml), carbendazim (100 ppm), triadimefon (100 ppm) and crude leaf extract of neem (25%). These teatments gave effects on viability of *Scelrotia* of *S. sclerotiorum* on pea, chickpea, cauliflower and cabbage to the tune of 14.25, 17.50, 10.00 and *Seterotia* 12.25 for *T. harzianum*; 16.50, 12.50, 10.50 and 14.50 for carbendazim; 12.50, 18.75, 21.25 and 19.25 for triadimefon and 13.00, 17.25, 19.50 and 15.75 for crude leaf extract of neem as against control 53.50, 63,75, 48.75 and 40.50 on pea, chickpea, cauliflower and cabbage, respectively. The best treatments among the biocontrol agents, fungicides and plant extracts were tested in combinations on viability of sclerotia of the above crops. The fungicides were used at sublethal dose 10 ppm whereas biocontrol agents and the plant extract were used at the same concentrations. *T. harzianum* + neem extract combination gave the percentage germination of sclerotia from pea, chickpea, cauliflower and cabbage as 3.75, 6.50, 9.75 and 5.50 while triadimefon + neem leaf extract gave 6.25, 5.25, 9.50 and 8.75 as against control 51.50, 53.75, 60.00 and 57.50, respectively. Thus the above combinations were effective treatments.

Ushamalini *et al.* (1997) did experiments to control cowpea wilt by non-chemical methods. The efficacy of biocontrol agents viz. *T. harzianum* and *T. viride*, FYM and plant products viz. *Abutilon indicum, Delonix regia* and neem cakes against cowpea (cv. Co-2) wilt caused by *F. oxysporum* f.sp. *tracheiphilum* was evaluated in field trials conducted during rabi at Coimbatore, India. Neem cake application to soil at 150 kg/ha resulted in minimum wilt incidence and recorded the highest yield of 833.7 kg/ha. Seed treatment with *T. harzianum* registered 98.8 percent seed germination and was equal to that with the standard seed treatment fungicide, carbendazim, compared with 93.2 percent seed germination in the control. The rhizosphere and non-rhizosphere population of fungi, bacteria and actinomycetes was greater in neem cake applied plots whereas the population of *Trichderma* was greater in plots with seed treated with talc based products of *T. viride* and *T. harzianum*. With respect to *Trichoderma*, the minimum root to shoot ratio of 1.55 was reocrded with neem cake at 20 days after sowing while at 40 and 60 days after showing, FYM and *D. regia* recorded the minimum root to shoot ratios of 1.30 and 1.33, respectively.

Karthi Keyan and Bhaskaran (1998) observed the effect of *Trichoderma harzianum* and neem cake in the control of basal stem rot of coconut caused by *Ganoderma lucidum* (Leys.) Karst. Field experiments in RBD were conducted with the application of farm yard manure (FYM) (50 kg/tree/year), neem cake (5 kg), tank silt (200 kg), coir dust (50 kg), *Pleurotus* decomposed coir dust (50 kg), poultry manure (10 kg), green leaves (50 kg) and FYM + 40 lt of Bordeaux mixture (BM) (1.0%) to basal stem rot affected palms with 500 g inoculum of *T. harzianum*. The antagonist was multiplied in respective organic amendments for a period of one month before field application. Eleven treatments were given once in a year with five replications. Disease intensity, nut yield and soil microbial population were recorded in various treatments. The data revealed that soil application of *T. harzianum* along with neem cake gave effective control of the disease showing 32.4 disease index. Neem cake treatment recorded a higher mean yield of 77 nuts/tree/year as compared to 14 nuts in control. Application of neem cake has been reported to control soil borne diseases of various crops.

4. Viral diseases

Tripathi and Tripathi (1982) observed that crude leaf extract of *Azadirachta indica* was to be most potent in reducing bean common mosaic virus infectivity.

Mariappan *et al.* (1982) studied the effect of custard apple oil and neem oil on the life span and rice tungro transmission. Studies in the Philippines showed that the treatment of rice seedlings with oil extracted from the seeds of *A. indica* and *Annona* sp. significantly shortened the life-span of *Nephotettix virescens* (Dist.) carrying rice tungro virus when they were confined on the plants and significantly reduced the incidence of transmission of the virus. The oils were emulsified with 0.1 percent liquid detergent and tested at 5 concentrations (5, 10, 20, 30 and 50%). The reduction in life-span was greater as the oil concentration increased, but 5 percent oil was generally at least as effective as higher concentrations in preventing virus transmission.

Rao *et al.* (1985) observed that neem induced local and systemic resistance of potato virus. Aqueous flower extracts of *Argemone mexicana, A. indica, Euphorbia milii, Jasminum sambac, Lantana indica, Nerium indicum* and *Vinca rosea* [*Catharanthus roseus*] induced resistance to PVX in the hypersensitive host *Chenopodium amaranticolor*. Resistance varied from 79.2 to 99.2 percent and remained active for up to 96 hr after leaf sprays. In the non-hypersensitive host, tomato, the extracts induced varying degrees of resistance, the highest being 87.2 percent with *A. indica*. The induced resistance was sensitive to actinomycin D; resistance induced by the extracts in basal inoculated and upper uninoculated leaves of both hosts was reversed by actinomycin D when applied simulaneously with the extract.

Srivastava *et al.* (1986) observed that inhibition of cucumber mosaic virus (CMV) by crude oil from *A. indica*. This oil, at 0.5 percent wate emulsion, inhibited the transmission of CMV by a single individual of *Aphis gossypii* from cucumber to cucumber to the extent of 40, 33.3, 23.3 and 30 percent when the aphids were allowed to come into contact with the oil before acquisition, during acquisition, between acquisition and inoculation, and at inoculation, respectively. No effect on trasmission was recorded when plants were sprayed with oil emulsion after inoculation. Oil did not affect the biological activity of the virus as evidenced by a mechanical transmission test using a mixture of oil and virus.

Aiyanathan and Narayanasamy (1987) studied the effect of neem oil on tungro (RTV) infection. It was observed that neem oil significantly reduced infection rates in both resistant and susceptible rice cultivars when inoculated with 1, 3 or 5 viruleiferous *Nephotettix virescens*/plant.

Saxena *et al.* (1987) observed the reduction of tungro virus transmission by *N. virescens* in neem cake treated rice seedlings. Rice seedlings grown in soil teated with neem seed cake at rates of _ 150 kg/ha had a significantly lower incidence of infection of rice tungro bacillidorm (RTBV) and spherical viruses transmission (RTSV) by the leafhopper *N. virescens* than did untreated seedlings. It was observed that infection by RTBV plus RTSV occurred in 58 percent of the untreated seedlings. However, RTBV had only 32 percent infection. It is, therefore, suggested that in developing countries where the cost of insecticide is very high, neem cake application on seed beds and fields can protect rice from tungro infection. Neem cake is safe for natural enemies of rice leaf hoppers and plant hoppers. Protection with neem seed cakes at 250 kg/ha was comparable with that provided by application of carbofuran 3G at 0.75 kg a.i./ha. Electronic monitoring showed that the insect fed mainly on the phloem of untreated plants but changed to xylem ingestion on neem-treated plants. Reduction in the incidence of phloem-specific tungro viruses in neem-treated plants was thus attributed to a shift in insect feeding from phloem to xylem vessels.

Saxena *et al.* (1987) reported neem cake at 250 kg/ha and carbofuran at 0.75 kg a.i./ha equally effective in controlling rice virus. Narasimhan and Mariappan (1988) sprayed rice plant with neem seed oil in water emulsified with 1 percent teepol, crude seed extract and cake extract and found that the plants had lower disease incidence. Kareem *et al.* (1988) observed that NSKP and carbaryl are the most effective treatment in reducing RTV infection up to 2 percent.

Mariappan *et al.* (1988) conducted greenhouse study on the effect of nonedible oils extracted from tree seeds on transmission of RTV by the cicadellid *N. virescens*. Oils extracted from seeds of *Pongamia pinnata, Madhuca longifolia* var. *latifolia, Calophyllum inophyllum* and *Annona squamosa* were more effetive than neem (*A. indica*) oil in reducing survival of *N. virescens* and transmission of the virus. Insect mortality was 100 percent after 4 days on rice plants sprayed with 5 percent oils and virus infection was 17-35 percent, compared with respective values of 69 percent for control plants.

Singh *et al.* (1988) studied the antiviral activity and physical properties of the extracts of *A. indica*. Leaf and bark extracts of *A. indica* inhibited infection of *Chenopodium amaranticolor* by cowpea mosaic comovirus. The inhibitor was heat stable, non-dialysable through cellophane, remained active after 25 d storage, was not destroyed by desiccation for 50 days. This showed appreciable loss of activity on dilution and was destroyed by activated charcoal and induced localised resistance. Antiviral activity was lowered in strong acidic and basic solutions and was reveresed by simultaneous application of Actinomycin D. All the extracts did not show inhibitory effect when applied on virus inoculated plants but were very effective when applied to the leaves prior to inoculation.

Zaidi *et al.* (1988) observed the inhibition of spinach mosaic virus (SMV) by extracts of some medicinal plants. The culture of SMV was maintained on *Nicotiana glutinosa* L. by periodic inoculations. The standard extract of SMV and plant parts were mixed in the ratio of 1:0.5 and 1:1. The number of local lesions evoked after 5-7 days of inoculation was counted. The result indicated that extracts of all 10 plants tested inhibited SMV to varying degrees. Inhibition was directly correlated with concentration of the extract. Maximum inhibition was given by extracts of *Ocimum sanctum* roots, followed by *Glycyrrhiza glabra* roots, *Anagallis arvensis* leaves, *Aconitum heterophyllum* roots, *Azadirachta indica* leaves, *Catharanthus roseus* leaves and *Rauvolfia serpentina* roots.

Aiyanathan and Narayanasamy (1988) studied the effect of neem oil in RTV infection in rice varities with different levels of resistance. The rice varieties TNAU831520 (resistant to RTV), ACM9 and IR50 (moderately resistant) and ADT31 (susceptible) were inoculated with the virus using 1, 3 or 5 green leafhoppers (GLH; *N. virescens*) per plant. Neem oil (5%) was applied as both a pre and post-inoculation spray. Increasing the number of insects for inoculation increased infection incidence from 50 to 95 percent in ADT31, from 20 to 60 percent in ACM9 and IR50, and from 0 to 35 percent in TNAU831520. Pre-inoculation application of neem oil to ADT31, ACM9 and IR50 reduced infection from 50 to 15, from 20 to 5 and from 25 to 10 percent, respectively, when one GLH was used for inoculation; similar results were obtained with 3 or 5 percent GLHs. All treated plants of TNAU831520 were free from infection when GLH was used for infection, and infection was reduced from 25 and 35 to 10 percent in treated plants inoculated using 3 or 5 GLHs, respectively. Post-inoculating treatment with neem oil also reduced infection but was less effective than pre-inoculating treatment.

Srinivasulu and Jeyarajan (1988) observed the effect of seed oil on tungro virus transmission and survival of green leaf hopper of rice crop. In greenhouse experiments in Tamil Nadu, India, pre-inoculation sprays with seed oils extracted from cotton, neem, pongamia, mustard and kapok (*Ceiba pentandra*) reduced RTV transmission and survival of the vector *N. virescnes*, and increased the incubation peirod of the tungro virus complex in rice plants. Reduction in the number of infected plants was greatest following treatment with cotton seed oil (88.8% over the untreated variant), and the incubation period was greatest in plants treated with neem oil (78.8%). On plants treated with cotton seed oil, neem oil and pongamia oil, all the insects died on the 6th day after application. Neem oil and pongamia oil reduced tungro transmission for up to 5 days after spraying.

Mishra and Rao (1988) observed inhibition of tobacco mosaic virus (TMV) by root extracts of higher plants. Crude aqueous root extracts of various higher plants were screened for antiviral properties against TMV. The root extracts of *Psidium guajava* and *Prunus persica* completely inhibited the virus. The activity of root extracts of both these plants survived up to 1:500 dilution and 72 hrs storage. Significant inhibition (90-100%) was also obtained with extracts of *Cassia fistula, Acacia arabica* [*A. nilotica*], *Azadirachta indica* annd *Ficus elastica* roots, showing the presence of potent inhibitor in them.

Saxena and Boncodin (1988) observed the effect of neem seed bitters (NSB) on GLH survival and RTV transmission. The ability of *N. virescens* to transmit RTV, and survival of the leafhopper, were evaluated on rice plants treated with NSB. Two-week 0ld seedlings of the susceptible variety TN1 were treated by 24-hrs root immersion or sprayed with NSB at 2500 ppm. Newly emerged adult females reared on virus-free TN1 plants were allowed a 3-day inoculation feeding period on source plants and transferred to freshly treated seedlings. Successive inoculation feeding continued for up to 5 days. Survival of *N. virescens* decreased significantly after 2 days' exposure to systemically treated plants but not to plants treated with foliar sprays. However, ability to transmit the virus was reduced significantly by both treatments after 1 day of inoculation feeding. After 2 days' feeding, the transmission efficiency of the vector became negligible in both treated and untreated plants.

Narasimhan ad Mariappan (1988) tested the effect of plant derivatives on GLH and RTV transmission. Leaf extracts of *Vitex negundo, Croton sparsiflorus, Aegle mamelos, Ocimum sanctum* and *Tribulus terrestris* in water, seed oils of neem, *Calophyllum inophyllum, Madhuca longifolia* var. *latifolia* and *Anona squamosa* in water emulsified with 1 percent teepol, a crude seed extract of neem and cake extract of neem in water, and the insecticides phosphamidon and carbofuran were tested for their effects on the survival of *N. virescens* on rice seedlings and on transmission of RTV. The seed oils gave higher mortality of *N. virescens* than the leaf extacts, neem oil giving the highest mortality. Mortality on seedlings treated with leaf extracts did not differ significantly from that on untreated ones. The percentage of virus-infected seedlings was much lower in all treatments than in the untreated variant.

Kareem *et al.* (1988) tested neem seed kernel (NSK) or neem cake powder (NCP) and carbofuran G mixture for controlling GLH and RTV. NSK and NCP mixed with carbofuran at a rate of 1 kg a.i./ha (33.3 kg neem powder and 33.3 kg granular carbofuran [Furadon]) were tested for their effectiveness in reducing the incidence of *N. virescens* and RTV on potted rice plants, in comparison with carbofuran alone at 2 kg a.i./ha (66.6 kg Furadon). Mortality of the pest was 98-100 percent with carbofuran alone, 96-100 percent with the neem-carbofuran mixture and zero percent in the untreated variant. Carbofuran mixed NSKP was the most effective treatment against RTV, reducing infection to 2 percent as compared with 100 percent in the untreated variant.

In rice crop, RTV transmitted by *N. virescens* was checked by 5 percent neem oil spraying or by soil basal application of neem cake (Saxena *et al*., 1989).

Narashimha Rao and Narayanaswamy (1990) observed that neem oil at 0.1 percent is effective in controlling rice yellow dwarf disease. Neem cake at 5 percent is effective against yellow vein mosaic of okra (Murthy and Reddy, 1992). Lal (1993) found that in tobacco, neem leaf extract at 5 percent was quite effective in controlling mosaic virus.

Rao and Narayanasamy (1990) studied the effect of plant extracts and oil on rice yellow dwarf (RYD) infection. Leaf extracts of young and mature leaves of *Azadirachta indica, Mirabilis jalapa, Catharanthus roseus*, maize and *Nerium oleander* were applied to rice cv. TN1 plants 1-72 hrs before inoculation with the mycoplasma-like organism causing this disease by means of infective GLH. *A. indica* reduced infection from 82.2 percent (control) to 40 percent and increased the incubation period from 28.5 d to 40.3 days. *A. indica* oil gave the least infection (13.3%) and increased the incubation period.

Nagarajan *et al*. (1990) utilized leaf extracts of many plants against tobacco mosaic virus (TMV). A large number of botanicals possess antiviral substance which when sprayed on tobacco early in the season gave good protection against TMV infection. The TMV culture was maintained on *Nicotiana tabacum* L. var. *jayasri*. Extract of the plants was obtained by macerating the green leaves in water (1 g/1 ml) and the juice was filtered and centrifuged. Equal quantities of this plant extract and TMV sap were mixed and thoroughly shaken for 5 minutes. Leaves of the test plant *N. glutinosa* were inoculated and the observations on the number of local lesions developed were recorded 96 hrs after inoculation. Percentage inhibition of virus was calculated. Aqueous extracts of *Acacia arabica* [*A. nilotica*], *Prosopis spicigera* [*P. cineraria*], *Boerhavia diffusa, Telanthera ficoidea, Lawsonia inermis, Azadirachta indica, Ficus religiosa* and *Nerium odorum* [*N. oleander*] possessed strong antiviral activity. The aqueous extrct of *A. indica* showed 86.8 percent inhibition. Aqueous leaf extracts of *Ricinus communis, Achyranthes aspera, Nymphaea pubescens, Artemesia parviflora* and *Cas-*

sia occidentalis increased TMV infection. Partial purification of the antiviral substance from *B. alba* and *C. inerme* indicated that the active principle is a macromolecule of glycoprotein containing 74 percent proteins and 26 percent carbohydrates.

Ragupathi and Doraiswamy (1990) laid out a field experiment in a RBD replicated thrice for the management of soyabean yellow mosaic virus (SYMV) by plants and other products. During rabi (winter) 1988, field trials were carried out using plant products, animal products and chemicals for the control of SYMV. Neem oil (1%) led to lower disease incidence (7.34%) on a par with monocrotophos (6.51%), compared with the control (25.87%). Methyl demeton gave the next best control with 9.54 percent disease incidence followed by buttermik, sorghum extracts, *Prosopis chilensis* extracts and acetyl salicylic acid. In the neem oil and monocrotophos treated plots, the whitefly population was 1.8/plant and 1.2/plant, respectively, compared with 8.9/plant in the control. Yield was not significantly affected by any of the treatments except the neem oil sprayed plot which gave a yield of 566.5 kg/ha compared with 438.5 kg/ha in the control.

Ganapathy and Narayanasamy (1990) studied the effect of plant products on the incidence of major diseases of groundnut. In laboratory and field trials, neem oil from *Azadirachta indica* and leaf extract from *Nerium odorum* reduced the incidence of late leaf spot (*Phaeoisariopsis personala* [*Mycosphaerella berkeleyi*]), rust (*Puccinia arachidis*) and ring mosaic (tomato spotted wilt virus). In addition, neem oil increased pod yield by 62.3 percent.

Gupta and Naqvi (1991) studied the inhibitory activities of plant extracts on virus infection. Extracts (from leaves, roots, green stems, bark, seeds, green fruits and rhizomes) of 7 flowering plants (*Aconitum heterophyllum, A. indica, Capsicum annuum, Chenopodium amaranticolor, Datura metel, Glycyrrhiza glabra* and *Rauvolfia serpentina*) markedly decreased the number of local lesions and systemic infection caused by brinjal necrotic mosaic virus on aubergines.

Atiri *et al.* (1991) studied the effects of natural and 5 synthetic chemicals on the incidence and sevirity of okra mosaic tymovirus (OMV) in okra. The effects of 3 synthetic chemicals at recommended dosages on disease incidence, severity and total damage by the beetle-transmitted OMV in okra (*Abelmoschus esculentus*) were compared with those of natural extracts from the seeds of neem, *A. indica*. Only treatments with the synthetic pyrethroid, lambdacyhalothrin, at 15 g a.i./ha and aqueous neem solution at 467 lt/ha significantly reduced incidence, severity and total damage. Treatments with a cypermethrin + diemthoate mixture (3:25) at 280 g a.i./ha apparently had the same effect on disease incidence and severity, but it had no effect on total damage relative to the untreated control. In all cases, the effects of neem oil at 60 lt/ha and carbaryl at 85 g a.i./ha were not significantly different from the control. Lambdacyhalothrin and aqueous neem solution are easily degraded by metabolizing systems.

Hunter and Ullman (1992) studied the effect of the neem product, RD-Repelin, on settling behaviour and transmission of zucchini yellow mosaic virus (ZYMV) by the pea aphid *Acyrthosiphon pisum* (Harris). The botanical product RD-Repelin a combination of extracts from seeds of *A. indica, Pongamia glabra, Ricinus communis, Madhuca indica* and *Sesamum indicum* was highly repellent to *A. pisum* at concentration of 1, 4 or 10 percent. Aphids were repelled prior to zucchini (marrow) leaf contact at all 3 concentrations. RD-Repelin delayed symptom expressión of ZYMV in 81 percent of plants treated with 1 percent concentration, although virus transmission was not prevented.

Kannan *et al.* (1993) did experiments on the management of cowpea aphid borne mosaic virus (CAMV) by plant products. Field experiments were conducted to investigate the efficacy of plant extracts from *Prosopis chlensis, Vitex negundo, A. indica* and *Madhuca lognifolia* against cowpea aphid-borne virus (black cowpea mosaic potyvirus) on cowpeas. All the extracts reduced the incidence of disease and increased the yield, however, incidence of the disease was most reduced by application of a 1 percent emulsion of *A. indica* and this resulted in a significant increase in yield (890 kg/ha compared with 540 kg/ha for the control.).

Eppler (1993) studied the inhibitory effect of neem on plant virus infections on seeds. Aqueous (AE) and methanolic (ME) of neem seeds were tested for their ability to inhibit tobacco mosaic tobamovirus infections of tobacco. ME gave greater inhibition than AE, but results given by ME were in general not much better than those given by methanol alone. Tests with azadirachtin showed that this substance was not responsible for the inhibition observed with AE and ME. In field experiments, soil amendment with crushed neem seeds, neem cake and neem leaf powder did not prevent infection of marrows with watermelon mosaic 2 polyvirus, or prevent yield losses under high infection pressure, but a delay in symptom appearance was observed, especially with neem seeds.

Epper (1993) studied the inhibitory effect off neem on plant virus infections on bark, cortex and leaves. Neem extracts were tested for their ability to inhibit tobacco mosaic tobamovirus and tobacco necrosis necrovirus on various hosts. In general, bark extracts were more inhibitory than those from the cortex or leaves.

Singh and Korpraditskul (1996) studied the efficacy of some plant extracts on aphids, a key vector of chilly virus complex. As a substitute for insecticide, the efficacy of neem, garlic and *Rhinocanthus* extracts was compared with malathion on chilly against aphids at different time intervals. The effect of malathion was best observed 4 days after spraying, when no aphids were recorded on the plants. After that, the population of aphids increased 400-fold (P<0.001), 4 weeks after spray. Neem extract 5000 ppm generally recorded low aphid populations when compared with unsprayed control, malathion and other plant extracts, especially at 8 and 12 days after spray. This extract also recorded an average of fewer plants infected with virus symptoms when compared with other treatments.

Lowery *et al*. (1997) observed that neem seed oil inhibits aphid transmission of potato virus Y to pepper. Transmission of potato Y potyvirus to sweet pepper (*Capsicum annum*) cv. California Wonder by *Myzus persicae* was inhibited by foliar applications of 1.0 or 2.0 percent neem seed oil to infected source plants or to uninfected recipient plants. Neem seed oil interfered with virus acquisition and inoculation in a manner comparable to that of a commercial horticultural oil, while an oil-free neem seed extract did not reduce rates of transmission compared with controls. The finding that neem seed oil inhibits virus transmission, while oil-free neem seed extracts does not, suggests that the presence of the oil rather than biologically active limonoids such as azadirachtin interfere with virus transmission. None of the treatments affected rates of infection when potato virus Y was transmitted mechanically, or the resulting virus titre and symptom expression. It is concluded that in addition to direct control of insect pests, formulated neem oils may help to reduce or delay the spread of non-persistent plant viruses.

Kumar *et al*. (1997) studied antiviral property of neem (*Melia azadirachta*) and tulsi (*Ocimum sanctum*) against newcastle disease virus (NDV)F1. NDV-F1 strain was incubated with alcoholic extracts of leaves of *M. azadirachta* [*Azadirachta indica*] or *O. sanctum* at 37oC for 4 hrs. Viral activity of the incubated suspensions were tested on chick chorioallantoic membrane (CAM). Haemagglutination tites for tulsi treated NDV-CAM was significantly lower than controls and neem-treated NDV.

Nagaraju *et al*. (1997) studied the effect of exogenously applied plant products on pepper vein banding virus transmission in bell pepper. Two neem based products, Neemark (5%) and NSKE (4%), and leaf extracts from 7 plant species (*A. indica, Pongamia glabra, Ocimum sanctum, Vinca rosea,*

Phyllanthus niruri, Tagetes erectas and spinach) were tested against the transmission of pepper vein banding virus on *Capsicum* under galsshouse conditions. Spinach was inhibitory, resulting in only 25 percent transmission compared with 86.6 percent in the inoculated control. *O. sanctum* had no inhibitory effect on the virus recording 100 percent transmission.

Somasekhara *et al*. (1998) evaluated neem products and insecticides against whitefly (*Bemisia tabaci*), a vector of tomato leaf curl, Geminivirus disease. High mortality of *B. tabaci* was observed with neem products NSKE (100) and RD 9 Repelin at 4 and 1 percent, respectively but the efficacy persisted for 48 hrs only. The neem products Achook, Nimbicidin, Repelin 555 and Field Marshal were moderately effective. Triazophos mixed with Neemark, RD 9 Repelin or aphidin was very effective. Tomato leaf curl virus transmission was lower in neem treatments as well as in those in mixture with triazophos.

Heuvel *et al*. (1998) studied the effects of NSKE and azadirachtin on the ability of *Myzus persicae* to transmit potato leafroll luteovirus (PLRV). It was also determined whether treatments with these compounds would exert an effect on larval growth and mortality, and on the aphid intracellular symbionts. Endosymbiotic bacteria play an essential role in the performance of aphids, and in luteovirus transmission by aphids. NSKE and azadirachtin were offered to one-day-old *M. persicae* nymphs via a membrane feeding system. The neem metabolites displayed a 100 percent mortality at concentrations higher than 2560 ppm. At intermediate concentrations ranging between 320 and 2560 ppm, larval growth and mortality were affected in a concentration-dependent manner. The transmission of PLRV by *M. persicae* was inhibited by 55-90 percent. The endosymbiont population of the aphid was clearly affected by a treatment with neem metabolites as the release of their most abundant protein. At concentrations lowerr than 160 ppm, NSKE or azadirachtin, the endosymbiont population of *M. persicae*, and mortality, growth and feeding behaviour, were similar to that of the untreated groups of aphids. However, PLRV transmission was still inhibited by 40-70 percent.

References

Achimu, P. and Schlosser, E. (1992). Effect of neem seed extracts (*Azadirachta indica* A. Juss) against downy mildew (*Plasmopara viticola*) of grape-vine. *Mededelingen van de Faculteit Landbouwwetenschappen*, Universiteit Gent, **57**(2b) : 423-431.

Aiyanathan, K.E.A. and Narayanasamy, P. (1988). Effect of neem (*Azadirachta indica*) oil on rice tungro virus infection in rice varieties with different levels of resistance. *Madras Agricultural Jounral*, **75**(7-8) : 259-266.

Aiyanathan, T.E.A. and Narayanasamy, P. (1987). Effect of neem oil on tungro (RTV) infection in susceptible and resistant varieties. *International Rice Research Newsletter*, **12**(6) : 23-24.

Ali, B.H. and El Sanousi, S. (1989). Sensitivity of different aerobic bacterial species to azadirachtin. *Journal of Arid Environments*, **17**(1) : 117.

Ali, T.E.S.; Nasir, M.A. and Shakir, A.S. (1992). *In vitro* evaluation of certain neem products as mould inhibitors against post-harvest fruit rotting fungi of tomato. *Pakistan Journal of Phytopathology*, **4**(1-2) : 58-61.

Alice, D.; Ebenezar, E.G. and Sivaprakasam, K. (1996). Biocontrol of *Macrophomina phaseolina* causing root rot of jasmine. *Journal of Ecobiology*, **8**(1) : 17-20.

Anandaraj, M. and Leela, N.K. (1996). Toxic effect of some plant extracts on *Phytophthora capsici*, the foot rot pathogen of black pepper. *Indian Phytopathology*, **49**(2) : 181-184.

Atiri, G.I.; Ivbijaro, M.F. and Oladele, A.D. (1991). Effects of natural and synthetic chemicals on the incidence and severity of okra mosaic virus in okra. *Tropical Agriculture*, **68**(2) : 178-180.

Banerjee, S.; Bhattacharya, I. and Mukherjee, N. (1989). Sensitivity of three sclerotial rice pathogens to plant oils. *International Rice Research Newsletter*. **14**(6) : 23.

Bansal, R.K. and Sobti, A.K. (1990). An economic remedy for the control of two species of *Aspergillus* on groundnut. *Indian Phytopathology*, **43**(3) : 451-452.

Bhaskaran, R.; Suriachandraselvan, M. and Ramachandran, T.K. (1990). Ganoderma wilt disease of coconut - a threat to coconut cultivation in India. *Planter*, **66**(774) : 467-471.

Bhatnagar, D. and McCormick, S.P. (1988). The inhibitory effect of neem (*Azadirachta indica*) leaf extracts on aflatoxin synthesis in *Aspergillus parasiticus*. *Journal of the American Oil Chemists' Society*, **65**(7) : 1166-1168.

Bhowmick, B.N. and Vardhan, V. (1981). Antifungal activity of some leaf extracts of medicinal plants on *Curvularia lunata*. *Indian Phytopathology*, **34** : 385-386.

Chandrasekar, V.; Narayanaswami, R. and Ramabadran, R. (1994). Effect of foliar spray of potash and neem seed extract on the Tikka leaf spot of groundnut. *Indian Phytopathology*, **47**(2) : 188-189.

Chauhan, V.B. and Singh, V.B. (1998). *In vitro* effect of some neem products on mycelial growth of *Phytophthora drechsleri* f.sp. *cajani*. *Abs. ICPPMSA*, 11-13 Dec., 1998 at CSAU A&T, Kanpur : 148.

Chaurasia, S.C. and Jain, P.C. (1978). Antibacterial activity of essential oils of four medicinal plants. *Indian J. Hosp. Pharm.*, **15**(6) : 166-168.

Chourasia, H.K. and Roy, A.K. (1991). Effect of temperature, relative humidity and light on aflatoxin B1 production in neem and Datura seeds. *International Journal of Pharmacognosy*, **29**(3) : 197-202.

Coventry, E.; Allan, E.J.; Kleeberg, H. (ed.) and Zebitz, CPW. (1997). The effect of neem-based products on bacterial and fungal growth. *Proceedings 5th Workshop Wetzlar*, Germany, 22-25 Jan. 1996 : 237-242.

Damayanti, M.; Susheela, K. and Sharma, G.J. (1996). Effect of plant extracts and systemic fungicide on the pineapple fruit-rotting fungus, *Ceratocystis paradoxa*. *Cytobios*, **86**(346) : 155-165.

Das, S.R.; Pani, B.K. and Kar, S. (1997). Evaluation of some plant extracts against *Sclerotium rolfsii* causing stem rot in tuberose. *Environment and Ecology*, **15**(4) : 975-976.

D'Aulerio, A.Z. and Zambonelli, A. (1997). Fungal diseases of officinal plants : spread, effects and defence. *Informatore Agrario*, **53**(1) : 91-93.

D'Aulerio, A.Z.; Zambonelli, A. and Bianchi, A. (1996). Natural products in the protection of medicinal plants. *Colture Protette*, **25**(11) : 99-102.

D'Aulerio, A.Z.; Zambonelli, A.; Bianchi, A.; Caterllani, P.L. and Biffio, S. (1998). Natural products against rusts of mint and tarragon. *Informatore Agrario*, **54**(45) : 89-90.

Daya Ram (1997). Fungitoxicity of some plants extract against *Alternaria brassicae*. *Annals of Agricultural Biological Reserach*, **2**(1) : 25-26.

Doshi, A. and Thakore, B.B.L. (1991). Effect of oil cakes on the development of opium poppy mildew under field condition. *Plant Disease Research*, **6**(2) : 77-81.

Dwivedi, R.S. and Dubey, R.C. (1986). Effects of volatile and non-volatile fractions of two medicinal plants on germination of *Macrophomina phaseolina* sclerotia. *Transactions of the British Mycological Society*, **87**(2) : 326-328.

Dwivedi, T.S. and Chaube, H.S. (1985). Effect of oil cakes on survival of *Macrophomina phaseolina* in soil. *Journal of Soil Biology and Ecology*, **5**(2) : 86-91.

Ehteshamul Haque; Zaki, M.J.; Vahidy, A.A. and Abdul Ghaffar. (1998). Effect of organic amendments on the efficacy of *Pseudomonas aeruginosa* in the control of root rot diseases of sunflower. *Pakistan Journal of Botany*, **30**(1) : 45-50.

Ekpo, E.J.A. and Banjoko, K.M. (1994). Efficacy of Fernasan D and wood ash in the control of seed-borne fungi, pre-emergence mortality and seedling blight of maize. *Discovery and Innovation*, **6**(1) : 84-88.

Enikuomehin, O.A.; Ikotun, T. and Ekpo, E.J.A. (1998). Evaluation of ash from some tropical plants of Nigeria for the control of *Sclerotium rolfsii* Sacc. on wheat (*Triticum aestivum* L.). *Mycopathologia*, **142**(2) : 81-87.

Eppler, A. (1993). Studies on the inhibitory effect of neem on plant virus infections. I. Bark, cortex and leaves. *Mededelingen van de Faculteit Landbouwwetenschappen*, Universiteit Gent, **58**(3a) : 1125-1134.

Eppler, A. (1993). Studies on the inhibitory effect of neem on plant virus infections. II. Seeds. *Mededelingen van de Faculteit Landbouwwetenschappen*, Universiteit Gent, **58**(3a) : 1135-1144.

Eswaramurthy, S.; Mariappan, V.; Muthusamy, M.; Alagianagalingam, M.N. and Subramaniam, K.S. (1993). Efficacy of neem products in controlling bacterial blight of paddy. *World Neem Conference*, Bangalore, India : 33.

Fabry, W.; Okemo, P. and Ansorg, R. (1996). Fungistatic and fungicidal activity of East African medicinal plants. *Mycoses*, **39**(1-2) : 67-70.

Fajardo, J.E.; McCollum, T.G.; McDonald, R.E. and Mayer, R.T. (1998). Differential induction of proteins in orange flavedo by biologically based elicitors and challenged by *Penicillium digitatum* Sacc. *Biological Control*, **13**(3) : 143-151.

Fiume, F. (1997). Therapeutical or protective effects of two new fungicides against powdery mildew on sweet pepper and their physiological implications. *Proceedings of the 49th International Symposium on Crop Protection*, Gent, Belgium, 6 May, 1997, Part IV, Mededelingen Faculteit Landbouwkundige en Toegepaste Biologische Wetenschappen Universiteit Gent., **62**(3b) : 1141-1150.

Ganapathy, T. and Narayanasamy, P. (1990). Effect of plant products on the incidence of major diseases of groundnut. *International Arachis Newsletter*, **7** : 20-21.

Ganguly, L.K. (1994). Fungitoxic effect of certainn plant extracts against rice blast and brown spot pathogen. *Environment and Ecology*, **12**(3) : 731-733.

Garcia, R.P. and Garcia, M.L. (1990). Laboratory evaluation of neem derivatives against *Aspergillus* growth and aflatoxin formation. *Philippine Agriculturist*, **73**(3-4) : 333-342.

Ghewande, M.P. (1989). Management of foliar diseases of groundnut (*Arachis hypogaea*) using plant extracts. *Indian Journal of Agricultural Sciences*, **59**(2) : 133-134.

Ghewande, M.P.; Desai, S.; Prem Narayan; Ingle, A.P. and Narayan, P. (1993). Integrated management of foliar diseases of groundnut (*Arachis hypogaea* L.) in India. *International Journal of Pest Management*, **39**(4) : 375-378.

Govindachari, T.R.; Suresh, G.; Geetha Gopalakrishnan; Balaganesan Banumathy and Masilamani, S. (1996). Identification of antifungal compounds from the seed oil of *Azadirachta indica*. *Phytoparasitica*, **26**(2) : 109-116.

Govindachari, T.R.; Suresh, G. and Gopalakrishnan, G. (1996). Bioassay-guided isolation and identification of anti-fungal compounds from seed oil of *Azadirachta indica* A. Juss. *Neem Newsletter*, **13**(2) : 18-19.

Gunasekaran, M.; Ramadoss, N.; Ramiah, M.; Bhaskaran, R. and Ramanathan, T. (1986). Role of neem cake in the control of Thanjavur wilt of coconut. *Indian Coconut Journal*, **17**(1) : 7-12.

Gupta, V.P.; Govindaiah, and Datta, R.K. (1996). Plant extracts : a non-chemical approach to control *Fusarium* diseases of mulberry. *Current Science*, **71**(5) : 406-409.

Gupta, V.P. and Naqvi, Q.A. (1991). Inhibitory activities of plant extracts on virus infection. *Acta Botanica Indica*, **19**(1) : 18-24.

Haque, E.S.; Abid, M. and Ghaffar, A. (1995). Efficacy of *Bradyrhizobium* sp. and *Paecilomyces lilacinus* with oil cakes in the control of root rot of mungbean. *Tropical Science*, **35**(3) : 294-299.

Haque, E.S.; Zake, M.J.; Vahidy, A.A.; Abdul Ghaffar and Ghaffar, A. (1998). Effect of organic amendments on the efficacy of *Pseudomonas aeruginosa* in the control of root rot disease of sunflower. *Pakistan Journal of Botany*, **30**(1) : 45-50.

Hashmi, R.Y.; Saleem Shahzad; Khanzada, A.K.; Aslam, M. and Kazmi, S.A.R. (1992). Studies on the seed mycoflora of lentil and its control. *Pakistan Journal of Scientific and Industrial research*, **35**(9) : 345-347.

Hebte, M.; Muruleedhara, B.N. and Ikawa, H. (1993). Response of neem (*Azadirachta indica*) to soil P concentration and mycorrhizal colonization. *Arid Soil Research and Rehabilitation*, **7**(4) : 327-333.

Heuvel, J.F. J.M. van den; Hogenhout, S.A.; Verbeek, M.; Wilk and F. van der (1998). *Azadirachta indica* metabolites interfere with the host-endosymbiont relationship and inhibit the transmission of potato leafroll virus by *Myzus persicae*. *Entomologia Expermentalis et Applicata*, **86**(3) : 253-260.

Hooda, K.S. and Srivastava, M.P. (1998). Impact of neem coated urea and potash on the incidence of rice blast. *Plant Disease Research*, **13**(1) :

28-34.

Hundekar, A.R.; Anahosur, K.H.; Patil, M.S.; Kalapannavar, I.K. and Chattannavar, S.N. (1998). *In vitro* evaluation of organic amendments against stalk rot of sorghum. *Journal of Mycology and Plant Pathology*, **28**(1) : 26-30.

Hulloli, S.S.; Singh, R.P. and Verma, J.P. (1998). Management of bacterial blight of cotton induced by *Xanthomonas axonopodis* pv. *malvacearum* with the use of neem based formulations. *Indian Phytopathology*, **51**(1) : 21-25.

Hunter, W.B. and Ullman, D.E. (1992). Effects of the neem products, RD-Repelin, on settling behaviour and transmission of zucchini yellow mosaic virus by the pea aphid, *Acyrthosiphon pisum* (Harris) (Homoptera : Aphididae). *Annals of Applied Biology*, **120**(1) : 9-15.

Iyer, S.R. and Williamson, D. (1991). Efficacy of some plant extracts to inhibit the protease activity of *Trichophyton* species. *Geobios Jodhpur*, **18**(1) : 3-6.

Jagannathan, R. and Sivaprakasam, K. (1996). Effect of botanicals on managing sheath rot of rice. *International Rice Research Notes*, **21**(1) : 49-50.

Jain, P.P.; Suri, R.K.; Deshmukh, S.K. and Mathur, K.C. (1987). Fatty oils from oilseeds of forest origin as antibacterial agents. *Indian Forester*, **113**(4) : 297-299.

Jamadar, M.M. and Desai, S.A. (1996). Management of charcoal rot caused by *Macrophomina phaseolina* (Tassi.) Goid in sorghum, *Sorghum bicolor* (L.) Moench. *Journal of Biological Control*, **10**(1-2) : 93-96.

Jeyalakshmi, C.; Durairaj, P. and Seetharaman, K. (1998). Neem oil to tackle chilli fruit rot caused by *Colletotrichum capsici* (Syd) Butler and Bisby. *Neem Newletter*, **15**(2) : 21.

Jeyalakshmi, C. and Seetharaman, K. (1998). Biological control of fruit rot and die-back of chilli with plant products and antagonistic microorganisms. *Plant Disease Research*, **13**(1) : 46-48.

Jeyarajan, R.; Sabitha Doraiswamy; Bhaskaran, R. Jayaraj, S.; Doraiswamy, S. and Schmutterer, S. (ed.) (1987). Effect of neem [*Azadirachta indica*] and other plant products in the management of plant diseases in India. *Natural Pesticides from the Neem Tree Azadirachta indica A. Juss and other Tropical Plants* : 635-644.

Kannan, N.R. and Sabitha Doraiswamy (1993). Management of cowpea aphid borne mosaic virus (CAMV) using plant products. *Madras Agricultural Journal*, **80**(7) : 393-395.

Kareem, A.A.; Boncodin, M.E.M. and Saxena, R.C. (1988). Neem seed kernel or neem cake powder and carbofuran granule mixture for controlling green leafhopper (GLH) and rice tungro virus (RTV). *International Rice Research Newsletter*, **13**(3) : 35.

Kazmi, A.R.; Niaz, I. and Jilani, G. (1993). Evaluation of some plant extracts for antifungal properties. *Pakistan Journal of Phytopathology*, **5**(1-2) : 93-97.

Khan, M.I. and Shah, N.H. (1992). Antifungal activityof leaf extract of neem on seed mycoflora of wheat. *Bioved.*, **3**(2) : 209-210.

Kareem, A.A.; Boncodin, M.E.M. and Saxena, R.C. (1988). Neem seed kernel or neem cake powder and carbofuran granule mixture for controlling green leaf hopper and rice tungro virus (RTV). *International Rice Research Newsletter*, **13**(3) : 35.

Karthikeyan, A. (1996). Effect of organic amendments, antagonist *Trichoderma viride* and fungicides on seed and collar rot of groundnut. *Plant Disease Research*, **11**(1) : 72-74.

Karthikeyan, A. and Bhaskaran, R. (1998). Effect of *Trichoderma harzianum* and neem cake in the control of basal stem rot of coconut. *Neem Newsletter*, **15**(3) : 29.

Karthikeyan, A. and Karunanithi, K. (1996). Influence of organic amendments on the intensity of *Fusarium* wilt of banana. *Plant Disease Research*, **11**(2) : 180-181.

Khan, M.I.; Ashaq, M.; Bhat, M.Y. andd Fazal, M. (1996). A comparative study of the efficacy of some plant extracts and fungicides on mycoflora of chickpea seed. *Annals of Plant Protection Sciences*, **4**(1) : 85-87.

Khan, M.W.; Alam, A.M.; Khan, A.M. and Saxena, S.K. (1974b). Effect of water soluble fractions of oil cakes and bitter principles of neem on some fungi and nematodes. *Acta Bot. Indica*, **2** : 120-128.

Khan, M.W.; Khan, A.M. and Saxena, S.K. (1973). Influence of certain oil cake amendments on nematodes and fungi in tomato fields. *Acta. Bot. Indica*, **1** : 49-54.

Khan, M.W.; Khan, A.M. and Saxena, S.K. (1974a). Rhizosphere fungi and nematodes of egg palnt as influenced by oil cake amendments. *Indian Phytopathology*, **27** : 480-484.

Khilare, V.C. and Gangawane, L.V. (1997). Application of medicinal plant extracts in the management of thiophanate methyl resistant *Penicillium digitatum* causing green mold of Mosambi. *Journal of Mycology and Plant Pathogloy*, **27**(2) : 134-137.

Khune, N.N.; Patil, B.G.; Kale, K.B.; Newsakar, V.B.; Wangikar, P.D. and Moghe, P.G. (1985). In vitro studies on the effect of fungicides and plant extracts against the fungus of sooty mould of Nagpur oranges. *PKV*

Research Journal, **9**(1) : 95-97.

Korpraditskul, V.; Ratanakreetakul, C. and Korpraditskul, R. (1991). Control of anthracnose of mango fruit by plant crude extract. *The 29th Kasetsart University Annual Conference*, 4-7 Feb., 1991, Bankok : 95-96.

Kurucheve, V.; Ezhilan, J.G. and Jayaraj, J. (1997). Screening of higher plants for fungitoxicity against *Rhizoctonia solani in vitro*. *Indian Phytopathology*, **50**(2) : 235-241.

Lakshmanan, P. and Mohan, S. (1989). Antifungal properties of some plant extracts against collar rot of *Phaseolus aureus*. *Madras Agricultural Journal*, **76**(5) : 266-270.

Lal, R.J. (1993). Grow neem tree to sustain ecological balance. *Kisan World*, **20**(4) : 15-17.

Lakshmanan, P.; Mohan, S. and Jeyarajan, R. (1990). Antifungal properties of some plant extracts *Thanatephorus cucumeris*, the causal agent of collar rot disease of *Phaseolus aureus*. *Madras Agricultural Journal*, **77**(1) : 1-4.

Lehmann, W. and Ibenthal, W.D. (1993). Fungicidal effects of neem extracts against seedling blight of sugarbeet. *Symposium on Useful Tropical Plants*, held in Hamburg, Germany on 22-24 September, 1993. *Angewandte Botanik Berichte*, 1995, No. **5** : 214-219.

Lokhande, N.M.; Lanjewar, R.D. and Newaskar, V.B. (1998). Effect of different fungicides and neem products for control of leaf spot of groundnut. *Journal of Soils and Crops*, **8**(1) : 44-46.

Louis, V.; Balakrishnan, S. and Mathew, J. (1996). *In vitro* screening of some medicinal plant extracts against *Phytophthora capsici*, incitant of foot rot of black pepper (*Piper nigrum* L.). *Journal of Tropical Agriculture*, **34**(1) : 72-73.

Lowery, D.T.; Eastwell, K.C. and Smirle, M.J. (1997). Neem seed oil inhibits aphid transmission of potato virus Y to pepper. *Annals of Applied Biology*, **130**(2) : 217-225.

Lukade, G.M. (1992). Effect of organic soil amendments on root rot incidence of safflower. *Madras Agricultural Journal*, **79**(3) : 179-181.

Maharishi, R.P. (1993). Management of chilli (*Capsicum annuum* L.) diseases by neem based preparations. *World Neem Conference*, Bangalore, India : 34.

Mariappan, V.; Jayaraj, S. and Saxena, R.C. (1988). Effect of nonedible seed oils on survival of *Nephotettix virescens* (Homoptera : Cicadellidae) and on transmission of rice tungro virus. *Journal of Economic Entomology*, **81**(5) : 1369-1372.

Mariappan, V.; Saxena, R.C. and Ling, K.C. (1982). Effect of custard apple oil and neem oil on the life span of and rice tungro transmission by *Nephotettix virescens*. *International Rice Research Newsletter*, **7**(3) : 13-14.

Mishra, A.S. and Rao, G.P. (1988). Inhibition of tobacco mosaic virus by root extracts of higher plants. *Phytophylactica*, **20**(1) : 93-94.

Moline, H.E. and Locke, J.C. (1993). Comparing neem seed oil with calcium chloride and fungicides for controlling postharvest apple decay. *HortScience*, **28**(7) : 719-720.

Manasi Mishra and Tewari, S.N. (1990). Ethanolic extract toxicity of three botanicals against five fungal pathogens of rice. *National Academy Science Letters*, **13**(11) : 409-412.

Meena, S.S. and Mariappan, V. (1993). Effect of plant products on seed borne mycoflora of sorghum. *Madras Agricultural Journal*, **80**(7) : 383-387.

Murthy, S.K.V.V. and Reddy, R.R.D. (1992). Chemical control of yellow vein mosaic disease of bhendi. *Indian Journal of Plant Protection*, **20** : 198-201.

Nagarajan, K.; Murthy, N.S. and Reddy, T.S.N. (1990). Utilization of botnaicals possessing antiviral principle agaist tobacco mosaic virus. *Proceedings of National Symposium* held on January, 21-22, 1990 at Central Tobacco Research Institute, Rajahmundry, India : 407-412.

Nagarajan, K. and Reddy, T.S.N. (1980). Preliminary evaluation of Ridomil against damping off and leaf blight disease of tobacco in seed beds. *Tob. Res.*, **6** : 127-131.

Nagaraju. M.; Reddy, H.R. and Ravi, K.S. (1997). Effect of exogenously applied plant products on pepper vein banding virus transmission, multiplication and symptom production in bell pepper (*Capsicum annuum* L.). *Indian Jounral of Virology*, **13**(2) : 161-163.

Naik, S.T. and Shivanna, H. (1997). Management of powdery mildew of *Acacia auriculiformis*. *Indian Forester*, **123**(9) : 868-869.

Nair, P.K.U.; Mammootty, K.P.; Sasikumaran, S. and Pillai, V.S. (1993). Phtophthora foot rot of black pepper (*Pepper nigrum* L.) a management study with organic soil amendments. *Indian Cocoa Arecanut and Spices Journal*, **17**(1-2) : 1-2.

Narashimha Rao, G. and Narayanasamy, P. (1090). Effect of plant extracts and oils on rice yellow dwarf infection. *Madras Agricultural Journal*, **77** : 197-201.

Narasimhan, V. and Mariappan, V. (1988). Effect of plant derivatives on green leafhopper (GLH) and rice tungro (RTV) transmission. *International Rice Research Newsletter*, **13**(1) : 28-29.

Narasimhan, V.; Rajappan, K.; Ushamalini, C. and Kareem, A.A. (1998). Efficacy of new EC formulations of neem oil and pungam oil for the management of sheath rot disease of rice. *Phytoparasitica*, **26**(4) : 301-306.

Narasimhan, V.; Sridhar, V.V. and Kareem, A.A. (1993). Efficacy of botanicals in managing sheath rot (ShR) of rice. *Internatinal Rice Research Notes*, **18**(3) : 33.

Nevill, R.J. and Safranyik, L. (1996). Effects of neem seed extract on the growth of fungal associates of the mountain pine beetle. *Journal of the Entomological Society of British Columbia*, **93** : 129-130.

Nwosu, M.O. and Okafor, J.I. (1995). Preliminary studies of the antifungal activities of some medicinal plants against Basidiobolus and some other pathogenic fungi. *Mycoses*, **38**(5-6) : 191-195.

Osnaya Gonzalez, M.; Steinhauer, B. and Schlosser, E. (1997). Effect of alternative plant protection materials on blackspot of roses. *Proceedings of the 49th International Symposium on Crop Protection*, Gent, Belgium, 6 May, 1997, Part IV, Mededelingen Faculteit Landbouwkundige en Toegepaste Biologische Wetenschappen Universiteit Gent., **62**(3b) : 1041-1048.

Pan, S. (1990). Additive effects of plant oils and fungicides. *Journal of Mycopathological Research*, **28**(1) : 57-61.

Panda, R.N.; Tripathy, S.K.; Kar, J. and Mohanty, A.K. (1996). Antifungal efficacy of homeopathic drugs and leaf extracts in brinjal. *Environment and Ecology*, **14**(2) : 292-294.

Pandey, K.R. and Prashad, J. (1992). Effect of neem products on *Xanthomonas campestris* pv. *vesticatoria* incitant of bacterial leaf spot of green pepper *in vitro*.Paper presented in *World Neem Conference*, Bangalore, India, 24-28 Feb., 1992.

Pandey, R.S.; Bhargava, S.N.; Shukla, D.N. and Dwivedi, D.K. (1983). Control of Pestalotia fruit rot of guava by leaf extracts of two medicinal plants. *Revista Mexicana de Fitopatologia*, **2**(1) : 15-16.

Pani, B.K. and Patra, A.K. (1997). Utilization of some phyto-extracts of *Sclerotium rolfsii* during paddy straw mushroom (*Volvariella volvacea*) cultivation - a new approach. *Mushroom Research*, **6**(1) : 37-41.

Panneerselvam, A. and Saravanamuthu, R. (1996). Studies on the saprophytic survival of *Fusarium moniliforme* J. Sheld in soil under treatment of oil cakes. *Indian Journal of Agricultural Research*, **30**(1) : 12-16.

Pasini, C.; D'Aquila, F.; Curir, P.; Gullino, M.L.; Albajes, R. (ed.) and Carnero, A. (1997). Alternative strategies to control rose powdery mildew. *Bulletin OILB, SROP*, **20**(4) : 84-87.

Patil, M.R. and Ghoderao, B.N. (1997). Evaluation of some medicinal and aromatic plants against cotton bacterial blight infection. *PKV Research Journal*, **21**(2) : 179.

Phavaphut Anon, L.; Davies, F.T. Jr. and Duray, S.A. (1996). Growth, root alteration, and nutrient uptake of neem tree (*Azadirachta indica* A. Juss) seedlings in resoponse to vesicular-arbuscular mycorrhizal fungi and phosphorus nutrition. *International Tree Crops Journal*, **9**(1) : 59-67.

Philip, T. and Sharma, D.D. (1997). *In vitro* evaluation of leaf and oil cake extracts of *Azadirachta indica* and *Pongamia glabra* on mulberry root rot pathogens. *Indian Journal of Sericulture*, **36**(2) : 150-152.

Pierce, R. (1981). Natrual products repel cucumber beetle. *Agricultural Research (USA)*, **30**(2) : 12.

Pramanick, T.C. and Phookan, A.K. (1998). Effect of plant extracts in the management of sheath rot of rice. *Journal of the Agricultural Science Society of North East India*, **11**(1) : 85-87.

Prasad, M.S.; Lakshmi, B.S. and Shaik Mohiddin (1998). Survival of sclerotia of *Rhizoctonia solani* in rice soils amended with oil cakes and green leaf manures. *Annals of Agricultural Research*, **19**(1) : 44-48.

Ragupathi, N. and Doraiswamy, S. (1990). Management of soybean yellow mosaic virus by plant and other products. *Proceedings of National Symposium* held on January 21-22, 1990 at Central Tobacco Research Institute, Rajahmundry, India : 469-472.

Rai, M.K. (1996). *In vitro* evaluation of medicinal plant extracts against *Pestalotiopsis mangiferae*. *Hindustan Antibiotics Bulletin*, **38**(1-4) : 53-56.

Rajan, F.S.; Vedamuthu, P.G.B.; Khader, M.D.A. and Jeyarajan, R. (1991). Management of root-disease of fenugreek. *South Indian Horticulture*, **39**(4) : 221-223.

Rajappan, K.; Kareem, A.A.; Narasimhan, V. and Manickam, K. (1996). Management of three major rice (*Oryza sativa* L.) disease by neem (*Azadirachta indica* A. Juss) formulations. *Neem Newsletter*, **13**(4) : 64-65.

Rajesh Kumar; Singh, D.P.; Chaturvedi, V.K. and Pathak, R.C. (1997). A note on antiviral property of neem (*Melia azadirachta*) and tulsi (*Ocimum sanctum*) against Newcastle disease virus. *Indian Journal of Comparative Microbiology, Immunology and Infectious Diseases*, **18**(2) : 192-193.

Ramanujam, B.; Nambiar, K.K.N. and Rohini Iyer (1997). Management of stem bleeding disease of coconut with fungicides and a biocontrol agent. *Journal of Plantation Crops*, **25**(2) : 175-179.

Rao, G.P.; Singh, R.K.; Pandey, A.K. and Singh, R. (1985). Induction of local and systemic resistnace of potato virus X by flower extracts. *Indian Journal of Virology*, **1**(2) : 174-182.

Rao, R.N. and Narayanasamy, P. (1990). Effect of plant extracts and oils on rice yellow dwarf infection. *Madras Agricultural Journal*, **77**(5-6) : 197-201.

Rao, V.K. and Ratnasudhakar, T. (1992). Effect of some plant powders on grain microflora of paddy during storage. *Indian Phytopathology*, **45**(1) : 55-58.

Reddy, A.B. and Sethi, M.S. (1986). Preliminary studies on the antiviral effects of some indigenous (to India) plant extracts (*Berberis floribunda, Azadirachta indica, Moringa oleifera, Cynodon dactylon*) on vaccinia virus *in vivo*. *Annali della Facolta di Medicina Veterinaria*, Meesina, 1984, publ, 1986, **21** : 65-76.

Redondo, E.E.; Varon, deAgudelo, F. and Agudelo, D.O. (1996). Evaluation of different treatments for inhibition of growth and sporulation of *Cylindrocladium scoparium in vitro*. *ASCOLFI Informa*, **22**(4) : 51-53.

Rettenassababady, C. and Ramadoss, N. (2000). Effect of neem products in the control of powdery mildew of ricefallow blackgram (*Erysiphe polygoni* DC.). *Neem Newsletter*, **17**(1) : 12.

Rizvi, G.; Saxena, P.; Faruqui, S.A. and Chandra, N. (1998). The effect of botanicals pesticides on the germination of rust spores of lucerne. *Abs. ICPPMSA*, 11-13 Dec., 1998 at CSAU A&T, Kanpur : 141.

Rovesti, L.; Marco, S. Di and Pancaldi, D. (1992). Effect of neem kernel extract on some phytopathogenic fungi under greenhouse conditions. *Zeitschrift fur Pflanzenkrankheiten und Pflanzenschutz*, **99**(3) : 293-296.

Roy, A.K.; Prasad, M.M.; Nira Kumari and Chourasia, H.K. (1988). Studies on association of mycoflora with drug-plant and aflatoxin producing potentiality of *Aspergillus flavus*. *Indian Phytopathology*, **41**(2) : 261-262.

Sadanandan, A.K. and Iyer, R. (1986). Effect of organic amendments on rhizome rot of ginger. *Indian Cocoa, Arecanut and Spices Journal*, **9**(4) : 94-95.

Sajid, M.N.; Ihsan, J.; Nasir, M.A. and Shakir, A.S. (1995). Comparative efficacy of neem products and Baytan against leaf rust of wheat. *Paki-*

stan *Journal of Phytopathology*, **7**(1) : 71-75.

Sarvamangala, H.S.; Govindaiah and Datta, R.K. (1993). Evaluation of plant extracts for the control of fungal diseases of mulberry. *Indian Phytopathology*, **46**(4) : 398-401.

Saxena, R.C. and Boncodin, M.E.M. (1988). Effect of neem seed bitters (NSB) on green leafhopper (GLH) survival and rice tungro virus (RTV) transmission. *International Rice Research Newsletter*, **13**(1) : 25-26.

Saxena, R.C.; Kareem, A.; Palanginan, E.L. and Malayba, M.T. (1989). Systemic and foliar applications of neem seed bitter to control green leaf hoppers and rice tungro virus (RTV) disease. *International Rice Research Newsletter*, **14**(1) : 31.

Saxena, R.C.; Khan, Z.R. and Bajet, N.B. (1987). Reduction of tungro virus transmission by *Nephotettix virescens* in neem cake treated rice seedlings. *Journal of Economic Entomology*, **80**(5) : 1079-1082.

Sharma, B.K. and Basandrai, A.K. (1997). Effect of biocontrol agents, fungicides and plant extracts on sclerotial viability of *Sclerotinia sclerotiorum*. *Indian Journal of Agricultural Sciences*, **67**(3) : 132-133.

Shivpuri, A.; Sharma, O.P. and Jhamaria, S.L. (1997). Fungitoxic properties of plant extracts against pathogenic fungi. *Journal of Mycology and Plant Pathology*, **27**(1) : 29-31.

Singh, A.K. and Singh, A.K. (1988). Antiviral activity and physical properties of the extracts of *Azadirachta indica* L. *Indian Journal of Virology*, **4**(1-2) : 76-81.

Singh, H. and Korpraditskul, V. (1996). Efficacy of some plant extracts on aphids, a key vector of chilly-virus complex. *Neem Newsletter*, **13**(4) : 56.

Singh, H. and Vichai Korpraditskul (1996). Efficacy of some plant extracts on control of *Colletortichum capsici* (Syd.) Butler & Bisby, the causal agent of chilly anthracnose. *Neem Newsletter*, **13**(4) : 58.

Singh, H. and Vichai Korpraditskul (1999). Evaluation of some plant extracts for the control of *Colletotrichum capsici* (Syd.) Butler & Bisby, the causal agent of chilly anthracnose. ***Azadirachta indica*** **A. Juss.** Ed. Singh and Saxena : 131-138.

Singh, H.B. and Singh, U.P. (1981). Effect of volatiles of some plant extracts and their oils on conidia of *Erysiphe polygoni* DC. *Australasian Plant Pathology*, **10**(4) : 66-67.

Singh, H.N.P.; Prasad, M.M. and Sinha, K.K. (1993). Efficacy of leaf extracts of some medicinal plants against disease development in banana. *Letters in Applied Microbiology*, **17**(6) : 269-271.

Singh, N. and Singh, R.S. (1982). Effect of oil-cake amendment soil atmoshpere on pigeon pea wilt pathogen. *Indian Phytopathology*, **35**(2) : 300-305.

Singh, N. and Singh, R.S. (1983). Inhibition of *Fusarium udum* (pigeon pea wilt pathogen) by ether distillate of margosa cake amended soil. *Indian Journal of Mycology and Plant Pathology*, **13**(3) : 329-330.

Singh, R.K.; Shukla, R.P. and Dwivedi, R.S. (1989). Studies on fungitoxicity of oils against *Sclerotium rolfsii* Sacc. and soil mycoflora. *National Academy Science Letters*, **12**(6) : 183-185.

Singh, S.N.; Yadav, B.P.; Sinha, S.K. and Ojha, K.L. (1997). Efficacy of plant-extracts in inhibition of radial growth and spore germination of *Colletotrichum capsici*. *Journal of Applied Biology*, **7**(1-2) : 58-61.

Singh, U.P. and Prithiviraj, B. (1996). Azadirachtin, a product of neem, induces resistance in pea (*Pisum sativum*) against powdery mildew. *Neem Newsletter*, **13**(3) : 37.

Singh, U.P. and Prithiviraj, B. (1997). NeemAzal, a product of neem (*Azadirachta indica*), induces resistance in pea (*Pisum sativum*) against *Erysiphe pisi*. *Physiological and Molecular Plant Pathology*, **51**(3) : 181-194.

Singh, U.P.; Singh, H.B. and Singh, R.B. (1980). The fungicidal effect of neem (*Azadirachta indica*) extracts on some soil-borne pathogens of gram (*Cicer arietinum*). *Mycologia*, **72**(6) : 1077-1093.

Sobti, A.K.; Sharma, O.P. and Bhargava, A.K. (1995). A comparative study of fungicidal compounds and plant extracts against three pathogens of *Arachis hypogaea*. *Indian Phytopathology*, **48**(2) : 191-193.

Somasekhara, Y.M.; Nateshan, B.M. and Muniyappa, V. (1998). Evaluation of neem products and insecticides against whitefly (*Bemisia tabaci*), a vector of tomato leaf curl Geminivirus disease. *Neem Newsletter*, **15**(2) : 16.

Srinivasan, D.; Muthukumar, T. and Udaiyan, K. (1996). Influence of effluent irrigation on VA mycorrhizae. *Indian Journal of Microbiology*, **36**(1) : 37-40.

Srinivasulu, B. and Jeyarajan, R. (1988). Effect of seed oils on tungro transmission and survival of green leaf hopper of rice crop. *Indian Journal of Plant Protection*, **16**(2) : 171-175.

Srivastava, A.K. and Bihari Lal (1997). Studies on biofungicidal properties of leaf extract of some plants. *Indian Phytopathology*, **50**(3) : 408-411.

Srivastava, A.K. and Singh, R.B. (1991). Use of organic amendment against *Fusarium solani* and *Meloidogyne incognita* on *Phaseolus vulgaris*. *New Agriculturist*, **2**(1) : 63-64.

Srivastava, K.M.; Gupta, P.K.; Tripathi, Y.C. and Sarvate, R. (1997). Antifungal activity of plant products on spermoplane fungi of *Azadirachta indica* (neem) seeds. *Indian Forester*, **123**(2) : 157-161.

Srivastava, K.M.; Rana, N.S.; Dwadash Shreni, V.C. and Singh, B.P. (1986). Mechanism of inhibition of CMV by crude oil from margosa (*Azadirachta indica*) during inoculation with single apterous *Aphis gossypii*. *Indian Phytopathology*, **39**(1) : 20-25.

Srivastava, M.P. and Hooda, K.S. (1996). Neem in control of blast (*Pyricularia oryzae* Cav.) of rice. *Neem Newletter*, **13**(3) : 40-41.

Suresh, G.; Narasimhan, N.S.; Masilamani, S.; Partho, P.D. and Gopalakrishnan, G. (1997). Antifungal fractions and compounds from uncrushed green leaves of *Azadirachta indica*. *Phytoparasitica*, **25**(1) : 33-39.

Thakore, B.B.L.; Sneh Mathur; Singh, R.B. and Chakravarti, B.P. (1987). Soil amendment with oil cakes in ginger field for rhizome rot control. *Korean Journal of Plant Protection*, **26**(4) : 267-268.

Tripathi, R.K.R. and Tripathi, R.N. (1982). Reduction in bean common mosaic virus (BCMV) infectivity vis-a-vis crude leaf extract of some higher plants. *Experientia*, **38**(3) : 349.

Ushamalini, C.; Rajappan, K. and Kousalya Gangadharan (1997). Control of cowpea wilt by non-chemical means. *Plant Disease Research*, **12**(2) : 122-129.

Ushamalini, C.; Rajappan, K. and Kousalya Gangadharan (1997). Mangaement of charcoal rot of cowpea using biocontrol agents and plant products. *Indian Phytopathology*, **50**(4) : 504-507.

Vir, D. and Sharma, R.K. (1985). Efficacy of fungicide : XXXII. Evaluation of neem oil for control of plant pathogens. *Asian Farm Chemicals*, **1**(7-8) : 23-24.

Vir, D. and Sharma, R.K. (1985). Efficacy of fungicides XXIX. Studies on the fungicidal properties of neem oil. *Indian Journal of Plant Pathology*, **3**(2) : 241-242.

Vasishtha, A.K.; Pathak, J.P. and Lehri, A. (1992). Studies on solvent extraction of neem seed (*Azadirachta indica*). *Journal of the Oil Technologists' Association of India*, **24**(3) : 75-80.

Viswanathan, R. and Kandiannan, K. (1990). Effect of urea applied with neem cake on disease intensity and insect population in rice fields. *International Rice Research Newsletter*, **15**(5) : 20.

Volf, O. and Steinhauer, B. (1997). Fungicidal activity of neemleaf extracts. *Proceedings of the 49th International Symposium on Crop Protec-*

tion, Gent, Belgium, 6 May, 1997, Part IV, Mededelingen Faculteit Landbouwkundige en Toegepaste Biologische Wetenschappen Universiteit Gent., **62**(3b) : 1027-1033.

Wan, M.T.; Rahe, J.E. and Wong, C. (1997). Persistence of azadirachtin-A in two biological systems used for culturing mycorrhizal fungus, *Glomus intraradices*. *Journal of Environmental Science and Health, Part-B, Pesticides, Food Contaminants and Agricultural Wastes*, **32**(6) : 929-953.

Wan, M.T. and Rahe, J.E. (1998). Impact of azadirachtin on *Glomus intraradices* and vesicluar-arbuscular mycorrhiza in root inducing transferred DNA transformed roots of *Daucus carota*. *Environmental Toxicology and Chemistry*, **17**(10) : 2041-2050.

Zaidi, Z.B.; Gupta, V.P.; Samad, A. and Naqvi, Q.A. (1988). Inhibition of spinach mosaic virus by extracts of some medicinal plants. *Current Science*, **57**(3) : 151-152.

Zaman, M.A.; Saleh, A.K.M.; Rahman, G.M.M. and Islam, M.T. (1997). Seed-borne fungi of mustard and their control with indigenous plant extracts. *Bangladesh Journal of Plant Pathogloy*, **13**(1-2) : 25-28.

Zeringue, H.J.Jr and Bhatnagar, D. (1990). Inhibition of aflatoxin production in *Aspergillus flavus* infected cotton bolls after treatment with neem (*Azadirachta indica*) leaf extracts. *Journal of the American Oil Chemists' Society*, **67**(4) : 215-216.

Chapter 7

Neem products in controlling plant parasitic nematodes

Plant parasitic nematodes are the serious pests of agricultural and horti cultural crops, they are tiny mainly inhabiting in the soil. To control these, synthetic nematicides are under use. However, the high cost of these nematicides pose a serious problem in their wider application in under developed or developing countries. These nematicides are highly toxic and causes environmental hazzards and led to ground water contamination. They even leave residues in our under ground crops. To overcome these problems neem products have been found promising for the management of nematodes. Neem cake has been found to be relatively toxic to the plant parasitic nematodes and hence majority of experiments have been done with neem cake. The wormicidal nature of neem is known in India since medieval times. The major plant parasitic nematodes are *Anguina tritice* on wheat, *Heterodera avenae* in wheat and barley *Globodera* sp. on potato, *Meloidogyne* spp. on vegetable and pulse crops, *Hoplolaimus* sp., *Pratylenchus* sp. and *Rotylenchulus* sp. The aqueous extract of the neem cake inhibited hatching of second stage juveniles from treated egg masses. These extracts were also found toxic to many species of nematodes. Neem cake alone or in combination with NPK or sublethal doses of synthetic nematicide, has been reported to reduce the population of nematodes. The details of research work conducted are given below :-

Sitaramaiah (1977) tested the effect of temperature on decomposition of margosa cake and on the survival of *Tylenchorhynchus elegans*. Field soil with the addition of powdered margosa oil-cake (*Azadirachta indica*) was inoculated with 100 *Tylenchorhynchus elegans*/200 gram soil in pots and incubated at 15 to 35°C. After 1, 2 or 3 weeks the nematodes were extracted and counted. Nematode population were significantly reduced in the amended soil at 20°C and above but the length of time allowed after application of the cake had little effect on survival. Survival was least in the treatment with the largest concentration (1.0%) of margosa cake.

Saxena et al (1977), studied the effect of certain soil amendments and nematicides on the population of nematode infesting grapevines. Mustard-oil cake (1500 kg/ha) and margosa (*Azadirachta indica*) leaves (300 kg/ha) were effective in controlling *Tylenchorhynchus* sp., *Helicotylenchus* sp. and *Meloidogyne* sp. on grape vines at Ludhiana, India. DBCP (24 lt a.i./ha) and fensulphothion 30 kg/ha) were more effective than fensulphothion (20 kg/ ha), carbofuran (6 kg/ha) and aldicarb (9 kg/ha). All treatments were effective for 3 to 4 months.

Gowda and Setly (1978), studied the comparative efficacy of neem cake and methomyl on the growth of tomato and root-knot development. Pot-grown tomato seedlings, cv. Marglobe, were inoculated with *Meloidogyne incognita*, and 1 week later either neem (*A. indica*) cake at 2 g/500 g of soil was added or methomyl at 10 p.p.m. was applied as a soil drench. Average shoot height and weight were increased by neem cake. Gall numbers were reduced by both treatments, methomyl being the more effective.

Sitaramaiah and Singh (1978), studied the effect of organic amendment on phenolic content of soil and plant and response of *Meloidogyne javanica* and its hosts of related compounds. Tomato seedlings were grown in soil amended with margosa (*A. indica*) cake or mixed wood sawdust from *Shorea robusta* and *Dalbergia sisoo* plus NPK. The phenolic content of the amended soils increased in relation to the quantity of the amendment added and the length of the decomposition period. There were more phenols in soil amended with margosa cake than in sawdust-amended soil, and more in tomato roots grown in the amended soil than in the non-treated soil. Significantly fewer larvae of *Meloidogyne javanica* invaded tomato roots exposed to phenols than untreated roots. There were fewer galts and egg masses/root system when roots were exposed to the aromatic acids, benzoic, phenyl butyric, phenyl acetic and cinnamic. The aromatic acids also suppressed larval movement and nematode egg laying. Studies with phenols e.g. phenyl acetic-, benzoic-, phenyl butyric- and cinnamic acids showed that all these compounds reduced egg-laying capacity and larval motility in *M. javanica*. Exposure of tomato roots to these compounds also imparted some resistance to invasion by the nematode, and in treated plants fewer larvae penetrated the roots to develop into mature females, and fewer eggs were produced.

Haseeb et al. (1978) tested the extracts of shoots and roots of *Chenopodium ambrosioides, Canabis sativa, Papaver rhoeas, Nicotiana tabacum, Withania somnifera, Solanum hispidum* or *Melia azadirachta* for nematicidal activity against *Tylenchorhynchus brassicae, Rotylenchulus reniformis* and *Hoplolamius indicus*. Extracts collected after 24 hrs were named as standard (S). Highest mortality of *H. indicus* was obtained in 'S' root extracts of *Chenopodium ambrosioides* and shoot extracts of *S. hispidum, M.*

azadiractha and *Canabis sativa*. Mortality of *R. reniformis* was highest in the 'S' concentration of *S. hispidum* and *C. sativa* and shoot extracts of *S. hispidum, M. azadirachta, Chenopodium ambrosiodies, N. tabacum* and *Canabis sativum*. Root extracts of *Chenopodium ambrosioides* and shoot extracts of *N. tabacum* and *Canabis sativa* were most toxic to *T. brassicae*.

Khan *et al.* (1979) obtained that, In pot and field trials the application of oil cakes of neem (*Azadirachta indica*), groundnuts (*Arachis hypogea*), mahua (*Madhuca indica*) or castor (*Ricinus communis*) significantly suppressed several nematode speices, including *Tylenchorhynchus brassicae* and *Meloidogyne incognita*, in the okra rhizosphere. Although there was an increase in the population of saprophytic fungi, the oil cake amendments adversely affected several parasitic fungi, including *Rhizocotonia solani* and *Fusarium oxysporum*.

Gowda and Setty ((1979) observed in vitro studies of neem cake extract on hatching of eggs and survival of larvae of root knot nematode (*M. incognita* Cehitwood). Hatching of larvae from fresh eggs of *M. incognita* soaked for 24 hrs in water extract of neem cake was significantly (at the 5% level) reduced by a 1:10 dilution of the extract. Survival of hatched larvae in the 1:10 and 1:100 dilutions was significantly less (at the 1% level), than in the control.

Singh *et al*. (1980) treated naturally infested soil having moderately high papulation of plant parasitic nematodes with oilcakes of mustard, castor or neem at the rate of 20 g/kg soil or the nematicide 5% Rogor G (2 g/kg soil) or Furadan 3G (3 g/kg soil), either singly or in various combinations. Highest increase in tomato growth was found with mustrad cake followed by castor and neem cakes. Furadan 3G gave the greatest reduction in nematode populations. When nematicides were mixed with oilcakes, the efficacy of oilcakes in reducing plant nematodes was increased being more effective than the nematicides in lower dose. It is suggested that an integration of these 2 methods gives best control.

Singh *et al.* (1980) compared soil amendment with oilseed cakes from *Ricinus communis, Brassica campestris, Azadirachta indica* (all at 20 g/kg soil), *Madhuca indica* (33.3 g/kg) or *Arachis hypogaea* (14.29 g/kg) was compared for nematode control with the granular nematicides Furadan 3G [carbofuran] at 3.3 g, 5% rogor G [dimethoate] at 2 g/kg and Thimet 10G [phorate] at 1 g/kg soil. The preparations were either applied to the soil surface or thoroughly incorporated. Seedlings of the tomato cv. Marglobe infested with *M. incognita* were planted in pots of treated or untreated soil, after which watering was applied at 3 levels: (1) equivalent to 21% moisture holding capacity (MHC), (2) 3 times daily at 21% MHC, or (3) on 3 occasions, once every 3 days, at 21% MHC. Of the oilseed cakes, *A. indica*, enhanced plant growth. All nematicides were equally effective against the nematodes and 5% Rogor G gave the best results in terms of plant growth; control was slightly better than with *A. indica* cake. The heaviest watering, (2) greatly enhanced the beneficial effects of all treatments. Soil incorporation was more effective than surface application for all the preparations.

Mohammad *et al.* (1980) observed that: the addition of *Azadirachta indica*, sesame, olive, soyabean or cotton seed cake to the soil improved root and shoot growth of potted *Citrus reticulata* seedlings and considerably reduced *Tylenchulus semipenetrans* populations.

Ram and Gupta (1981) studied the control of *Meloidogyne javanica* on chickpea. In pot trials with seedlings of *Cicer arietinum* infected with *M. javanica* at 1000 larvae/kg soil, the soil was mixed with neem or datura leaves (200 or 400 quintals/ha), aldicarb or carbofuran 1.5 or 3.0 kg a.i./ha or phorate (3.0 or 6.0 a.i./ha). Aldicarb and neem (400 q/ha) were the most effective in increasing growth and reducing galling.

Netscher (1981) observed in pot tests that on 20 cm high *Azadirachta indica, Anacardium occidentale* and *Eucalyptus camaldulensis* inoculated with 2000 *Meloidogyne* juveniles from 11 populations from vegetable gardens at 4 months showed all 3 trees to be resistant to root-knot.

Prot and Kornprobst (1983) observed the effect of *Azadirachta indica, Hannao undulata* and *Hannao klaineana* seed extract on *M. javanica* juveniles to penetrates tomato roots. Pretreatment with crude seed extracts of *A. indica, H. undulata* and *H. klaineana* inhibited the penetration of *M. javanica* juveniles into tomato roots. The degree of inhibition was correlated with the concentrations used: full inhibition occurred with concentrations of 20 000, 10 000 and 200 ppm for the 3 extracts, respectively.

Husain *et al.* (1984) observed that root-dip treatment of *Solanum melongena* seedlings with margosa or marigold leaf extract, aldicarb, Decaris, mustard cake or carbofuran considerably reduced development of *M. incognita* as compared to treatment with piperazine citrate, *Chenopodium* oil and ground nut cake.

Hasan & Jain (1984) observed that incorporation of 5 oil cakes or urea, each at 70 kg N/ha, and 2.5 kg phenamiphos/ha into nematode-infested soil showed that *Brassica campestris* cake was the most effective in increasing fodder production and photosynthetic pigment (chlorophylls and carotene) content of *T. alexandrinum*, while *A. indica* cake was the most effective in reducing nematode populations. In the subsequent season, fresh fodder yield of *P. typhoides* [*P. americanum*] was increased and nematode populations were decreased by the residual effects of the oil cakes.

Jagdale *et al.* (1985) observed the effects of nematicidal plants (Tagetes erecta or Vinca rosea [Catharanthus roseus]), oil cakes (neem or karanj cakes), sawdust or aldicarb on populations of *M. incognita* on *Piper betle* was studied under field conditions in Maharashtra, India. Maximum reduction in nematode populations and numbers of galls on roots was observed with aldicarb treatment followed by neem cake.

Jagdale *et al.* (1985) applied organic amendments in the field in India to *P. betle* infested with *M. incognita*. Highest reduction in nematode populations and number of galls 180 days after application was observed in the neem cake treatment applied at the rate of 1.0 t/ha followed by the sawdust treatment applied at 2.0 t/ha. These treatments also significantly increased the number and weight of leaves.

Gowda *et al.* (1985) compared pressmud tank silt, neem (*A. indica*) oil cake and farmyard manure with aldicarb 10G for the control of *M. incognita* in a tobacco at 5 kg/m2 and neem cake at 2 kg/m^2 were the most effective in increasing the germination count of tobacco and were at a par with aldicarb at 4 g/m^2, in giving more healthy transplantable seedlings at 60 days than the control. Aldicarb was the most effective in reducing the number of root-knot infested seedlings and galls/plant followed by pressmud, neem cake, farmyard manure and tank silt (both at 5 kg/m^2). All the amendments gave better results than the controls.

Dev Kumar *et al.* (1985) screened neem kernel fractions against *M. incognita*. Four fractions of neem oil were tested for their effect on larval emergence and mortality of *M. incognita*. Pure oil extracts were inactive in both cases, whereas limonoids were highly active.

Mani *et al.* (1986) studied toxic plant products against citrus nematode *Tylenchulus semipenetranse* Cobb. The effect of leaf extracts from 14 different plant species on the mobility of 2nd-stage juveniles of *T. semipenetrans* is shown. Juvenile mobility decreased as the concentration of extracts and the exposure period was increased. Among the effects of oil cake extracts, neem (Azadiractha indica) showed the highest toxicity, with only 19.27% mobility at the 5% concentration. Study on the effect of root exudates revealed 61.8% juvenile mobility in the exudate of marigold (*Tagetes erecta*) followed by 74.8% and 80.6% for mustard (*Brassica campestris*) and periwinkle (*Vinca rosea*) [*Catharanthus roseus*], respectively.

Sundararaju and Koshy (1986) tested neem oil cake in the control of *Radopholus similis* in arecanut palm. The effect of fensulfothion (50 g a.i./palm), aldicarb (10g a.i./palm), DBCP (10 ml a.i./palm) or neem oil cake (1.5 kg/palm) on the control of *R. similis* was studied in yellow disease-affected *Areca catechu*. All treatments significantly reduced nematode populations compared with controls, with fensulfothion treatment giving maximum number of nuts and total nut weight/palm.

Ahuja and Mukhopadhyaya (1987) observed that application of neem (*A. indica*) or Persian lilac (*Melia azedarach*) leaves to aubergine crops in the absence of fruit borer suppressed *M. incognita* populations. In fields heavily infested with *Meloidogyne* spp., carbofuran or aldicarb at 1 kg a.i/ha were not effective. Higher doses and split applications of these nematicides were effective.

Mahmood and Saxena (1987) tested the effect of water extracts of oil cakes and deoiled cakes on the mortality of *Rotylenchulus reniformis*. *R. reniformis* juveniles were exposed to oil-cakes and deoiled cakes [*Ricinus communis*], neem [*A. indica*] and groundnut [*A. hypogaea*]. Morality decreased with an increase in dilution of the extracts and increased with exposure time. Groundnut cake extract was the most toxic.

Sharma and Raj (1987) observed the effect of nematicides and organic amendments on root-knot nematodes infecting potato. The efficacy of aldicarb (Temik 10G) at 5kg a.i./ha, ethoprop (Mocap 10G) at 15kg a.i./ha, sawdust and neem-cake (*A. indica*) both at 26 quintals/ha for controlling *M. incognita* and improving yield of potato cv. Kufri Jyoti was evaluated in randomised-block experiment on an infested field in India. All treatments were applied in furrow at planting, with untreated plots as control. There was little or no tuber infestation and reduced nematode populations in soil after all treatments. Aldicarb gave greatest yield dincrease (24%), ethoprop (92% reduction) and sawdust reduced the nematode population by 56% but neither organic amendment increased yield. Overall, aldicarb gave best results followed by ethoprop, neemcake and sawdust.

Bhattacarya & Goswani (1987) tested comparative efficacy of neem and groundnut oil cakes with aldicarb against *M. incognita* in tomato. In pot experiments, powdered oil-cakes containing 4% neem (*A. indica*) or 1 percent groundnut [*A. hypogaea*] were added to the soil; just before transplantation of tomato seedling, a treatment with aldicarb at 2 kg a.i./ha was applied 7 days after planting. The *M. incognita* juveniles/g soil were introduced. Plant growth response to neem and aldicarb was better than to groundnut or in unamended controls. Generally the number of juveniles penetrating the roots increased significantly with time and decreased with treatment, with least penetration in aldicarb treatment, followed by neem, groundnut and control. However, when egg masses obtained from different treatments were kept for hatching, egg sacs from groundnut and neem treatments gave a smaller hatch than those from aldicarb treatments.

Siddiqui and Alam (1987) studied the efficacy seed dressing with extracts of neem and persian lilac against *M. incognita* and *Rotylenchulus reniformis*. Seeds of tomatoes, aubergines and okra (*Hibiscus esculentus*) were coated with water-soluble extracts of *A. indica* ad *M. azedarach* and the resultant seedlings were inoculated with different doses of *M. incognita* or *R. renifiormis*. Both seed treatments significantly reduced galling and R. reniformis population build-up. The effect was more pronounced in plants grown from seed coated with fruit extracts than in those treated with extracts of leaves, flowers, barks or with gum.

Rao *et al.* (1987) studied the effect of different nematicide, oil cakes and urea in the control of *R. reniformis* on okra. Pot culture studies were conducted to find out the effect of nematicides (aldicarb, carbofuran, ethoprophos or UC-54229 at 2 kg a.i./ha), oil cakes (*A. indica*), karanj (*Pongamia pinnata*) or polang (*Beta vulgaris* var. *orientalis*), at 1 ton/ha) and urea at 30 kg N/ha singly and in combinations (using half of the dose) against *R. reniformis* infecting okra (*Hibiscus esculentus*) cv. Pusa Sawani. Application of aldicarb (1 kg a.i./ha) + neem or karanj oilcake (0.5 ton/ha)+ urea (15 kg N/ha) proved the most effective in reducing the nematode populations while carbofuran or aldicarb (1 kg a/i./ha) + neem oilcake (0.5 ton/ha) + urea (15 kg N\ha) were the most effective in increasing plant growth.

Goswani and Vijayalakshmi (1987) studied the effects of period of decomposition of oilseed cakes in soil on *M. incognita* juveniles. Pot experiments using *M. incognita* J2 (2nd-stage juveniles) were used to find out a correlation between the period of decomposition of neem (*A. indica*) and undi (*Calophyllum inophyllum*) oilseed cakes (at 1 to 3% w/w) in soil and their nematicide effect. The toxicity of exudate from the oilseed cakes increased up to 3 weeks and then decreased until the 5th week.

Bhattacharya and Goswani (1987) studied the role of microorganisms in the decomposition of 4 percent neem (*A. indica*) and 1 percent groundnut (*A. hypogaea*) (w/w) oilcakes and their effect on the penetration, development and population build up of *M. incognita* on tomato. The effectiveness of the cakes was better in unsterilized soil than in sterilized soil, providing increased plant growth response and low nematode populations.

Padhi and Misra (1987) studied the control of *Rotylenchulus reniformis* on french bean. *P. vulgaris* cv. Premier seedlings were sown in pots each containing 2000 individuals of *R. reniformis*/kg soil and treated with one of 7 control treatments. Neem (Azadirachta indica) oil cake (at 2 ton/ha) was the most effective in reducing the nematode population by 86.2 percent and increasing the dry weight of *P. vulgaris* shoots and roots 208.3 and 11.7 percent, respectively. Karanj (*Pongamia pinnata*) oilcake (2 ton/ha), carbofuran (4 kg/ha a.i.), aldicarb (4 kg/ha a.i.) and prawn exoskeleton (2 ton/ha) were also effective. Ethopropos (4 kg/ha a.i.) and organic manure (2 ton/ha) were less effective.

Vijayalkshmi and Goswami (1987) gave root-dip treatments in water extracts of oilseed cakes of *Arachis hypogaea, Azadirachta indica, Sesamum indicum, Brassica rapa* and *Madhuca indica* [*M. ongifoila*] and then exposed to *Meloidogyne incognita* (1000 juveniles/pot). After 45 days, there was a significant reduction in nematode infestation due to all treatments and a marked increase in plant growth.

Jain and Bhatti (1988)tested the effect of neem leaves on root-knot nematode. Neem (*A. indica*) leaves were added at the rates of 2.5 and 5.0 g/kg soil, to potted soil at 0, 1, 2 ,4 and 6 weeks before transplanting 4-week-old tomato seedlings. Each pot was then inoculated with 1000 Juveniles *M. javanica*. Maximum shoot weight and length and minimum galling were observed when *A. indica* leaves were first allowed to degrade over a period of 6 weeks before transplanting plants.

Siddiqui and Alam (1988) studied the control of root-knot and reniform nematodes by bare-root-dip in leaf extracts of margosa and persian lilac. In pot experiments, bare-root-dip in leaf extracts of margosa/neem (*A. indica*) and a related species of persian lilac/bakain (*Melia azedarch*) caused significant inhibition in the penetration of *M. incognita* larvae and the subsequent root-galling and the population build-up of *R. reniformis* on tomato cv. Pusa Ruby and aubergine cv. Pusa Purple Long. The damaging effects of the nematodes were marked by the root-dip treatment. The root-galling and the population of *R. reniformis* were gradually decreased with an increase in the concentration of the extracts and the dip duration.

Schauer *et al*. (1988) observed the effect of extraction residues of neem seed kernels on the nematode *Pratylenchus prenetrans*. The effect of de-oiled, crushed seeds of neem (*A. indica*) and residues of a MeOH/MTB seed kernal extraction (AZT-residue) on *P. penetrans* was investigated. Both extraction-residues were applied in concentrations between 500 and 4170 mg/kg as soil amendments in pot trials. All neem amendments resulted in a marked reduction in numbers of *P. penetrans* in roots of winter wheat. In the oilcake and AZT-residue applications the nematode populations was reduced to 7 and 20 percent of the control, respectively. The soil amendments resulted in increased numbers of microbivorous nematodes but decreased numbers of obligate ad facultative plant parasitic nematodes compared to controls. These effects were more apparent in pots.

Haseeb *et al.* (1988) studied the comparison of nematicides and oil seed cakes for control of *Meloidogyne incognita* on *Ocimum basilicum*. Three-week-old seedlings of O. baslicum (sweet basil) were transplanted to pots containing soil treated 10 days previously with oil cakes of mahua (*Madhuca*) or neem (*A. indica*), or with aldicarb at 0.002 g a.i., carbofuran at 0.0015 g a.i. or bavistin at 0.001 g a.i./kg. Each plant was then inoculated with 5000 (2nd-stage juvenile) and eggs of *M. incognita*. Neem oil cake provided the greatest improvement in plant growth and oil yield, followed by aldicarb, bavistin, mahua oil cake and carbofuran. The greatest root galling (76-100% roots galled) was observed in utreated plants whereas galling ranged from 12.5-75% in treated plants. Neem oil cake was the most effective in reducing root-knot galling and nematode popluations, followed by mahua oil cake and aldicarb. Bavistin and carbofuran were less effective. Mahua oil cake caused phytotoxicity.

Khanna *et al.* (1988) screened 15 plant extracts for nematicidal action against *Aphelenchoides composticola* infesting mushroom. Out of 15 plant extracts tested for their toxic effect against *A. composticola, Calotropis procera* and *Ricinus communis* were most effective, giving 100 percent mortality after 72 hrs *Chrysanthemum indicum* [*Dendranthema indicum*], *Azadirachta indica, Cannabis sativus* and *Eucalyptus hybrida* [*E. tereticornis*] all gave 70 percent or higher mortality.

Salawu (1988) studied the effectiveness of neem leaf extracts in controlling sugarcane root-knot nematode. The comparative effectiveness of neem (*A. indica*) leaf extract used as soil amendment and ethoprop [ethoprophos] singly and in combination in controlling *Meloidogyne incognita* was investigated. Neem extract amendment alone considerably suppressed root-knot development. Neem extract in combination with ethoprophos at the rate of 2-4 kg a.i./ha was as effective as ethoprophos at the rate of 5-8 kg a.i./ha in controlling *M. incognita*. Neem leaf extract amendment was not toxic to sugarcane.

Ravindra (1988) reported the effect of neem cake, antibiotic KT-199 and aldicarb to control *M. incognita* on green gram. The effect of various larval densities of *M. incognita* were tested on greengram under glass house conditions. There was reduction in the plant growth as the larval density increased. Screening of 6 popular varieties for their reaction to the nematode revealed that none of the varieties were free from infection. The variety Pusa-103 was resistant and recorded the least number of galls, egg masses and eggs per egg mass. Control studies showed that aldicarb was better than the antibiotic KT-199 and neem cake in pre-application with respect to increase in shoot height, weight and number of leaves and in reducing the number of galls and eggs per egg mass. The number of pods were greater in plants treated with neem cake in pre-application, KT-199 was found to be superior in increasing shoot height, weight, number of leaves, number of pods and root weight and root length in post-application when compared with aldicarb and neem cake. Neem cake was superior than aldicarb and KT-199 in reducing galls and egg masses in post application.

Gokte & Swarup (1988) observed the effect of neem seed kernel powder treatment in penetration and development of *Anguina tritici* on wheat. Wheat seeds cv. Sonalika were treated with kernal powder at 400, 600 or 800 mg powder/100 g seed. Three nematode galls per seed (4200 larvae/gall) were introduced into each pot. There was a significant reduction in the numbers of larvae able to penetrate in the treated seed in comparison with untreated seed. The 400 mg dose was less effective than the other two dosages.

Goswami *et al*. (1988) compared karanj (*P. pinnata*) and castor (*Ricinus communis*) oil cakes and a commercial organic manure, Hind-O-Meal (containing oilseed of *Melia azadiracht* [*Azadirachta indica*], N, P, K and bone meal) with carbofuran (2 kg a.i./ha) for control of *Meloidogyne incognita* on *Vigna ungiculata* in a pot experiment. Maximum suppression of *M. incognita* populations in roots occurred in plants treated with karnaj and Hind-O-Meal followed by carbofuran treated plants.

Rehman & Khan (1988) tested the effect of oilcakes from castor (*R. communis*), neem (*A. indica*) and groundnut (*Arachis hypogaea*) on seed germination of *C. album* as well as on nematodes and fungi infesting soils of vegetable plots. The water soluble fraction of the oilcakes was inhibitory for seed germination of *C. album*. Oilcakes in different states of decomposition in soil also suppressed the seed germination. Populations of plant nematode ware adversely affected by incorporation of oilcakes in soil.

Dhawan and Kaushal (1988) tested the effect of neem emulsion in the hatching of cereal cyst nematode, *Heterodera avenae. H. avenue* cysts were placed in 500, 1000 or 2000 p.p.m. neem emulsions in petri dishes. Maximum reduction (67.58%) in emergence was observed at 2000 p.p.m. and minimum (43.12%) at 500 ppm.

Bhattacharya and Goswani (1988) studied the effect of different dosages of neem and groundnut oil cakes on plant growth characters and population of *M. incognita* in tomato. The most significant increases in host plant growth and reduction of root galling as well as root and soil populations of *M. incognita* were obtained with neem (*A. indica*) oil cake at 4% w/w. Dosages above 4 percent were phytotoxic.

Pathak *et al*. (1988) studied the effect of water hyacinth and neem leaves for the control of root-knot nematode on brinjal. Fresh, chopped, dried and ground leaves and plant parts and compost of water hyacinth (*Eichhornia crassipes*) and dried and ground neem (*A. indica*) leaves at 30 q/ha were added to pots of aubergines infested with *M. incognita*. Results were compared with those obtained from aldicarb at 2 kg a.i./ha. Aldicarb and dried and ground neem leaves gave maximum yield. All treatments significantly decreased numbers of galls and egg masses.

Jain *et al*. (1988) gave in a field experiment in India with natural infestation of *M. javanica* on tomato seedlings the following chemical treatments: aldicarb, carbofuran, ethoprophos, oncol [benfuracarb], quinalphos, phorate, UC 54229, malathion or BHC (HCH) dust at 4 kg a.i./ha or neem (*Azadirachta indica*) or *Calotropis procera* leaves each at 100 q/ha. These treatments were given alone or combined with carbofuran at 1kg a.i./ha at transplanting. All nursery treatments except oncol and *C. procera* leaves gave increased shoot length compared with control at transplanting. After transplanting there was significant increase in tomato yield when nursery treatments were supplemented with carbofuran. Aldicarb and UC 54299 were the most effective in increasing tomato yield in *M. javanica*-infested plots.

Sundaram and Velayatham (1988) observed the comparative efficacy of insecticides and neem cake in the control of *Rotylenchulus reniformis* and *Helicotylenchus dihystera* affecting garden bean. Aldicarb, phorate, phosphamidon or dimethoate alone or aldicarb + neem cake or phorae + neem cake were applied to *Dolichos lablab* [*Lablab purpureus*] for control of *R. reniformis* and *H. dihystera*. Aldicarb at 0.5 kg a.i./ha + neem cake at 250 kg/ha or dranching with phosphamidon at 1.25 litres/ha were effective in reducing the population of both nematodes up to the ninetyth day.

Khan and Husain (1988) treated *Vigna unguiculata* seeds with aldicarb, bavistin [carbendazim], groundnut (*A. hpogaea*) cake, neem-leaf (*A. indica*) paste or a culture filtrate of *P. lilacinus*. Seed treatment with aldicrarb or bavistin significantly reduced individual or combined adverse effects of *R. reniformis, M. incognite* and *Rhizoctonia solani* on plant growth and nodulation. Neem cake seed treatment caused a significant improvement in plant growth only when inoculated with *M. incognita* or *R. solani* individually or with all 3 pathogens.

Chhobra *et al.* (1988) tested the efficacy of some plant extracts on root-knot nematodes. Six plant species were tested for their leaf extract toxicity to *M. incognita*. Leaf extract of *Ricinus communis* was most toxic followed by *Leucaena leucocephala, Populus deltiodes, Azadiractha indica, Lantana camera* and *Eucalyptus hybrida* [*E. tereticornis*], respectively. Fresh extracts were more toxic than boiled ones except in *E. teretiornis*.

Singh *et al.* (1988) tested the effect of organic soil amendment on rhizosphere fungi and root-knot nematode on egg plant. In a pot experiment aubergine seedlings were grown in soil heavily infested with *Meloidogyne* spp. and then treated with castor (*R. communis*), mustard, neem (*A. indica*), Mahua (*M. longifolia*) or groundnut (*A. hypogaea*) oil-cakes alone or in combination. All oil-cake treatments reduced the nematode population.

Prasad *et al.* (1988) studied the control of root lesion nematode, *Pratylenchus indicus*. Crude extracts of *Lippia germinata*, leaves of *Cleistanthus collinus* and *Azadiractha indica* oil were used as foliar sprays (at 0.5 kg a.i.) and as seed treatment (at 1 kg a.i./100 kg seed) against *P. indicus* on rice var. MW-10 but were ineffective in controlling the nematode. Carbofuran was given as seed treatment at 0.5 kg/100 seed or incorporated into soil in furrows between the rows at 1 kg a.i./ha. The latter treatment was the most effective in controlling P. indicus.

Acharaya and Padhi (1988) tested the effect of neem oil cake and sawdust against *M. incognita* on betelvine (*Piper betle* L). Neem (*A. indica*) oilcake at 1 ton/ha applied to soil in trenches near the root zone at the time of lowering of betelvine cv. Gudibangla, was the most effective treatment in controlling *M. incognita* and reducing the number of root galls 10-12 months after application. The maximum number of marketable leaves was produce by the same treatment followed by sawdust treatment at 2 tonnes/ha.

Jain and Bhatti (1988) studied the integrated control of root-knot nematode. *M. javanica* infecting tomato. In a field naturally infested with *M. javanica* in Haryana, India, transplanting of 8-week-old tomato seedlings alone reduced root-knot index and final populations of the nematodes resulting in better yield compared with transplanting 4-week-old seedlings. However, the results were not significantly different. Spot applications of aldicarb at 1 kg a.i./ha at transplanting in tomato seedlings of either age gave significantly lower nematode populations and higher yields compared with other treatment combinations. Bare-root dip treatment with phosphamidon at 1000 p.p.m. for 8 h at transplanting was effective against the nematode but phytotoxic. Spot applications of neem (*A. indica*) leaves (50 qts/ha) also reduced nematode populations compared with untreated plants but were not as effective as the 2 nematicides.

Darekar *et al.* (1989) studied the management of nematode infesting pomegranate. Carbofuran 3G at 6 kg a.i./ha was the most effective treatment in reducing nematode (*Helicotylenchus* spp., *Xiphinema insigne* and *Meloidogyne incognita*) populations (55.95%) and increasing the yield (33.27%) of pomegranate (*Punica granatum* var. G.B.G. No.1). Treatment with neem (*A. indica*) cake at 2.5 t/ha was the next most effective treatment followed by karanj (*Pongamia pinnata*), mahua (*Bassia latifolia= Madhuca longifolia*) and castor (*Ricinus communis*) cakes.

Zaik and Bhatti (1989) studied the phytotherapeutic effect of some plant leaves in *M. javanica* on tomato. The nematicidal effect of leaves of three species (*Calotropis procera, Ricinus communis* and *Azadirachta indica*) and of their paired combinations on the growth of tomato and infection by *M. javanica* showed that plant growth (height and weight) in all treated plants was significantly higher than that of the control with a subsequent reduction in the nematode population. *R. communis* (40 g/kg soil) was the most effective treatment in reducing *M. javanica* infection.

Siddiqui and Alam (1989) tested the effect of root exudates of neem and persian lilac on plant parasitic nematodes. Root exudates of *Azadirachta indica* and *Melia azedarach* caused considerable mortality of *Hoplolaimus indicus, Helicotylynchus indicus, Tylenchus* sp, *Tylenchorhynchus brassicae, Rotylenchulus reniformis* and *Meloidogyne incognita* in a laboratory experiment. There was a linear relationship between concentration of root exudate and percentage mortality. *A. indica* exudates were more toxic than those of *M. azedarach*. Root exudates of both plants were also found inhibitory to larval hatching of *M. incognita*.

Mishra and Gaur (1989) studied the effect of nematicides, and neem cake on the control of nematode infesting mung. Field microplots initially infested with *Tylenchorhynchus* spp., *Helicotylenchus* sp., *R. reniformis* and *M. incognita* were treated with aldicarb (4 kg/ha) or neem (Azadirachta indica) at 17 q/ha, 10 days prior to sowing *Vigna radiata*. In another experiment, *V. radiata* seeds were treated with fensulfothion (1.0%). Aldicarb gave maximum reduction in nematode population. Fensulfothion gave the greatest increase in grain yield. Neem was not very effective.

Patel and Thakar (1989) studied organic amendments in the control of stunt nematode, *Tylenchorhynchus vulgaris*. Results of pot experiments revealed that fresh and dry azolla mustard oil cake (both at 800 g/pot) effectively reduced growth of *T. vulgaris* populations and increased wheat cv. Sonalika growth. Castor, neem, karanj and famyard manure did not significantly reduce nematode populations over the control. Urea reduced nematode populations but did not effectively improve plant growth.

Akhtar and Alam (1989) studied evaluation of nematicidal properties in some medicinal plants. The incorporation of chopped leaves of *Azadirachta indica, Calotropis procera, Clerodendrum inerme, Eucalyptus citirodora, Lantana indica, Melia azedarach, Ricinus communis* or *Thuja orientialis* at 50 or 100 g/pot significantly suppressed build-up of *Hoplolaimus indicus, Helicotylenchus indicus, Tylenchorhynchus brassicae, Rotylenchulus reniformis* and *Tylenchus filiformis* on *Capsicum annum*. Higher doses gave better results.

Govindaiah *et al*. (1989) studied the effect of mulching of green leaves for the control of root-knot nematode in mulberry. In 4 trails, mulching with *Pongamia pinnata* leaves was found to be superior over *Azadirachta indica* leaves in controlling *Meloidogyne incognita* and increasing growth of mulberry (*Morus alba*) var. Kanva-2. The leaf yield was increased by 28.44.41.71% using pongamia and by 16.13-38.67 percent using neem leaves, (both at 1 tonne/acre) over the control. The numbers of root-knots and egg masses of *M. incognita* were lowest in pongamia and neem.

Qamar *et al*. (1989) applied leaf extracts from *Azadirachta indica, Vinca rosea* (*Catharanthus roseus*) and Jatropha curcas in microplots to evaluate their efficacy against *Anguina tritici*. After 12 weeks, the number of cockles was zero in the plants treated with *A. indica* and *C. roseus*. In the plants treated with *J. curcas*, the number of cockles was 10/plant whereas in control their number was 13/plant.

Gaur and Mishra (1989) studied the integrated control of nematodes in lentil with aldicarb, neem cake. Microplots of *Lens culinaris* in India initially infested with *Tylenchorhynchus* spp., *Helicotylenchus* sp., *Hoplolaimus indicus, Rotylenchulus reniformis* and *Hemicriconemoides* sp. when treated with a combination of seed treatment of phorate (1.0% wet weight), aldicarb at 4.0 kg/ha and neem (*A. indica*) cake at 17 q/ha gave the greatest reduction in nematode populations and the greatest increase in lentil yield. This combination gave a 90.7 percent increase in the grain yield of the subsequent *Vigna radiata* crop.

Krishnamurthy *et al.* (1990) tested the effect of oil seed cakes on the control of *Meloidogyne javanica* in tobacco. Castor, cotton, gingelly, groundnut, mustard, neem and pongamia oil seed cakes were each applied at 100 and 200 g/m^2 before sowing of tobacco in *M. javanica*-infested soil. Observations were taken of tobacco seed germination, nematode population in beds before and after treatment and at the end of nursery growth as well as the root-knot index after 2 and 3 months and yield of transplantable and non-transplantable seedlings. None of the cakes influenced tobacco seed germination, nematode population and seedling yield. The cakes did not offer any significant control of galling on roots in the tobacco nursery at the doses tested.

Facknath and Jadunundun (1990) observed that neem and marigold (*Tagetes erecta*) both suppressed nematode juvenile in the soil and reduced significantly the number of galls on *Meloidogyne*-intested tomato plants.

Das and Padhi (1990) observed that tomato cv. Pusa Ruby seeds with 4 neem products-neem bitter (0.25, 0.5%), Repelin (0.5, 1.0%) neem seed kernel extract (1.25, 2.5%) and Welgrow (0.5, 1.0%) for control of *M. incognita*. All compounds at the higher concentration significantly reduced nematode populations and increased plant growth with neem seed kernel extracts (2.5%) and Welgrow (1%) being the most effective.

According to Azmi (1990), seeds of *L. leucocephala* were sown in polybags containing nursery soil infested with about 400 nematodes/[litre] of soil. Seedlings were thinned to one per bag after germination and the study started in July when seedlings were 10-15 cm tall. Treatments were carried out by excavating the soil round each seedling to 5 cm (*Madhuca latifolia* [*M. lingifolia*]), and returning it to the bag. Nematode populations were reduced by 40-66% and seedling growth (length and dry weight of shoots and roots) increased in all treatments compared with control. Overall, the best performance was with neem cake.

Reddy *et al.* (1990) observed the effect of oilcakes on root-knot nematode and yield of okra. Castor (*Ricinus communis*), groundnut, karanj (*Pongamia pinnata*) and neem oil cakes each at 1t/ha and carbofuran at 2 kg a.i/ha were tested on okra infested with *Meloidogyne incognita* under field conditions. Finely powdered oil cakes were applied 2 weeks prior to sowing okra seeds (cv. Pusa Savani). The nematicide was applied a day prior to sowing. All the oil cakes and the nematicide tested significantly reduced the root-knot index and increased the okra fruit yield compared to control. However, carbofuran gave the least root galling which was on par with karanj cake. The highest fruit yield was obtained with carbofuran and karanj which were on par with the other 3 oil cakes.

Srivastava and Singh (1990) studied the effect of organic amendment on interaction of *Macrophomina phaseolina* and *Meloidogyne incognita* on french bean (*Phaseolus vulgaris*). A pot trail was initiated to see the effect of organic amendments, neem (*Azadirachta indica*) and mustard cake on *M. phaseolina* and *M. incognita* infesting French bean (*Phaseolus vulgaris*). Mustard cake was effective in reducing the disease incidence significantly.

Darekar *et al.* (1990) tested the effect of non-edible oilseed cakes on the control of root-knot nematode on tomato. Neem (*A. indica*), karanj (*P. pinnata* [*P. glabra*]), mahua (*M. indica* [*M. longifolia*]) and castor (*R. communis*) oilsed cakes were tested in a field experiment in furrows and spots for control of *M. incognita* on tomato. All treatments at 400 kg/ha reduced *M. incognita* populations and gall index, with neem and karanj cake being most effective. Spot applications were superior to in-furrow applications.

Akhtar and Alam (1990) evaluated some nematicidal potential in some plants against root knot nematode on tomato and chilli. Three-weak-old seedlings of tomato cv. Pusa Ruby and chilli (*Capsicum anum*) cv. NP-46-A were transplanted singly into clay pots, one week after 50 and/or 100g chopped leaves of a test plant had been added to the soil. The plants were then inoculated with 5000 freshly hatched larvae of *M. incognita*. Three months later, suppressed root galling by the nematode and enhanced plant growth were observed in all amendment treatments. *Calotropis procera* was the most effective followed by *Azadirachta indica, Ricinus communis, Melia azedarach* and *Lantana indica*. *Eucalyptus citriodora* and *Thuja orientalis* gave poorer results while *Clerodendrum inerme* was the least effective.

Akhtar *et al.* (1990) studied the effects of organic amendments to soil as nematode suppressant. Chopped leaves of *Melia azedarach, Calotropis procera, Ricinus communis, Eichhornia crassipes* and *Azadiractha indica* were applied to a field infested with plant parasitic nematodes. All chopped leaves suppressed nematode reproduction.

Siddiqui and Alam (1990) studied sawdusts as soil amendments for the control of nematode infesting some vegetables. Sawdusts of neem (*A. indica*) and mango (*Mangifera indica*) reduced populations of *Rotylenchulus reniformis* on tomatoes and aubergines and *Tylenchorhynchus brassicae* on cabbages and cauliflowers. Nematode control increased with the increasing dose of sawdust, neem being more effective than mango. The combined effect of sawdust and ammonium sulfate was greater than for either of the separate treatments.

Pandey and Singh (1990) tested the effect of organic amendments on soil microflora and nematode (*M. incognita*). Neem cake (*A. indica*), mustard cake, linseed cake (*Linum usitatissimum*), mahua cake (*M. lingifolia*), compost, gober (callte dung), gas product (biogas) and sawdust were added to soil in micro plots. The plots were sown with seeds of *Cicer arietinum* (chick pea) 15 days after the addition of amendment. Neem cake caused maximum reduction in the number of root-knot nematodes (*M. incognita*) followed by sawdust.

Khan (1990) studied the effect of medicinal plants against *Pratylenchus zeae*. Of the 6 plant species tested in pot experiments, incoporation of leaves of *Aburus precatorius* allowed the highest multiplication of *P. zeae* population on *Capsicum frutescens* cv. PL 2289 while *Azadirachta indica* allowed the lowest. *Acacia albida, Albizia adianthifolia, Cnestis ferruginea* and *Tamarindus indica*, exhibited nematode toxicity to varying extents. There was a direct relationship between the dose of amendment (25, 50 and 100 g/kg soil) and plant growth.

Darekar and Mhase (1990) did trials to manage grapevine nematodes in the field with neem cake and nematicide. Neem cake (*A. indica*) alone or in combination with carbofuran or benfuracarb was used to control *Meloidogyne incognita, Rotylenchulus reniformis* and *Xiphinema insgine* on grape in Maharashtra, India. All treatments reduced the nematode population and increased the yield of crops. Carbofuran and benfuracarb both at 6 kg a.i.\ha were the most effective, reducing the nematode population by 60.18 and 56.93% respectively and and increasing yields by 30.00 and 27.94 percent.

Kaushal (1990) tested the effectiveness of neem products against *Heterodera avenue. Azadiracha indica* concentrate and powder were compared with carbosulfan, sebufos or padan for control of *H. avenae* on wheat in pot trials. Sebufos gave the greatest reduction in cyst population followed by carbosulfan. A. indica products were least effective.

Siddiqui and Alam (1990) Worked on the control of root-knot reniform and stunt nematodes by nimbin dressing. The application of nimbin, a triterpenoid from *Azadirachta indica*, as seed dressing significantly reduced the root-knot development/nematode population, with a consequent improvement in plant growth at all levels of nematode inoculation on tomato, eggplant (aubergine) and okra attacked by *M. incognita* and *R. reniformis* and on cabbage and cauliflower attacked by *Tylenchorhynchus brassicae*.

Jonathan *et al.* (1991) studied effect of organic amendments on the control of sugarcane nematodes. Press mud (25 t/ha), farmyard manure (20 t/ha), poultry manure (20 t/ha), neem cake (2 t/ha), coir pith (25 t/ha) and sugarcane bagasse (25 t/ha) were tested for their effect on a population of sugarcane rhizosphere nematodes, *M. incognita, Pratylenchus coffee* and *Helicotylenchus dihystera*. Neem (*A. indica*) cake and press mud were the most effective, giving populations 81-82% lower than control after 90 days and 60-62% lower after 180 days.

Reddy and Khan (1991) Studied integrated management of root knot nematodes infecting okra. Castor (*R. communis*), groundnut, karanj (*P. pinata*) and neem oilseed cakes at 1.0 ton/ha in single applications and 0.5 ton/ha in combined applications and carbofuran at 2 kg/ha in single applications and 1 kg/ha in combined treatments were evaluated for control of *M. incognita* on okras in India. Plots treated with karanj cake, carbofuran and groundnut cake + carbofuran gave significantly higher okra fruit yield compared to controls. All treatments significantly reduced the root gall index.

Mojumder and Mishra (1991) tested the effect of agueous extracts of neem seeds on hatchabilitiy of eggs and penetrability of hatched juveniles of *Meloidogyne incognita* into roots of mungbean. *M. incognita* egg masses were exposed to different concentrations of aqueous extracts of neem kernel and neem seed coat for 12, 24 and 48 h. Both extracts reduced hatching of eggs with neem seed kernel being more effective than neem seed coat. The penetrability of hatched juveniles into roots of *Vigna radiata* seedlings after 15 days of inoculation was also reduced in both treatments.

Akhtar and Alam (1991) studied integrated control of plant parasitic nematodes on potato with organic amendments nematicide and mixed cropping. Increase in the population of plant parasitic nematodes (mainly *Meloidogyne incognita*) on potato was significantly arrested when mustard (*Brassica juncea*) was also grown with potatoes in alternate rows. There was further suppression of the population of parasitic nematodes when the soil was also treated with oil-seed cakes and leaves of neem (*A. indica*) and castor (*R. communis*) and carbofuran. As a consequence of reduction in the nematode population there was an increase in the yield of potato. The beneficial effects of all the above treatments persisted in the next okra (*Abelmoschus esculentus*) crop.

Reddy *et al.* (1991) studied integrated management of the citrus nematode *Tylenchulus semipenetrans* using oil cakes and *Paecilomyces lilacinus*. Oil-cakes of castor (*R. communis*), karanj (*P. pinnata*) and neem (*A. indica*) at 20 g per plant and the parasitic fungus, *P. lilacinus* at 8 g infected rice seeds per plant alone and in combination at half the dosages were evaluated for the management of *T. semipenetrans* infecting acid lime under glasshouse conditions. Neem cake + *P. lilacinus* gave maximum shoot length and shoot weight, while castor cake + *P. lilacinus* gave maximum root length and root weight. Maxmum reduction in soil nematode population was obtained with neem cake + *P. lilacinus*, while castor cake + *P. lilacinus* gave the least nematode population in the roots.

Sharma (1991) applied mustard or neem cakes (2 kg/tree), carbofuran or phorate (5 kg a.i./ha) to plum trees to determine losses due to plant parasitic nematodes. All treatments reduced nematode populations. Haseeb and Butool (1991) evaluated different nematicides and oil cake on the control of *Meloidogyne incognita*. The efficacy of aldicarb, bavistin, carbofuran (all at 4 kg a.i./ha), ethoprophos (6 kg a.i./ha), Onchol [benfuracarb], Rubby [cadusafos] (both at 4 kg a.i./ha) and neem (*A. indica*) cake (1 g N/kg soil) was evaluated for control of *M. incognita* on *Artemisia pallens*. Best results in terms of improvement in plant growth and reduction in root-knot population was obtained with aldicarb followed by ethoprophos and neem cake.

Khan *et al*. (1991) studied the effect of water soluble extracts of certain oil cakes on the development of root knot disease of okra. Okra seedlings were given 25 or 50 minutes root dips in water soluble extracts of neem (*A. indica*), mustard and mahua oil cakes. All treatments reduced damage due to *M. incognita*. The longer dip period was more effective irrespective of the oil cake extract used. Neem-cake extract treatments give more beneficial results than the other two extracts.

Khanna (1991) studied *in vitro* some plant extracts as nematicides aganist *Meloidogyne incognita*. The efficacy of leaf extracts of 8 plants was tested against *M. incognita*. *Azadiractha indica, Calortopis procera* and *Ricinus communis* were the most effective with more than 50 percent of nematodes being killed even at the lowest concentration after 24 hrs.

Alam (1991) evaluated oil seed cakes against plant parasitic nematodes on vegetables. In field trials oilseed cakes of mahua, castor, mustard, neem (*A. indica*) and groundnut each at 110 kg N/ha, singly or in different combinations significantly controlled populations of plant parasitic nematodes on tomato (*Lycopersicon esculentum*), eggplant (aubergine), chilli (*Capsicum annum*), okra (*Abelmoschus escultentus*), cabbage (*Brassica oleracea* var. *capitata*) and cauliflower (*Brassica oleracea* var. *botrytis*). Mahua oilseed cake was phytotoxic to all test crops except eggplant. All other oilseed cakes improved plant growth significanlty.

Ahmad *et al*. (1991) studied the effect of extracts of various plants on egg hatching and larval mortality of root knot nematode. Leaf extracts of *Azadirachta indica, Melia azedarach, Datura alba* and *Ricinus communis* were highly toxic to *M. incognita, A. indica* gave maximum (100%) inhibition of egg hatching and larval mortality followed by *M. azedarach, D. alba* and *R. communis*. Inhibition of egg hatch and larval mortality were significantly affected by the concentration of the leaf extract and the exposure time.

Jain and Bhatti (1991) studied on the evaluation of the effective integrated methods for the control of root knot nematode. Nursery treatment of tomato using aldicarb at 4 kg/ha or neem leaves at 50 qt/ha alone or in combination with spot application of aldicarb at 1 kg/ha or neem leaves applied in the field were evaluated for control of *M. javanica*. Aldicarb treated nursery plants + spot application of aldicarb at transplanting showed maximum reduction in *M. javanica* popluations (62.3%). Highest tomato yields were obtained with uninfected tomato plants + aldicarb at transplanting.

Abid and Magbool (1991) studied the effect of oil cakes and neem leaf extract on the root knot nematode development. Bare-root dip treatment in the extracts of oil-cake of castor (*R. communis*) and mustard and neem leaf significanlty reduced root-knot infection caused by *Meloidogyne javanica* on tomato cv. Rutgers and eggplant [aubergine] cv. Round purple. Pandey *et al.* (1991) investigated the influence of neem (*A. indica*), mustard, linseed (*Linum usitatissimum*) or mahua cakes, cattle manure gas product, compost and sawdust on population of *Meloidogyne incognita* attacking *Vigna radiata* in microplot experiments. Neem cake followed by sawdust gave greatest control of *M. incognita*.

Borah and Phukan (1992) studied effect of organic amendments for the management of *M. incognita* on green gram. Poultry manure, mustard cake or neem cake (*A. indica*) were applied at 1 or 2 tonnes/ha for the control of *M. incognita* on *Vigna radiata* in the field in Assam, India. There was a significant reduction in number of galls, eggmasses and nematode populations in both doses of organic amendments with the higher dose of poultry manure being the most effective.

Hassan *et al.* (1992) studied the effect of plant exudates and by products on development of root-knot nematode. Root exudates of 5 plant species (*Argemone mexicana, Ricinus commuis, Allium cepa, Azadirachta indica* and *Solanum xanthocarpum*) significantly increased the percentage of 2nd and 3rd-4th stage juveniless of the root-knot nematode *Meloidogyne incognita*, and reduced the percentage of adult females associated with tomato roots compared with control crop plants. The percentage of adult males was unaffected. Root galling of tomato was reduced by all the test plants, with *Azadirachta indica* the most effective. Four plant products (*Azadirachta indica* and *Madhuca longifolia* oilcakes, sawdust and pressmud) also increased the occurrene of nematode juveniles on tomato, but reduced the percentage of adult females with *A. indica* oilcake the most effective.

Salawu (1992) tested the effect of neem leaf extract and ethoprop singly and in combination with *Meloidogyne incognita*. The effect of neem leaf extract (*A. indica*) used as soil amendment and ethoprop [ethoprophos], singly and in combination on *M. incognita* activity and growth of sugarcane cv. Co 1001 was investigated. Neem extract amendment alone considerably suppressed root-knot development. Neem extract in combination with ehtoprophos at the rate of 2 to 4 kg a.i./ ha was effective as ethoprophos alone at the rate of 5 to 8 kg a.i./ha. Neem extract amendment was not toxic to sugarcane.

Awan *et al.* (1992) studied the effect of leaf extract of the four plant species on larval mortality of citrus nematode, *Tyenchulus semipenetrans*. The effect of leaf extract of *Azadiractha indica, Calotropis procera, Nerium indicum* [*N. oleander*] and *Datura alba* [*D. metel*] on larval mortality of *T. semipenetrans* was investigated. *A indica* gave maximum larval mortality followed by *D. metal, N. oleander* and *C. procera*. Larval mortality increased with increase in exposure time and concentration of extract.

Agbakli *et al.* (1992) studied the extracts of neem against *Meloidogyne* sp. on leaf vegetable. Different concentrations of neem (leaf extracts and neem cake) were applied as protective treatments in soil infested with *Meloidogyne*, before sowing jute (*Corchours oiltorius*), lettuce and celosia (*Celosia argentea*) as test crops. The foliar extracts had no effect on nematode damage but neem cake at 0.5 and 1 kg/ha/m2 increased yields and reduced the disease rate, often more effectively than carbofuran.

Ganai *et al.* (1992) studied the effect of leaf extracts of neem and other plants against *Ditylenchus myceliophagus*. The effect of leaf extracts of neem *Azadirachta indica, Mentha pipertia* and *Parthenium hysterophours* on *D. myceliophagus* was investigated *in vitro*. All extracts caused significant nematode mortality with the maximum occurring with *A. indica*. The same trend were observed by Wani (1992).

Mojumdar and Mishra (1992) did management studies of *Meloidogyne incognita* on chickpea. Powdered *Azadirachta indica* seed coat was applied to soil infested with *M. incognita* or given as seed treatments to chickpeas in pot trials. Soil and seed treatments were given separately or in combination. All treatments significantly reduced the number of root galls, with combined soil and seed treatments having the greatest reduction. Rao and Reddy (1992) studied the effect of plant leaves in the management of *M. incognita* on tomato. The efficacy of the incorporation of leaves of *Brassica campestris, Catharanthus roseus, Pedilanthus tithymaloides, Bougainvillea spectabilis, Azadirachta indica, Pongamia pinnata, Calotropis procera* and *Ricinus communis* at 80 g/kg soil was compared with carbofuran application at 2 kg a.i./ha for control of *M. incognita* on tomato. *B. campestris, P. tityhmaloides, R. communis, A. indica* and *C. procera* were as effective as carbofuran for control.

Kathirvel *et al.* (1992) studied the effect of seed treatment with plant products and chemicals for the control of *M. incognita*. Carbosulfan 6 percent seed dressing was the most effective treatment for reducing *M. incognita* populations on okra in pot experiments. Of the other nematicides, fenamiphos at 3, 6 or 12 g a.i. w/w seed soak was more effective than monocrotophos at the same concentrations. Of the plant extracts, neem (*A. indica*) oil, mahua oil and pinnai (*Calophyllum inophyllum*) oil and neem seed kernel extract all at 3 percent, and neem leaf extract at 0.5 kg/litre water were effective. However, the neem leaf extract and mahua oil were the most effective but to a lesser extent than carbosulfan.

Abid *et al.* (1992) observed in a pot experiment, complete control of *Meloidogyne incognita* infection in mungbean (*Vigna radiata*) in plants grown in 3% w/w cotton mixed with 25 ml of *Bradyrhizobium* sp. (3.5 X 109 cells/ ml). The combined use of *Bradyrhizobium* sp., *P. lilacinus* (3 X 109 conidia/ml) or Furadan (carbofuran) at 55 kg/ha with cotton and neem (*A. indica*) cakes was more effective than individual applications.

According to Pandey *et al.* (1992), in Uttar Pradesh (India), during the cropping season of Japanese mint (*Mentha arvensis* cv. MAS-1) the crop was severely infested with *Meloidogyne incognita*. In addition to *Meloidogyne* larvae, populations of *Tylenchus* sp., *Tylenchorhynchus* sp., *Hoplolaimus* sp. *Helicotylenchus* sp., *Pratylenchus* sp., *Rotylenchulus* sp., *Aphelenchoides* sp., *Longidorus* sp., *Xiphinema* sp., and larvae of *Heterodera* sp. were encountered in the rhizosphere of *M. arvensis* cv. MAS-1. *Meloidogyne* larvae and *Tylenchorhynchus* sp. were found most abundant as compared to other phytoparasitic nematodes. Green-house experiments were carried out to determine the effect of graded populations densities of *M. incognita* and the use of different oil-seeds cakes neem (*A. indica*), castor (*R. communis*), mahua (*M. longifolia*) on the growth, yield of the crop and reproduction of the nematode. Results indicated that an increase in initial population densities of nematode, caused a considerable reduction in root-shoot length., fresh and dry weight, chlorophyll a, b and total rate of photosynthesis and oil yield of the crop. The nematode reproduction (Rf=Pf/Pi) was found to be density dependant. Neem oil-seed cake at 1 g N/kg soil was found best for the significant enhancement of the crop yield followed by castor and mahuna cakes.

Khan (1992) tested neem leaves and carbofuran an *Pratylenchus thornei*. The effect of carbofuran at 10 kg a.i./ha and crushed dry neem (*A. indica*) leaves at 833 or 1666 kg/ha, on *P. thornei* were examined. All the three treatments resulted in increased root weight as well as the thousand grain seed weight. The population level of *P. thornei* was reduced the most by the higher dose of neem.

Dwivedi and Pandey (1992) studied the effect of organic amendments on root knot nematode. Neem (*A. indica*) cake, mustard (*Brassica juncea*) cake, linseed (*Linum usitatissimum*) cake, mahuat (*M. longifolia*) cake, compost, cattle gas manure or sawdust were applied to microplots prior to sowing *V. radiata* seeds. Each plant was inoculated with 1000 *Meloidogyne incognita* juveniles. All treatments significantly reduced the total number of nematodes, egg sacs, eggs, males and females/plant with the greatest reduction occurring in the neem cake application.

Murthy and Rao (1992) controlled *Meloidogyne incognita* with certain oil cake amendment. Aldicarb, carbofuran and phorate all at 3.75, 6.25 and 8.75 kg a.i./ha and the oil cakes of neem (*A. indica*), castor (*R. communis*), gingelly (*Sesamum indicum*) and coconut at 100 or 200 kg N\ha were effective in suppressing the effect of *M. incognita* and improving the yield of betelvine (*Piper betle*). Neem was the most effective treatment.

Mishra and Gupta (1992) applied five oilseed cakes i.e. mustard (*Brassica juncea*), neem (*A. indica*), mahua (*M. longifolia*), linseed (*Linum usitatissimum*) and castor (*R. commuis*) or sawdust to microplots and then sown with soyabean seeds. All treatments reduced populations of plant parasitic nematode compared with control at 120 days after sowing.

Mojumdar and Mishra (1992) tested the effect of seed soaking in aqueous extracts of neem seed on germination on mungbean and penetration of second stage juveniles of *M. incognita*. Seeds of *Vigna radiata* cv. PS-16 were soaked in 3 concentration of aqueous exacts of *Azadirachta indica* kernels and seed coats for 3, 6, 12 and 24 h in a pot trial. The number of *M. incogntia* penetrating roots declined with increasing concentration and soaking time.

Hasseb and Butool (1993) studied some pesticides and oil seed cakes for the control of *Meloidogyne incognita* on *Trachyspermum ammi*. Use of aldicarb, carbendazim, carbofuran, ethoprophos and ebufos at 4 kg a.i./ha and oil cakes of linseed (*Linum usitatissiumum*), mustard (*Brassica juncea*) and neem (*Azadiractha indica*) at 1 g N/kg soil, for the management of *M. incognita* parasitizing *T. ammi* was investigated. In general, all the treatments were able to reduce the root-knot development and nematode population and to increase the plant fresh and dry weights. The best result in terms of reduction in root-knot index and improvement in plant growth was achieved by treating plants with ethoprophos followed by carbofuran, neem cake, carbendazim, ebufos, mustard and linseed cakes.

Akhtar & Mahmood (1993) studied the fallen leaves as soil amendment for the control of plant parasitic nematode. Leaves of *Melia azedarach, Ricinus communis* or *Azadiractha indica* applied to pots containing tomato or *Capsicum annuum*, significantly suppressed development of *M. incogntia* and populations of other plant parasitic nematodes *Azadirachta indica* was the most effective treatment.

The authors again in 1993 studied the effect of Nimin and some plant oils for controlling plant parasitic nematode. Bare-root dip treatment with 'Nimin' containing neem triterpenes and some plant oils namely neem (*A. indica*), castor (*R. communis*), rocket-salad (*Eruca sativa* [*E. vesicaria*]) and mustard (*Brassica juncea*) showed significant inhibition of root-knot development caused by *M. incogntia*, and population build-up of commonly occurring plant-parasitic nematodes on tomato in pot experiments. 'Nimin' was highly effective in reducing the nematode population.

Farahat *et al*. (1993) tested margosan against reniform nematode infecting sunflower. Margosan (active ingreadient azadirachtin) and sincocin (cytokinin and fattly acids) were evaluated in the control of *Rotylenchulus reniformis* infecting sunflower. Results indicated the application of margosan either once at nematode inoculation time, one week after nematode inoculation or twice at and after nematode inoculation at the rates of 100, 500, and 1000 p.p.m, significantly reduced numbers of eggmasses/root and hatched juveniles. A significant increase in shoot fresh and dry weights was observed with increasing concentrations. The application of sincocin (5, 10 and 15%) once at nematode inoculation or twice (at and after nematode inoculation) significantly decreased numbers of the hatched juveniles and eggmasses with the double treatment of sincocin giving the best results.

Akhtar and Mahmood (1993) tested the effect of organic amendment to conrtol root-knot nematode on chilli. Root-knot development caused by *M.*

incognita on chilli (*Capsicum annuum*) cv. Jawala significantly declined in the presence of *Mononchus aquaticus*. The serverity of root-knot infection was greatly reduced when chopped leaves of neem (*A. indica*) and castor (*R. communis*) were incorporated into the soil.

Mojumdar and Mishra (1993) tested the effect of neem seed kernal and neem seed coat against *M. incognita*. *Vigna radiata* seeds were soaked in aqueous extracts of *Azadirachta indica* seed kernels and seed coats at 100, 50, or 25 percent. All extracts reduced the number of root-knot galls in plants and the juvenile of *M. incognita* in the soil. Significant effects were obtained with the 100 and 50 percent concentrations for 3 hrs soaking or the 50 and 25 percent concentrations for 6 hours soaking. Mojumdar and Mishra (1993) did experiment with soil application of aqueous extracts of neem products for the management of *M. incognita* in chickpea. Aqueous extracts of *Azadirachta indica* cake, seed kernal or seed coat were applied as a single full dose or split doses to soil naturally infested with *M. inconita* in a pot experiment. Chickpea seeds were then sown in the pots. All treatments significantly reduced the number of root galls, *A. indica* seed kernels applied at the full dose being most effective.

Khan and Shaukat (1993) tested the effect of neem and carbofuran against population of three nematodes associated with wheat varieties. Carbofuran at 12 g/miniplot or neem leaves at 1 kg (single dose) or 2 kg (double dose)/miniplot were applied to wheat cultivars. Faisalabad-85, Khyber 87 and Sarhad-82 for control of *Bitylenchus goffarti, Helicotylenchus indicus* and *Hoplolaimus seinhorsti*. All treatments reduce nematode populations with carbofuran being most effective against *H. indicus* and the double dose of neem against *B. goffarti* and *H. seihorsti*.

Vats and Nandol (1993) tested the effect of leaves extracts of neem and *Eucalyptus* in root dip treatment of tomato seedlings aganist *M. javanica*. Healthy tomato seedlings were treated with 5, 10, 20, 30 or 40 percent w/v concentration of leaf extracts of *Azadirachta indica* or *Eucalyptus tereticornis* and then grown in pots containing soil infested with *M. javanica*. The 20 and 30 percent concentrations of *A. indica* and *E. teretiocrnis* were applied in pots containing soil infested with *M. javanica*. The 20 and 30 percent concentrations of *A. indica* and *E. tereticornis*, respectively, with a dipping period of 6 hrs, were most effective in controlling *M. javanica*.

Goswani & Devi (1993) tested comparative efficacy of neem and mahua oil seed cakes aganist root knot nematode. A preliminary study was conducted in pots under greenhouse conditions to compare the efficacy of non-edible oilseed cakes of mahua (*M. longifolia*) and neem (*A. indica*) with attapulgite based clay dust (ABCD) individually and in combination against *M. incognita* on pea. Observations recorded 30 days after inoculation showed better plant health in pots treated with neem cake and mahua cake. Maximum reduction of *M. incognita* population was recorded in ABCD treatment, followed by neem and mahua. Combined treatment with neem mahua and ABCD showed improvement in plant growth and reduced *Meloidogyne incognita* infestation.

Haidar *et al*. (1993) tested the effect of organic soil amendments on plant parasitic nematode. Effect of application of castor (*R. communis*), karanj (*P. pinnata*), mustard (*B. campestris*) and *Azadirachta indica* cakes, pressmud and carbofuran in soil was examined on nematode population yield and juice quality of sugarcane. All the treatments reduced nematode papulation in the field.

Vats and Nandal (1993) studied the efficiency of plant extracts against *M. javanica* . Eighteen plant extracts were screened for their efficacy as bare-root-dip treatments of tomato seedlings against *M. javanica*. The *Azadirachta indica* and *Eucalyptus tereticornis* leaf extracts were observed most effective in reducing the root galls.

Sahoo and Sahu (1994) Studied the effect of different oil cakes and two nematicides in the control of lesion nematode *Pratylenchus zeae* in rice. Neem (*A. indica)*, karanj (*P. pinnata*), mustard (*B. juncea*), polanga (*Calophyllum inophyllum*), til (*Sesamum indicum*), groundnut (*Arachis hypogaea*), mahua (*M. longifolia*) and cotton (*Gossypium*) oil cake were compared with carbofuran or phorate for control of *P. zeae* in rice in pot trials. All treatments significantly reduced the number of nematodes in roots with carbofuran. Neem cake gave the greatest reductions.

Mukhtar *et al*. (1994) studied the control of root knot desiase of tomato with organic oils amendment. Addition of organic soil amendments proved effective in controlling *M. incogntia* in tomato plants, in pot experiments. Castor-oil cake at 25 g/kg of soil amended before sowing give better results than neem leaf (*A. indica*) cake and sawdust at the same dosage, whereas a combination of these amendments gave maximum control of nematodes.

Mehta *et al*. (1994) observed the effect of the oil cakes on the control of *Pratylenchus zeae* in sugarcane. Application of 5 oil cakes viz., groundnut, sesame, neem, cotton seed and coconut significantly increased yield and quality of sugarcane. Population of *P. zeae* in soils and roots was significanlty suppressed by all the treatments. Rathore (1994) tested the effect of neem, mustard and karanj cakes against *Rotylenchulus reniformis* on cowpea. Pot experiments indicated that karanj cake and neem (*A. indica*) cake at 25 g/ha were effective improving plant growth and reducing *R. reniformis* infection on cowpeas. Akhtar & Mahmood (1994) tested the effect of oil cakes and leaves of neem and castor extracts for the control of root knot nematode on chilli. Bare-root treatment of chilli (*Capsicum annuum*) seedlings with extracts of decomposed and undecomposed oil-cakes and levaes of neem (*A. indica*) and castor (*R. communis*) provided protection against root-knot disease cuaused by *M. incognita*.

Rao and Reddy (1994) observed that the addition of neem cake extracts (5 and 10%) mixed with spores of *Pinus* to the soil was effective in the management of *M. incognita* on egg plant. Results also indicated that the neem cake extracts (5 and 10%) support the growth of *P. lilacinus* and *M. incognita* on egg plant.

Mukhtar *et al.* (1994) tested the extracts of leaves of some plants on *M. incognita*. Leaf extracts of *Azadirachta indica, Melia azedarach, Datura alba* (*D. metel*) and *Ricinus communis* were applied to tomato plants inoculated with *M. incognita* in a pot trial. All extracts except *R. communis* enhanced plant growth and reduced disease severity. The *A. indica* extracts was most effective.

Khan *et al.* (1994) tested the effect of three plant extracts on nematode population and on growth parameters of wheat. Leaf extracts of *Azadiractha indica, Eucalyptus citriodora* and tobacco decreased the population density of *Pratylenchus thornie* in soils, while only *A. indica* and tobacco extracts reduced the population density of *Hoplolaimus seinhorsti* [*Basirolaimus seinhoristi*]. Of the leaf extracts, only *A. indica* significantly increased the number of spikes and root weight of wheat cv. Pirsabak-85.

Walia *et al.* (1994) observed that the growth of tomato (cv. HS-101) plants was significanlty incresed in soil amended with subabool (*Leucaena leucocephala*) and neem (*A. indica*) levaes compared with the unamended control of number of galls and egg-masses of *M. javanica* popluation was also reduced in amended soil. Neem decreased the incidence of both *Rhizoctonia batatiocla* and *R. solani* in the treatment where nematodes were inoculated one week earlier than the fungus.

Tiyagi and Alam (1995) tested the effect of oil seed cakes aganist plant parasitic nematode on mungbean and chickpea. The efficiency of oil-seed cakes of neem (*A. indica*), castor (*R. communis*), mustard (*Brasica campestris*) and duan (*Eruca sative* [*Eruca vesicaria*]) was evaluated against plant-parasitic nematodes and soil-inhabiting fungi infesting mungbean (*Phaseolus aureus*

[*Vigna radiata*]) and the subsequent crop, chickpea, in field trials. The population of plant-parasitic nematodes, including *Meloidogyne incognita, Rotylenchulus reniformis, Tylenchorhynchus brassicae* and *Helicotylenchus indicus*, and the frequency of pathogenic fungi, including *Marcophomina phaseolina, Rhizoctonia solani, Phyllostica phaseolina* and *Fusarium oxysporum* f. *ciceri*, were significantly reduced by these treatements, but the frequency of saprophytic fungus was increased. Depth of ploughing also influenced the population of plant-parasitic nematodes and the frequency of fungi.

Alagumalai *et al.* (1995) used acid extract of neem cake at different dilutions to control the root knot nematode *M. incognita* on potted seedlings of *Vigna unguiculata*. The extracts enhanced the growth of the host plants and also reduced the population build up of the nematode. The extract of lowest dilution (1/10) was most effective in controlling the nematode population.

Vijayalakshmi *et al.* (1995) did management experiment of root knot nematode through neem products. Pot experiment was conducted to reduce the damage of *M. incognita* in mung bean byadding decoposed neem seed, seed kernel and seed coat in nematode infested soil in two doses. Observations recorded on the nematode multiplication and plant growth revealed drastic reduction in root-knot galling and nematode population in soil and significant increase in plant growth in all the treatments in either doses. The grain yield was also increased in respect of all the treatments. The grain yield in case of neem seed and seed kernel was comparable to uninfested check. Among the treatments, neem seed kernel provided maximum reduction in nematode infestation.

Thakur and Darekar (1995) did experiments on the effect of non-edible oil seed cakes against *M. incognita* and growth parameters of brinjal. The effectiveness of four non-edible oilseed cakes viz., neem, karanj, castor and mustard was tested against *M. incogntia* infesting brinjal (aubergine). The results indicated that neem cake @ 35 g\plant and karanj cake @ 44 g/plant were most effective in controlling *M. incognita* both under pot culture and field conditions. These oilseed cakes had recorded reduced root galling, egg/egg masses, root-knot index and increased shoot and root length as well as dry shoot and root weight and yield of aubergine. Again in the same year, they tested the relative efficacy of decomposed non-edible oilseeds cakes against *M. incognita* juveniles. The studies conducted on the decomposition of 4 non-edible oilseeds cakes indicated that the leachates obtained from decomposed neem oil cake had a promising effect on mortality of *M. incognita* juveniles at 3 percent w/w dose until the end of the third week in which it was maximum.

Thakur & Darekar (1995) studied the period of decomposition of some non-edible oilseed cakes for the control of root-knot nematode. A higher percentage of polyphenols was released at the higher dose (3% w/w) of neem, karanj, castor and mustard cakes than the lower (1% w/w) doses of these cakes. The percentage polyphenols were observed to be increased up to the third weak and gradually became very low with the time interval. The results indicated that the planting/sowing of the crops should be done immediately after three weeks from the addition of cakes into the soil for the effective control of *Meloidogyne* spp.

Reddy *et al.* (1995) evaluated castor (*R. communis*), karanj (*P. pinnata*) and neem (*A. indica*) oil cakes each at 20 g/plant and vesicular arbuscular mycorrhiza, *Glomus fasciculatum* at 4000 spores per plant either individually or in combination at half the dosages for the management of *Tylenchulus semipenetrans* infecting acid lime under greenhouse conditions. The neem cake in combination with VA mycorrhiza gave maximum shoot length, shoot weight, root length and root weight. The highest reduction in nematode population in both roots and soil was obtained with neem cake + *G. fasciculatum* which was on par with karanj cake + *G. fasciculatum*, neem cake and karanj cake.

Abid *et al.* (1995) tested the effect of neem cake in the control of root-knot nematode in mung bean. Soil amendment with neem cake, cotton cake, *Datura fastuosa* and *Calotropis procera* at 0.1, 0.5 and 1 percent w/w reduced infection of *M. incongnita* on mungbean (*Vigna radiata*) with no significant differences at different dosages. *Carlotropis* at 1 percent showed greater reduction in root-knot index. Mustard cake at low dosages (at 0.1 & 0.5% w/w) effectively reduced the gall formation but was phytotoxic at 1 percent. Organic amendment significantly increased plant height but reduced root nodulation.

Sobita and Gupta (1995) observed larval emergence from egg sacs of *Heterodera cajani* in extracts of cakes and their effect on cowpea. Inhibition of larval emergence from egg-sacs of *H. cajani* was observed when they were dipped in various extracts of neem [Azadiractha indica], mahua (*Madhuca longifolia*), mustard, linseed cakes and sawdust for 24 or 48 h. These organic amendments were also found to decrease the cyst population and increase the plant growth characters. The neem and sawdust being the most effective.

Akhtar (1995) observed that the root-knot symptom caused by *M. incognita* was significantly suppressed by the predatory nematode *Mononchus aquatidcus*. There was further suppression on the development of root-knot symptoms when neem leaves (*A. indica*) and castor leaves (*R. communis*) were incorporated into the soil. The addition of these leaves to the soil increased the population of *M. aquaticus* to some extent. Enchanced plant growth of the susceptible tomato was positively correlated with inhibiting of root-knot incidence.

Vats and Nandal (1995) tested plant extracts as bare root dip treatment of tomato seedlings against *M. javanica*. Evaluation of variously prepared plant extracts revealed that root dipping in decomposed chopped and whole leaves extracts of *Azadirachta indica* and *Eucalyputs tereticornis* at 1:20 and 1:10 dilutions, respectively, for 5 hrs was more effective than fresh leaves extracts. Root dipping in the decomposed chopped and whole leaves extracts (40% w/v) of *A. indica* and *E. tereticornis* for 13 hrs dip durations, respectively, was the most effective among the different dipping durations of the 3 variously prepared plant extracts.

Akhtar and Mehmood (1995) observed the effect of seed coating with Achook and neem oil against *M. incognita* in tomato. The incidence of *Meloidogyne incognita* significanlty declined when seeds of tomato were coated with Achook, a neem based product, and neem, oil in different concentration and for various exposure durations. As a result of a reduction in the severity of root-knot infection, plant growth improved and there was a positive correlation between treatment and plant growth.

Sheela *et al*. (1995) found *Meloidogyne incognita, Radopholus similis, Rotylenchulus reniformis, Helicotylenchus multicinctus, Praytlenchus* sp., *Tylenchorhynchus* sp., *Hoplolaimus indicus* (*Basirolaimums indicus*), *Criconemoides* sp. and *Xiphinema* sp. in the rhizosphere of ginger (*Zingiber officinale*) in Kerala, India. *M. incognita* and *R. similis* were the major nematode species. *M. incognita* was the most widley distributed species with maximum frequency of occurrence (100 and 34%, repsectively, in soil and root). The avoidable yield loss due to *M. incognita* was 43 percent at an initial population level of 166 larvae per 250 g soil sample. The overall assessment of various treatments in an integrated management revealed that neem cake at 2.5 t/ha at planting + carbofuran 1 kg a.i. at 45 days after planting was effective in reducing the nematode population in soil, root and root-knot index and increasing the yield.

Sreenivasa and Reddy (1996) Studied effective use of neem in sustainable bio-intensive management of root knot nematode on tomato. The aqueous suspensions (5 & 10%) of neem leaves and neem cake was given to support and encourage the growth of certain promising fungal biocontrol agents such as *Paecilomyces lilacinus* (Thom.) Samson and *Verticillium lacanni* Zim. Neem leaf and cake suspensions (5 &10%) mixed with the spore suspension of biocontrol agents were used for the seed treatment, nursery bed drenching and seedling root dip, treatment before transplanting in the main field. These methods resulted in the significant reduction of *M. incognita* infestation on tomato plants and significant reduction of nematodes in nursery beds and in the main field.

Integration of neem with biocontrol agents decreased the requirement of biocontrol agents and increased their efficacy under field conditions. These results suggest the term "bio-intensive management" where all components of management are of biological origin and utilised in an intensive manner. These investigations on the features of neem could form the basis for commercial exploitation.

Parveen and Alam (1996) tested the efficacy of neem products for management of root knot nematode. The root-knot nematode, *M. incognita* caused significant reductions in the growth parameters of tomato, *Lycopersicon esculentum* (Linnaeus) Karsten. The organic materials included in the present study were oil-seed cakes and leaves of neem (*A. indica* A. Juss) and castor (*Ricinus communis* Linnaeus) (Geraniales: Euphorbiaceae) and rice polish, a by-product of rice milling. All the above treatments significanlty controlled the root-knot development and improved plant growth parameters. Neem cake caused greatest improvement in plant growth parameters followed by castor cake, neem leaf, castor leaf and rice polish.

Rai & Singh (1996) Observed the efficacy of certain oilseeds amendments on *Heterodera cajani*. Oilcakes of neem, mustard, mahua, coconut, linseed and sesame were tested at different concentrations (0.25, 0.5, 1.0 and 2.0%) against radial growth of *F. udum*. Neem, mustard and mahua oilcakes were found most effective in reducing frungal growth and were used in pot culture to test their efficacy on *H. cajani, F. udum* and associated wilt of pigeon pea, with carbofuran as standard nematicide. The best growth of pigeon pea plants was recorded with mahua oilcake but the neem oilcake was most effective in controlling wilit incidence, caused by *F. udum* alone and in combination with *H. cajani*. Further, all treatments adversely affected total nematode population with carbofuran the most effective, followed by neem and mustard oilcake.

Mojumdar and Raman (1996) observed that seed treatments with crude neem products such as neem seed kernel, seed shell (seed coat) and seed cake were effective in reducing nematode attack in cowpea and chickpea. Seeds were treated in two ways seed coating, with gum as adhesive and chalk powder as drier and seed soaking in aqueous extracts. A new formulation series, "NEEMA-S" was made with the most effective of the crude neem products (powdered seed kernel). The formulations in this series are suitable for seed treatments only. The formulations combine both methods of seed coating and seeds soaking as adhesive for making aqueous formulations. The present studies tested the efficacy of formulation "NEEMA-S1" against pigeon pea cyst nematode, *Heterodera cajani* in cowpea and root-knot nematode, *Meloidogyne incognita* in chickpea. The observations on *H. cajani* in cowpea and *M. incognita* in chickpea showed a reduction in penetration and development in both nematodes in their respective host crops.

Mojumdar and Raman (1996) studied the comparative efficacy of neem based products for the management of nematode pests of pulse crops. The present studies werre conducted to test the efficacy of seed coating with powdered neem products (neem seed kernel, seed shell and cake) in two important pulse crops of India viz., mungbean (*Vigna radiata*) and chickpea (*Cicer arietinum*). Greenhouse trails involved 25 cm pots with *M. incognita* and *R. reniformis* infestation @ 3J2/g and 2 premature females/g of soil, respectively. Field trials involved 4 sq.m microplots with major phytonematode species: *M. incognita, Helicotylenchus inducus, R. reniformis, Tylenchorhynchis mashoodi* and *Hoplolaimum indicus*. In the first set of experiment conducted with mungbean (cv. Pusa-105) during Kharif season (July-September), seeds were coated @ 20% w/w with gum as an adhesive, chalk powder as a drier and uncoated seed as checks. Observations taken in respect of root-knot and reniform nematode multiplications in the pot trial showed that all treatments reduced galling due to *M. incognita* and population of both nematodes. In case of field trials, all treatments reduced all species of phytonematodes and increased grain yield, with neem seed kernel most effective. The percentage decrease in total plant parasitic nematode population was 78, 67 and 52 and increase in yield was 74, 58 and 28 in neem seed kernel, cake and seed shell, respectively.

Rao *et al*. (1996) conducted experiments to observed the effective utilisation of neem (*A. indica* A. Juss) plant products along with the endomycorrhiza *Glomus mossae* (Niclo. and Gerd) in the integrated management of root-knot nematode *M. incognita* (Kofoid et White Chitw.) on egg-plant *Solanum melongena* Linn. cv. Pusa Purple Round under field conditions. Inoculum of *G. mossae* was added in the nursery beds amended with either neem leaf or neem cake @ 400 g/l mt bed. Seedlings raised in the nursery beds integrated with *G. mossae* and neem leaf or neem cake were highly vigorous with least levels of infestation by *M. incognita*. Further significant increase in the consolidation of endomycorrhiza was observed on the roots of the egg plant seedlings indicating the synergistic effect of neem amendement on the growth and multiplication of *G. mossae*. Since these transplants were heavily colonised by *G. mossae* the egg plants were least infected by *M. incognita* in the main field after transplanting. Significant reduction in the infestation of *M. incognita* and increased yield of crop could be due to the combined and complementary interactive effect of neem and *G. mossae*. The number of eggs per egg mass was significantly smaller on the plants on the treatments where neem and *G. mossae* were integrated. Supplements of neem in the nursery beds facilitated the reduction of nematode population and increased the colonisation of *G. mossae* on the roots leading to the creation of desirable rhizospheric changes for sustainable management of *M. incognita* under the field conditions.

Shukla and Haseeb (1996) did experiments on the evaluation of some nematicides (aldicarb, carbofuran, ethoprophos) and oil cakes of linseed (*Linum ustitatissimum*), mustard (*Brassica capmpestris*), neem (*Azarirachta indica*) aganist *Pratylenchus thornie* infesting *Mentha citrata, M. piperita* and *M. Spicata* in glasshouse experiments. All the treatments were effective in increasing herb weight and oil yield of all the test species of mint as compared to untreated-inoculated plants. However, herb weight and oil yield of plants treated with aldicarb, mustard and neem cakes were significantly higher than those of untreated-uninoculated plants. All the treatments were effective in minimizing nematode reproduction in all the species of mint. Neem cake was most effective in reducing the reproduction rate of *P. thornei* and increased the herb weight and oil yield, followed by mustard cake, aldicarb, ethoprophos, carbofuran and linseed cake, respectively.

Haque *et al.* (1996) observed that use of *Verticillium chlamydosporium, Paecilomyces lilacinus, Rhizobium meliloti* or soil amendment with *Sotechospermum marginatum* seaweed extract controlled the infection of *Meloidogyne javanica* and *Macrophomina phaseolina, Rhizoctonia solani* and *Fusariu solani* on okra in pot experiments. Neem (*A. indica*) cake and datura (*Datura fastuosa*) powder were also effective against *M. javanica* and *M. phaseolina* infection. The combined use of *V. chlamydosporuim* and neem cake completely controlled *M. javanica*. The greatest height and fresh weight of shoot were obtained when *R. meliloti* was used with seaweed followed by *V. chlamydosporium* used with neem cake.

Sharma *et al.* (1996) tested in field trials between the efficacy of soil solarization, neem (*A. indica*) cake and carbofuran against a naturally occurring mixed population of *M. javanica, Rotylenchulus reniformis, Hoplolaimus seinhorsiti, Helicotylenchus retusus, Tylenchorhynchus vulagaris* and *Tylenchus* sp. on tomato cv. Pusa Ruby. Soil solarization trreatments, alone and in combination with carbofuran (0.5 kg a.i./ha) and neem cake (0.75 t/ha) applications were highly effective in reducing the nematode populations. Soil solarization + carbofuran (1 kg a.i/ha) increased the tomato yield by 96 percent. Soil application of neem cake (1.5 t/ha) was more effective.

Sebastian and Gupta (1996) evaluated some oil cakes against *Pratylenchus thornei* on chickpea. Groundnut, linseed, mustard and *Azadirachta indica* oil seed cakes were evaluated for control of *P. thornie* on chickpeas in field trials in India. Mustard and linseed cakes were most effective in reducing root populations of *P. thornei* at 120 days after sowing (DAS). Greatest reduction in soil nematode populations occurred with groundnut cake. Significant increase in shoot and root mass of chickpea at 90 DAS occurred with mustard and *A. indica* cakes.

Akhtar & Mehmood (1996) tested the effect of biopesticides against root-knot nematode *M. incognita*. All root dip treatments of the neem based product azadirachtin (5 g at 1500 , 750, 150 ug/ml for 20, 40 and 80 min.) singificantly reduced *M. incognita* development on tomato and chilli seedlings in pot tests. No phytotoxicity was recorded.

Ahmad *et al.* (1996) observed the efficacy of same organic amendments against reniform nematode, *R. reniformis* on blackgram. Oil-seed cakes and chopped leaves of neem (*A. indica*) and castor (*R. communis*) and rice were found to be effective against *R. reniformis*, resulting in improved plant growth of black gram, *Phaseolus* (*Vigna*) *mungo*. Neem cake was the most effective giving 41.9 percent reduction over untreated control followed by castor cake. rice polish, neem leaf and castor leaf.

Anver and Alam (1996) observed that oilseed cakes of *Azadiractha indica, Ricinus communis, Brassica campestris* and *Eruca sativa* (*E. vesicaria*) at 20 g/kg soil reduced single and concomitant inculations with *M. incognita* and *R. reniformis* on okra cv. Pusa Sawani, in pot experiments.

Akhter and Mahmood (1996) observed that populations of plant parasitic nematodes on potato were significantly suppressed when rocket-salad (*Eruca vesicaria*) was grown with potato in alternate rows. There was a further decline in the nematode population when the soil was treated with oilcakes or leaves of neem (*A. indica*) and castor (*R. communis*) or with carbofuran. Reduction in nematode populations led to an increase in yield. The beneficial effects of all the above treatments persisted into the next growing season when okra (*Abelmoschus esculentus*) was grown.

Rao *et al.* (1996) conducted field trials with soil solarization, before planting *Piper betle* crop and the experiment showed a reduction of 43.8 and 24.3% in the population of *Rotylenchulus reniformis* and *H. indicus* respectively. Application of *Calotropis* leaves at 80 t.ha in the first year was found significantly better than neem (*A. indica*) and castor (*R. communis*) leaves and was proved as effective as carbofuran granules 3G at 1.5 kg a.i./ha, in reducing nematode populations. In the second year, minimum root gall incidence (27 galls/5 g root) was recorded when neem cake was applied at 2 t/ha compared to check plots (70 galls/5 g root).

Reddy *et al.* (1996) observed that *Trichoderma harzianum* (2 and 4g dosages) in combination with neem (*A. indica*), karanj (*P. pinnata*) and castor oil (*R. communis*) cakes (20/40g dosages) was effective in increasing the growth of acid lime (*Citrus aurantifolia*) seedlings and reducing the population of *Tylenchulus semipenetrans* both in soil and roots in pot experiments. The parasitization of citrus nematode females with *T. harzianum* increased in the presence of the oil cakes.

Vats *et al.* (1996) tested camparative efficacy of leaf extracts of plant products against *M. javanica* in tomato nursery. Of 3 different methods of extract preparation of *Azadirachta indica* and *Ecualyputs tereticornis* leaves, 1:40 and 1:20 dilutions of decomposed chopped leaves extracts were the most effective against *M. javanica* in the tomato nursery. These dilutions also helped in promoting growth parameters and in reducing number of root-knot galls and egg masses of *M. javanica*.

Reddy *et al.* (1996) evaluated *Verticillium chlamydosporium* and *V. lecanii* each at 4 g/plant and oil cakes of castor (*R. communis*), karanj (*P. pinnata*) and neem (*A. indica*) each at 20 g/plant either individually or in combination (at half the dosages) for the management of *T. semipenetrans* infecting acid lime (*Citrus aurantiifolia*) under glasshouse conditions. *V. lecanii* in combination with the oil cakes was effective in increasing the plant growth and in reducing the citrus nematode population both in soil and roots. The parasitization of nematode eggs was also maximum with *V. lecanii* when used in combination with the oil cakes.

Firoza and Maqbool (1996) observed that soil amendment with leaves of neem (*A. indica*), datura (*D. fastuosa*) and calotropis (*Calotropis procera*) reduced the population of *Helicotylenchus dihystera* and improved growth of tomato. The population of nematode decreased with increase in the rate of amendments. Datura was found most effective in reducing the nematode population build-up followed by calotropis and neem.

Kumar and Vadivelu (1996) evaluated neem cake (*A. indica*), castor cake (*R. communis*), mahua (*Madhuca longifolia*) cake and press-mud each at 500 kg/ha for the managment of *M. incognita* and *R. reniformis* infecting bringjal (aubergine) under field conditions and their effects were compared with carbofuran. Carbofuran and neem cake were effective in increasing plant growth parameters, while carbofuran, neem cake and castor cake reduced root-knot and reniform nematode populations.

Umrao and Goswani (1996) compared the efficacy of organic amendment with groundnut, karanj, mahua, mustared and neem (*Azadirachta indica*) oil-seed cakes and an inorganic amendment attapulgite based caley dust (ABCD) with a standard nematicide (carbofuran) on root-knot development and growth characters of cowpea. Reduction in root knot development caused by *Meloidogyne incogntia* was significantly high with the use of mustard, neem, carbofuran and ABCD, the least effect being with groundnut. The plant growth characters were greatly improved in mustard and neem amended soil followed by karanj, mahua and ABCD.

Sharma and Singh (1996) observed that plant drenching with Neemark and Nimbecidine (*Azadirachta indica* extracts) resulted in nematode gall and egg mass reduction in pot experiments. There was no significant effect on okra plant growth.

Akhtar and Mahmood (1996) tested the effect of plant based product Nimin and soil plant oils on nematodes. Nimin, an industrial product containing neem triterpenes, and oil from castor (*R. communis*), mustard (*B. juncea*), neem (*A. indica*) and rocket-lettuce (*Eruca sativa* [*E. vesicaria*]) significantly suppressed populatoins of some ecto-parasitic nematodes and reduced the incidence of *M. incognita* in chilli (*Capsium annuum*) cv. Jawala. Nimin gave the best results in tems of reduced root-knot incidence and improved plant growth , followed by oils of neem, castor, rocket-lettuce and mustard.

Barman and Das (1996) found seed treatment with carbofuran 36 at 3 percent w/w and orangic amendments neem (*A. indica*) cake, poultry manure and mustard oil cake each at 2 t/ha alone and in combined application of seed dressing followed by organic amendments at 1 t/ha each, very effective in improving plant growth characters and yield of green gram (*Vigna raidata*). The number of galls, egg masses and final nematode populations of *M. incognita* were reduced over the untreated control. The best result was obtained with poultry manure applied 2 t/ha followed by the combined application of seed dressing + neem cake at 1/ha.

Ajith and sheela (1996) studied the utilization of green leaves of neem for the management of soil organisms in Bhindi & cowpea. Additoin of chopped green leaves of neem (*A. indica*) and *Eupatorium* (15 t/ha) effectively reduced *M. incognita, R. reniformis* and *Helicotylenchus* spp. on bhindi (okra) and cowpea in pot experiments. The lower dose of neem (7.5 t/ha) suppressed *Helicotylenchus* spp. in okra.

Rehman (1996) observed that the management of ufra disease (casued by *Ditylenchus angustus*) with or without stubble cleaning, $ZnSO_4$, neem (*A. indica*) cake, neem seed dust and Furadan 5G (carbofuran) was tested in transplanted Aman and irrigated Boro rice. Neem seed dust was as effective as carbofuran in controlling ufra disease with 72.2-91.0% healthy panicles producing 5.3 to 6.7 t/ha yield. Results were at par when carbofuran was applied alone or in combination with the other treatments. Yields in neem seed dust and carbofuran treatments were almost equal, or even sometimes higher.

Rajeshwari *et al.* (1997) conducted studies in a greenhouse to assess the effects of organic amendments, *Trichoderma viride* and carbofuran on the population of *Arthrobotrys cladodes* var. *macroides* and plant parasitic nematodes, and their interaction effects on the growth of tomato. Application of organic amendments enhanced the population of the nematophagous fungus. Increase in density of *A. cladodes* var. *macroides* was highest in loam soil amended with 2% farmyard marnure (FYM) or neem (*Azadirachta indica*) cake. Carbofuran at 15 mg a.i./pot gave the maximum reduction in the population of *Helicotylenchus dihystera, Tylenchorhynchus capitatus* and *Xiphinema basiri* and was followed by FYM, neem cake and *T. viride*. The plant growth was significanlty increased in combination treatments consisting of FYM + carbofuran followed by neem cake + carbofuran. Application of the organic amendments reduced the efficacy of carbofuran.

Akhtar and Mahmood (1997) studied control of *M. incognita* by seed coating with neem. Development of *M. incognita* infection and juveniles was significantly reduced in tomato plants whose seeds had been previously treated wtih Suneem (azadirachtin 80% a.i.) or neem oil, in pot experiments. Suneem showed no phytotoxicity and improved plant growth whereas neem oil was slightly phytotoxic.

Khan and Saxena (1997) recorded amemdments of oil-cakes, bone and horn meals in soil increased tomato growth and reduced the nematode multiplication. Highest improvement in plant growth and reduced reproduction factor and root galling in neem (*A. indica*) cake amended soil was observed. The least effect was with sesame cake. The integration of oil-cakes except mahua (*Madhuca longifoia*) cake, bone and horn meals with *P. lilacinus*, resulted in increased plant growth and reduced population build-up of nematodes and root gallings.

Mishra and Gupta *et al.* (1997) studied the effect of oil cakes on root knot nematodes. In pot experiments, the application of sawdust or oil cakes of mustard, *Azadirarchta indica, Linum usitatissimum, Madhuca longifolia* or *Ricinus communis* increased growth of soyabean and reduced nematode damage.

Khanna (1997) tested the effcet of some wild plants on *M. incognita* on tomato. finely crushed fresh leaves of some wild plant including *Cannabis sativa, Ficus dioica, Melia azadirachta* (*Azadirachta indica*) an *Urtica dioica* were mixed in potted soil having tomato transplants inoculated with *M. incognita*. Though all the test plant species checked the nematode multiplication, the plants treated with *F. dioica* and *U. dioica* were more vigorous as compared to those treated with *C. sativa*. However, plants with *Azadiractha indica* exhibited poor growth when compared with the uninoculated untreated control.

Devappa *et al.* (1997) evaluated carbofuran at 2 kg a.i. per ha, neem cake at 2.5 t percent and *Pasteuria penetrans* at 924 g of inoculum per m2 either singly or in combination, for the management of *M. incognita* infecting sunflower, under field conditions. A combination of carbofuran, neem cake and *P. penetrans* significantly increased the plant growth characteristics like shoot height, shoot weight, root weight and grain yield, and reduced the nematode population in soil and root galling.

Rao *et al.* (1997) attempted integration of *P. lilacinus* with neem (*A. indica*) leaf aqueous suspension was for the management of *M. incognita* on aubergine. Results revealed the suitability of neem leaf suspensions (5 and 10%) for mass multiplication of *P. lilacinus*. Further, significant reduction in root-knot index and final population of *M. incognita* was observed in the aubergine seedlings which were given root-dip treatment in neem leaf suspensions mixed with *P. lilacinus* spores. In addition, significant increases were observed on the colonization of *P. lilacinus* on the root of egg plant and parasitisation of egg of *M. incgonita*, indicating the complementary interaction between these 2 components for the sustainable managment of root-knot nematodes on aubergine.

Paruthi *et al.* (1997) studied the effectiveness of some available formulations of neem (*A. indica*) pesticides (Achook, Neemguard, Neemark and Margosa) against *M. javanica* under laboratory conditions. The results of the study revealed that mortality of the larvae increased with the increase in time and concentration of the pesticides in all the formulations. However, comparatively higher mortality was observed in Achook than the other formulations . With respect to doses, comparatively high (more than 50%) juvenile mortality was recorded at 1.0% concentration. Juvenile hatching was also reduced with an increase in concentration of the pesticides. Maximum reduction in hatching was observed at 1 percent concentration in all the pesticides, with respect to formulations. Maximum reduction in hatching was observed in Achook, whereas it was minimum in Margoside.

Pathak and Kumar (1997) observed that application of various organic amendments in reniform nematode infested soil had a significant effect on plant growth of *S. trichophilla*. Carbofuran (2 kg a.i.), neem (*A. indica*) cake (250 kg/ha) and neem cake + carbofuran (1 kg a.i. + 100 kg/ha) soil application gave the best plant growth with significant nematode inhibitory effects.

Nagesh and Reddy (1997) studied : The use of oil cakes (at 15 g/plant) with *G. mosseae* (1000 chlamydospores/plant) in pot experiements, promoted better plant growth, reduced nematode multiplication rate, and increased root colonization by the mycorrhiza and the number of mycorrhizal propagules. The treatments were effective in the order: *G. mosseae* plus *Azadirachta indica* plus *Pongamia pinnata*> plus *Ricinus communis*> *G. mosseae* alone.

Jain and Trivedi (1997) observed the nematicidal activity of certain plants against root knot nematode, *M. incognita* infecting chickpea. Leaf powders of locally available plants were used as organic amendments at 1g, 2 g and 3 g doses, 15 days prior to sowing of chickpea. Ten-day-old seedlings were inoculated with 1000 freashly hatched nematode juveniles and examined after 60 days. The average root and shoot length, dry and fresh weight tended to increase in amended soil. Soil amended with *Calotropis procera* showed the highest increase followed by *Azadirachta indica*, *Lawsonia inermis, Ocimum sanctum* and *Parthenium hysterophorus*. These amendments showed a marked decline in number of root galls, eggs per eggmass and final nematode populations indicating a reduction in severity of disease.

Sankaranarayanan and Sundara babu (1997) tested effect of oil cakes on root knot nematode Post soil was either treated with *Azadirachta indica, Madhuca longifoila, Ricinus communis* or groundnut oil cakes, 15 days prior to sowing of *V. mungo* cv. Co5, or incoulation with mycorrhizal fungi at 10 g inoculum/plant, 22 days prior to sowing. Thirty seven days later, nematodes were inoculated at one J2/cc soil. One week following nematode application phorate 10G was applied. All treatments wee effective but the greatest reduction in nematode population was found in the oilcake treatments.

Poornima and Vadivelu (1997) tested effect of some organic oils and their cakes on the infectivity of root knot nematode on tomato. Higher concentrations of organic oils of karanj (*Pongamia pinnata*), neem (*Azadirachta indica*), mahua (*Madhuca longifoila*) and castor (*Ricinus communis*) proved effective in preventing larval penetration and gall production in the roots of tomato infected by *Meloidoygne incogntia*. Cakes of these plants also gave significant reduction in nematode gal numbers. Treatment with oil was relatively more effective than soil treatment with cakes in controlling root-knot nematode infection.

Govindaiah *et al.* (1997) observed that among the different treatments used against *M. incognita* infecting mulberry, sebufos (cadusafos) (1-3 kg a.i./ha) was found the most effective. It has increaseed leaf yield by 10-14 percent and reduced root-galls and egg mass production by 80-90 percent. Neem (*A. indica*) and *Pongamia* oilcakes and mulched green leaves at 2 ton/ha, were equally effective. While poultry manure, sewage manure, saw dust and mulching with *Parthenium, Vinca rosea, Calotropis* and *Euphorbia* spp. were less effective.

Ramakrishan *et al.* (1997) observed the control of *M. Incognita* on okra through botanical as soil amendments. A study under glasshouse conditions showed that *Melia azedarach* (80 g/pot) recorded maximum shoot length, fresh shoot weight and pod yield of okra, infected with *M. incognita*. The highest reduction in root-knot index and nematode population was also recorded with *M. azedarach*, followed by *Azadirachta indica* leaves (80 g/pot).

Sebastian and Gupta (1997) observed that phorate followed by carbofuran and carbosulfan, at 100 ppm or 1000 ppm stock solutions made by soaking 10 g a.i. in 100 ml water for 48 hrs, were effective against *P. thornei* in laboratory tests. Oilcakes of groundnut, linseed, *Madhuca longifolia*, mustard and *A. indica* were also all effective and gave 70-100% nematode mortality.

Anita and Vadivelu (1997) observed that fenamiphos at 2 kg a.i./ha and *Paecilomyces lilacinus* at 10 g/plant were the most effective treatments for increasing geranium plant growth and oil yield. Nematicides were effective in reducing nematode populations for a period of 60 days while the fungus, neem cake and farmyard manure were effective for 90-120 days.

Siddiqui (1997) Observed the effect of organic amendments on the population of plant parasitic nematode, the highest nematode reduction of *T. alexandrinum* was achieved with *Azadirachta indica* cake at 110 g/ha, followed by *Ricinus commuins* cake at the same dose, chopped leaves of *Eichhornia crassipes* at 8 ton/ha and organic fertilizers (urea, phosphate and potash). With the *A. indica* treatment, populations of *Hoplolaimus indicus, Helicotylenchus indicus* and *Tylenchus filiformis* were arrested after the 3rd clipping.

Choudhary and Siva Kumar (1997) observed neem (*A. indica*), castor (*R. communis*), mahua (*M. longifolia*) and groundnut cakes; and farmyard manure @ 5 g/pot emcouraged the multiplication of *Mylonchulus minor* on tomato in pot experiments. Along with *M. minor*, the population of saprohytic nematodes was also increased significanlty by the addition of different orgnaic amendments, while the population of *Helicotylenchus dihystera* declined significantly.

Reddy *et al*. (1997) evaluated castor, karanj and neem oilcakes each at 400 g per plant and *G. mosseae* at 200 g inoculum per plant (containging 25 chlamydospores per g of inoculum) either individually or in combination (at half the dosage) for the management of *R. similis* under field conditions. Karanj cake + G. mosseae was the most effective treatment in increasing plant growth parameters (plant height, pseudostem girth, leaf area and number of leaves and fruit yield) while neem cake + *G. mosseae* was the most effective in reducing the nematode populations both in soil and roots. Mycorrhizal root colonization and number of chlamydospores of *G. mosseae* were maximum in neem cake amended soil.

Nagesh *et al*. (1997) conducted a field experiment for the management of *M. incognita* infecting *Polianthes tuberosa*, by integrating the use of the antagonistic fungus, *Paecilomyces lilacinus*, with 5 percent leaf extracts of castor (*R. communis*) and neem (*A. indica*) as bulb treatments and soil drenches. Combination of *P. lilacinus* (at 2 X 104 spores/ml) with 5 percent neem leaf extract resulted in significantly higher fresh plant weight and flower yield. Root gall index was least under *P. lilacinus* plus neem leaf extract (5%) combination followed by *P. lilacinus* plus castor leaf extract (5%) treatment. Although the per cent egg and egg mass parasitization by *P. lilacinus* was higher when integrated with the leaf extracts, neem leaf extract improved the parasitization by *P. lilacinus* more than that with castor leaf extract.

Akhtar and Mahmood (1997) observed that application of poultry and cattle manures, inorganic fertilizer in the form of urea, neem based urea coated agent 'Nimin' and neem based nematicide 'Suneem G' singificantly reduced population of *Meloidogyne incognita* and other plant parasitic nematodes and their symptoms on the plant and increased the populations of microbivorous nematodes. Enhanced growth of tomato in treated soil was directly correlated with the suppression of the plant parasitic nematode populations.

Kumar *et al*. (1997) observed that seed extracts of *Acacia eburnea, Azadirachta indica, Cassia* sp., *Parakinsonia aculeata, Sesbania sesban, Poinciana regia* (*Delonix regia*) had the highest nematicidal potential aganist *M. incognita* juveniles while maximum potential aganist *M. javanica* was shown by *A. eburnea, Cassia* sp., *Melia azedarach, S. sesban* and *Tribulus terrestris*, as indicated by strong activity in all concentrations. Similarly, *Calotropis procera* and *S. sesban* were the best in preventing egge hatch of *M. incognita* and *C. procera*, *S. sesban* and *Chenopodium album* were the most effective against hatch of *M. javanica* eggs.

Verma and Anwar (1997) observed that oil cakes incorporated into the soil at the rate of 250 kg/ha significantly suppresses the incidence of *M. incognita* in *Trichosanthes dioica*. Neem (*A. indica*) cake was more effective than mahua (*M. longifoila*) or mustard cake in reducing root galling and in protecting fruit yield and compared favourably with carbofuran applied at 2 kg a.i./ha.

Rao *et al*. (1997) observed the effective use of neem cake extract for the management of root knot nematode. The combination of the use of 5 or 10% aqueous extract of neem (*A. indica*) cake for seed treatment and soil drenching under field conditions was found as effective as an application of carbofuran at 2 kg a.i./ha or neem cake at 2 tonnes/ha for the management of *M. incognita* on okra.

Aly Khan *et al*. (1997) observed that the effect of neem leaves, wheat husk and sugarcane bagasse on the population growth of *H. dihystera* and *P. thornei* in soil, and the growth of wheat var. ZA-77. All 3 organic amendments reduced the nematode population. Wheat husk singificantly increased root and shoot length of wheat, while neem leaves significantly increased the root weight and grain weight over the control.

Verma *et al*. (1997) observed that application of decomposed composts of neem (*A. indica*), subabul, mustard and water hyacinth (*Eichhornia crasspies*) at 25 q/ha to nematode infested plots, increased fruit yield of bottle gourd (*Lagenaria siceraria*) and reduced nematode damage. However carbofuran at 2.0 kg a.i./ha was more effective then the composts.

Chakaraborty and Dutta (1997) tested the efficacy of neem (*A. indica*) and garlic (*Allium sativum*) against floral malady of *P. tuberosa* var. *single* caused by *Aphelenchoides besseyi* and *Meloidogyne incognita* in field experiments conducted in West Bengal, India. A combined treatment of bulb dipping either in neem seed kernel extract or in garlic clove extract along with neem cake as an organic amendment significanlty controlled the disease, 82% and 77%, respectively. A positive correlation between reduction in disease incideence and increase in flower yield was observed.

Pandey (1997) studied eco-friendly management organic soil application in the form of neem (*A. indica*) oil, seed cake and dried leaf powder in combination with *Adhatoda vasica, Mentha arvenisis* or *Murrahya koenigiii* extracts caused a significant reduction in the population of *M. incognita* on *H. niger* in pot experiments. The most effective application for reducing nematode population and increasing plant growth was neem cake in conjunction with *M. koeniggi*. This was followed by neem cake and *A. vasica*, and neem cake and *M. arvensis* combinations. These different treatments also inhibited root galling, whereas, higher doses of neem cake alone showed some phytotoxic effect on the growth of the host plant.

Reddy *et al*. (1997) evaluated neem (*A. indica*) cake at 1 kg/m^2, *Pasteuria penetrans* at 28 X 10^7 spores/m^2 and *Paecilomyces lilacinus* at 20 g inoculum/m^2 either singly or in combination at half the dose for the management of *M. incognita* infecting tomato under nursery and field conditions. Integration of all the 3 components gave maximum increase in plant growth parameters and the number of seedlings per bed. Parasitism of *M. incognita* females was maximum when neem cake was integrated with *P. penetrans*, while egg parasitism was highest when neem cake was integrated with *P. lilacinus*. Under field conditions, planting of tomato seedlings (rasied in nursery beds amended with neem cake + *P. penetrans*) in pits incorporated with *P. lilacinus* at 0.5 g inoculum per plant gave the least root galling and nematode multiplication rate. The above treatment was also effective in increasing fruit weight and yield of tomato.

Deka and Phukan (1997) observed that castor bean, mustard and neem (*A. indica*) oilcakes each at 125 g per furrow and 15 g per spot application were found to be effective for the management of *M. incognita* on tomato, and increasing the plant growth. Spot application of castor and mustard cakes was found superior over the furrow application in respect to plant height and yield of tomato.

Rao *et al*. (1997) evaluated neem cake (*A. indica*) and a biocontrol fungus, *T. harzianum*, either singly or in combination for the management of *M. incognita* on tomato. Significant increase in plant growth, and reduction in root galling and final population of *M. incognita* were observed in tomato seedlings trasplanted in neem cake amended soil incorporated with *T.harzianum*. Increase in colonization of *T. harzianum* on roots of tomato was also observed in the above treatments which indicated favourable effect of neem cake amendment on the growth of *T. harzianum*.

Jain & Gupta (1997) tested the efficacy of neem cakes in the management of root knot nematode. Neem cake at 80 q/ha was more effective then carbofuran applied at 2 kg a.i./ha, in reducing nematode population and improving tomato yield in pot experiments. Lower doses of neem, however, showed less reduction in nematode numbers compared to the control.

Sobita Devi (1997) studied the use of various neem products for the management of *Heterodera cajani* on pegem pea. Neem products namely the cake, seeds, bark, leaves and flowers were obtained and properly dried and powdered. Each of the respective products was applied in 2.5 kg of soil @ 400 kg/ha and mixed thoroughly. The soil was filled in 6 inches diameter pots. The neem products were allowed to decompose in soil for 15 days before sowing the pigeonpea seeds. Twenty days after decomposition of neem products, the local variety of pigeonpea seeds was sown. One hundred cysts/pot (equivalent to 5 larvae and 4 eggs/g of soil) were inoculated at the age of 7 day-old 2 seedlings of pigeonpea. Observations were recorded on 45th day for plant height after inoculation. Observations on shoot length and weight, root length and weight, cyst populations, and the rate of multiplication were made at 90 days after inoculation of *H. cajani*. Thirty gram of each neem product was finely powdered and soaked in 100 ml sterilized water for 24 hrs and filtered. The filterate of each product as a pure solution and its 50 percent dilution were used. With each 5 ml solution of test material, 5 egg sacs of *H. cajani* were introduced in the cavity block of plants. After these treatments, the larval emergence was recorded after 24 and 48 hrs exposure and the percent viability of emerged larvae calculated.

Results indicated that various neem products inhibited completely the larval emergence from the egg sacs of *H. cajani* at 24 and 48 hrs exposure to both the concentrations. In case of neem bark and flowers, the emergence of larvae from egg sacs was significantly reduced as compared to control at both 24 and 48 hrs of exposure. The results obtained from pot experiments indicate that population of *H. cajani* on pigeonpea was reduced with the application of neem products. The results also showed that final cyst populations per gram of root and rate of cyst multiplication from the initial inoculum level were also reduced with all the neem treatments as compared to control.

Haseeb *et al.* (1998) studied the influence of hydrodistillations wastes of medicinal plants, oil cakes and pesticidies on root kont nematode infesting *Ocimum canum*. Evaluation of eight hydrodistillation wastes of medicinal plants @ 1500 kg/ha), five oil cakes @ 1500 kg/ha) and five pesticides (@ 3 kg a.i./ha) was done to manage the infestation of *M. incognita* on *O. canum* in glasshouse pot studies. All the treatments were effective to increase the growth and oil yield of *O. canum* and to reduce nematode reproduction as compared to untreated inoculated pots. All the pesticides used and neem cake were highly effective in restricting the final nematode population below half of the initial inoculum level, and increased shoot weight and percent oil yield as compared to untreated plants. In general, carbofuran was most effective followed by aldicarb, neem cake, phorate, carbosulfan, cadusafos, mustard cakes, groundnut cakes, *Ocimum sanctum* (*O. americanum*) cv. *Krishan wastes, O. sanctum* cv. *Shyama-wastes*, castor cakes, linseed cakes, *O. basilicum* cv. *Indian, O. basilicum* cv. *French, O. canum, Mentha arvensis* cv. *Gomti, O. gratissiumum* and *O. gratissimum* and *O. kilimandscharicum* wastes, respectively.

Sreeja and Charles (1998) initiated a study to screen 20 plant extracts for their efficacy against *R. similis* attacking banana. The leaf extracts of *Glycosmis pentaphylla, Azadiractha indica, Kalanchoe pinnata*, the fruit extract of *Melia azedarach* (*A. indica*) and the rhizome extract of *Cyperus rotundus* exhibited a high degree of nematicidal action against the adults and larvae of *R. similis*.

Akhtar (1998) tested various products prepared from neem (*A. indica*) such as leaf powder, sawdust and oilseed cake, and urea for their activities against plant parasitic nematodes (*Hoplolaimus indicus, Helicotylenchus indicus, Rotylenchulus reniformis* and *Meloidogyne incognita* juveniles), a predatory nematode (*Dorylaimus elongatus*) and free-living nematodes. Oilcake was most effective, though all the neem products and urea markedly suppressed plant-parasitic nematodes. However, leaf powder increased populations of predatory and free-living nematodes. All treatments resulted in increased fresh and dry weights and the height and number of pods on chick-pea plants.

Rao *et al.* (1998) evaluated aqueous extracts of neem (*A. indica*), castor

(*R. communis*) and pongamia (*P. pinnata*) cakes as substrates for mass production of *T. harzianum* which was used in the management of *M. incognita* in eggplant (aubergine) under field conditions. Castor cake extract at 10 percent gave maximum growth of mycelial mat and spore production of *T. harzianum* compared with moderate growth in 10 percent pongamia cake and 5 percent castor cake extracts. Application of plant-based formulations of *T. harzianum* to nursery beds of aubergine was effective in producing vigorous seedlings with the least root galling. The above treatments also increased root colonization and parasitization of *M. incognita* females by *T. harzianum*.

Mareggiani *et al.* (1998) studied the toxicity of secondary metabolites extracts from different organs of neem to root-knot nematode. The moratlity of *M. incognita* larvae due to flower, fruit and leaf extract of *Melia azedarach* (*A. indica*) was determined 1 and 48 hrs after their application.

Basu and Mojumdar (1998) studied the effect of agueous extracts of neem based-pesticides on the moratlity of Juveniles of *M incognita*. The standard aqueous extracts of 5 neem (*A. indica*)-based pesticidal formulations neem seed kernel, neem seed coat, Achook, Neemark and Nimbecidine were investigated for their toxicity aganist *M. incognita*. All the neem products were toxic, with mortality being directly correlated with the concentration of the extract and the period of exposure. Neem seed kernel was the most effective at all concentrations and exposures, and Achook was the most effective amongst the commercially available neem pesticides tested.

Verma *et al.* (1998) studied effect of organic amendments on sprout emergence of pointed gourd in *M. incognita.* Organic amendments (250 g/m^2 of oilseed cakes of *Madhuca longifoila, Azadiractha incdica* and mustard and chopped leaves of *Ricinus communis, Calotropis procera, Leucaena leucocephala* or *Melia azedarach* (at 500, 1000 or 1500 g/m^2) were incorporated into soil infected with *M. incognita*. Soil amendments increased sprout emergence of pointed gourds (*Trichosanthes diocia*) grown in the infected soil. Of the oilseed cakes investigated, neem cake resulted in the highest sprouting, and *M. longifolia* resulted in the lowest. Chopped leaves decreased infection by *M. incognita* and increased sprouting.

Reddy *et al.* (1998) studied the management of *M. incognita* on tomato. The effects of integration of the mycorrhizal fungus *G. mosseae* and oil cakes (of castor (*R. communis*), karanj (*P. pinnata*) and margosa (*A. indica*) were evaluated for the management of *M. incognita* infecting tomatoes under nursery and field conditions in Karnataka, India. In nursery beds, integration of *G. mosseae* with margosa cake significantly reduced *M. incognita* populations in soil and reduced root galling, egg mass production and fecundity, and produced vigorous tomato seedlings with increased root colonization by *G. mosseae*. When planted in the field, these seedlings had reduced root galling, increased fruit yields and increased root colonization with *G. mosseae*.

Butool *et al*. (1998) studied the management of root knot nematode *M. incognita* by the use of nematicides and oil cakes. Studies were carried out to evaluate the efficacy of various nematicides: aldicarb and carbofuran at 3 kg a.i./ha ethoprophos at 5 kg a.i./ha and oilcakes of castor (*R. communis*, 4.2% N), groundnut (*Arachis hypogaea,* 7% N), linseed (*Linum usitatissimum*, 4.7% N), neem (*A. indica*, 5.4% N) and mustard (*Brassica campestirs*, 4.8% N) at 1 g N/kg of soil in the form of soil amendments for the management of *M. incognita* infecting *H. muticus*. All the treatments resulted in a significant increase in length, fresh and dry weights of plants as compared with untreated inoculated plants. The maximum improvement in plant growth parameters was observed in plants treated with mocap followed by neem cake, aldicarb, carbofuran, castor, mustard, groundnut and linseed cakes, respectively.

Kumar and Shiva Kumar (1998) studied the management of white tip nematode *Aphelenchoides besseyi*. Monocroptophos sprayed at 1000 ml/ha at the boot leaf stage, reduced the white tip incidence and grain chhaffiness, and increased the yield of rice in field experiments. Phosphamidon, malathion, neem oil (*A. indica*) copper oxychloride were also tested and were effective to a lesser degree.

Vadhera *et al*. (1998) studied integrated management of *M. incognita* in ginger amending the soil with different oil cakes was found effective for reducing the soil population of *M. incognita*, as well as for reducing root gall formation in ginger grown. The highest yields and the maximun reduction in soil populations was recorded in neem cake (*A. indica*)-treated plots.

Pandey *et al*. (1998) studied the comparative efficacy of some neem formulations and pesticides against root knot nematode in tomato. The results revealed that among the four neem formulations tried nimbicidine proved significantly most effective in minimising the infestation through producing 1.09 and 33.75 galls/plant when observed at 15 days and after harvesting the crop. The tomato seedlings treated and planted with this formulation also gave the maximum yield (232.44 g/ha) of tomato fruits. However, each treatment has given significantly better yield than control but Carbofuran @ 1 kg a.i./ha gave maximum yield and least infestation of nematode, i.e., 185.67

g/ha and 71.79 galls/plant, respectively and thus proved most effective among all the chemical pesticides and also to Suneem, which comparatively provided lower yield (181.33 g/ha) and higher gall formation (81.18/plant) than it. Thus, it could be concluded that Nimbicidine is the best replacement of chemical pesticides including Carbofuran and safer to use just by dipping the roots in 1 percent solution in water prior 1 hour of transplanting the seedlings in field.

Sharma and Ezhil (1998) studied root dip treatment with azadirachtin for nematode management in brinjal. For the field trial, 192 Usa Purple cluster were dipped for 30 minutes in a a solution of 30 ppm azadirachtin prepared from a commercial formulation 1500 ppm azadirachtin and all 192 seedling were dipped in water for 30 min. The seedlings were transanted into a field naturally infested with root-knot (*Meloidogyne* spp.) and reniform nematicide (*Rotylenchulus reniformis*). All the treatments oil-based formulation were found phytotoxic. Around 40 percent reduction in height along with reduction in root shoot mass was observed. Though average root gall index and egg sac were reduced by 73 and 38 percent, but probably more due to reduction in mass.

Mojumder (1998) studied the effect of seed soaking in aqueous extracts of neem seed kernel, neem seed coat and neem cake against root knot and reniform nematodes on chickpea. The results (Table 1) indicated that both root knot and reniform nematodes severely damaged chickpea crop as shown by the significantly less plant growth with nematode infestation than in checks without nematodes. Seed soaking in aueous extracts of all the three neem products at either dosage significantly reduced root knot nematode gall and soil population of both *M. incongnita* and *R. reniformis*. Neem seed kernel and neem cake were on par at higher dosage and were the most effective while at lower dosage neem seed kernel was more effective reducing the nematode infestations. The lower dosage of neem seed coat extract was the least effective against either of the nematodes. The data also revelaed that all treatments improved plant growth over check. The shoot length, shoot weight and root weight in case of neem seed kernel extract at higher dosage was better than that of unifested check. In lower dosage of neem seed kernel and neem cake both dosages were on par with uninfested control while neem seed coat was less effective. Root length and root weight did not show much difference. Seed soaking of chickpea for 6 hrs in aqueous extracts of neem products was found to be an effective technique for the management of two major nematode pests.

Table-1 : Effect of seed soaking of chickpea in aqueous extracts of neem products against *Meloidogyne incognita* and Rotylenchulus reniformis*

Treatment	Dose	Shoot length (cm)	Shoot wt. (g)	Root length (cm)	Root wt.(g)	No. of galls/plant	Soil population/ 100cc soil	
						M. incognita		*R. reniformis*
Neem cake	S	41.2	5.84	11.2	0.70	10.0	300	350
	S/2	40.0	5.00	9.2	0.64	26.0	425	550
Neem seed	S	42.5	6.52	12.8	0.96	10.0	275	400
Kernel	S/2	39.6	4.84	10.0	0.80	20.0	300	600
Neem seed	S	39.2	5.60	11.4	0.70	26.0	300	600
Coat	S/2	36.0	4.34	9.2	0.60	30.0	525	800
Check (infested)	--	35.2	4.34	8.4	0.60	38.0	950	875
Check (uninfested)	--	40.0	5.52	10.0	0.80	--	--	--
CD at 5%	--	2.6	0.60	4.2	0.20	2.8	57.9	70.9

**Mean of four replication

Mojumdar (1998) tested the effect of neem products and neem based pesticides on seed coating of mungbean for the management of *M. incognita* and *R. reniforms*. The data presented in Table 2 indicated that all the treatments significantly reduced root-knot galling and soil population of *M. incognita* and *R. reniformis*. Neem seed kernel was the most effective with minimum numbers of root-knot galls, juveniles of *M. incognita* and all stages of *R. reniformis* in soil. This was followed by neem cake and achook. Neemark and nimbicidine were at par in reducing infestation of both nematodes and were the least effective. Neem seed coat was at par with neemark in respects of number of root knot galls while soil population of both nematodes were less than that of neemark and nimbicidine.

Table-2 : Effect of seed coating of mungbean with crude and commercially available neem products for the management of *Meloidogyne incognita* and *Rotylenchulus reniformis**

		Nematode population per 100 g soil	
Treatment	No. of root-knot galls	*M. incognita*	*R. reniformis*
Neem cake	23.00	250.0	300.0
Neem seed kernel	21.60	225.0	250.0
Neem seed coat	29.60	375.0	450.0
Achook	25.00	300.0	350.0
Neemark	31.30	425.0	350.0
Nimbecidine	32.60	450.0	500.0
Check	54.30	650.0	780.0
CD at 5%	2.70	39.4	30.2

*Average of 3 replications

Mojumdar and Raman (1999) tested the nematicidal efficacy of Neem-S1, against *Heterodera cajani* and *Meloidogyne incognita* in cowpea and chickpea. A new siries of neem-based pesticidal formulation of Neem-S series were made. These were prepared from neem seed kernel powder and crystralline gum from *Acacia arabica* in aqueous from. In the first experiment, cowpea seed was tested with Neem S1 against *H. cajani*. The plastic pots containing cyst infested soil @ 6 J2 of *H. cajani* per g soil penetration of juveniles was recorded at 10 days after germination. In the second experiment, chickpea seeds was treated with Neema S-1 against *M. incotnita* in the plastic pots filled with root knot nematode infected soil @ 4 J2/of *M. incognita* per g soil. At 10 days after germination, the number of second stage juveniles that penetrates the plants was recorded. It was observed that Neema -S1 reduced penetration of second stage juvenile of *H. cajani* in cowpea and *M. incognita* in chickpea. The treatment also reduce the number of root knot galls in chickpea. These treatmetns were as effective on the application.. of powdered neem seed kernel @ 20% w/w of seed. These treatments restrict the infestation of store grain pests for six months in cowpea and chickpea.

Mojumdar and Raman (1999) studied the efficacy of neem-based products against nematode pests of pulse crops. Seed treatments with neem-based laboratory formulations of neem seed kernel, neem seed coat (seed shell) and neem seed cake @ 20% w/w reduced root-knot and reniform (*Meloidogyne incognita* and *Rotylenchulus reniformis*) nematode multiplication in two important pulse crops, mungbean and chickpea in greenhouse trials. All these treatments reduced major phytonematode species in the field, viz., *M. incognita, R. reniformis, Ttylenchorhynchus mashoodi, Helicotylenchus indicus* and *Hoplolaimus indicus* in both the crops in field trials in 4 sq. m microplots. In either case, neem seed kernel was the most effective followed by neem seed cake and neem seed coat. There was 78, 67 and 52 percent reduction in total plant parasitic nematode populations in the mungbean crop and the increase in the grain yield was 74, 58 and 28 percent neem seed kernel, neem seed cake and neem seed coat (seed shell), respectively. In the case of chickpea crop, the percentagae reduction in the total plant parasitic nematode was 87, 76 and 25 and the increase in the grain yield was 122, 78 and 48, respectively in neem seed kernel, neem seed cake and neem seed coat.

Parveen and Alam (1999) tested the efficacy of neem products for the management of root-knot nematode on tomato. The organic materials included in the present study were oilseed cakes and leaves of neem (*A. indica*) and castor (*R. communis*) and rice polish, a by-product of rice milling. All the above treatments significantly controlled the root-knot development and improved plant growth parameters. Neem cake resulted in the greatest improvement in plant growth parameters followed by castor cake, neem leaf, castor leaf and rice polish. Both the cakes gave better results than inorganic fertilizers (urea + superphosphate + murate of potash) used as control treatment. Efficacy of the organic additives was also evaluated in the presence of heavy metal pollutants, e.g. cadmium (Cd) and lead (Pb). The heavy metals were found to be more damaging to plant growth parameters than *M. incognita*, irrespective of the treatments. Even though cadmium and lead showed nematotoxic, yet it was noted that both the heavy metals and the nematode together caused more plant damage than that caused by either of them alone. It was interesting to note that barring the oilseed cakes of neem and castor, the treatments could not check the synergistic effect of the metal-nematode combine. Therefore, it is safe to conclude that these oilseed cakes are beneficial to tomato plants under the attack of root-knot nematode, even in the soils polluted with cadmium and lead. This study is the first of its kind in formulating a suitable nematode control strategy.

References

Abid, M. and Maqbool, M.A. (1991). Effect of bare-root dip treatment in oil-cakes and neem leaf on the root-knot development and growth of tomato and eggplant. *Pakistan Journal of Nematology*, **9**(1) : 13-16.

Abid, M.; Haque, S.E.; Maqbool, M.A. and Ghaffar, A. (1992). Effect of oilcakes, *Bradyrhizobium* sp., *Paecilomyces lilacius* and Furadan on root nodulation and root-knot nematode in mungbean. *Pakistan Journal of Nematology,* **10**(2) : 145-150.

Abid, M.; Ehteshamul Haque, S.; Sultana Ara, J.; Ghaffar, A. and Maqbool, M.A. (1995). Comparative efficacy of neem cake and other organic amendments in the control of root-knot nematode in mungbean. *Pakistan Journal of Nematology*, **13**(2) : 103-107.

Acharya, A. and Padhi, N.N. (1988). Effect of neem oilcake and sawdust against root-knot nematodes, *Meloidogyne incognita* on betelvine (*Piper betle* L.). *Indian Journal of Nematology*, **18**(1) : 105-106.

Agbakli, R.; Maraite, H. and Colin, J.E. (1993). Use of neam extracts (*Azadirachta indica*) to cotrol knot nematodes (*Meloidogyne* sp.) in leafy vegetable crop in South Benin Mededelingen van de Faculteit Landbouwetenshappen, Rijksuniversiteit Gent, **57**(3A) : 913-918.

Ahmad, R.; Tariq, M.; Javed, N.; Inam, Ul and Haq, M. (1991). Effect of extract of various plants on egg hatching and larval mortality of root knot nematode, *Meloidogyne incognita*. *Pakistan Journal of Phytopathology*, **3**(1-2) : 38-42.

Ahuja, S. and Mukhopadhayaya, M.C. (1987). Split application of nematicides to control root-knot (*Meloidogyne incognita*) and shoot and fruit borer (*Leucinodes orbonalls* Guen) in egg plant (*Solanum melongena*) seed crop. *Capsicum Newsletter*, **6** : 81-82.

Ajith, K. and Sheela, M.S. (1996). Utilization of green leaves of neem and eupatorium for the management of soil organism in bhindi and cowpea. *Indiain Journal of Nematology*, **26**(2) : 139-143.

Akhila, A. (1996). Biogenesis of azadirachtin in *Azadirachta indica* A. Juss. *Neem Newsletters*, **13**(2) : 24.

Akhila, A. and Kumar, S. (1996). Biosynthesis of limonoids in *Azadirachta indica*, meliantriol and nimbocinone. *Neem Newsletter*, **13**(2) : 19-20.

Akhtar, M. (1995). Biological control of the root-knot nematode *Meloidogyne incognita* in tomato by the predatory nematodes *Mononchus aquaticus*. *International Pest Control*, **37**(1) : 18-19.

Akhtar, M. (1998). Biological control of plant-parasitic nematodes by neem products in agricultural soils. *Applied Soil Ecology*, **7**(3) : 219-223.

Akhtar, M. and Alam, M.M. (1990). Evaluation of nematicidal potential in some medicinal plants. *International Nematology Network Newsletter*, **6**(1) : 8-10.

Akhtar, M. and Alam, M.M. (1990). Evaluation of nematicidal potential in some plants against root-knot nematode on tomato and chilli. *International Nematology Network Newsletter*, **7**(3) : 10-12.

Akhtar, M. and Alam, M.M. (1991). Integrated control of plant-parasitic nematodes on potato with organic amendments, nematicides and mixed cropping with mustard. *Nematologia Mediterranea*, **19**(2) : 169-171.

Akhtar, M. and Mahmood, I. (1996). Effect of plant based products-Nimin and some plant oils on nematodes. *Nematologia Mediterranea*, **24**(1) : 3-5.

Akhtar, M. and Mahmod, I. (1996). Evaluation of nematicidal properties of a biopesticide against root-knot nematrode, *Meloidogyne incognita*. *Tests of Agrochemicals and Cultivars*, **17** : 104-105.

Akhtar, M. and Mahmood, I. (1993). Control of plant -parasitic nematodes with 'Nimin and some plant oils by bare-root dip treatment. *Nematology Mediterranea*, **21**(1) : 89-92.

Akhtar, M. and Mahmood, I. (1993). Effect of *Mononchus aquaticus* and organic amendments on *Meloidogyne incognita* development on chilli. *Nematologia Mediterranea*, **21**(2) : 251-252.

Akhtar, M. and Mahmood, I. (1993). Utilization of fallen leaves as amendments for the control of plant-parasitic nematodes. *Pakistan Journal of Nematology*, **11**(2) : 131-138.

Akhtar, M. and Mahmood, I. (1994). Prophylactic and therapeutic use of oilckaes and leaves of neem and castor extracts for the control of root-knot nematode on chilli. *Nematologia Mediterranea*, **22**(2) : 127-129.

Akhtar, M. and Mahmood, I. (1995). Control of root-knot nematode *Meloidogyne incognita* in tomato plants by seed-coating with Achook and neem oil. *International Pest Control*, **37**(3) : 86-87.

Akhtar, M. and Mahmood, I. (1996). Integrated nematode control in potato, *Solanum tuberosum*. *International Pest Control*, **38**(2) : 62-64.

Akhtar, M. and Mahmood, I. (1997). Control of root-knot nematode *Meloidogyne incognita* in tomato plants by seed coating with suneem and neem oil. *Journal of Pesticides Science*, b, **22**(1) : 37-38.

Akhtar, M. and Mahmood, I. (1997). Impact of organic and inorganic management and plant-based products on plant parasitic and microbivorous nematodes communities. *Nematologia Mediterranea*, **25**(1) : 21-23.

Akhtar, M. Mahmood, I. (1994). Control of root-knot nematode by bare-root dip in uncomposed and decomposed extracts of neem and leaf. *Neamtologia Mediterranea*, **22**(1) : 55-57.

Akhtar, M.; Anver, S. and Yadav, A. (1990). Effect of organic amendments to soil as nematode suppressants. *International Nematology Network Newsletter*, **7**(3) : 21-22.

Alagumalai, K.; Tiruvalluvan, M.; Nagendra, N. and Panjaly Ramraj (1995). Effect of acid extract of neem cake on the population build up of *Meloidogyne incognita. Environmemt and Ecology*,.**13**(2) : 275-277.

Alam, M.M. (1991). Control of plant-parasitic nematodes with oilseed cakes on some vegetables in field. *Pakistan Journal of Nematology*, **9**(1) : 21-30.

Aly Khan; Shaukat, S.S. and Aslam, M. (1997). Effect of organic amendments on population of *Helicotylenchus dihystera* and *Pratylenchus thornei* in soil and growth parameters of wheat var.ZA-77. *Sarhad Journal of Agriculture*, **13**(5) : 425-430.

Anita, B. and Vadivelu, S. (1997). Management of root knot nematode *M. hapla* on scented geranium, *Pelargonium graveolens* L. *Indiain Journal of Nematology*, **27**(1) : 123-125.

Anjum Ahmad and Alam, M.M. (1996). Efficacy of some organic amendments against reniform nematode, *Rotylenchulus reniformis* on blackgram. *Annals of Plant Protection Sciences*, **4**(2) : 122-125.

Anver, S. and Alam, M.M. (1996). Effect of some oilseed cakes against *Meloidogyne incognita* and *Rotylenchulus reniformis* infecting okra. *Annlas of Plant Protection Science*, **4**(2) : 176-178.

Ara, I.; Siddiqui, B.S.; Faizi, S. and Siddiqui, S. (1989). Structurally novel diterpenoid constituents from the stem bark of *Azadirachta indica* (Meliaceae). *Journal of the Chemical Society*, Perkin Transactions-I, 1989, **No. 2** : 343-345.

Awan, M.N.; Javed, N.; Ahmed, R. and Inam-ul Haq, M. (1992). Effect of leaf extract of fourplant species on larval mortality of citrus nematodes (*Tylenchulus semipenetrans* Cob) and citrus plant growth. *Pakistan Journal of Phytopathology*, **4**(1-2) : 41-45.

Azmi, M.I. (1990). Control of plant parasitic nematode(s) incorporation of deolied cakes of tree seeds of subabul, *Leucaena leucocephala* (Lam) de wit. *Indian Foreser*, **116**(10) : 812-814.

Balasubramanian, C.; Mohan, P.S.; Arumugasamy, K. and Udaiyan, K. (1993). Flavonoid from resin glands of *Azadirachta indica*. *Phytochemistry*, **34**(4) : 1194-1195.

Barman, M. and Das, P. (1997). Effect of chemical seed dressing and organic amendment alone and in combination for the management of root-knot nematode, *Meloidogyne incognita* in green gram. *Indian Journal of Nematology*, **26**(1) : 72-76.

Bhattacharya, D. and Goswami, B.K. (1987). A study of the comparative efficacy of neem and groundnut oil cakes against root-knot nematode, *Meloidogyne incognita* as influenced microorganism on sterilized and unsterilized oil. *Indian Journal of Nematology*, **17**(1) : 81-83.

Bhattacharya, D. and Goswami, B.K. (1989). Effect of different dosage of neem and groundnut oil-cakes on plant growth character and population of root-knot nematode, *Meloidogyne incognita* in tomato. *Indian Journal of Nematology*, **18**(1) : 125-127.

Bhattacharya, D. and Goswami, B.K. (1987). Comparative efficacy of neem and groundnut soil-cakes with aldicarb against *Meloidolgune incognita* in tomato. *Revue de Nematologie*, **10**(4) : 467-470.

Bidmon, J.H.S.; Kavser, K.; Mobus, P. and Koolman, J. (1987). Effect of azadirachtin on blowfly larvae and pupae. In natural pesticides from the neem tree (*Azadeirchtin indica* A. Juss) and other tropical plants, H. Schmutterer and K.R.S. Ascher, eds. *Proc 3rd Int. Neem Conference*, Nairobi Ereborn : GT2 : 141-150.

Borah, A. and Phukan, P.N. (1992). Effect of different organic amendments for the management of *Meloidogyne incogita* on green gram. *Current Nematology*, **3**(1) : 49-50.

Butool, F.; Haseeb, A. and Shukla, P.K. (1998). Management of root-knot nematodes, *Meloidogyne incognita*, infesting Egyptian henbane, *Hyoscyamus muticus* L., by the use of nematicides and oilcakes. *International Journal of Pest Management*, **44**(4) : 199-200.

Chaiyapat Chirathamjaree; Klaus Ermel and Arom Sanwanich (1997). Azadirachtin content of neem seed kernels from selected location in Thailand. *Proceedings of First International Symposium on Biopesticide*, held of Narvesuan University Phitsanulok, Thailand, Oct. 27-31, 1996 : 192.

Chakraborty, H.S. and Dutta, P. (1997). Phytochemical control of floral malady of *Polianthes tuberosa* L. *Journal of Mycopathological Research*, **35**(2) : 111-113.

Chhabra, H.K.; Grewal, P.S. and Singh, A. (1988). Efficacy of some plant extracts on root-knot nematodes (*Meloidogyne incognita*). *Journal of Tree Sciences*, **7**(1) : 24-25.

Choudhary, B.N. and Sivakumar, C.V. (1997). Influence of organic amendmens on a predatory mononch *Mylonchulus minor* and its associated nematodes. *Journal of the Agricultural Science Society of North East India*, **10**(2) : 245-248.

Cohen, E.; Quistad, G.B. and Casida, J.E. (1996). Cytotoxiciy of nimbolide, epoxyazadiradione and other limonoids from neem insecticide. *Life Sciences*, **58**(13) : 1075-1081.

Cohen, E.; Quistad, G.B. and Casida, J.E. (1998). Cytotoxicity of nimbolide, epoxyazadiradione and other limonoids from neem insecticide. *Neem Newsletter*, **15**(2): 55-56.

Darekar, K.S. and Mhase N.L. (1990). Management of grapevine nematodes in field with neem cake and granular nematicides *Bioved*, **1**(1) : 101-103.

Darekar, K.S.; Mhase, N.L. and Shelke, S.S. (1989). Management of nematodes infesting pomegranate. *International Nematology Network Newsletter*, **6**(3) : 15-17.

Darekar, K.S.; Mhase, N.L. and Shelke, S.S. (1990). Effect of placement of oilseed cakes on the control of root-knot nematodes on tomato. *International Nematology Network Newsletter*, **7**(1) : 5- 7.

Dash, B.K. and Padhi, N.N. (1990). Response of root-knot nematode infecting tomato to seed treatment by some neem products. *Orrisa Journal of Agricultural Research*, **2**(3 & 4) : 148-150.

Deka, U. and Phukan, P.N. (1997). Management of root-knot nematodes, *Meloidogyne incognita* on tomato through soil amendments. *Journal of the Agricultural Science Society of North East India*, **10**(1) : 91-94.

Devakumar, C.; Goswami, B.K. and Mukherjee, S.K. (1985). Nematicidal principles from neem (*Azadirachta indica* A. Juss) part I. Screening of neem kernel fractions against *Meloidogyne incognita*. *Indian Journal of Nematology*, **15**(1) : 121-124.

Devapa, V.; Krishnappa, K. and Reddy, B.M.R. (1997). Management of rot-knot nematode *Meloidogyne incognita* in sunflower. *Mysore Journal of Agricultural Science*, **31**(2) : 155-158.

Dhawan, S.C. and Kaushal, K.K. (1988). Effect of neem (*Azadirachta indica*) emulsion on the hatching of cereal cyst nematodes, *Heterodera avanae*.

Indian Journal of Nematology, **18**(2) : 364.

Dwivedi, B.K. and Pandey, G. (1992). Effect of organic amendments on root-knot nematodes with its development of egg sac, egg juvenile male and female in green gram (*Vigna radiata*). *Current Nematology*, **3**(2) : 209-210.

Ehteshamul Haque, S.; Abid, M.; Sultana, V.; Ara, J. and Ghaffar, A. (1996). Use of organic amendments on the efficacy of biocontrol agents in the control of root rot and root knot disease complex of okra. *Nematologia Mediterrranea*, **24**(1) : 13-16.

El Nagar, H.I.; Hendy, H.H.; Osman, A.A. and Farhat, A.A. (1994). The combined effect and sequence of application of different concentrations of some biocides on reproduction of the reniform nematodes. *Bulletin of Faculty of Agricultural*, University of Cairo, **45**(3) : 739-751.

Faccknath, S. and Jadunundun, J.K. (1990). Potential of neem and tagetes as nematicides against *Meloidogyne incognita* infesting tomato. *Revue Agricole et Sucriere de Ilie Maurice*, **69**(1-3) : 75-85.

Farahat, A.A.; Osman,AA; E.L.; Nagar, H.I. and Hendy, H.H. (1993). Evaluation of margosan and sincocin as biocides of the reniform nematodes infecting sunflower. *Bulletin of Faculty of Agriculture*, University of Cario, **44**(1) : 191-204.

Firoja, K. and Maqbool, M.A. (1996). Effect of plant extract in the control of *Helicotylenchulus dihystera* (Cobb, 1893) Sher, 1961. *Pakistan Journal of Nematology*, **14**(1) : 61-66.

Firoza, K. and Maqbool, M.A. (1996). Nematicidal properties of leaves of some plant species against *Helicotylenchus dihystera* (Cob, 1893) Sher, 1963 on tomato. *Pakistan Journal of Nematology*, **14**(2) : 107-110.

Ganai, G.H.; Kaul, V.K. and Chhabra, H.K. (1992). Nematicidal action of neem, Mentha and parthenium against *Ditylenchus myceliophagus*. *Plant Disease Research*, **7**(2) : 279-281.

Gaur, H.S. and Mishra, S.D. (1990). Integrated control of nematodes in lentil with aldicarb, neem cake and seed treatment thimet and its residual effect on the subsequent mung crop. *Indian Journal of Entomology*, **51**(3) : 283-287.

Ghazala Parveen and Alam, M.M. (1996). Efficacy of neem products for management of root-knot nematode on tomato in soil polluted with cadmium. *Neem Newsletter*, **13**(3) : 38-39.

Ghazala Parven and Alam, M.M. (1999). Efficacy of neem products for the management of root-knot nematode on tomato, in soil polluted with heavy metals, cadmium and lead. ***Azadirachata indica* A juss**, Ed. Singh and Saxena : 235-246.

Gokte, N. and Swarup, G. (1989). Effect of neem (*Azadirachta indica*) seed kernel powder treatment on penetration and development of *Anguina tritici* on wheat. *Indian Journal of Nematology*, **18**(1) : 149.

Goswami, B.K. (1993). Effect of different soil amendments with neem cake on root-knot nemtaode and soil mycoflora in cowpea rhizosphere. *Indian Journal of Plant Protection*, **21**(1) : 87-89.

Goswami, B.K. and Devi, T.P. (1993). Comparative efficacy of neem and mahua oilseed cakes and a clay dust "ABCD" against root-knot nematode infecting pea. *Indian Journal of Plant Protection*, **21**(2) : 245-246.

Goswami, B.K. and Vijayalakshmi K. (1987). Effect of period of decomposition of oilseed cakes in soil on *Meloidogyne incognita* juveniles. *Indian Journal of Nematology*, **17**(1) : 84-86.

Goswami, B.K.; Haque, M.; Mishra, S.D. and Majumdar, V. (1990). Comparative efficiency of Hind-O-Meal, karanj, caster oil cakes and carbofuran against root-knot nematode in cowpea. *Annals of Agricultural Research*, **10**(3) : 316-318.

Govindachari, T.R.; Geetha Gopala Krishnan; Raghunathan, R., and Rajan, S.S. (1994). Crystallisation of azadirachtin-A. *Current Science*, **66**(4) : 295-297.

Govindachari, T.R. and Geetha Gopalakrishnan (1997). 13,14-Desepoxyazadirachtin-A, a tetranotriterpenoid from *Azadirachta indica*. *Phytochemistry*, **45**(2) : 397-399.

Govindaiah Sharma, D.D.; Himanthraj, M.T. and Bajpai, A.K. (1997). Nematicidal efficacy of organic manures, intercrops, mulches and nematicide against root-knot nematodes in mulbery. *Indian Journal of Nematology*, **27**(1) : 28-35.

Govindaiah Suryanarayana; Sharma, N. and and Gargi, D.D. (1989). Effect of mulching of green leaves for the control root-knot nematode in mulberry. *Indian Journal of Nematology*, **19**(1) : 25-28.

Gowda, D.N. and Setty, K.G.H. (1978). Comparative efficacy of neem cake and methomyl on the growth of tomato and root-knot development. *Current Research*, **7**(7) : 118-120.

Gowda, D.N. and Sety, K.G.H. (1979). *In vitro* studies of neem cake (*Azadirachta indica* A. Juss) extract on hatching of eggs and survival of larvae of rootknot nematodes (*Meloidogyne incognita* Chitwood). *Current Reseacrh*, **8**(5) : 78-89.

Hadhera, I.; Tiwari, S.P. and Dave, G.S. (1998). Integrated management of root-knot nematode, *Meloidogyne incognita* in ginger. *Indian Phytopathology*, **51**(2) : 161-163.

Haidar, M.G.; Samad, M.A. and Waris, M.L. (1993). Efficacy of organic soil amendments and carbofuran on plant parasitic nematodes, yield and juice quality of sugarcane. *Indian Sugar*, **42**(12) : 925-929.

Hasan, N. and Jain, R.K. (1984). Effect of soil amendments on fodder production, photosynthetic pigments and nematodes associated with berseem (*Trifolium alexandrinum* L.) followed by bajra (*Pennisetum typhoides* L.). *Agricultural Science Digest*,□**4**(1) : 12-14.

Haseeb, A and Butool, F. (1991). Evaluation of different pesticides and oil cake on the control of root-knot nematodes (*Meloidogyne incognita*) infesting on davana (*Artemisia pallens*). *Current Nematology*, **2**(2) : 210-204

Haseeb, A. and Butool, F. (1993). Use of some pesticides and oil seed cakes for the control of *Meloidogyne incognita* on *Trachyspermum ammi*. *Afro Asian Journal of Nematology*, **3**(1) : 112-114.

Haseeb, A.; Birpal Singh; Khan, A.M.; Saxena, S.K. and Singh, B. (1978). Evaluation of nematicidal property in certain alkaloid bearing plants. *Geobios*, **5**(3) : 116-118.

Haseeb, A.; Pandey, R. and Husain, A. (1988). A comparison of nematicides and oilseeds cakes for the control of *Meloidogyne incognita* on *Ocimum basilicum*. *Nematropica*, **18**(1) : 65-23.

Haseeb, A.; Shukla, P.K.l. and Butool, F. (1998). Influence of hydrodistillation wastes of medicinal plant of oil cakes and pesticides on *Meloidogyne incognita* Infesting *Ocimum canum*. *International Journal of Nematology*, **8**(1) : 77-80.

Hassan, A.; Tauro, P. (ed); Narwal, S.S. (ed) (1992). Effect of certain plants and by products on development of root-knot nematode. *Proceeding First National Symposium Allelopathy in Agroecosystems (Agriculture & Forestry)*, February 12-14, 1992, held at CCS Haryana Agricultural University, Hisar, India : 184-186.

Husain, S.L.; Kumar, R.; Khan, T. and Titov, A. (1984). Effect of root-dip treatment of egg plant seedlings with plant extracts, nematicides, oil-cake extracts and anthelmintic drugs on plant growth and root-knot development. *Pakistan Journal of Nematology*, **2**(2) : 79-83.

Jacobson, M. (1986). Pharamcological and toxicological effects of neem and chinaberry on warm-blooded animals. *Neem Newsletter*, **3**(4) : 39-43.

Jagdale, G.B.; Pawar, A.B. and Darekar, K.S. (1985). Studies on control of root-knot nematode on betelvine. *International Nematology Network Newsletter*, **2**(1) : 10-13.

Jagdale, G.B.; Pawar, A.B. and Darekar, K.S. (1985). Effect of organic amendments on root-knot nematodes of betelvine. *International Nematology Network Newsletter*, **2**(3) : 7 -10.

Jain, C. and Trivedi, P.C. (1997). Nematicidal activity of certain plants against root-knot nematode, *Meloidogyne incognita* infecting chickpea, *Cicer arietinum*. *Annals of Plant Protection Sciences*, **5**(2) : 171-174.

Jain, R.K. and Bhatti, D.S. (1988). Effect of degredation of neem leaves on incidence of root-knot nematode in tomato. *International Nematology Network Newsletter*, **5**(1) : 7-9.

Jain, R.K. and Bhatti, D.S. (1990). Integrated control of root-knot nematode, *Meloidogyne javanica* infecting tomato. *Indian Journal of Nematology*, **18**(1) : 30-34.

Jain, R.K. and Bhatti, D.S. (1993). Evaluation of the effective integrated methods for the control of root-knot nematodes (*Meloidogyne javanica*) in tomato. *Indian Journal of Nematology*, **21**(2) : 107-112.

Jain, R.K. and Gupta, D.C. (1997). Efficacy of neem (*Azadirachta indica*) cake as nursery bed treatment in the management of root knot nematode (*Meloidogyne javanica*) infecting tomato. *Indian Journal of Nematology*, **27**(2) : 249-251.

Jain, R.K.; Paruthi, I.J. and Gupta, D.C. (1988). Control of root-knot nematode (*Meloidogyne javanica*) in tomato through nursery bed treatment alone and in combination with application of carbofuran at transplanting. *Indian Journal of Nematology*, **18**(2) : 340-341.

Jarvis, A.P.; Johnson, S. and Morgan, E.D. (1998). Stability of the natural insecticide azadirachtin in aqueous and organic solvents. *Pesticide Science*, **53**(3) : 217-222.

Jarvis, A.P.; Johnson, S.; Morgan, E.D.; Simmonds, M.S.J. and Blaney, W.M. (1997). Photooxidation of nimbin and salannin, tetranotriterpenoids from the neem tree (*Azadirachta indica*). *Journal of Chemical Ecology*, **23**(12) : 2841-2860.

Jarvis, A.P.; Morgan, E.D. and Edwards, C. (1999). Rapid separation of triterpenoids from neem seed extracts. *Phytochemical Analysis*, **10**(1) : 39-43.

John, D. Stark (1997). Risk assessment of neem insecticides: Persistence in the environment and potential impact on non-target organisms. *Proceedings First International Symposium on Biopesticides*, held of Naresuan University Phitsanulok, Thailand, Oct. 27-31, 1996 : 69-74.

Johnson, S. and Morgan, E.D. (1997). Supercritical fluid extraction of oil and triterpenoids from neem seeds. *Neem Newsletter*, **15**(4) : 48.

Jonathan , E.I.; Krishnamoorthy, S.V.; Mandharan, M.L. and Muthukrisnan, K. (1991). Effect of organic amendments on the control of sugarcane nematodes. *Bhartiya Sugar*, **16**(6) : 37, 39-49.

Kathiervel, M.; Balasubrmanium, M.; Gopalan, M. and Sivakumar, C.V. (1992). Effect of seed treatment with botanical and chemicals for the control of root-knot neamatode, *Meloidogyne incognita* infesting okra, *Abelmoschus esculentus* L. *Indian Journal of Plant Protection*, **20**(2) : 191-194.

Kaushal, K.K. (1990). On the effectiveness of neem products against the cereal nematode *Heterodera avenae*. *Indian Journal of Nematology*, **20**(2) : 218-219.

Khan, A. (1992). Effect of neem leaves and carbofuran on *Pratylenchus thornei* associated with three wheat varieties. *Pakistan Journal of Scientific and Industrial Research*, **35**(11) : 459-460.

Khan, A. and Shaukat, S.S. (1993). Effect of neem and carbofuran against population of three nematodes associated with wheat varieties. *Sarhad Journal of Agriculture*, **9**(2) : 157-163.

Khan, A.; Shaukat, S.S.; Qamar, F.; Jaffry, A.H. and Hakro, A.A. (1994). Effect of three plant extracts on nematode population on growth parameteres of wheat (var. Pirsabak-85). *Sarhad Journal of Agriculturae*, **10**(4) : 415-418.

Khan, F.A. (1990). Nematicidal potential of some naturally growing medicinal plants against *Pratylenchus zeae*. *Revue de Nemtologia*, **13**(4) : 463-465.

Khan, M.W.; Khan, A.M. and Saxena, S.K. (1979). Suppression of phytophagous nematodes and certain fungi in the rhizosphere of orka due to oil cake amendments. *Acta-Botania Indica,* **7**(1) : 51-56.

Khan, T. and Husain, S.I. (1988). Studies on the efficacy of seed treatments with pesticides, oil cakes neem-leaf and culture filtrate of *Paecilomyces lilacinus* for the control of disease caused by the presence of *Rotylenchulus reniformis*, *Meloidogyne incognita* and *Rhizoctonia solani* either individually or concomitantly on cowpea. *Indian Journal of Nematology*, **18**(2) : 192-198.

Khan, T.A. and Saxena, S.K. (1997). Integrated management of root knot nematode Meloidogyne javanica infecting tomato using organic material and *Paecilomyces lilacinus*. *Bioresource Technology*, **61**(3) : 247-250.

Khan, T.A.; Nisar, S. and Husain, S.I. (1991). Effect of water soluble extracts of certain oil-cakes on the development of root-knot disease of okra, *Abelmoschus esculentus* var. *sevendhari*. *Current Nematology*, **2**(2) : 167-170.

Khanna, A.S. (1991). *In Vitro* studies on some plants as nematicides against *Meloidogyne incognita*. *Current Nematology*, **2**(2) : 199-200.

Khanna, A.S. (1997). Effect of some wild plants on *Meloidogyne incognita* and plant growth of tomato. *Annals of Agricultural Research*, **18**(1) : 48-50.

Khanna, A.S.; Grewal, P.S.; Sharma, N.K. and Sohi, H.S. (1988). Screening of plant extract for nematicidal action *Aphelenchoides composticola* infesting mushroom beds. *Indian Journal of Hill Farming*, **1**(2) : 41-44.

Khurma, U.R. and Archana Singh (1997). Nematicidal potential of seed extracts *in vitro* effects on juvenile mortality and egg hatch of *Meloidogyne incognita* and *M. javanica*. *Nematologia Mediterranea*, **25**(1) : 49-54.

Klaus Ermel; Chaiyapat Chirathamjaree and Arom Sangwanich (1997). Processing of Thai Neem (*Azadirachta siamensis* Valenten) and the quality problem it. *Proceedings First International Symposium on Biopesticide*, held at Naresuan University Phitsanulik, Thailand, Oct. 27-31, 1996 : 171-184.

Kraus, W.; Gutzeit,H. and Bokel, M. (1989). 1,3-Diacetyl- 11,19-deoxa-11-oxo-meliacarpin, a possible precursor of azadirachtin, from *Azadirachta indica* A. Juss. (Meliaceae). *Tetrachedron Letters*, **30**(14) : 1797-1798.

Krishnamurthy, G.V.C.; Chari, M.S. (ed) and Ramaprasad, G. (1993). Effect of oil seed cakes on the control of root-knot nematode, *Meloidogyne javanica* in tobacco nursery. *Botanical pesticides in integrated pest management : Proceeding National Symposium* held on January 21-22, at Central Tobacco Research Institute, Rajahmundry, India : 434-437.

Kumar, C.S.S.R.; Malladi, Srinivas, and Shirish, Yakkundi (1996). Limonoids from the seeds of *Azadirachta indica*. *Phytochimstry*, **43**(2) : 451-455.

Kumar, S. and Sivakumar, C.V. (1998). Management of rice white-tip nematode, *Aphelenchoides besseyi*. *Indian Journal of Nematology*, **28**(1) : 85-87.

Kumar, S. and Vadivelu, S. (1996). Evaluation of organic amendments for the management of root-knot and reniform nematodes infesting brinjal as compared with carbofuran. *Pest Management in Horticultural Ecosystems*, **2**(2) : 71-74.

Ley, S.V.; Anderson, J.C.; Blaney, W.M.; Lidert, Z.; Morgan, E.D.; Robinson, N.G. and Simmonds, M.S.J. (1988). Chemistry of insect antifeedants from *Azadirachta indica* (part 3): reactions on the C-22,23 enol ether double bond of azadirachtin and conversion to 22,23-dihydro-23-beta-methoxyazadirachtin. *Tetrahedron Letters*, **29**(442) : 5433-5436.

Li, Xiao Dong and Zhao Shan Huan (1996). The toxic effects and mode of azadirachtin on insects. *Journal of South China Agricultural University*, **17**(1) : 118-122.

Mahanta, B. and Phukan, P.N. (1992). Effect of different organic amendments for the management of *Meloidogyne incognita*. *Indian Journal of Hill Farming*, **5**(1) : 45-47.

Mahmood, I. and Saxena, S.K. (1987). Effect of water extract of oil-cakes and deoiled cakes on the mortality of *Rotylenchulus reniformis* Linford and Oliveira. *Indian Journal of Plant Pathology*, **5**(2) : 184-187.

Maki, J.; Kofi Tselkpo, M.W.; Fujimaki, Y.; Mitsui, Y.; Ito, Y. and Aoki, Y. (1997). Non-effectiveness of *Azadirachta indica* (neem tree) leaf extract against the larvae of *Angiostrongylus cantonensis* and *Trichinella spiralis*. *Tropical Medicine Nagasaki*, **39**(2) : 65-68.

Mani, A. and Anandam, R.J. (1989). Evaluation of plant leaves, oil cakes and agro-industrial wastes as substrates for mass multiplication of the nematophagous fungus, *Paecilomyces lilacinus*. *Journal of Biological Control*, **3**(1) : 56-58.

Mani, A.; Ahmed, S.N.; Rao, P.K. and Dakshinamurti, V. (1986). Plant products toxic to the citrus nematodes *Tylenchulus semipenetrans* Cobb. *International Nematology Network Newsletter*, **3**(2) : 14-15.

Mareggiani, G.; Leicach, S. and Laner, P. (1998). Toxicity of secondary metabolites extracts from different organs of *Melia azedarach* to the root-knot nematode. *Fitopatologia*, **33**(2) : 122-126.

Mehta, U.K.; Sundararaj, P. and Natesan, N. (1994). Effect of five oil cakes on control of *Pratylenchulus zeae* in sugarcane. *Nematologia Mediterranea*, **22**(2) : 219-220.

Mishra, S.D. and Gaur, H.S. (1989). Control of nematodes infesting mung with nematicidal seed treatment and field application of aldicarb and neem cake. *Indian Journal of Entomology*, **51**(4) : 422-426.

Mishra, S.M. and Gupta, P. (1992). Effect of organic amendments on phytonematodes and *Rhizobium* population associated with soyabean. *New Agriculturist*, **3**(2) : 161-166.

Mishra, S.M. and Pramila Gupta (1997). Effect of some oil cakes and saw dust application on root-knot (*Meloidogyne incognita*) infection and root nodulation of soyabean. *Indian Phytopathology*, **50**(1) : 131-132.

Mohammad, H.J.; Husain, S.I. and Al Zarari, A.J. (1980). Effect of oil cakes on the growth of *Citrus recticulata* L. and the control of citrus nematodes in Iraq. *Phytopathologia Mediterranea*, **19**(2-3) : 153-154.

Mojumder, V. (1998). Effect of seed coating of mungbean with crude neem products and neem based pesticidal for mulations for the management of *Meloidogyne incognita* and *Rotylenchulus reniformis*. *Neem Newsletter*, **15**(2) : 12-13.

Mojumder, V. (1998). Effect of seed soaking in aquesous extracts of neem seed kernel, neem coat and neem cake against root-knot and reniform nematodes infesting chickpea. *Neem Newletter*, **15**(2) : 14-15.

Mojumder, V. and Mishra, S.D. (1991). Effect of aqueous extracts of neem seeds on hatchability of eggs and penetrability of hatched juveniles of *Meloidogynes incognita* into roots of mungbean. *Current Nematology*, **2**(2) : 117-120.

Mojumder, V. and Mishra, S.D. (1992). Effect of seed soaking in aqueous extracts of neam seed on germination on mungbean and penetration of second stage juveniles of *Meloidigyne incognita*. *Annals of Agricutural Research*, **13**(3) : 297-299.

Mojumder, V. and Mishra, S.D. (1992). Management of root-knot nematode *Meloidogyne incognita* infecting chickpea with neem seed coat. *Annals of Agricutural Research*, **13**(4) : 388-390.

Mojumder, V. and Mishra, S.D. (1993). Management of *Meloidogyne incognita* infecting mungbean by seed-soaking in aqueous extracts of neem seed kernel and neem seed coat. *Current Nematology*, **4**(1) : 25-28.

Mojumder, V. and Mishra, S.D. (1993). Soil application of aqueous extracts of neem products for the management of *Meloidogyne incognita* in chickpea. *Current Nematology*, **4**(1) : 105-107.

Mukhtar, K.; Ahmad, R.; Javed, N. and Khan, S.H. (1994). Control of root-knot disease of tomato with organic soil amendments. *Pakistan Journal of Phytopathology*, **6**(2) : 152-154.

Mukhtar, T.; Ahmad, R.; Inam-ul Haq, M. and Javed, N. (1994). Effect of leaf extracts of some plants on *Meloidogyne incognita*. *Pakistan Journal of Phytopathology*, **6**(1) : 35-37.

Murthy, M.M. and Rao, K.T. (1992). Control of root knot nematode, *Meloidogyne incognita* on betelvine with certain oil cake amendments and non voliatile nematicides. *Indian Journal of Plant Protection*, **20**(2) : 171-173.

Nagesh, M. and Reddy, P.P. (1997). Maanagement of *Meloidogyne incognita* on *Crossadra undulaefolia* using vesicular arbuscular mycorrhiza *Glomus mosseae* and oil cakes. *Mycorrhiza News*, **9**(1) : 12-14.

Nagesh, M.; Reddy, P.P. and Rao, M.S. (1997). Integrated management of *Meloidogyne incognita* on tuberose using *Paecilomyces lilacinus* in combination with plant extracts. *Nematlogia Mediterranea*, **25**(1) : 3-7.

Nair, M.S.; Sindhu Gopal; Deepa Issac; GOP's, Issac,D. (1997). Optimised isolation procedure for biologically active compounds nimbolide and 28-deoxonimbolide from *Azadirachta indica* leaves. *Phytochemistry*, **46**(7) : 1177-1178.

Nanje Gowda, D.; Shivanandappa, N.; Karegowda, C. and Gowda, D.N. (1985). Efficacy of soil amendments in control of root-knot nematodes on tobacco nursery. *Journal of Soil Biology and Ecology* , **5**(2) : 131-133.

Netscher, C. (1981). Trees resistant to *Meloidogyne* spp.: their use as wind-break in Senegal L'. *Agronomie Tropicale*, **36**(2) : 175-177.

Niranjan Ramji; Krishnamachari Venkatakrisshan and Madyastha, K.M. (1998). 11-Epi-azadirachtin D: an epimeric azadirachtin analogue from *Azadiractha indica*. *Phytochemisty*, **49**(1) : 265-267.

Niranjan Ramji; Venkatakrishnan, K.; Madyastha, K.M. and Ramji, N. (1996). 11-epi-Azadirachtin H from *Azadirachta indica*. *Phytochemistry*, **42**(2) : 561-562.

Padhi, N.N. and Mishra, R.P. (1987). Control of *Rotylenchulus reniformis* of french bean (*Phaselous vulgaris* L). *Indian Journal of Nematology*, **17**(1) : 130-131.

Pandey, G. and Singh, K.P. (1990). Effect of organic amendments on soil microflora fauna with special reference to *Meloidogyne incognita* and nematode in chickpea. *New Agriculturist*, **1**(1) : 65-70.

Pandey, G. and Singh, K.P. (1990). Effect of organic and soil microflora and nematode fauna with special reference to *Meloidogyne incognita* on chick pea. *New Agriculturist*, **1**(1) : 65-70.

Pandey, G.; Dwivedi, B.K. and Singh, R.B. (1991). Effect of organic amendments on the root-knot disease of greengram and its rhizosphere microflora. *New Agriculturalist*, **2**(2) : 203-206.

Pandey, R. (1994). Comparative performance of soil, seed cakes and pesticides in the managements of root knot disease of davana. *Nematologia Mediterranea*, **22**(1) : 17-19.

Pandey, R. (1997). Ecofriendy management of *Meloidogyne incognita* on

Hyoscyamus niger. Indian Journal of Nematology, **27**(2) : 175-178.

Pandey, R.; Haseeb, A. and Husain, A. (1992). Distribution, pathogenicity and management of *Meloidogyne incognita* on *Mentha arvensis* cv. MAS-1. *Afro Asin Journal of Nematology*, **2**(1-2) : 27-34.

Pandey, R.C.; Tripathi, M.K. and Ahmad, F. (1998). The comparative efficacy of some neem formulations and pesticides against root-knot nematodes in tomato crop. *ICPPMSA*, 11-13 Dec., 1998, CSAU A&T, Kanpur (India) : 162-163.

Paruthi, I.J.; Kanwar, R.S. and Gupta, D.C. (1996). *In vitro* evaluation of achook-A, neem based pesticides against root-knot nematodes, *Meloidogye incognita. Indiain Journal of Nematology*, **26**(2) : 270-272.

Paruthi, I.J.; Pannu, A. and Banerjee, M.K. (1997). *In vitro* evaluation of some neem pesticides against root-knot nematode, *Meloidogyne javanica. Annals of Agri. Bio. Research*, **2**(2) : 137-139.

Parvatha Reddy, P.; Khan, R.M. and Rao, M.S. (1991). Integrated management of the citrus nematodes, *Tylenchulus semipenetrans* using oil-cakes and *Paecilomyces lilacinus. Afro Asian Journal of Nematology*, **1**(2) : 221-222.

Patel, P.N. and Thakar, N.A. (1989). Organic amendments in control of the stunt nematodes, *Tylenchorhynchus vulgaris* on wheat. *Indian Journal of Nematology*, **19**(1) : 81-82.

Pathak, A.K.; Yadav, B.S. and Brar, J.S. (1988). Water hyacinth and neem leaves for the control of root knot nematode *Meloidogyne incognita* on brinjal. *Plant Disease Research*, **3**(1) : 74-76.

Pathak, K.N. and Kumar, P. (1997). Efficacy of organic amendments and carbofuran in the control of *Rotylenchulus reniformis* infesting Kochia (*Scoparia trichophillla*). *Journal of Research*, Bisra Agricultural University, **9**(2) : 159-161.

Poornima, K. and Vadivelu, S. (1997). Effect of some organic oils and their cakes on the infectivity of root-knot nemartode, *Meloidogyne incognita* affecting tomato. *Indiain Journal of Nematology*, **27**(1) : 70-73.

Pradhan, S.; Goswami, M.G. and Rai, B.K. (1962). Neem seed different to locusts. *Indian Farming*, **12**(8) : 7-11.

Prasad, J.S.; Panwar, M.S.; Rao, Y.S. and Rajamani, S. (1987). Control of root lesion nematode *Pratylenchus indicus* Das in rainfed upland rice. *Rice Research Newsletter*, **8**(1) : 3.

Prasad, J.S.; Panwar, M.S.; Rao, Y.S. and Rajamani, S. (1988). Control of the lesion nematode, *Pratylenchus indicus* Das in reinfed upland rice.

Indian Journal of Nematology, **18**(1) : 117-118.

Prot, J.C. and Kornprobst, J.M. (1983). Effect of *Azadirachta indica, Hannoa undulata* and *Hannoa klaineana* seed extracts on the ability of *Meloidogyne javanica* juveniles to penetrates totamo roots. *Revue de Nematology*, **6**(2) : 330-332.

Qamar, F.; Saeed, M.; Khan, A. and Seema , N. (1989). Control of anguinosis of wheat using plant exatracts. *Pakistan Journal of Scientific and Industrial Research*, **32** : 4.

Rahman, M.L. (1996). Ufra disease management in rainfed lowland and irrigated rice. *Bangladesh Journal of Botany*, **25**(1) : 13-18.

Rahman, R. and Khan, M.W. (1988). Effect of oilcake amendments on seed germination of *Chenopodium album* and population of fungi and nematodes. *Acta Botanica Indica*, **16**(1) : 106-110.

Rai, P.K. and Singh, K.P. (1996). Efficacy of certain oilcake amendments on *Heterodera cajani, Fusarium udum* and associated wilt on pigeonpea. *International Journal of Tropical Plant Diseases*, **14**(1) : 51-58.

Rajeshwari, S.; Santhi, A. and Sivakumar, C.V. (1998). Effect of organic soil amendments *Trichoderma viride* Pers. ex. Fr. and carbofuran on the nematode trapping fungus, *Anthrobotrys cladodesd* var. *macroides* (Drechsler, 1944) and plant parasitic nematodes. *Journal of Biological Control*, **11**(1-2) : 49-52.

Ram, K. and Gupta, D.C. (1981). Studies on the control of *Meloidogyne javanica* chickpea (*Cicer arietinum*). *Haryana Agriculturtal University Journal of Research*, **11**(1) : 77-81.

Ramakrishnan, S.; Gunashekaran, C.R. and Sivagami Vadivelu (1997). Control of *Meloidogyne incognita* on okra through botanicals as soil amendments. *Indiain Journal of Nematology*, **27**(1) : 24-27.

Rangasamy, S. and Parmar, B.S. (1997). Azadirachtin-A contents of seeds of neem ecotypes inrelation to the agro-ecological region of India. *Neem Newsletter*, **14**(1) : 19.

Rao, A.B.V.K.; Padhi, N.N.; Krishna and Rao, A.B.V. (1987). Effect of different nematicides, oilcakes and urea in the control of *Rotylenchulus reniformis* on okra. *Indian Journal of Nematology*, **17**(2) : 171-173.

Rao, D.V.S.; Sitaramaiah, K. and Maiti, S. (1997). Management of plant parasitic nemtodes in betelvine gardens trough non-chemical methods. *Indiain Journal of Nematology*, **26**(1) : 93-97.

Rao, M.S. and Reddy, P.P. (1992). Studies on the comparative efficacy of certain plant leaves and carbofuran in the management to *Meloidogyne incognita* on tomato. *Current Nematology*, **3**(1) : 5-6.

Rao, M.S. and Reddy, P.P. (1994). A method for conveying *Paecilomyces lilacinus* to soil for the management of root knot nematodes on egg plant. *Nematologia Mediterranea*, **22**(2) : 265-267.

Rao, M.S.; Reddy, P. and Nagesh, M. (1997). Management of root-knot nematode, *Meloidogyne incognita* on tomato by integration of *Trichoderma harzianum* with neem cake. *Zeitschrift fur Pflanzenkrankheiten und Pflanzenschutz*, **104**(4) : 423-425.

Rao, M.S.; Reddy, P. and Sukhandmohandas (1996). Bioefficacy of natural pesticides from neem and intensive use of neem in integrated pest management of root-knot nematode on egg plant. *Neem Newsletter*, **13**(3) : 37-38.

Rao, M.S.; Reddy, P.P. and Nagesh, M. (1997). Effective use of cake extracts for the management of root-knot nematodes infecting okra (*Abelmosochus esculentus*). *Pest Management in Horticultural Ecosystems*, **3**(2) : 95-99.

Rao, M.S.; Reddy, P.P. and Nagesh, M. (1997). Integration of *Paecilomyces lilacinus* with neem leaf suspension for the management of root-knot nematodes on agg plant. *Nematologia Mediterranea*, **25**(2) : 249-252.

Rao, M.S.; Reddy, PP.. and Nagesh, M. (1998). Evaluation of plant based formulations of *Trichoderma harzianum* for the management of *Meloidogyne incognita* on egg plant. *Nematologia Mediterranea*, **26**(1) : 59-62.

Rathor, S. (1994). The effect of neem, mustard and karanj cakes against reniform nematode, *Rotylenchulus reniformis* on cowpea. *Indian Journal of Mycology and Plant Pathology*, **24**(1) : 51-52.

Ravindra, H. (1988). Studies on root-knot nematodes *M. incognita* on green gram (*Vigna radiata* L.) Wilezeck. *Thesis Mysore Journal of Agricultural Sciences*, **22**(1) : 201-202.

Reddy, P.P. and Khan, R.M. (1991). Integrated management of root-knot nematodes infecting okra. *Current Nematology*, **2**(2) : 115-116.

Reddy, P.P.; Khan, R.M.; Chari, M.S. (ed) and Ramaprasad, G. (1990). Effect of oil cakes on root-knot nematode and yield of okra. *Botanical pesticides in integrated pest management* : *Proceeding National Symposium* held on January 21-22, 1990 at Central Tobacco research Institute, Rajahmundry : 424-426.

Reddy, P.P.; Nagesh, M. and Devappa, V. (1997). Effect of integration of *Pasteuria penetrans, Paecilomyces lilacinus* and neem cake for the management of root-knot nematodes infecting tomato. *Pest Management in Horticultural Ecosystems*, **3**(2) : 100-104.

Reddy, P.P.; Nagesh, M.; Divappa, V. and Kumar, M.V.V. (1998). Management of *Meloidogyne* on tomato by integrating endomycorrhiza, *Glomus mosseae* with oil cakes under nursery and field conditions. *Zeitschrift fur Pflanzenkrankheiten un Pflanzenschutz*, **105**(1) : 53-57.

Reddy, P.P.; Nagesh, M.; Rao, M.S. and Devappa, V. (1997). Integrated management of the burrowing nemartode, *Radopholus similis*, using endomycorrhiza, *Glomus mosseae* and oil cakes. *Pest Management in Horticultural Ecosystems*, **3**(1) : 25-29.

Reddy, P.P.; Rao, M.S. and Nagesh, M. (1996). Management of the citrus nematodes, *Tylenchulus semipenetrans*, by integration of *Trichoderma harzianum* with oil cakes. *Nematologia Mediterrnea*, **24**(2) : 265-267.

Reddy, P.P.; Rao, M.S. and Nagesh, M. (1996). Management of the citrus nematodes on acid lime by integration of parasitic fungi and oil cakes. *Pest Management in Horticultural Ecosystems*, **2**(1) : 15-18.

Reddy, P.P.; Rao, M.S.; Mohandas S. and Nagesh, M. (1995). Integrated management of the citrus nematode, *Tylenchulus semipenetarns* Cobb using VA mycorrhiza *Glomus fasciculatum* (Thaxt.) Gerd & Trappe and oil cakes. *Pest Management in Horticultural Ecosystems*, **1**(1) : 37-41.

Rembold, H.; Forster, H.; Czoppelt, C.; Schmutterer, H. (ed) and Ascher, K.R.S. (1987). Structure and biological activity of azadirachtins A and B. Natural pesticides from the neem tree (*Azadirachta indica* A. Juss) and other tropical plants. *Proceedings of the 3rd International Neem Conference*, Nairobi, Kenya, 10-15 July 1986 : 149-160.

Renu Sharma and Ezhil, P.N. (1998). Root dip treatment with azadirachtin for nematode management in brinjal cv Pusa purple cluster. *Neem Newsletter*, **15**(2) : 21.

Reshmi Basu and Vijaylakshmi Mojumder (1998). Effect of aqueous extracts of neem based pesticides on the mortality of second stage juveniles of *Meloidogyne incognita*. *Annals of plant Protection Science*, **8**(1) : 53-55.

Rojatkar, S.R.; Bhat,V.S.; Kulkarni, M.M.; Joshi, V.S. and Nagasampagi, B.A. (1989). Tetranortriterpenoids from *Azadirachta indica*. *Phytochemistry*, **28**(1) : 203-205.

Rojatkar, S.R.; Sawaikar, D.D. and Nagasampagi, B.A. (1996). New tetra and pentanortriterpenoids from *Azadirachta indica* A. Juss. *Neem and Environmental*, **1** : 203-206.

Sahoo, C.R. and Sahu, S.C. (1994). Studies on the effect of different oilcakes and two nematicides in the control of lesion nematodes, *Pratylechus zeae* in rice. *Annals of Plant Protection Sciences*, **2**(1) : 30-32.

Salawu, E.O. (1988). Effectiveness of neem leaf extract and ethoprop in controling sugarcane root-knot nematode. *Bangladesh Journal of Sugarcane*, **10**(48) : 75-80.

Salawu, E.O. (1992). Effect of neam leaf and ethoprop singly and in combination on *Meloidogyne incoginta* and growth of sugarcane. *Pakistan Journal of Nematology,* **10**(1) : 15- 56.

Sankaranarayanan, C. and Sundarababu, R. (1997). Effect on oil cakes and nematicides on the growth of blackgram (*Vigna mungo*) inoculated with VAM fungus (*Glomus fasciculatum*) and root-knot nematodes (*Meloidogyne incognita*). *Indian Journal of Nematology*, **27**(1) : 128-130.

Saxena, P.K.; Chhabra, H.K. and Jasail, K. (1977). Effect of certain soil amendments on the population of nematodes infesting grape-vines. *Zeitscrift fur Angewantde Zoologie,* **64**(3) : 325-330.

Schauer, Blume, M. (1988). Effect of extraction-residues of neem seed kernels (*Azadirachta indica*) on the nematode (*Pratylenchus penetrans*). *Gesunde Pflanzen*, **40**(6) : 229-233.

Schoonhoven, L.M. (1981). Perception of azadirachtin by some lepidopterons larvae. *Proceedings of First International Neem Conference*, RottachEgrn, 1980 GT2 Eschborn : 105-108.

Sebastian, S. and Gupta, P. (1997). Laboratory screening of some nematicides and oil-cake extracts against *Pratylenchulus thornei*. *Indiain Journal of Nematology*, **27**(1) : 133-134.

Sebastian, S. and Gupta, P. (1996). Evaluation of some oil-cake against *Pratylenchulus thornei* on chickpea. *International Chickpea and Pigeonpea Newsletter*, **3**(40-41) : 7.

Selvi, S.T. and Mohan, P.S. (1997). Chemical examination of the resinous exudate from *Azadirachta indica* A. Juss. *Indian Journal of Chemistry Section-B, Organic including Medicinal*, **36**(11) : 1082-1083.

Selvi, S.T.; Balasubramanian, C. and Mohan, P.S. (1996). Chemical examination of the resinous exudate from *Azadirachta indica* A. Juss. *Neem Newsletter*, **13**(2) : 21.

Sen, K. and Dasgupta, M.K. (1989). Control of root-knot nematodes (*Meloidogyne* spp.) with organic amendments and soil nematicides on some successive crop. *Horticultural Journal*, **2**(1) : 48-54

Senniappan Thamarai, Selvi; Chinnusamy Balasubramanian and Palathurai

Subramaniam Mohan (1996). Chemical examination of the resinous exudate from *Azadirachta Indica*. A. Juss. *Neem Newsletter*, **13**(2) : 21.

Sharma, G.C. (1991). Yield losses in plum (*Prunus domestica*) due to nematodes. *Current Nematology*, **2**(2) : 171-174.

Sharma, H.K. and Singh, R.V. (1997). Drench application of Neemark and Nimbecidine against root-knot nematode, *Meloidogyne incognita* of okra. *Indian Journal of Nematology*, **26**(1) : 124-125.

Sharma, R.; Rai, R.R. and Farhat, S. (1996). Interaction of soil solarization carbofuran and neem cake application for nematode management on tomato. *Indian Journal of Plant Protection*, **24**(1-2) : 110-114.

Sharma, R.K. and Raj, D. (1987). Effect of nematicides and organic amendments of root-knot nematodes infecting potato. *International Nematology Network Newsletter*, **4**(1) : 8-10.

Sheela, M.S.; Bai, H.; Jiji, T. and Kuriyan, K.J. (1995). Nematodes associated with ginger rhizosphere and their management in Kerala. *Pest Management in Horticultural Ecosystems,* **1**(1) : 43-48.

Shukla, P.K. and Haseeb, A. (1996). Effectiveness of some nematicides and oil cakes in the management of *Pratylenchus thornei* on *Mentha citrata, M. piperita* and *M. spicata*. *Bioresource Technology*, **57**(3) : 307-310.

Siddiqui, A.M. (1997). Effect of organic amendments together with clippings on the population of plant parasitic nematodes associated with berseem (*Trifolium alexandrinum*). *Indian Journal of Nematology*, **27**(1) : 117-120.

Siddiqui, B.S.; Ghiasuddin; Faizi, S. and Siddiqui, S. (1992). Triterpenoids from the fresh fruit coats of *Azadirachta indica*. *Phytochemistry*, **31**(12) : 4275-4278.

Siddiqui, M.A. and Alam, A.A. (1990). Sawdust as oil amendments for control nematodes infesting some vegetables. *Biological Wastes*, **33**(2) : 123-129.

Siddiqui, M.A. and Alam, M.M. (1987). Efficacy of seed dressing with extracts of neem and Persian lilac against *Meloidogyne incognita* and *Rotylenchulus reniformis*. *Nematologia Mediterranea*, **15**(2) : 399-403.

Siddiqui, M.A. and Alam, M.M. (1988). Control of root-knot reniform nematodes by bare-root dip in leaf extract of margosa and Persian Iiliac. *Zeitschrift fur Pflanzenkrankheiten und Pflanzenschutz,* **95**(2) : 138-142.

Siddiqui, M.A. and Alam, M.M. (1989). Effect on root-exudates of neem and

Persian lilac on plant parasitic nematodes. *Anzeiger fur Schadlingskunde*, Pflanzenschutz, Umweltschutz, **62**(2) : 33-35.

Siddiqui, M.A. and Alam, M.M. (1990). Control of root-knot, reniform and stunt nematodes by nimbin seed dressing. *Nematologia Mediterranea*, **18**(1) : 19-22

Siddiqui, M.A. and Saxena, S.K. (1987). Effect of interculture of margosa and Persian liliac with tomato and eggplant on root-knot and reniform nematodes. *International Nematology Network Newsletter*, **4**(2) : 5-8.

Siddiqui, M.A. and Saxena, S.K. (1987). Studies on control of the stunt nematodes *Tylenchorhynchus brassicae* by interculture of Margosa and Persian lilac. *International Nematology Network Newsletter*, **4**(4) : 27-29.

Siddiqui, S.; Ara, I.; Faizi, S.; Mahmood, T. (1988). Phenolic tricyclic diterpenoids from the bark of *Azadirachta indica*. *Phytochemistry*, **27**(12) : 3903-3907.

Siddiqui, S.; Mahood, T.; Siddiqui, B.S. and Faizi, S. (1986). Two new tetranortriterpenoids from *Azadirachta indica. Journal of Natural Products*, **49**(6) : 1068-1073.

Siddiqui, S.; Tariq Mahmood; Siddiqui, B.S. and Shaheen Faizi (1986). Isolation of a triterpenoid from *Azadirachta indica. Phytochemistry*, **25**(9) : 2183-2185.

Sidhu, O.P. and Behl, H.M. (1996). Seasonal variation in azadirachtins in seeds of *Azadiractha indica. Current Science*, **70**(12) : 1084-1086.

Singh , S.P.; Khan, A.M. and Saxena, S.K. (1980). Effect of watering and mode of application of oil cakes and nematicides on their efficacy in controlling root-knot nematode on tomato. *Acta Botanica Indica*, **8**(2) : 193-195.

Singh, R.P; Dev Kumar, C. and Dhingra, C. (1988). Activity of neem seed kernel extract against the mustard aphid. *Phytoparasitica*, **16**(3) : 225-229.

Singh, S.K.; Khan, M.R. and Khan, A.A. (1988). Effect of organic soil amendments on rhizosphere fungi and root-knot nematode on aggplant cv. PPL. *Indian Journal of Applied and Pure Biology*, **3**(2) : 103-106.

Singh, S.K.; Khan, M.R. and Khan, A.A. (1990). Control of root-knot nematode (*Meloidogyne incognita*) by organic soil amendment on tomato cv. Pusa Ruby. *Indian Journal of Applied and Pure Biology*, **5**(1) : 21-23.

Singh, S.P.; Khan, A.M. and Saxena, S.K. (1980). Effect of nematicides of oilscakes separately and in mixture on plant parasitic nematodes. *Geobios*,

7(3) : 111-113.

Sitaramaiah (1979). Effect of temperature on decompositions of margosa cake and on survival of *Tylenchorhychus elegans*. *Indian Jornal of Nematology,* **7**(2) : 167-169.

Sitaramaiah, K. and Singh, R.S. (1978). Effect of organic amendment on phenolic content of soil and plant and response of *Meloidogyne javanica* and its host to related compounds. *Plant and Soil*, **50**(3) : 671-679.

Sobita Devi and Pramila Gupta (1995). Larval emergence from egg sacs of *Heterodera cajani* in extracts of various media and their effect on cowpea. *Indian Journal of Nematolog*,.**25**(2) : 190-193.

Sobita Devi, L. (1997). Use of of various neem, (*Azardichta indica* A. Juss) products for the management of *Heterodera cajani* on pigeonpea and a water screening test on the emergence of the nematode larvae from egg sacs. *Neem Newsletter*, **14**(1) : 1-6.

Sombastsiri; Ermel, K. and Schmutterer, H. (1995). The Thai neem tree *Azadirachta siamensis* (Val.) pp. 585-597. In H. Schmutterer (ed.). The Neem Tree *Azadirachta indica* A. Juss and Other Meliaceous Plants, VCH Verlagsgeellschagh, MBH, Weinhiim, German.

Sreeja, P. and Charles, J.S.K. (1998). Screening of botanical against the burrowing nematode, *Radopholus similis* (Cobb. 1893), Thorne, 1949. *Pest Management in Horticultural Ecosystems*, **4**(1) : 36-39.

Sreenivasa Rao, M.S. and Reddy, P.P. (1996). Effective use of neem in sustainable biointensive management of root-knot nematode on tomato. *Neem Newsletter*, **3**(3) : 1996.

Srivastava, A.K. and Singh, R.B. (1990). Effect of organic amendment on interaction of *Macrophomina phaseolina* and *Meloidogyne incognita* on French bean (*Phaseolus vulgaris*). *New Agriculturist,* **1**(1) : 99-100.

Sundaram, K.M.S. and Curry, J. (1996). Effect of some UV light absorbers on the photostabilization of azadirachtin, a neem-based biopesticide. *Chemosphere*, **32**(4) : 649-659.

Sundaram, R. and Velayutham, B. (1988). Relative efficiency of some inscticides and neem (*Azadirachta indica*) cake in the control of *Rotylenchulus reniformis* and *Helicotylenchus dihystera* affecting garden bean. *Indian Journal of Nematology*, **18**(2) : 329-331.

Sundararaju, P. and Koshy, P.K. (1986). Effect of different nematodes and neem oil cake in the control of *Radopholus similis* in yellow leaf disease affected arecanut palms.*Indian Journal of Nematology*, **16**(1) : 44-47.

Suresh, G.; Narasimhan, N.S. and Palani, N. (1997). Structure of nimonol from fresh whole green leaves of *Azadirachta indica. Phytochemistry*, **45**(4) : 807-810.

Thakur, S.G. and Darekar, K.S. (1995). Determination of the waiting period (period of decompositions) of some non-edible oilseed cakes for the control of root-knot nematode. *Current Nematology*, **6**(1) : 1-4.

Thakur, S.G. and Darekar, K.S. (1995). Effect of some non-edible oilseed cakes against root-knot nematode, *Meloidogyne incognita* and growth parameters of brinjal. *Current Nematology*, **6**(1) : 21-26.

Thakur, S.G. and Darekar, K.S. (1995). Relative efficacy of leachates of decomposed non-edible oilseed cakes against root-knot nematode juveniles. *Current Nematology*, **6**(1) : 31-34.

Tiyagi, S.A. and Alam, M.M. (1995). Efficacy of oil-seed cakes against plant parasitic nematodes and soil-inhabiting fungi on mungbean and chickpea. *Bioresource Technology*, **51**(2-3) : 233-239.

Tonthubthimthong, P and Tasakorn, P. (1987). Concentration of crude azadirachtin extract by adsorption on active carbon. *Proceeding of First International Symposium on Biopesticide* held at Naresuan University, Phitsanulok, Thailand, Oct. 27-32, 1996 : 169.

Tuticorin Raghavachari Govindachari; Geetha Gopalakrishanan and Govindaraghavan Suresh (1996). Isolation and quantification of azadirachtins and other major triterpenoids in the seed oil of *Azadirachta indica* A. Juss. *Neem Newsletter*, **13**(2) : 21-22.

Umrao and Goswami, B.K. (1996). Comparative efficacy of soil amendments with carbofuran against root-knot nematode, *Meloidogyne incognita* on cowpea. *Pesticide Research Journal*, **8**(1) : 87-89.

Vats, R. and Nandal, S.N. (1993). Effect of different concentrations of leaves extracts of neem and *Eucalyptus* used as bare-root-dip treatment of tomato seedlings against *Meloidogyne javanica. Current Nematology*, **4**(1) : 15-18.

Vats, R. and Nandal, S.N. (1995). Evaluation of differently prepared plant extracts as bare-root dip treatment of tomato seedlings against *Meloidogyne javanica. Crop Research Hisar*, **9**(1) : 123-128.

Vats, R. and Nandal, S.N. (1993). Preliminary studies on efficiency of aqueous plant extract as bare-root dip treatment of tomato seedlings against *Meloidogyne javanica. Current Nematology*, **4**(2) : 181-185.

Vats, S. Nandal, S.L. and Dalal, M.R. (1996). Comparative efficacy of different preparations of leaf extracts of *Azardirachta indica* and *Eucalyptus tereticornis* against *Meloidogyne javanica* in tomato nursery. *Haryana Agricultural University Journal of Research*, **26**(2) : 99-102.

Verma , R.D.; Mahendra Singh; Samar, R.; Sharma, G.L. and Singh, M. (1997). Effect of soil amendment against root-knot nematode (*Meloidogyne incognita*) on bottlegourd. *Indian Journal of Nematology*, **27**(2) : 255-256.

Verma, A.C. and Ali Anwar (1998). Effect of organic amendments on sprout emergence of pointed gourd, in root-knot nematodes, *Meloidogyne incognita* infested field. *Annals of Plant Protection Science*, **6**(1) : 102-104.

Verma, A.C. and Anwar, A. (1997). Control of *Meloidogyne incognita* on pointed gourd. *Nematologia Mediterranea*, **25**(1) : 31-32.

Vijaylakshmi Mojumder and Rajagopal Raman (1996). Comparative efficacy of neem-based nematicidal products for management of nematode pests of pulse crops. *Neem Newsletter*, **13**(3) : 45-46.

Vijaylakshmi Mojumder and Rajagopal Raman (1996). Nematicidale efficacy of "Neema"-S1" - an experimental formulation of neem for seed treatments against *Heterodera cajani* and *Meloidogyne incognita* in cowpea and chickpea. *Neem Newsletter*, **13**(3) : 41-42.

Vijaylakshmi Mojumder and Rajagopal Raman (1999). Comparative efficacy of neem based nematicidal products for management nematode pests of pulse crops. ***Azadiracchta indica* A Juss,** Ed. Singh & Saxena : 223-234.

Vijaylakshmi Mojumder and Rajagopal Raman (1999). Nematicidal efficacy of neema-S1, experimental formulation for neem seed treatments against *Heterodera cajani* and *Meloidogyne incognita* in cowpea and chickpea respectively. Azadirachata indica A Juss, Ed. Singh & Saxena : 217-222.

Vijaylakshmi, K. and Goswami, B.K. (1987). Effect on root dip treatment seedlings of aquesous extracts of some oil-seed cakes on root-knot nematode infestation. *Annals of Agricultural Research*, **8**(1) : 168-171.

Vijaylakshmi, Mojumder and Mishra, S.D. (1995). Management of root-knot nematode, *Meloidogyne incognita* through decomposed neem products. *Current Nematology*, **6**(2) : 139-141.

Walia, K.K.; Mehta, N. and Gupta, D.C. (1994). Effect of green manuring on *Rhizoctonia* and root-knot nematode complex on tomato. *Nematologia Mediterranea*, **22**(2) : 131-133.

Wani, A.H. (1992). Control of root-knot nematodes on okra with seed soaking in neam leaf extract. *Current Nematology*, **3**(1) : 39-49.

Zaki, F.A. and Bhatti, D.S. (1989). Phytotherapeutic effect of some plant leaves on *Meloidogyne javanica* infecting tomato plants. *Nematologia Mediterranea*, **17**(2) : 1-72.

Chapter 8

Neem products in controlling phytophagous mites

The mites are of great economic importance. They are found as serious pest of crops, stored grain and grain products, in humus breakdown as biological control agents and also as parasite of various insects, animals including man. Some mites are exclusively phytophagous, while others include external and internal parasites. The small insects suck the sap from the leaves. In severe cases, the whole plantation looks yellowish and cobwebs are seen on the leaves.

Mansour and Ascher (1983) carried out studies in Israel to determine the effect of NSKE against *Tetranychus cinnabarinus* (Boisduval). Females were repelled by the treatment and oviposition was reduced. The effect of solvent was assessed and on the basis of the EC50s for both parameters, the order of decreasing effectiveness of the solvents was pentane, chloroform, butanol, acetone and methanol; aqueous extracts were inactive. The activity decreased with increasing dielectric constant of the solvent. The chloroform and butanol extracts were phytotoxic. Seven days after spraying, leaf discs sprayed with pentane or acetone extracts still caused mortality and reduced the fecundity of the females, but were not repellent, when sprayed directly onto females on bean leaf discs, these solvents were repellent and toxic to the mites and reduced fecundity.

Chiu (1985) obtained good results with 0.5 percent neem oil against citrus spider mite *Panonychus citri* but there was no adverse effect on the predatory mite *Amblyseius* sp.

Mansour *et al*. (1987) studied the toxicity effect of neem extracts to predaceous mite *Phytoseiulus persimilis*. The extracts (2%) had no effect for the mite. However, extracts were found more toxic to the prey of *P. persimilis*. Mansour *et al*. (1993), however, again observed that RD9-Replin was highly toxic to the predaceous mite *Tetranychus athiosiae* and its prey *T. cinnabarinus*.

Wel *et al*. (1989) found that neem seed oil was effective against the citrus mite *P. citri* (LC50=0.27%) and innocuous to its predator, *Amblyseius newsami* (LC50=1.63%). They also observed that in the field experiment,

0.5 percent seed oil emulsion gave 93-96 percent control.

Sanguanpong (1992) carried out investigations in the laboratory, greenhouse and field in Germany on the effects of neem extracts (from *A. indica* and *A. excelsa*) on *Tetranychus urticae* Koch. and its predator *Phytoseiulus persimilis*. The laboratory studies indicated that azadirachtins had only minor potential for the control of the tetranychid, and extracts at concentration up to 1 percent were harmless to the phytoseiid. In the field, 2-3 percent extracts against *T. urticae* increased yields of field beans (faba beans), and neem extracts had no effects on the population dynamics of *P. persimilis*.

Sanguanpong and Schmutterer (1992) observed that laboratory studies with *T. urticae* showed that NSKE prepared with pentane were considerably more toxic than an extract prepared with solvents (AZT extract), cold-pressed neem oil (CPNO) or commercial neem oil (CNO). LD50s of topically applied extracts were calculated as 1.33, 1.86, 2.06 and 3.86 percent for pentane extract (PE), CNO, CPNO and AZT extract, respectively. PE and CPNO reduced the fecundity of the mites on treated plants and the survival of nymphs hatched from treated eggs. Application of PE or NO in sublethal concentration caused growth disrupting effects on the nymphal stages and ovicidal effects. Nonpolar components of the oil and PE were more effective than the polar components. Quantification of the insecticidal substance azadirachtin in the extracts revealed that this compound was not the most active principle against the mites.

Serra (1992) did trials with neem oil and neem seed water extracts against *Aculops lycopersici* on tomato and observed no obvious harm to unidentified predatory mites.

Kumar and Sharma (1993) conducted studies on the population dynamics and control of *Tetranychus ludeni* on okra at Samastipur in Bihar, India, in 1990-91. the mite appeared in the 1st week of April. The population peak was recorded in June, and a sharp decline in mite density was observed in July. No mites were observed later in the year than September. Predatory *Amblyseius* spp. (including *A. tetranychivorus* and *A. multidentatus*) appeared in the 2nd week of June. Sulfur, dicofol, tetradifon and NSKE were effective for controlling the pest.

Patel et al. (1993) compared the effectiveness of the botanical insecticides Neemark (0.5%), Repelin (1.0%), Margocide (CK 0.1%, of unstated composition) and Margocide (OK 0.8%, of unstated composition) in laboratory studies with that of conventional insecticides against *T. cinnabarinus* and *T. macfarlanei* in brinjal, okras and indian bean (*Lablab purpureus*) in laboratory studies. The effectiveness of the botanical insecticides was comparable with that of conventional insecticides/acaricides, but varied against the same species on different food plants.

Dimetry *et al*. (1993) tested 2 commercial preparations of NSKE for their effect as a deterrent, toxicant and growth inhibitor against *T. urticae*. Both formulations exhibited a strong deterrent effect against females of *T. urticae*. Mites feeding on leaf discs treated with different concentrations of the 2 products showed increased mortality and retarded development. A significant reduction in the total number of eggs laid as well as a decrease in the percentage of egg hatchability was observed.

Sundram *et al*. (1995) applied a commercial neem formulation containing azadirachtin-A (AZ-A) to the soil around the root system of potted aspen (*Populus tremuloides*) plants. The uptake, translocation, persistence and dissipation of the chemical in the plants were studied. The effect of foliar residues of AZ-A on *T. urticae* populations was also evaluated. The compound was taken up by the root system within 3 hrs and translocated in the stem and foliage within 3 days, confirming that AZ-A is systemic. The peak concentration (μg/g, fresh weight) of AZ-A occurred at 10 days post-treatment, and were distributed in roots, stem and foliage in the ratio of 8.1:1, 0:2.3, respectively. The rate of dissipation of AZ-A from the matrices was moderately rapid, and the residual concentration on the last day of samplling (50 days post-treatment) in roots, stem and foliage were in the ratio 2.7:1, 0:1.2, respcetively. Control of mites by AZ-A residues in foliage was statistically significant, and the bioactivity declilned within 30 days.

Sundram and Sloane (1995) tested pure AZ-A and 4 neem-based formulations [Margosan-O (RH), Azatin (MO), RH-9999 (PT) and Neem PTI-EC4 (AT)] containing the insecticide isomer for their repellency, toxicity and oviposition deterrence against *T. urticae*. Mites were placed on treated and untreated aspen (*P. tremuloides*) leaf discs and their fecundity, feeding rate and mortality were assessed. Results indicated that the effects of AZ-A on *T. urticae* varied with formulation type and AZ-A concentrations. Significant reductions in feeding and oviposition, which correlated with AZ-A concentration, were recorded with a significant increase in repellency. Leaf discs treated with increasing concentrations of the formulations showed increased mortality of *T. urticae*, reduction in the total of number of eggs laid, reduced proportion of eggs hatched and reduced survival of emerged mites. The deterrent and biological effects decreased in the order AT > PT > MO > RH > AZ-A.

Nandagopal and Gedia (1996) studied Residual toxicity of neem oil alone and in combination with some pesticides against 2 spider mites (*Tetranychus hypogaea* and *T. cinnabarinus*) on groundnuts in India. Residual toxicity of synthetic insecticides (phosphamidon and endosulfan) persisted for a few hours and gave 45-50 percent mortality of *T. hypogaea*. The crude neem oil (2%) gave the highest LT50 values of 156 hrs for *T. hypogaea* and 182.4 hrs for *T. cannabarinus*. Mixture of synthetic insecticides and kelthane with neem oil resulted in higher LT50 than alone.

Cadatal and Cadapan (1996) studied regulation of spider mite (*Tetranychus* sp.) population on roses. Water blasting (25-30 psi) was as effective as formetanate hydrochloride spray (0.33 gm a.i./2.5 m^2 plot) in reducing spider mite infestation on roses. Neem extract, detergent (Tide powder) and rice hull mulching were not as effective against the pest but were generally better than the untreated control.

Mohapatra (1996) studied the field efficacy of dicofol (0.05%), ethion (0.05%), endosulfan (0.05%), fenpropathrin (0.01%), monocrotophos (0.01%), dimethoate (0.05%), phosalone (0.05%), sulfur WP (0.25%) and NSKE (2%) agianst *Polyphagotarsonemus latus* infesting jute in Orissa, India. Dicofol was the most effective treatment, followed by endosulfan.

Sterk *et al.* (1997) carried out field and pots studies in Belgium to compare the effectiveness of 19 acaricides for the control of *Tarsonemus* [*Phytonemus*] *pallidus* in strawberries (cvs. Elsanta and Evita). All insect growth regulators and Apollo (clofentezine) and Nissorun (hexythiazox) were ineffective, while the results of mitochondrial electron transport inhibitors and Intrepid (dicamba + dichlorprop + MCPA) were not consistent. Organotin compounds and Neoron (bromopropylate) had moderate effects. Azadirachtin (neem seed extract), Polo (diafenthiuron), Talstar (bifenthrin) and Sirbon (halfenprox) failed to control *P. pallidus*.

Mironova *et al.* (1997) tested the effect of NeemAzal-T/S on *T. urticae*. Laboratory sutdies showed that NeemAzal-T/S caused 100 percent mortality of adults of *T. urticae* when they fed on fig leaves treated with a 0.5 percent concentration. Direct treatment of adults caused 98 percent mortality, but this was observed 4 days earlier than in the leaf treatment variant. Nymphal mortality was 100 percent when they fed on leaves treated with 0.5 percent. Treatment with all concentrations tested (0.1, 0.3 and 0.5%) significantly reduced the reproductive capacity of mites, and the survival of progeny of treated females was greatly reduced in comparison with the untreated control.

Jhala *et al.* (1998) tested 10 synthetic pesticides viz. bromopropylate (0.05 and 0.1%), diafenthiuron (0.05, 0.0625 and 0.075%), dicofol (0.03%), dimethoate (0.03%), endosulfan (0.07%), ethion (0.1%), fenpropathrin (0.005%), monocrotophos (0.07%), phosphamidon (0.03%) and wettable sulfur

(0.25%); one azadirachtin rich ready to use botanical pesticide viz. Gronimm X (0.04%); 2 neem products viz. neem oil (1.0%) and NSKE for their acaricidal activity against red mite, *T. cinnabarinus* infesting okra in field condition during May, 1995. The pooled data on percent mortality revealed that bromopropylate, dicofol, diafenthiuron and fenpropathrin with more than 75 percent mortality were most effective as acaricides. Among all treatments, bromopropylate at higher dose exhibited the higest acaricidal effect. All the three neem-based products caused mortality between 40 to 50 percent, were at par with each other and less effective than synthetic acaricides.

Singh and Singh (1998) tested a neem-based formulation Nimbecidine (AZT 0.03%) alone and with combination of established acaricides like dicofol, ethion and sulphur for its toxicity in managing two spotted spider mite (TSM), *T. urticae* on okra in the field condition during 1994-95. TSM has been identified as most detrimental mite species on okra crop during summer months. Water sprays was included as check for evaluating efficacy of different treatments. Neem oil extract was found to be effective either alone at its recommended field condition or in combination with recommended acaricides. The doses of the acaricides could be reduced to half of the recommended doses when used in combination with Nimbecidine and had given good performance. Phytopesticides were comparable with promising chemicals. In addition, Nimbecidine worked as good acaricide too and safer to predators.

Jhala *et al.* (1998) evaluated vairous plant extracts viz. Arduso (*Alianthus excelsa* Roxburgh) leaf extract (AALE), Ardusi (*Adhatoda vasica* L.) leaf extract (AILE), castor seed extract (CSE), Damara (*Ocimum bacilum* L.) leaves extract (DLE), Pudina (*Mentha piperita* L.) leaves extract (FLE), Naffatiya (*Ipomoea carnea* L.) leaves extract (NFLE), neem leaves extract (NLE), NSKE and Tulsi (*Ocimum sanctum*) leaves extract (TLE) in the laboratory during April, 1995 for their ovicidal and adulticidal action against *T. cinnabarinus* on okra. Water extracts of all above botanicals were evaluated at 5 percent concentration (w/v). The descending order of above plant extracts for their ovicidal action based on percent egg mortality given in brackets was : NSKE (71.4) > NLE (56.9) > DLE (52.6) > NFLE (50.3) > AILE (46.1) > AALE (35.0) > TLE (27.1) > FLE (26.8) > CSE (19.8). While descending order for their adulticidal action based on adult percent mortality given in bracket was NSKE (64.77) > NLE (46.39) > NFLE (43.63) > AALE (43.56) > AILE (36.60) > TLE (35.05) > FLE (32.28) > DLE (30.93) > CSE (20.98).

Patel *et al.* (1998) tested the effeciveness of neem-based botanical pesticides against red spider mite, *T. cinnabarinus* on okra in laboratory during March, 1995 for their ovicidal and adulticidal action. On the basis of percent egg mortality given in brackets, the order of effectiveness of various neem-based formulations for their ovicidal action was Neemark (86.4 with 1%) > Repelin (77.7 with 1%) > neem oil (64.4 with 1%) > Gronim (50.3 with 0.4%) > Margocide CK (44.5 with 0.1%) > Nimbecidine (36.2 with 0.2%) > Parasmani (30.8 with 0.3%) > Achook (28.7 with 1%). While the descending order for their adulticidal action based on adult percent mortality after 3 days exposure was Gronim (87.67 with 0.4%) > Neemark (81.95% with 1%) > Achook (76.38 with 1%) > Margocide CK (72.20 with 0.1%) > Repelin (69.41 with 1%) > Nimbecidine (59.71 with 0.2%) > neem oil (45.80 with 1%) > Parasmani (44.41 with 0.3%). Thus, Neemark (1%) showed highest acaricidal (ovicidal + adulticidal) effect. Gronim (0.4%) was found best adulticide but moderate ovicide. Achook (1%) showed higher adulticidal effect but poor ovicidal action. Parasmani (0.3%) was lees effective as ovicide and adulticide.

Singh *et al*. (1998) did experiment on the treatment of stored pigeonpea seed with certain oils for the protection from *Suidasia nesbitti* Hughes. Neem, castor, mustard, groundnut and sesame oils were tested for their protection against *S. nesbitti* infesting stored pigeonpea seeds. All 3 dosages tested (0.25, 0.50 and 0.75 ml/kg) effectively controlled the mites and protection increased with an increase dosage, for all the oils tested. The maximum population of mites was recorded on sesame oil at 0.25 ml/kg. The minimum number of mites was observed at 0.75 ml/kg level of neem oil. It is concluded that neem oil was the most effective in controlling the population of mite pest followed by castor, groundnut and mustard oil.

Singh *et al*. (1998) tested the effect of different oils on the oviposition and fecundity of an acarid mite *S. nesbitti* infesting stored pigeonpea seeds. The efficacy of neem, castor, groundnut, Indian mustard and sesame oils at three different levels, i.e. 0.25, 0.50 and 0.75 ml/kg, was assessed in terms of fecundity, discrimination quotient (ovipositional preference) and percentage hatchability of *S. nesbitti*. The largest reduction in egg laying was noticed with neem oil at 0.75 ml/kg level, when only 70 eggs were laid compared with 148 eggs in the control. All the treatments deterred oviposition as DQ ranged from 0.07-0.38. However, deterrence was maximum in neem oil at all three dosages 0.25, 0.50 and 0.75 ml/kg as indicated by DQ 0.238, 0.33 and 0.35, respectively.

Gulate (1998) studied inhibitory action of neem products on *Tyrophagus putrescentiae* on stored wheat grains during storage. Neem products (neem oil, Neemark and neem cake) were tested for their effect on the fecundity and development of *T. putrescentiae*. The number of eggs oviposited was reduced from 147.6 in untreated wheat to 10.2, 21.4 and 27.6 following treatment with 0.4 percent neem oil, Neemark and neem cake, respectively. Other stages were affected similarly. Irrespective of concentration and neem product, all four life stages of *T. putrescentiae* were equally susceptible. Neem oil was more efficacious, followed by Neemark and neem cake.

References

Ascher, K.R.S.; Rembold, H. (ed.) (1986). Natural pesticides from the neem tree (*Azadirachta indica* A. Juss). *Proceedings of 3rd International Neem conference*, Nairobi, Kenya : 577-587.

Cadatal, T.D. and Cadapan, E.P. (1996). Regulation of spider mite (*Tetranychus* sp.) population on roses. *Philippine Entomologist*, **10**(2) : 195-201.

Chiu, S.F. (1985). Recent research findings on Metiaceae and other promising botanical insecticides in China. *Z. pflkrank Pflschutz*, **92** : 310-319.

Dimetry, N.Z.; Amer, S.A.A. and Reda, A.S. (1993). Biological activity of two neem seed kernel extracts against the two-spotted spider mite *Tetranychus urticae* Koch. *Journal of Applied Entomology*, **116**(3) : 308-312.

Gulati, R. (1998). Inhibitory action of neem products on *Tyrophagus putrescentiae* (Schrank) (Acarina : Acaridae) in wheat during storage. *Annals of Agri. Bio. Research*, **3**(2) : 227-230.

Jhala, R.C.; Patel, B.D. and Bharpoda, T.M. (1998). Ovicidal and adulticidal efficacy of various plant-extracts against red spider mite, *Tetranychus cinnabarinus* on okra. *National Seminar in Entomology in 21st Century*, April 30-May 2, 1998, Udaipur : 125.

Jhala, R.C.; Patel, B.D. and Patel, J.J. (1998). Field efficacy of synthetic and botanical pesticides against red spider mite, *Tetranychus cinnabarinus* (Boisduval) on okra. *Neem Newsletter*, **15**(2) : 20.

Kumar, V. and Sharma, D.D. (1993). Bio-ecology and chemical control of spider mite, *Tetranychus ludeni* Zecher on okra. *Indian Journal of Plant Protection*, **21**(1) : 68-71.

Mansour, F.; Ascher, K.R.S. and Abo-Moch, F. (1993). Effects of Margosan OTM, Azatin TM and RD-9 RepelinR on spiders and on predaceous and phytophagous mites. *Phytoparasitica*, **21** : 205-211.

Mansour, F.; Ascher, K.R.S. and Omari, N. (1987). Effects of neem seed kernel extracts from different solvents on the predaceous mite, *Phytoseiulus persimilis* and the phytophagous mite, *Tetranychus cinnabarinus* as well as on the predatory spider, *Chiracanthium mildei*. In : Natural Pesticides from the Neem Tree (*Azadirechta indica* A. Juss) H. Schmutterer and KRS Ascher (eds) Proceeding of third Internation Neem Conference Nairobi, Keyna, 1986), pp. 577-587.

Mansour, F.A. and Ascher, K.R.S. (1983). Effects of neem (*Azadirachta indica*) seed kernel extracts from different solvents on the carmine spider mite, *Tetranychus cinnabarinus*. *Phytoparasitica*, **11**(3-4) : 177-185.

Mironova, M.K.; Khorkhordin, EG.; Kleeberg, H. (ed.) and Zebitz, CPW. (1997). Effect of NeemAzal-TS on *Tetranychus urticae* Koch. *Proceedings 5th Workshop Wetzlar*, Germany, 22-25 January, 1996 : 129-137.

Mohapatra, L.N. (1996). Chemical control of yellow mite, *Polyphagotarsonemus latus* Banks (Acarina : Tarsonemidae) on jute. *Indian Journal of Plant Protection*, **24**(1-2) : 15-17.

Nandagopal, V. and Gedia, M.V. (1996). Residual effect of pesticides against two spider mites of groundnut. *Karnataka Journal of Agricultural Sciences*, **9**(4) : 601-605.

Patel. C.B.; Rai, A.B.; Patel, M.B.; Patel, A.J. and Shah, A.A. (1993). Acaricidal tests of botanical pesticides in comparison to conventional acaricides/pesticides against red spider mites (Acari : Tetranychidae) okra, brinjal and Indian bean. *Indian Journal of Entomology*, **55**(2) : 184-190.

Patel, B.D.; Jhala, R.C. and Pattel, N.C. (1998). Effectiveness of neem based botanical pesticides against red spider mite, *Tetranychus cinnabarinus* on okra. Abstract *National Seminar in Entomology in 21st Century*, April 30-May 2, 1998, Udaipur : 128.

Sanguanpong, U. (1992). The effect of products containing oils of neem and marrango seeds on the two-spotted spider mite, *Tetranychus urticae* Koch and side-effects on its natural enemy *Phytoseiulus persimilis* Athias Henriot. *Bibliographic citation*, pp. 102.

Sanguanpong, U. and Schmutterer, H. (1992). Laboratory studies on the effect of neem oil and neem seed extracts on the spider mite, *Tetranychus urticae* Koch. (Acarina : Tetranychidae). *Zeitschrift fur Pflanzenkrankheiten und Pflanzenschutz*, **99**(6) : 637-646.

Serra, A.C. (1992). Untersuchungen Zum Einastz Von Niemsamenextrakten im Rahmen Integrierter Ansotze zur Bekampfung Von Tomatenschadlingen inder Dominikanischen Republik. *Doctor Thesis*, University of Giessen, Germany : 109-121.

Singh, R.N. and Singh, J. (1998). Evaluation of azadirachtin and some conventional acaricides against two spotted spider mite, *Tetranychus urticae*. *Natioal Seminar in Entomology in 21st Century*, April 30-May 2, 1998, Udaipur : 121.

Singh, Manju; Raheja, Akshma; Singh, M. and Raheja, A. (1998). Treatment of stored pigeonpea seed with certain oils for protection from *Suidasia nesbitti* (Hughes). *Crop Research Hisar*, **16**(3) : 382-388.

Singh, Manju; Raheja, Akshma and Singh, M. (1998). Effect of different oils on the oviposition and fecundity of an acarid mite, *Suidasia nesbitti* (Hughes). *Annals of Agricultural Biological Research*, **3**(1) : 49-53.

Sterk, G.; Bal, E.; Goossens, W. and Bylemans, D. (1997). Semi-field and field experience in the control of the strawberry mite, *Tarsonemus pallidus* (Banks) (Acari : Tarsonemidae). *Parasitica*, **53**(1) : 25-33.

Sundram, K.M.S. and Sloane, L. (1995). Effects of pure and formulated azadirachtin, a neem-based biopesticide, on the phytophagous spider mite, *Tetranychus urticae* Koch. *Journal of Environmental Science and Health*, Part-B, Pesticides, Food Contaminants, and Agricultural Wastes, **30**(6) : 801-814.

Sundram, K.M.S.; Campbell, R.; Sloane, L. and Studens, J. (1995). Uptake, translocation, persistance and fate of azadirachtin in aspen plants (*Populus tremuloides* Michx.) and its effect on pestiferous two-spotted spider mite (*Tetranychus urticae* Koch). *Crop Protection*, **14**(5) : 415-421.

Wel, X. L.; Chiu, S.F. and Huang, Z.X. (1989). Studies on the control of citrus red mite *Panonychus citri* with chinaberry seed oil. *Journal of Science*, China Agricultural University, **10** : 48-55.

Chapter 9

Effect of neem products on Honey bees

Sharma *et al*. (1980) observed insect growth disruption by neem seed fractions in *Apis mellifera* L. and *Ephestia kuehniella* Zell. During laboratory studies carried out in the German Federal Republic, 4 partially purified neem (*Azadirachta indica*) seed fractions (NN18-701, NN18-705, NN 18-79a and NN 18-79b) were tested for growth disruption activity against the holometabolous insects *E. kuehniella* and *A. mellifera*. The neem fractions were applied topically to the spinning stage of the last-instar larva of *E. kuehniella* and to the 3rd-instar larva of *A. mellifera*. Fraction NN 18-79a was the most effective, followed by NN 18-705, in causing metamorphic disturbances in both species; NN 18-79b was relatively ineffective, while NN 18-701 was almost inactive. However, fraction NN 18-705 showed higher antifeedant activity than fraction NN 18-79a. The treatment of *E. kuehniella* larvae with the neem fractions caused delayed larval development, and after treatment with NN 18-79a many died as larval-pupal intermediates, pupal and adult malformations were also observed. In *A. mellifera*, maximum mortality occurred immediately before larval-pupal ecdysis. The growth disruption activity of neem seed fractions NN 18-705 and NN 18-79a appeared to be independent of feeding inhibition.

Rambold and Czoppelt (1981) studied IGR effect of neem on honeybees. Azadirachtin has recently been recognized as a potent insect grwoth regulator which may be an interesting candidate for plant protection purposes. Evaluation of its toxicity to honeybee brood has, therefore, become important. The compound was purified from neem seeds by chromatography of a defatted methanol extract on silica gel and subsequent high-pressure liquid chromatography. Azadirachtin was tested for its growth disrupting activity with honeybee larvae, which were reared on royal jelly/yeast extract in an incubator. Various amounts of azadirachtin (0-1.0 ppm), dissolved in methanol, were topically applied to 3rd-instar larvae. Weight gains of treated larvae (approximate larval weight 5 mg) were comparable to controls, thereby indicating no feeding inhibition. At doses of 0.25 µg/larva and higher, a disturbance of larval and prepupal development was initiated. High larval mortality and poor weight gain were observed, compared to controls.

Rambold *et al.* (1982) evaluated the effect of azadirachtin, a potent insect growth regulator from neem tree (*A. indica*) on three holometabolous species *Epilachna varivestis* Muls., *Ephestia kuehniella* Zell. and *Apis mellifera* L. under laboratory conditions in the German Federal Republic. Various amounts of the compound were fed to 4th-instar larvae of *E. varivestis* and topically applied to final-instar larvae of *E. kuehniella* (spinning stage) and 3rd-instar larvae of *A. mellifera*. The number of larvae developing to the adult stage was greatly reduced by azadirachtin treatment, this reduction being dose-dependent. Pupal mortality was also observed. However, weight gains were comparable with those observed in untreated larvae, indicating that the compound did not cause feeding inhibition. The strong effect of azadirachtin on larval-pupal and pupal-adult moults is interpreted as an interference with moulting hormone pools.

Rembold *et al.* (1982) treated larvae of *Apis mellifera* with fractions of extracts from NSKE and also pure azadirachtin. They observed metamorphic disturbances leading to abnormal larvae. Topical applications of third instar larvae with doses of 0.25 and 0.5 μ g/larva of azadirachtin reduced the survival rate of adult stage to 58 and 22 percent, respectively.

Naumann *et al.* (1994) studied repellent effect of neem products against foraging honey bees and other pollinators. A standardized, oil-free neem (*A. indica*) seed extact (NSE) was tested for repellency to *Apis mellifera* using a feeding dish choice bioassay, and to *A. mellifera* and other pollinators in field applications on blooming canola (rape) in Manitoba. Foraging workers were able to discriminate between untainted sugar syrup and syrup containing formulated NSE at concentration as low as 0.1 ppm azadirachtin, the principal active ingredient of NSE. However, there were no significant differences in the numbers of foraging *A. mellifera* collected in neem-treated, solvent-treated or untreated rape plots. Other pollinators species present were similarly unaffected. The results suggested that *A. mellifera* may be successfully utilized in blooming crops that have been treated with doses of NSE sufficient to control phytophagous insect pests. The results of this study also sugested that using feeding-dish bioassays to screen potential repellent compounds, although demonstrating the ability of *A. mellifera* to detect compounds, may not yield results in the field.

Schmutterer and Holst (1987) tested the effect of neem product AZT-VR-K on the behaviour and developmet of small bee colonies. The neem product was sprayed at 2 ml/lt water at a rate of 60 ml/m^2. No damage was caused to bee colonies consisting of a queen and 3000 workers but in small colonies of a queen and 200-300 workers do have some side effects. Some young bees were unable to hatch from the cells after biting off the lids. It was also observed that egg laying was not reduced by the treatment with neem products. It was also recommended to avoid spraying of higher concentrations of neem products on flowering plants which are attended by numerous honey bees.

Larson (1989) reported that the neem product Margosan-O was non-toxic to honey bees in concentrations up to 4478 ppm a.i./ha.

Malaipan *et al.* (1992) found the numbers of honeybees (*Apis mellifera*) and thrips (*Scirtothrips dorsalis*) of flowers of pummelos in open plots in Thailand and in cages without any insecticide applications were twice as high as for those with spray applications of phosalone, endosulfan or neem extract. Pummelos pollinated by honeybees produced numerous fruit sets. In open pollination and cages with bees, fruit set was 54 and 67 percent, respectively, while treatment with the pesticides gave only 44, 34 and 36 percent fruit set, respectively. Fruit drop was higher for treatment with chemical insecticides than with neem extract, and damage by thrips was higher in untreated than treated crops.

Bunsen (1992) tested compatibility of neem oil and other plant oils to honeybees and also their repellency to a parasitic mite of bees *Varroa jacobsoni*. The neem oil in a petridish was placed under the breeding bee colonies in hives and were allowed to evaporate. It was observed that there was no significant adverse effect on the bees of any instars but a strong repellent effect of the oil vapours on the mite was observed.

Sadakathulla (1994) studied at Veppankulam, India to evaluate the effect of neem derivatives on the safety of three honeybees viz. Indian bee, *Apis cerana indica* F. little bee *Apis florea* F. and dammer bee *Melipona iridipennis* D. In this experiment, 8 treatments were taken viz. 3 and 6 percent neem oil emulsion (NOE), 5 and 10 percent neem seed extracts, 5 and 10 percent neem seed kernel extracts, water spray and no spray (control). It was recorded that after the spraying of neem derivatives, the mean percent of bees that visited was 55.0, 31.9 and 13.1 for the three bees, respectively. The mean number of bees that visited the inflorescence, ranged from 7.3 to 10, 3.0 to 6.6 and 1.7 to 2.6 for the three bees, respectively, and there was no significant difference in the number of different honeybees species among various treatments. The bees were observed visiting the inflorescence normally without any inhibition, even on the next day of spraying. This gives clear indication that neem sprays had no side effects on the three species of honeybees.

Liu (1995) fed azadirachtin from *A. indica* [*Melia azadirachta*] to honeybee (*A. mellifera*) colonies as 1 ml or 2 ml Margosan-O (3000 ppm a.i./lt of sugar syrup). These colonies produced fewer chalk brood (*Ascosphaera apis*) mummies, had lower levels of *Nosema apis* spores, produced more honey and collected more pollen than controls. Azadirachtin added to a growth medium inhibited the growth and development of *A. apis*.

Naumann and Isman (1996) injected neem-based insecticide at various concentrations of azadirachtin (a.i.) into the food in cells of 1st and 4th instar *A. mellifera* maggots. Mortality increased with dose, and workers removed all dead larvae. Estimated LD50 values were 37 µ/g body wt of 1st instar larvae and 61 micron/g for 4th instar larvae; these are higher than reported for most other insect species. Leaves of flowering *Brassica campestris* were sprayed with the insecticide; in analysis of nectar and pollen samples on the following 6 days, no azadirachtin was found (detection limit below 5 ppm), indicating that translocation from leaves to flowers probably does not occur.

Thapa and Wongsiri (1997) sprayed neem (*A. indica*)-based biopesticides; azadirachtin-A (Neemix - 50cc/20 lt) and azadirachtin-B (Advantage - 80cc/20 lt) and synthetic pyrethroids; lambda-cyhalothrin (Karate - 16cc/20lt) and permethrin (Ambush - 10cc/20lt) at the rate of 200 ml/minute during the peak foraging hours of *A. cerana* on 99 m^2 caged field of flowering oilseed rape (*B. chinensis*) to determine the toxicity to honeybees. Observations were made 3 days before and 3 days after the application of the four pesticides. No mortality of the honeybees was observed after spraying azadirachtin-A and -B, with no significant difference ($P>0.05$) in mortality between control and these treatments. In contrast, a large number of dead bees accumulated in front of the hive immediately after the application of lambda-cyhalothrin and permethrin. Foraging activity of this species declined for 1-1 and 1/2 hr immediately after the application of azadirachtin-A and -B. A slightly more prolonged decline in foraging activity occurred after the application of lambda-cyhalothrin and permethrin. Flight activity also stopped for 3 to 5 hrs after the application of the four pesticides. However, the mortality caused by lambda-cyhalothrin and permethrin was quite low compared with some other insecticides.

Sithanantham *et al*. (1997) carried out a field study during April-July, 1995 in Kenya to test *Azadirachta indica* seed kernel extract sprays (5, 10 and 20%) for controlling some insect pests on cowpea (*Vigna unguiculata*). Observations were made on the visitation frequently of honeybees (*Apis mellifera littorea*) and pitfall traps were used to capture 2 groups of predatory arthropods (spiders and ants). On the first day after spraying, the number of honeybee visits to *A. indica*-sprayed plots was reduced at all doses, as compared to the "no-spray" plots. On 3 and 5 days, such a reduction was only seen at the highest dose. It is concluded that the use of *A. indica* sprays in cowpea ecosystems appears to be relatively safe to beneficial arthropods.

References

Bunsen, J.D. (1992). [Experimentelle Untersuchungen Zur. Bekampfung der Milbe Varroa gacobsoni Oud. einem Ektoparasit der Honigbiene (*Apis mellifera* L.) mit Stoffen naturlicher Herkunft.] *Doctor Thesis*, University of Giessen, Germany.

Larson, R.O. (1989). The commercialization of neem. In Jacobson, M. (ed.), 1988, *Focus on Phytochemical Pesticides* **Vol. I**. **The Neem Tree**, CRC Press, Bocaraton, FL : 155-168.

Liu, T.P. (1995). A possible control of chalkbrood and Nosema disseases of the honey bee with neem. *Canadian Beekeeping*, **18**(5) : 107-109.

Malaipan, S.; Kongpitak, P. and Sujitratanan. S. (1992). Comparative study on the effect of insecticides and neem extracts to the thrips and honey bee pollinators of pomelo cultivar Tongdee. *Kasetsart Journal of Natural Scienice*, **26**(2) : 146-150.

Naumann, K.; Currie, R.W. and Isman, M.B. (1994). Evaluation of the repellent effects of a neem insecticide on foraging honey bees and other pollinators. *Canadian Entomologist*, **126**(2) : 225-230.

Naumann, K. and Isman, M.B. (1996). Toxicity of a neem (*Azadirachta indica* A. Juss) insecticide to larval honey bees. *American Bee Journal*, **136**(7) : 518-520.

Rambold, H. and Czoppelt, C. (1981). Evaluation of insect growth regulators from *Azadirachta indica* (neem) using rearing tests on honeybee larvae. *Mitteilungen der Deutschen Gesellschft fur Allgemeine und Angewandte Entomologie*, **3**(1-3) : 196-198.

Rambold, H.; Sharma, G.K.; Czoppelt, C.H. and Schmutterer, H. (1982). Azadirachtin : a potent insect growth regulator of plant origin . *Z. Angew. Entomol.* **93** : 12-17.

Sadakathulla, A. (1994). evaluation of safety of neem derivatives on three honey bee species. *Neem Newletter* **11**(3) : 31-32.

Schmutterer, H. and Holst, H. (1987). [Untersuchungen uberdee wirkung des angereicherten, formulierten Niemsamon extrakts AZT-VR-K aug die Honigbiene *Apis mellifera* L.] *Journal of Applied Entomology*, **103** : 208-213.

Sharma, G.K.; Czoppelt, C. and Rambold, H. (1980). Further evidence of insect growth disruption by neem seed fractions. *Zeitschrift fur Angewandte Entomologie,* **90**(5) : 439-444.

Sithanantham, S.; Otieno, Z.N. and Nyarko, K.A. (1997). Preliminary observations on effects on neem sprays on some non-target arthropods in cowpea ecosystem. *Quatrieme Conference Internationale sur les Ravageurs en Agriculture*, 6-7-8 janvier 1997, le Corum, Montpellier, France. Tome 2, 1997, 643-648.

Thapa, R. and Wongsiri, S. (1997). Toxicity of azadirachtin derivatives and synthetic pyrethroids on oilseed rape to *Apis cerana* (Hymenoptera : Apidae). *Proceedings of the First International Symposium on Biopesticides*, Phitsanulok, Thailand, October, 27-31, 1996 : pp. 82.

Chapter 10

Effect of Neem products on bioagents

The use of pesticides for the cotrol of pests and diseases of major food and cash crops both in field and under storage has increased enor mously during the past 5-10 years. The dramatic increase in pesticide usage is predominant among some and most of the medium and large scale farmers, respectively. Due to the prohibitive costs and erratic supply and manufacturing of substandard pesticides, the majority of farmers and small scale holdings have been adopting alternative methods of pest and disease control. Some of which involve the application of neem products, vegetable oils, essential oils, ashes and heat and smoke treatment etc. Mulching and soil organic amendments are used for disease control. Normally we use these plant products against our targeted insects, but at the same time, it is necessay to know the effect of these plant products etc. on our beneficial insects i.e. parasites, predators, insect pathogens, honey bees etc. The toxicity effects of neem to vertebrates appears to be negligible but we do get variable effects of neem products on beneficial arthropods. Such studies are discussed in this chapter.

1- Effect of neem products against predators

Schmutterer (1981) observed in the laboratory by topical application of 2 percent neem seed kernel extract and 3 percent neem oil on the adult of *Coccinella septempunctata*, a predator of aphids and jassids and observed negative influence. However, a few larvae do had some disturbances in metamorphosis.

Chelliah and Rajendran (1984) tested the toxicity of 7 insecticides to the rice pest predator *Cyrtorhinus lividipennis* in which the mirid was confined on previously sprayed rice plants, the least toxic of the sprays was 0.07 percent endosulfan, which is effective against rice hoppers, followed by 5 percent neem oil, the corrected mortality percentages were 34.72-36.90 mortality on the 1 day after spraying, 7.40-38.11 on the 2nd day and 20.01-20.03 on the 3rd day. The other sprays tested (0.075% quinalphos, 0.04% chlorpyrifos, 0.07% phosalone, 0.08% monocrotophos and 1.0% carbaryl) were highly toxic to the bugs giving 36.07-53.96% mortality on the 3rd day.

The toxicity of all the sprays diminished 5 days after spraying. However, Saxena *et al.* (1984) observed that neem oil was slightly harmful to the mirid byg, *Cyrtorhinus lividipennis* which predates on plant hoppers of rice crop. Fernandez *et al.* (1992) did trial in greenhouse against *C. lividipennis* by taking four treatments viz. 3 percent neem oil, 5 percent aqueous NSKE, endosulfan and water. They observed that there was no mortality in case of neem oil and aqueous NSKE while in case of endosulfan cent percent mortality was obtained.

Saxena *et al.* (1984) and Saxena (1987) observed that the predatory wolf spider *Pardosa pseudoannulata* [*Lycosa pseudpannulata*] of rice leaf hoppers can tolerate high concentations of neem products with out being harmed. They found that even the concentration of 50 μg does not have bad effect on the spiders.

Schmidt and Pesel (1987) found in a laboratory trial the effect of azadirachtine-enriched neem products AZT-VR-K and MTB/H2O-K-NR on the fecundity of beneficial red ant. *Formica polyctena*. The neem products were fed to the ant through liquid food. It was observed that low concentration of neem products stimulate egg production where as higher concentrations especially of MTB/H2O-K-NR decrease egg production.

The predator cricket, *Metioche vittaticollis* were observed to feed on the eggs of leaf folder, *Cnaphlocrosis medinalis* in Philippines. It was observed that three sprayings of neem seed bitters had not bad effect on the predator (Lamb and Saxena, 1988).

Schauer (1985) observed in a laboratory test that application of 100 ppm of azadirachtine-rich product MTB/H2O-VR-K + sesame oil in the ratio of 1:4 resulted in the death of most treated syrphid flies.

Hellpap (1985) found that a laboratory trial the neem product AZT-VR-K fed larvae of *Spodoptera frugiperda* was supplied in the ant colonies of Hymenopteran ant, *Ectatomma vuridum*. The neem treated larvae were accepted by the ants. According to Hellpap (1985), the corn leaves treated with AZT-VR-K were exposed to the larvae of armyworm, *Spodoptera frugiperda* and kept for 2-4 days. The predatory earwigs *Doru laeniatum* were also exposed to the armyworm larvae. It was observed that after 7 days of exposure, there was significant difference in mortality between earwings fed with treated larvae of armyworm.

Schauer (1985) and Eisemlohr *et al.* (1992) observed that the earwig *Forficula auricularia* had no side effects by neem products.

Wu (1986) and Serva (1992) observed that the neem products are not at all toxic to spider predators. Mansour *et al.* (1987) studied the effect of neem seed kernel extracts from different solvents on the predaceous mite *Phytoseiulus persimilis*, the phytophagous mite *Tetranychus cinnabarinus* and the predatory spider *Cheiracanthium mildei*. All the extracts tested were more toxic to the pest than to the predator. The toxicity indexes of LC50 *P. persimilis* : LC50 *T. cinnabarinus* were 3 for methanol, 4 for ethanol, 23 for acetone and 58 for pentane extracts. The EC50 for fecundity reduction was similar in the 2 species. The order of toxicity of 4 percent extracts for the predatory spider *Cheiracanthium mildei* was pentane<right arrow>acetone<right arrow>ethanol>methanol=water (non toxic). The 2.5 percent extracts were also found non-toxic.

Kareem *et al.* (1988) tested the effect of neem seed kernel extract on population of predatory mirid and Araneae in rice and compared with those of monocrotophos (0.75kg a.i./ha). It was revealed that the populations of mirids and Araneae were also lower in plots treated with monocrotophos than in plots treated with neem, 48 days after treatement. Yadav and Patel (1990) tested the efficacy of different botanical insecticides on oviposition of *Chrysopa scelestes* and their ovicidal action. The results revealed that neem products viz. 0.2 percent neemark, 1.0 percent Repelin, 2.0 percent neem seed kernel suspension, 0.1 peercent neem-rich repelled the female from egg laying as greater number of eggs were deposited on the sites other than treated surface. It was found that there was no effect on the hatching of *Chrysopa* eggs.

Hoelmer *et al.* (1990) studied the effect of neem product Margasan-O against predators *Delphastus pusillus* predating on *Bemisia tabci* and *Scymnus* sp. predating on *Aphis gossypii* and *Myzus persicae*. The predators do not have any adverse effect.

Kaethner (1990) studied the effect of AZT-VR-K (at 1000 ppm) and the mixture of AZT-VR-K + neem oil (at 250-30000 ppm) on *Chrysoperla carnea*. The eggs and adults were given topical treatment but there was no adverse effects. There was no reduction in fecundity and the emergence rate of larvae was normal.

Kaethner (1990) tested the effect of AZT-VR-K (at 1000 ppm) and a combination of AZT-VR-K + neem oil (at 250 to 30000 ppm) against eggs, larvae and adults of *Coccinella septumpunctata* in lab and field conditions. It was observed that there was no effect on the rate of emergence from the eggs, on larvae also no effect was observed. However, when sprayed on adults, there was no adverse effect on the fecundity and activity of the adult beetles.

Kaethner (1991) found neem extracts with or without neem oil were hermless to the eggs, larvae or adults of *Chrysoperla carnea* and *Coccinella septempunctata*, and no effects on mortality or fecundity were observed. Under conditions proposed by the IOBC/WRRS working group 'Pesticides and beneficial organisms' the neem products can be considered harmless, but according to Banken and Stark (1997), it is stated that although we did not do separate studies to determine the toxicity of neem when ingested only or after exposure to residues only. They further concluded that susceptibility of *C. septempunctata* after exposure to 3 routes to neemix was much greater than if this species was only exposed to direct spray.

Hough-Goldstein and Keil (1991) found moult and morphogenetic defects in the 3rd instar nymphs of the predator *Perillus bioculatus*, when sprayed with Margosan-O.

Forster (1991) studied the effect of Margosan-O, the neem product against the predatory Carabid beetle *Platynus dorsalis*. The soil where the beetle habitate was treated with the neem products but the studies revealed that there was no adverse effect on the predator.

Saleem and Matter (1991) observed that the neem oil act as temporary repellent against the predatory beetle, *Paederus alfierii*, *Coccinella undecimpunctata* and the lacewig, *Chrysoperla carnea* in cotton but otherwise neem oil has no adverse effect on these predators of *Spodoptera littoralis*.

Fernandez *et al*. (1992) treated the adult of *Pardosa pseudoannulata* with 3 percent neem oil, aqueous neem seed kernel extract (NSKE) and 5 percent endosulfan. They observed that only one adult died in case of neem oil treatment while in other case none of the spider died. However in endosulfan treatment, cent percent mortality was observed on 5th day, while Karim *et al*. (1992) observed that *Oxyopes javanus* was less sensitive to neem oil 50 percent EC than *P. pseudoannulata*.

Mohapatra *et al*. (1991) reported that neem oil at 24 peercent a.i. had no significant adverse effect on the coleopteran predators in rice.

Krishnaiah and Kalode (1992) tested the efficacy of neem oil against predaceous mirid black bug *Tytthus parviceps* and *Nephotettix virescens*. The LC50 values calculated were 2.876 and 1.308 percent, respectively. Serra (1992) observed in laboratory malformed symptoms of the few nymphs of the predator *Nesidiocoris* sp. after the spraying of neem seed water extract and 2 percent neem oil. However, it was observed that these neem products had no significant effect on the field population of these bug population.

Sontakke (1993) tested field efficacy of insecticide alone and in combination with neem oil against insect pests of rice and their predators viz. spiders and the mirid *Cyrtorhinus lividipennis*. It was observed that neem oil was safest to mirid bug and spider population (4.2 and 3.1/clump, respectively). Chlorpyriphos alone as well as combined with neem oil proved very toxic to both mirid bug and spiders with least population (mirid bug 0.9 and 0.8 and spiders 6.5 and 1.0, respectively). Monocrtophos + neem oil seems to be little safer to predators.

Matter *et al.* (1993) tested neem oils against the predator *Coccinella undecimpunctata*. Neem oil had residual activity for up to 6 days. It was observed that neem oil had no effect on survival or behaviour of larvae of *C. undecimpunctata* except for a prolongation of the fourth instar larva. Consumption of the aphids by these predator was unaffected.

Nanda *et al.* (1993) tested the bioefficacy of neem derivatives against the predatory spiders, wolf spiders *Lycosa pseudoannulata* and jumping spider *Phidippus* sp., Lynx spider *Oxyopes* sp. dwarf spider *Callitrichia formosana*, orb spider *Argiope* sp. Damseflies *Agriocnemis* sp. and mirid bug *Cyrtorhinus lividipennis*. It was observed that the neem kernel extract and oil were found relatively safer than the insecticides to predators like the spider *L. pseudoannulata* and *Phidippus* sp. and the mirid bug *C. lividipennis* in field conditions.

Dimetry *et al.* (1994) did laboratory trials of two neem seed extracts on the predatory mites *Amblysius barkeri* (Hughes) and *Typhlodromus richteri* (Karg). The effect of 2 commercial products of neem seed extracts (Neem Azal-S and Margosan-O) on the predatory mites *A. barkeri* and *T. richteri* was studied in the laboratory. Both products were considered to be safe for *A. barkeri* as no mortality occurred at the concentration tested (0.05 and 0.2%). However, the treatements had an adverse effect on the rate of oviposition and food consumption of *A. barkeri*, the degree varying with concentration tested. Margosan-O had no adverse effects on *T. richteri* at 0.05 percent and slight adverse effects (10% mortality) at 0.2 percent, while Neem Azal-S caused 30 and 50 percent mortality at 0.05 and 0.2 percent, respectively. Neither product affected the sex ratio of the progeny of *A. barkeri*, but the percentage of males of *T. richteri* was increased greatly in all the treatments.

Srinivas *et al.* (1994) studied the behavioural responses of *Gambusia affinis* the predator of mosquito larvae in ralation to leaf extracts of *Azadirachta indica*. It is observed that leaf extract at 0.0005-0.1 percent showed a clear repellent action in a Jones Gradient apparatus. Lowery and Isman (1994) found no effect on survival of *C. undecimpuntata* L. larvae topically treated with neem seed oil at concentration up to 1.0 percent but when larvae of predatory beetle were exposed to treated foliage and allowed to feed on contaminated aphids, neem seed oil reduced eclosion rates to 100 percent.

Roger *et al.* (1995) studied mortality and predation efficiency of *Coleomegilla maculata* Lengi Timb (Coleoptera : Coccinellidae) following applications of neem extracts. The toxicity of the 2 neem (*A. indica*) extracts and of the chemical insecticide malathion was evaluated on adults of the predacious coccinellid *C. maculata*. Bioassays were carried out with both neem formulations at 1, 5, 10, 20 and 50 percent concentrations. The azadirachtin quantities contained in neem oil (v/v) and neem seed kernels (w/v) were 13.7 and 91.0 ppm, respectively. Malathion was tested at the field rate of 2.85 g a.i./lt. The mortality rate was evaluated in the laboratory after topical applications on the ventral side of the coccinellids. Adult mortality rate after 72 hr was 100 percent following malathion treatments. Only 1 dose of neem oil (10%) resulted in significantly greater mortality than in the control. No toxicity was observed after the treatments with the aqueous suspension of ground neem seeds. The predation efficiency of *C. maculata* was also evaluated after topical application of these 3 insecticides at sublethal doses. Fifteen minutes after treatments, adult coccinellids were presented 30 aphids for 24 hr. The aqueous suspension of ground neem seeds caused a 50 percent reduction in the number of aphids consumed.

Subba Rao and Mohan (1996) tested the efficacy of neem seed kernel extracts on pegion pea pod borer *Helicoverpa armigera*. The neem products namely, 0.2 percent Nimbecidine, 0.5 percent Achook, and 5.0 percent NSKE were evaluated. Endosulfan (0.07%) and monocrotophos (0.04%) were also taken as insecticide. The data reported that 5.0 percent NSKE is the best.

Hermann *et al.* (1996) studied different Neem Azal formulations on the larvae of green lacewig *Chrysoperla carnea* Steph. The effects of three Neem Azal formulations (NA-T/S, NA-A1 and NA-GMS/SO/F10) on 1st- and 2nd-instar larvae of *C. carnea* were tested under laboratory and semi-field conditions. In the laboratory tests, larvae were exposed to a dried pesticide film on glass plates. Semi-field tests were performed on potted apple trees in gauze tents. Larvae were released 24 hr after treatment of the trees and recaptured with bait cards acontaining eggs of *Sitotroga cerealella*. The number of cocoons was counted. All Neem Azal formulations caused disturbances to growth and moulting. High mortalities in laboratory experiments after a test period of 3-4 weeks led to the classification "harmful" for all formulations. In further tests Neem Azal-T/S yielded lower mortalities and was, therefore, classified "slightly harmful". In the semi-field test, no significant differences between pupation rates on treated and untreated trees were found, therefore, no contact toxicity is to be expected in field.

Mansour *et al*. (1997) studied the efffects of Neemguard on predaceous mite *Phytoseiulus persimilis* and the predatory spider *Cheiracanthium mildei*. It was observed that Neemguard was highly toxic to phytophagous mite *Tetranychus cinnabarinus* but had no effect on *P. persimilis* and *C. mildei*.

Schmutterer (1997) studied the side effects of neem products against natural enemies of spider mites. He found that neem products are safe to spiders, adults of numerous beneficial insects and eggs of many predators such as coccinellids. He emphasized that due to their relative selectivity, neem products can be recommended for many programmes of IPM as it is unlikely that they will cause severe disturbances in ecosystems. These neem products can be s substantial contribution towards preservation of biodiversity in ecosystems inspite of the fact that they are not completely safe to all stages of beneficial nematodes, mites and insects.

Momen *et al*. (1997) tested the Neem Azal-F on *Tetranychus urticae* and three predaceous mites. NeemAzal-F, a commercial preparation of neem seed kernel extract, was tested for its repellency, toxicity and oviposition deterrence to the phytophagous mite *T. urticae*. Significant reductions in oviposition correlated with Neem Azal-F concentrations were recorded, with a significant increase in repellency. Leaf discs treated with increasing concentrations of the product showed increased mortality of *T. urticae* and a reduction in the total number of eggs laid, but percentage egg hatch was 100 percent. The effect of NeemAzal-F on the predatory mites *Amblyseius barkeri*, *A. swirskii* and *A. zaheri* was also stuied. Neem Azal-F decreased the food consumption rate at the two concentrations used (0.2 and 0.05%) for all the predatory mites, as well as egg-laying at both concentrations in *A. swirskii* only. Neem Azal-F was highly toxic to *A. swirskii*. In contrast, the two tested concentrations were considered to be safe for *A. barkeri* and *A. zaheri*.

Sundrababu *et al*. (1997) tested residual toxicity of some insecticides and neem seed kernel extracts against the predatory beetle *Brumoides suturalis* (Fabricius). It was found that neem seed kernel extract and endosulfan exhibited low toxicity to *B. suturalis* larvae and adults.

Drescher *et al*. (1997) studied the effect of neem product Neem Azal-T/S on the anthocorid *Orius majusculus* Reuter. Laboratory tests were carried out on the effect of different concentrations of the product Neem Azal-T/S (1:200, 1:100, 1:50, Neem Azal-T/S:water; active ingredients:1% azadirachtin, 4% Neem Azal) on *O. majusculus*, a predator of Aphidoidea, Tetranychidae and Thysanoptera. The effects of these different concentrations as mist, wet solution, dried solution, and of an oral ingestion of the solution on the predator were studied in relation to mortality, fecundity, sex ratio, rate of emergence, and behaviour. The product Neem Azal-T/S showed little noticeable influence on the predator. Only the highest concentration (1:50) of the wet solution which was used and oral ingestion of the same increased the mortality of the test population significantly above that of the control population. No negative effect on fecundity, sex ratio, rate of emergence or behaviour was observed.

Mani *et al*. (1997) studied the effect of 5 percent neem seed kernel extracts on the predator *Cryptolaemus montrouzieri* Mulsant. No detrimental effect was observed on the progeny production of *C. montrouzieri*.

Banken and Stark (1997) studied the stage and age influence on the susceptibility of *Coccinella septumpunctata* after direct exposure to neem products 'Neemix'. The effects of Neemix 4.5 EC on larvae of the predator *C. septempunctata* were determined after directspray exposure. First instars were treated by direct application with 0, 40, 100, 200, 400, 600, amd 1000 ppm and 4th instars were treated withh 400, 600, 800 and 1000 ppm azadirachtin, the active ingredient in Neemix. Survivorship and development of the larvae were affected in a dose-dependent manner. Symptoms of exposure included delay or prevention of pupation, blackening of the pupal case, formation of pupal-adult intermediates, and deformation of wings and elytra in adults. To account for the slow action of the pesticide, probit analysis of survivorship data was performed after all surviving larvae from each replicate had emerged as adults: day 16 for 1st instars and day 13 for 4th insrtars. The LC50 values and 95 percent FL for 1st and 4th instars were estimated to be 1120 (719-3677) ppm and 520 (405-600) ppm azadirachtin, respectively. These values were much higher than the recommended rates for control of aphids (3 weekly applications of 20 ppm), suggesting that Neemix might be used in integrated pest management programs because application of rates that control aphids should not result in appreciable mortality of predators. Fourth-instar larvae of *C. septempunctata* were innately more sensitive to the growth disrupting effects of acute exposureto Neemix than 1st instrars. It may be possible for early instars to sustain the effects of Neemix as long as the pesticide is detoxified before the onset of pupation. These results suggest that it is extremely important to examine more than 1 life stage of a species to estimate the total effect of pesticides. The results presented here build on the argument presented by Stark and Wennergren (1995). According to them, it appears that estimating toxicity of pesticides to bioagents may not be straightforward as we think unless the susceptibility of various life stages is estimated, erroneous conclusions can be drawn. It is, therefore, suggested that toxocity should be estimated for one life stage to obtain better conclusions.

Bunken and Stark (1997) studied the exposure and the risk of neem products against seven spotted lady bird beetle *Coccinella septumpunctata*. Microcosms were used to illuminate community-level interactions among *Acyrthosiphon pisum* and *C. septempunctata* after a spray exposure to a commercial neem pesticide, Neemix. The instantaneous rate of increase (ri) of aphid pipulations was used as a response variable in a 2 x 3 factorial experiment. Microcosms were treated with 100 ppm (mg/lt) or 600 ppm azadirachtin, the active ingredient in the commercial neem insecticide, or with water in the presence or absence of adult or 4th instar coccinellids. The results were compared with those of a previous study where *C. septempunctata* was exposed to only 1 route of exposure, direct application. Results from the previous study indicated that 100 ppm of the neem insecticide was sublethal (<LC1) and 600 ppm was equivalent to the LC2 for 4th instars. Both concentrations caused no mortality to adults based on direct application. However, in the current study, *C. septempunctata* was exposed to direct sprays, residues on leaves, and pesticide-contaminated prey. Population growth rates (ri values) of the aphid populations 4 days after treatment were compared withh 2-way analysis of variance. The pesticide alone and the predator alone caused a significant decrease in aphid pipulation growth rates. However, no significant (P>0.05) interaction between the predator the pesticide was detected, indicating that the chemical and biological control agents were not working synergistically. Furthermore, exposure to the pesticide in mocrocosms significantly reduced or completely eliminated oviposition in adult *C. septempunctata*, and all of the larvae exposed to 100 or 600 ppm died within 10 days of treatment. Realistic exposure of beneficial species is, therefore, critcal when estimating the potential risk of a pesticide. Microcosm toxicity tests can defect subtle, sublethal effects that might be overlooked in traditional laboratory bioassays. Although survivorship of adult lady bird beetles was unaffected, exposure to Neemix resulted in a severe reduction in fecundity or complete sterility depending on the concentration. The results have demonstrated that pesticides that appear to be harmless to organisms when tested with conventional tests are not necessarily harmless to the same species when they are exposed within terrestrial microcosms. The influence of natural routes of pesticide exposure and protection provided by foliage may have subsequent effects on reproduction and survivorship that would not be detected in the laboratory tests.

Lakshmi *et al.* (1997) studied the safety of neem formulations and insecticides to *Microveliadouglasi atrolineata* Bergroth, a predator of plant hopper in rice ecosystem. The observation revealed that Neemix (2 and 4%) and Rakshak (0.2 and 0.5%) were the safest neem formulations where as phorate and carbofuran (1 kg a.i./ha) granular application and quinalphos spray at 0.5 percent were the least toxic to the predator.

Babu *et al*. (1998) observed the efficacy of neem oil and neem cake for the control of green leaf hoppers, brown plant hoppers and their effect on predators viz. the mirid *Cyrtorhinus lividipennis* and the lycosid *Lycosa pseudoannulata*.

Prashad and Logiswaran (1998) compared the toxicity of different insecticids to the adults of *Chilomenes sexmaculata* F. Adult *C.` sexmaculata* were exposed to insecticides and toxicity was established. The insecticides used were : monocrotophos (0.07%), endosulfan (0.07%), dimethoate (0.03%), quinalphos (0.05%), carbaryl (1.0%), phosalone (0.07%), phosphamidon (0.085%), fenvalerate (0.01%), deltamethrin (0.0014%) and neem oil (3.0%). Mortality percentages were recorded at 3-hr intervals over a 30 hr exposure period. Based on LT50 values, neem oil was the safest insecticide, while the synthetic pryrethroids were highly toxic. It is concluded that the less toxic phosalone, monocrotophos or neem oil could be integrated with the release of *C. sexmaculata* in the field.

Sundari (1998) observed that azadirachtin and dichlorvos showed the lowest toxicity to the predator *Cryptolaemus montrouzieri*.

Mishra and Mishra (1998) studied the effect of biopesticides Biotox and neem formulation Neemix and multineem. They observed that the population of predatory coccinellids was least effected by biopesticides compared to Neemix and multineem which gave similar net return as that of malathion treatment.

Singh and Singh (1998) tested different neem based formulations *vis-a-vis* to synthetic insecticides. On aphidophagous coccinellids on *Brassica juncea* to find out suitable and safer insecticide for the predators. Results revealed that the neem products, Achook (WSP), RD-9 Repelin, Neemazal T/S, Neemgold, Neemta 2100, Nimbecidine 0.05 percent, proved safer to coccinellids as the synthetic insecticides, i.e. endosulfan 35EC, fenvalerate 20EC, dimethoate 30EC and Chess 25. The order of safety was found maximum in Achook followed by RD-9 Repelin, Neemazal T/S, Neemgold, Neemta 2100, Nimbecidine 0.03 percent, endosulfan, chess 25, fenvalerate and dimethoate during the first year (1994-95) of experiment and Neemgold followed by Achook, Annona EC, Neemta 2100, Achook EC, Neemazal T/S, endosulfan, Nimbecidine, Chess 25, fenvalerate, dimethoate during second year (1995-96) of experiment.

Nirmala and Balasubramaniam (1999) studied the effects of insecticides and neem based formulations on the predatory spiders of rice-ecosystem. It was observed that feeding efficiency of *L. pseadoannulata* was higher when compared to *T. javana* in all the treatments except in NSKE with green leaf hopper, *Nephotettix virescens* as prey whereas rise in body weight was obtained in both predator species when they were treated with neem products indicating the safety of neem to spiders.

2. Neem in relation to parasitoids

It has been observed that use of chemical insecticides have been the main mode to control insect pests and these are still in use on large scale as preventive and curative measures. These toxic chemicals not only kill our targeted insects but also kill large number of natural enemies viz. parasites, predators and insect pathogens.

It is, therefore, desirable to have insecticides that will selectively control the targeted pests but having no side effects on bioagents.

Neem is one tree whose various parts do have insecticidal properties and in the process, controls many crop pests. Lot of work has been done by various workers to test the side effects of neem on bioagents of the crop pests. Many useful references are available which clearly indicate no or little side effects of neem on bioagents and at the same time effective in controlling noxious pests. The important findings are being listed here.

Saxena *et al*. (1981) sprayed 30 percent neem oil against the rice folder *Cnaphalocrosis medinalis*. The pest larvae were parasitised by Ichneumonid, braconids and encyrtid groups of parasitoids. Surprisingly the parasitization of the leaf folder larvae in neem oil treated plots were double than control. This is due to the fact that the larvae of the host could not spin the leaves together due to high toxicity of neem oil, theryby gave enough opportunity for parasitization. However, the neem oil do not have side effects on parasitoids.

Joshi *at al*. (1982) demonstrated that the application of a 2 percent neem (*Azadirachta indica*) seed kernel suspension to eggs of *Spodoptera litura* (F.) parasitised by *Telenomus remus* Nixon did not adversely affect the emergence of the parasite, nor did it repel ovipositing female parasties. When the treatment was applied to the host eggs before they were parasitised, it significantly accelerated parasite development but shortened its adult life-span. When the treatment was applied to eggs that had already been parasitised, it significantly increased the adult life-span and thus favoured the parasite. *T. remus* is being reared and released in tobacco nurseries in India and these results indicate that the spray can safely be included in integrated programmes for the control of *S. litura* in these nurseries.

The effect of neem oil + acetone + triton X-100 was studied by Tewari and Moorthy (1985) on the degree of parasitization by eulophid parasitoid *Pediobus foreolatus* of the phytophagous pest *Epilachna vigintioctopunctata*. When the pest larvae were exposed to neem oil (0.075 and 0.05%) spray and then exposed to parasitoids, the rate of parasitization was very low but exposure after one day treatment had no reduction in the rate of parasitization.

Li *et al*. (1986) tested the toxicity of many formulations of *Bacillus thuringiensis* (*B.t.*) and neem oil to the egg parasitoid, *Trichogramma japonicum*. It was concluded that both these biopesticides are safest to the parasitoids.

Wu, H.L. *et al*. (1986) tested the toxicity of various insecticides to the lycosid predator *Lycosa pseudoannulata* and the braconid parasitoid *Apanteles cypris* in the laboratory. The LC50 of Shachongsuang [thiosulfuric acid ($H_2S_2O_3$) S,S'-[2-(dimethylamino)-1,3-propanediyl] ester disodium salt], chlordimeform hydrochloride and acephate were 34.5, 113.6 and 447.3 ppm, respectively, to *L. pseudoannulata* and 22.4, 56.8 and 50.4 ppm to *A. cypris*. The mortality of *A. cypris* 2 hr after treatment with the 3 insecticides was 12.3, 93.2 and 16.97, respectively, and 72, 8.7 and 19.5 percent after 5 hrs. Of the insecticides tested, neem seed oil was the safest to the2 natural enemies.

Prashad *et al*. (1987) tested the effect of treatment of chickpea (*Cicer arietinum*) seed with aldrin, chlorpyrifos, neem oil (*A. indica*), monocrotophos and karanj (*Pongamia pinnata*) oil at 15 g/kg seed, and 4, 20, 3 and 20 ml/kg seed on populations of the noctuid *Heliothis* [*Helicoverpa*] *armigera* and its ichneumonid parasitoid *Campoletis chlorideae* in the field in Bihar, India, In 1983. The larval population of *H. armigera* was highest (2.2-2.5 larvae/m row) in plots teated with monocrotophos and karanj oil and untreated plots. Populations of *C. chlorideae* were highest in plots treated with aldrin (1.8 pupae/m row compared to <1 pupa/row in the other plots). Parasitism was 37.0-42.7 percent following treatment with aldrin, chlorpyrifos and neem oil, and 12.4 and 12.0 percent following treatment with monocrotophos and karanj oil.

Velayuddan *et al*. (1988) studied the effect of host species on the sex ratio and parasitism rate of *Anastatus ramakrishnai* in the laboratory and field in Tamil Nadu, India, with the pentatomid *Halys dentata* (on *Cassia marginata, Azadirachta indica* and *Casuarina equisetifolia*) and the coreid *Homoeocerus prominulus* (on *Cassia marginata, Prosopis spicigera* and *Acacia leucophloea*) as hosts, from April 1986 to March, 1987. Development from egg to adult took about 14 days and adult longevity was 11-34 days. With

Halys dentata, an average of 36.56 progeny/female was produced with a male : female ratio of 1:6.67. The intrinsic rate of increase was 0.207/ female per day and the population multiplied 60.45 times in a generation time of 19.82 days. With *H. prominulus*, the male : female ratio was 1:3.56 and the intrinsic rate of increase was 0.178/female/day. The population multiplied 57.93 times in a generation time of 21.98 days. The order of parasitism was *C. marginata* > *A. indica* > *C. equisetifolia* with *H. dentata*, and *C. marginata* > *P. spicigera* > *A. leucophoea* with *H. prominulus*. Higher rates of parasitism were recorded with *H. dentata* than *H. prominulus*, throughout the year.

Lamb and Saxena (1988) gave topical treatment to the females of ectoparasite *Goniozus triangulifer* at doses from 5-50 micron g of a 0.1 micron lt solution of neem seed bitters. The result indicated decreased fecundity at 50 micron g per female. When the rice plants were sprayed with 1000 ppm neem seed bitters, very few larvae of leaf folder *Marasmia patnalis* sustained the development of *Goniozus traingulifer* up to pupation stage. However, when 1000 ppm neem seed bitters were sprayed three times, there appears to be negative influence on parasitization by *G. traingulifer*. The authors further observed that when *Tetrastichus howardi* [*T. ayyari*] parasitized pupae of *M. patnalis* were dipped in 1000 ppm of neem seed bitters, the adult emergence decreased significantly. It was also observed that topical application of 1000-10000 ppm of the neem seed bitters had no effect on *T. howardi*.

Shilke *et al.* (1990) conducted bioassays to determine the toxicity of oil/ extracts to *Chelomus blackburni* Cam. and thereby to explore the possobility of using parasite alongwith oils/extracts in integrated control programmme of potato tuber worm, *Phthorimoea operculella* (Zell). Seven oils/extracts viz. neem oil from seeds of *Azadirachta indica* J., Ratnajoti oil from seeds of *Jatropha cureas* L., Mohwa oil from seeds of *Pongamia glabro* Vent, kagzi lime extract from fruit peel of *Citrus aurantifolia* Swing, Dodi extract from leaves of *Leptaenia reticulata* Weight and Arn and sadabahar extract from leaves of *Ipomea carnea* Jacg. The data were statistically analysed and it was concluded that all the vegetable oils including neem oil are safe to *C. blackbruni*, an egg-larval parasite of *P. operculella*.

Chandra Mohan and Nanjan (1990) observed higher population of the parasitoid *Diadegma semiclausum* than in control after the treatment of neem based product 'Biosol' in cabbage plots. However, Schneider and madel (1992) observed no side effects with residues of fumes of NSKE (0.1-5%) on the adults of *D. semiclausum* after exposure for 3 days or throughout their life time. It is also found that the life span of the parasitoid wasp was prolonged after the NSKE exposure. The female parasitoid of *D. semiclausum* obtained from the neem treated larvae of *Plutella xylostella* did not show reduced fecundity or activity.

Hoelmer *et al.* (1990) did experiments with parasitoids of *B. tabaci* and *Aphis gossypii* with the neem product Margosan-O. It was found that the aphid parasitoids namely *Lysiphlebus testaceipes* and *Aphelinus asychis* were more sensitive to neem treated surface whereas the survival of the aphid parasitoid *Eretmocerus californicus* was same on treated and untreated hibiscus foliage. The *E. californicus* pairs in sealed petridishes in laboratory with treated and untreated foliage survived for 5 days. It was also observed that dipping of aphid mummies parasitized by *L. testaceipes* and also dipping the parasitized puparia of *B. tabaci* by *Encarsia formosa* and *E. transversa*, did not affect the emergence of the parasitoids. However, when *E. californicus* parasitized white fly puparia was dipped, the emergence of parasitoids reduced by more than 50 percent. However, similar observations were recorded by Schauer (1985) who found that neem products i.e. MeO H-NR (0.1%), AZT (0.05%) and MTB (0.01%) + sesam oil were treated on parasitized nymphs and mummies of the aphid *Myzus persicae*. It was observed that the parasitoids *Diaeretiella rapae* and *Ephedrus cerasicola* developed normally.

Stark *et al.* (1990) observed that a highly purified and cocentrated neem extract prevented adults of fruit flies from emerging form puparia. However, the parasitoid *Opius* sp. emerged.

Dutta *et al.* (1992) harvested cocoons of *Antheraea mylitta* from 2 plots of natural forest *Terminalia arjuna, T. tomentosa, Shorea robusta, Lagerostromia* [*Lagerstroemia*] spp. and *Ziziphus mauritiana*) in Bihar, India, during July-August, September-October and January-February. *A. indica* was present in one plot but absent from the other. Parasitism of cocoons by *Blepharipa zebina* was 8-20 percent greater in plots without *A. indica*. Fewer pupae of *B. zebina* were found in soil from plots with *A. indica*.

Schneider and Madel (1992) studied the fecundity and vitality of females of the ichneumonid parasitoid *Diadegma semiclausum* by exposing them in short- and long-term trials to different surfaces treated with an aqueous NSKE. Following the short-term exposure to treated cabbage leaves, parasitoids were kept on untreated cabbage and offered host larvae of *Plutella xylostella* for parasitization. During long-term exposure, ichneumonids were kept over the whole adult lifesapn in 5.0 percent NSKE-treated cages and were allowed to parasitize host larvae periodically on untreated cabbage. None of the treatments showed a significant reduction in parasitization rate or fecundity. The treatment also had no influence on the longevity of the parasitoids. Aqueous NSKE had no adverse effects on *D. semiclausum* following direct contact.

Schmutterer (1992) found in laboratory experiments, concentrations of 10 and 20 ppm of azadirachtin, of an azadirachtin-free fraction and of 50 and 100 ppm of an enriched, formulated seed kernel extract of *A. indica* were only slightly harmful to *Apanteles glomeratus* [*Cotesia glomerata*], provided they were applied against the 5th instar of *Pieris brassicae*. Under these circumstances, numerous larvae of *C. glomerata* emerged from their hosts, pupated and hatched as normal adults. However, higher concentration (40 ppm) of azadirachtin and of the azadirachtin-free fraction as well as 50 and 100 ppm of the enriched product reduced the number of parasitoids considerably. The latter were mainly killed by lack of food and died within their hosts. Larvae of *P. brassicae* under the influence of metamorphosis-disturbing neem products did not die immediately after uptake of active principles, but reduced food uptake, leading to an increase of intraspecific competition among the gregarious larvae of *C. glomerata*. Direct growth-regulation effects of neem products against *C. glomerata* were not observed. Application of neem products against young (1st- 3rd) larval instars of *P. brassicae*, which may be typical under practical conditions, lead to the death of the caterpillars together with the larvae of the parasitoid. However, Osman and bradley (1993) gave an opposite view and explained highh mortality of larvae and morphogenetic defects of adult parasitoids *Apanteles glomeratus* developed from hosts treated withh NSKE. They further expressed that adult parasitoids were not affected by neem products even after spraying with higher concentration i.e. AZT-VR-K, 2000 ppm.

Beitzen Heineke and Hofmann (1992) studied the side effects of neem product AZT-VR-NR on the endoparasitic tachnid fly *Drino inconspicua* when the fly was exposed for 7 days with residues of neem product (-45 g ai/ha), it did not harm the adult flies but fecundity was reduced by 18.5 percent in comparision to control.

Fernandez *et al.* (1992) conducted experiments with the eggs of yellow stem borer of rice *Tryporyza incertulus* by dipping for 30 seconds in Bordan neem products and water. The eggs were exposed for 40 hrs for parasitization by the parasitoid *Telenomus rowani*. The data revealed highest mean number of parasitoids emergence i.e. 65.8 percent in water treatment, 59.90 percent in 5 percent aqueous NSKE, 31.4 percent in 3 percent neem oil and 38.2 percent in Bordan treatment.

Serra (1992) observed no or only side effects of neem products on the parasitoids of the genera *Ganaspidium, Desorygma* and *Opius* emerged from the tomato leaf miner *Lireomyza sativae*. He also obtained simmilar results on the genera *Pseudapanteles* and *Glyptapanteles* emerged form tomato pinworm *Keiferia lycopersicella*.

Stark *et al*. (1992) studied the survival, longevity and reproduction of the three braconid parasites namely *Psystallia incisi* and *Diachasmimorpha longicaudata* [*Biosteres longicaudatus*] form *Bactrocera dorsalis* and *Diachasmimorpha tryoni* from *Ceratitis capitata*. They also studied the effect of azadirachtin concentration to these three parasites. Two experiments were conducted, the test I, in which azadirachtin concentrations were used in a range of low fly emergence (1:0, 1, 5 and 10 ppm azadirachtin) and test II with higher concentrations where no fly emergence would occur (2:20, 50 and 100 ppm azadirachtin). Results of the test I are in confirmatory to Stark *et al*. (1990). All larvae that were exposed to sand treated with azadirachtin, pupated. Adult eclosion was concentration-dependent in both fly species, with little or no fly eclosion at concentration of 10 ppm. However, P. incisi and D. longicaudata successfully enclosed form fliles treated with ≤ 10 ppm azadirachtin.

In all the cases after the expossure of azadirachtin, the adult eclosion is inhibited. Even life spans of parasitoids that emerged form treated flies were not different significantly from controls. The azadirachtin had no effect on the longevity of parasitoid species tested in this study, indicating that the parasitoids were less sensitive to this chemical than were their hosts. The reproduction of *P. incisi* that developed in flies exposed to azadirachtin concentration of >20 ppm was reduced by 63.88 percent. The reproduction of *D. longicaudatus* and *D. tryoni* was unaffected.

Loke *et al*. (1992) observed cocoons of *Apanteles* [*Cotesia*] *plutellae* were topically treated with neem (*A. indica*) oil at 1.25, 2.5, 5.0 and 10.0 percent in the laboratory. Adult eclosion was significantly inhibited by 2.5 percent neem oil and no adult emergence was observed at 10 percent neem oil. Treated cocoons yielded adults with reduced longevities but no morphological deformities were observed.

Klemm and Schmutterer (1993) tested the effects of neem seed kernel water extracts (25 or 30 g seed kernels/lt of water) on the diamond back moth *Plutella xylostella* and its egg parasites *Trichogramma pretiosum* and *T. principium* in the laboratory and in the fields of Taiwan. It was observed that parasitization of host's eggs by the two trichogrammatids was not effected by neem products treatment. In the laboratory, however, neem treated host eggs were less parasitized (17.7 eggs laid per parasitoid) than untreated eggs (33.63 eggs/parasitoid). It was also observed that after two treatments, neem treated *Trichogramma* pupae in host eggs did not develop in to adults. A decrease of 13.33 eggs/female parasitization was observed after treatment of eggs with 0.2 percent neem oil. However, an increase of 7.42/ female was observed when the eggs were treated with AZT-VR-K.

Feldhege and Schmutterer (1993) dipped the parasitized puparia of the whitefly, *Trialeurodes vaporariorum* in Margosan-O solution containing 10 and 20 ppm azadirachtin. It was observed that the lower concentration was relatively non-toxic to the puparia of the parasitoid *Encarsia formosa* where as higher concentration resulted in a slight but significant reduction of activity thereby parasitization capacity was reduced by 60-70 percent.

McCloskey *et al.* (1993) tested the effects of azadirachtin on the parasitoid *Diadegma terebrans* parasite on the European corn borer, *Ostrinia nubilalis*. The azadirachtin 0.1 and 0.3 ppm were used in the diet of second instar larvae of the borer. It was observed that both the concentrations did not have any side efffects. Further it was found that the mortality of unparasitized hosts was higher when treated with azadirachtin than in controls.

Osman and Bradley (1993) applied neem (*A. indica*) seed extract orally and topically to 4th instar larvae of *Pieris brassicae* which also contained larvae of the parasitoid, *Pholetestor* [*Apanteles*] *glomeratus* [*Cotesia glomerata*], and also topically to the cocoons of *C. glomerata*. Heavy parasitoid mortality of *C. glomerata* due to premature host death, suffered increased mortality at all subsequent life stages and deformities in emerged adults. Neem seed extract was capable of having direct negative effects on *C. glomerata* populations. It is suggested that care should be taken in the use of such extracts, and that an assessment of the potential impact on parasitoid populations should take place beforehand.

Lafleur (1994) tested the effects of carbofuran isofenphos and neem on the principal pests of rice and the two parasites of *Orseolia oryzivora* namely *Platygaster* sp. and *Tetrastichus* sp. It was observed that under the combined action of heavy pest pressure and washing neem subsequently proved ineffective. It was also observed that the synthetic insecticides reduce natural parasitism of *O. oryzivora* by the two parasites by 15-30 percent.

Dash *et al.* (1994) observed in field trials in Orissa, in 1989 and 1990, some neem (*A. indica*) derivatives (alone or in combination with synthetic organic insecticides) produced no effective control of the cecidomyiid rice pest *Orseolia oryzae*. Attack by the parasitoid *Platygaster oryzae* was not adversely affected by neem derivatives alone or in combination with monocrotophos or chlorpyriphos [chlorpyrifos]. There was no linear relationship between the percentage of silver (damaged) shoots and the extent of parasitism. Maximum parasitism was observed in an untreated control plot (53.2%), followed by a plot treated with a 3 percent neem oil spray (51%). The percentage parasitism was lower in insecticide-treated plots.

Moser (1994) observed no side effects of aqueous NSE(2.5 and 5.0%) among natural enemies (coccinellids, syrphids, chrysopids and braconids) of *Aphis gossypi* on okra fields in Dominian Republic, but slight harm viz. morphogenetic defects, delay of larval and pupal development was caused in the laboratory.

Lowery and Isman (1996) tested the effects of extracts from neem on aphids and their natural enemies. In field trials, populations of aphid natural enemies (predators and parasitoids) were not affected by application of neem insecticides, suggesting compatibility of neem with biological control agents.

Nand Kumar and Saradamma (1996) observed the activity of natural enemies in cucurbit fields in Kerala, India, where neem-based pesticides are applied for the control of *Henosepilachna* [*Epilachna*] *vigintioctopunctata*. Natural enemies observed in considerable numbers were *Tetrastichus* sp., *Chrysocharis johnsoni, Tetragnatha* sp., *Oxyopes* sp. and orb-web spiders which had no side effects by neem products.

Mani (1996) observed that neem products were safe to parasitoids and predators of citrus insect pests.

Schmutterer (1996) described the varying sensitivity of bioagents to neem products. According to him, eggs of predators such as coccinellids and chrysopids are not sensitive. However, ecoparasitic gregarious larvae of *Bracon* sp., *Colpoclypeus* sp. show high mortality after contact with neem. Endoparasitic solitary or gregarious hymenopterans larvae are less endangered as they are protected by their hosts. Often, lack of food in neem treated hosts resulted in the death of parasitoids due to hunger.

Borah and Basit (1996) tested in laboratory experiments the effects of insecticides on the emergence of *Trichogramma japonicum* from eggs of *Corcyra cephalonica* on the 3rd or 6th days after parasitization were investigated. Chlorpyrifos, quinalphos, monocrotophos, cypermethrin, dimethoate, phosphamidon, fenvalerate, Biolep and Bioasp (both *Bacillus thuringiensis* subsp. *kurstaki*) and Neemazal-F and Fortune Aza (both neem-based) were tested. The *B. thuringiensis* and neem products had the least effect on emergence of parasitoids. Of the other insecticides, fenvalerate and monocrotophos had the least effect while quinalphos had the most. Adult emergence was relatively less when eggs were sprayed on the 6th day after parasitization compared with when they were sprayed on the 3rd day after parasitization.

Barry Lyons *et al.* (1996) tested the effects of azadirachtin based insecticides on the egg parasitoid *Trichogramma minutum* Riley. Effects of azadirachtin formulations on reproduction and survival of the egg parasitoid *T. minutum* were examined to assess their compatibility in integrated pest management programmes. A novel laboratory bioassay was developed for this purpose which could be used as a model ecotoxicological system to assess the potential effects of natural products and other chemicals on a representative beneficial insect parasitoid through its life cycle. Radiation-killed eggs of Mediterranean flour moth, *Ephestia kuehniella* Zeller (Lepidoptera : Pyralidae) were attached to the sticky end of 25 x 6 mm strips of Post-it notes. Each strip with about 200 of these factitious host eggs was treated with 20 ml of an acetone solution. The strip was placed in a shell vial with one *T. minutum* female to parasitize the teated eggs. These eggs were then held until any parasitoids completed development and emerged from the eggs.

Azatin EC (3% aza., AgriDyne Inc., Salt Lake City, USA), "Neem EC" (experimental formulation, 4.6% aza.., Phero Tech, Delta, Canada) and 100 percent azadirachtin were tested at 50 g azadirachtin/ha (standard dosage) and 500 g/ha. At 50 g/ha, no significant effects were observed with these treatments. AT 500 g/ha, female survival after one day was reduced to 64 percent by Azatin and 40 percent by Neem EC compared to controls but no reduction was evident with azadirachtin suggesting that other components of the formulatins were responsible for the toxicity to females. Likewise, at 500 g/ha, the number of eggs parasitized was reduced to 89 percent by Azatin, 29 percent by Neem EC but not reduced by azadirachtin. These reductions in egg parasitism were probably indirectly due to the effects on females. At 500 g/ha, parasitoid development success was reduced by all treatments including azadirachtin (Azatin - 79%, Neem EC - 45%, Azadirachtin - 37%). These results indicate that azadirachtin is compatible with *T. minutum* during egg parasitism at standard dosages.

Bandara and Kudagamage (1996) studied the potential of neem formulations for controlling leaf eating caterpillar complex in cabbage. They also studied the effect of neem formulations and insecticide against parasitoid *Apenteles plutellae* Kurdjumov. It was observed that there was no adverse effect of NSKWE (25g/lt) on the adult of *A. plutellae*. However, Methamidophos and chlorfluazuran drastically reduced their longevity and induced rapid mortality with in 24 hrs after introduction. It was also observed that after the spraying of the NSKWE and the two insecticides on the cocoons, there was no significant reduction in the adult emergence. Thus, NSKWE (25 g/lt) does not have any adverse effect on *A. plutellae*.

Dobelin *et al.* (1996) studied the side effects of Neem Azal-T/S (1.0% azadirachtin) against two parasitoids of aphids viz. *Aphidius colemani* and *Aphidoletes aphidimyza*. It was observed that neem product has no effects on these natural enemies, however, the number of aphids was reduced to nearly zero by the neem product.

Srivastava et al. (1997) reported that alcohol and hexane extracts of 17 neem (*A. indica*) ecotypes in India were found to be toxic to the egg, larval and pupal stages of the cosmopolitan, external gregarious larval parasitoid *Bracon brevicornis*. In general, the hexane ectracts showed higher toxicity against the egg and pupal stages, whereas the alcohol extracts were more toxic against the larvae. Azadirachtin content of the neem ecotypes revealed no apparent correlation with the observed toxicity against different stages of the parasitoid.

According to Michelakis and Vacante (1997), *Phyllocnistis citrella* is now widely distributed throughout the citrus-growing regions of Crete and Greece. A variety of control measures have been adopted over time to reduce populations of *P. citrella* on citrus. Initial control attempts were focused at preventing the spread of *P. citrella* by adopting quarantine measures. Other control measures included using insecticides and cultural control. Presently, there is emphasis on using insecticides which do not harm natural enemies of *P. citrella* (such as Pirate [chlorfenapyr], Agrimek [abamectin], Cascade [flufenoxuron], Neemark [extacts of *Azadirachta indica*], and mineral oils). Indigenous parasitoids of *P. citrella* include *Pnigalia* sp., which is very abundant. A biological control programme has been implemented using *Ageniaspis citricola, Citrostichus phyllcnistoides* and *Semielacher petiolatus*, which were all introduced during 1996.

Stansy and Liu (1997) observed *Encarsia pergandiella* is the most abundant parasitoid of Bemisia argentifolii in south Florida vegetable fields and can contribute significantly to natural biological control of this and other whitefly species. However, quality standards, costs, and risks in commercial tomato production are high, resulting in frequent recourse to chemical control. Therefore, successful utilization of biological control could depend on compatibility of the parasitoid with selective insecticides. A wide range of potentially selective insecticides was tested, represented by a mineral oil, an insecticidal soap, a neem seed extract and synthetic and natural sugar esters, against eggs, 1st- and 3rd instar larvae, pupae and adults, using a pyrethroid, Capture (20 g bifenthrin/lt EC), for comparison. Bifenthrin residues on nymphs exposed to adult parasitoids reduced parasitization more than other materials tested, and were most toxic to all parasitoid stages. Mineral oil caused high mortality of immature parasitoids, and residues reduced parasitization of *E.*

pergandiella. Neem extract, insecticidal soap and both sugar esters tested had little or no effect on *E. pergandiella*. These latter materials could be considered selective in respect to *E. pergandiella* and might be used to suppress *B. argentifolii* without decimating parasitoid populations.

Lakshmi *et al*. (1997) carried out laboratory experiments to study the effect of neem formulations as compared with insecticides on *Trichogramma japonicum*, based on rate of parasitization and emergenc of adults from parasitized eggs. The results revealed that Econeem and Neem Azal-T/S (0.1-1.0%) were safer compared to insecticides, particularly quinalphos and chlorpyrifos which had adverse effects on parasitization. Some neem formulations such as Nimbecidine (0.25-4.0%), Neemgold (2.0-4.0%) and Rakshak (1.0%) also had adverse effects on parasitism.

Sreenivasa and Patil (1998) observed that neem products were fairly safe to *Trichogramma* egg parasitoid.

Chanda and Chakravorty (1998) observed the effects of neem oil nimbecidine (azadirachtin 0.03%), a paralytic toxin of Bracon Hebetor, and of starvation on heart beat of the final instar of *C. cephalonica* were explored. The stress factors affected post-embryonic divelopment. Neem oil feedig in four doses 0.25, 0.50, 0.75 and 1.00 ml per 20 g of jowar-dust food, throughout the larval life of *C. cephalonica* caused 60 percent larval and 42 percent pupal mortality. Deformed adults unable to emerge were also found. The rate of heart beat in *Bracon* toxin injected larvae was significantly reduced and the reduction rate was more than that of neem-fed larvae. The rate, however, showed a sudden rise (109/min compared with 86/min in control) just after toxin injection and declined to 69/min after 12 hrs.

Raguraman and Singh (1998) observed an aqueous suspension and an ethanolic extract of neem seed kernel (NSK) at 0.3, 0.6, 1.2, 2.5 and 5.0 percent concentrations were tested for ovipositinal deterrency, feeding deterrency, toxicity, sterility and insect growth regulatory effects on a larval parasitoid, *Bracon hebetor*. Neither NSK extracts delivered in the food or by contact influenced the *B. hebetor* oviposition (parasitization) in *C. cephalonica*. They also did not cause parasitoid sterility through feeding, but they showed feeding deterrent effects for a limited period. parasitoid eggs and pupae also were unaffected by the extracts tested. The parasitoid larvae, however, were killed by feeding on contaminated host larvae and also through contact with neem extracts. Thus, use of a minimum safety period is suggested for inundative release of *B. hebetor* in integrated pest management.

3. Neem and insect microorganisms

Bacteria

Rao and Rao (1979) observed the effectiveness of spray residues of 0.25-1.0 percent suspensions of Thuricide HP (a preparation of *Bacillus thuringiensis*) and aqueous extracts of the roots or shoots of *Eclipta erecta* (alba) and of the leaves of neem (*Azadirachta indica*) applied to potted rice seedlings for the control of adults of *Nilaparvata lugens* Stal. released on the plants 1 hr after treatment in laboratory tests in India. Mortality counts were made 24 and 48 hrs after release. The data showed that all the Thuricide HP treatments gave 100 percent mortality at both times. All the plant extracts gave high mortality after 48 hrs.

Hellpap and Zebitz (1986) found that during a laboratory study in the German Federal Republic, 4- and 10-day-old larvae of *Spodoptera frugiperda* were fed for 1-3 days on castor leaves treated with either a neem seed kernel extract, Dipel (*B. thuringiensis* subsp. *kurstaki*) or a mixture of both. Fourth-instar larvae of *Aedes togoi* were kept in water treated with a neem seed kernel extract, *B. thuringiensis* subsp. *israelensis* (Teknar) or a mixture of both until adult emergence. The combinations of the extract and the *B. thuringiensis* formulations caused increased mortality as compared with that caused by each alone. The combination also caused a reduction in the pupal weights of the surviving test insects and reduced the time required to kill 50 percent of larvae of *S. frugiperda*. In the tests with *A. togoi*, more pupae than larvae were killed by the combination than by each component alone. The combinations had an additive effect in most cases, but in some a synergistic effect was observed, but according to Tillmann (1992), the use of mixture of NSKE + *B.t.t.* (*B. thuringiensis* subsp. *tenebrionis*) at the sametime was not successful against *Leptinotarsa decemlineata*. However, when *B.t.t.* was applied first and the neem products one day later, the results were observed satisfactory. Moar and Trumble (1987) observed antagonestic effect when NSKE and *B. thuringiensis Kurstaki* were applied against *Spodoptera exigua*.

Joshi, *et al.* (1993) tested relative efficacy of botanicals and biocides alone and in combination against tobacco caterpillar *Spodoptera litura* F. in tobacco nursery. The treatments were 2 percent neem seed kernel suspension (NSKS), *B.t.* (Bactospeine) (16 x 10^3 1 micron/mg, 2 and 1 g/lt), PCW (Pongamia cake water) extract 1:5 (w/v), NPV (nuclear Polyhedraous Virus) 250 LE/ha, NSB (neem seed bitters) 0.25 and 0.12 percent, NSKS 1 percent + *B.t.* 1 g/INSKS 1 percent + NPV 125 LE/ha, NSB (0.12%) + *B.t.* 1 g/lt and untreted beds as control. It was observed that the two botanicals NSKS and PCW extracts gave superior protection than the two biocides *B.t.*

and NPV. The NSKS combined well with both the biocides but mixing the two had no special advantage. Simillar observations were made by Krisch (1986) in Philippines against *Plutella xylostella* L. It was also found that there was no difference between two concentrations of NSB though higher concentraions had simillar effect as NSKS, and combination of NSB with *B.t.* and combination of PCW with *B.t.* Combined analysis revelaed that botanicals alone were more effective than biocides alone. Combined treatments of NSKS with either *B.t.* or NPV were as effective as botanicals alone. However, at seventh day after spraying the two botanicals continues to be more effective than *B.t.* and NPV.

Kroschel and Koch (1996) tested the effect of the pyrethroid fenvalerate and of the growth regulators diflubenzuron, fenoxycarb and teflubenzuron, of botanicals (seed extracts of chinaberry (*Melia azedarach*), neem (*A. indica*), garlic (*Allium staivum*), covering the tubers with leaves of the tree species(*Schinus molle* and *Eucalyptus* sp.) and of *B. thuringiensis* (Bt mixed with fine sand and dusted) to determine their potential to control *Phthorimaea operculella*. In bioassay on tubers, products were tested either when the potaoes were inoculated with eggs of the pest after treatment, or when the larvae were already in the tubers. Fenvalerate, diflubenzuron and Bt prevented the development of the larvae once eggs had hatched. Fenoxycarb stopped developmet to the adult stage, but 50 percent of the eggs developed to the larval and pupal stages and larvae casued damage tothe tubers. The effectiveness or the water extract of neem was 93.8 percent, and even sand without Bt was 94.6 percent effective. The sand's action was attributable to the high proportion of quartz which caused damage to the very sensitive cuticle of newly hatched larvae. With larvae already in tubers, the Bt/sand mixture was extremely effective, with a success rate of 96 percent. However, the effect must be attributed to Bt since sand alone was only 21.6 percent effecitve in reducing the development of adults. In store trials, the effectiveness of fenvalerate and Bt was confirmed. Many of the other products tested including neem also indicated a good potential for control.

Gupta (1996) conducted field trials with upland cotton *Gossypium hirsutum* var. Pusa-31 to evaluate the bioefficacy of neem products and their possible combinatins with synthetic insecticides in a spray schedule for the management of pest complex in cotton viz. *Amrasca devastans* Dist., *Bemisia tabaci* Genn, *Earias Vittella* Fab., *E. insulana* Boisd., *Pectinophora gossypiella* and *Heliothis armigera*. The neem products, Neemrich-I (Margocide OK), Neemrich-II (Margocide CK), Neemguard and Nimbecidine were included in the experiments. On the basis of three years data, it was concluded that the application of neem products alone or in combination with biopesticides (Bt), or conventional synthetic pyrethroids were encouraging in managing pest complex in cotton and increasing good quality yield. The seed cotton yield in such a schedule was more (1905 kg/ha and 2476 kg/ha) as compared to recommended schedule (1861 kg/ha and 2405 kg/ha) during 1991 and 1992, respectively. The findings show that neem products can replace conventional insecticides in a spray schedule thereby minimise the exposure of environment to toxic pesticides and will also postpone the development of resistance. Simillar observations were taken by Schmutterer (1990) who recommended the use of neem products alternatively or mixed with other products to prevent or at least postpone the development of possible resistance.

Basedow *et al.* (1997) conducted field experiments during 1995-96 in potato fields in Germany on the biological control of *Leptinotarsa decemlineata*. Double treatments were carried out in 1995 and a single treatment in 1996. The effects of treatments were measured by the degree of defoliation. Double treatments of 0.5 percent Neem Azal-TS (2 lt/ha) or *B. thuringiensis* subsp. *tenebrionis* (3 lt/ha as Novodor) significantly increased potato yields, while Neem Azal-TS increased crop yields in 1996. The economic output was highest in the Neem Azal-treatment (net gain DM 4450/ha in 1996).

Nagesh and Verma (1997) conducted a field trial during the year 1995 and 1996 in cabbage variety Goldenacre to determine the comparative efficacy of certain eco-friendly pesticides viz. neem, *B. thuringiensis*, diflubenzuron, lufenuron and cartap and synthetic organic insecticides against diamond back moth, *Plutella xylostella*. On considering the overall efficacy, cartap at 0.05 percent concentration was found to be the most effective in controlling DBM. The percent larval reduction over the untreated check was 93.3, 98.2 and 64.8, respectively after 1, 7 and 14 days of each spray. It was closely followed by lufenuron (0.006%), Bt (0.2%) and then Neem Azal-TS (0.002%). However, diflubenzuron (0.025%) was least effective. Hence, sequential spraying with these chemicals with different modes of action can be recommended to solve the problem of development of resistance in the above pests.

Babu and Krischnayya (1998) conducted field studies during the rabi season of 1995-96 in Andhra Pradesh, India, to determine the efficacy of quinalphos at 250 g a.i./ha, cartap at 500 g a.i./ha, neem oil at 1 percent and *B. thuringiensis* subsp. *kurstaki* (Btk) (0.15%), alone and in combination at half the recommended concentration, for the control of *S. litura*, *P. xylostella* and *Crocidolomia binotalis* [*C. parvonana*] on cauliflowers. Individual treatments of cartap and quinalphos and their combinations were highly effective and required fewer sprays (2 to 3) at much longer intervals (12 to 15 days). Toxicity was high resulting in a low mean insect damage score (0.46 to 0.85) and a high marketable curd yield (7058-8305 kg/ha) and input cost benefit ratio (1:3.9-1:6.0). Neem oil, Btk and their combinations with other insecticides were relatively less effective but were superior to the untreated control.

Men and Thakare (1998) conductded the field trials in Maharashtra, India, in kharif 1990-91 and 1991-92, to test efficacy of NPV, Bt and Neemark alone and in cimbination with endosulfan against *Thysanoplusia orichalcea* on sunflowers. Considerably better results were obtained with endosulfan (0.06%), and combination of a sub-lethal dose of endosulfan (0.03%) with a reduced dose of either Bt (250 ml/ha) or NPV (150 LE [Lethal Equivalents]/ ha). Treatments composed of Bt (500 ml/ha), NPV (300 LE/ha) and Neemark (0.25%) with either Bt (250 ml/ha) or NPV (150 LE/ha) also produced good results.

Murugan *et al*. (1998) investigated the combined effect of *B. thuringiensis* subsp. *kurstaki* (Dipel) and certain botanical insecticides on growth, feeding and nutritional efficiency of *Helicoverpa armigera* in the laboratory at 27±1°C, LD 14:10 and 75 percent RH. The antifeedant activity of neem oil and *Vitex negundo* leaf extract was greater than that of *Pongamia* oil. However, growth was severely inhibited by Pongamia oil. Mortality was significantly increased by *Pongamia* oil with Btk at different concentrations considerably inhibited the food consumptin of *H. armigera*. Approximate digestibility was decreased at higher concentraions of Btk. The efficiency of conversion of ingested and digested food was significantly reduced and the profiles of degestive enzymes were decreased by Btk treatment. A significantly greater reduction in growth of rate was observed when Btk was mixed with *Pongamia* oil than with the *V. negundo* extracts or neem oil. The complete inhibition of nutritional efficiency measures and digestive enzymes in the midgut of *H. armigera* suggested that Btk and plant extracts greatly affect the gut homeostasis and suppress the feeding physiology.

Tomar (1998) conducted field studies in the kharif season of 1995 and 1996 at Bilaspur, Madhya Pradesh, India to determine the efficacy of *B. thuringiensis* subsp. *Kurstaki* (Dipel) mixed with lower concentrations of multineem (neem extract) and chemical insecticides for the control of *Earias vittella* infesting okra. The results indicate that Dipel + endosulfan and Dipel + fenvalerate were very effective, reducing the percentage shoot and fruit infestation. The Maximum yield of healthy fruits was obtained using Dipel + fenvalerate, which also gave the highest profit and cost benefit ratio, followed by Dipel + acephate and Dipel + endosulfan. However, Dipel + multineem and Dipel alone were least effective in reducing percentshoot and fruit infestation by number and also by weight. Maximum yield (2168.75 kg/ha) of healthy fruits was obtained from the plot treated with Dipel + endosulfan. The Dipel and multineem combination recorded less yield (1465.58 kg/ha) which is at par with untreated control (999.75 kg/ha).

Gupta and Sharma (1998) tested 8 different spray schedule against bollworms in cotton system during 1993 and 1994 and their impact was studied on the build up of whitefly population. Neem alone or when used alternately with Bt or with conventional synthetic insecticides failed to check bollworms, however, no build-up in whitefly population was noted. Alternate use of conventional insecticides with synthetic pyrethroids proved effective against bollworms but whitefly resurgence was evident. However, spray schedule in which neem was used alternately with *B.t.* and atleast one spray of synthetic pyrethroid proved as effective as previous treatment in managing bollworm complex without resurgence of whitefly. Further studies showed satisfactory control of bollworms and cotton yield response by alternate use of Neem, Bt and 84 percent reduced rate of synthetic pyrethroid. This spray schedule has no adverse effect on environment.

Murli Krischna *et al.* (1998) found in his field study carried out during 1993-94 and 1994-95, that the application of *B. thuringiensis* var. *kursatki* and neem alone was ineffective against the three major insect pests viz., leafhopper (*Amrasca biguttula biguttula* Ishida), whitefly (*Bemisia tabaci* Genn.) and shoot and fruit borer (*Leucinodes orbonalis* Guen.) of brinjal.

Nuclear Polyhedrosis viruses (NPV)

Shapiro *et al.* (1994) observed that neem seed extract inhibited growth and development of gypsy moth, *Lymantria dispar* L. larvae. Untreated control larvae increased their weight by = 4-fold by day 14, whereas insects treated with 0.10 percent neem and 1.00 percent neem weight = 30 percent and = 40 percent, respectively, of the average weight of the untreated larvae. By day 14, 99 percent of the controls were in the fifth stage and 1 percent were prepupae. After treatment with 0.10 percent neem almost one-third of those larvae were still in the fourth stage, whereas larvae treated with 1.00 percent neem were still in the second and third stages. Neem extracts had little, if any, effect upon vital activity as measured by LC50. The addition of neem extract to the gypsy moth NPV, however, resulted in faster virus-caused mortality.

Sarode *et al.* (1996) tested neem seed kernel extract (NSKE) and NPV for the control of *H. armigrea* on cotton in Akola, India, during 1992-93. NSKE was applied at 3, 4, 5 and 6 percent and the virus at 200, 300, 400 and 500 LE/ha. The treatments were applied 3 times at 15-day intervals starting at square formation. The smallest infestaion (2.38% of green fruiting bodies was recorded with 500 LE virus/ha, 7 days after treatment. This was followed by 6 percent NSKE (3.78%) and 400 LE virus/ha (4.19%). Fouteen days after treatment, 400 LE virsu/ha and 6 percent NSKE were the most effective treatments (5.54 and 5.78% infestation, respectively). Virus at 500 and 400 LE/ha and 6 percent NSKE gave the minimum infestations of open bolls (6.91, 7.28 and 7.64%, respectively). The maximum yield was obtained from plots treated with 6 percent NSKE (8.14 qt/ha), followed by 500 and 400 LE virus/ha (7.26 and 7.25 qt/ha, respectively).

Patil and Sarode (1996) conducted field studies in India during 1993-94 to determine the efficacy of combinations of *B. thuringiensis*, neem seed extract, NPV, and half and full concentrations of the recommended insecticide, for their effect on the quality and yield of cotton. Neem seed extract + a half concentration of insecticide gave the highest seed yield. Applying microbial pesticides alone did not give adequate crop protection or produce the desired yield of cotton.

Rajasekhar *et al.* (1996) conducted the laboratory and field studies (in cotton fields in Andhra Pradesh, India) to determine the efficacy of 5 insecticidal additives (sasame oil as synergist, diflubenzuron as insect growth regulator, NPV as biocide, neem oil as repellent and *Parthenium* leaf extract as botanical insecticide) in combination with endosulfan and cypermethrin for the control of *Helocoverpa armigera*. Sesame oil (0.5%) reduced the impact of *H. armigera*. Diflubenzuron, NPV and neem oil enhanced the action of endosulfan and cypermethrin. Treatment with sesame oil as an additive gave the highest seed cotton yield. The efficacy of this treatment was also confirmed in laboratory tests.

Cook *et al.* (1996) placed second-instar larvae of *Lymantria dispar* on semi-synthetic diet and white oak (*Quercus alba*) seedlings that had been surface-treated with azadirachtin and *L. dispar* NPV. Both treatments affected larval development (weight gain and moulting) and survival. When consumed together, larvae died significantly faster compared than larvae that consumed only azadirachtin or virus. The combination also resulted in reduced larval survival compared with only 1 material was consumed. The combination of azadirachtin and virus should result in good foliage protection if used against *L. dispar* larvae. However, the addition of azadirachtin to viral formulations could also result in less virus being produced within the larval cadaver and released into the environment because the affected larvae are smaller.

Rabindra *et al.* (1997) did laboratory experiments and observed that neem bitter (0.1%), an ethanol-soluble concentrate of neem oil cake caused a significantly higher mortality of *Spodoptera litura* larvae when combined with NPV (5 x 10^5 POBs/ml) and crude sugar (1.0%). Without crude sugar, the NPV-neem bitter combination was not effective in increasing mortality. The larval weight and growth rate were significantly reduced in the NPV-neem combination.

Salamouny *et al*. (1997) studied increasing susceptibility to NPV by synergistic additives such as fluorescent brighteners, granulosis virus (GV) enhancin protein and neem (*A. indica*) extract for potential effects on susceptibility of various Lepidoptera to a range of baculoviruses. Susceptibility was increased with the addition of fluorescent brightener to hosts challenged by heterologous rather than homologous viruses, when up to 460-fold increases in potency were demonstrated. Increases in potency of up to 3.4-fold and 2.9-fold were demonstrated with *Trichoplusia ni* GV enhancin and neem extract (Neem Azal-T), resepectively.

Bhanukiran *et al*. (1997) conducted field studies at Bapatala, Andhra Pradesh, India. during the rabi season of 1995-96, to determine the effect of triazophos (0.05%) and methomyl (0.05%) mixed with either diflubenzuron (0.025%). *S. litura* NPV (SL) at 250 LE/ha or neem oil (1.0%), on predatory coccinellids of *S. litura* on groundnuts. High populations of coccinellids were observed in the untreated control, neem and NPV treatments and their combinations, followed by diflubenzuron (0.025%). A mixture of conventional insecticides at half the recommended concentrations with neem or NPV had little effect on predatory coccinellids.

Sarode et al. (1997) conducted field studies in Akola, Maharashtra, India, during 1993-94 to determine the efficacy of combined applilcations of Heliothis NPV (HNPV) and NSKE for the control of *H. armigera* infesting pigeon peas. The results showed that NPV and NSKE were more effective when applied as combinations than when applied alone. HNPV at 500 LE/ha + 5 percent NSKE recorded the highest percentage control (66.50 and 55.73) at 7 and 14 days after spraying, respectively, and the highest grain yield (1110 kg/ha).

Chaudhari and Singh (1997) tested bioefficacy of NPV in combination with a neem formulations against tobacco catetpillar, *S. litura* Fab. The neem formulation Nimba was tested in combination with NPV. It is observed from the data that NPV caused 60 percent mortality when used alone. Neem formulation Nimba (0.1%) when used alone caused 40 percent mortality. However, when both are mixed together, it gives additive effects on mortality (96%). Thus, mixture of neem + NPV had joint effects thereby giving better mortality of larvae than when used alone. The joint action has additional advantage as it will considerably delay the development of possible resistance.

Sonalkar *et al*. (1998) evaluted adjuvants Ranipal, indigo, Sandovit, urea, neem (*A. indica*) seed extract, cotton seed extract and Vitex negundo leaf extract for increasing the virulence of NPV against second instar larvae of *H. armigera*. Ranipal (0.5%0, indigo (0.2%) and urea (1.0%) were effective in increasing the larval mortality and reducing LT50 to 112.9, 124.12 and 149.12 hr, respectively, compared to virus alone (175.25 hr).

Wanjari *et al*. (1998) evaluated 8 treatments comprising of HaNPV, a commercial formulation of Btk (Dipel), neem seed extract and endosulfan either alone or in a combination of two products against *H. armigera* in chickpea in rabi 1994-95 in India. It was shown that all the treatments were effective in reducing larval populations as well as pod damage due to this pest and produced greater grain yields compared to the untreated control. Overall, the treatment with Btk alternated with endosulfan proved to be the best.

Murugan and Jeyabalan (1998) carried out the laboratory trial and observed that NSKE at 2.5 percent enhanced the activity of NPV at 102 polyhedral inclusion bodies/ml and Btk at 50 mg/ml against *H. armigera* on cotton leaves, suppressing overall consumption, fecundity and survival.

Shukla *et al*. (1999) evaluated neem products viz. NSKE, neem oils (NO) and Neemark along with NPV and synthetic insecticides at different concentrations against chickpea pod borer, *Heliothis armigera* Hubner (Noctuidae : Lepidoptera). In 1992-93, the neem products and NPV were at par, for larval population observed per 10 plants ranging from 13.33 to 21.00 larvae, whereas synthetic insecticides were at par with each other with 4.3 (endosulfan, 0.07%) to 6.0 larvae (cypermethrin, 0.002%). Significantly low pod damage was observed in neem products on par with synthetic insecticides.

Entomogenous fungi

Aguda *et al*. (1986) observed that in laboratory trials, 5 percent neem oil reduced the production of the conidia of the entomogenous fungus, *Metarhizium anisopliae*. It is also stated that normally the oil used for the plant protection were at lower concentrations, i.e. 0.5-3.0 percent. Hence, the neem oil may not have adverse effect on the conidia of the fungus.

Wilps *et al*. (1992) investigated in cage experiments, the effects of teflubenzuron/normolt [teflubenzuron], of extracts of *A. indica* and *Melia volkensii* and of spores of *Metarhizium flavoviride* and *Beauveria bassiana* on adults of *Schistocerca gregaria*. Teflubenzuron caused about 70 percent mortality. The other treatments caused 40-100 percent mortality. Of the survivors of the various treatments, flight was reduced by 40-70 percent and the provision of energy for musculature was reduced to 30-50 percent of controls. Jani *et al*. (1992) found that the entomogenous fungi, *Beauveria bronginartii* and *M. anisopliae* var. *anisopliae* tolerate the effect of neem product Neemark.

Peveling and Weyrich (1992) studied the effects of neem (*A. indica*) oil (0.1 and 0.2% azadirachtin), *B. bassiana* (1 x 10^{13} spores/ha) and dieldrin (18%) on the tenebrionid *Pimelia angulata tschadensis* in cage trials in Niger in 1990. Neem oil and *B. bassiana* were also tested in small scale field trials in typical habitats of *S. gregaria*. Dieldrin and 0.1 percent neem oil caused high mortalities in these trials (78 and 56%, resp.). Neem (0.2%) increased and dieldrin decreased the activity of tenebrionids. Infections by *B. bassiana* were not observed in *P. angulata tschadensis*. Only topically inoculated individulas of *Trachyderma* [*Ocnera*] *hispida* were infected by *B. bassiana* in laboratory trials. Field applications did not appear to have any side-effects on the tenebrionid fauna.

Vyas *et al*. (1992) observed in laboratory trials, in vitro application of nicotine sulfate, RD-9 Repelin (extracts of *A. indica*, *Pongamia glabra* [*P. pinnata*] and *Madhuca indica* [*M. longifolia*]) and indiara (diallyl disulfide and allyl propyl disulfide) inhibited the growth of *B. bassiana* and *M. anisopliae*. Phytoallexin (an unspecified herbal extract with a biostimulant) had an even greater inhibitory effect. Neemark (azadirachtin) did not inhibit the growth of either fungus.

Nasseh *et al*. (1992) carried out field trials in cages in Tamesna area of northern Niger, during 1990-91, to test various spray and bait treatments against *S. gregaria*. The growth regulator teflubenzuron (50 g a.i./ha) was compared with a standard dieldrin treatment (20 g a.i./ha) by direct application to adults and nymphs of *S. gregaria* and adults of *Anacridium wernerellum* and by application to the food plant *Schouwia thebaica*. Sprays of pure and enriched neem (*A. indica*) oil (10 lt/ha) and oil formulations of *B. bassiana* (1 x 10^{13}) conidia/kg applied at 320 and 960 g/ha) and baits containing *B. bassiana* and *Nosema locustae* (2.5 x 109/kg bran/ha) were also tested. Teflubenzuron caused less adult mortality than dieldrin but prevented development of 3rd- and 4th-instar nymphs. Neem oil caused no mortality by topical application but had strong antifeedant effects. However, severe burning of *S. thebaica* foliage by neem oil was also observed. None of the treatments involving spraying with *B. bassiana* or baiting with *B. bassiana* or *N. locustae* caused any mortality.

Stathers *et al*. (1993) assessed the viability of freshly harvested conidia of *M. flavoviride* stored at 25°C in peanut (groundnut) oil, cottonseed oil, shellsol, palm oil, codacide, soybean oil, molasses, rape oil, olive oil, liquid paraffin, kerosene, neem oil, cod liver oil or wheatgerm oil over a period of 4 months. Viability was reduced to <40 percent in all but in kerosene and soybean oil only after 4 weeks storage. Kerosene, soybearn and peanut were the only oils that gave conidial viability of >30 percent after 6 weeks. In a second experiment, storage at 5 or 15°C resulted in enhanced conidial viability of up to 71.9 percent with kerosene and 92.1 percent with soybean oil after 14 weeks storage.

Devi and Prashad (1996) tested seed kernel extracts from *A. indica, Melia azedarach* and *P. pinnata*, whole plant extracts from *Tephrosia purpurea, Parthenium hysterophorus* and *Cleome viscosa*, and vegetable oils from sunflowers, safflower, groundnut, rapeseed, sesame, coconut and cottonseed for their compatibility with the entomogenous fungus, *Nomuraea rileyi* applied to larvae of *S. litura*. It is concluded that none of the oils were determintal to the fungus.

Ranaivo *et al.* (1996) studied the effect of low doses of *Metarhizium flavoviride* and neem on walking activity, food consumption and weight of *Locusta migratoria*. Blastospore suspension was applied at 1 x 10^4, 5 x 10^4 and 1 x 10^5 spores/ml (LC10, LC40 and LC50, resp.). Neem oil was applied at 5 and 12.5 percent, corresponding to LC10 and LC50. Differences in walking activity were recorded for nymphs treated with increasing blastospore concentrations. On the 2nd day after treatment, activity increased, but decreased from the 4th day. Walking activity was reduced to 24 and 7 peercent of controls, 10 days after teatment with 5 and 12.5 precent neem oil. Food intake was reduced for all blastospore and neem oil treatments.

Roo *et al.* (1996) tested the efficacy of aqueous suspensions of oilseed cakes of neem (*A. indica*), castor (*Ricinus communis*) and pongamia (*Pongamia pinnata*) to support the growth of *Verticillium lecanii*. Results showed that autoclaved suspensions of neem and castor cake increased the growth of fungal mycelium and sporulation.

Gonzalez *et al.* (1996) tested in an attempt to use the entomopathogenic fungus *M. anisopliae* as part of an integrated pest management program against the coffee pest *Hypothenemus hampei*, the commercial (CD) and half commercial doses (1/2 CD) of the following susbtances to determine their effect on the fungus; the insecticides endosulfan, chlorpyrifos, pirimiphos-methyl, fenitrothion, isazofos and a neem extract: the fungicides benomyl, cyproconazole, hexaconazole, triadimefon and copper oxychloride; the herbicides oxyfluorfen, paraquat, glyphosate and glyphosate + terbuthylazine; the adjuvants Agral, Agrotin, Mezclafix and Mixel; and the foliar fertilizer Tottal. The inhibition measured in percentage at 24 hrs after germination for the CD and 1/2CD was, respectively, for the insecticides; pirimiphos-methyl (98.16 and 57.43%), chlorpyrifos (97.42 and 58.95%), fenitrothion (100 and 100%) and endosulfan (96.94 and 95.11%); for the fungicides : cyproconazole (98.79 and 96.37%), hexaconazole (58.08 and 36.56%) and triamidefon (100 and 95.11%); for the herbicides : oxyfluorfen (100 and 100%) and paraquat (99.06 and 98.80%) and for the adjuvants : Mixel (43.06 and 28.93%). The other susbstances tested had a lower inhibitory effect (<30%). The fungistatic effect of the pesticides on the fungus decreased 48 hrs after germination.

Hazarika and Puzari (1997) conducted a field trial for 4 years (1989-93) in different rice seasons in various areas endemic to rice hispa, *Dicladispa armigera* in Assam, India, to assess the efficacy of a mycopathogen, *B. bassiana* for controlling the insect. Multi-locational field trials with *B. bassiana* (10 million spores/ml dilution), 1 percent neem-seed oil and a conventional insecticide (0.072% monocrotophos) revealed that the mycopesticide was superior to neem-seed oil but at par with monocrotophos in controlling the rice hispa, leading to an increase in yield.

Rodriguez et al. (1997) carried out a tiral for evaluation of *B. bassiana* with aqueous extracts of *A. indica* seeds (Easmar), to observe their compatibility. Doses of Easmar of 0.5, 2.5 and 5.0 percent did not inhibit mycelial growth or spore viability of the entomopathogen.

Bajan *et al.* (1998) reared the isolates of *B. bassiana* from dead larvae of *Achroia grisella* from pine forests in zones of different degrees of SO2 and NOx pollution in the laboratory on pure substrate or substrate contaminated with field equivalent doses of Fastak [alpha-cypermethrin] or BioNEEM (extract of *A. indica*). Addition of Fastak to the substrate considerably inhibited the growth of an isolate from a less contaminated zone, did not affect an isolate from a zone with moderate contamination, but stimulated growth of an isolate from a heavily contaminated zone. Fastak also significantly reduced the pathogenicity of the isolate from the heavily contaminated zone. Addition of BioNEEM caused substantial inhibition of colony growth, regardless of the ecotype. It is concluded that, although contamination of the substrate with BioNEEM or Fastak affects the isolate in different ways and altered their biological properties to some extent, it does not change the role the fungus plays in the habitat. It is also suggested that preparations of BioNEEM or Fastak and appropriate *B. bassiana* isolates can be applied simultaneously in integrated pest control.

Grimm and Guharay (1998) compared and assessed the pathogenicity of three isolates of *B. bassiana* and one isolate of *M. anisopliae* in the laboratory against adults of *Leptoglossus zonatus* and *Pachycoris klugii*, the two most frequent pest species in physic nut (*Jatrophs curcas*) plantations in Nicaragua. In a dipping bioassay, the median lethal concentration (LC50) of the most efficient strain, *M. anisopliae* NB, was determined as 4.34 x 10^6 conidia/ml for adult *P. klugii*. In a field trial, a scheduled high-volume spray regime using *B. bassiana* increased fruit yield by 28 percent, and was more effective than malathion or an aqueous extracct of ground neem seeds. The effectiveness of *M. anisopliae* was further tested in field cages covering entire trees and containing a predetermined number of insects. Mineral oil-based ultra-low volume controlled droplet applications of *M. anisopliae* at a rate of 1 x 10^{10} conidia/tree were made using hand-held Micron ULVA + sprayers. The corrected mortalities ranged from 65 percent in P. klugii to 94 percent in L. zonatus. It is concluded that effect of *B. bassiana* is better than neem products.

Entomophilic nematode

Rovesti and Deseo (1989) exposed infective J3 of *Heterorhabditis bacteriophora* (1H 145 strain), *H. heliothidis*, *Neoaplectana carpocapsae* (195 strain), *N. glaseri* and *N. bibionis* to different concentrations (2% w/v at 1/2, 1/4, 1/8, 1/16 and 1/32) of neem kernel extract. Nematodes were affected at the higher concentraions with *Heterorhabditis* spp. more adversely affected than the *Neoaplectana* spp. It is suggested that entomophilic nematodes and neem should not be applied together in control programmes.

Gill and Raupp (1994) observed bagworm, *Thyridopteryx ephemeraeformis*, damages a wide range of evergreen and deciduous plants in the USA. Trials were conducted in Maryland during 1992-93 to control mid- to late-instar bagworm larvae on arborvitae (*Thuja occidentalis*) grown in large containers. Treatments tested included 2 formulations of neem, carbaryl, acephate, cyfluthrin, formulations of the entomopathogenic nematodes *Steinernema carpocapsae* and *S. feltiae*, and *B. thuringiensis* var. *kurstaki*. In 1992, neem gave the least control (36-56% reduction). The nematodes, either alone or with oil or antidessicant, gave 91-100 percent control of larvae. The synthetic pyrethroid cyfluthrin gave 100 percent control; carbaryl and acephate also gave acceptable control (83 and 86%, respectively). In 1993, cyfluthrin gave the greatest control of both mid- and late-instar larvae (95-97%). The biological control agents, *Bacillus* and the nematodes, provided intermediate levels of control. Both species of *Steinernema* were effective.

Pezowiez *et al*. (1997) observed a high mortality (>50%) of eco-friendly juveniles of *Steinernema feltiae* after 5 days of exposure to neem seed shell (*Azadirachta indica*) extract (5 g/100 ml water), but not after exposure to seed kernel extracts. The botanical inseticide, Neemix (azadirachtin), had a slight adverse effect on the survival of infective juveniles. Bancol 50 WP 0.75 percent (bensultap) resulted in 100 percent mortality after 3 days. The exposure of nematode infective to aqueous extracts did not affect their ability to infect larvae of *Galleria mellonella*, but limited their establishment in the insect. The mean number of established nematodes per *G. mellonella* decreased with increasing Neemix concentrations.

Stark (1997) studied the effects of Margosan-O, a commercial neem-based insecticide, on the entomopathogenic nematode speices *Steinernema carpocapsae* Steiner, *S. feltiae* Filipjev, and *S. glaseri* Steiner. Both acute and chronic toxicity were estimated, as were also effects on nematode infectivity after incubation with the insecticide. Margosan-O was toxic to all 3 species, but only at much higher concentrations than the recommended field rate of 20 mg azadirachtin/lt water. The acute LC50 and 95% FL for *S. carpocapsae* exposed to Margosan-O was 425 mg azadirachtin/lt. Chronic toxicity was not significantly different from acute toxicity until the 15th day of incubation. The acute LC50 and 95 percent FL for *S. feltiae* increased was 380 mg azadirachtin/lt. Chronic toxicity values were not significantly different until day 15. The susceptibility of *S. feltiae* increased greatly between 12 and 15 days of incubation with Margosan-O. The acute LC30 and 95 percent FL for *S. glaseri* was 351 mg azadirachtin/lt. Unlike the other 2 species, no difference was detected in susceptibility among incubation times for *S. glaseri*. When the 3 species were compared, susceptibility to Margosan-O at LC50 at each incubation time interval was not significantly different until day 15. *S. feltiae* was significantly more susceptible than the other species after exposure to Margosan-O for 15 days. Infectivity measured as the ability to kill greater waxmoth larvae, *Galleria mellonella* L., contained in well plates was reduced in all 3 species only when they had been incubated with Margosan-O at concentrations >200 mg azadirachtin/lt. When Margosan-O and nematodes were applied together at field rates to soil, infectivity was not reduced.

References

Aguda, R.M.; Rombach, M.C. and Shepard, B.M. (1986). Effect of neem oil on germination and sporulation of the entomogenous fungus, *Metarhizium anisopliae*. *International Rice Reseach Newsletter*, **11** : 34-35.

Babu, B.R. and Krishnayya, P.V. (1998). Bioinsecticides to mitigate the load of quinalphos against cauliflower caterpillars. *Pesticide Research Journal*, **10**(2) : 231-234.

Babu, G.R.; Rao, G.M. and Rao, P.A. (1998). Efficacy of neem oil and neem cake for the control of green leafhoppers, brown planthoppers and their effect on predators of brown plant hoppers. *Shashpa*, **5**(1) : 91-94.

Bajan, C.; Kmitowa, K. and Popowska-Nowak, E. (1998). Reaction of various ecotypes of entomopathogenic fungus *Beauveria bassiana* to the botanical preparation NEEM\trade mark\and pyrethroid Fastak. *Archives of Phytopathology and Plant Protection*, **31**(4) : 369-375.

Bandara, K.A.N.P. and Kudagamage, C. (1996). Potential of neem for controlling leaf-eating caterpillar complex on cabbage. *Neem and Environment*, **1** : 547-558, Ed. R. P. Singh *et al.*

Banken, J.A.O. and Stark, J.D. (1997). Stage and age influence on the susceptibility of *Coccinella septempunctata* (Coleoptera : Coccinellidae) after direct exposure to Neemix, a neem insecticide. *Journal of Economic Entomology*, **90**(5) : 1102-1105.

Banken, J.A.O. and Stark, J.D. (1998). Multiple routes of pesticide exposure and the risk of pesticides to biological controls : a study of neem and the seven spotted lady beetle (Coleoptera : Coccinellidae). *Journal of Economic Entomology*, **91**(1) : 1-6.

Basedow, T.; Lucknhard, J. and Peters, A. (1997). Results of two experiments on the control of the Colorado potato beetle *Leptinotarsa decemlineata* Say by *Bacillus thuringiensis tenebrionis* and Neem Azal-TS. *Proceedings of the Entomology Conference* in Bayreuth, Germany, 18-22 March, 1997. **11** : 1-6.

Beitzen Heineke, I. and Hofmann, R. (1992). [Versuchezur wirkung des azadirachtin haltigen, formulierten neemsamenextraktes, AZT-VR-NR aug *Hylobius obietis* L., *Lymantria monacha* L. und *Drino inconspicua* Meig.] *Z. Pflkrunkh, Pflschutz*, **99** : 337-348.

Bhanukiran, Y.; Rao, P.A. and Rao, K.T. (1997). Effect of certain chemicals and their combinations on natural enemies of *Spodoptera litura* (F.) in groundnut. *Pesticide Research Journal*, **9**(2) : 238-240.

Borah, M. and Basit, A. (1996). Effect of certain insecticides on the emergence of *Trichogramma japonicum* Ashmead. *Journal of the Agricultural Science Society of North East India*, **9**(2) : 224-225.

Chanda, S. and Chakravorty, S. (1998). Effect of stress on heart beat and post embryonic development in *Corcyra cephalonica* larvae. Indian *Journal of Experimental Biology*, **36**(8) : 796-799.

Chandra Mohan, N. and Nanjan, K. (1990). Effect of neem oil against diamond back moth in cabbage. *Neem Newletter*, **7**(3) : 35-36.

Chandrababu, A.; Gautam, R.D. and Garg, A.K. (1997). Residual toxicity of some insecticides against the predatory beetle, *Brumoides suturalis* (Fabricius). *Annals of Plant Protection Sciences*, **5**(2) : 131-136.

Chaudhari, S. and Singh, R.P. (1997). Bio-efficacy of nuclear polyhedrosis virus (NPV) in combination with a neem formulations against tobacco caterpillar, *Spodoptera litura* (Fab.). *Neem Newsletter*, **14**(1) : 12-14, New Delhi.

Chelliah, S. and Rajendran, R. (1984). Toxicity of insecticides to the predatory mirid bug, *Cyrtorhinus lividipennis* Reuter. *International Rice Research Newsletter*, **9**(4) : 15-16.

Cook, S.P.; Webb, R.E. and Thorpe, K.W. (1996). Potential enhancement of the gypsy moth (Lepidoptera : Lymantriidae) nuclear polyhedrosis virus with the triterpene azadirachtin. *Environmental Entomology*, **25**(5) : 1209-1214.

Dash, A.N.; Senapati, B. and Tripathy, M.K. (1994). Effect of neem derivatives alone and in combination with synthetic insecticides on gall midge and its parasitization. *Environment and Ecology*, **12**(3) : 581-583.

Devi, P.S.V. and Prasad, Y.G. (1996). Compatibility of oils and antifeedants of plant origin with the entomopathogenic fungus *Nomuraea rileyi*. *Journal of Invertebrate Pathology*, **68**(1) : 91-93.

Devi, P.S.V.; Prasad, Y.S. and Rajeswari, B. (1996). Effect of *Bacillus thuringiensis* var. *Kurstaki* and neem on castor defoliators - *Achaea janata* (Linnaeus) and *Spodoptera litura* (Fabricius). *Journal of Biological Control*, **10**(1-2) : 67-71.

Dimetry, N.Z.; Amer, S.A.A. and Momen, F.M. (1994). Laboratory trials of two neem seed extracts on the predatory mites, *Amblyseius barkeri* (Hughes) and *Typhlodromus richteri* Karg. *Bollettino di Zoologia Agraria e di Bachicoltura*, **26**(1) : 127-137.

Dobelin, H.; Kleeberg, H. (ed.) and Zebitz, C.P.W. (1997). Aphid control of cut anemones with Neem-Azal T/S taking into account its effect on beneficials. Practice oriented results on use and production of neem-ingredients and pheromones. *Proceedings 5th Workshop Wetzlar*, Germany, 22-25 January, 1996 : 115-121.

Drescher, K.; Madel, G.; Kleeberg, H. (ed.) and Zebitz, C.P.W. (1997). Effect of Neem Azal-T/S on the anthocorid predator *Orius majusculus* Reuter. Practice oriented results on use and production of neem ingredients and pheromones. *Proceedings 5th Workshop Wetzlar*, Germany, 22-25 January, 1996 : 177-181.

Dutta, R.N.; Patnaik, S.K.; Johri, S.C.; Ramanjaneyulu, Y.V.; Santha, P.C.; ChannaBasavanna, G.P. (ed.); Veeranna, G. (ed.) and Dandin, S.B. (1993). Role of vegetation on uzi fly control in tasar culture. *Proceedings of the National Seminar on Uzi Fly and Its Control*, 16-17 January, 1992 : 247-251.

Eisenlohr, U.; Domange, A.L.; Lenfant, C. and Sauphanor, B. (1992). Effect of NeemAzal-F on aphids and beneficial insects in peach orchards in France. In Kleeberg, H. (ed.). *Proceedings in 1st Workshop Practice - Oriented Results on Use and Production of Neem-Ingredients*, Druck and Graphic, Giessen, Germany : 27-40.

Feldhege, M. and Schmutterer, H. (1993). Investigations on side effects of Margosan-O on *Encarsia formosa* Gah. (Hym., Aphelinidae), parasitoid of the greenhouse whitefly, *Trialeurodes vaporariorum* Westw (Hom., Aleyrodidae). *Journal of Applied Entomology*, **115** : 37-42.

Fernandez, J.J.; Palanginan, E.L.; Soon, L.L. and Bottrell, D.G. (1992). Impact of neem on nontargeted organisms. *Proceedings of Final Workshop on Botanical Pest Control Programme, Phase 2* (IRRI, Los Banos Philippines), : 117-121.

Fernandez, N.J.; Palanginan, E.L.; Soon, L.L. and Bottrell, D.G. (1992). Impact of neem on nontarget organisms. *Proc. Final Workshop Botanical Pest Control Project Phase 2* (IRRI, Los Banos, Philippines) : 117-121.

Froster, P. (1991). EinfluB Von Pflanzenschutzmitteln auf-larven und Adulte Von *Platymus dorsalis* (Pont) (Coleoptera : Carabidae) Und auf Adulte Von *Tachyporus hyphorum* (L.) (Coleoptera : Staphylinidae) in Labor-Halbfreiland Und Freilandversuchen. *Doctor Thesis*, University of Giessen, Germany.

Gill, S.A. and Raupp, M.J. (1994). Using entomopathogenic nematodes and conventional and biorational pesticides for controlling bagworm. *Journal of Arboriculture*, **20**(6) : 318-322.

Gonzalez, D.M.E.; Valbuena, P.B.F.; Rivera, M.A.; Bustillo, P.A.E. and Chaves, B. (1996). Viability of the fungus *Metarhizium anisopliae* in mixture with agrochemicals. *Revista Colombiana de Entomologia*, **22**(1) : 31-36.

Grimm, C. and Guharay, F. (1998). Control of leaf-footed bug *Leptoglossus zonatus* and shield-backed bug *Pachycoris klugii* with entomopathogenic fungi. *Biocontrol Science and Technology*, **8**(3) : 365-376.

Gupta, G.P. (1996). Use of neem in management of pest complex in cotton. *Neem and Environment*, **1** : 469-484.

Gupta, G.P. and Sharma, K. (1998). Neem based pest management strategy in cotton system. *Neem Newsletter*, **15**(1) : 6.

Hazarika, L.K. and Puzari, K.C. (1997). Field efficacy of white muscardine fungus (*Beauveria bassiana*) on rice hispa (*Dicladispa armigera*). *Indian*

Journal of Agricultural Sciences, 67(10) : 463-465.

Hellpap, C. (1985). Populationsokolagie und biologischbiotechnische Bekampfung Von *Spodoptera* spp. in Nicaragua. *Doctor Thesis*, University of Frankfurt and Giessen, Germany.

Hellpap, C. and Zebitz, C.P.W. (1986). Combined application of neem seed extracts with *Bacillus thuringiensis* products in the control of *Spodoptera frugiperda* and *Aedes togoi*. *Journal of Applied Entomology*, **101**(5) : 515-524.

Hermann, P.; Zebitz, C.P.W.; Kienzle, J.; Kleeberg, H. (ed.) and Zebitz, C.P.W. (1997). Effects of different Neem Azal-formulations on larvae of the green lacewing *Chrysoperla carnea* (Steph.) (Neuroptera, Chrysopidae) in laboratory and semi-field. Practice oriented results on use and production of neem ingredients and pheromones. *Proceedings 5th Workshop Wetzlar*, Germany, 22-25 January, 1996 : 183-188.

Hoelmer, K.A.; Osborne, L.S. and Yokomu, R.K. (1990). Effects of neem extracts on beneficial insect in green house culture. *Proceedings USDA Neem Workshop on Neem's Potential in Pest Management Programs* (Beltsville, MD, USA) : 100-105.

Hough Goldstein, J. and Keil, C.B. (1991). Prospects of integrated control of the colorado potato beetle (Coleoptera : Chrysomelidae) using *Perillus bioculatus* (Heteroptera : Pentatomidae) and various pesticides. *Journal of Economic Entomology*, **84** : 1645-1651.

Jani, J.J.; Vyas, R.V. and Yadav, D.N. (1992). Compatibility of some botanical pesticides with muscardine fungi. *National Seminar on Changing Pest Scenario in Pests and Pest Management in India* (Hyderabad, A.P., India), Abstr. p. 9. *Neem Newsletter*, **9**(2) : pp. 17.

Joshi, B.G.; Ramaprasad, G. and Sitaramaiah, S. (1982). Effect of a neem seed kernel suspension on *Telenomus remus*, an egg parasite of *Spodoptera litura*. *Phytoparasitica*, **10**(1) : 61-63.

Joshi, B.G.; Sriprakash, K. and Murthy, P.S.N. (1993). Ralative efficacy of botanicals and biocides alone and in combination against tobacco caterpillar *Spodoptera litura* F. in tobacco nursery. *Botanical Pesticides in Integrated Pest Management* : 154-158.

Kaethner, M. (1990). Wirkung Von Niemsamenprodukten anf die. Reproduktionsfahigkeit und Fitneb Von *Leptinotarsa decemlineata* Say, *Melolontha hippocastani* F. and *Melolontha melolontha* L. Sowie Nebenwirkungen auf die aphidophagen Nutzlinge *Coccinella septempunctata* L. Und *Chrysoperla carnea* (Stephens). *Doctor Thesis*, University of Giessen, Germany.

Kaethner, M. (1991). No side effects of neem extracts on the aphidophagous predators *Chrysoperla carnea* (Steph.) and *Coccinella septempunctata* L. *Anzeiger fur Schadlingskunde, Pflanzenschutz, Umweltschutz*, **64**(5) : 97-99.

Kareem, A.A.; Saxena, R.C. and Malayba, M.T. (1988). Effect of sequential neem (*Azadirachta indica*) treatment on green leafhopper (GLH), rice tungro virus (RTV) infection, and predatory mirid and spiders in rice. *International Rice Research Newsletter*, **13**(6) : 37.

Kareem, R.A.N.M.; Chowdhury, M.M.A. and Hoque, M.N.M. (1992). Current research on neem in rice in Bangladesh. *Proc. Final Workshop Botanical Pest Control Project Phase 2* (IRRI, Los Banos, Philippines) : 30-34.

Klemm, U. and Schmutterer, H. (1993). Effects of neem preparations on *Plutella xylostella* L. and its natural enemies of the genus *Trichogramma. Zeitschrift fur Pflanzenkrankheiten und Pflanzenschutz*, **100**(2) : 113-128.

Krisch, K. (1986). Studies on the efficacy of neem extracts in controlling major insect pests in tobacco and cabbage. *Proceedings of Third International Neem Conference*, Nairobi, Kenya, 1986 : 495-515.

Krishnaiah, N.V. and Kalode, M.B. (1992). Environmental impact studies on neem. *Proceedings of Final Workshop on Botanical Pest Control Project Phase 2* (IRRI Los Banos, Philippines) : 50-53.

Kroschel, J. and Koch, W. (1996). Studies on the use of chemicals, botanicals and *Bacillus thuringiensis* in the management of the potato tuber moth in potato stores. *Crop Protection*, **15**(2) : 197-203.

Kuldeep Singh and N.N. Singh (1998). Effect of different neem formulations on aphidophagous coccinellids. Abst. *ICPPMSA*, 11-13 Dec. at CSAU A & T, Kanpur : 150.

Lafleur, G. (1994). The effects of carbofuran, isofenphos and neem on the principal pests of rice and parasites of the rice cecidomyiid, in Burkina Faso. *Sahel PV Info*., No. **68** : 9-14.

Lakshmi, V.J.; Katti, G. and Krishnaiah, N.V. (1998). Safety of neem formulations and insecticides to *Microvelia douglasi atrolineata* Bergroth (Heteroptera : Veliidae), a predator of plant hoppers in rice ecosystem. *Journal of Biological Control*, **11**(1-2) : 33-36.

Lakshmi, V.J.; Katti, G.; Krishnaiah, N.V. and Lingaiah, T. (1998). Laboratory evaluation of commercial neem formulations vis-a-vis insecticides against egg parasitoid, *Trichogramma japonicum* Ashmead (Hymenoptera : Trichogrammatidae). *Journal of Biological Control*, **11**(1-2) :

29-32.

Lamb, R. and Saxena, R.C. (1988). Effects of neem seed derivatives on rice leaf folders (Lepidoptera : Pyralidae) and their natural enemies. *Proc. Final Workshop, IARI, ADB-EWC, Proj. on Botanical Pest Control in Rice Based Cropping Systems* (IRRI, LosBanos, Philippines) : pp. 47.

Li, K.H.; Xu, X.Li, Y.F.; Meng, V.Z. and Zhou, L.C. (1986). Determination of toxicity of 29 chemicals to *Trichogramma japonicum* at various developmental stages [In Chinese]. *Natural enemies of Insects*, **8** : 187-194, 199.

Loke, W.H.; Heng, C.K.; Basirun, N.; Mardi, A.R.; Ooi, P.A.C. (ed.); Lim, G.S. (ed.) and Teng, P.S. (1992). Non-target effects of neem (*Azadirachta indica* A. Juss) on *Apanteles plutellae* Kurdj., cabbage, sawi and padi. *Proceedings of the 3rd International Conference on Plant Protection in the Tropics*, **2** : 108-110.

Lowery, D.T. and Isman, M.B. (1994). Effects of neem and azadirachtin on aphids and their natural enemies. [In P.A. Hedin (ed) Bioregulators for Crop Protectioin and Pest Control, American Chem. Society Symposium Seires-557] : 78-91.

Lowery, D.T. and Isman, M.B. (1996). Effects of extracts from neem on aphids (Homoptera : Aphididae) and their natural enemies. *Neem and Environment*, **1** : 253-264. Ed. R. P. Singh *et al*.

Lyons, B.; Helson, B. and Bourchier, R. (1996). Effects of azadirachtin-based insecticides on the egg parasitoid, *Trichogramma minutum* Riley [Hymenoptera : Trichogrammatidae]. *Neem Newsletter*, **13**(4) : 50-51.

Mani, M. (1996). Safety of neem and other plant products to parasitoids and predators of citrus insect pests. *Journal of Insect Science*, **9**(1) : 89-90.

Mani, M.; Lakshmi, V.J. and Krishnamoorthy, A. (1997). Side effects of some pesticides on the adult longevity, progeny production and prey comsumption of *Cryptolaemus montrouzieri* Mulsant (Coccinellidae, Coleoptera). *Indian Journal of Plant Protection*, **25**(1) : 48-51.

Mansour, F.A.; Ascher, K.R.S. and Abo Moch, F. (1997). Effects of Neemguard on phytophagous and predacious mites and on spiders. *Phytoparasitica*, **25**(4) : 333-336.

Mansour, F.A.; Ascher, K.R.S. and Omari, N. (1986). Effect of neem seed kernel extracts from different solvents on the predacious mite, *Phytoseiulus persimilis* and the polyphagous mite, *Tetranychus cinnabarinus* as well as on predatory spider, *Chiracanthium mildei*. In Schmutterer, H.; Ascher, K.R.S.; Rambold, H. (eds) : Natural Pesticides From the Neem Tree (*Azadirachta indica* A. Juss). *Proc. 3rd International Neem Conference*, Nairobi, Kenya, 1986 : 577-587.

Mansour, F.A.; Ascher, K.R.S. and Omari, N. (1986). Toxicity of neem (*Azadirachta indica*) seed kernel extracts prepared with different solvents, on the spider *Cheiracanthium mildei*. *Phytoparasitica*, **14**(1) : 73-76.

Mansour, F.A.; Ascher, K.R.S.; Omari, N. and Schmutterer, H. (ed.) (1987). Effect of neem seed kernel extracts from different solvents on the predacious mite *Phytoseiulus persimilis* and the phytophagous mite *Tetranychus cinnabarinus* as well as on the predatory spider *Cheiracanthium mildei*. *Proceedings of the 3rd Internatinal Neem Conference*, Nairobi, Kenya, 10-15 July, 1986 : 577-587.

Matter, M.M.; Marei, S.S.; Moawad, S.M. and El Gengaihi,S. (1993). The reaction of *Aphis gossypii* and its predator, *Coccinella undecimpunctata* to some plant extracts. *Bulletin of Faculty of Agriculture*, University of Cairo, **44**(2) : 417-432.

Mecloskey, C.; Arnason, J.T.; Donskov, N.; Chenier, R.; Kaminski, J. and Philogene, B.J.R. (1993). Third tropic level effects of azadirachtin. *Canadian Entomologist*, **125** : 163-165.

Men, U.B. and Thakare, H.S. (1998). Efficacy of biocides alone and in combination with insecticides against sunflower semilooper, *Thysanoplusia orichalcea* F. *PKV Research Journal*, **22**(1) : 57-59.

Michelakis, S.E. and Vacante, V. (1997). The citrus leafminer status in Greece. *Bulletin OILB-SROP, 1997*, **20**(7) : 81-82.

Mishra, N.C. and Mishra, S.N. (1998). Impact of biopesticides on the useful insect pests and defenders of okra. *Neem Newsletter*, **15**(2) : pp. 18.

Moar, W.J. and Trumble, J.T. (1987). Toxicity joint action and mean time of mortality of Dipel QX, avermectin BI, neem and thuringiensin against beet army worm (Lepidoptera : Noctuidae) *Journal of Economic Enotmology*, **80** : 588-592.

Mohapatra, R.N.; Patnaik, N.C.; Mishra, B.K. and Patnaik, H.P. (1991). Comparative toxicity of some insecticides and neem oil against the brown planthopper, *Nilaparvata lugens* (Staal) and its predators. *Orissa Journal of Agricultural Research*, **4**(3-4) : 239-240.

Momen, F.M.; Reda, A.S. and Amer, S.A.A. (1997). Effect of Neem Azal-F on *Tetranychus urticae* and three predacious mites of the family Phytoseiidae. *Acta Phytopathologica et Entomologica Hungarica*, **32**(3-4) : 355-362.

Mosar, G. (1994). [Nutzlingskomplex um Aphis gossypii Versuchemd Niems menkernwasser extrakten, an Beispeel von okra.] In der Dominikanischen Republik, *Diploma Thesis*, University of Giessen, Germany.

Murli Krishnan, T.; Lal, O.P.; Singh, V.S.; Dhingra, S. and Srivastava, Y.N.S. (1998). Effectiveness of *Bacillus thuringiensis* var. Kurstaki and neem alone as well as in combinatin with other insecticides against three major insect pests of egg plant. Abstr. *ICPPMSA*, 11-13 Dece., 1998, CSAUA and T, Kanpur : 268-269.

Murugan, K. and Jeyabalan, D. (1998). Neem enhances the activity of microbial pesticides. *Insect Environment*, **4**(1) : 3-4.

Murugan, K.; Babu, R.; Pandey, A. (ed.); Soccol, C.R. (ed.) and Josh, V.K. (1998). Impact of certain plant products and *Bacillus thuringiensis* Berliner subspecies *Kurstaki* on the growth and feeding physiology of *Helicoverpa armigera* Hubner (Insecta : Lepidoptera : Noctuidae). *Journal of Scientific and Industrial Research*, **57**(10-11) : 757-765.

Nagesh, M. and Verma, S. (1997). Bioefficacy of certain insecticides against diamondback moth (*Plutella xylostella*) on cabbage. *Indian Journal of Entomology*, **59**(4) : 411-414.

Nanda, U.K.; Parija, B.; Pradhan, N.C.; Nanda, B. and Dash, D.D. (1996). Bioefficay of neem derivatives against the insect pest complex of rice. *Neem and Environment*, **1** : 517-528.

Nandakumar, C. and Saradamma, K. (1996). Botanical pesticides are benign to the natural enemies of epilachna beetle. *Insect Environment*, **2**(1) : 27.

Nasseh, O.M.; Freres, T.; Wilps, J.; Kirkilionis, E. and Krall, S. (1992). Field cage trials on the effects of enriched neem oil, insect growth regulators and the pathogens *Beauveria bassiana* and *Nosema locustae* on desert locusts in the Republic of Niger. *Biological Control of Locusts and Grasshoppers : Proceedings of a Workshop* held at the International Institute of Tropical Agriculture, Cotonou, Republic of Benin, 29 April-1 May, 1991 : 311-320.

Nirmala, R. and Balasubramaniam, G. (1999). Residual toxicity effects of insecticides and neem based formulations on the predatory spiders of rice-ecosystem. *Neem Newsletter*, **16**(4) : 37-38.

Osman, M.Z. and Bradley, J. (1993). Effects of neem seed extracts on *Pholeastor [Apanteles] glomeratus* L. (Hym., Braconidae), a parasitoid of *Pieris brassicae* L. (Lep., Pieridae). *Journal of Applied Entomology*, **115**(3) : 259-265.

Patil, P.P. and Sarode, S.V. (1996). Impact of bio-rational insecticides combined with chemical insecticides on quality and yield of cotton. *Indian Journal of Entomology*, **58**(2) : 175-181.

Peveling, R. and Weyrich, J. (1992). Effects of neem oil, *Beauveria bassiana* and dieldrin on non-target tenebrionid beetles in the desert zone of the Republic of Niger. *Biological Control of Locusts and Grasshoppers : Proceedings of a Workshop* held at the International Institute of Tropical Agriculture, Cotonou, Republic of Benin, 29 April-1 May, 1991 : 321-336.

Pezowicz, E.; Ignatowicz, S. and Kamionek, M. (1997). Comparative effect of water extracts of seeds of the Indian neem tree, *Azardirachta indica* A. Juss, a plant derived insecticide, Neemix TM and a synthetic insecticides, Bancol 50 WP, on the efficacy of the entomongenous nematode, *Steinernema feltiae* (Filipjev). *Annals of Warsaw Agricultural University SGGW, Horticulture*, **18** : 31-39.

Prasad, D.; Chand, P. and Haque, M.F. (1987). Parasitization of *Heliothis armigera* (Hub.) in chickpea field as affected by pesticidal seed treatment. *Research and Development Reporter*, **4**(2) : 236-238.

Prasad, G.S. and Logiswaran, G. (1998). Toxicity of different insecticides to the adult of *Cheilomenes sexmaculata* F. Coccinellidae : Coleoptera). *Insect Environment*, **3**(4) : 117-118.

Rabindra, R.J.; Rajasekaran, B. and Jayaraj, S. (1997). Combined action of nuclear polyhedrosis virus and neem bitter against *Spodoptera litura* (Fabricius) larvae. *Journal of Biological Control*, **11**(1-2) : 5-9.

Raguraman, S. and Singh, R.P. (1998). Behavioural and physiological effects of neem (*Azadirachta indica*) seed kernel extracts on larval parasitoid, *Bracon hebetor*. *Journal of Chemical Ecology*, **24**(7) : 1241-1250.

Rajasekhar, P.; Rao, N.V. and Venkataiah, M. (1996). Utility of insecticide additives for managing *Helicoverpa armigera* on cotton. *Pesticide Research Journal*, **8**(2) : 119-123.

Ranaivo, F.; Welling, M.; Zimmermann, G. and Schmutterer, H. (1996). Fitness reduction by tee African migratory locust, *Locusta migratoria*, after application of low concentrations of *Metarhizium flavoviride* blastospores and neem oil. *Insect Pathogens and Insect Parasitic Nematodes. Proceedings of the first Joint Meeting Bulletin-OILB-SROP*, **19**(9) : 236-239.

Rao, M.S.; Reddy, P.P. and Chawla, G. (1996). Potential use of botanicals for the development of formulation of the bio-control fungus *Verticillium lecanii* (Zimm.) Viegas. *Zeitschrift-fur-Pflanzenkrankheiten-und-Pflnzenschutz*, **103**(4) : 436-439.

Rao, P.R.M. and Rao, R.S.P. (1979). Note on the use of *Bacillus thuringiensis* and extracts of *Eclipta alba* (Linn.) Hassak. and *Azadirachta indica* A. Juss. for the control of rice brown planthopper. *Indian Journal of Agricultural Sciences*, **49**(11) : 905-906.

Rodriguez, A.-L.D.; Lagunes Tejeda, A.; Riestra Diaz, D.; Rodriguez Maciel, C.; Velazquez Mendoza, J.; Becerril Roman, E. Rodriguez Colorado, S. and Pacheco Velasco, E. (1997). Compatibility of *Beauveria bassiana* and aqueous extracts of neem (*Azadirachta indica*) to control coffee borer (*Hypothenemus hampei*). *Manejo Integrado de Plagas*, **44** : 14-19.

Roger, C.; Vincent, C. and Coderre, D. (1995). Mortality and predation efficiency of *Coleomegilla maculata lengi* Timb. (Coleoptera : Coccinellidae) following application of neem extracts (*Azadirachta indica* A. Juss., Meliaceae). *Jouranl of Applied Entomology*, **119**(6) : 439-443.

Rovesti, L. and Deseo, K.V. (1989). Effect of neem [*Azadirachta indica*] kernel extract on steinernematid and heterorhabditid nematodes. *Nematologica*, **35**(4) : 493-496.

Salamouny, E.I.; Huber, J.; E.I Nagar S., and E.I. Sheikh, M.A.K. (1997). Increasing the susceptibility to nuclear polyhedrosis viruses by synergistic additives. Microbial Insecticides, *Novelty or Necessity? Proceedings of a Symposium* held at the University of Warwick, Coventry, UK, 16-18 April, 1997 : 289-292.

Saleem, S.A. and Matter, M.M. (1991). Relative effects of neem seed oil and Deenate on the cotton leaf worm, *Spodoptera littoralis* Boisd. and the most prevalent predators in cotton fields in Menoufyia Governorate. *Bulletin Faculty of Agricultural University*, Cairo, 42 : 941-952.

Sarode, S.V.; Jumade, Y.S. and Borkar, S.L. (1997).Efficacy of Heliothis nuclear polyhedrosis virus and neem seed kernel extract combinations against *Helicoverpa armigera* (HB) on pigeonpea. *PKV Research Journal*, **21**(2) : 227-229.

Sarode, S.V.; Jumde, Y.S.; Deotale, R.O. and Thakare, H.S. (1996). Bioefficacy of neem seed kernel extract (NSKE) and Heliothis nuclear polyhedrosis virus (HNPV) against *Helicoverpa* [*Heliothis*] *armigera* (Hb) on cotton. *PKV Research Journal*, **20**(1) : 96-97.

Saxena, R.C. (1987). Neem seed derivatives for management of rice insect pests - a review of recent studies. In Schmutterer, H.; Ascher, K.R.S. (eds) : Natural Pesticides From the Neem Tree and Other Tropical Plants. *Proc. 3rd International Neem Conference*, Nairobi, Kenya, 1986 : 81-93.

Saxena, R.C.; Epino, P.B.; Chang-Wen, T. and Puma, B.C. (1984). Neem, China berry and custard apple : antifeedant and insecticidal effects of seed oils on leaf hoppers and plant hoppers, pest of rice. In Schmutterer,

H.; Ascher, K.R.S. (eds) : Natural Pesticides From the Neem Tree and Other Tropical Plants. *Proc. 3rd International Neem Conference* (Raueschholzhausen, Germany, 1983) : 403-412.

Saxena, R.C.; Waldbouer, G.P.; Liguido, N.J. and Puma, B.C. (1981). Effects of neem seed oil on the rice leaf folder *Cnaphalocrosis medinalis.* In Schmutterer, H.; Ascher, K.R.S. and Rambold, H. (eds.) - **Natural Pesticides from the Neem Tree (*Azadirachta indica* A. Juss).** *Proc. First Int. Neem Conf.* (Rottach Egern Germany, 1980) : 198-203.

Schauer, M. (1985). [Die wirkung von nieminhalt sstoffen auf, Blattlause und die Rubenblattwanze]. Doctor, Thesis, Univ. of Giessen, Germany.

Schmidt, G. and Pesel, E. (1987). Studies on the sterilizing effect of neem extracts in ants. In Schmutterer, H.; Ascher, K.R.S. (eds) : Natural Pesticides From the Neem Tree and Other Tropical Plants. *Proc. 3rd International Neem Conference*, Nairobi, Kenya, 1986 : 361-373.

Schmutterer, H. (1981). Ten years of neem research in the Federal Republic of Germany. In Schmutterer, H.; Ascher, K.R.S. (eds) : Natural Pesticides From the Neem Tree (*Azadirachta indica* A. Juss). *Proc. 3rd International Neem Conference* (Rottach Egern, Germany, 1980) : 21-31.

Schmutterer, H. (1990). Properties and potential of natural pesticides from the neem tree *Azadirachta indica. Annual Review of Entomology*, **35** : 271-297.

Schmutterer, H. (1992). Influence of azadirachtin, of an azadirachtin-free fraction of an alcoholic neem seed kernel extract and of formulated extracts on pupation, adult emergence and adults of the braconid, *Apanteles glomeratus. Journal of Applied Entomology*, **113**(1) : 79-87.

Schmutterer, H. (1996). Side effects of neem products on insect pathogens, predators and parasitoids. *Neem Newsletter*, **13**(4) : 54-55.

Schmutterer, H. (1997). Side-effects of neem (*Azadirachta indica*) products on insect pathogens and natural enemies of spider mites and insects. *Journal of Applied Entomology*, **121**(2) : 121-128.

Schneider, A. and Madel, G. (1992). Fecundity and vitality of adult parasitoids following exposure to neem (*Azadirachta indica*) treated surfaces. [Fekunditat und vitalitat adulter schlupfwespen nach exposition auf Niem (*Azadirachta indica*) behandelten Flachen]. *Mitteilungen der Deutschen Gesellschaft fur Allgemeine und Angewandte Entomologie*, **8**(1-3) : 273-278.

Serra, A.C. (1992). [Untersuchungen zum einsatz von Niemsamenextracten,

Im Rahmen Integrierter Ansatze zur Bekampfung von tomatenschadlingen in der Dominikanischen Republik]. *Doctor Thesis*, Univ. of Giessen Germany.

Shapiro, M.; Jacqueline, L.; Robertson, and Ralph, E.W. (1994). Effect of neem seed extract useful upon the gypsy moth (Lepidoptera : Lymantriidae) and its nuclear polyhedrosis virus. *Neem Newsletter*, **11**(2) : 28-29.

Shelke, S.S.; Jadhav, L.D. and Salunkhe, G.N. (1990). Toxicity of some vegetable oils/extracts to *Chelonus blackburni* Cam., an egg-larval parasite of *Phthorimaea operculella* (Zell.). *Indian Journal of Entomology*, **52**(4) : 709-711.

Shukla, B.C.; Kaushik, U.K. and Gupta, R. (1998). Efficacy of neem products for useful management of chickpea pod borer. Neem Newletter, **15**(2) : pp.19.

Singh, Kuldeep and Singh, N.N. (1998). Effect of different neem formulations on aphidophagous coccinellids. Abst. *ICPPMSA*, 150, 11-13 Dece., CSAUA & T, Kanpur.

Sonalkar, V.U.; Deshmukh, S.D.; Satpute, U.S. and Ingle, S.T. (1998). Efficacy of nuclear polyhedrosis virus in combination with adjuvants against *Helicoverpa armigera* (HBN). *Journal of Soils and Crops*, **8**(1) : 67-69.

Sontakke, B.K. (1993). Field efficacy of insecticide alone and in combination with neem oil against insect pests and their predators in rice. *Indian Journal of Entomology*, **55**(3) : 260-266.

Sreenivasa, A.G. and Patil, B.V. (1998). Role of commercial neem products in cotton insect pest management. *Neem Newsletter*, **15**(2) : pp. 17.

Srinivas, S.D.; Jeyasuresh, B.; Edwin, S. and Dwarakanath, S.K. (1994). The behavioural responses of *Gambusia affinis* in relation to leaf extracts of *Azadirachta indica*. *Journal of Ecotoxicology and Environmental Monitoring*, **4**(1) :75-76.

Srivastava, M.; Paul, A.V.N.; Rengasamy, S.; Kumar, J. and Parmar, B.S. (1997). Effect of neem (*Azadirachta indica* A. Juss) seed kernel extracts on the larval parasitoid *Bracon brevicornis* Wesm. (Hym., Braconidae). *Journal of Applied Entomology*, **121**(1) : 51-57.

Stansly, P.A. and Liu, T.X. (1997). Selectivity of insecticides to *Encarsia pergandiella* (Hymenoptera : Aphelinidae), an endoparasitoid of *Bemisia argentifolii* (Hemiptera : Aleyrodidae). *Bulletin of Entomological Research*, **87**(5) : 525-531.

Stark, J.D. and Wennergren, U. (1995). Can population effects of pesticides be predicted from demographic toxicological studies. *Journal of Economic Entomology*, **88** : 1089-1096.

Stark, J.D.; Vargas, R.I. and Thalman, R.K. (1990). Azadirachtin effects on metamorphosis, longevity and reproduction of three tephritid fruitfly species (Diptera : Tephritidae). *Journal of Economic Entomology*, **83**(2) : 2168-2174.

Stark, J.D.; Vargas, R.I. and Wong, T.Y. (1990). Effects of neem seed extracts on tephritid fruiflies (Diptera : Tephritidae) and their parasitoids in Hawaii. In Lock, Jc, Lawson, R.H. (eds.), **Neem Potential in Pest Management Programme**, U.S. Deptt. of Agri. Res. Serv., ARS-**86** : 106-111.

Stark, J.D.; Wong, T.T.Y.; Vargas, R.I. and Thalman, R.K. (1992). Survival, longevity and reproduction of tephritid fruitfly parasitoids (Hymenoptera : Braconidae) reared from fruit flies exposed to azadirachtin. *Journal of Economic Entomology*, **84**(4) : 1125-1129.

Stark, John D. (1997). Entomopathogenic nematodes (Rhabditida : Steinernematidae) : Toxicity of neem. *Neem Newsletter*, **14**(1) : 18.

Stathers, T.E.; Moore, D. and Prior, C. (1993). The effect of different temperatures on the viability of *Metarhizium flavoviride* conidia stored in vegetable and mineral oils. *Journal of Invertebrate Pathology*, **62**(2) : 111-115.

Sundari, M.S.N. (1998). Residual toxicity of insecticides against *Cryptolaemus montrouzieri* Mulsant (Coleoptera : Coccinellidae). *Insect-Environment*, **4**(2) : 45.

Tewari, G.L. and Moorthy, K.P.N. (1985). Plant extracts as antifeedant against *Hemosepilachna vigintioctopunctata* (F.) and their effects on their parasites. *Indian Journal of Agricultural Science*, **55** : 120-124.

Tillmann, M. (1992). [Kombincerte Anwendung der praparte *Bacillus thuringiensis* var. *tenebrionis* und Neem-Azal-S gegen laren von *Leptinotarsa decemlineata* Say (Coleopetra : Chrysomelidae) und *Epilachna varivestis* Muls. (Coleoptera : Coccinellidae)]. *Diploma Thesis*, University of Giessen, Germany.

Tomar, R.K.S. (1998). Efficacy and economics of biopesticide and insecticide combinations against okra shoot and fruit borer. *Indian Journal of Entomology*, **60**(1) : 25-28.

Velayudhan, R.; Senrayan, R. and Rajadurai, S. (1988). Parasitoid - host interactions with reference to *Anastatus ramakrishnae* (Mani) (Hymenoptera : Eupelmidae) in relation to pentatomid and coreid hosts. *Proceedings of the Indian National Science Academy, Biological Sciences*, **54**(2-3) : 145-153.

Vyas, R.V.; Jani, J.J. and Yadav, D.N. (1992). Effects of some natural pesticides on entomogenous muscardine fungi. *Indian Journal of Experimental Biology*, **30**(5) : 435-436.

Wanjari, R.R.; More, G.D.; Supare, N.R.; Turkar, K.S. and Agarkar, V.K. (1998). Management of *Helicoverpa armigera* (Hub.) on chickpea with some herbal, chemical and bio-pesticides. *Journal of Soils and Crops*, **8**(1) : 34-37.

Wilps, H.; Nasseh, O.; Krall, S.; Kabo, O. and Nasseh, O.M. (1992). Control of *Schistocerca gregaria* adults in the area of new preparations and methods. *Sahel PV Info.*, **44** : 7-18.

Wu, H.L. (1986). Determination of toxicity of some insecticides to *Lycosa pseudoannulata* and *Apanteles cypris*. *Natural Enemies of Insects*, **8**(4) : 230-231.

Yadav, D.N. and Patel, A.R. (1990). Effect of some botanical insecticides on oviposition of *Chrysopa scelestes* and their ovicidal action. *Proceedings Symposium on Botanical Pesticides in IPM*, Rajahmundry : 166-169.

Chapter 11

Neem in Integrated Pest Management

The concept of integrated pest management (IPM) is very encouraging as it provides an appropriate alternative control measures to the pest problems which have been caused due to widespread introduction of monogenic resistant cultivars, killing of natural enemies, resistance problem among insects to insecticides, residue hazards of pesticides, upsetting of pest balance in nature and also due to abandonment of cultural controls. The need to phase out the use of chemical pesticides is now taking momentum and the concept of integrated pest management has now been broadened to include use of plant products, biological control agents, cultural methods, mechanical methods, sterile male techniques, use of chemosterilants and sex attractants etc. Among plant products, neem, tobacco and pyrethrum were under use and the simple plant part extracts or aqueous extracts were generally applied on our crops to control pests.

The pesticidal quality of neem was known to human beings for centuries ago and in those days, the dried neem leaves were used to protect our food grains in stores and to protect our cloths. However, due to rapid development in the field of research, these methods are seldom used by the users and they easily switched on to the use of chemical pesticides. These pesticides indue course of time developed side effects and thus the use of neem products again gain momentum in IPM strategy. The neem seed kenrel and cake powder are now in use to control several pests including stem borer pests. The seed oil is very effective against a large number of sucking insects and others. The aqueous extract of neem seed kernel powder is under use to control many pests including locusts. Leaf extracts and leaf powder do have several triterpenoids having insecticidal properties. The insecticidal compounds of neem gained importance only during the last two decades when various neem extracts viz. alcoholic, ether, acetone extracts etc. were prepared having repellent, antifeedant, growth disruptant, insecticidal and sterilizing properties against insects of economic importance.

The overall idea of IPM is to either minimize the use of chemical pesticides for pest control strategy or to develop zero chemical pesticides strategy.

Neem products thus have the potential to take care of both aspects. The research work conducted with mixing of sublethal doses of chemical pesticides and neem formulations gave in many cases synergistic effect in controlling pests. This, therefore, would be an additional advantage for pest control strategy.

The research work conducted in the IPM of crop pests in which neem is one of the component is being discussed below :

Siddiq *et al*. (1987) tested the application of neem extracts at different sowing dates in the Sudan for the control of insect pests of potatoes. Other cultural means of control were evaluated against soil pests. The control of postharvest infestations of the gelechiid *Phthorimaea operculella* by neem sprays and other means was also studied. The application of neem extacts, ground fruits and leaves soaked in water at 1 kg/40 litres, controlled foliage pests, including the aleyrodid *Bemisia tabaci*, the cicadellid *Jacobiasca lybica* and *Aphis gossypii*, and increased yield by 0.5 tonnes/ha. Spraying tubers with neem and then placing them in jute sacks reduced postharvest loss. Early sowing on 15 Novermber reduced foliage pests and increased the marketable yield by 3.2 tonnes/ha compared with late sowing in December. Increase of sowing depth from 2.5 cm to 7.6 cm reduced damage by the noctuid *Agrotis ipsilon* and *P. operculella* and increased marketable yield by 3.7 tonnes/ha.

Schmutterer (1988) gave his views on the potential of neem products in integrated pest management in developing and industrialized countries. According to him, azadirachtin and azadirachtin-containing neem (*Azadirachta indica*) seed extracts act as antifeedants, growth regulators and sterilants in insects. The effect upon insect development is most important from the viewpoint of practical insect pest control. Several hypotheses exist on the mode of action. An interference with the neuroendocrine system controlling ecdyson and juvenile hormone synthesis is suggested. In addition, azadirachtin is a chitin synthesis inhibitor. Good results in insect control are obtained with azadirachtin-containing seed extracts under field conditions. Aqueous, alcoholic and enriched (azadirachtin-rich) extracts are used. The residual effect usually lasts about 4-8 days, depending on the environmental conditions and the plant species treated. Systemic effects last somewhat longer. Ultraviolet light, rainfall and perhaps high acidity on treated surfaces of plants cause a fast degradation or loss of the active material. Consequently, much higher concentrations of azadirachtin have to be used in the field to obtain comparable to those in the laboratory. Several problems have to be overcome to develop reliable pesticides form neem-seed extracts. In spite of the sensitivity of insects of most orders to azadirachtin, neem products are selective as they do not harm important natural enemies of pests. They are also non-toxic to warm-blooded animals. Neem-seed extracts have, therefore, a considerable potential for integrated pest control measures, provided certain strategies are considered for their application.

Kumar *et al*. (1989) tested kerosene and neem extracts in the laboratory and in the field in Karnataka against *Planococcus citri* and *P. lilacinus* on coffee. Kerosene was tested alone at 1.5 percent and with 0.025 percent a.i. methyl parathion [parathion-methyl]. The combination of kerosene and parathion-methyl was more effective than parathion-methyl alone (99.96 and 96.70% mortality, respectively, on the 4th day), while kerosene alone was also effective (64.10% mortality). Effective control was also obtained with 3, 5 and 7 percent neem oil against *P. citri*. It is advocated that kerosene, neem oil are important in the integrated pest management of scale insect.

Kumar *et al*. (1989) observed that 1 percent kerosene alone or in combination of 4 percent neem extract were effective in the integrated pest management of *Coccus viridis* on coffee. This combination causes about 80-90 percent mortality in 13-19 days. Pascual *et al*. (1990) observed that azadirachtin induced imaginal moult deficiencies in *Tenebrio molitor* L. Azadirachtin was injected into newly moulted pupae of *T. molitor* at doses ranging from 0.01 to 10 µg/insect. It induced a delay of the development and an inhibition of apolysis and acdysis, which was dose dependent. Low doses of azadirachtin suppressed adult ecdysis and high doses also prevented apolysis and secretion of adult cuticle. Its potential for integrated control of stored food pests is emphasized.

Abdul Kareem *et al*. (1989) studied in greenhouse experiments, survival of *Nephotettix virescens* was significantly lower on potted TN1 rice plants treated with carbofuran (1 kg a.i./ha) mixed with neem seed kernel or cake powder (1:1, w/w) or carbofuran alone (1 or 2 kg a.i./ha) than on untreated plants. Rice tungro virus (RTV) infection was 28 percent in plants teated with carboruran (1 kg a.i./ha), while it was 100 percent in untreated plants. Plants treated with the neem cake-carbofuran mixture or with carbofuran (2 kg a.i./ ha) had 8 and 10 percent RTV infection, respectively. Plants treated with the neem kernel-carbofuran mixture had the lowest rate of RTV infection (2%). No nymphs emerged on plants treated with the neem-carbofuran mixtures or with carbofuran at 2 kg a.i..ha.

Krishnaiah *et al*. (1990) reported RD-9 Repelin in combination with monocrotophos and ethofenpox at lower concentrations resulted in reduced survival of BPH and GLH nymphs than the application of insecticide alone.

In field trials against cotton pests viz. boll worm *Helicoverpa armigera*, aphid *Aphis gossypii* and mite *Tetranychus telarius*, RD-9 Repelin at 1.5

percent sprayed 4-6 times and at 1 percent sprayed 6-8 times was found significantly superior to application of insecticides (RARS, 1989). Further studies revealed that 0.5 percent Repelin in combination wtih quinalphos (0.05%) was significantly superior to lone use of quinalphos (0.05%) in controlling *H. armigera*. In the field experiments conducted against aphid population and american boll worm in okra, it was found that RD-9 Repelin (0.5%) alone or in combination with carbaryl (0.15%) reduced the population as compared to carbaryl (0.15%) alone suggesting synergistic action (RARS, 1993).

In oil seed crops, RD-9 Repelin was tested against ground nut aphid *Aphis craccivora* Koch and hopper *Empoasca kerri*, the best treatments were 0.25 percent dimethoate alone or in combination with reduced dose of RD-9 Repelin i.e. 0.1 percent. The data indicate that RD-9 Repelin at 0.1 percent or 1.5 percent alone and at 1.0 percent in combination with dimethoate or spraying RD-9 Repelin alternated with them gave significantly higher yields than the untreated control.

Seek (1991) described the biology and damage caused in Senegal by *Collosobruchus maculatus* on stored cowpeas, *Sitophilus zeamais, S. oryzae, Rhizopertha dominica, Tribolium castaneum, Sitotroga cerealella, Ephestia cautella* [*Cadra cautella*] and *Corcyra cephalanica* on strored cereals. An integrated control approach against the insect pests of stores of maize millet and cowpea have been formulated by the use of insecticides, varietal resistance, use of neem extacts and bioagents.

Shivayogeshwara *et al*. (1991) did field experiments in 1985-87 in Shimoga, Karnataka, India, a system of integrated pest management (IPM) was compared to a recommennded regime of insecticide applications for control of *Spodoptera litura* on tobacco. In the IPM system, castor beans (*Ricinus communis*) were grown as a trap crop, 200 litres/ha of 0.05 percent quinalphos was sprayed on the 30th day after transplanting, a nuclear polyhedrosis virus spray was applied on the 45th day an 400 litres/ha 1.0 percent neem kernel suspension was applied on the 60the day. In the chemical control system, 0.05 percent quinalphos spray was applied at 200, 300 and 400 litres/ha on the 30th, 45th and 60th day, respectively. The numbers of eggs removed from the *R. communis* were monitored and it is concluded that it served effectively as an effective ovipositional crop. Average leaf damage in the tobacco grown in the IPM system was 6.56 percent, instead of 10.98 percent in insecticide-treated crops and 31.26 percent in untreated controls.

Temurde *et al*. (1992) conducted field trials of fenvalerate (0.01%), cypermethrin (0.01%), endosulfan (0.05%) and carbaryl (0.2%) alone and at half concentration mixed with Neemark (extract of *A. indica*) (0.5%) against *Leucinodes orbonalis* on brinjal were carried out in Maharashtra, India, in 1990-91. All the systhesized insecticides on their own gave better control of the infestation and larger yields than Neemark on its own. Mixing Neemark with cypermethrin or fenvalerate gave better control of the pest than did Neemark alone.

Krishnaiah *et al.* (1993) studied the effects of seed oils of *Calophyllum inophyllum* and *Bassia latifolia* [*Madhuca longifolia*] on the brown planthopper (*Nilaparvata lugens*) and the green leafhopper (*Nephotettix virescens*) were compared with those of neem (*A. indica*) in the greenhouse. The effect of *C. inophyllum* oil on orientation and settling of *N. lugens* was similar to that of neem oil. *C. inophyllum* oil had the greatest antifeedant effects against *N. lugens* and the greatest growth disruptant effects against both pests. Survival of nymphs of *N. virescens* and *N. lugens* was lower with repelin than with neemta 2100 and neemguard. A mixture of neem oil and monocrotophos was more toxic to *N. lugens* than monocrotophos alone showing synergistic effect. A mixture of repelin with monocrotophos or etofenprox gave a greater reduction in the survival of nymphs of *N. lugens* than insecticides alone. In field studies on rice in Andhra Pradesh, India, incorporation of neem cake (150 kg/ha) followed by spraying with neem oil reduced incidence of leaf folder (*Cnaphalocrocis medinalis*). Combinations of repelin and neemta 2100 with monocrotophos at half the recommended concentrations gave control of stem borer and *C. medinalis* equal to monocrotophos at the recommended concentrations and greater to that with botanical insecticides alone.

Manoharan and Uttamasamy (1993) observed that addition of neem oil to endosulfan and phosalone increases the mortality of larvae by 16.66 and 25.00 percent, respectively compared with insecticide alone.

Katole *et al.* (1993) conducted field trials in Nagpur, Maharashtra, India, during 1989-90 to determine the effectiveness of monocrotophos and phosalone (each 0.025 and 0.05%), butacarboxim 50EC (0.075 and 0.1%), fish oil resin soap (3.33%), neem (*A. indica*), karanj (*Pongamia pinnata*) oil, mahua (*Madhuca longifolia*) oil and castor (*Ricinus communis*) oil (all 1.00%) for the control of *Aleurocanthus woglumi* infesting mandarins. Pooled analysis revealed that monocrotophos (0.05%) was the most effective treatment. There was no significant difference between the effectiveness of the reamaining insecticides and the untreated control (castor oil was the least effective treatment). Treatment effects were not influenced by seasonal variations. It is concluded that the application of any insecticide at low concentration or fish oil resin soap can be recommended during the early nymphal stage, when the population of natural enemies is generally low, followed by neem oil sprays.

Dhawan and Simwat (1993) observed that cotton crop in Punjab is attacked by pink bollworm and spotted bollworms. Three botanical insecticides viz. RD-9 Repelin, Neemark and Ind-Ne 20EC were tested alone and in combination with conventional insecticide (quinalphos) and sysnthetic pyrethroid (cypermethrin) for the control of bollworm complex on cotton in three diferent experiments. Botanical insecticides when used alone, significantly reduced the bollworm damage and increased the seed cotton yield than control. All except RD-9 Repelin were significantly inferior to synthetic insecticides. RD-9 Repelin was as effective as quinalphos for reducing bollworm damage and increasing seed cotton yield. Botanical insecticides gave better bollworm control and higher yield when used in combination than quinalphos and cypermethrin alone. Mixing of these botanicals with conventional insecticides make them as effective as synthetic pyrethroids. These combinations can be substituted for synthetic pyrethroids for bollworm integrated pest management.

Rao and Rosaiah (1993) tested neem formulations against insect pest complex of okra alone and in combination with carbaryl. Botanical insecticides such as Neemguard (0.5%), Repelin (0.5%), neem seed kernel extract (5.0%) and nicotine sulphate (0.5%) were evaluated in the field individually and in combination with carbaryl (0.15%) against the major pest complex of okra. Nicotine sulphate alone or in combination with carbaryl (0.15%) was effective in suppressing jassids. Similarly, Repelin followed by nicotine sulphate with and without carbaryl was equally effective in reducing aphid populations and fruit borer incidence due to *Heliothis*. Significant seed yields were realized with the use of nicotine sulphate and Repelin both with and without carbaryl, carbaryl alone, Neemguard + carbaryl, and NSKE + carbaryl. The botanical insecticides did not exert any adverse effect on the natural beneficial predatory fauna such as coccinellids, syrphids, the spiders and hence can be safely used for the IPM of the key pest problems on okra.

Saucke (1996) studied integrated control of *Plutella xylostella* and *Crocidolomia pavonana* in different cabbage growing regions of Papua, New Guinea. The parasitoids *Diadegma semiclausum* and *Cotesia plutellae* and effect of neem products and microbial pesticides based on *Bacillus thuringiensis* were studied in laboratory and field conditions during 1992-95 in three regions. Mahapatra and Nanda (1996) observed that the treatment of neem derivatives alone and in combination with conventional insecticides against the yellow stem borer of rice *Scirpophaga incertulus* is effective in controlling the pest. Maximum protection treatment where full protection umbrella was extended to crop registered the lowest level of incidence i.e. 0.19 percent and 0.28 percent dead hearts in Jaya and Lalot rice, respectively. Crop protection schedules where only neem oil or neem seed extracts were provided to the plots at 20 DAT arrested the activities of stem borers and the incidence level ranged form 0.5 to 1.57 percent dead hearts in both the rice varieties. A moderate level of infestation was observed at 50 DAT. The control plots recorded as high as 8.09 percent and 5.48 percent dead hearts as against 4.67 and 3.71 percent in maximum protection plots of two varieties, respectively. Crop protection treatments with neem products alone or along monocrotophos exhibited moderate level of effectiveness recording 4.76 to 6.30 percent dead hearts and 1.13 to 3.28 percent dead hearts in two rice varieties. It is observed that Lalot rice exhibited a moderate level of varietal resistance and this can be introduced as a component in IPM of stem borer. Neem oil and neem seed extract (both at 2%) alone or in combination with monocrotophos (0.4 kg a.i./ha produced a better level of control on Lalot rice. More or less similar type of findings have been repoted by several workers like Balasubramanian *et al*. (1987), Gunathilagaraj and Sundrababu (1987) and Nanda *et al*. (1993).

Sinha (1996) reported monitoring of the male *Helicoverpa* [*Heliothis*] *armigera* Hubn., from 1986 to 1998 at Karnal (Haryana), using synthetic sex pheromone traps. Maximum activity was recorded during the spring and summer (7th to 28th standard week) while a minor peak activity was observed in autumn (42nd to 50th standard week). Moth catches in pheromone traps were found positively correlated with the larval count of *H. armigera* on chickpea made after one, two and three weeks of trap catches.

In a four year field efficacy trial, neem oil, neem seed kernel extract (NSKE), neem emulsion (NE) and neem wettable powder (NWP) along with insect growth regulators (IGR) and synthetic insecticides were evaluated against *H. armigera* infesting chickpea seed crop. At a level of 20.5 percent (1986-87) and 12.5 percnet (1988-89) overall infestation, diflubenzuron (0.05%), neem oil and NSKE each at 5.0 percent sprayed twice at an interval of 10 days, gave two-fold reduction in the pod borer infestation over untreated control and were found comparable to endosulfan (0.07%) an insecticide. All these treatments gave significantly higher yield of seed in both years.. A 39 to 47 percent increase in seed yield was registered with diflubenzuron (0.05%) and endosulfan (0.07%) over untreated control.

During 1989-90 and 1990-91, some new neem-products and chitin biosynthetic inhibitors were evaluated along with endosulfan. A high level of infestation, 35.6 percent (1989-90) and 37 percent (1990-91) of the pest was recorded. Flufenoxuron irrespective of its dosages (0.0125 and 0.025%) proved to be the most effective treatment and reduced pod borer infestation by 45-50 percent in 1989-90 and 77-78 percent in 1990-91, over control, respectively. Neem emulsion and neem wettable powder afforded 40-60 percent whereas fenvalerate and endosulfan gave 72-73 percent reduction in the infestation over untreated control (1990-91). The NSKE at 5.0 percent gave 40 percent reduction in the infestation and was comparable to endosulfan at 0.07 percent (1989-90). Neemguard was found ineffective. There was no significant difference in the seed yield obtained from NE (0.125%), NSKE (5.0%) flufenoxuron and endosulfan (0.07%) treatments. The findings can be incorporated in the IPM schedule of *H. armigera*.

Srivastava (1996) conducted two fold experiments to evaluate 2, 3, 4 and 5 percent neem seed kernel in water, 0.1 percent cartap hydrochloride, 0.07 percent endosulfan and 1 percent endosulfan and 1 percent of neem oil and repelin against major pest complex of cabbage. These were evaluated both under standard spray regime at 10 days internal and in the IPM involving Indian mustard as trap crop. Five sprays were required as against three in the IPM plot until the cabbage growing period. In both these trials, neem seed kernel extract (NSKE) at 4 and 5 percent consistently recorded significant reduction of diamond back moth (DBM), *Plutella xylostella* larvae with a consequent increase in percent marketable heads as compared to rest of the treatments. Sprays with NSKE did not affect the dry matter and size of cabbage heads. In addition, DBM larvae sampled from these plots, had higher levels of parasitism by *Cotesia plutellae*, a dominant parasitoid. Long term use of neem seed kernel extract for the management of insecticide resistant strain of *P. xylostella* under the IPM programme is indicated.

Siddiq and Khalafalla (1996) formulated IPM of potato pests by experimenting in farmer's fields. Based on research, an IPM package in light soil and heavy soil conditions were prepared.

(i) IPM for light soil conditions

  (a) Three sprays at 10 days intervals against early larval stages and leaf hoppers with neem seed-water extract at 1 kg seed powder/ 40 lit water.

  (b) Potato seed cut in two longitudinal halves, each sown at a spacing of 20 cm as a control measure against older larval stages of cutworms.

(ii) IPM for heavy soil conditions

  (a) Spraying leaf hoppers with neem seed water extract.

  (b) Control of tuber worms through early sowing of tubers in November, deep sowing at 3 inch depth and early harvesting of crop.

Subramaniam *et al.* (1996) studied an effective botanical RD-9 Repelin (neem-based formulation) for the IPM of rice, cotton, vegetable and oil seed crops. The RD-9 Repelin was tested for its bioefficacy against rice pests namely *Scirpophaga incertulus, Nilaparvata lugens* Stol. and gall midge *Orseolia oryzae* Wood Mason, leaf folder *Cnaphalocrosis medinalis* Guinea, white backed plant hopper *Sogatella furcifera* Horvath, green leaf hopper *Nephotettix virescens* and rice tungro virus (RTV). Repelin at 1 percent concentration and half the dose of monocrotophos at 0.03 percent significantly reduced the population of the insects.

Katole *et al*. (1996) tested the efficacy of neem oil and monocrotophos alone and in combination against *Aleurocanthus woglumi* on 3 flush broods on Nagpur Mandarins during 1991-93 in Akola, India. The results revealed that a spray of 0.025 percent monocrotophos plus 1.0 percent neem oil (emulsion) at an early nymphal stage followed by a spray of 1.25 percent neem oil alone after 15 days was the most effective treatment. Insecticide usage was reduced by 75 percent compared to using 2 sprays of monocrotophos.

Swaran Dhingra (1996) carried out the studies to enhance the susceptibility level of *Mylabris pustulata* to cypermethrin in combination with certain non-toxic vegetable oils (karanj [*Pongamia pinnata*] oil, neem [*Azadirachta indica*] oil, sesame oil and citronella [*Cymbopogon nardus*] oil along with the standard synergist piperonyl butoxide). Neem oil and citronella oil proved synergistic at both the ratios, i.e., 1:1 and 1:5, whereas sesame oil and the well known synergist PBO produced only in an additive effect at a 1:1 ratio. Karanj oil exhibited antagonism with cypermethrin at both ratios. It was concluded that cypermethrin + neem oil and cypermethrin + citronella oil provided the desired combination to overcome resistance in *M. pustulata* and gave significantly greater control than cypermethrin alone.

Gupta and Sharma (1997) tested 8 different spray schedules against boll worms (Lepidoptera) in cotton fields in Delhi, India, during 1993-94, and their impact on the population build-up of *Bemisia tabaci* was studied. Neem used alone or alternately with *Bacillus thuringiensis* or conventional synthetic insecticides failed to control boll worms. However, no population build-up of *B. tabaci* was noted. The alternate use of conventional insecticides with synthetic pyrethroids was effective against boll worms, but resurgence of *B. tabaci* was evident. However, a spray schedule in which neem was used alternately with *B. thuringiensis*, and at least one spray with a synthetic pyrethroid, was effecitve in managing the bollworm complex without a resurgence of *B. tabaci*. Further studies indicated a satisfactory level of control of bollworms and good cotton yield following the alternate use of neem. *B. thuringiensis*, and an 84 percent, reduction in the use of synthetic pyrethroids.

This spray schedule had no adverrse effect on the environment.

Gopal and Senguttuvan (1997) conducted field trials in India during 1989-92 to determine the efficacay of insecticides (endosulfan and diflubenzuron), neem products and nuclear polyhedrous virus (NPV) alone or in combination for the control of fruit borer, *Helicoverpa armigera*, on tomatoes. Neem seed kernel extract (3.0%) + endosulfan (0.035%) + NPV at 250 equivalents (LE) ha-1 applied three times at 45, 55 and 65 days after planting gave the highest larval mortality, reduced fruit damage, and the highest fruit yield, followed by neem oil (3.0%) + endosulfan (0.035%) + NPV at 250 LE ha-1, and endosulfan (0.07%). Endosulfan (0.07%) gave the highest cost-bebefit ratio, followed by NSKE (3.0%) + NPV at 250 LE ha-1, and NSKE (3.0%) + endosulfan (0.035%) + NPV at 250 LE ha-1.

According to Katole *et al*. (1997), a survey was carried out among 150 cotton growers at Akola, Maharashtra, India, to investigage the adoption of plant protection measures for the cotton hybrid AHH-468. The majority (55.33%) of the respondents adopted cultural practices of IPM (burning residues, deep ploughing and crop rotations), followed by chemical, mechanical (hand picking of eggs and larvae, cutting affected shoots), 5 percent neem seed extract, and physical control (sun drying of seeds) measures. Only a negligible percentage of the respondents used biological control measures.

Garg *et al*. (1997) found that in an improved Basmati rice variety Taraori, a neem based pesticide, the parasitoid, *Trichogramma japonicum* and insecticides were effective against leaf folder *Cnaphalocrosis medinalis* and yellow stem borer *Scirpophaga incertulas* in an IPM approach during 1994 and 1995 in Haryana, India. The result revealed that the IPM technology is very effective, environmentally safe and economically viable.

Das and Singh (1998) conducted field experiments for the integrated management of jute pests by integrating cultural practices and minimum use of pesticides. Among the pesticides, three chemical pesticides and three non-chemical pesticides were sprayed twice against stem weevil *Apion corchori*, yellow mite *Polyphagotarsonemus latus* and jute semilooper *Anomis sabulifera*. Endosulfan (0.075% a.i.) was most effective against all th major pests of jute. Among the non-chemical treatments, neem oil and *Bacillus thuringiensis* could control jute semilooper better than chirata (*Swertia chirata*) extract.

Parthiban and Anonthan (1998) conducted a field trail on cotton at Theni, Madurai district, Tamil Nadu, India, during the Kharif season of 1996. Crops were sprayed with alternate applications of a neem formualtion, containing 1500 ppm azadirachtin (1 lt/ha), and a biopesticide formulation, based on

Bacillus thuringiensis subsp. *kurstaki* (1 lt/ha), at 7, 10 and 14 day intervals. The standard treatment consisted of endosulfan 35EC (2 lt/ha) and cypermethrin 10EC (0.6 lt/ha) alternated at 10-day intervals. The alternate application of neem and bacterial formulations significantly reduced the incidence of sucking pests compared to the untreated control, but the insecticide treatment recorded significantly lower levels of pests. The larval populations of *Spodoptera litura* (0.24 to 0.46 larvae/plant) and *Helicoverpa armigera* (0.25 to 1.75 larvae/plant) were controlled at all spray intervals, and were comparable to the insecticides treatment (0.17 larvae/plant). The insecticide treatment gave a significantly higher yield (66.56 q/ha) than neem and bacterial formulations (52.58-54.31 q/ha) or the untreated control (31.22 q/ha). It is concluded that an integrated pest management programme can offer adequate protection against sucking pests and bollworms on cotton.

Sanap and Pawar (1998) conducted a field experiment in Maharashtra during 1993-96 for controlling *H. armigera* infesting gram (chickpea). Integrated pest management treatment comprising endosulfan (0.07%), neem-seed-kernel extract (5%) and nuclear polyhedrosis virus (250 LE/ha) were evaluated. The field trial conducted during winter 1993-96 revealed that 3 spray applications starting from initiation of flowering and subsequent 2 sprays at fortnightly intervals with first 2 sprays either with NPV (250 LE/ha) or NSKE followed by a third spray with endosulfan (0.07%) were most effecitve in controlling the pest and resulted in a 26.94 and 27.29 percent increase in yield, respectively.

Webb *et al*. (1998) compared 5 biological insecticide treatments (1 *B. thuringiensis* [BT] and 4 gypsy moth nuclear polyhedrosis virus treatments) and 2 reduced-risk insecticides (azadirachtin from neem [*Azadirachta indica*] and tebufenozide) against 2 commonly used standard insecticides (diflubenzuron and cyfluthrin) for gypsy moth (*Lymantria dispar*) management. The trial was conducted on oak (primarily *Quercus palustris*) trees on the edge of a forest clearing near Smyrna, Delaware. Significant details pertaining to the use of the specific control materials were clarified in support of a decision matrix around which an IPM system can be formulated for use by arborists and nurserymen. The standard insecticides and the ecdysone against tebufenozide, demonstrated excellent activity against all instars in both a 1-hour residue study and a 35-day residue study. BT was clearly more effective against second instar larvae than against fourth instar larvae and lost significant activity against the gypsy moth after 7 days. The efficacy of the neem product was similar to that of BT in that it controlled younger instars better than older instars. Although its speed of action was slower than that of BT, it remained highly active against gypsy moth 21 days after treatment. NPV without the activity enhancer Blankophor BBH gave significant levels of control for all 4 larval instars fed, although mortality was higher for younger instars than for older instars. Residual effectiveness was significantly reduced after 1 day. The addition of Blankophor BBH to the NPV tank mix led to improved kill in the 1-hour study and vastly improved residual activity up to 35 days.

Rodriguez *et al.* (1998) studied the potential use of neem seeds (*A. indica*) as a natural insecticide to control the coffee berry borer (*Hypothenemus hampei*). The stability of aqueous extracts of neem seeds and the ability of different doses to protect coffee fruits were evaluated. Analysis of the results showed that aqueous extracts of neem seeds, prepared 24 and 48 h before application, remain stable. Of the evaluated doses, concentrations of 2.0 and 5.0 percent were most effective, protecting 73 and 75 percent of the coffee fruits, respectively. The level of control provided by neem seed extracts exceeded than by commercial insecticide control. The authors propose the inclusion of the use of neem seeds in the programme of integrated pest management to control *H. hampei* in Mexico.

Vaish (1998) studied the efficacy of neem formulation nimbecidine (azadirachtin 0.03%) alone and mixed with other insecticides against *Helicoverpa armigera* for their efficacy. Pooled data of four years (1991-92 to 1994-95) revealed that nimbecidine (azadirachtin 0.03%) at the rate of 1.0 and 2.0 lit. when mixed with endosulfan 35EC at 1.2 lit or fenvalerate 20EC at 0.4 lit/ha recorded low pod damage of 8.7 and 8.2 percent and 8.5 and 7.7 percent, respectively. Pod damage was at par with endosulfan at the rate of 1.5 lt/ha (7.86%) and fenvalerate at 0.4 lt/ha (8.57%) in two sprays commencing at 50 percent flowering and repeated after 15 days. Yield was at par in treatments and ranged from 30.730 q/ha in nimbecidine @ 2.0 lt + fenvalerate at 0.4 lt with C : B ratio 1:5.09 and 28.906 /ha in nimbecidine @ 1.0 lt + endosulfan @ 1.2 lt/ha with C:B ratio 1:4.86. Standard treatment recorded 10.14 percent pod damage and 18.39 q/ha grain yield whereas untreated control recorded 18.39 pod damage and 25.60 q/ha grain yield. Hence, such formulations can be utilized in the IPM programme of *H. armigera*.

Dhingra *et al.* (1998) evaluated the efficacy of cypermethrin, fenvalerate, fenpropathrin and deltamethrin alone and in combination with neem oil in ratios of 1:1 and 1:4 against adults of *Bagrada cruciferarum* [*B. hilaris*] in the laboratory at 27±1°C and 50-60 percent RH. On the basis of LC50s, deltamethrin, fenpropathrin and cypermethrin were 3.99, 2.60 and 2.05 times as toxic as fenvalerate, respectively. However, amongst the combinations tested, fenvalerate in combination with neem oil showed a synergistic effect in both ratios. Combinations of other systhetic pyrethroids exhibited antagonism with both ratios. Such combinations having synergistic effect can be used in the IPM programme. Sarode and Chaudhary (1998) observed that neem seed extracts in combination with half dose of insecticide was found equally effective as insecticide schedule against bollworms. It is expected that this can be fitted in IPM programme.

Durairaj *et al.* (1999) studied IPM strategy to control *Sciota divisella* (Pyralidae) with biocides viz. *B. thuringiensis, Beauveria bassiana*, NeemAzal, Forture Aza and chemical pesticide monocrotophos.

References

Abdul Kareem, A.; Saxena, R.C. and Boncodin, M.E.M. (1989). Neem-carbofuran mixture against *Nephotettix virescens* (Distant) (Homoptera, Cicadellidae) and its transmission tungro associated viruses. *Journal of Applied Entomology*, **108**(1) : 68-71.

Balasubramanian, G.; Jayraj, S. and Balasubramanian, M. (1987). Effect of slow release fertilizer nitrogen and neem products in the control of stem borer of rice. *Indian Journal of Entomology*, **15**(2) : 132-135.

Das, L.K. and Singh, B. (1998). Integrated management of jute pests. *Environment and Ecology*, **16**(1) : 218-219.

Dhawan, A.K. and Simwat, G.S. (1996). Field evaluation of some botanical insecticides alone and in combination with other insecticides for management of bollworm complex on cotton. *Neem and Environment*, **1** : 485-492.

Durairaj, S; Srinivasan, G.; Balu, A.; Deeparaj, B. and Rajarishi, R. (1999). Ecofriendly methods to control the Jatropha shoot and fruit borer, *Sciota divisella. Neem Newsletter*, **16**(4) : 38.

Garg, D.K.; Baranwal, V.K.; Dhaliwal, G.S.; Arora, R.; Randhawa, N.S. and Dhawan, A.K. (1998). An integrated approach for pest management in Basmati rice in Haryana. *Ecological Agriculture And Sustainable Development* Vol. **2**, *Proceedings of International Conference on Ecological Agricultural : Towards Sustainable Development*, Vol. **2** held at Chandigarh, India, 15-17 Novemsber, 1997 : 216-221.

Gopal, S. and Senguttuvan, T. (1997). Integrated management of tomato fruit borer with insecticides, neem products and virus. *Madras Agricultural Journal*, **84**(2) : 82-84.

Gunathilagaraj, K. and Sundrababu, P.C. (1987). Role of plant products in pest management in rice. *Neem Newsletter*, **4**(3) : pp.35.

Gupta, G.P. and Sharma, K. (1997). Neem based pest management strategy in cotton system. *Pesticide Research Journal*, **9**(2) : 190-197.

Katole, R.T.; Nikhade, D.M.; Ingle, P.O. and Wankhade, P.P. (1997). Extent of adoption of plant protection measures in hybrid cotton AHH-468. *Journal of Soils and Crops*, **7**(2) : 160-162.

Katole, S.R.; Mahajan, R.K.; Tayde, G.S. and Gawande, R.B. (1996). Neem oil : an environmentally safer biocide in the management of citrus blackfly (*Aleurocanthus woglumi* Ashby). *PKV Research Journal*, **20**(1) : 28-30.

Katole, S.R.; Thakare, H.S. and Mahajan, R.K. (1993). Management of citrus blackfly nymphs with some insecticides and plant oils. *Journal of Maharashtra Agricultural Universities*, **18**(1) : 68-71.

Krishnaiah, N.V.; Kalode, M.B. and Pasalu, I.C. (1990). New approaches in utilization of botanicals in rice insect pest control. *Proceedings Symposium on Botanical Pesticides in IPM*, Rjamundry : 114-117.

Krishnaiah, N.V.; Kalode, M.B. and Pasalu, I.C. (1990). New approaches in utilization of botanicals in rice insect pest control. *Proceedings of Symposium on Botanical Pesticides in Integrated Pest Management*, Rajamundry : 203-216.

Kumar, M.G.; Bhat, P.K. and Ramaiah, P.K. (1989). Potential role of kerosene and neem derivatives in integrated management of mealybugs on coffee. *Journal of Coffee Research*, **19**(1) : 17-29.

Kumar, M.G.; Bhat, P.K. and Vijayalakshmi, C.K. (1989). Screening of kerosene and neem extract against coffee green scale. *Journal of Coffee Research*, **19**(1) : 51-55.

Mahapatra, G.K. and Umakanta Nanda (1996). Integrating neem in yellow stem borer management in Kharif rice. *Indian Journal of Entomology*, **58**(4) : 369-373.

Manoharan, T and Uthamasamy, S. (1993). Evaluation of plant products for synergistic action with insecticides for control of gram-podborer (*Helicoverpa armigera*). *Indian Jouranl of Agricultural Sciences*, **63**(11) : 759-761.

Nanda, U.K.; Parija, B.; Nanda, B.; Dash, D.D. and Prahan, N.C. (1993). Bioeffecacy of neem derivatives against the paddy insect pest complex. Abstract *World Neem Conference*, 24-28th February, 1993, Bangalore, India : pp.14.

Parthiban, M and Ananthan, G. (1998). Alternate application of neem and biopesticide in cotton pest management. *Insect Environment*, **3**(4) : 97-98.

Pascual, N.; Marco, M.P. and Belles, X. (1990). Azadirachtin induced imaginal moult deficiencies in *Tenebrio molitor* L. (Coleoptera : Tenebrionidae). *Journal of Stored Products Research*, **26**(1) : 53-57.

Rao, S.V. and Rosaiah, B. (1996). Evaluation of botanical insecticides alone and in combination with carbaryl against insect complex of okra. *Neem and Environment*, **1** : 493-496. Ed. R. P. Singh *et al*.

RARS (1989). Studies on the evaluation of Repelin on the control of cotton. A.P. Agricultural University Regional Agricultural Research Station, Lam Report.

RARS (1993). Affect of RD-9 Repelin in the management of okra pests. Regional Agricultural Research Station, Lam Report.

Rodriguez Lagunes, D.A.; Lagunes Tejeda, A.; Riestra Diaz, D.; Rodriguez Maciel, C.; Velazquez Mendoza, J.; Becerril Roman, E. and Pacheco Velasco, E. (1998). Aqueous extracts of neem for the control of coffee berry borer. *Manejo Integrado de Plagas*, **49** : 73-77.

Sanap, M.M. and Pawar, V.M. (1998). Integrated management of *Helicoverpa armigera* on gram (*Ciceer arietinum*). *Indian Journal of Agricultural Sciences*, **68**(3) : 162-164.

Sarode, S.V. and Chaudhary, A.E. (1998). Efficacy of neem products against cotton bollworms. Abstract in *ICPPMSA*, Chandra Shekhar Azad University of Agriculture and Technology, Kanpur (India) : 149-150.

Saucke, H. (1996). Biological and integrated control of *Plutella xylostella* (Lepidoptera, Yponomeutidae) and *Crocidolomia pavonana* (Lepidoptera, Pyralidae) in different cabbage-growing regions of Papua, New Guinea. Bibliographic citation, pp. 131.

Schmutterer, H. (1988). Potential of azadirachtin-containing pesticides for integrated pest control in developing and industrialized countries. *Journal of Insect Physiology*, **34**(7) : 713-719.

Seck, D. (1991). Economic importance and development of an integrated control approach against insect pests of stores of maize, millet and cowpea in farming areas. *Sahel PV Information*, **33** : 15-20.

Shivayogeshwara, B.; Mallikharjunaiah, H.; Krishnaprasad, N.K. and Shetty, M.V.N. (1991). Integrated management of *Spodoptera litura* Fabricius (Noctuidae : Lepidoptera) in FCV tobacco crop. *Tobacco Research*, **17**(2) : 59-61.

Siddiq, S.A. and Khalafalla, A. (1996). The performance of an IPM package including neem for the control of potato pests in farmers fields. *Neem and Environment*, **1** : 541-546.

Siddiq, S.A.; Schmutterer, H. (ed.) and Ascher, K.R.S. (1987). A proposed pest management program including neem treatments for combating potato pests in the Sudan. *Proceedings of the 3rd International Neem Conference*, Nairobi, Kenya, 10-15 July, 1986 : 449-459.

Sinha, S.N. (1996). Neem in integrated management of *Helicoverpa armigera*

Hubn. infesting chickpea. *Neem and Environment*, **1** : 439-446. Ed. R. P. Singh *et al*.

Srinivasan, K. and Krishna Moorthy, P.N. (1996). Evaluation of neem products and other standard chemicals for the management of major pest complex on cabbage : Comparison between standard spray regime and I.P.M. involving mustard as a trap crop. *Neem and Environment*, **1** : 447-458.

Subramaniam, T.S.; Sivaprasad, B. and Prasad, V.V.L.N. (1993). RD-9 Repelin : An effective botanical for integrated pest management. *Neem and Environment*, **2** : 857-864.

Swaran Dhingra (1996). Effect of different vegetable oils on the toxicity of cypermethrin in mixed formulations to the adults of *Mylabris pustulata* Thunb. *Journal of Entomological Research*, **20**(1) : 19-22.

Swaran Dhingra; Seema Dureja, P. and Sudhakar, K. (1998). Effect of neem oil on the toxicity of different synthetic pyrethroids in mixed formulation against the adults of *Bagrada cruciferarum* Kirk. *Journal of Entomological Research*, **22**(2) : 153-156.

Temurde, A.M.; Deshmukh, S.D.; Nemade, S.B. and Khiratkar, S.D. (1992). Efficacy of neemark and its combinations with other groups of insecticides against the shoot and fruit borer of brinjal. *Journal of Soils and Crops*, **2**(1) : 29-31.

Vaish, O.P. (1998). Effect of a neem formulation alone and in combination with insecticides on *Helicoverpa armigera* on chickpea. Abstract in *National Seminar in Entomology in 21st Century* held at Udaipur, India, 30 April-2 May, 1998 : pp.125.

Webb, R.E.; Peiffer, R.; Fuester, R.W.; Thorpe, K.W.; Calabrese, L. and McLaughlin, J.M. (1998). An evaluation of the residual activity of traditional, safe and biological insecticides against the gypsy moth. *Journal of Arboriculture*, **24**(5) : 286-293.

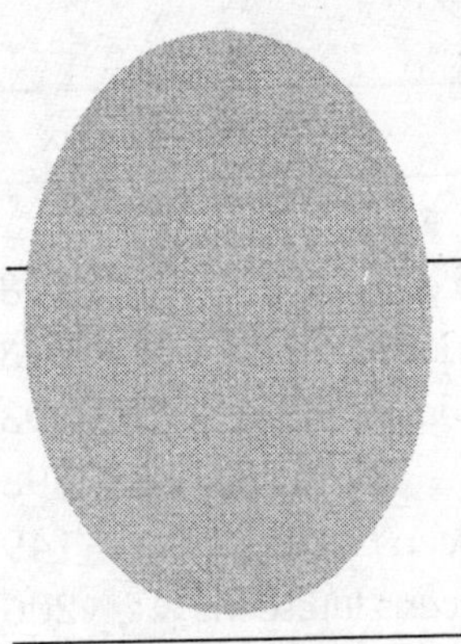

Species Index

A

B

C

D

E

F

G

H

N

O

P

Q

R

S

T

U

V

W

X

Z